18 Pasture L.
Bughkeepsie, NY
12605

STEWART

STEWART

STEWART

$$\text{Power (H/p)} = \frac{\Delta P \text{ in" water} \times CFM}{6350 \times \text{Efficiency (.6)}} \quad \text{(page 97)}$$

$$\text{Power} = \frac{\# \text{ force}}{ft^2} \times \frac{ft^3}{sec} \times \frac{1}{550}$$

McGRAW-HILL SERIES IN CHEMICAL ENGINEERING

SIDNEY D. KIRKPATRICK, *Consulting Editor*

UNIT OPERATIONS OF CHEMICAL ENGINEERING

McGRAW-HILL SERIES IN CHEMICAL ENGINEERING

SIDNEY D. KIRKPATRICK, *Consulting Editor*

BUILDING FOR THE FUTURE OF A PROFESSION

Fifteen prominent chemical engineers first met in New York more than thirty years ago to plan a continuing literature for their rapidly growing profession. From industry came such pioneer practitioners as Leo H. Baekeland, Arthur D. Little, Charles L. Reese, John V. N. Dorr, M. C. Whitaker, and R. S. McBride. From the universities came such eminent educators as William H. Walker, Alfred H. White, D. D. Jackson, J. H. James, J. F. Norris, Warren K. Lewis, and Harry A. Curtis. H. C. Parmelee, then editor of *Chemical & Metallurgical Engineering*, served as chairman and was joined subsequently by S. D. Kirkpatrick as consulting editor.

After several meetings, this Editorial Advisory Committee submitted its report to the McGraw-Hill Book Company in September, 1925. In it were detailed specifications for a correlated series of more than a dozen text and reference books, including a chemical engineers' handbook and basic textbooks on the elements and principles of chemical engineering, on industrial applications of chemical synthesis, on materials of construction, on plant design, on chemical-engineering economics. Broadly outlined, too, were plans for monographs on unit operations and processes and on other industrial subjects to be developed as the need became apparent.

From this prophetic beginning has since come the McGraw-Hill Series in Chemical Engineering, which now numbers about forty books. More are always in preparation to meet the ever growing needs of chemical engineers in education and in industry. In the aggregate these books represent the work of literally hundreds of authors, editors, and collaborators. But no small measure of credit is due the pioneering members of the original committee and those engineering educators and industrialists who have succeeded them in the task of building a permanent literature for the chemical-engineering profession.

THE SERIES

Unit Operations of Chemical Engineering

WARREN L. McCABE

Administrative Dean
Polytechnic Institute of Brooklyn

JULIAN C. SMITH

Professor of Chemical Engineering
Cornell University

McGRAW-HILL BOOK COMPANY, INC.

New York   Toronto   London

1956

UNIT OPERATIONS OF CHEMICAL ENGINEERING

Library of Congress Catalog Card Number 56-8181

III

PREFACE

This book is a beginning text on the unit operations. It is written for undergraduate students in the junior or senior years who have the usual training in mathematics, physics, chemistry, and mechanics. An elementary knowledge of material-balance calculations and thermodynamics is helpful but not essential.

We have continued the traditions of earlier books on this subject. During the past generation the unit-operation approach has demonstrated its usefulness, both in education and in engineering practice. Although certain operations—notably gas absorption, distillation, and extraction—have tended to fuse together, we believe it wise to protect the integrity of the individual operation by giving it separate treatment, as each one has a definite task to perform.

The limits of the text are fixed by space and by level of treatment. Many of the less conventional operations, such as adsorption, dialysis, expression, colloid milling, ion exchange, freezing, sublimation, and specialized methods of mechanical separation, are omitted because of space. Other important topics have been left out both because of space and because any adequate treatment of them would raise the level of the book to a point inaccessible to the undergraduate. For these reasons, multicomponent separations, transient phenomena, and operations at extremes of temperature, pressure, or velocity have been omitted or touched on but lightly.

The reasons why we have not made further cuts are several: (1) some schools give a three-semester sequence in unit operations, and this text contains sufficient material for such courses; (2) both to treat properly the scientific foundations and to describe adequately the equipment used in the conventional operations requires considerable space, and to cut the manuscript further would mean the omission of standard operations, important equipment, or valuable matter on foundations and theory; and (3) the extensive additions made during the recent years to the knowledge of unit operations should, we believe, be included in texts.

For the usual two-semester course, a selection of topics is necessary. We believe this to be an advantage rather than otherwise, as courses may be arranged to meet the needs of individual situations. We do not believe that it is the job of textbook writers to dictate course contents. Every effort has been made to keep a uniform level of treatment throughout the book. The key chapters are Chap. 2, Fluid Mechanics; Chap. 8, Flow of Heat; and Chap. 10, Mass Transfer. Any selection of material for a specific course should include these chapters, as they contain most of the basic concepts of all unit operations.

An effort has also been made to carry the treatment to a point where the student will find an easy transition to more advanced and specialized chemical engineering handbooks, texts, and monographs.

Two ideas have influenced the choice of equipment for discussion in the text. Most of the standard "bread-and-butter" types are shown, and a selection of newer and more specialized devices has also been included. The choice of the latter is quite arbitrary, but those illustrating important engineering principles have been favored.

Most of the apparatus shown in the figures is generalized and is not exactly like that of individual makers. Where specific units of definite manufacturers are shown, appropriate credits are given.

We express our thanks to the following colleagues who have reviewed various portions of the manuscript, and who have contributed many helpful suggestions: Robert F. Benenati, C. F. Bonilla, Ju Chin Chu, H. B. Caldwell, T. B. Drew, P. Harriott, H. S. Kemp, J. C. Whitwell, and C. C. Winding.

WARREN L. McCABE
JULIAN C. SMITH

CONTENTS

Chapter 1

INTRODUCTION

Chemical engineering has to do with industrial processes in which raw materials are changed or separated into useful products. The chemical engineer must develop, design, and engineer both the complete process and the equipment used in it. He must choose the proper raw materials; he must operate his plants efficiently, safely, and economically; and he must see to it that his products meet the requirements set by his customers. Consistent with engineering generally, chemical engineering is both an art and a science. Whenever science helps the engineer to solve his problems, he should use science. When, as is usually the case, science does not give him a complete answer, he must use experience and judgment. His professional stature depends on his skill in combining all sources of information to reach practical solutions to processing problems.

The range and variety of processes and industries that call for the services of chemical engineers are both great. The field is not one that is easy to define. The processes described in standard treatises on chemical technology and the process industries give the best idea of the field of chemical engineering.[6] †

Because of the variety and complexity of modern processes, it is not practicable to cover the entire subject matter of chemical engineering under a single head. The field is divided into convenient, but arbitrary, sectors. This text covers that portion of chemical engineering known as the unit operations.

UNIT OPERATIONS

The most economical method of organizing the subject matter of chemical engineering is based on two facts: (1) Although the number of individual processes is great, each one may be broken down into a series of steps, called operations, each of which in turn appears in process after process. (2) The individual operations have common techniques and are based on the same scientific principles. For example, in most processes solids and fluids must be moved, heat or other forms of energy must be transferred from one substance to another, and tasks like drying, size reduction, distillation, and evaporation must be performed. The unit oper-

† Superior numerals in the text correspond to the numbered references at the end of each chapter.

1

ation concept is this: by studying systematically these operations themselves—operations which clearly cross industry and process lines—the
treatment of all processes is unified and simplified.

The unit operations are as applicable to processes that are essentially
physical in nature as to those that are chemical. For example, the process
used to manufacture common salt consists of the following sequence of the
unit operations: transportation of solids and liquids, transfer of heat, evaporation, crystallization, drying, and screening. No chemical reaction appears in these steps. On the other hand, the cracking of petroleum, with
or without the aid of a catalyst, is a typical chemical reaction conducted on
an enormous scale. Here the unit operations—transportation of fluids and
solids, distillation, and various mechanical separations—are vital, and the
cracking reaction could not be utilized without them. Even the chemical
steps themselves are conducted by controlling the flow of material and
energy to and from the reaction zone.

Because the unit operations are a branch of engineering, they are based
on both science and experience. Theory and practice must combine to
yield designs for equipment that can be fabricated, assembled, operated,
and maintained. A balanced discussion of each operation requires that
theory and equipment be considered together. An objective of this book
is to present such a balanced treatment.

Scientific Foundations of Unit Operations. A number of scientific
principles and techniques are basic to the treatment of the unit operations.
Some are elementary physical and chemical laws, and others are special
techniques particularly useful in chemical engineering. It is assumed that
the reader is familiar with the elementary principles, which are summarized here for reference. The special techniques are discussed more fully
later in this chapter. Other areas in engineering science are considered at
the proper places in the text.

BASIC LAWS

The following physical and chemical laws will be used frequently in
later chapters.

Material Balances. The law of conservation of matter states that
matter cannot be created or destroyed. This leads to the concept of mass,
and the law may be stated in the form that the mass of the materials taking part in any process is constant. It is known now that the law is too
restricted for matter moving at velocities near that of light or for substances undergoing nuclear reactions. Under these circumstances energy
and mass are interconvertible, and the sum of the two is constant, rather
than only one. In ordinary engineering, however, this transformation is
too small to be detected, and in this book it is assumed that mass and
energy are independent.

Conservation of mass requires that the materials entering any process
must either accumulate or leave the process. There can be no loss or gain
during the process. In this book the only kinds of process considered will

involve neither accumulation nor depletion, and the law of conservation of matter takes the simple form that input equals output. The law is often applied in the form of material balances. The process is debited with everything that enters it and is credited with everything that leaves it. The sum of the credits must equal the sum of the debits. Material balances must hold over the entire process or equipment and over any part of it. They must apply to all the material that enters and leaves the process and to any single material that passes through the process unchanged.

Material balances may often be simplified by a judicious choice of basis. The typical problem is one in which a stream of material flows through a unit of equipment and one or more constituents in the stream are augmented or depleted while other constituents are unchanged. Then it is convenient to express the concentration of the active constituent as a ratio to the inert constituent. When such a choice of basis is made, the terminal concentrations can be subtracted directly to give the quantity of the active constituent added or removed.

In some processes a given stream does not vary in total mass during its flow through the equipment, and the loss of one or more components is compensated for by a gain of an equal mass of other components, so the concentrations of all constituents are affected. The material-balance calculations are then simplified if concentrations are expressed in terms of the total mass of the stream, rather than as a ratio to one component. Differences may then be taken between the terminal concentrations to give the mass of substance transferred from stream to stream during the process.

Molal Units. In material balances where chemical reactions are involved, or in the use of relationships like the ideal-gas law, molal units are simpler than the usual units of mass. A mole of any pure substance is defined as the quantity of that substance whose mass is numerically equal to its molecular weight. The mole is a mass unit, and it can be used in material balances just like mass in pounds or grams. From the definition of mole the meanings of the terms pound mole and gram mole follow. In engineering calculations the pound mole is used. The average molecular weight of a mixture of substances is defined by the equation

$$\frac{m_A + m_B + m_C + \cdots}{m_A/M_A + m_B/M_B + m_C/M_C + \cdots} = \bar{M} \tag{1-1}$$

where $m_A, m_B, m_C, \cdots$ = masses of individual pure components A, B, C, $\cdots$ in mixture

$M_A, M_B, M_C, \cdots$ = molecular weights of pure components

$\bar{M}$ = average molecular weight

Mole Fraction. It is often convenient to express compositions not as mass fraction or mass per cent but rather as mole fraction or mole per cent. The mole fraction is the ratio of the number of moles of one component to the total number of moles in the mixture. For example, using the same symbols as in Eq. (1-1), the mole fraction x_A of component A is

$$x_A = \frac{m_A/M_A}{m_A/M_A + m_B/M_B + m_C/M_C + \cdots} \tag{1-2}$$

Similarly, the mole fraction of component B, x_B, is

$$x_B = \frac{m_B/M_B}{m_A/M_A + m_B/M_B + m_C/M_C + \cdots}$$

It is apparent that the sum of the mole fractions of all components in a given mixture must be unity. Mole per cent is, of course, mole fraction multiplied by 100.

Example 1-1. An evaporator is fed continuously with 25 tons/hr of a solution consisting of 10 per cent NaOH, 10 per cent NaCl, and 80 per cent H_2O. During evaporation, water is boiled off, and salt precipitates as crystals, which are settled and removed from the remaining liquor. The concentrated liquor leaving the evaporator contains 50 per cent NaOH, 2 per cent NaCl, and 48 per cent H_2O.

Calculate: (a) the pounds of water evaporated per hour, (b) the pounds of salt precipitated per hour, and (c) the pounds of concentrated liquor produced per hour.

Solution. (a) Because the NaOH is the only constituent passing through the evaporator unchanged, it is convenient to base the concentrations of the H_2O and the NaCl on the NaOH. The total input of NaOH is $25 \times 2{,}000 \times 0.10 = 5{,}000$ lb/hr. The ratio of H_2O to NaOH in the feed is $80{:}10$, and that in the concentrated liquor is $48{:}50$. Since these ratios are on a common basis, they may be subtracted to give the mass of water evaporated per pound of NaOH. The total evaporation is then

$$5{,}000(\tfrac{80}{10} - \tfrac{48}{50}) = 35{,}200 \text{ lb/hr}$$

(b) By the same method, the ratios of NaCl to NaOH in feed and concentrated liquors are $10{:}10$ and $2{:}50$, respectively. The NaCl precipitated is then

$$5{,}000(\tfrac{10}{10} - \tfrac{2}{50}) = 4{,}800 \text{ lb/hr}$$

(c) The concentrated liquor may be calculated in two ways. The first method uses a NaOH balance. The ratio of concentrated liquor per pound of NaOH is $100{:}50$, and the total liquor is then

$$5{,}000 \times \tfrac{100}{50} = 10{,}000 \text{ lb/hr}$$

The second method uses an over-all balance, which states that the mass of the concentrated liquor equals that of the feed liquor less the sum of the masses of the salt precipitated and the water evaporated. The feed liquor is $25 \times 2{,}000 = 50{,}000$ lb/hr. The concentrated liquor is then

$$50{,}000 - (4{,}800 + 35{,}200) = 10{,}000 \text{ lb/hr}$$

The two methods are not independent, and must give identical results.

Example 1-2. In the distillation technique called rectification (see Chap. 12), condensing vapor and boiling liquid streams are passed through a rectifying column in intimate countercurrent contact. The more volatile components tend to transfer from the liquid to the vapor and the less volatile components from the vapor to the liquid. The energy relations are such that the number of moles of both liquid and vapor is constant during the process, and for every mole of liquid vaporized there is a mole of vapor condensed. In a specific rectification process, operating on ethanol-and-water mixtures, the analyses of the vapor and liquid streams entering and leaving the column are those given in Table 1-1.

TABLE 1-1. DATA FOR EXAMPLE 1-2

	Ethanol, mole %	Water, mole %
Liquid entering top............	75.1	24.9
Liquid leaving bottom.........	25.1	74.9
Vapor entering bottom.........	40.2	59.8
Vapor leaving top............	77.3	22.7

Calculate the number of moles of vapor flowing up the column per mole of liquid flowing down the column.

Solution. In this problem, since the moles of both vapor and liquid are constant, concentrations are best expressed in mole fractions of the total stream. The mole fractions are found by dividing the mole percentages by 100. Choosing a basis of 1 mole of liquid, and letting V_m represent the number of moles of vapor, the ethanol lost by the liquid is $0.751 - 0.251$ mole, and the ethanol gained by the vapor is $V_m(0.773 - 0.402)$ mole. Equating these quantities gives

$$V_m(0.773 - 0.402) = 0.751 - 0.251$$

This gives $V_m = 1.35$ moles of vapor per mole of liquid.

The transfer of $0.751 - 0.251 = 0.500$ mole of ethanol from liquid to vapor is balanced by a transfer of 0.500 mole of water from vapor to liquid.

Ideal-gas Law. The characteristics of gases at low pressures are the basis for a useful law called the ideal-gas law. The law applies strictly to actual gases only at very low pressures, but for many gases and vapors it is sufficiently precise for engineering calculations. The law is usually written in the form

$$pV = nR_oT \tag{1-3}$$

where p = pressure, lb force/ft^2
V = volume, ft^3
T = absolute temperature, °R †
R_o = gas constant, equal for all gases to 1,545 ft-lb force/(lb mole)(°R)
n = number of pound moles of gas

Equation (1-3) is seldom used in the above form. It states three facts, however: (1) The volume of a gas is directly proportional to the number of moles; (2) the volume is directly proportional to the absolute temperature; and (3) the volume is inversely proportional to the pressure. These

† Absolute temperatures are expressed in degrees Kelvin (°K), which is temperature in degrees centigrade, or Celsius, plus 273.1, or in degrees Rankine (°R), which is temperature in degrees Fahrenheit plus 459.6.

facts are valuable in converting a gas volume from one temperature and pressure to another temperature and pressure. For example, V_a ft^3 of an ideal gas at an absolute temperature of T_a and a pressure of p_a will occupy V_b ft^3 at an absolute temperature of T_b and a pressure of p_b, where V_b is given by the equation

$$V_b = V_a \frac{p_a T_b}{p_b T_a} \tag{1-4}$$

Molal Volume. Equation (1-3) shows that a mole of gas under definite conditions of temperature and pressure always occupies a definite volume, regardless of the nature of the gas. The gas law applies to a mixture of gases as well as to a pure gas. Thus a gram mole of ideal gas, or a mixture of ideal gases, occupies 22.41 liters at 0°C and 760 mm Hg. This volume is called the gram-molal volume. A pound mole of gas, either pure or a mixture, occupies 359 ft^3 at 32°F and 29.92 in. Hg. The pound-molal volume at any other combination of temperature and pressure may be calculated by Eq. (1-4).

The mass of a gas mixture that occupies the molal volume is called the *average molecular weight* of the gas mixture.

Partial Pressures. A useful quantity for dealing with the individual components in a gas mixture is partial pressure. The partial pressure of a component in a mixture, e.g., component A, is defined by the equation

$$\bar{p}_A = px_A \tag{1-5}$$

where $\bar{p}_A$ = partial pressure of component A in mixture

x_A = mole fraction of component A in mixture

p = total pressure on mixture

Note that this definition does not involve the concept of ideal gases in any way.

Each other component in the mixture has its own partial pressure. For example, for components B and C

$$\bar{p}_B = x_B p \qquad \bar{p}_C = x_C p$$

If all the partial pressures for a given mixture are added together, the result is

$$\bar{p}_A + \bar{p}_B + \bar{p}_C + \cdots = p(x_A + x_B + x_C + \cdots)$$

Since the sum of the mole fractions is unity,

$$\bar{p}_A + \bar{p}_B + \bar{p}_C + \cdots = p \tag{1-6}$$

All partial pressures in a given mixture add to the total pressure. This applies to mixtures of both ideal and nonideal gases.

Dalton's Law. A relationship concerning partial pressures that is restricted to ideal gases is Dalton's law. This states that in a mixture of ideal gases each component gas fills the entire volume of the mixture at the temperature of the mixture and at the partial pressure of that component.

Then, for component A, for example,

$$\bar{p}_A = \frac{x_A R_o T n}{V} \tag{1-7}$$

Equivalent equations may be written for all the other components.
A generally useful expression of the ideal gas laws is

$$\text{Volume per cent} = \text{pressure per cent} = \text{mole per cent} \tag{1-8}$$

For example, since air contains 79 per cent nitrogen and 21 per cent oxygen by volume, 1 ft^3 of air under a pressure of p atm may be considered to be a mixture of 0.21 ft^3 of oxygen and 0.79 ft^3 of nitrogen, both measured at p atm and the temperature of the mixture. It may also be considered to be 1 ft^3 of oxygen at 0.21 atm and 1 ft^3 of nitrogen at 0.79 atm, so that 21 per cent of the pressure is from oxygen and 79 per cent from nitrogen. Finally, 1 mole of air contains 0.21 mole of oxygen and 0.79 mole of nitrogen at all temperatures and pressures.

Example 1-3. A solvent-recovery system delivers a gas saturated with benzene vapor (C_6H_6) that analyzes on a benzene-free basis to 15 per cent carbon dioxide, 4 per cent oxygen, and 81 per cent nitrogen. This gas is at 70°F and 750 mm pressure. It is compressed to 5 atm and cooled to 70°F after compression. How many pounds of benzene are condensed by this process, per 1,000 ft^3 of the original mixture? The vapor pressure of benzene at 70°F is 75 mm Hg.

Solution. Since volume per cent equals pressure per cent, the volume of inert gas is $1,000(750 - 75)/750 = 900$ ft^3. This volume may be converted to moles as follows. The pound-molal volume at 750 mm and 70°F is, based on that for 760 mm and 32°F,

$$\frac{359(460 + 70)760}{(460 + 32)750} = 392 \text{ ft}^3/\text{lb mole}$$

The inert gas is, then, $900/392 = 2.30$ lb moles. The ratio of moles of benzene vapor to moles of inert gas before and after compression is

Before compression:

$$\frac{75}{750 - 75} = 0.1111$$

After compression:

$$\frac{75}{760 \times 5 - 75} = 0.0201$$

Therefore, per mole of inert gas, $0.1111 - 0.0201 = 0.0910$ mole of benzene is condensed. The molecular weight of benzene is 78, and the total condensation is, then,

$$0.091 \times 2.30 \times 78 = 16.3 \text{ lb}$$

Example 1-4. A mixture of 25 per cent ammonia gas and 75 per cent air is passed upward through a vertical scrubbing tower, to the top of which water is pumped. Scrubbed gas containing 0.5 per cent ammonia leaves the top of the tower, and an aqueous solution containing 10 per cent ammonia by weight leaves the bottom. Both entering and leaving gas streams are saturated with water vapor. The gas enters the tower at 100°F and leaves at 70°F. The pressure of both streams and throughout the tower is 15 lb force/in.^{2} gauge. The air-ammonia mixture enters the tower at a rate of 1,000 ft^3/min, measured as dry gas at 60°F and 1 atm. What percentage of the ammonia entering the tower is not absorbed by the water? How many gallons of water per minute are pumped to the top of the tower?

Solution. Because the problem concerns mixtures of gases, the use of Eq. (1-8) is indicated. Also, the temperatures and pressures vary, and the calculations are simplified if pound moles are used in the material balances. Again, since the air in the entering gas passes through the absorber unchanged in mass, the dry, ammonia-free air provides a convenient basis for calculating the moles of ammonia absorbed.

The molal volume at 60°F and 1 atm is 359(460 + 60)/492 = 379 ft^3/lb mole, and the total ammonia and air entering the tower is 1,000/379 = 2.64 lb moles/min. Gas analyses are always by volume and are automatically on the dry basis. By Eq. (1-8) the volume analyses can be used directly as molal analyses. The ammonia entering the absorber is, therefore, 0.25 × 2.64 = 0.660 mole/min, and the dry air is 0.75 × 2.64 = 1.98 moles/min. The molal ratio of ammonia to air in the effluent gas is 0.005:0.995, and the unabsorbed ammonia is 1.98 × 0.005/0.995 = 0.0099 mole/min. The fraction of the entering ammonia unabsorbed is

$$\frac{0.0099}{0.66} \, 100 = 1.5 \text{ per cent}$$

The vapor pressure of water at 100°F and at 70°F is 0.949 and 0.363 lb force/in.^{2}, respectively (see Appendix 6). The total pressure of the gas is 14.7 + 15.0 = 29.7 lb force/in.^{2} The partial pressure of water in the entering gas is 0.949 lb force/in.^{2}, and that of the dry gas is 29.7 − 0.949 = 28.75 lb force/in.^{2} Because a partial-pressure ratio equals a molal ratio, the water in the inlet gas is (0.949/28.75)2.64 = 0.0872 mole/min. Likewise, since the air and ammonia in the effluent gas are 1.98 + 0.0099 = 1.99 moles/min, the water in this stream is 1.99 × 0.363/(29.70 − 0.363) = 0.0246 mole/min. The water condensed from the gas stream is (0.0872 − 0.0246)18 × 60 = 67 lb/hr.

The ammonia liquor leaving the tower is 10 per cent by weight, and carries 0.660 − 0.0099 = 0.650 mole of ammonia per minute. The total water in the liquor is, therefore, 0.650 × 17 × 0.90/0.10 = 99.4 lb/min. Of this, 67 lb/hr, or 1.1 lb/min, is absorbed from the entering air, and 99.4 − 1.1 = 98.3 lb/min is pumped to the tower. The mass of 1 gal water is 8.33 lb, and the volume of water to the tower is 98.3/8.33 = 11.8 gal/min.

Note that the water condensed from the gas is approximately 1 per cent of the total and could be neglected in the calculation of the water consumption of the absorber.

Law of Motion. The fundamental law of mechanics, and one of the truly basic laws of engineering, is Newton's correlation of momentum and force, which may be expressed by the following proportionality:

$$F \propto \frac{d(mu)}{d\theta} \tag{1-9}$$

The quantity m is the mass of the body, u is its linear velocity, and F is the resultant of all forces acting on the body. The product mu is called the momentum of the body, and Newton's law of motion states that the resultant, or net, force acting on a body is measured quantitatively by the time rate of change of the momentum of that body.

For the important special case of constant mass, Eq. (1-9) can be written, after introduction of the proportionality factor k_n, as

$$F = k_n m \frac{du}{d\theta} = k_n ma \qquad (1\text{-}10)$$

where a is the acceleration, defined as $du/d\theta$.

Equation (1-10) may be interpreted qualitatively. It states that, when a body is at rest or in uniform motion, so the acceleration is zero, the resultant force on the body is also zero; conversely, when the forces on a body all cancel, the body is not accelerating and is either at rest or moving in a straight line at constant velocity.

Forces, accelerations, and momenta are vector quantities; i.e., they have both direction and magnitude. In Eqs. (1-9) and (1-10) the vectors involved must all be acting in the same direction.

Energy Balances. The law of conservation of energy expresses the same fact with regard to the energy input and output of a process or unit of equipment as does the law of conservation of mass for materials. To be valid an energy balance must include all types of energy that are involved in the process, whether these energies be heat, mechanical energy, electrical energy, radiant energy, chemical energy, or other forms.

Equilibria. Systems that are undergoing change spontaneously do so in a definite direction. If left to themselves, they eventually reach a state where apparently no further action takes place. Such a state is called an equilibrium state. For example, when a piece of hot iron is placed in contact with a piece of cold iron, the temperature of the hot iron falls, and that of the cold iron rises, until an equilibrium point is reached, when both pieces are at the same temperature. Again, a handful of salt placed in a beaker of water at a definite temperature dissolves until, if there is an excess of salt, the concentration of the salt in the solution reaches a definite value. Here, again, the process apparently stops when the equilibrium state is reached, and the solution is called a saturated solution. Such instances are examples of universal behavior. Equilibrium states represent end points of naturally occurring processes. An equilibrium cannot be shifted without making some change in the conditions governing the system.

The Phase Concept. In considering equilibria, the idea of phase is useful. A phase may be defined as a homogeneous substance viewed independently of its shape and size. Thus, raindrops, water in a tank, or water in a river are all the same from a phase point of view: all are water in liquid phase. A mixture of salt and saturated brine consists of two phases, one solid and one liquid, regardless of how much of either is present. A phase may be either a pure substance or a homogeneous solution of two or more substances. Liquid water, saturated steam, and ice may all exist together at the triple point of water, where the temperature is 0.010°C and the pressure

0.0060 atm. Here there are three phases: solid, liquid, and gas. The utility of the phase concept comes from the fact that, at equilibrium, the temperature, pressure, and concentrations are independent of the geometric shape and size of the phases.† For example, the temperature and pressure at the triple point of water do not change if the amounts of any or all of the three phases are changed, provided some of each phase is present.

The complete treatment of chemical and physical equilibria lies in the field of thermodynamics. The equilibrium relationships encountered in the study of the unit operations are simple. In heat transfer, uniformity of temperature in each phase and equality of temperature between phases exist at equilibrium; in processes with masses of fluids, a hydrostatic-pressure distribution (see page 37) represents equilibrium; and in processes having to do with changes in concentration, uniformity of concentration within each phase denotes equilibrium. When two or more phases are in equilibrium, all phases must be at the same temperature and pressure, but the concentrations in the individual phases usually differ. The important specific situations are considered at appropriate points in the text.

Process Kinetics. More important in practice than the end point of a process is the question of how rapidly a process not in equilibrium is moving toward its end point. Systems far from equilibrium but moving toward it slowly are, for practical purposes, at equilibrium. When processes are expected to function, it is important that the rates at which they take place be at least reasonably fast, because time is an all-important factor in practical operations. If the speed of the desired change is slow, large and expensive equipment is needed to produce a given amount of product, and investment costs are large. If the speed is great, small and cheaper equipment may be used. The study of rates of change in physical and chemical systems is called kinetics.

Kinetics may be divided into two parts, chemical and physical. The first is outside the scope of this book. Those parts of physical kinetics that are important in unit operations are developed at the appropriate points in the text. One general rule is, however, of sufficient importance to be discussed at this point.

The rate of most physical processes is proportional to a quantity called a driving force or driving potential. The nature of a driving potential depends on the kind of process. In heat transfer it is a temperature difference; in fluid flow through equipment it is a pressure difference; and in transfer of material through a phase it is a concentration difference.

Opposing a driving force is always a resistance, and the rate of the process at any instant is proportional to the driving force and inversely proportional to the resistance. This may be expressed mathematically by the equation

$$\frac{dG}{d\theta} = \frac{\Delta F}{R} \tag{1-11}$$

† If a phase is very finely divided, the curvatures and areas of its boundaries will affect its equilibrium properties. In this discussion, it is assumed that the phases are large enough to render this surface-energy effect unimportant.

where G = quantity being transferred, which may be a fluid or other matter, heat, or electricity

ΔF = driving force

θ = time

R = resistance

Three points are important in interpreting Eq. (1-11): (1) The equation is written as a differential equation because ordinarily ΔF changes with time, and the equation must then be integrated. (2) A proportionality constant has been absorbed into R, and Eq. (1-11) is, in a sense, a definition of R, as one can always divide a measured driving force by a measured rate and call the quotient a resistance; however, it is a fact that the resistance so defined often is *independent of* ΔF, and it is this fact that makes Eq. (1-11) useful. For example, in the flow of a steady current of electricity through a metallic conductor, Eq. (1-11) becomes the familiar Ohm's law

$$\frac{dQ_e}{d\theta} = I = \frac{\Delta E}{R_e}$$

where Q_e = quantity of electricity

I = current

ΔE = potential difference

R_e = resistance

Here R_e is independent of ΔE and of I and depends on the temperature and geometric size and shape of the metal. It is shown later that analogous relations hold for the transfer of heat and of matter. (3) An equilibrium state is a special case of Eq. (1-11), where ΔF is zero. Then the rate is also zero, and the process stops. The farther the process is from equilibrium, the faster it goes, and the rate at any time is measured by the distance from equilibrium the system is at that time.

UNITS AND DIMENSIONS

The fundamentals of the units and dimensions of physical quantities are important to the chemical engineer for several reasons. He must use with equal facility both scientific and technical units. Many of his basic data are in the metric system, but he puts the results of his calculations and measures most of his quantities in English units. He often must convert a quantity from one system to another or from one unit to an equivalent unit in the same system. He must think straight in force, weight, and mass. Many of his problems are simplified by use of a technique called dimensional analysis.

Physical Quantities. Any physical quantity consists of two parts: a unit, which tells what the quantity is and gives the standard by which it is measured, and a number, which tells how many units are needed to make up the quantity. For example, the simple statement that the distance between two points is 8 ft means all this: a definite length has been measured; to measure it a standard length, called the foot, has been chosen as a unit;

and eight 1-ft units, laid end to end, are needed to cover the distance. If an integral number of units is either too many or too few to cover a given distance, the standard unit is divided into fractions, so a measurement may be made to any degree of precision in terms of a decimal number of units. No physical quantity is completely defined until both the number and the unit are given, nor is any equation containing physical quantities correct unless it is correct both numerically and dimensionally.

Primary and Secondary Quantities. Physical quantities are divided into two groups. First, a comparatively small number of them are chosen; and second, the remainder is expressed in terms of them. The quantities of the first kind are called primary or fundamental quantities, and those of the second kind are called secondary or derived quantities. The choice of primary quantities is quite arbitrary, both in number and in type, and is based largely on convenience. A satisfactory minimum list of primary quantities for all engineering is length, mass, time, temperature, and quantity of electric charge. In unit operations, electric charge is not needed, and it is convenient, but not essential, to add force and heat to this list.

Dimensions and Dimensional Formulas. Each primary quantity may be represented by a letter, which symbolizes that quantity generally. This letter is called the dimension of the quantity it represents. Thus, let $\bar{L}$ represent the idea or concept of distance, without regard to any specific unit or numerical magnitude. The distance L differs from the dimension $\bar{L}$ in that L stands for an actual distance, the unit for which has been chosen and the magnitude determined. Similarly, let the dimension of mass m be $\bar{M}$; that of time θ be $\bar{\theta}$; that of force F be $\bar{F}$; of temperature t be $\bar{T}$; and of heat Q be $\bar{H}$.

The dimensions of a secondary quantity show how that quantity is constructed from primary quantities. Thus, any velocity, regardless of unit or of magnitude, is found by dividing a length by a time, and the dimensions of velocity are, then, $\bar{L}/\bar{\theta}$, or $\bar{L}\bar{\theta}^{-1}$. Likewise, the dimensions of acceleration are $\bar{L}/\bar{\theta}^2$, or $\bar{L}\bar{\theta}^{-2}$. The dimensions of pressure p (force divided by area) are $\bar{F}\bar{L}^{-2}$.

Dimensional Formulas. The dimensions of any quantity may be shown by the use of square brackets, thus

$$[L] = \bar{L} \qquad [p] = \bar{F}\bar{L}^{-2}$$

The second equation, for example, says, "The dimensions of pressure are force times length to the minus 2."

Once the primary quantities have been chosen, the dimensional formula for any secondary quantity can be found from its definition, just as those for velocity, acceleration, and pressure were written above. The general result for any quantity G may be written as

$$[G] = \bar{L}^{\alpha}\bar{M}^{\beta}\bar{\theta}^{\gamma}\bar{F}^{\delta}\bar{T}^{\epsilon}\bar{H}^{\zeta} \tag{1-12}$$

The exponents α, β, γ, δ, ϵ, and ζ are always positive or negative integers, small positive or negative integral fractions, or zero. Dimensional formulas for a number of secondary quantities are shown in Appendix 1, where values of the exponents are given for each quantity.

Force, Mass, and Weight. The units and dimensions of three common physical quantities, force, mass, and weight, are especially important technically. The engineer must deal with forces and pressures. Mass is the quantity conserved in chemical reactions, and material balances are in terms of mass. It is also the basic quantity in buying and selling. "Weight" is often used for both "force" and "mass." The relations existing among these quantities should be understood, and their units and dimensions must be watched closely in formulas and calculations.

Force, mass, and acceleration are correlated by Eq. (1-10). Several systems of units and dimensions have been defined and used for the quantities in this equation.[3] The various systems differ in the number and choice of the fundamental quantities and in the choice of standard units for the fundamental quantities. Two systems are especially important to the chemical engineer. One is the centimeter-gram-second (cgs) system, in which most scientific work is reported and in which most basic data are given. The other is the foot-pound-second (fps) system, in which engineering and industrial quantities are expressed. These fps units are also called English units. In the fps system, two units, mass and force, have the same name: the pound.

In the cgs system, only three fundamental mechanical quantities, mass, length, and time, are used. Force is then a derived unit. It is defined, first, by making the constant k_n in Eq. (1-10) dimensionless, and second, by choosing its numerical value as unity. Note that these choices are separate decisions. Then, to complete the system, mass is measured in grams, length in centimeters, and time in seconds.†

In the fps system, as used in this text, force is a fundamental quantity, along with mass, length, and time. Four, not three, mechanical units are used. The constant k_n then becomes a dimensional quantity. Also, its numerical value instead of being fixed at unity is given a special designation, $1/g_c$, and the numerical size of g_c is fixed arbitrarily at 32.174. Then in this system—which should be called the pound mass, foot, second, pound force system—Eq. (1-10) becomes

$$F = \frac{ma}{g_c} \tag{1-13}$$

In this system, the unit of mass is the avoirdupois pound, the unit of length the foot, the unit of time the second, and the unit of force the pound force. The dimensions of g_c are, from Eq. (1-13),

$$[g_c] = \left[\frac{ma}{F}\right] = \frac{\overline{M}\overline{L}}{\overline{F}\overline{\theta}^2} = \overline{M}\overline{L}\overline{F}^{-1}\overline{\theta}^{-2}$$

† Another unit system, closely related to the cgs system, is the mks system, in which mass is measured in kilograms, length in meters, and time in seconds. Constant k_n is dimensionless and equal to unity. This system has been accepted internationally for electrical science and technology.

The complete specification of the dimensional constant g_c is, then, 32.174 ft-lb mass/lb force-sec^2. It is called the Newton's-law conversion factor.

The number 32.174 is chosen because it equals the numerical value of the average acceleration of gravity at sea level, in feet per second per second. The force of gravity varies by 0.1 or 0.2 per cent from locality to locality. Within this precision, gravity exerts a pound force on a pound mass. It must be remembered, however, that numerical equality does not mean dimensional equality. The dimension of mass is $\bar{M}$, and that of force is $\bar{F}$. Also, the constant g_c is not an acceleration. This is obvious by comparing their dimensions. If g is the acceleration of gravity, the numerical value of g/g_c is practically unity, but its dimensions are $\bar{F}\bar{M}^{-1}$. A numerical value of 32.17 is used in ordinary calculations for both g and g_c.

To differentiate between the two pound units in this book the mass pound is denoted by lb without qualification, and the force pound is denoted by lb force. Pressures, for example, are in pounds force per square foot (lb force/ft^2) or in pounds force per square inch (lb force/in.^{2}).

As examples of the use of the cgs and fps units, consider two common forms of mechanical energy: the kinetic energy of translation of a body and the potential energy of a body because of its distance above the earth's surface or above any other arbitrary datum plane. Energies in the cgs system are in ergs and in the fps system in feet times pounds force. Let m be the mass of the body, Z its distance above the datum plane, u its velocity, and g the local acceleration of gravity. The kinetic and potential energies of mass m in the two systems of units are given in Table 1-2.

TABLE 1-2. POTENTIAL AND KINETIC ENERGIES

Energy	Cgs units	Fps units
Potential........	mgZ	mgZ/g_c
Kinetic.........	$mu^2/2$	$mu^2/2g_c$

In order to avoid erroneous omission of the constant g_c, it is advisable to derive equations in the fps system. Then this constant will appear correctly. Equations in the fps system may be easily converted to the cgs system by deleting g_c.

Weight. The term "weight" is used as a synonym for both mass pounds and gravitational forces. It has the same ambiguity as the term pound when the latter is not differentiated between mass and force. For careful work, the term weight should not be used without specifying its meaning.

Conversion of Units. To convert the magnitude given in one set of units to magnitude in another set conversion factors are needed. A conversion factor is the ratio of the size of a unit in one system to that of a corresponding unit in another. Since the two units must have the same dimensions, conversion factors are pure numbers without dimensions.

Because of the great variety of units used in engineering, the number of possible conversion factors is very large. It is convenient, therefore, to be able to calculate conversion factors from a relatively small number of fundamental factors, which have been established by arbitrary definition or by official action. An example of a defined conversion factor is that between temperature intervals on the centigrade and Fahrenheit scales, namely, $1°C/1°F = 1.8$. Examples of factors established by official action are the legal conversion factors in the United States for converting mass and length between English and cgs units, which are

$$\frac{1 \text{ yd}}{1 \text{ m}} = \frac{3,600}{3,937} \text{ †}$$

$$\frac{1 \text{ avdp lb}}{1 \text{ g}} = 453.5924277 \text{ ‡}$$

Obviously, official factors are often carried to precision far beyond that required in engineering practice. The factor for mass conversion is usually shortened to 453.6 g/lb. A list of basic conversion factors for use in this book is given in Appendix 2.

To use conversion factors, the quantity to be converted is written in terms of its dimensions, and each fundamental unit is converted to the desired unit by direct algebraic substitution in the proper factor.

Example 1-5. The dimensional formula for a heat-transfer coefficient h is

$$[h] = \bar{H}\bar{L}^{-2}\bar{\theta}^{-1}\bar{T}^{-1}$$

In experimental work on heat transfer, a coefficient

$$h = 450 \text{ Btu}/(\text{ft}^2)(\text{hr})(°F)$$

was obtained. What is the conversion factor for converting this unit to cal/(cm²)(min)(°C), and what is the magnitude of the observed coefficient in these metric units?

Solution. The basic conversion factors required are

$$\frac{1 \text{ hr}}{1 \text{ min}} = 60$$

From Appendix 2,

$$\frac{1 \text{ ft}}{1 \text{ cm}} = 30.48 \qquad \frac{1 \text{ Btu}}{1 \text{ cal}} = 252 \qquad \frac{1°C}{1°F} = 1.8$$

Then the required conversion factor is

$$\frac{1 \text{ Btu}}{(\text{ft}^2)(\text{hr})(°F)} = \frac{252 \text{ cal}}{(30.48 \text{ cm})^2(60 \text{ min})(°C/1.8)}$$

$$= 0.00814 \text{ cal}/(\text{cm}^2)(\text{min})(°C)$$

† Act of Congress, 1866.
‡ *Natl. Bur. Standards (U.S.) Circ.* **47.**

The observed coefficient is

$$\frac{450 \text{ Btu}}{(\text{ft}^2)(\text{hr})(°\text{F})} = \frac{0.00814 \times 450 \text{ cal}}{(\text{cm}^2)(\text{min})(°\text{C})}$$

$$= 3.66 \text{ cal}/(\text{cm}^2)(\text{min})(°\text{C})$$

Dimensionless Equations and Consistent Units. Equations that have been derived mathematically from basic laws consist of terms that have the same dimensions, because these same basic laws are used to define the secondary quantities. Equations in which all terms have the same dimensions are called *dimensionless* or *dimensionally homogeneous* equations. If such an equation is divided by one of its terms, all terms in the resulting equation are dimensionless.

A dimensionally homogeneous equation can be used as it stands with any set of units provided that the same units for the fundamental quantities are used throughout. Units meeting this requirement are called consistent units. No conversion factors are needed when consistent units are used.

For example, consider the ordinary equation for the vertical distance traversed by a freely falling body during time θ when the initial velocity is u_0:

$$Z = u_0\theta + \frac{g\theta^2}{2} \tag{1-14}$$

If the dimensional formulas of the terms in this equation are inspected, it is seen that the dimension of each term is $\overline{L}$. If the equation is divided by Z,

$$1 = \frac{u_0\theta}{Z} + \frac{g\theta^2}{2Z} \tag{1-15}$$

A check of the dimensions of each term of Eq. (1-15) shows that the dimensions of each term cancel and each term is dimensionless. A combination of variables such that all dimensions cancel in this manner is called a *dimensionless group*. The numerical value of a dimensionless group for given values of the quantities contained in it is independent of the units used, provided they are consistent. Both terms on the right-hand side of Eq. (1-15) are dimensionless groups.

If consistent units are used for all terms in Eqs. (1-14) and (1-15), either cgs or fps units may be used, and no factors converting units in one system to those in another need be used. In this text all equations are dimensionless *unless otherwise noted*. For convenience, however, units in the pound force, foot, second (or hour), pound mass, degree Fahrenheit and Btu system are given in nomenclature tables.

Dimensional Equations. Equations derived by empirical methods, in which experimental results are correlated by empirical equations without regard to dimensional consistency, usually are not dimensionally homogeneous, and contain terms of various dimensions. Equations of this type are dimensional equations, or dimensionally nonhomogeneous equations. In these equations there is no advantage in using consistent units, and two or more length units, such as inches and feet, or two or more time units, like seconds or minutes, may appear in the same equation. For example, a formula for the rate of heat loss from a horizontal pipe to the atmosphere by conduction and convection is

$$\frac{q}{A} = 0.50 \frac{\Delta t^{1.25}}{(D_o')^{0.25}} \tag{1-16}$$

where q = rate of loss of heat, Btu/hr

A = area of pipe surface, ft^2

Δt = excess of temperature of pipe wall over that of ambient (surrounding atmosphere), °F

D_o' = outside diameter of pipe, in.

Obviously, the dimensions of q/A are not those of the right-hand side of Eq. (1-16), and the equation is dimensional. Quantities substituted in Eq. (1-16) must be expressed in the units given, or the equation will give a wrong answer. If other units are to be used, the coefficient must be changed. If it is desired to express Δt in degrees centigrade, for example, the numerical coefficient must be changed to $0.50 \times 1.8^{1.25} = 1.04$, since there are 1.8 Fahrenheit degrees in one centigrade degree.

TOTAL-ENERGY EQUATION OF STEADY FLOW

The most important processes in unit operations are flow processes, in which materials flow into, through, and out of, pieces of equipment. Useful general equations, based on the laws of conservation of mass and energy, have been derived for flow systems.[2,7] The following equations are restricted to the class of processes characterized by steady flow.

Steady-flow Process. In a steady-flow process, the flow rates and the properties of the flowing materials, such as temperature, pressure, composition, density, and velocity, at each point in the apparatus, including all entrance and exit ports, are constant with time. These quantities can, and usually do, vary from point to point in the system, but at any one location they do not change. Because of this constancy of local conditions, there is no accumulation or depletion of either mass or energy within the apparatus, and all material and energy balances are of the simple type

Input = output

Energy Balance for Single-stream Process. As an example of steady-flow process in which a single stream of material is treated, consider the process shown in Fig. 1-1. The equipment is any device through which the material is passing. Assume the material is flowing through the system at a constant mass rate. Consider the flow of m lb of material. The entering stream has a velocity of u_a ft/sec, and is Z_a ft above the horizontal datum above which heights are measured. Its enthalpy (a quantity discussed later) is i_a Btu/lb. The corresponding quantities for the leaving stream are u_b, Z_b, and i_b. Heat in the amount of Q Btu is being transferred through the boundaries of the equipment to the material flowing through it during the time m lb of fluid enters the equipment. If the equipment includes a turbine or engine, it may do work, usually by means of a turning shaft, on the outside. If the unit includes a pump, work from the outside must be done on the material, again through the agency of a turning shaft. Work effects of this kind are called *shaft work*. Assume that shaft work equal to W_s ft-lb force is being done on the outside by the equipment. For

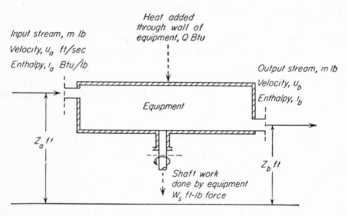

Fig. 1-1. Diagram for steady-flow process.

this process, the following equation, which is derived in standard texts on technical thermodynamics, applies.[2,7]

$$m\left[\frac{u_b^2 - u_a^2}{2g_cJ} + \frac{g(Z_b - Z_a)}{g_cJ} + i_b - i_a\right] = Q - \frac{W_s}{J} \qquad (1\text{-}17)$$

where J is the mechanical equivalent of heat, in feet times pounds force per Btu, and g and g_c have their usual meanings.

Discussion of Eq. (1-17). Several limitations and interpretations of Eq. (1-17) must be understood.

1. Changes in electrical, magnetic, surface, and mechanical-stress energies are not taken into account. Except in rare situations, these are absent or unimportant.

2. To apply Eq. (1-17) to a specific situation, a precise choice of the boundaries of the equipment must be made. The inlet and outlet streams must be identified, the inlet and outlet ports located, and rotating shafts noted. All heat-transfer areas between the equipment and its surroundings must be located. The boundaries of the equipment and the cross sections of all shafts and inlet and outlet ports form the *control surface*. This must be a closed envelope, without gaps. Equation (1-17) really applies to the equipment and material inside the control surface. For ex-

ample, the control surface of the process of Fig. 1-1 is bounded by the walls of the equipment and the cross sections of the shaft and inlet and outlet ports, as shown by the dotted lines.

3. The constant J is a universal constant, the value of which depends only on the units chosen for heat and work. The precise value of J is 778.26 ft-lb force/Btu. For ordinary calculations, a value 778 ft-lb force/Btu is used.

4. The heat effect Q is, by convention, positive when heat flows from the outside of the control surface into the equipment and negative when heat flows in the opposite direction. The shaft work W_s is taken as positive when the work is done on the outside of the control surface by the equipment and is negative when the work is supplied to the equipment from outside the control surface. Thus work required by a pump located within the control surface is negative.

Both Q and W_s are net effects. If there is more than one heat flow or shaft work, the individual values are added algebraically, and the net values of Q and W_s used in Eq. (1-17).

5. No term appears in Eq. (1-17) for friction. Friction is an internal transformation of mechanical energy into heat and occurs inside the control surface. Its effects are included in the other terms of the equation.

Enthalpy. The quantities i_a and i_b in Eq. (1-17), the enthalpies of the inlet and outlet streams, respectively, are physical properties of the material. The enthalpy of a unit mass of a pure substance is a function of pressure and temperature. Tables and diagrams provide numerical values of this property at various temperatures and pressures. These are given in texts and handbooks. Enthalpies of liquid water and saturated steam are given in the steam table, a shortened form of which is given in Appendix 6.

Absolute enthalpies are not obtainable, and numerical values of this property for a given substance are based on an arbitrarily defined datum, or standard state, for that substance. The method is analogous to that of specifying heights above sea level. The datum chosen in the steam table, for example, is liquid water at 32°F and in equilibrium with its own vapor at that temperature. Steam-table enthalpies for liquid water, saturated steam, and superheated steam at other temperatures and pressures are excess values over that of water at the datum condition. An independent choice of datum must be made for each substance for which numerical enthalpies are desired. Because all enthalpies are relative to an arbitrary datum, only enthalpy differences have physical significance.

The enthalpy change accompanying the vaporization or condensation of a pure substance at constant pressure (and therefore at constant temperature) is the ordinary latent heat of vaporization λ. Then,

$$i_y - i_x = \lambda \tag{1-18}$$

where i_y and i_x are the enthalpies of vapor and liquid, respectively. Latent heats are also given in tables of properties of substances.

Although enthalpy is, in general, a function of both temperature and pressure, two special cases are met where the pressure effect may be neglected: (1) The enthalpy of an ideal gas is independent of pressure, and the pressure effect can be ignored when the ideal-gas law is used; (2) the enthalpies of liquids and solids are not greatly affected by pressure, and under ordinary conditions, unless pressures of several atmospheres are involved, the effect of pressure on the enthalpy of a liquid or solid may be neglected.

At constant pressure, or under conditions where the effect of pressure on enthalpy may be neglected, the enthalpy difference over the temperature range from t_a to t_b is given by the equation

$$i_b - i_a = \int_{t_a}^{t_b} c_p \, dt = \bar{c}_p (t_b - t_a) \tag{1-19}$$

where c_p = specific heat (at constant pressure), Btu/(lb)(°F)

$\bar{c}_p$ = mean specific heat over the temperature range t_a to t_b

Equation (1-19) cannot be used if a phase change occurs in the temperature range covered by the equation. For temperature ranges of less than 100 to 200°F it is satisfactory to use a constant value of c_p in place of $\bar{c}_p$ and to choose the value of c_p at a temperature midway between t_a and t_b.

Flow Equation for Several Streams. Equation (1-17) can be generalized for use in treating steady-flow processes where there are several streams entering and leaving the equipment. For each stream calculate the quantity

$$E = m \left(\frac{u^2}{2g_cJ} + \frac{gZ}{g_cJ} + i \right) \tag{1-20}$$

Put the quantity E for each leaving stream on the left-hand side of Eq. (1-17), using a plus sign, and put the quantity E for each entering stream on the same side of the equation with a minus sign; then equate this algebraic sum of the E's to $Q - W_s/J$. This may all be expressed mathematically by the equation

$$\Sigma E = \Sigma m \left(\frac{u^2}{2g_cJ} + \frac{gZ}{g_cJ} + i \right) = Q - \frac{W_s}{J} \tag{1-21}$$

where the operator Σ means to add algebraically all the E values, using positive values for leaving streams and negative values for entering streams. The over-all mass balance written this way is

$$\Sigma m = 0 \tag{1-22}$$

Since the process is one of steady flow, Eq. (1-22) must hold as well as Eq. (1-21).

It is seldom that all the terms for all streams appear in a specific problem. Also, Q is zero if the process is adiabatic, and W_s is zero if no shaft work is done. The most common special case in unit operations is that where the kinetic energies $u^2/2g_cJ$ and potential energies gZ/g_cJ are all negligible in comparison with Q and W_s is zero. Then Eq. (1-21) becomes simply

$$\Sigma mi = Q \tag{1-23}$$

Equation (1-23) is the simple "heat balance."

Example 1-6. Air is flowing steadily through a horizontal heated tube. The air enters at 40°F and at a velocity of 50 ft/sec. It leaves the tube at 140°F and 75 ft/sec. The average specific heat of air, from Appendix 12, is 0.24 Btu/(lb)(°F). How many Btu per pound of air are transferred through the wall of the tube?

Solution. The quantities for use in Eq. (1-17) are

$$Z_a = Z_b \qquad W_s = 0 \qquad u_a = 50 \text{ ft/sec} \qquad u_b = 75 \text{ ft/sec}$$

$$i_b - i_a = 0.24(140 - 40) = 24 \text{ Btu/lb}$$

Taking a basis of 1 lb of air ($m = 1.0$), Eq. (1-17) gives

$$\frac{75^2 - 50^2}{2 \times 32.2 \times 778} + 24 = Q$$

$$Q = 24 + 0.06 = 24.1 \text{ Btu/lb}$$

It is clear that the effect of the kinetic-energy terms is negligible at these velocities when gas is heated. In high-speed flow where gas velocities approach the speed of sound, the kinetic-energy change is important.

DIMENSIONAL ANALYSIS

Many important problems in chemical engineering cannot be solved completely by theoretical or mathematical methods. Problems of this type are especially common in fluid-flow, heat-flow, and diffusional operations. One method of attacking a problem for which no mathematical equation can be derived is that of empirical experimentation. For example, the pressure loss from friction in a long, round, straight, smooth pipe depends on all these variables: the length and diameter of the pipe, the flow rate of the liquid, and the density and viscosity of the liquid. If any one of these variables is changed, the pressure drop also changes. The empirical method of obtaining an equation relating these factors to pressure drop requires that the effect of each separate variable be determined in turn by systematically varying that variable while keeping all others constant. The procedure is laborious, and it is difficult to organize or correlate the results so obtained into a useful relationship for calculations.

There exists a method intermediate between formal mathematical development and a completely empirical study.[1,3] It is based on the fact that, if a theoretical equation does exist among the variables affecting a physical process, that equation must be dimensionally homogeneous. Because of this requirement it is possible to group many factors into a smaller number of dimensionless groups of variables. The numerical values of these groups, in any given situation, are independent of the dimension system used, and the groups themselves rather than the separate factors appear in the final equation.

Dimensional analysis does not yield a numerical equation, and experiment is required to complete the solution of the problem. The result of a dimensional analysis is valuable in guiding experiments and is useful in pointing a way to correlations of experimental data suitable for engineering use.

A dimensional analysis cannot be made unless enough is known about the physics of the situation to decide what variables are important in the problem and what basic physical laws would be involved in a mathematical solution if one were possible. The basic laws are important because such laws introduce dimensional constants that must be considered along with the list of variables. In the applications of dimensional analysis considered in this book two such constants may appear. One is g_c, which must be introduced whenever Newton's law [Eq. (1-13)] is involved; the other is J, the mechanical equivalent of heat, which must be used when the heat originating in the conversion of mechanical energy to heat by friction is important to the problem. The decision as to the dimensional factors and variables that enter the problem is the definitive step in a dimensional analysis. The method is illustrated by the following example.

Example 1-7. A steady stream of liquid is heated by passing it through a long, straight, heated pipe. The temperature of the wall of the pipe is assumed to be greater by a constant amount than the average temperature of the liquid. The conversion of mechanical energy into heat by friction is negligible in comparison with the heat transferred to the liquid through the wall of the pipe. It is desired to find a relationship that can be used to predict the rate of heat transfer from the wall to the liquid, in Btu per square foot of tube area in contact with the liquid per hour.

Solution. The mechanism of this process is discussed in Chap. 8. From the known characteristics of the process it may be expected that the rate of heat

TABLE 1-3. QUANTITIES AND DIMENSIONAL FORMULAS FOR EXAMPLE 1-7

Quantity	Symbol	Dimensions
Heat flow per unit area	q/A	$H L^{-2} \bar\theta^{-1}$
Diameter of pipe (inside)	D	L
Average velocity of liquid	$\bar V$	$L\bar\theta^{-1}$
Density of liquid	ρ	$M L^{-3}$
Viscosity of liquid	μ_F	$F L^{-2}\bar\theta$
Specific heat, at constant pressure, of liquid	c_p	$H M^{-1} \bar T^{-1}$
Thermal conductivity of liquid	k	$H L^{-1}\bar\theta^{-1}\bar T^{-1}$
Newton's-law conversion factor	g_c	$M L F^{-1}\bar\theta^{-2}$
Temperature difference between wall and fluid	Δt	$\bar T$

transfer per unit area q/A depends on a number of quantities, which are listed with their dimensional formulas in Table 1-3.

It is known that viscosity forces and forces needed to accelerate and decelerate eddies are involved; so Newton's law enters the situation. The dimensional constant g_c is therefore included in the list of factors. Since conversion of mechanical energy to heat is negligible, J is not needed.

If a theoretical equation for this problem exists, it can be written in the following general form

$$\frac{q}{A} = \psi(D,\overline{V},\rho,\mu_F,g_c,c_p,k,\Delta t) \tag{1-24}$$

where ψ means "function of." The form of this function is completely unknown.

If Eq. (1-24) is a relationship derivable from basic laws, all terms in the function ψ must have the same dimensions as those of the left-hand side of the equation, q/A. Then any term in the function must conform to the dimensional formula

$$\left[\frac{q}{A}\right] = [D]^a[\overline{V}]^b[\rho]^c[\mu_F]^d[g_c]^e[c_p]^f[k]^g[\Delta t]^h \tag{1-25}$$

Substituting the dimensions from Appendix 1 gives

$$\overline{H}\overline{L}^{-2}\overline{\theta}^{-1}$$

$$= \overline{L}^a\overline{L}^b\overline{\theta}^{-b}\overline{M}^c\overline{L}^{-3c}\overline{F}^d\overline{\theta}^d\overline{L}^{-2d}\overline{M}^e\overline{L}^e\overline{F}^{-e}\overline{\theta}^{-2e}\overline{H}^f\overline{M}^{-f}\overline{T}^{-f}\overline{H}^g\overline{L}^{-g}\overline{\theta}^{-g}\overline{T}^{-g}\overline{T}^h \tag{1-26}$$

Since Eq. (1-24) is assumed to be dimensionally homogeneous, the exponents of the individual primary units on the left-hand side of Eq. (1-26) must equal those on the right-hand side. This gives the following set of equations:

Exponents of $\overline{H}$: $\qquad 1 = f + g$ $\hfill$ (1-27)

Exponents of $\overline{L}$: $\qquad -2 = a + b - 3c - 2d + e - g$ $\hfill$ (1-28)

Exponents of $\overline{\theta}$: $\qquad -1 = -b + d - 2e - g$ $\hfill$ (1-29)

Exponents of $\overline{M}$: $\qquad 0 = c + e - f$ $\hfill$ (1-30)

Exponents of $\overline{F}$: $\qquad 0 = d - e$ $\hfill$ (1-31)

Exponents of $\overline{T}$: $\qquad 0 = -f - g + h$ $\hfill$ (1-32)

Here there are eight unknowns but only six equations. Six of the unknowns may be found in terms of the remaining two. Arbitrarily, two letters must be retained. The final result is equally valid for all choices, but for this problem it is customary to retain the exponents for the velocity $\overline{V}$ and the specific heat c_p. The letters b and f will be retained, and the remaining six eliminated. One method of doing this is as follows: From Eq. (1-27),

$$g = 1 - f, \tag{1-33}$$

From Eqs. (1-32) and (1-33),

$$h = f + g = 1 \tag{1-34}$$

From Eq. (1-31),

$$d = e \tag{1-35}$$

From Eqs. (1-29) and (1-35),

$$2e - d = 1 - b - g = e = d \tag{1-36}$$

From Eqs. (1-33) and (1-36),

$$d = e = 1 - b - 1 + f = f - b \tag{1-37}$$

From Eqs. (1-30) and (1-37),

$$c = f - e = f - f + b = b \tag{1-38}$$

From Eqs. (1-28), (1-33), (1-37), and (1-38),

$$
\begin{aligned}
a &= -2 - b + 3c + 2d - e + g \\
&= -2 - b + 3b + 2f - 2b - f + b + 1 - f \\
&= b - 1
\end{aligned}
\tag{1-39}
$$

Equation (1-25) becomes, by substituting values from Eqs. (1-33) to (1-39) for letters a, c, d, e, g, and h,

$$\left[\frac{q}{A} \right] = [D]^{b-1}[\overline{V}]^b[\rho]^b[\mu_F]^{f-b}[g_c]^{f-b}[c_p]^f[k]^{1-f}[\Delta t]$$

By collecting all factors having integral exponents in one group, all factors having exponents b into another group, and those having exponents f into a third,

$$\left[\frac{qD}{Ak\,\Delta t} \right] = \left[\frac{D\overline{V}\rho}{\mu_F g_c} \right]^b \left[\frac{c_p \mu_F g_c}{k} \right]^f \tag{1-40}$$

The dimensions of each of the three bracketed groups in Eq. (1-40) are zero, and all groups are dimensionless. Any function whatever of these three groups will be dimensionally homogeneous, and the equation will be a dimensionless one. Let such a function be

$$\frac{qD}{Ak\,\Delta t} = \Phi\left(\frac{D\overline{V}\rho}{\mu_F g_c}, \frac{c_p \mu_F g_c}{k} \right) \tag{1-41}$$

or

$$\frac{q}{A} = \frac{k\,\Delta t}{D}\,\Phi\left(\frac{D\overline{V}\rho}{\mu_F g_c}, \frac{c_p \mu_F g_c}{k} \right) \tag{1-42}$$

The relationship given in Eqs. (1-41) and (1-42) is the final result of the dimensional analysis. The form of function Φ must be found experimentally, by determining the effects of the groups in the brackets on the value of the group on the left-hand side of Eq. (1-41). The correlations that have been found for this are given in Chap. 8.

It is usual, in Eqs. (1-41) and (1-42), to use the so-called absolute viscosity μ in place of its equal $\mu_F g_c$. Viscosity is further discussed in Chap. 2.

It is clear that to correlate experimental values of the three groups of variables of Eq. (1-41) is simpler than to attempt to correlate the effects of each of the individual factors of Eq. (1-24).

Formation of Other Dimensionless Groups. If a pair of letters other than b and f is selected for retention, three dimensionless groups are again obtained, but one or more differ from the groups of Eq. (1-41). For example, if b and g are kept, the result is

$$\frac{q}{A\overline{V}\rho c_p\,\Delta t} = \Phi_1\left(\frac{D\overline{V}\rho}{\mu}, \frac{c_p \mu}{k} \right) \tag{1-43}$$

Other combinations may be found. However, it is unnecessary to repeat the algebra to obtain such additional groups. The three groups in Eq.

(1-41) may be combined in any desired manner, by multiplying and dividing them, or reciprocals or multiples of them, together. It is necessary only that each original group be used at least once in finding new groups and that the final assembly contain exactly three groups. For example, Eq. (1-43) is obtained from Eq. (1-41) by multiplying both sides by $(\mu/D V \rho)(k/\mu c_p)$

$$\frac{qD}{Ak\,\Delta t}\frac{\mu}{D\overline{V}\rho}\frac{k}{\mu c_p} = \left[\Phi\left(\frac{D\overline{V}\rho}{\mu},\frac{c_p\mu}{k}\right)\right]\frac{\mu}{D\overline{V}\rho}\frac{k}{\mu c_p} = \Phi_1\left(\frac{D\overline{V}\rho}{\mu},\frac{c_p\mu}{k}\right)$$

and Eq. (1-43) follows. Note that function Φ_1 is not equal to function Φ. In this way any dimensionless equation may be changed into any number of new ones. This is often useful when it is desired to isolate a single factor in one group. Thus, in Eq. (1-41) c_p appears in only one group, and in Eq. (1-43) k is found in only one. It is shown in Chap. 8 that Eq. (1-43) is more useful for some purposes than Eq. (1-41).

"Named" Dimensionless Groups. A number of dimensionless groups of the type given in Eqs. (1-41) and (1-43) have been found by various methods, including dimensional analysis, and are important enough to justify names and symbols. A list of the more important ones is given in Appendix 3. In view of the ease with which dimensionless groups are made, there must be better reasons for naming a group than the fact that it popped out of a dimensional analysis. Actually, the more important groups have definite physical significance as ratios of certain kinds of forces or energy effects.[4,5] This interpretation of dimensionless groups is beyond the scope of this book.

The numerical value of a dimensionless group for a given case is independent of the units chosen for the primary quantities provided consistent units are used within the group. The units chosen for one group need not be consistent with those for another. For example, it is quite customary to choose the second as the unit of time in group $D\overline{V}\rho/\mu$, and the hour as the unit in group $c_p\mu/k$.

One word of warning must be given. No equation derived by dimensional analysis should be accepted without question until experiment shows that the variables used in the original list are all necessary and that no essential variables have been omitted.

USEFUL MATHEMATICAL METHODS

Two simple mathematical devices are especially useful in treating the subject matter of this text. The first is graphical integration, and the second is logarithmic plotting.

Graphical Integration. By definition, the value of the definite integral

$$\int_{x_a}^{x_b} y\,dx$$

may be represented by the area bounded by the curve of x vs. y, the ordinates $x = x_a$ and $x = x_b$, and the x axis. The formal method of evaluating this integral is to express y as a function of x, substitute this function for y in the integral, and integrate by using an integral table. This is the preferred method when it can be easily used, but there are two situations when it is impractical. The first is where the relationship between y and x is known only as a table of corresponding values of x and y, and the second is where the functional relationship between x and y is so complicated that formal integration cannot be accomplished. One method of treating the first situation is to fit an empirical equation to the data and use the equation in a formal integration. Here, however, fitting the equation may be difficult, and when found, the equation may still be unintegrable. Both cases may be handled graphically. It is necessary only to plot y vs. x over the range of x values between x_a and x_b and to determine the area under the curve.

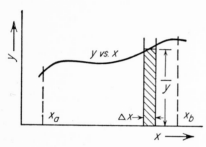

FIG. 1-2. Principle of graphical integration.

Using a planimeter, which is a mechanical device for measuring areas, is the preferred method for doing this, but if this rather expensive device is not available, a reasonably precise value for the integral may be obtained by the following technique.

Divide the axis of the abscissa between $x = x_a$ and $x = x_b$ into a series of short segments. Draw ordinates at the points defining the segments, and extend these ordinates to just beyond the curve. Then, by eye, draw a short horizontal line for each segment, so adjusting the height of the line that the little triangular area below the curve and above the line equals the area of the triangle above the curve and below the line. Figure 1-2 shows two such ordinates and the short line drawn at the top. If the short line has been accurately located, the area of the rectangle formed by the x axis, the two ordinates, and the line at the top equals the area under the curve between the ordinates. The crosshatched rectangle in Fig. 1-2 is an area balance of this kind. The area of a single rectangle is $\bar{y} \, \Delta x$, where $\bar{y}$ is the height of the rectangle and Δx is the segment of the x axis covered by the same rectangle. By adding the areas of all rectangles between $x = x_a$ and $x = x_b$, the total area is found, and this is the value of the integral.

To evaluate the integral of a complicated function over a definite range of the independent variable x, the principle is to plot the function as the ordinate against x as the abscissa and to determine the area under the curve between the upper and lower limits of the independent variable in the manner given above. This may be done no matter how complicated the function, as long as it can be plotted.

Example 1-8. Table 1-4 shows the instantaneous rate of flow of crude oil through a pipeline. The flow rate is measured by an indicating flowmeter and is given in thousands of pounds per hour. Readings are given at 12-min intervals.

(a) How much oil flows through the line in the 2-hr period covered by the data? (b) What is the average rate of flow, in pounds per hour, during the period? (c) Plot a curve showing the cumulative flow at each time between noon and 2 P.M.

TABLE 1-4. OIL FLOW RATES FOR EXAMPLE 1-8

Time	Flow rate, 1,000 lb/hr	Time	Flow rate, 1,000 lb/hr
12:00 noon	6.1	1:12 P.M.	5.7
12:12	4.7	1:24	6.8
12:24	4.0	1:36	8.0
12:36	4.1	1:48	9.3
12:48	4.4	2:00	10.5
1:00	5.0		

Solution. If w is the instantaneous flow rate in pounds per hour at time θ hr from the start of the observation period, then during the next very short time $d\theta$ the flow is $w\,d\theta$ in pounds. Call this flow dm. Then $dm = w\,d\theta$, and the cumulative flow up to time θ is

$$m = \int_0^\theta w\,d\theta$$

The total flow for the entire time period θ_T hr is m_T, and this is found by integrating over the entire time, or

$$m_T = \int_0^{\theta_T} w\,d\theta$$

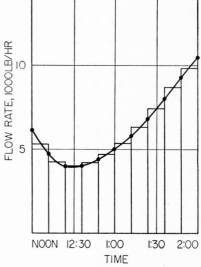

FIG. 1-3. Graphical integration for Example 1-8.

Solution of the problem calls for a graphical integration of flow rate over a time period of 2 hr, from noon to 2 P.M.

The first step is to plot the data of instantaneous flow rate vs. time given in Table 1-4. The flow rates as given are values of $w/1,000$, and the time intervals from noon on provide values of θ. These points are plotted in Fig. 1-3. A smooth curve is then drawn through the points. In drawing this curve, the assumption is made that changes in flow rates between the actual points of observation were continuous and that there were no surges or other discontinuities.

The next step is to choose the increments Δx. These should be sufficiently small to ensure that the changes in curvature of the y vs. x curve between ordinates are not so severe as to prevent an accurate location of the area-equalizing lines. An increment of 0.2 hr is satisfactory in this problem. Then ordinates are drawn for each 0.2 hr as shown in Fig. 1-3, and the short lines at the tops of the rectangles are placed as shown. The remaining calculation is best done tabularly, as shown in Table 1-5. In this table, column 1 is the time; column 2 is the increment of time for each individual segment (0.2 hr in this problem); column 3 gives values of $\bar{y}$,

TABLE 1-5. CALCULATIONS FOR EXAMPLE 1-8

Time	Δx	$\bar{y}$	$\bar{y}\,\Delta x$	$\bar{y}\,\Delta x$
(1)	(2)	(3)	(4)	(5)
Noon	....	...	....	0
12:12	0.20	5.3	1.06	1.06
12:24	0.20	4.2	0.84	1.90
12:36	0.20	4.0	0.80	2.70
12:48	0.20	4.2	0.84	3.54
1:00	0.20	4.7	0.94	4.48
1:12	0.20	5.3	1.06	5.54
1:24	0.20	6.3	1.26	6.80
1:36	0.20	7.4	1.48	8.28
1:48	0.20	8.7	1.74	10.02
2:00	0.20	9.8	1.96	11.98

the height of the rectangles, read after the horizontal lines have been located; column 4 is the area of the individual rectangles $\bar{y}\,\Delta x$; and column 5 is the cumula-

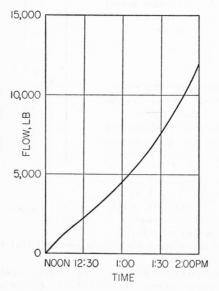

Fig. 1-4. Integral curve for Example 1-8.

tive sum of the areas up to time θ, obtained by adding all the incremental areas in column 5 up to that time. The quantities in column 5 are, then, the total flow in thousands of pounds, at the corresponding times given in column 1. The total at 2 P.M. is 11.98, so the total flow during the test period is 11,980 lb, or 12,000 lb. This is the answer to question (a). The average flow rate during the period is 12,000/2 = 6,000 lb/hr, and this is the answer to part (b). A plot of the data in column 5 against time in column 1 is given in Fig. 1-4. This curve is the relationship called for in part (c). Mathematically, Fig. 1-4 is an example of an *integral curve*, as it is the value of a definite integral shown as a function of a variable upper limit.

Logarithmic Plots. Ordinarily, when a relationship between two variables is shown graphically, one variable is plotted as the ordinate vs. the other as the abscissa, using rectangular, or cartesian, coordinates. In many situations, a plot more convenient than that of y vs. x on rectangular coordinates is one of $\log y$ vs. $\log x$. Such a plot, called a log-log plot, has two advantages over rectangular plots in some circumstances: (1) an equation of the form

$$y = ax^n \qquad (1\text{-}44)$$

where a and n are constants, gives a straight line on log-log coordinates; (2) when both variables cover a very wide range of magnitude, a single plot may be used for the entire range.

Taking these advantages in order, the first use of log-log plots is based on the fact that, if logarithms are taken of Eq. (1-44),

$$\log y = n \log x + \log a \qquad (1\text{-}45)$$

Then, letting $\log y = y'$, $\log x = x'$, and $\log a = a'$,

$$y' = nx' + a' \qquad (1\text{-}46)$$

Equation (1-46) is, of course, that of a straight line of slope n and an intercept a where $x' = 0$ $(x = 1)$. Equations of the form of Eq. (1-44) are common, especially in correlations of dimensionless groups.

To illustrate the second use of log-log coordinates, assume that a variable has an experimental range of 10^{-3} to 10^3. Plotting this variable on a single rectangular plot is not possible without either losing the lower end or requiring an enormous plot for the larger end of the range. On log-log coordinates, however, the plot for this variable would cover only a range equivalent in actual distance from -3 to $+3$, and the entire range can be easily accommodated in a single plot. Many plots covering several logarithmic cycles are used in this text.

To plot log-log relationships easily, special plotting paper called log-log paper is available and is convenient to use. The scales on this paper are logarithmic, just like those of the A, B, C, and D scales of a slide rule. The numbers given on log-log paper, like those engraved on the scales of a slide rule, are actually laid out on a linear scale of their logarithms.

Plotting data on log-log paper automatically gives a straight line if the data follow the exponential law given by Eq. (1-44). The actual geometric slope of the straight line is the value of the exponent n. The value of a can be found by reading the coordinates of any convenient point on the line and, knowing n, calculating a directly from Eq. (1-44). If the plot covers a range such that it crosses the ordinate $x = 1$, a may be read directly, as it is the y coordinate where $x = 1$, since $\log 1 = 0$ and $\log y = \log a$.

Example 1-9. A calibration of an orifice meter, which is a device for measuring the rate of flow of a fluid in a pipeline, gave the data shown in Table 1-6.

If the flow rate through an orifice follows the exponential equation

$$\overline{V} = aR_m^n \qquad (1\text{-}47)$$

where $\overline{V}$ is the velocity and R_m is the reading, determine the values of constants a and n.

TABLE 1-6. DATA FOR EXAMPLE 1-9

Average velocity of water in pipe $\overline{V}$, ft/sec	Orifice manometer reading R_m, mm Hg
3.42	30.3
4.25	58.0
5.25	75.5
5.88	93.5
7.02	137.5
7.30	148.0
10.05	261.0

Solution. The data of Table 1-6 are plotted in Fig. 1-5. A straight line fits the data with considerable accuracy. The slope of the line is found by measuring the

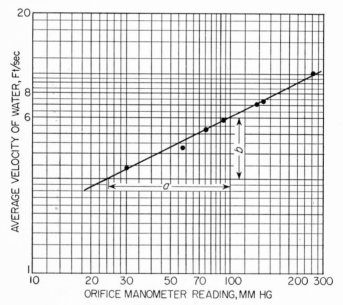

FIG. 1-5. Solution of Example 1-9.

distances shown in Fig. 1-5 as a and b. The slope is the ratio b/a, and this is the value of n. The result is $n = 0.50$.

The line does not reach the value of $R_m = 1$, so the simple use of the ordinate for $R_m = 1$ cannot be used to obtain a. However, the ordinate for $R_m = 100$ is 6.05, so, by Eq. (1-47),

$$a = \frac{\bar{V}}{R_m^n} = \frac{6.05}{100^{0.5}} = 0.605$$

The desired equation is, then,

$$\bar{V} = 0.605 R_m^{0.5} = 0.605 \sqrt{R_m}$$

It is shown in Chap. 2 that the theoretical value of the exponent n is 0.5.

SYMBOLS

A Area of heating surface, ft^2

a Coefficient in exponential equation; $a' = \log a$

c_p Specific heat at constant pressure, Btu/(lb)(°F); $\bar{c}_p$, average value of c_p

D Diameter, ft; also inside diameter; D_o', outside diameter, in.

E Total-energy term, ft-lb force/lb, defined by Eq. (1-20)

F Force, lb force

G General quantity

g Acceleration of gravity, ft/sec^2

g_c Newton's-law conversion factor, 32.174 ft-lb/lb force-sec^2

I Electric current, amp

i Enthalpy, Btu/lb; i_a, enthalpy of entering stream; i_b, enthalpy of leaving stream; i_x, enthalpy of saturated liquid; i_y, enthalpy of saturated vapor

J Mechanical equivalent of heat, 778.26 ft-lb force/Btu

k Thermal conductivity, Btu-ft/(ft^2)(hr)(°F)

k_n Constant in Newton's law, general form

L Length, ft

M Molecular weight; M_A, for component A; M_B, for component B; M_C, for component C; $\bar{M}$, average molecular weight

m Mass, lb; m_A, for component A; m_B, for component B; m_C, for component C; m_T, total mass of fluid at time θ_T

n Number of pound moles; exponent in exponential equation

p Pressure, lb force/ft^2; p_a, initial pressure or pressure of entering fluid; p_b, final pressure or pressure of leaving fluid

$\bar{p}$ Partial pressure, lb force/ft^2; $\bar{p}_A$, for component A; $\bar{p}_B$, for component B; $\bar{p}_C$, for component C

Q Quantity of heat, Btu; Q_e, quantity of electricity, coulombs

q Rate of heat transfer, Btu/hr

R Resistance, general; R_e, electrical resistance, ohms

R_m Reading of manometer of orifice meter

R_o Gas-law constant, 1,545 ft-lb force/(lb mole)(°R)

T Absolute temperature, °R; T_a, initial temperature; T_b, final temperature

t Temperature, °F; t_a, initial temperature or temperature of entering fluid; t_b, final temperature or temperature of leaving fluid

u Linear velocity, ft/sec; u_a, velocity of entering stream; u_b, velocity of leaving stream; u_0, initial velocity of falling body

V Volume, ft^3

$\bar{V}$ Average velocity of fluid stream, ft/sec

V_m Flow of vapor stream, moles

W_s Shaft work, ft-lb force

w Mass flow rate, lb/hr

x Mole fraction; x_A, for component A; x_B, for component B; x_C, for component C; general independent variable and abscissa; x_a, lower limit; x_b, upper limit; $x' = \log x$

y General dependent variable and ordinate; $y' = \log y$

Z Height above datum plane, ft; Z_a, for entering fluid; Z_b, for leaving fluid

Greek Letters

α Exponent in dimensional equation of general quantity

β Exponent in dimensional equation of general quantity

γ Exponent in dimensional equation of general quantity

ΔE Electric potential difference, volts

ΔF General driving potential

Δt Temperature difference between wall and fluid, °F

Δx Increment on x axis

δ Exponent in dimensional equation of general quantity

ϵ Exponent in dimensional equation of general quantity

ζ Exponent in dimensional equation of general quantity

θ Time, sec or hr; θ_T, total time

λ Latent heat, Btu/lb

μ Absolute viscosity, lb/ft-sec; μ_F, gravitational viscosity, lb force-sec/ft^2

ρ Density, lb/ft^3

Σ Operator, meaning "algebraic sum of"

Φ, Φ_1, ψ Operators, meaning "function of"

Dimensions of Primary Units

$\bar{F}$ Force

$\bar{H}$ Heat

$\bar{L}$ Length

$\bar{M}$ Mass

$\bar{T}$ Temperature

$\bar{\theta}$ Time

PROBLEMS

1-1. Dry gas containing 75 per cent air and 25 per cent ammonia vapor enters the bottom of an absorbing column, which is constructed as follows.

A welded steel cylinder, 20 ft high and 2 ft in diameter, is filled with lumps of coke from $1\frac{1}{2}$ to 2 in. in diameter. This coke tower filling is supported by a grid, positioned 2 ft from the bottom of the tower, and the inlet gas enters below the grid. The depth of the filling is 16 ft. In the top of the tower nozzles distribute fresh water over the coke. A solution of ammonia and water is drawn from the bottom of the column, and scrubbed gas leaves the top.

The gas enters at 80°F and 760 mm Hg pressure. It leaves at 60°F and 730 mm. The leaving gas contains, on the dry basis, 1.0 per cent ammonia.

(*a*) If the entering gas flows through the empty bottom of the column at an average velocity (upward) of 1.5 ft/sec, how many cubic feet of entering gas are treated per hour?

(*b*) How many pounds of ammonia are absorbed per hour?

1-2. Prepare a freehand sketch of the top of the tower showing the connections for water and gas and the distribution device for the water.

1-3. A suspension of calcium carbonate in water was pumped through a plate-and-frame filter press. The rate of filtrate flow was determined from time to time during the process by reading a flowmeter. The following data were obtained:

Time since start, min	Filtrate flow rate, gal/min
0	8.0
1	4.5
2.0	2.8
2.9	2.4
6.0	1.63
8.33	1.29
11.25	1.03
14.50	0.93
18.33 (end of run)	0.70

What is the total volume of filtrate collected during the entire run, in gallons?

1-4. Repeat Example 1-7 on the assumption that the velocity is so great that the conversion of mechanical energy into heat through the action of friction must be taken into account.

1-5. Air flows steadily and adiabatically through a horizontal straight pipe. The air enters the pipe at an absolute pressure of 100 lb force/in.^{2}, a temperature of 100°F, and a linear velocity of 10 ft/sec. The air leaves at 2 lb force/in.^{2} abs. What are the temperature and velocity of the leaving air?

1-6. Water enters the bottom of a vertical evaporator tube 20 ft long at a temperature of 130°F and a velocity of 1.5 ft/sec. The tube is 2.00 in. OD, and has a wall 0.065 in. thick. Heat amounting to 284,000 Btu/hr flows through the wall of the tube and is absorbed by the water. The pressure at the exit of the tube is 4.0 lb force/in.^{2} abs. Assuming the liquid-and-vapor mixture leaving the tube is in equilibrium, what fraction of the liquid water entering the bottom of the tube is vaporized?

1-7. The definition of the standard atmosphere is in terms of a column of mercury 760 mm high acted upon by standard gravity. The density of mercury is taken as 13.59504 g/cm^3. From this definition, show that 1 atm = 14.696 lb force/in.^{2} and 1 atm = 29.92 in. Hg.

1-8. Given that 1 cal/1 joule = 4.1873, show that J = 778.26 ft-lb force/Btu.

1-9. It is believed that the velocity of fall of a ball in a liquid depends on the diameter of the ball, the density of the ball, the density and viscosity of the liquid, and the acceleration of gravity. What equation would be predicted by dimensional analysis for the velocity of fall in terms of these variables?

REFERENCES

1. Bridgman, P. W.: "Dimensional Analysis," rev. ed., Yale University Press, New Haven, Conn., 1931.
2. Dodge, B. F.: "Chemical Engineering Thermodynamics," pp. 308–312, McGraw-Hill Book Company, Inc., New York, 1944.
3. Focken, C. M.: "Dimensional Methods and Their Application," Edward Arnold & Co., London, 1952.
4. Hunsaker, J. C., and B. G. Rightmire: "Engineering Applications of Fluid Mechanics," pp. 113–121, McGraw-Hill Book Company, Inc., New York, 1947.
5. Klinkenberg, A., and H. H. Mooy: *Chem. Eng. Progr.*, **44:** 17 (1948).

6. Shreve, R. H.: "The Chemical Process Industries," 2d ed., McGraw-Hill Book Company, Inc., New York, 1956
7. Smith, J. M.: "Introduction to Chemical Engineering Thermodynamics," pp. 35–39, McGraw-Hill Book Company, Inc., New York, 1949.
8. Weber, Ernst, in O. W. Eshbach (ed.): "Handbook of Engineering Fundamentals," 2d ed., p. 3-05, John Wiley & Sons, Inc., New York, 1952.

Chapter 2

FLUID MECHANICS

The behavior of fluids is important in many unit operations. An understanding of the elements of the mechanics of fluids is essential, not only in accurately treating problems of fluid flow through pipes, pumps, and other process equipment but also in the study of heat transfer and of the operations of absorption, distillation, and extraction. In this chapter the elements of fluid mechanics are discussed, and important applications to chemical engineering are considered. Fluids include liquids, gases, and vapors.

Nature of Fluids. A fluid may be defined as a substance that does not permanently resist distortion. An attempt to change the shape of a mass of fluid results in layers of fluid sliding over one another until a new shape is attained. During the change in shape, shear stresses exist, the magnitudes of which depend upon the viscosity of the fluid and the rate of sliding, but when a final shape has been reached, all shear stresses will have disappeared. A fluid in equilibrium is free from shear stresses.

At a given temperature and pressure, a fluid possesses a definite density, which in engineering practice is usually measured in pounds per cubic foot. Although the density of a fluid depends on temperature and pressure, the variation of density with changes in these variables may be large or small. If the density is but little affected by moderate changes in temperature and pressure, the fluid is said to be *incompressible,* and if the density is sensitive to changes in these variables, the fluid is said to be *compressible.* Liquids are considered to be incompressible and gases compressible. The terms are relative, however, and the density of a liquid can change appreciably if pressure and temperature are changed over wide limits. Also, gases subjected to small percentage changes in pressure and temperature act as incompressible fluids, and density changes under such conditions may be neglected without serious error.

The science of fluid mechanics includes two branches that are of importance in the study of the unit operations: fluid statics and fluid dynamics. Fluid statics treats of fluids in the equilibrium state of no shear stress, and fluid dynamics treats of fluids under conditions where portions of fluid are in motion relative to others.

FLUID STATICS

In a stationary mass of a single static fluid, the pressure is constant in any cross section parallel to the earth's surface but varies from height to height. Consider the vertical column of fluid shown in Fig. 2-1. Assume the cross-sectional area of the column is S ft^2. At a height Z ft above the base of the column let the pressure be p lb force/ft^2 and the density be ρ lb/ft^3. Because the fluid is at rest, it follows from Eq. (1-10) that the

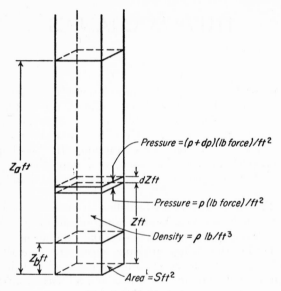

Fig. 2-1. Fluid statics.

resultant of all forces on the small volume of fluid of height dZ and cross-sectional area S must be zero. Three vertical forces are acting on this volume: (1) The force from pressure p acting in an upward direction. This is pS. (2) The force from pressure $p + dp$ acting in a downward direction. This is $(p + dp)S$. (3) The force of gravity acting downward. This is $(g/g_c)\rho S\,dZ$. By Eq. (1-10),

$$-pS + (p + dp)S + \frac{g}{g_c}\rho S\,dZ = 0$$

which, after simplification and division by S, becomes

$$dp + \frac{g}{g_c}\rho\,dZ = 0 \qquad (2\text{-}1)$$

Strictly, Eq. (2-1) cannot be integrated for compressible fluids unless the variation of density with pressure is known throughout the column of fluid. In chemical engineering practice, however, Eq. (2-1) is nearly always applied only where ρ is essentially constant. The density is constant for incompressible fluids and, except for large changes in height, is nearly so for compressible fluids. For example, at a temperature of 70°F and a pressure of 1 atm, a height of 100 ft of air is equivalent to a density difference of 0.36 per cent. Integration of Eq. (2-1) on the assumption that ρ is constant gives

$$\frac{p}{\rho} + \frac{g}{g_c} Z = \text{const} \tag{2-2}$$

or, between the two definite heights Z_a and Z_b shown in Fig. 2-1,

$$\frac{p_b}{\rho} - \frac{p_a}{\rho} = \frac{g}{g_c}(Z_a - Z_b) \tag{2-3}$$

Equation (2-2) expresses mathematically the condition of hydrostatic equilibrium.†

It is common, in using Eqs. (2-2) and (2-3), to assume that g/g_c is unity. Numerically this assumption is justified because the local value of g varies but slightly from place to place. Equation (2-3) can then be written in the dimensionally inconsistent form

$$p_b = p_a - \rho(Z_b - Z_a) \tag{2-5}$$

To calculate, from the known pressure at a definite level in a column of a single fluid, the pressure at another level, the procedure is to multiply the change in level by the density and to add the product to, or subtract it from, the known pressure, noting that the lower the level, the higher the pressure.

Manometers. The manometer is an important device for measuring pressure differences. Figure 2-2 shows the simplest form of manometer. Assume that the shaded portion of the U tube is filled with liquid A, having a density ρ_A lb/ft^3, and that the arms of the U tube above the liquid are filled with fluid B, having a density ρ_B lb/ft^3, which is both immiscible with liquid A and lighter than A.

A pressure p_a lb force/ft^2 is exerted in one arm of the U tube and a pressure p_b in the other. As a result of the difference in pressure $p_a - p_b$, the meniscus in one branch of the U tube is higher than in the other, and the vertical distance between the two meniscuses, R_m ft, may be used to measure the difference in pressure. To derive a relationship between $p_a - p_b$ and R_m, start at the point 1, where the pressure is p_a; then, as shown in the last section, the pressure at point 2 is $p_a + (Z_m + R_m)\rho_B$.

---

† In hydraulic practice, Eq. (2-2) is often written as

$$\frac{p}{\gamma} + Z = \text{const} \tag{2-4}$$

where $\gamma = (g/g_c)\rho$, and γ is called the specific weight. In Eq. (2-4) p/γ is known as the pressure head and Z as the static head. Both heads are measured in feet.

By the principles of hydrostatics, this is also the pressure at point 3. The pressure at point 4 is less than that at point 3 by the amount $R_m\rho_A$, and

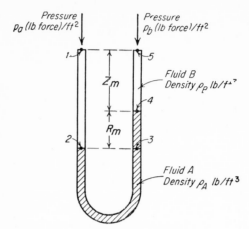

FIG. 2-2. Simple manometer.

the pressure at point 5, which is p_b, is still less by the amount $Z_m\rho_B$. These statements can be summarized by the equation

$$p_a + (Z_m + R)\rho_B - R_m\rho_A - Z_m\rho_B = p_b$$

Simplification of this equation gives

$$p_a - p_b = R_m(\rho_A - \rho_B) \tag{2-6}$$

It will be noted that this relationship is independent of the distance Z_m and of the dimensions of the tube provided that p_a and p_b are measured in the same horizontal plane.

Example 2-1. A manometer of the type shown in Fig. 2-2 is used to measure the pressure drop across an orifice (see Fig. 2-36). Liquid A is mercury (specific gravity 60°F/60°F = 13.6), and fluid B, flowing through the orifice and filling the manometer leads, is brine (specific gravity 60°F/60°F = 1.26). When the pressures at the taps are equal, the level of the mercury in the manometer is 3 ft below the orifice taps. Under operating conditions, the pressure at the upstream tap is 2.0 lb force/in.² gauge,† and that at the downstream tap is 10 in. Hg below atmospheric. What is the reading of the manometer in millimeters?

Solution. Calling atmospheric pressure zero, the numerical data for substitution in Eq. (2-6) are:

$$p_a = 2 \times 144 = 288 \text{ lb force/ft}^2$$

From Appendix 14, the density of water at 60°F is 62.37 lb/ft³.

$$p_b = -\tfrac{10}{12} \times 13.6 \times 62.37 = -707 \text{ lb force/ft}^2$$

$$\rho_A = 13.6 \times 62.37 = 848.2 \text{ lb/ft}^3$$

$$\rho_B = 1.26 \times 62.37 = 78.6 \text{ lb/ft}^3$$

† Gauge pressure is pressure measured above the prevailing atmospheric pressure.

Substituting in Eq. (2-6),

$$288 + 707 = R_m(848.2 - 78.6)$$

$$R_m = 1.29 \text{ ft}$$

Since 1 ft = 304.8 mm, the reading of the manometer is $1.29 \times 304.8 = 393$ mm.

For measuring small differences in pressure, the *inclined manometer* shown in Fig. 2-3 may be used. In this type, one leg of the manometer is inclined in such a manner that, for a small magnitude of R_m, the meniscus in the inclined tube must move a considerable distance along the tube.

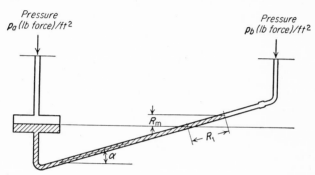

FIG. 2-3. Inclined manometer.

This distance is R_m divided by the sine of α, the angle of inclination. By making α small, the magnitude of R_m is multiplied into a long distance R_1, and a large reading becomes equivalent to a small pressure difference; so

$$p_a - p_b = R_1(\rho_A - \rho_B) \sin \alpha \qquad (2\text{-}7)$$

In this type of pressure gauge, it is necessary to provide an enlargement in the vertical leg so the movement of the meniscus in the enlargement is small within the operating range of the instrument.

FLOW OF FLUIDS

Fluids in motion, through channels or pipes or past solid shapes, are constantly encountered in chemical engineering practice, and therefore the mechanics of fluids in flow are basic in chemical engineering.

Laminar and Turbulent Flow. Depending upon conditions, a fluid may move in either of two contrasting types of flow at any given point in a flowing stream. The first is called laminar,† and the second, turbulent. The distinction between the two flow patterns was first demonstrated by a classic experiment performed by Osborne Reynolds. The equipment used

† Laminar flow is also called viscous or streamline flow.

by Reynolds is shown in Fig. 2-4. A horizontal glass tube was immersed in a glass-walled tank full of water. A controlled flow of water could be drawn through the tube by opening a valve. The entrance of the tube was flared, and provision was made to introduce a fine filament of colored water from the overhead tank into the stream at the tube entrance. Reynolds found that, at low flow rates, the jet of colored water flowed intact along with the main stream and no cross mixing occurred. The behavior of the color band showed clearly that the water was flowing in parallel straight lines. This type of fluid motion is laminar flow. When the flow rate was

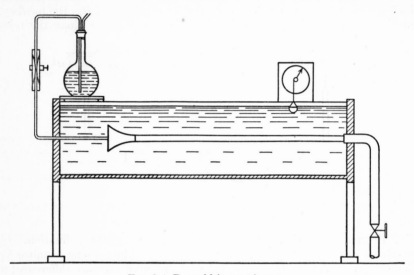

Fig. 2-4. Reynolds's experiment.

increased, a velocity, called the *critical velocity*, was reached at which the thread of color disappeared and the color diffused uniformly throughout the entire cross section of the flowing water. This behavior of the colored water showed that the water no longer flowed in laminar motion but moved erratically in the form of crosscurrents and eddies. This type of fluid motion is turbulent flow.

Deviating Velocities in Turbulent Flow. Because of its importance in many branches of engineering, turbulent flow has been extensively investigated in recent years. Refined methods of measurement have been used to follow experimentally the eddies and velocity fluctuations that occur in fluids in turbulent flow. An example of these measurements is shown in Fig. 2-5. The measurements show that, in a fluid in turbulent flow, the instantaneous velocity at a given point varies rapidly with time, both in direction and magnitude. Since the fluid has a net flow in a definite direction, the component of the instantaneous velocity in the direction of flow can be separated into two parts, a constant part that equals the net velocity of flow and a fluctuating part, called the deviating velocity, which represents the component of the variable velocity in the direction of flow. This can be shown by the equation

$$u_i = u + u' \qquad (2\text{-}8)$$

where u_i = instantaneous total velocity component in direction of flow

u = constant net velocity in direction of flow

u' = deviating velocity in direction of flow

Over a sufficient time interval, the deviating velocity u' passes through a long succession of positive and negative values, and its time average is zero.

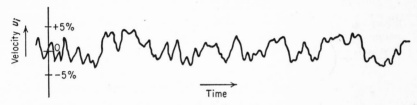

Fig. 2-5. Velocity fluctuations in turbulent flow. The percentages on the axis of ordinates are based on the constant net velocity. (*Courtesy of California Institute of Technology.*[16])

The time average of the instantaneous velocity u_i is, then, equal to the net velocity u.

In a direction perpendicular to the direction of flow the instantaneous velocity is a deviating velocity only, which is denoted by u''. The time average of u'' is zero, otherwise there would be a net flow through the plane parallel with the direction of flow, which is not possible.

In ordinary pipes and channels, the deviating velocities of u' and u'' are in magnitude not more than a few per cent of the net velocity u. They fluctuate in sign many times in a second. In rivers and in atmospheric currents, eddies may be many feet in size, and such eddies possess large periods of change.

In laminar flow there are no eddies, the deviating velocities do not exist, and the total velocity in the direction of flow is, at all times, u.

Steady Flow. When the net velocity u at any point is constant, the flow is called steady. Unsteady flow, where the velocity u changes with time at a given locality, will not be considered in this text. In steady flow the velocity u can, and often will, vary from place to place in the fluid.

Streamlines, Stream Tubes, and Continuity. Discussions of fluid-flow phenomena are facilitated by visualizing, in the stream of fluid, fluid paths called streamlines. A streamline is an imaginary curve in a mass of flowing fluid so drawn that at every point on the curve the net-velocity vector u is tangent to the streamline. No net flow takes place across such a line. In turbulent flow, eddies do cross and recross a streamline, but, as shown in the last section, the net flow from such eddies is zero.

A stream tube, or stream filament, is a tube, of small or large cross section and of any convenient cross-sectional shape, that is entirely bounded by streamlines. A stream tube can be visualized as an imaginary pipe in the mass of flowing fluid through the walls of which no net flow is occurring.

The law of conservation of matter yields an important relation concerning flow through a stream tube. The rate of mass flow into the tube must equal the rate of mass flow out of the tube. Consider the stream tube shown in Fig. 2-6. Let the fluid enter at a point where the area of the cross section of the tube is dS_a and leave where the area of the cross section is dS_b. Let the velocity and density at the entrance be u_a and ρ_a, respectively, and let the corresponding quantities at the exit be u_b and ρ_b, re-

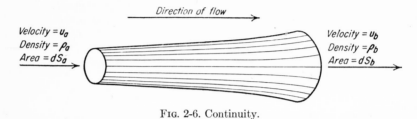

FIG. 2-6. Continuity.

spectively. Then the mass of fluid entering and leaving the tube in unit time is

$$dw = \rho_a u_a \, dS_a = \rho_b u_b \, dS_b \tag{2-9}$$

where w is the rate of flow, in mass per unit time. From Eq. (2-9) it follows, for a stream tube,

$$dw = u\rho \, dS = \text{const} \tag{2-10}$$

Equation (2-10) is called the equation of continuity. Assuming ρ is constant throughout the total cross section S, integration of this equation gives

$$w = \rho \int_0^S u \, dS \tag{2-11}$$

Average Velocity. When the cross section of a stream tube is large, the local velocity u may not be the same at all points in a single cross section. This is especially true when the boundary of the tube is a solid wall, where the local velocity is zero at the wall and changes rapidly with distance from the wall. For example, Fig. 2-13 shows how the local velocity in a tube can vary with distance from the wall. In such situations, it is important to distinguish between local and average velocities.

If the fluid is being heated or cooled, the density of the fluid also varies from point to point in a single cross section. In this text, density variations in a single cross section of a stream tube are neglected.

The average velocity $\overline{V}$ across an entire stream tube is defined by the equation

$$\overline{V} = \frac{w}{S\rho} = \frac{q}{S} \tag{2-12}$$

$w = q\rho$

where q = volumetric flow rate through stream tube
S = cross-sectional area of tube

Comparison of Eqs. (2-11) and (2-12) shows that the relation between the average velocity and the local velocity is

$$\overline{V} = \frac{w}{S\rho} = \frac{\displaystyle\int_0^S u \, dS}{S} \tag{2-13}$$

The density has been canceled in Eq. (2-13) on the assumption that ρ is constant throughout the cross section. Velocities $\overline{V}$ and u are equal if, and only if, the local velocity is the same at all points in cross-sectional area S.

The continuity equation for flow through a finite stream tube is

$$w = \rho_a \overline{V}_a S_a = \rho_b \overline{V}_b S_b = \rho \overline{V} S = \text{const} \tag{2-14}$$

For the important case where the flow is through a tube or pipe of circular cross section

$$w = \rho_a \overline{V}_a \frac{\pi}{4} D_a^2 = \rho_b \overline{V}_b \frac{\pi}{4} D_b^2$$

so

$$\rho_a \overline{V}_a = \left(\frac{D_b}{D_a}\right)^2 \rho_b \overline{V}_b \tag{2-15}$$

The average velocity can be expressed in any convenient units. It is usually expressed in either feet per second or in feet per hour. The cross-sectional area is expressed in square feet, the mass flow rate in pounds per second or in pounds per hour, and the volumetric flow rate in cubic feet per second or in cubic feet per hour.

Mass Velocity. Equation (2-13) can be written

$$\overline{V}\rho = \frac{w}{S} = G \tag{2-16}$$

This equation defines the mass velocity G, which is calculated by dividing the mass flow rate by the cross-sectional area of the channel. In practice, the mass velocity is expressed either in pounds per square foot per second or in pounds per square foot per hour. The advantage of using G is that it is independent of temperature and pressure when the flow is steady (constant w) and the cross section is unchanged (constant S). This fact is especially useful when compressible fluids are considered, where both $\overline{V}$ and ρ vary with temperature and pressure. Also, certain relationships appear later in this book in which $\overline{V}$ and ρ are associated together as their product, and the mass velocity represents the net effect of both variables.

Example 2-2. Crude oil, specific gravity 60°F/60°F = 0.887, flows through the piping shown in Fig. 2-7. Pipe A is 2-in. Schedule 40, pipe B is 3-in. Schedule 40, and each of pipes C is $1\frac{1}{2}$-in. Schedule 40. An equal quantity of liquid flows through each of the pipes C. The flow through pipe A is 30 gal/min. Calculate: (a) the

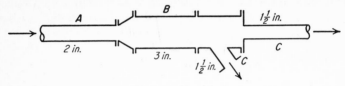

Fig. 2-7. Example 2-2.

mass flow rate in each pipe, in pounds per hour, (b) the average linear velocity in each pipe, in feet per second, and (c) the mass velocity in each pipe, in pounds per square foot per second.

Solution. Dimensions and cross-sectional areas of standard pipe are given in Appendix 4. Cross-sectional areas needed are for 2-in. pipe, 0.0233 ft²; for 3-in. pipe, 0.0513 ft²; for $1\frac{1}{2}$-in. pipe, 0.01414 ft².

(a) The density of the fluid is

$$\rho = 0.887 \times 62.37 = 55.3 \text{ lb/ft}^3$$

Since there are 7.48 gal in 1 ft³ (Appendix 2), the total volumetric flow rate is

$$q = \frac{30 \times 60}{7.48} = 240.7 \text{ ft}^3/\text{hr}$$

The mass flow rate is the same for pipes A and B and is the product of the density and the volumetric flow rate, or

$$w = 240.7 \times 55.3 = 13,300 \text{ lb/hr}$$

The mass flow rate through each of pipes C is one-half the total, or $13,300/2 = 6,650$ lb/hr.

(b) Use Eq. (2-12). The velocity through pipe A is

$$\overline{V}_A = \frac{240.7}{3,600 \times 0.0233} = 2.87 \text{ ft/sec}$$

through pipe B is

$$\overline{V}_B = \frac{240.7}{3,600 \times 0.0513} = 1.30 \text{ ft/sec}$$

and through each of pipes C is

$$\overline{V}_C = \frac{240.7}{2 \times 3,600 \times 0.01414} = 2.36 \text{ ft/sec}$$

(c) Use Eq. (2-16). The mass velocity through pipe A is

$$G_A = \frac{13,300}{0.0233} = 571,000 \text{ lb/ft}^2\text{-hr}$$

through pipe B is

$$G_B = \frac{13,300}{0.0513} = 259,000 \text{ lb/ft}^2\text{-hr}$$

and through each of pipes C is

$$G_C = \frac{13,300}{2 \times 0.01414} = 470,000 \text{ lb/ft}^2\text{-hr}$$

Prandtl Boundary Layer. The local velocities in a fluid stream are profoundly affected when the stream is brought into contact with a solid object or with the wall of a conduit. Thus, consider the flow of fluid parallel with a thin plate, as shown in Fig. 2-8. The velocity of the fluid up-

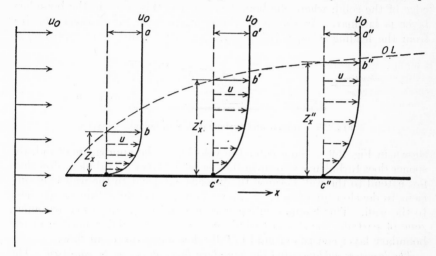

Fig. 2-8. Prandtl boundary layer: x, distance from leading edge. u_0, velocity of undisturbed stream. Z_x, thickness of boundary layer at distance x. u, local velocity. abc, $a'b'c'$, $a''b''c''$, velocity vs. distance-from-wall curves at points c, c', c''. OL, outer limit of boundary layer.

stream from the leading edge of the plate is uniform across the entire stream of fluid. The fluid in immediate contact with the plate adheres to it, and therefore the velocity of the fluid at the interface between the solid and fluid is zero. The velocity increases with distance from the plate shown in Fig. 2-8. Each of these curves corresponds to a definite value of x, the distance from the leading edge of the plate. Each curve changes slope rapidly near the plate and shows that the local velocity approaches asymptotically the velocity of the bulk of the fluid stream. In Fig. 2-8, the dotted line OL is so drawn that the velocity changes are confined between this line and the trace of the wall. Because the velocity lines are asymptotic with respect to distance from the plate, it is assumed, in order to locate the dotted line definitely, that the line passes through all points where the velocity is 99 per cent of the bulk fluid velocity. Line OL represents an imaginary surface which separates the fluid stream into two parts, one in which the fluid velocity is constant, and the other in which the velocity varies from zero at the wall to a velocity substantially equal to that of the undisturbed fluid. The surface separates the fluid that is directly affected by the plate from that where the local velocities are constant and which is unaffected by the plate. The relatively narrow zone, or layer, between the dotted line and the plate is called the Prandtl boundary layer. The thickness of the layer increases with the distance from the leading edge of the plate. It is zero at the leading edge and is a maximum at the trailing edge. Any solid body in contact with a flowing fluid will develop a boundary layer, regardless of the shape of the body.

Laminar and Turbulent Flow in the Boundary Layer. Toward the leading edge of the solid, where the boundary layer is thin, flow in the boundary layer is laminar. As the layer thickens, however, at distances farther from the leading edge, a point is reached where turbulence appears, as

FIG. 2-9. Laminar and turbulent flow in boundary layer.

shown in Fig. 2-9. The greater the velocity of the approaching fluid, the sooner does turbulence appear in the boundary layer. Turbulent flow does not extend to the wall because the velocity adjacent to the wall is insufficient to develop turbulence. Laminar flow persists in a thin region next to the wall. This laminar sublayer is overlaid by a transition, or buffer, zone of partially developed turbulence, and the remaining portion of the boundary layer consists of fluid in fully developed turbulent flow.

The laminar sublayer and the boundary layer must not be confused. The former refers only to that part of the boundary layer, immediately next to the solid, that remains in laminar flow. The boundary layer includes all the fluid in which there is a velocity variation in a plane perpendicular to the solid.

Tractive Force and Velocity Gradient. The velocity change across a boundary layer is a result of fluid layers sliding past one another and is, therefore, accompanied by shear. *All actual fluids resist shear*, both in laminar and in turbulent flow. The shear force per unit shear area is called the tractive force. Thus, consider the curve of u vs. y shown in Fig. 2-10. The curve represents the change of velocity with thickness of layer at a definite point in a typical boundary layer. The dotted line aa is the trace of an imaginary plane parallel with the solid wall. Let F_s be the total shear force acting on the plane, and let A_s be the area of the plane. The tractive force τ is defined for both laminar and turbulent flow by the equation

$$\tau = \frac{F_s}{A} \qquad (2\text{-}17)$$

The units of τ are pounds force per square foot.

The slope of the u vs. y curve at any given value of y is called the velocity gradient at y, and is, of course, du/dy. Thus, the slope of the tangent drawn at point M in Fig. 2-10 is the gradient at the point $y = y_A$. Inspection of Fig. 2-10 shows that the gradient is a maximum at the wall and zero outside of the boundary layer, where the velocity u_m is a maximum.

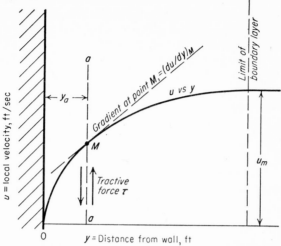

FIG. 2-10. Velocity gradient near solid wall.

Newtonian and Non-Newtonian Fluids. The relation between the tractive force τ and the velocity gradient du/dy is clearly of importance in the behavior of flowing fluids. In Fig. 2-11, four typical curves of τ vs. du/dy are shown. In all cases the temperature and pressure are constant, and laminar flow is assumed. The line for fluid A is a straight line passing through the origin. Gases, noncolloidal liquids, and true solutions follow this simple law. Fluids of this type are called Newtonian fluids. The line for liquid B is linear, or nearly so, but it intersects the τ axis at a finite magnitude τ_0. A definite minimum traction τ_0 is required before a velocity gradient will form. At tractive forces less than τ_0, the material acts like a solid, deforms but does not flow, and permanently resists shear stresses. The flow type shown by curve B is called plastic flow. Certain colloidal

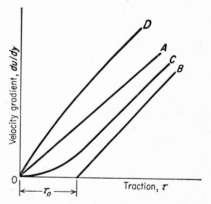

FIG. 2-11. Velocity gradient vs. traction: A, Newtonian. B, plastic. C, pseudo-plastic. D, dilatant.

clays and muds exhibit plastic flow. The curve for material C passes through the origin and is not linear at low values of du/dy but becomes so at higher velocity gradients. Liquids of this type are called pseudo-plastic. Rubber latices are of this type. The curve for liquid D is the reverse of that for material C. The ratio of traction to gradient increases with τ. Liquids of this type are called dilatant. Only Newtonian fluids will be considered further in this text.

Viscosity. The constant slope of the τ vs. du/dy line of a Newtonian fluid is used to define an important fluid property called the viscosity. The *absolute* viscosity μ is defined by the equation

$$\mu = \frac{\tau g_c}{du/dy} = \frac{g_c(F_s/A)}{du/dy} \qquad (2\text{-}18)$$

The factor g_c is introduced so F_s may be expressed in pounds force. The quantity μ/g_c is the slope of the straight τ vs. du/dy line. The viscosity of a Newtonian fluid is independent of both τ and du/dy. It is a quantitative measure of the tendency of the fluid to resist shear. Thus, the viscosity of a lubricating oil is greater than that of brine or water, and that of water is much greater than that of a gas, such as air.

The dimensions of viscosity may be found by substituting the dimensional formulas for the various terms into Eq. (2-18).

$$[\mu] = (\overline{F}\overline{L}^{-2})(\overline{M}\overline{L}\overline{\theta}^{-2}\overline{F}^{-1})\overline{L}(\overline{L}^{-1}\overline{\theta}) = \overline{M}\overline{L}^{-1}\overline{\theta}^{-1}$$

The scientific unit for viscosity is the *poise*, which is named after the French scientist Poiseuille, who performed fundamental experiments on flow through tubes. The poise is 1 g/cm-sec. This unit is inconveniently large for many practical purposes, and viscosities are more commonly expressed in *centipoises*, 1 centipoise being $\frac{1}{100}$ poise. Water at 68.6°F, for example, has a viscosity of 1 centipoise.

In English units, viscosity is expressed either as pounds per foot per second or as pounds per foot per hour. From standard conversion factors, the relations between the poise and the English units are

$$1 \text{ poise} = 0.0672 \text{ lb/ft-sec} = 242 \text{ lb/ft-hr}$$

$$1 \text{ centipoise} = 6.72 \times 10^{-4} \text{ lb/ft-sec} = 2.42 \text{ lb/ft-hr}$$

The so-called *gravitational* viscosity μ_F, which was used in dimensional analysis in Chap. 1, is often encountered in the engineering literature. It is defined by the equation

$$\mu_F = \frac{\tau}{du/dy} \qquad (2\text{-}19)$$

Comparison of Eqs. (2-18) and (2-19) shows that

$$\mu = \mu_F g_c \qquad (2\text{-}20)$$

The engineering units of μ_F are pound force-seconds per square foot. A viscosity of 1 poise is 2.089×10^{-3} lb force-sec/ft^2.

In certain fluid-flow problems the viscosity is associated with density as the ratio μ/ρ. For convenience, this ratio has been given a name, the *kinematic viscosity*, and a symbol, ν. From its definition, the dimensions of kinematic viscosity are $\bar{L}^2\bar{\theta}^{-1}$. The usual unit of ν is that obtained by expressing the viscosity in poises and the density in grams per cubic centimeter. This unit is the *stoke*, which is 1 cm^2/sec. The centistoke is $\frac{1}{100}$ stoke. In English units the kinematic viscosity is in square feet per second.

Viscosities of Gases and Liquids. The viscosity of a fluid depends primarily on temperature and to a lesser degree on pressure. The viscosity of a gas increases with temperature in accordance with an equation of the type

$$\frac{\mu}{\mu_0} = \left(\frac{T}{492}\right)^n \tag{2-21}$$

where μ = viscosity at absolute temperature T,°R

μ_0 = viscosity at 32°F, or 492°R

n = constant

The constant n ranges between 0.65 and 1.00.

At high pressures, the viscosities of gases increase with pressure, especially in the vicinity of the critical point.

The viscosities of liquids are sensitive to temperature, and decrease with increasing temperature. For example, the viscosity of water at 0°C is more than six times that at 100°C. The viscosities of liquids increase moderately with large increases in pressure, although water is an exception in that its viscosity first decreases and then increases as the pressure increases.

Appendixes 7 and 8 show the effect of temperature on the viscosities of various gases and liquids. Additional data, showing the effect of both temperature and pressure, are given in standard tables of physical data.[12a]

Boundary-layer Formation in Straight Tubes. Perhaps the most important single case of fluid flow in chemical engineering is that through long, straight tubes of constant cross section.

Consider a tube having a bell-shaped entrance, such as that used in the Reynolds experiment shown in Fig. 2-4. As shown in Fig. 2-12, a boundary layer begins to form at the entrance, and as the fluid moves through the

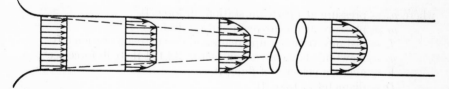

FIG. 2-12. Development of boundary-layer flow in pipe. Dotted line shows limit of boundary layer.

first part of the straight tube, the layer thickens. During this stage, the boundary layer occupies only part of the cross section of the tube, and the total stream consists of a core of fluid flowing in a rodlike manner at a constant velocity and an annular boundary layer between the wall and the core. In the boundary layer, the velocity increases from zero at the wall to that in the core. As the stream moves farther down the tube, the boundary layer occupies an increasing portion of the cross section. Finally, at a point well downstream from the entrance, the boundary layer reaches the center of the tube, the rodlike core disappears, and the boundary layer occupies the entire cross section. At this point, the velocity distribution in the tube reaches its final form, and it remains unchanged through the remaining length of the tube. The velocity gradient is zero at the center of the tube. The attainment of the final velocity distribution is shown by the last curve at the right of Fig. 2-12.

The length of tube necessary for the boundary layer to reach the center of the tube and for the final velocity distribution to be completely established is called the *transition length*. The transition length may be demonstrated in the equipment of Fig. 2-4.

If the average velocity of the fluid is above the critical, the bulk of the stream is in turbulent flow. The laminar sublayer, next to the wall, and the buffer layer, between the laminar layer and the turbulent core, are permanent characteristics of the flow and persist beyond the transition length. The thicknesses of the laminar and buffer layers are small fractions of the diameter of the tube, and most of the cross section is occupied by fluid in fully developed turbulent flow. This complex pattern is called simply *turbulent flow*.

Hagen-Poiseuille Equation. An important equation can be derived for the laminar flow of a Newtonian fluid through a long, straight tube. The assumptions made in the derivation are: (1) the tube is accurately bored and its diameter is accurately known, (2) the fluid adheres to the wall of the tube, (3) the temperature of the fluid is constant, so the density and viscosity of the fluid are constant, (4) the velocity distribution is fully developed at both upstream and downstream stations of pressure measurement. Under these assumptions, the pressure difference between upstream and downstream stations is given by the Hagen-Poiseuille equation

$$p_a - p_b = \frac{32L\overline{V}\mu}{g_c D^2} \tag{2-22}$$

where p_a = pressure at upstream station, lb force/ft^2
p_b = pressure at downstream station, lb force/ft^2
L = distance between upstream and downstream stations, ft
g_c = Newton's-law conversion factor, 32.174 ft-lb/lb force-sec^2
$\overline{V}$ = average velocity of fluid, ft/sec
D = diameter of tube, ft
μ = viscosity, lb/ft-sec

If the viscosity is known, this equation is used to calculate the pressure drop from friction in laminar flow. The Hagen-Poiseuille equation is also used as a basis for the experimental determination of viscosity, by measuring a flow rate through a tube of known dimensions under a measured pressure drop.

Measurement of Viscosity. The practical use of Eq. (2-22) as it stands is difficult because of the length of tube necessary to build a fully developed velocity distribution at the upstream station and so to conform to assumption (4). A device used to measure viscosity is called a viscometer. Capillary viscometers are so constructed that the fluid starts at rest and enters a capillary tube from a reservoir. One type consists of a glass capillary with a bulb at one end. The bulb is calibrated at two points with a known volume between them. The bulb and capillary are filled with the liquid to be investigated, the tube is immersed in a thermostat, and a known pressure head is applied. The predetermined volume in the reservoir is forced through the capillary, and the time of flow is measured.

In this method, energy must be supplied to accelerate the liquid from rest to the average velocity $\overline{V}$ and to build the final velocity distribution for laminar flow by the time the liquid leaves the capillary. As shown on page 60, this energy calls for an additional pressure drop of $\rho\overline{V}^2/g_c$. If the pressure drop $p_a - p_b$ is kept constant during the flow so that the velocity $\overline{V}$ is constant, Eq. (2-22) becomes, for a capillary viscometer,

$$p_a - p_b = \frac{32L\overline{V}\mu}{g_c D^2} + \frac{\rho\overline{V}^2}{g_c} \tag{2-23}$$

The average velocity $\overline{V}$ is related to the volume of flow and the time of flow by the equation

$$\overline{V} = \frac{4V}{\pi\theta D^2} \tag{2-24}$$

where V is the volume of fluid flowing through the apparatus in time θ. The units of V and θ are cubic feet and seconds, respectively. Substitution of $\overline{V}$ from Eq. (2-24) into Eq. (2-23) gives

$$p_a - p_b = \frac{16\rho V}{\pi g_c D^4 \theta}\left(\frac{8\mu L}{\rho} + \frac{V}{\pi\theta}\right) \tag{2-25}$$

Since

$$\nu = \frac{\mu}{\rho} \tag{2-26}$$

and

$$q = \frac{V}{\theta} \tag{2-27}$$

Eq. (2-25) can also be written in the form

$$8L\nu = \frac{(p_a - p_b)g_c\pi D^4\theta}{16\rho V} - \frac{V}{\pi\theta} = \frac{(p_a - p_b)g_c\pi D^4}{16\rho q} - \frac{q}{\pi} \tag{2-28}$$

where q is the flow rate in cubic feet per second. All terms in this equation except ν are known, and the kinematic viscosity can be calculated from measurements of V and θ. In these equations ν is in square feet per second.

For liquids having viscosities comparable in magnitude to that of water, it is necessary to use fine capillaries and rather elaborate apparatus. For viscous liquids such as oils, the diameter of the tube may be larger so that the liquid can flow under gravity head. Viscometers of this type are quite common. Such an instrument consists of a vessel with a short capillary in the bottom and surrounded by a constant-temperature bath. A definite volume of liquid, the viscosity of which is to be determined, is put in the vessel, and a calibrated receiver is placed below the tube. By removing the stopper from the tube and measuring the time required to fill the calibrated receiver, a number, in seconds, that is a function of the kinematic viscosity is obtained. An empirical form of Eq. (2-28) is used to calibrate viscometers of this type. The quantity $(p_a - p_b)/\rho$ is proportional to the gravity head causing the flow through the instrument, and this quantity depends on the design of the viscometer and is independent of the fluid. The terms L, D, and V are also constants of the equipment, and Eq. (2-28) can be written in the form of the dimensional equation

$$\nu = A\theta - \frac{B}{\theta} \qquad (2\text{-}29)$$

where A and B are constant for a given viscometer.

Two common viscometers are the Engler and the Saybolt viscometers. For readings in seconds, and kinematic viscosities in stokes, the standard constants for these instruments are as shown in Table 2-1.

TABLE 2-1. CONSTANTS FOR VISCOMETERS

Instrument	A	B
Saybolt Universal........	0.0022	1.80
Engler................	0.00147	3.74

Many other types of viscometers are available. Some of these are designed to determine the τ vs. du/dy curves of non-Newtonian liquids.

Example 2-3. A light lubricating oil is draining by gravity from a reservoir through a vertical $\frac{1}{4}$-in. pipe 10 ft long. The level of the oil in the reservoir is constant and is 2 ft above the entrance to the pipe. The oil is at 100°F, and at this temperature its specific gravity is 0.89, referred to water at 100°F, and its viscosity is 200 SSU. Calculate the rate of flow of the oil, in gallons per minute.

Solution. From Eq. (2-29) the kinematic viscosity is

$$\nu = 0.0022 \times 200 - \frac{1.80}{200} = 0.431 \text{ cm}^2/\text{sec}$$

$$= \frac{0.431}{30.48^2} = 4.64 \times 10^{-4} \text{ ft}^2/\text{sec}$$

Equation (2-25) is used. The quantities required are

$$L = 10 \text{ ft}$$

From Appendix 14, the density of water at 100°F is 62.0 lb/ft³, so

$$\rho = 0.89 \times 62.0 = 55.2 \text{ lb/ft}^3$$

$$p_a - p_b = 55.2(10 + 2) \frac{g}{g_c} = 662.4 \text{ lb force/ft}^2$$

$$D = \frac{1}{4 \times 12} = \frac{1}{48} \text{ ft}$$

Substituting into Eq. (2-25) and using Eq. (2-27),

$$662.4 = \frac{16 \times 55.2q}{32.17\pi(1/48^4)} \left(8 \times 10 \times 4.64 \times 10^{-4} + \frac{q}{\pi} \right)$$

As a first approximation, neglect the kinetic energy term q/π and solve for q.

$$q = \frac{662.4 \times 32.17\pi(10^4)}{48^4 \times 16 \times 55.2 \times 8 \times 10 \times 4.64} = 3.85 \times 10^{-4} \text{ ft}^3/\text{sec}$$

The term in the parentheses is

$$8 \times 10 \times 4.64 \times 10^{-4} + \frac{3.85}{\pi} 10^{-4} = 10^{-4}(371.2 + 1.2)$$

The kinetic-energy term can be neglected, and the first approximation accepted. The flow is $3.85 \times 10^{-4} \times 7.48 \times 60 = 0.172$ gal/min.

Reynolds Number and Transition from Laminar to Turbulent Flow. The concept of viscosity and the development of methods for measuring it have led to an important numerical factor for treating flow problems.

Reynolds, in an extension of his work on laminar and turbulent flow, studied the conditions under which one type of flow changes into the other. He found that the critical velocity, at which laminar flow changes into turbulent flow, depends on four quantities: the diameter of the pipe, the viscosity and density of the fluid, and the average linear velocity of the fluid. Furthermore, he found that these four factors can be combined into one group, and that the change in type of flow occurs at a definite magnitude of the group. The grouping of variables found by Reynolds was

$$N_{\text{Re}} = \frac{D\overline{V}\rho}{\mu} = \frac{D\overline{V}}{\nu} = \frac{DG}{\mu} \tag{2-30}$$

where D = diameter of pipe, ft
 $\overline{V}$ = average linear velocity of fluid, ft/sec
 μ = viscosity of fluid, lb/ft-sec
 ρ = density of fluid, lb/ft³
 ν = kinematic viscosity of fluid, ft²/sec
 G = mass velocity of fluid, lb/ft²-sec

The group of variables defined by Eq. (2-30) is called the Reynolds number. It is denoted by N_{Re}, and has appeared in Example 1-7. It will be met many more times in this text.

Since Reynolds's time additional observations have been made on the transition from laminar to turbulent flow. It has been found that the transition actually occurs over a wide range of Reynolds numbers. Laminar flow is always encountered at Reynolds numbers below 2,100, but laminar flow can persist up to Reynolds numbers of several thousand under special conditions of well-rounded tube entrances and very quiet water in the tank. Under ordinary conditions of flow, the flow is turbulent at Reynolds numbers above about 4,000. Between 2,100 and 4,000, a transition, or "dip," region is found, where the type of flow may be either laminar or

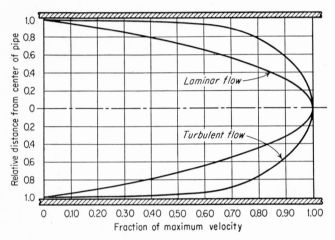

FIG. 2-13. Distribution of velocity across pipe, fully developed flow.

turbulent, depending upon conditions at the entrance of the tube and on the distance from the entrance.

It has been shown in Chap. 1 that the Reynolds number is one of the dimensionless groups that appear in dimensional analysis. Its magnitude is independent of the units chosen for the individual factors D, $\bar{V}$, ρ, and μ provided the units are consistent.

Velocity Distribution in Tubes for Laminar and Turbulent flow. Typical curves showing the final velocity distribution in flow through pipes are shown in Fig. 2-13. Distances are given as fractions of the radius of the tube measured from the center line of the tube, and velocities are expressed as fractions of the maximum velocity, which occurs at the center line of the tube. In both laminar and turbulent flow the curve of velocity vs. distance rises sharply from zero at the wall and has a rounded top in the neighborhood of the center line.

In laminar flow, theory and experiment agree that the curve of velocity vs. distance is a parabola and that the average velocity $\bar{V}$ is exactly one-half the maximum velocity u_m.

In turbulent flow, the velocity-distribution curve is flatter than in laminar flow, and most of the total velocity change occurs in the laminar and buffer layers near the wall. The relationship between N_{Re} and $\overline{V}/u_m$ for turbulent flow is shown in Fig. 2-14. The upper curve A is used to relate $\overline{V}/u_m$ to a Reynolds number based on the average velocity $\overline{V}$, and the lower curve B is used with a Reynolds number based on the maximum velocity u_m. In either case, the Reynolds number is calculated and used as the abscissa in Fig. 2-14, and $\overline{V}/u_m$ is read from the corresponding ordinate of a point on the appropriate curve. An average value of $\overline{V}/u_m$ in turbulent flow, often used in approximate calculations, is 0.8.

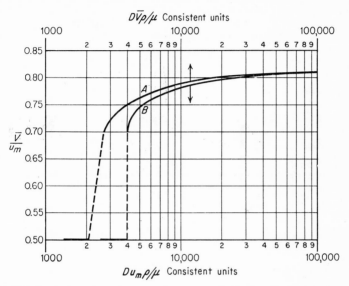

FIG. 2-14. Relation between maximum and average velocity.[10a] (By permission of author and publishers, from "Heat Transmission," 2d ed., by W. H. McAdams. Copyright by author, 1942. McGraw-Hill Book Company, Inc.)

The curves shown in Figs. 2-13 and 2-14 apply only when the fluid is being neither heated nor cooled by heat transfer to or from the tube wall, and in long, straight, smooth tubes well beyond any change in direction or cross section.

Example 2-4. Water at 100°F is flowing through a long, straight, 4-in. Schedule 40 pipe at a rate of 6.20 gal/min. What is the velocity at the center of the pipe?

Solution. From Appendix 4 the inside diameter and cross-sectional area of the pipe are 4.026 in. and 0.0884 ft², respectively. The average linear velocity is, by Eq. (2-12),

$$\overline{V} = \frac{6.20}{60 \times 0.0884 \times 7.48} = 0.156 \text{ ft/sec}$$

The density and viscosity of the water are, from Appendix 14, 62.0 lb/ft³ and 0.682 centipoise, respectively. The Reynolds number is

$$N_{Re} = \frac{0.156(4.026/12)62.0}{0.682 \times 6.72 \times 10^{-4}} = 7,080$$

From Fig. 2-14, when $N_{Re} = 7,080$, $\overline{V}/u_m = 0.775$, and the velocity at the center of the pipe is

$$u_m = \frac{\overline{V}}{0.775} = \frac{0.156}{0.775} = 0.20 \text{ ft/sec}$$

This flow rate is much lower than that usually specified for water in a pipe of this size. Average velocities of 4 to 8 ft/sec are common in practice.

Transition Length for Laminar and Turbulent Flow. The approximate length of straight pipe necessary for completion of the final velocity distribution is, for laminar flow,[14]

$$\frac{x_t}{D} = 0.05 N_{Re} \tag{2-31}$$

where x_t is the length required for the development of the final velocity distribution and D is the diameter of the pipe. Thus, for a 2-in.-ID pipe and a Reynolds number of 1,500, a transition length of $x_t = 0.05 \times 1,500 \times (2/12) = 12.5$ ft is necessary.

In turbulent flow, the transition length is nearly independent of the Reynolds number and is about 40 to 50 pipe diameters. There is but little difference between the distribution at 25 diameters and that at greater distances from the entrance. For a 2-in.-ID pipe, 6 to 8 ft of straight pipe is sufficient. The transition length for turbulent flow is less than that for laminar flow unless the Reynolds number is small.

Bernoulli Equation: Frictionless Flow. Because of the change in velocity with distance from the wall, the fluid in a boundary layer is subjected to shear stresses, and viscosity must therefore be taken into account in studying the mechanics of the boundary layer. In the stream of fluid outside a boundary layer, where the velocity does not vary from point to point in a given cross section, shear stresses do not appear, and viscosity is not a factor. An important equation, called the Bernoulli equation without friction, applies to stream tubes and streamlines lying outside a boundary layer. A derivation of this equation follows.

Consider a volume element of fluid flowing along a stream tube as shown in Fig. 2-15. Let the cross section of the element be dS, the density of the fluid in the element be ρ, the pressure and the velocity of the fluid at the upstream end of the volume element be p and u, respectively; and let the axis of the tube be inclined from the vertical by the angle β. Assume that the pressure and velocity at the downstream end of the element are $p + dp$ and $u + du$, respectively. Assume that the length of the element is dL and that $d\theta$ is the time required for the element to move a distance equal to its own length dL.

Consider the forces acting on the volume element and tending either to push it along the stream tube or to retard its movement.

No effective forces in the direction of flow are acting on the cylindrical bounding surface of the tube, because the pressure forces on this surface are perpendicular to the direction of flow and, by assumption, shear stresses do not exist. The forces that do act to accelerate or retard flow are (1) force $p\,dS$ in the direction of flow, (2) force $(p + dp)\,dS$ acting to oppose flow, (3) a component of the force of gravity acting to oppose the flow.† The force of gravity is the product of the mass of fluid in the element and the acceleration of gravity divided by the dimensional constant g_c. The mass of the element is the product of the density ρ and the volume of the element $dS\,dL$. The force of gravity is, then, $\rho(g/g_c)\,dS\,dL$, and the

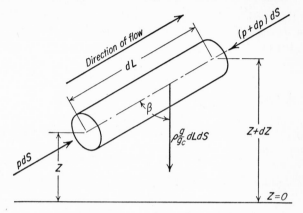

Fig. 2-15. Forces acting on volume element of stream tube.

component of this force along the axis of the tube is $\rho\cos\beta\,(g/g_c)\,dS\,dL$. The resultant force in the direction of flow is

$$p\,dS - (p + dp)\,dS - \frac{\rho g}{g_c}\cos\beta\,dS\,dL = F$$

By Eq. (1-13) the resultant force F is equal to the mass of the fluid in the volume element multiplied by the acceleration, which is $du/d\theta$, and divided by the constant g_c. Therefore,

$$p\,dS - (p + dp)\,dS - \frac{\rho g}{g_c}\cos\beta\,dS\,dL = \rho\,dS\,dL\frac{du/d\theta}{g_c} \qquad (2\text{-}32)$$

Simplifying Eq. (2-32) and dividing by $\rho\,dS$ gives

$$\frac{dp}{\rho} + \frac{g}{g_c}\cos\beta\,dL + \frac{du}{g_c}\frac{dL}{d\theta} = 0 \qquad (2\text{-}33)$$

† Gravity opposes motion because Fig. 2-15 is so drawn that the flow is in an upward direction, against the pull of gravity. If the flow were downward, the pull of gravity would favor the flow. In that case, however, the sign of the quantity $\cos\beta$ in Eq. (2-35) would be negative, and the final equation [Eq. (2-36)] would be unchanged.

Equation (2-33) can be further modified by noting that

$$\frac{dL}{d\theta} = u \qquad (2\text{-}34)$$

and
$$\frac{dZ}{dL} = \cos \beta \qquad (2\text{-}35)$$

where dZ is the vertical distance between the ends of the element, and substituting for $dL/d\theta$ from Eq. (2-34) and $\cos \beta$ from Eq. (2-35). The result of the substitutions is

$$\frac{dp}{\rho} + \frac{g}{g_c} dZ + d\left(\frac{u^2}{2g_c}\right) = 0 \qquad (2\text{-}36)$$

Equation (2-36) is the differential form of the Bernoulli equation without friction. It applies as written either along a streamline or along a stream tube in which there is no variation in velocity within any cross section.

For compressible fluids, where the density ρ depends on the pressure p, Eq. (2-36) cannot be integrated without further information. For incompressible fluids, or where the variation of ρ is negligible, Eq. (2-36) can be integrated to give

$$\frac{p}{\rho} + \frac{g}{g_c} Z + \frac{u^2}{2g_c} = \text{const} \qquad (2\text{-}37)$$

where Z is the height above an arbitrary datum plane. Between two definite cross sections of the stream tube, or between two points along a streamline, Eq. (2-37) can be written

$$\frac{p_a}{\rho} + \frac{gZ_a}{g_c} + \frac{u_a^2}{2g_c} = \frac{p_b}{\rho} + \frac{gZ_b}{g_c} + \frac{u_b^2}{2g_c} \qquad (2\text{-}38)$$

The subscript a refers to the inlet station and subscript b to the outlet station, chosen as the limits for the application of Eq. (2-37).

Discussion of Bernoulli Equation. Any consistent units can be used in the various forms of the Bernoulli equation. The usual choice is pressure in lb force/ft^2; density in lb/ft^3; height in ft; velocity in ft/sec; g in ft/sec^2; and g_c in ft-lb/lb force-sec^2. Hour units can be used, of course, instead of second units, in u, g, and g_c.

Equation (2-38) is the Bernoulli equation without friction for the flow of incompressible fluids. It is of great importance in treating the flow of incompressible fluids. The equation states that in the absence of friction, when the velocity u is reduced, either the height above datum Z or the pressure p, or both, must increase. When the velocity increases, it can do so only at the expense of either Z or p. If the height is changed, compensation must be found in a change in either pressure or velocity. The reason for this complete interconvertibility of pressure, height, and velocity as shown by the Bernoulli equation becomes more apparent when it is noted that all terms have the dimensions feet times pounds force per pound.† Each term represents a mechanical-energy effect based on 1 lb mass of flowing fluid. The terms $(g/g_c)Z$ and $u^2/2g_c$ are the mechanical-potential and mechanical-kinetic energy, respectively, of 1 lb of fluid, and p/ρ represents the mechanical work done by forces, external to the stream tube, on the fluid in pushing the fluid into the tube or the work recovered from the fluid when the fluid leaves the tube. For these reasons, Eq. (2-37) is essentially a special case of the law that energy input equals energy output.

To apply Eq. (2-38) to a specific problem it is essential to identify the streamline or the stream tube and to choose a definite upstream station a and a definite downstream station b before using the equation. Stations a and b are chosen on the basis of convenience and are usually taken at locations at which the most is known about pressures, velocities, and heights.

Example 2-5. Brine, specific gravity 60°F/60°F = 1.15, is draining from the bottom of a large open tank through a standard 2-in. Schedule 40 pipe. The drainpipe ends at a point 15 ft below the surface of the brine in the tank. Considering a streamline starting at the surface of the brine in the tank and passing through the center of the drain line to the point of discharge and assuming that friction along the streamline is negligible, calculate the velocity of flow along the streamline at the point of discharge from the pipe.

Solution. To apply Eq. (2-38), choose station a at the brine surface and station b at the end of the streamline at the point of discharge. Since the pressure at both stations is atmospheric, p_a and p_b are equal, and $p_a/\rho = p_b/\rho$. At the surface of the brine, u_a is negligible, and the term $u_a^2/2g_c$ is dropped. The datum for measurement of heights can be taken through station b, so $Z_b = 0$ and $Z_a = 15$ ft. Substitution in Eq. (2-38) gives

$$\frac{15g}{g_c} = \frac{u_b^2}{2g_c}$$

and the velocity on the streamline at the discharge is

$$u_b = \sqrt{15 \times 2 \times 32.17} = 31.1 \text{ ft/sec}$$

Note that this velocity is independent of density and of pipe size.

Bernoulli Equation: Correction for Effects of Solid Boundaries. The Bernoulli equation for frictionless flow applies to fluid streams that are undisturbed by solids. Most fluid-flow problems encountered in chem-

† In hydraulic practice, Eq. (2-37) is written as

$$\frac{p}{\gamma} + Z + \frac{u^2}{2g} = \text{const} \tag{2-39}$$

where γ is the specific weight, in pounds force per cubic foot, as in Eq. (2-4). The terms in this equation all have the dimension of length, and are called heads. The term p/γ is a pressure head, Z is a static head, and $u^2/2g$ is a velocity head. The same terminology has carried over in the use of Eq. (2-37), and the terms therein are often called heads.

ical engineering, however, do involve streams that are influenced by solid boundaries, and which, therefore, contain boundary layers. This is especially true in the flow of fluids through pipes and other equipment where, as shown above, the entire stream may consist of a boundary layer.

To extend the Bernoulli equation to cover these practical situations two modifications are needed. The first, usually of minor importance, is a correction of the kinetic-energy term for the variation of the local velocity u over the cross section of the stream, and the second, of major importance, is the correction of the equation for the effect of fluid friction. Friction appears wherever the flow is modified by a solid boundary.

Also, the utility of the corrected Bernoulli equation in solving problems of flow of incompressible fluids is enhanced if provision is made in the equation for the work effect of a pump.

Kinetic Energy of Total Stream. The term $u^2/2g_c$ in the Bernoulli equation for frictionless flow is modified to take account of the velocity distribution in the cross section of the stream by so defining a correction factor α that the true kinetic energy of the entire stream can be calculated from the equation.

$$E_k = \frac{\overline{V}^2}{2\alpha g_c} \tag{2-40}$$

where E_k is the true value of the kinetic energy of the stream in feet times pounds force per pound and the other terms have their usual meanings. The factor α is called the *kinetic-energy correction factor*. It is dimensionless and approaches unity at high velocities, where u becomes nearly constant at all points in the cross section.

In laminar flow when the final velocity distribution has been fully developed through a circular channel, α can be calculated from the parabolic velocity distribution. It is found to be exactly 0.5. In laminar flow in a pipe, then, the kinetic energy of 1 lb of fluid is $\overline{V}^2/g_c$. In turbulent flow, because of the flatness of the velocity-distribution curve, α is nearly unity. As shown in Fig. 2-16, α varies with the Reynolds number and covers a range of about 0.88 to 0.97, increasing with N_{Re}. It can be taken as unity without appreciable error unless the kinetic-energy term is important in comparison with the other terms in the Bernoulli equation.†

When Eqs. (2-36) to (2-39) are used for the entire cross section of a channel, each $u^2/2g_c$ term is replaced by the equivalent term $\overline{V}^2/2\alpha g_c$.

Correction of Bernoulli Equation for Fluid Friction. Friction manifests itself by the disappearance of mechanical energy. In frictional flow, the quantity

$$\frac{p}{\rho} + \frac{u^2}{2g_c} + \frac{g}{g_c} Z$$

is not constant along a streamline, as called for by Eq. (2-37), but always

† The term $\rho\overline{V}^2/g_c$ in Eq. (2-23) represents the kinetic energy of the fluid leaving the viscometer tube. The velocity in the reservoir at the entrance of the tube is zero.

decreases in the direction of flow, and, in accordance with the principle of conservation of energy, an amount of heat equivalent to this loss in mechanical energy is generated. Fluid friction can be defined as any conversion of mechanical energy into heat in a flowing stream.

For incompressible fluids, the Bernoulli equation is corrected for friction by adding a term to the right-hand side of Eq. (2-38). Thus, after also

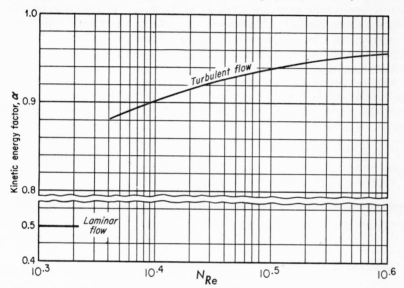

FIG. 2-16. Kinetic-energy factors. (*Kays.*[6])

introducing the kinetic-energy correction factors α_a and α_b, this equation becomes

$$\frac{p_a}{\rho} + \frac{g}{g_c} Z_a + \frac{\overline{V}_a^2}{2g_c\alpha_a} = \frac{p_b}{\rho} + \frac{g}{g_c} Z_b + \frac{\overline{V}_b^2}{2g_c\alpha_b} + H_f \qquad (2\text{-}41)$$

The term H_f represents all the friction (and therefore all the conversion of mechanical energy into heat) that occurs in the fluid between stations a and b. It differs from all the other terms in Eq. (2-41) in two ways: (1) The mechanical terms represent conditions *at* specific locations, namely, the inlet and outlet stations a and b, whereas H_f represents the loss of mechanical energy at all points *between* stations a and b. (2) The friction term differs from all others in Eq. (2-41) in not being interconvertible with the remaining quantities. The sign of H_f is always positive.

The practical physical significance of friction is apparent if Eq. (2-41) is written for the special case where $Z_a = Z_b$ and $\overline{V}_a = \overline{V}_b$. Then

$$\frac{p_a}{\rho} - \frac{p_b}{\rho} = H_f$$

If H_f is zero, $p_a = p_b$, and no pressure difference is needed to maintain flow. In actual equipment, where $H_f > 0$ and $p_a > p_b$, a pressure decrease in the direction of flow, $p_a - p_b$, is required to maintain flow, and this pressure drop must be available, either through a pump or some other source of pressure.

Pump Work in Bernoulli Equation. A pump is used in a flow system to increase the mechanical energy of the fluid. The increase in energy is used to maintain flow. The shaft work done by the pump is, by the sign convention of Chap. 1, negative, since it flows into the control surface rather than out of it. The shaft work of the pump $-W_s$ is inserted into the left-hand side of Eq. (2-41).

An ideal pump would operate without internal friction, and $-W_{sr}$, the shaft work supplied to the fluid, would equal $-W_s$, the shaft work used to drive the pump. In an actual pump, however, $-W_{sr}$ is less than $-W_s$ (or W_{sr} is greater than W_s). The difference is the friction generated in the pump. The over-all efficiency η is defined by the equation

$$\eta = \frac{-W_{sr}}{-W_s} = \frac{W_{sr}}{W_s} \tag{2-42}$$

The efficiency is always less than unity. It is a characteristic datum of a pump, and is discussed further in Chap. 3. The difference between $-W_s$ and $-W_{sr}$, or $-W_s(1 - \eta)$, is the mechanical energy transformed into heat in the pump.

Equation (2-41) with a term for pump work can be written

$$\frac{p_a}{\rho} + \frac{g}{g_c} Z_a + \frac{\overline{V}_a^2}{2\alpha_a g_c} - W_s\eta = \frac{p_b}{\rho} + \frac{g}{g_c} Z_b + \frac{\overline{V}_b^2}{2\alpha_b g_c} + H_f \tag{2-43}$$

Equation (2-43) applies when the density ρ can be considered constant. It is a final working equation for the treatment of problems on the flow of incompressible fluids.

Example 2-6. In the equipment shown in Fig. 2-17, a pump draws a solution, specific gravity = 1.84, from a storage tank through a 3-in. Schedule 40 steel pipe.

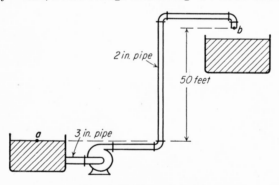

Fig. 2-17. Example 2-6.

The efficiency of the pump is 60 per cent. The velocity in the suction line is 3 ft/sec. The pump discharges through a 2-in. Schedule 40 pipe to an overhead tank. The end of the discharge pipe is 50 ft above the level of the solution in the feed tank. Friction losses in the entire piping system are 10 ft-lb force/lb. What pressure must the pump develop in pounds force per square inch? What is the horsepower of the pump?

Solution. Use Eq. (2-43). Take station a at the surface of the liquid in the tank and station b at the discharge end of the 2-in. pipe. Take the datum plane for elevations through the station a. Since the pressure at both stations is atmospheric, $p_a = p_b$. The velocity at station a is negligible because of the large diameter of the tank in comparison with that of the pipe. The kinetic-energy factor α can be taken as 1.0 with negligible error. Equation (2-43) becomes

$$-W_s\eta = \frac{g}{g_c} Z_b + \frac{\bar{V}_b^2}{2g_c} + H_f$$

By Appendix 4, the cross-sectional areas of the 3- and 2-in. pipes are 0.0513 and 0.0233 ft², respectively. The velocity in the 2-in. pipe is

$$\bar{V}_b = \frac{3 \times 0.0513}{0.0233} = 6.61 \text{ ft/sec}$$

then

$$-0.60W_s = 50 \frac{g}{g_c} + \frac{6.61^2}{64.34} + 10 = 60.68$$

and

$$-W_s = \frac{60.68}{0.60} = 101.1 \text{ ft-lb force/lb}$$

The pressure developed by the pump can be found by writing Eq. (2-43) over the pump itself. Station a is in the suction connection and station b in the pump discharge. The difference in level between suction and discharge can be neglected, so $Z_a = Z_b$, and Eq. (2-43) becomes

$$\frac{p_b - p_a}{\rho} = \frac{\bar{V}_a^2 - \bar{V}_b^2}{2g_c} - W_s\eta$$

The pressure developed by the pump is

$$p_b - p_a = 1.84 \times 62.37 \left(\frac{3^2 - 6.61^2}{2 \times 32.17} + 60.68 \right)$$

$$= 6,913 \text{ lb force/ft}^2, \text{ or } 6,913/144 = 48.0 \text{ lb force/in.}^2$$

The power used by the pump is the product of $-W_s$ and the mass flow rate divided by the conversion factor, 1 hp = 550 ft-lb force/sec. The mass flow rate is

$$w = 0.0513 \times 3 \times 1.84 \times 62.37 = 17.66 \text{ lb/sec}$$

and the power is

$$P = \frac{-wW_s}{550} = \frac{17.66 \times 101.1}{550} = 3.25 \text{ hp}$$

FLUID FRICTION

The calculation of friction in specific cases of fluid flow through equipment and the minimizing of friction where it is excessive are important engineering problems. In some situations, the calculation of friction can be made with considerable accuracy. In many cases, however, judgment and experience must be relied upon for practical estimates of friction. Then knowledge of the mechanism of friction and of the laws that apply to the flow of actual fluids is helpful.

Types of Fluid Friction: Skin Friction and Form Friction. Friction in a flowing fluid is caused by solid boundaries in contact with the stream. Two flow phenomena lead to fluid friction. These types of friction

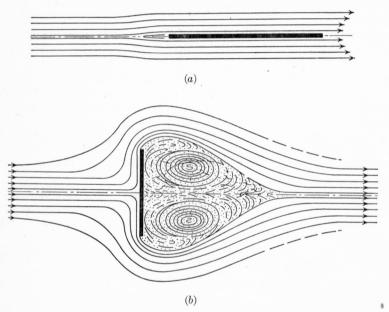

(a)

(b)

Fig. 2-18. Flow past flat plate: (a) Flow parallel with plate.   (b) Flow perpendicular to plate.

may be illustrated by considering flow past a thin plate, first with the plate parallel with the direction of flow and second with the plate perpendicular to the flow.

In the first situation, shown in Fig. 2-18a, a boundary layer forms on each side of the plate, and the friction originates in the tractive shear forces in the boundary layers. These forces act through the distance traveled by the fluid along the streamlines in the boundary layer and perform work. Just as in the case of mechanical friction that is generated when a solid body is pushed along a rough surface, the work performed by the shear forces is converted into heat. This kind of fluid friction, which occurs only in boundary layers in contact with solid boundaries, is called skin friction. In skin friction, the boundary layer does not break contact with the solid and is under the control of the solid.

In the second situation, as shown in Fig. 2-18b, the boundary layer is forced to flow across the plate outwardly to the edges, and because of its momentum, it is unable to maintain contact with the back of the plate. Accordingly, it breaks away or separates from the solid and forms a current that is no longer under the control of the solid. This phenomenon is called boundary-layer separation. Well downstream the separated currents rejoin.

Immediately behind the plate, in the space enclosed by the back of the plate and the separated currents, there exists a mass of fluid called the wake. The wake consists of large-scale swirling eddies, called vortices, which are kept in motion by the shear stresses between the wake and the separated current. The kinetic energy of the rotating vortices is replenished by absorbing kinetic energy from the current and is converted into heat in the wake by shear. As a result of this transfer of energy, wake formation corresponds to a large friction loss.

Friction resulting from boundary-layer separation and wake formation is called form friction, since its character and extent depend on the shape and position of the solid.

In a given situation, both skin and form friction may be active in varying degrees. In the case of Fig. 2-18a, the friction is entirely skin; in that of Fig. 2-18b, the friction is largely form friction because of the large wake, and skin friction is unimportant.

Friction in Long, Straight Pipe. Excepting a short distance at the entrance, the friction in a long, straight pipe is entirely skin friction, and, for incompressible fluids, the friction per unit length of tube is independent of length.

In laminar flow in circular pipes, the friction loss is given by the Hagen-Poiseuille relation [Eq. (2-22)] and is

$$H_{fs} = \frac{32\mu L \overline{V}}{g_c D^2 \rho} \tag{2-44}$$

The subscript s calls attention to the fact that only skin friction is involved. In turbulent flow, the relationship between H_{fs} and the conditions of flow is too complicated for simple mathematical treatment, although considerable progress has been made along theoretical lines. This is a situation where dimensional analysis and experiment have been combined to yield a satisfactory relationship.

For long, smooth pipes carrying an incompressible fluid, it is reasonable to assume that H_{fs}/L, the friction loss per unit length of pipe, might depend upon: (1) the viscosity μ of the fluid, (2) the density ρ of the fluid, (3) the average velocity $\overline{V}$ of the fluid, and (4) the diameter D of the pipe. Also, since Newton's law is involved in accelerating the eddies in the fluid, the conversion factor g_c should be added to this list of factors; or

$$\frac{H_{fs}}{L} = \mathfrak{F}(\mu,\rho,\overline{V},D,g_c) \tag{2-45}$$

where $\mathfrak{F}$ means "function of." The procedure described in Chap. 1 is applied. The equation of dimensions is

$$\left[\frac{H_{fs}}{L}\right] = [\mu]^a[\rho]^b[\overline{V}]^c[D]^d[g_c]^e \tag{2-46}$$

Substitution of dimensions from Appendix 1 gives

$$\overline{F}\overline{M}^{-1} = \overline{M}^a\overline{L}^{-a}\overline{\theta}^{-a}\overline{M}^b\overline{L}^{-3b}\overline{L}^c\overline{\theta}^{-c}\overline{L}^d\overline{M}^e\overline{L}^e\overline{\theta}^{-2e}\overline{F}^{-e}$$

Collecting the exponents of the fundamental units gives

Exponents of $\overline{F}$: $\qquad 1 = -e$ $\qquad\qquad\qquad\qquad\qquad\qquad$ (2-47)

Exponents of $\overline{L}$: $\qquad 0 = -a - 3b + c + d + e$ $\qquad\qquad$ (2-48)

Exponents of $\overline{M}$: $\qquad -1 = a + b + e$ $\qquad\qquad\qquad\qquad$ (2-49)

Exponents of $\overline{\theta}$: $\qquad 0 = -a - c - 2e$ $\qquad\qquad\qquad\qquad$ (2-50)

There are four equations and five unknowns. One unknown must be chosen arbitrarily and the other four calculated in terms of it. A valid result is obtained whichever choice is made. A common choice is a. From Eq. (2-47),

$$e = -1$$

From Eq. (2-49,

$$b = -1 - a - e = -1 - a + 1 = -a$$

From Eq. (2-50),

$$c = -a - 2e = -a + 2$$

From Eq. (2-48),

$$d = a + 3b - c - e = a - 3a + a - 2 + 1 = -a - 1$$

Equation (2-46) becomes

$$\left[\frac{H_{fs}}{L}\right] = [\mu]^a[\rho]^{-a}[\overline{V}]^{-a+2}[D]^{-a-1}[g_c]^{-1}$$

Collecting all factors containing a in one group and those containing integers in the other gives

$$\frac{H_{fs}Dg_c}{L\overline{V}^2} = \phi\left(\frac{D\overline{V}\rho}{\mu}\right) = \phi(N_{\mathrm{Re}}) \tag{2-51}$$

Again the Reynolds number appears as a controlling factor. The term $\phi(N_{\mathrm{Re}})$ means only "a function of the Reynolds number," as dimensional analysis tells nothing about the form of the function. It simply states that, if the variables have been properly chosen, for every numerical value of the Reynolds number there is a definite numerical value of the dimensionless quantity $H_{fs}Dg_c/L\overline{V}^2$.

It is customary in chemical engineering practice to define a friction factor f by the equation

$$f = \frac{H_{fs}Dg_c}{2L\overline{V}^2} = \frac{\phi(N_{\mathrm{Re}})}{2} = \Phi(N_{\mathrm{Re}}) \tag{2-52}$$

Then

$$H_{fs} = \frac{2fL\overline{V}^2}{Dg_c} = 4f\frac{L}{D}\frac{\overline{V}^2}{2g_c} \tag{2-53}$$

For smooth pipes, f depends only on N_{Re}. Equation (2-53) is called the Fanning equation, and the factor f is the Fanning friction factor.[†]

For laminar flow, a formula for f is obtained by substituting H_{fs} from Eq. (2-53) into Eq. (2-44), which gives

$$f = \frac{16\mu}{D\overline{V}\rho} = \frac{16}{N_{\mathrm{Re}}} \tag{2-54}$$

Friction-factor Chart. In industrial practice, turbulent flow is the rule. For turbulent flow, where the relation between f and N_{Re} is not simple, an experimental relationship has been found by hundreds of careful experiments in which the pressure drop from skin friction has been measured for a wide range of tube diameters, tube lengths, viscosities, densities, and flow rates. The friction factor, calculated from the observed data, has been plotted against the Reynolds numbers, as suggested by Eq. (2-52). The result is shown in Fig. 2-19. The range of Reynolds numbers is so great (from approximately 100 to 3,000,000) that it is necessary to plot the logarithm of the friction factor against that of the Reynolds number or, what is an equivalent method, to plot the data on log-log coordinates. All data for smooth pipe fall on a single curve, which is the lowest line on Fig. 2-19. The straight line with a slope of -1, covering the range of Reynolds numbers below 2,100, shows the friction factor for laminar flow in circular conduits and corresponds to Eq. (2-54). The lower curved line, covering Reynolds numbers greater than 3,000, applies to turbulent flow. The range between 2,100 and approximately 4,000 is a transition range, where the flow may be either laminar or turbulent, depending mainly on the conditions at the entrance of the channel.

The following empirical equation for the friction factor for turbulent flow in smooth pipes fits the data with a precision of about ± 5 per cent.[1]

$$f = 0.00140 + \frac{0.125}{N_{\mathrm{Re}}^{0.32}} \tag{2-55}$$

[†] Another friction factor, equal to $4f$, is found in many engineering references.

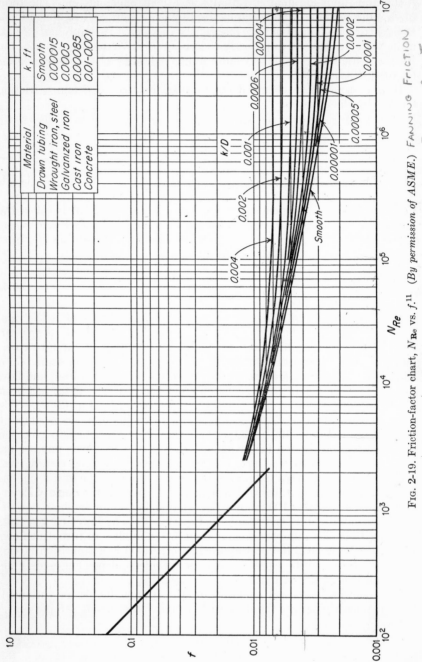

Fig. 2-19. Friction-factor chart, N_{Re} vs. f.[11] (*By permission of ASME.*) FANNING FRICTION FACTOR CHART.

It is apparent from Eq. (2-55) that as the Reynolds number increases, the friction factor f approaches asymptotically the magnitude 0.0014. At large Reynolds numbers, then, the friction factor becomes independent of viscosity, and, by Eq. (2-53), H_{fs} is proportional to $L\bar{V}^2/D$.

The friction H_{fs} calculated from Eq. (2-53) can be substituted for H_f in Eq. (2-43) if, and only if, the entire friction loss is the skin friction from flow through a straight pipe. Additional friction losses are encountered at the entrance and exit of each length of pipe, and unless the pipe is so long that these effects are unimportant relative to the skin friction, they must be included in the H_f term of Eq. (2-43).

Example 2-7. Brine, specific gravity 60°F/60°F = 1.18, viscosity = 2.5 centipoises, is to be pumped through 100 ft of smooth copper tube having an inside diameter of 0.995 in. The flow rate is to be 25 gal/min. What is the pressure drop from friction, in pounds force per square inch? How much power is required to overcome friction?

Solution. The cross-sectional area of the pipe is 0.995² $(\pi/4)$ = 0.777 in.² The average velocity is

$$\bar{V} = \frac{25 \times 231}{60 \times 0.777 \times 12} = 10.32 \text{ ft/sec}$$

The Reynolds number is

$$N_{Re} = \frac{(0.995/12) \times 10.32 \times 62.37 \times 1.18}{6.72 \times 10^{-4} \times 2.5} = 3.75 \times 10^4$$

From Fig. 2-19, $f = 0.0056$. From Eq. (2-53),

$$H_{fs} = \frac{2 \times 0.0056 \times 100 \times 10.32^2}{(0.995/12) \times 32.17} = 44.7 \text{ ft-lb force/lb}$$

The pressure drop corresponding to this friction is

$$p_a - p_b = H_{fs}\rho = 44.7 \times 1.18 \times 62.37$$

$$= 3{,}288 \text{ lb force/ft}^2, \text{ or } \frac{3{,}288}{144} = 22.8 \text{ lb force/in.}^2$$

The power needed to overcome friction can be calculated from the relationship

$$\text{Power (hp)} = \frac{\text{ft-lb force}}{33{,}000 \text{ min}} = \frac{\text{ft-lb force}}{33{,}000 \text{ lb}} \frac{\text{lb}}{\text{min}}$$

The flow rate is $25 \times 1.18 \times 62.37/7.48 = 246$ lb/min, and the power is $44.7 \times 246/33{,}000 = 0.33$ hp.

Effect of Roughness. The discussion has thus far been restricted to smooth channels, without defining smoothness. It has long been known that, in turbulent flow, a rough pipe leads to a larger friction factor for a given Reynolds number than does a smooth pipe. If a rough pipe is smoothed, the friction factor will be reduced. Continued polishing can get a tube so smooth that additional polishing does not further reduce the friction factor for a given Reynolds number. The tube is then said to be hydraulically smooth. The lowest curve in Fig. 2-19 is for a smooth pipe.

Figure 2-20 shows several idealized kinds of roughness. The height of a single unit of roughness is denoted by k and is called the *roughness param-eter*. From dimensional analysis, f is a function of both N_{Re} and the relative roughness k/D, where D is the diameter of the pipe. For any given type of roughness, e.g., that shown in Fig. 2-20a and b, it can be expected that a different curve of f vs. N_{Re} would be found for each magnitude of the relative roughness and also that, for other types of roughness, such as those shown in Fig. 2-20c and d, a different family of curves of N_{Re} vs. f would be found for each type of roughness. Experiments on artificially roughened pipe have confirmed these expectations. It has also been found

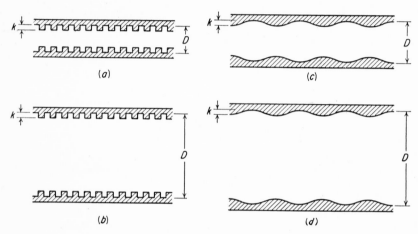

FIG. 2-20. Types of roughness. Parameter k is equal in all cases.

that all clean, new, commercial pipe seems to have the same kind of roughness and that each material of construction has its own characteristic roughness parameter. On Fig. 2-19 are plotted several curves, lying above that for smooth pipe, each curve drawn for a constant value of k/D. The parameters for several common metals are given in the figures. Clean wrought-iron or steel pipe, for example, has a roughness parameter of 0.00015 ft, regardless of the diameter of the pipe. The roughness of drawn copper and brass pipe is so small that such pipe can be considered smooth unless the Reynolds number is large.

Old, fouled, and corroded pipe can become very rough, and the character of the roughness differs from that of clean pipe, so the roughness curves of Figs. 2-19 and 2-21 do not apply to such pipe.

Roughness has no appreciable effect on the friction factor for laminar flow unless k is so large that the measurement of the diameter becomes uncertain.

Calculation of Flow Rate from Known Skin Friction. Figure 2-19 is useful for calculating H_{fs} from a known pipe size and condition of flow. It is not convenient for the calculation of the flow rate from a known H_{fs}, because

$\overline{V}$, the average linear velocity, appears in both N_{Re} and f. A trial-and-error solution can be used for this calculation, but a more convenient direct solution can be obtained by converting Fig. 2-19 into a plot in which the velocity appears in only one coordinate. One way of accomplishing this is to plot f vs. $N_{Re}\sqrt{f}$, as shown in Fig. 2-21. That $\overline{V}$ is thereby eliminated from the abscissa is shown by the following equation for $N_{Re}\sqrt{f}$.

$$N_{Re}\sqrt{f} = \frac{D\overline{V}\rho}{\mu}\sqrt{\frac{H_{fs}Dg_c}{2L\overline{V}^2}} = \frac{D\rho}{\mu}\sqrt{\frac{H_{fs}Dg_c}{2L}} \qquad (2\text{-}56)$$

The lines for the various roughness parameters are also plotted in Fig. 2-21.

To use Fig. 2-21, the abscissa is calculated from the known friction loss, the ordinate f is read, and the velocity $\overline{V}$ is calculated from f by means of Eq. (2-52).

An equation for the lowest line in Fig. 2-21 is

$$\frac{1}{\sqrt{f}} = 4.0 \log N_{Re}\sqrt{f} - 0.40 \qquad (2\text{-}57)$$

This is the von Kármán equation.[10c] It is consistent with a fundamental treatment of turbulent flow in pipes,[10b] and is recommended for extrapolation to large Reynolds numbers.

Shapes Other than Circular. For turbulent flow in closed channels that have cross sections other than circular, Figs. 2-19 and 2-21 can be used if the diameter D, in both coordinates, is replaced by an "equivalent diameter," which is four times an important quantity called the *hydraulic radius.* The hydraulic radius is the ratio of the area of the cross section to the wetted perimeter of that section. It is denoted by r_H, and its defining equation is

$$r_H = \frac{S}{L_p} \qquad (2\text{-}58)$$

where S = cross-sectional area, ft^2
  L_p = perimeter, ft
For example, the hydraulic radius for a circular channel is $(\pi D^2/4)(1/\pi D)$ = $D/4$, and that for a square conduit of width of side b is $b^2/4b = b/4$. In terms of r_H, $f = 2g_c r_H H_{fs}/L\overline{V}^2$, $N_{Re} = 4r_H\overline{V}\rho/\mu$, and $N_{Re}\sqrt{f} = (4r_H\rho/\mu)\sqrt{2H_{fs}r_Hg_c/L}$.

The simple hydraulic-radius rule does not apply in laminar flow through noncircular sections. Equations for various cross sections are given in standard handbooks.[12b]

Example 2-8. How much flow, in barrels per day, can be expected if crude oil, specific gravity = 0.85, viscosity = 5.1 centipoises, is pumped through 1,500 ft of 2-in. Schedule 40 steel pipe under a pressure drop of 50 lb force/in.²?

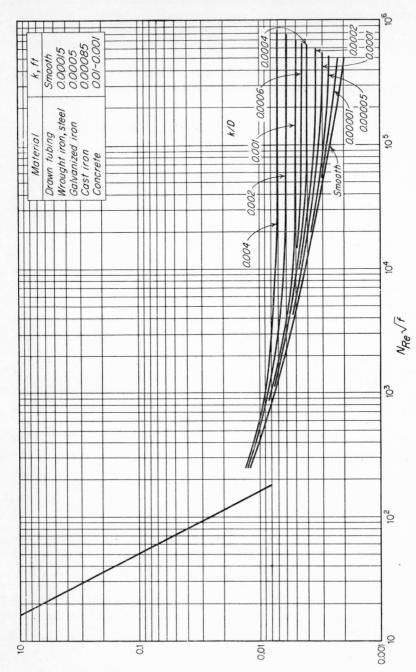

Material	k, ft
	Smooth
Drawn tubing	0.00015
Wrought iron, steel	0.0005
Galvanized iron	0.00085
Cast iron	0.001–0.001
Concrete	

Fig. 2-21. Friction-factor chart, $N_{Re}\sqrt{f}$ vs. f.[11] (By permission of A.S.M.E.)

Solution. The friction is

$$H_{fs} = \frac{p_a - p_b}{\rho} = \frac{50 \times 144}{0.85 \times 62.37} = 136.0 \text{ ft-lb force/lb}$$

The inside diameter and cross-sectional area of a standard 2-in. Schedule 40 pipe are 2.067 in. and 0.0233 ft², respectively. Then

$$D = \frac{2.067}{12} = 0.172 \text{ ft,}$$

and the abscissa for use with Fig. 2-21 is

$$N_{\text{Re}} \sqrt{f} = \frac{0.172 \times 0.85 \times 62.4}{5.1 \times 6.72 \times 10^{-4}} \sqrt{\frac{32.17 \times 0.172 \times 136.0}{2 \times 1,500}} = 1.33 \times 10^3$$

For clean steel pipe, $k = 0.00015$ ft, and the relative roughness k/D is $(0.00015 \times 12)/2.067 = 0.00087$. From Fig. 2-21, for $N_{\text{Re}}\sqrt{f} = 1.33 \times 10^3$ and $k/D = 0.00087$, the friction factor f is 0.074. From Eq. (2-52), the average velocity is

$$\overline{V} = \sqrt{2 \times 32.17 \frac{136.0}{1,500} \frac{0.0431}{0.074}} = 1.84 \text{ ft/sec}$$

Since 1 bbl is 42 U.S. gal, the flow is

$$1.84 \times 0.0233 \times 24 \times 3,600 \frac{7.48}{42} = 660 \text{ bbl/day}$$

Effect of Heat Transfer on Friction Factor. The methods for calculating friction described thus far apply only where there is no transfer of heat between the wall of the channel and the fluid. When the fluid is either heated or cooled by a conduit wall hotter or colder than the fluid, the friction factor can be corrected for the effect of heat transfer by the following method: (1) The Reynolds number is calculated on the assumption that the fluid temperature is the "mean bulk temperature," which is defined as the arithmetic average of the inlet and outlet temperatures. (2) The friction factor corresponding to this temperature is *divided* by a factor ψ, which, in turn, is calculated from the following equations.[10d] For $N_{\text{Re}} > 2,100$,

$$\psi = 1.0 \left(\frac{\mu}{\mu_s}\right)^{0.14} \tag{2-59}$$

For $N_{\text{Re}} < 2,100$,

$$\psi = 1.1 \left(\frac{\mu}{\mu_s}\right)^{0.25} \tag{2-60}$$

where μ = viscosity of fluid at mean bulk temperature
μ_s = viscosity at temperature of wall of conduit

Example 2-9. Water enters the tubes of a heater at a velocity of 8 ft/sec and a temperature of 100°F. The water leaves the tubes at a temperature of 180°F.

The temperature of the tube wall is 210°F. The tubes are 0.652 in. ID and are 12 ft long. What is the skin friction in feet times pounds force per pound?

Solution. The mean bulk temperature of the water is $(100 + 180)/2 = 140°F$. At this temperature, from Appendix 14, μ, the viscosity of water, is 0.470 centipoise, and the density of the water entering the tubes is 62.00 lb/ft³. The Reynolds number is

$$N_{Re} = \frac{(0.652/12)8 \times 62.00}{0.470 \times 6.72 \times 10^{-4}} = 8.53 \times 10^4$$

The friction factor, uncorrected for the heat-transfer effect, is, from Fig. 2-19, 0.00455. The viscosity μ_s of water at 210°F, the tube-wall temperature, is 0.287 centipoise. The correction factor is, by Eq. (2-59),

$$\psi = 1.0 \left(\frac{\mu}{\mu_s}\right)^{0.14} = \left(\frac{0.470}{0.287}\right)^{0.14} = 1.07$$

and the corrected friction factor is

$$f = \frac{0.00455}{1.07} = 0.00426$$

The friction loss is, by Eq. (2-53),

$$H_{fs} = \frac{2 \times 0.00426 \times 12 \times 8^2}{32.17 \times 0.652/12} = 3.74 \text{ ft-lb force/lb}$$

Friction from Changes in Velocity or Direction. Whenever the velocity of a fluid is changed, either in direction or magnitude, by a change in the direction or size of the conduit, friction additional to the skin friction from flow through the straight pipe is generated. These friction effects include form friction resulting from vortices which develop when the normal streamlines are disturbed and when boundary-layer separation occurs. In most situations, these effects cannot be calculated accurately, and it is necessary to rely on empirical data. It is often possible to estimate friction of this kind in specific cases from a knowledge of the losses in known arrangements.

Effect of Fittings and Valves. Fittings and valves disturb the normal flow lines and cause friction. In short lines with many fittings, the friction loss from the fittings may be greater than that from the straight pipe. A common method of estimating the friction from fittings is to use empirical factors that evaluate the effect of each fitting in terms of equivalent length of straight pipe. A list of factors is given in Table 3-3. For example, one 90° T, used as an elbow with the fluid entering the branch, has an equivalent length of 90 pipe diameters. In a 2-in. line, one T will add a friction equal to that of $90 \times 2/12 = 15$ ft of straight pipe. This length is added to the length of straight pipe to give the magnitude of L for use in Eq. (2-53).

The length of straight pipe, to which the equivalent lengths of the fittings are added, is the length measured from face to face of the fittings, as shown in Fig. 2-22.

Friction Loss from Sudden Expansion of Cross Section. If the cross section of the conduit is suddenly enlarged, the fluid stream separates from the wall and issues into the enlarged section as a jet. The jet then expands to fill

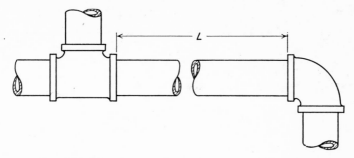

FIG. 2-22. Measurement of straight pipe between fittings.

the entire cross section of the larger conduit. The usual vortices, characteristic of separation, exist in the space between the jet and the conduit wall. Figure 2-23 shows this effect.

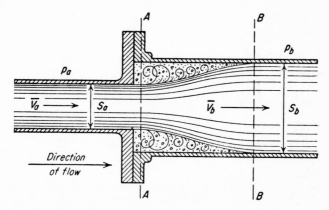

FIG. 2-23. Flow at sudden enlargement of cross section.

The friction loss H_{fe} from a sudden expansion of cross section is proportional to the velocity head of the fluid in the small conduit and can be written

$$H_{fe} = K_e \frac{\overline{V}_a^2}{2g_c} \tag{2-61}$$

where K_e is a proportionality factor called the *expansion-loss coefficient*, and $\overline{V}_a$ is the average velocity in the smaller, or upstream, conduit. The coefficient K_e depends primarily on S_a/S_b, which is the ratio of the area of the upstream cross section to that of the downstream, and secondarily on the upstream Reynolds number. The relation between K_e, S_a/S_b, and

N_{Re} is shown, for conduits of circular cross section, in Fig. 2-24, in which K_e is plotted against S_a/S_b for constant values of the Reynolds number for the smaller conduit. The relation between K_e and S_a/S_b for laminar flow in the smaller conduit is also shown. These relations are calculated from theoretical equations based on the application of the momentum law of

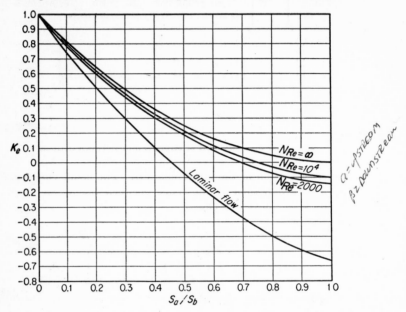

FIG. 2-24. Coefficient, loss from sudden enlargement of cross section.† (*By permission of ASME.*)

Eq. (1-9) and from the known velocity distribution in laminar and turbulent flows.[6]

For laminar flow, the equation plotted in Fig. 2-24 is

$$K_e = 1 - 2.667 \frac{S_a}{S_b} + \left(\frac{S_a}{S_b}\right)^2 \qquad (2\text{-}62)$$

For turbulent flow at large Reynolds numbers, it is

$$K_e = \left(1 - \frac{S_a}{S_b}\right)^2 \qquad (2\text{-}63)$$

The curves of Fig. 2-24 can be used for single tubes or for multiple tubes discharging into a common header. Equation (2-63) and the curve for large Reynolds numbers can be used for cross sections other than circular. The curves for lower Reynolds numbers in the turbulent-flow range apply approximately to cross sections other than circular, but the curve for laminar flow applies only to circular cross sections.

† Including effect of α of kinetic energy. This leads to negative values of K_e at large S_a/S_b.

Friction Loss from Sudden Contraction of Cross Section. When the cross section of the conduit is suddenly reduced, the fluid stream cannot follow around the sharp corner, and the stream breaks contact with the wall of the conduit. A jet is formed, which flows into the stagnant fluid in the smaller section. The jet first contracts and then expands to fill the smaller cross section, and downstream from the contraction the normal velocity distribution eventually is established. The cross section of minimum area, at which the jet changes from a contraction to an expansion, is called the

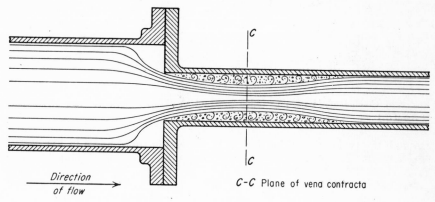

Direction of flow

C-C Plane of vena contracta

Fig. 2-25. Flow at sudden contraction of cross section.

vena contracta. The flow pattern of a sudden contraction is shown in Fig. 2-25. Section *CC* is drawn at the vena contracta. Vortices appear as shown in the figure.

The friction loss from sudden contraction is proportional to the velocity head in the smaller conduit, and can be calculated by the equation

$$H_{fc} = K_c \frac{\overline{V}_b^2}{2g_c} \tag{2-64}$$

where the proportionality factor K_c is called the *contraction-loss coefficient* and $\overline{V}_b$ is the average velocity in the smaller, or downstream, section. This coefficient is a function of the area ratio S_b/S_a and of the Reynolds number in the smaller conduit. The relation between K_c and S_b/S_a, at constant Reynolds numbers, is shown in Fig. 2-26. No simple theoretical equation applies to contraction losses.[6] The relations in Fig. 2-26 apply under the same conditions as those in Fig. 2-24, and they are subject to the same limitations.

Expansion and Contraction Losses in Bernoulli Equation. Expansion and contraction losses are incorporated in the H_f term of Eq. (2-43). They are combined with the friction loss through the straight pipe to give the total friction loss. Consider, for example, the flow of an incompressible fluid

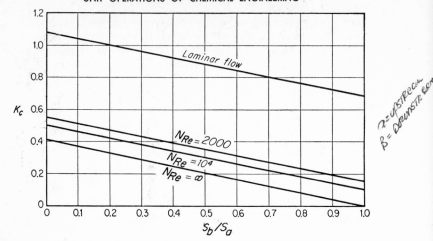

FIG. 2-26. Coefficient, loss from sudden contraction of cross section.† (*By permission of ASME.*)

through the two enlarged headers and connecting tube shown in Fig. 2-27. Let $\overline{V}$ be the average velocity in the tube, D the diameter of the tube, and

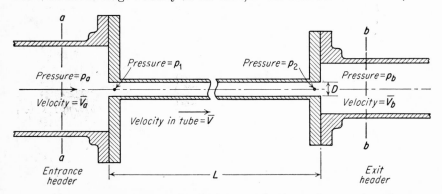

FIG. 2-27. Flow of incompressible fluid through headers and connecting tubes.

L the length of the tube. The friction loss in the straight tube is, by Eq. (2-53),

$$H_{fs} = 4f \frac{L}{D} \frac{\overline{V}^2}{2g_c} \qquad (2\text{-}65)$$

The contraction loss at the entrance to the tube is, by Eq. (2-64),

$$H_{fc} = K_c \frac{\overline{V}^2}{2g_c} \qquad (2\text{-}66)$$

† Includes effect of α of kinetic energy term.

The expansion loss at the exit of the tube is. by Eq. (2-61),

$$H_{fe} = K_e \frac{\overline{V}^2}{2g_c} \qquad (2\text{-}67)$$

Neglecting skin friction in the entrance and exit headers, the total friction is

$$H_f = H_{fs} + H_{fc} + H_{fe} = \left(4f\frac{L}{D} + K_c + K_e\right)\frac{\overline{V}^2}{2g_c} \qquad (2\text{-}68)$$

To write the Bernoulli equation for this case, take station a in the inlet header and station b in the outlet header. Because there is no pump between stations a and b, $W_s = 0$, and Eq. (2-43) becomes †

$$\frac{p_a - p_b}{\rho} + \frac{g}{g_c}(Z_a - Z_b) + \frac{\overline{V}_a^2 - \overline{V}_b^2}{2g_c} = \left(4f\frac{L}{D} + K_c + K_e\right)\frac{\overline{V}^2}{2g_c} \qquad (2\text{-}69)$$

It can be seen that the friction loss is expressible as a number of "velocity heads." For long pipes, the expansion and contraction losses, which are measured by the coefficients K_c and K_e, are negligible in comparison with the friction in straight pipe, which is measured by the coefficient $4f(L/D)$. The effect of valves and fittings is accounted for in the L term of the friction coefficient for straight pipe.

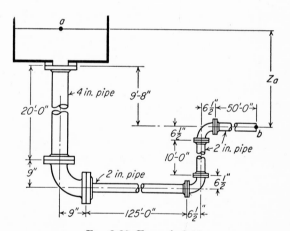

FIG. 2-28. Example 2-10.

Example 2-10. Figure 2-28 represents an elevated tank connected to a pipeline. The system contains water at 180°F. What must be the height of the water surface in the tank to produce a flow of 100 gal/min through the line?

Solution. The Bernoulli equation (2-43) may be written between station a, in the surface of the water in the tank, and station b, at the discharge from the 2-in.

† Note α not used in kinetic energy term because of inclusion in K_e and K_c.

pipe. The datum plane of elevations may be taken through point b, and $Z_b = 0$. There is no pump, and $W_s = 0$; the pressure at both stations is atmospheric, so $p_a = p_b$; the diameter of the tank may be assumed to be large enough to make $\overline{V}_a$ zero. Equation (2-43) becomes

$$Z_a = \frac{\overline{V}_b^2}{2\alpha_b g_c} + H_f$$

The friction H_f consists of: (1) the contraction loss at the exit of the tank, (2) the friction in the straight length of 4-in. pipe, (3) the contraction loss between the 4-in. pipe and the 2-in. pipe, and (4) the straight-pipe friction in the 2-in. pipe. Of these, (1) and (2) are associated with the 4-in. pipe, and (3) and (4) are associated with the 2-in. pipe.

FRICTION IN 4-IN. PIPE. The diameter and cross-sectional areas of the 4-in. pipe are, by Appendix 4, 4.026 in. and 0.0884 ft², respectively. The average velocity in the pipe is, then,

$$\overline{V} = \frac{100}{60 \times 7.48 \times 0.0884} = 2.52 \text{ ft/sec}$$

The viscosity and density of the water are, by Appendix 14, 0.347 centipoise, and 60.58 lb/ft³, respectively. The Reynolds number in the 4-in. pipe is

$$N_{Re} = \frac{(4.026/12)2.52 \times 60.58}{0.347 \times 6.72 \times 10^{-4}} = 2.2 \times 10^5$$

From Table 3-3, the equivalent length of one long-radius L is 20 pipe diameters, or $20 \times 4.026/12 = 6.7$ ft, and the total equivalent length of the 4-in. pipe is $20 + 6.7 = 26.7$ ft. The relative roughness is

$$\frac{k}{D} = 0.00015 \frac{12}{4.026} = 0.00045$$

From Fig. 2-19, the friction factor f, for $N_{Re} = 2.2 \times 10^5$ and $k/D = 0.00045$, is 0.0046. The friction coefficient for straight pipe is

$$4f\frac{L}{D} = \frac{4 \times 0.0046 \times 26.7}{4.026/12} = 1.46$$

CONTRACTION LOSS AT EXIT OF TANK. The area ratio S_b/S_a is zero, and, from Fig. 2-26, $K_c = 0.43$. The total friction associated with the 4-in. pipe is, then,

$$\frac{(0.43 + 1.46)2.52^2}{2 \times 32.17} = 0.19 \text{ ft-lb force/lb}$$

The diameter and cross-sectional area of the 2-in. pipe are, by Appendix 4, 2.067 in. and 0.0233 ft², respectively, and the average velocity is

$$\overline{V}_b = \frac{100}{60 \times 7.48 \times 0.0233} = 9.56 \text{ ft/sec}$$

The Reynolds number is

$$N_{Re} = \frac{(2.067/12)9.56 \times 60.58}{0.347 \times 6.72 \times 10^{-4}} = 4.3 \times 10^5$$

CONTRACTION LOSS BETWEEN 4- AND 2-IN. PIPES. The area ratio S_b/S_a is 0.0233/0.0884 = 0.26. From Fig. 2-26, $K_c = 0.35$.

FRICTION IN 2-IN. PIPE. The equivalent length of two L's is $2 \times 20 \times 2.067/12$ = 6.9 ft. The total equivalent length of pipe and fittings is $125.0 + 10.0 + 50.0 + 6.9 = 191.9$ ft. The relative roughness is $0.00015 \times 12/2.067 = 0.00087$. For $k/D = 0.00087$ and $N_{Re} = 4.3 \times 10^5$ the friction factor f is 0.0050, and the friction coefficient for straight pipe is

$$\frac{4fL}{D} = \frac{4 \times 0.0050 \times 191.9}{2.067/12} = 22.27$$

The total friction associated with the 2-in. pipe is

$$\frac{(0.35 + 22.27)9.56^2}{2 \times 32.17} = 32.15 \text{ ft-lb force/lb}$$

The total friction in the entire system is

$$H_f = 0.19 + 32.15 = 32.34 \text{ ft-lb force/lb}$$

From Fig. 2-16, the kinetic-energy factor for $N_{Re} = 4.3 \times 10^5$ is $\alpha = 0.95$. Then, substitution in the above equation gives

$$Z_a = \frac{9.56^2}{0.95 \times 2 \times 32.17} + 32.34$$

$$= 1.50 + 32.34 = 33.84 \text{ ft}$$

Since the discharge is 9 ft 8 in., or 9.67 ft, below the bottom of the tank, the water level is $33.84 - 9.67 = 24.2$ ft above the bottom of the tank.

Example 2-11. Water at 40°F is to flow through 1,000 ft of horizontal Schedule 40 pipe at the rate of 150 gal/min. A head of 20 ft is available. What size pipe must be used?

Solution. In Fig. 2-19, the abscissa and ordinate are both functions of the pipe diameter, and a trial-and-error solution must be used. A diameter is chosen at random, the friction loss calculated, and the result so obtained compared with the available head to determine whether the assumed pipe size was too large or too small. A second assumption should then give further information, on the basis of which the proper size can be chosen. Since standard pipe comes only in certain definite sizes, it is necessary only to "bracket" the known head to choose the pipe needed, which will be the smallest pipe that allows at least 150 gal/min of flow under the specified head.

The density and viscosity are, from Appendix 14, 62.43 lb/ft³ and 1.546 centipoises, respectively.

ASSUME 4-IN. PIPE. From Appendix 4, the diameter and cross-sectional area are 4.026 in. and 0.0884 ft², respectively. The average velocity is

$$\overline{V} = \frac{150}{60 \times 7.48 \times 0.0884} = 3.78 \text{ ft/sec}$$

The Reynolds number is

$$N_{Re} = \frac{(4.026/12)3.78 \times 62.43}{1.546 \times 6.72 \times 10^{-4}} = 7.6 \times 10^4$$

The friction factor f is, from Fig. 2-19, 0.0053, and the friction loss is

$$H_f = \frac{2 \times 1,000 \times 0.0053 \times 3.78^2}{(4.026/12)32.17} = 14 \text{ ft-lb force/lb}$$

The velocity head is negligible, and the friction is less than the head available; hence a 4-in. pipe may be larger than necessary.

ASSUME $3\frac{1}{2}$-IN. PIPE. This is the next smallest standard pipe. The diameter and cross section are 3.548 in. and 0.0687 ft², respectively. Repeating the above calculation shows that the velocity is 4.85 ft/sec, that the Reynolds number is 8.6×10^4, that the friction factor is 0.0052, and that the friction is 25.7 ft-lb force/lb. Since this is more than the available head, the $3\frac{1}{2}$-in. pipe is too small, and the 4-in. pipe must be used.

This type of problem can be solved directly by preparing a plot of $f/N_{Re} = H_{fs}\mu g_c/2\overline{V}^3 L\rho$ as the abscissa vs. f as the ordinate. Then the diameter appears only in the ordinate. This is an application of the principle used in preparing the plot of Fig. 2-21, namely, combining f and N_{Re} to give a new dimensionless group that does not contain the specific factor to be calculated.

Separation from Velocity Decrease. Separation can occur even where there is no sudden change in cross section if the cross section is constantly enlarged. For example, consider the flow of a fluid stream through the conical expander shown in Fig. 2-29. Because of the increase of cross sec-

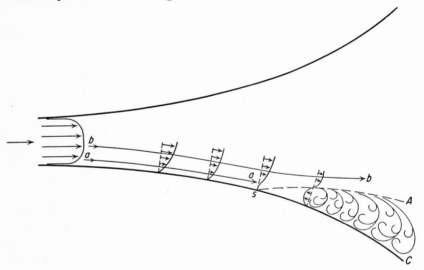

FIG. 2-29. Separation in converging channel: s, separation point. aa, bb, streamlines. AsC, wake.

tion in the direction of flow, the velocity of the fluid decreases, and, by the Bernoulli equation, the pressure must increase. Consider two stream filaments, one, aa, very near the wall, and the other, bb, a short distance from the wall. The pressure increase over a definite length of conduit is the same for both filaments, because the pressure throughout any single cross section is uniform. The loss in velocity head is, then, the same for both filaments.

The initial velocity head of filament aa is less than that of filament bb, however, because filament aa is the nearer to the wall. A point is reached, at a definite distance along the conduit, where the velocity of filament aa becomes zero but where the velocities of filament bb and of all other filaments farther from the wall than aa are still positive. This point is point s in Fig. 2-29. Beyond point s the velocity at the wall changes sign, a backflow of fluid between the wall and filament aa occurs, and the boundary layer separates from the wall. In Fig. 2-29, several curves are drawn of velocity u vs. distance from the wall y, and it can be seen how the velocity near the wall becomes zero at point s and then reverses in sign. The point s is called a *separation point*.

The vortices formed between the wall and the separated fluid stream, beyond the separation point, cause excessive form-friction losses.

Separation occurs in both laminar and turbulent flow. In turbulent flow, the separation point is farther along the conduit than in laminar flow.

Separation can be prevented if the angle between the wall of the conduit and the axis is made small. The maximum angle that can be tolerated in a conical expander without separation is 7°.

Minimizing Expansion and Contraction Losses. A contraction loss can be nearly eliminated by reducing the cross section gradually rather than suddenly. For example, if the reduction in cross section shown in Fig. 2-25 is obtained by a conical reducer or by a trumpet-shaped entrance to the smaller pipe, the contraction coefficient K_c can be reduced to approximately 0.05 for all values of S_b/S_a. Separation and vena-contracta formation do not occur unless the decrease in cross section is sudden.

An expansion loss can also be minimized by substituting a conical expander for the flanges shown in Fig. 2-23. The angle between wall and axis of the cone must be less than 7°, however, or separation may occur. For angles of 30° or more, the loss through a conical expander can become greater than that through a sudden expansion for the same area ratio S_a/S_b because of the excessive form friction from the vortices formed by the separation.

FLOW OF COMPRESSIBLE FLUIDS THROUGH STRAIGHT CONDUITS OF CONSTANT CROSS SECTION

When the fluid is compressible, its density is not constant in the direction of flow but changes with both temperature and pressure. The integrated equations [Eqs. (2-41) and (2-43)] used for incompressible fluids do not apply, and the basic differential equation [Eq. (2-36)] must be used, with suitable modifications, for problems in which the fluid density varies along the channel.

Equation (2-41) can be written over a short length of pipe in the following differential form

$$\frac{dp}{\rho} + d\left(\frac{\overline{V}^2}{2\alpha g_c}\right) + \frac{g}{g_c}dZ + dH_{fs} = 0 \qquad (2\text{-}70)$$

In this equation, the symbol dH_{fs} is the skin friction in a short length dL of pipe. From Eq. (2-53),

$$dH_{fs} = \frac{2\bar{V}^2 f}{g_c D} dL \qquad (2\text{-}71)$$

Eliminating dH_{fs} from Eqs. (2-70) and (2-71), substituting G/ρ for $\bar{V}$, and assuming α constant gives

$$\frac{dp}{\rho} + \frac{G^2}{2\alpha g_c} d\left(\frac{1}{\rho^2}\right) + \frac{g}{g_c} dZ + \frac{2G^2 f}{\rho^2 g_c D} dL = 0 \qquad (2\text{-}72)$$

An integration of Eq. (2-72) for a general case is not possible, and special cases must be considered separately.

Large Change in Temperature and Small Change in Pressure. For the first special case of the flow of a compressible fluid, consider turbulent flow of a fluid under such conditions that the effect of temperature on the density is relatively large and that of pressure is relatively small. Turbulent flow of a liquid through the tubes of a heat exchanger, where the temperature of the fluid is changed through a wide range, and flow of gas with a small relative pressure change through the same type of equipment are examples of this kind. The density ρ is, under these assumptions, independent of pressure but does change with temperature, which, in turn, varies with L, the length of the tube. If Eq. (2-72) is multiplied by ρ and the differentiation of $d(1/\rho^2)$ performed, the result is

$$dp + \frac{\rho g}{g_c} dZ + \frac{G^2}{\alpha g_c} d\left(\frac{1}{\rho}\right) + \frac{2G^2 f}{\rho g_c D} dL = 0 \qquad (2\text{-}73)$$

The Reynolds number DG/μ, and therefore the friction factor f, is independent of ρ and will change only with the viscosity μ, which in turn varies with temperature. The change of f with N_{Re} is, however, relatively small, and it is permissible to use an arithmetic average friction factor, $\bar{f} = (f_a + f_b)/2$ as a constant in the integration of Eq. (2-73). Also, the percentage change in ρ is small in the present case; the arithmetic mean average $\bar{\rho} = (\rho_a + \rho_b)/2$ can be used as a constant without serious error and Eq. (2-73) integrated term by term to give

$$p_b - p_a + \frac{\bar{\rho} g(Z_b - Z_a)}{g_c} + \frac{G^2}{\alpha g_c}\left(\frac{1}{\rho_b} - \frac{1}{\rho_a}\right) + \frac{2G^2 \bar{f} L_b}{g_c D \bar{\rho}} = 0 \qquad (2\text{-}74)$$

where $L = 0$ where $p = p_a$
$L = L_b$ where $p = p_b$

Flow of Ideal Gas. A second important special case of compressible fluid flow is that of an ideal gas through a long, straight conduit. For an ideal gas

$$p = \frac{R_0 \rho T}{M} \qquad (2\text{-}75)$$

where R_o = molal gas constant, 1,545 ft^3-lb force/(lb mole)(°R)

M = molecular weight

T = absolute temperature, °R

From Eq. (2-75),

$$dp = \frac{R_o}{M} d(\rho T) \tag{2-76}$$

In gas flow through conduits, the static term in Eq. (2-73), $(\rho g/g_c) \, dZ$, is negligible and can be omitted. Substituting p from Eq. (2-75) and dp from Eq. (2-76) into Eq. (2-73) and solving for dL gives

$$-dL = \frac{D}{2f} \left[\frac{g_c R_o \rho \, d(\rho T)}{G^2 M} - \frac{d\rho}{\alpha \rho} \right] \tag{2-77}$$

If the temperature changes over a range of not more than 10 per cent of the absolute-temperature level in the conduit, an arithmetic mean temperature, $\overline{T} = (T_a + T_b)/2$, where T_a and T_b are the absolute temperatures at the inlet and outlet, respectively, can be substituted for T in Eq. (2-77). The average friction factor $\bar{f}$ can also be used as a constant. Then, Eq. (2-77) can be integrated to give

$$-\int_0^{L_b} dL = \frac{D}{2\bar{f}} \left(\frac{R_o g_c \overline{T}}{G^2 M} \int_{\rho a}^{\rho b} \rho \, d\rho - \frac{1}{\alpha} \int_{\rho a}^{\rho b} \frac{d\rho}{\rho} \right)$$

or

$$L_b = \frac{D}{2\bar{f}} \left[\frac{R_o g_c \overline{T}}{2 G^2 M} (\rho_a^2 - \rho_b^2) - \frac{2.303}{\alpha} \log \frac{\rho_a}{\rho_b} \right] \tag{2-78}$$

Usually $\rho_a - \rho_b$ is small in comparison with $\rho_a + \rho_b$, and Eq. (2-78) can be transformed into a more convenient form by noting that

$$\rho_a^2 - \rho_b^2 = 2(\rho_a - \rho_b) \frac{\rho_a + \rho_b}{2} = 2\bar{\rho}(\rho_a - \rho_b) \tag{2-79}$$

Substituting from Eq. (2-79) into Eq. (2-78) and solving for $\rho_a - \rho_b$ gives

$$\rho_a - \rho_b = \frac{M G^2}{g_c \bar{\rho} R_o \overline{T}} \left(\frac{2\bar{f} L_b}{D} + \frac{2.303}{\alpha} \log \frac{\rho_a}{\rho_b} \right) \tag{2-80}$$

Isothermal Flow. When the temperature throughout the pipe is constant, $\bar{f} = f$, $\overline{T} = T$, and $\rho_a/\rho_b = p_a/p_b$. Equation (2-80) becomes

$$p_a - p_b = \frac{G^2}{g_c \bar{\rho}} \left(\frac{2 f L_b}{D} + \frac{2.303}{\alpha} \log \frac{p_a}{p_b} \right) \tag{2-81}$$

In Eqs. (2-78), (2-80), and (2-81), the length of conduit corresponding to a given flow rate and known inlet and outlet pressures can be calculated directly, as L_b occurs in only one term. In the more usual problem, the length, flow rate, inlet and outlet temperatures, and one pressure are known. The remaining pressure must be calculated by trial and error. For long pipes, the logarithmic term is relatively unimportant, and can be neglected in the first approximation. For large pressure drops, Eq. (2-78) can then be used directly. For lower pressure drops, Eqs. (2-80) and (2-81) are more precise, although a preliminary estimate of either ρ_a or ρ_b is needed to establish a trial value of $\bar{\rho}$.

Adiabatic Flow. The flow of a fluid through a conduit in the absence of heat transfer is essentially adiabatic even if the conduit is unlagged. Equations are available that cover the adiabatic flow of ideal gases through horizontal conduits.[7] Unless the conduit is short, however, the discharge rate for a given pressure drop is practically the same for adiabatic as for isothermal flow, and in practical calculations, Eq. (2-81) can be used for both types of flow.

Example 2-12. Air at 25 lb force/in.^{2} gauge (referred to a normal barometer) and 60°F enters a horizontal Schedule 40 3-in. pipe 600 ft long. The entering velocity is 50 ft/sec. What is the pressure at the end of the pipe?

Solution. Equation (2-81) is used. The viscosity of air at 60°F is, from Appendix 7, 0.018 centipoise. The inside diameter of Schedule 40 3-in. pipe is, from Appendix 4, 3.068 in. Other quantities for substitution into Eq. (2-81) are

$$\rho_a = \frac{29 \times 492(25 + 14.7)}{359(460 + 60)14.7} = 0.2064 \text{ lb/ft}^3$$

$$G = 50 \times 0.2064 = 10.32 \text{ lb/ft}^2\text{-sec} \qquad D = \frac{3.068}{12} = 0.256 \text{ ft}$$

$$N_{\text{Re}} = \frac{0.256 \times 10.32}{0.018 \times 6.72 \times 10^{-4}} = 2.2 \times 10^5$$

For 3-in. pipe, $k/D = 0.00015/(3.068/12) = 0.00059$. From Fig. 2-19, $f = 0.0047$.

$$p_a = 144(25 + 14.7) = 5{,}717 \text{ lb force/ft}^2$$

$$\frac{2fL_b}{D} = \frac{2 \times 0.0047 \times 600}{0.256} = 22.03$$

As a first approximation, assume $\bar{\rho} = 0.20$ lb/ft^3 and neglect log (p_a/p_b). Substitution into Eq. (2-81) gives

$$5{,}717 - p_b = \frac{22.03 \times 10.32^2}{0.20 \times 32.17} = 365$$

$$p_b = 5{,}717 - 365 = 5{,}352 \text{ lb force/ft}^2, \text{ or } \frac{5{,}352}{144} = 37.2 \text{ lb force/in.}^2$$

As a second approximation

$$\rho_b = \frac{29 \times 492 \times 37.2}{359 \times 520 \times 14.7} = 0.193 \text{ lb/ft}^3$$

$$\bar{\rho} = \frac{0.206 + 0.193}{2} = 0.200 \text{ lb/ft}^3$$

$$2.302 \log \frac{0.206}{0.193} = 0.065$$

$$5{,}717 - p_b = \frac{10.32^2}{0.200 \times 32.17} (22.03 + 0.065) = 366$$

$$p_b = 5{,}717 - 366 = 5{,}351 \text{ lb force/ft}^2$$

$$= \frac{5{,}351}{144} = 37.2 \text{ lb force/in.}^2$$

The discharge pressure is $37.2 - 14.7 = 22.5$ lb force/in.² gauge.

Maximum Discharge Rate. Consider a long conduit of constant cross section which discharges into a reservoir, as shown in Fig. 2-30a. Let a compressible fluid enter the pipe at constant pressure p_a and absolute temperature T_a. Assume also that the flow is adiabatic. Let the pressure in

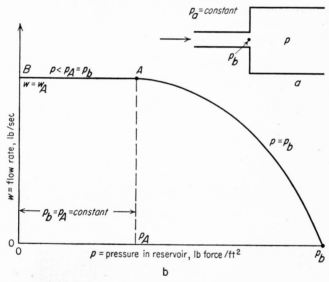

Fig. 2-30. Maximum discharge rate, adiabatic flow of compressible fluid: p_A, pressure at which Mach number = 1.0. A, point of maximum mass flow rate.

the reservoir be p, and let this pressure be steadily reduced toward zero. If the flow rate, in pounds per second, is plotted against the reservoir pressure, a curve such as that shown in Fig. 2-30b is obtained. As the pressure p is reduced, the flow rate increases, and, at a definite pressure, denoted by p_A in Fig. 2-30b, the flow rate reaches a maximum magnitude denoted by

w_A. Further reduction of reservoir pressure does not affect the flow rate. Point A of Fig. 2-30b corresponds to the reservoir pressure at which the maximum flow rate is attained.

The reason for the phenomenon of maximum flow rate is found when conditions at the end of the pipe are examined. The pressure at the pipe exit, which is denoted by p_b, is found to equal that in the reservoir p until the condition of maximum flow rate is attained. As p is further reduced, p_b remains constant at the value p_A, and over the pressure range of line AB of Fig. 2-30, $p < p_A$. Mathematically it can be demonstrated that at point A the velocity of the fluid equals that of sound under the conditions at the tube exit. This velocity is the *acoustical velocity*, and in an ideal gas is given by the equation

$$u_A = \sqrt{\frac{\kappa g_c p_A}{\rho_A}} \tag{2-82}$$

where κ is the ratio of the specific heat at constant pressure to that at constant volume.

The reason for the lack of effect on flow rate of further reduction of reservoir pressure is that sound itself is a pressure wave, and if the velocity of the fluid is equal to, or greater than, that of sound, a reduction in back pressure cannot influence the flow because the gas is moving faster than a change of pressure can be carried back into the conduit.

The ratio of the velocity of the fluid to that of sound in the fluid is the *Mach number* and is denoted by N_{Ma}.

In pipelines under normal conditions, the acoustical velocity is seldom approached, but plots and equations are available [7,12c] when needed for situations where the effect is important. Equations (2-78), (2-80), and (2-81) should not be used if N_{Ma} approaches unity.

RESISTANCE OF IMMERSED BODIES

The discussion thus far has centered about the laws of fluid flow and the factors that control changes of pressure and velocity of fluids flowing past solid boundaries and especially flow through closed conduits. Emphasis during the discussion has been placed on the fluid. In many problems, on the other hand, it is of interest to consider the effect of the flowing fluid (or, more generally, of relative motion between solid and fluid) on the solid. These questions are of especial importance when the solid shapes are immersed in the fluid. It is immaterial whether the fluid is considered to move past the solid or the solid to move through the fluid. Important technical examples include the settling of particles through fluids, the transfer of solid particles by fluid streams, the flow of fluids through beds of solid particles, and the resistance offered by the atmosphere to the flight of projectiles or aircraft.

Because of friction, a force is required to maintain fluid flow past solid boundaries or around objects immersed in the stream. For example, in the simple situation of incompressible flow through a horizontal conduit of constant cross section, the net force on the fluid is $S(p_a - p_b)$, where S is the cross-sectional area of the channel and p_a and p_b are the upstream and downstream pressures, respectively. If the channel also contains a bed of broken solids with channels between the lumps, additional friction will develop for the same flow rate, and a larger pressure drop will be required to maintain the flow. The force on the fluid is transmitted to the solid, and, by Newton's third law of motion, the force of the fluid on the solid is equal and opposite to the force of the solid on the fluid. Such forces, resulting from friction, between solid and flowing fluid are called drag forces or, simply, drag. Only in the ideal situation of frictionless flow is drag zero. The flow mechanisms leading to drag are the same as those causing friction, and large friction is accompanied by large drag.

Drag Coefficients. Consider a smooth sphere immersed in a flowing fluid and at a sufficient distance from the solid boundary of the stream to ensure that the boundary does not influence the flow around the sphere. Define the projected area of a solid body as the area obtained by projecting the body on a plane perpendicular to the direction of flow. Denote this area by A_p. For a sphere, the projected area is that of a great circle, or $(\pi/4)D^2$, where D is the diameter. If F_D is the drag, the drag per unit projected area is F_D/A_p.

For dimensional analysis, assume that F_D/A_p depends on (1) the velocity u_0 of the fluid approaching the sphere, (2) the density ρ of the fluid, (3) the viscosity μ of the fluid, and (4) the diameter D of the sphere. Since Newton's law is involved, the conversion factor g_c is added to the list of factors.

A dimensional analysis shows that the quantity $F_D g_c/A_p u_0^2 \rho$ is a function of the Reynolds number, $Du_0\rho/\mu$. This result leads to the definition of the drag coefficient C_D by the equation

$$C_D = \frac{F_D g_c}{(\rho u_0^2/2) A_p} \tag{2-83}$$

Solving for F_D/A_p gives

$$\frac{F_D}{A_p} = \frac{C_D u_0^2 \rho}{2 g_c} \tag{2-84}$$

The drag coefficient is analogous to f, the friction factor for flow through conduits. Like the friction factor, the drag coefficient must be measured experimentally, although a theoretical equation for spheres exists for small Reynolds numbers. Confirming the prediction of the dimensional analysis, C_D depends only on N_{Re} when the surface of the solid is smooth.

Particles of shapes other than spheres are defined by choosing one dimension for size measurement and specifying the ratios of other dimensions to the one chosen. Ratios of this type are called shape factors. The chosen dimension is used as D in the Reynolds number. Thus, a cylinder can be defined by specifying the diameter and the ratio of the length to the diameter. Also, the projected area A_p can be calculated for any given shape. The projected area of a cylinder having its axis parallel with the direction

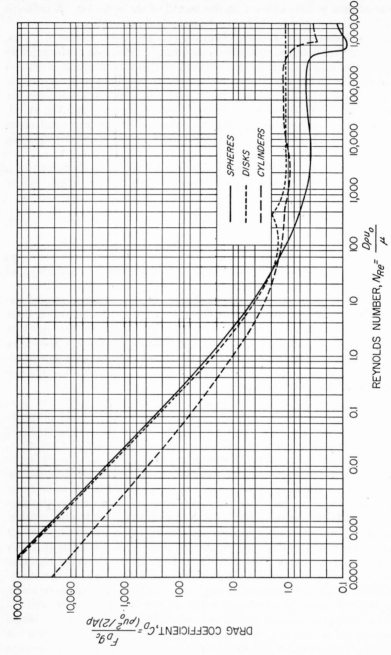

FIG. 2-31. Drag coefficients for spheres, disks, and cylinders. (By permission, from "Chemical Engineers' Handbook," 3d ed., by J. H. Perry. Copyright, 1950. McGraw-Hill Book Company, Inc.)

of flow is $(\pi/4)D^2$, and that of a cylinder oriented perpendicularly to the direction of flow is LD, where L is the length of the cylinder. A different C_D vs. N_{Re} relationship exists for each shape and orientation.

Drag Coefficients of Typical Shapes. In Fig. 2-31, curves of C_D vs. N_{Re} are shown for spheres, cylinders, and disks.[12f] The axis of the cylinder and the face of the disk are both perpendicular to the direction of flow of the fluid. It is clear that the variation of C_D with N_{Re} is more complicated than that of the friction factor f with N_{Re}, especially with spheres and cylinders. The variations in slope of the curves of C_D vs. N_{Re} at different Reynolds numbers are the result of changes in the flow mechanism in the several ranges of N_{Re}. These effects can be followed by discussing the case of the sphere.

For low Reynolds numbers, below about 0.06, the drag coefficient of a sphere follows a theoretical equation known as *Stokes' law*, which can be written

$$C_D = \frac{24}{N_{\text{Re}}} \tag{2-85}$$

At the low velocities at which this law is valid, the particle moves through the fluid by deforming it, and the resistance is the result of viscosity shear forces only. Forces necessary to accelerate the fluid, which are called inertia forces, are negligible. No boundary layer forms. The motion of the solid affects the fluid at considerable distances from the body, and if there is a solid wall within 20 or 30 diameters of the body, Stokes' law must be corrected for the wall effect. The law is useful for calculating the resistance of small particles, such as dusts or fogs, or for the motions of larger particles through liquids of large viscosity. The flow considered by this law is sometimes called creeping flow.

As the Reynolds number is increased beyond the range of Stokes' law, separation occurs at a point forward of the equatorial plane, as shown in Fig. 2-32a, and a wake, covering the entire rear hemisphere, is formed. It has been shown above that a wake is characterized by a large friction effect, and the drag must also increase accordingly. In a wake, the velocity of the fluid in the vortices is large, and the kinetic energy is greater than that of the separated current enclosing the wake. The pressure in the wake is, by the Bernoulli principle, less than that elsewhere in the stream, and a suction exists at the back of the sphere, which contributes considerably to drag.

At larger Reynolds numbers, a boundary layer forms, which is in contact with the sphere in front of the separation point and which flows freely around the wake after separation. At first, the boundary layer is in laminar flow, both when in contact with the sphere and when free. As the Reynolds number is further increased, transition to turbulent flow takes place, first in the free boundary layer and then in the boundary layer still attached to the sphere. When turbulence occurs in the latter, the separation point moves toward the rear of the body, and the wake shrinks, as shown in Fig. 2-32b. Both friction and drag decrease, and the remarkable drop in drag coefficient from 0.45 to 0.20 at a Reynolds number of about 300,000 is a result of the shift in the separation point when the boundary layer attached to the sphere becomes turbulent. At Reynolds numbers above 300,000, the drag coefficient is nearly constant.

The curve of C_D vs. N_{Re} for a cylinder is much like that for a sphere, as the same phenomenon occurs in both cases. Disks do not show the drop in drag coefficient at a large Reynolds number, because once the separation occurs at the edge of the disk, the separated stream never returns to the

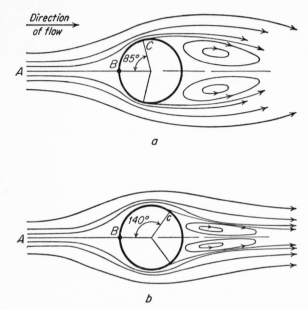

FIG. 2-32. Flow past single sphere, showing separation and wake formation: (a) Laminar flow in boundary layer. (b) Turbulent flow in boundary layer. B, stagnation point. C, separation point. (*Hunsaker and Rightmire.*[4])

back of the disk, and the wake does not shrink when the boundary layer becomes turbulent.

Drag resulting from separation and wake formation is called *form drag*. Form drag can be minimized by forcing separation toward the rear of the body. This is accomplished by streamlining. The usual method of streamlining is to so proportion the rear of the body that the increase in pressure in the boundary layer, which is the basic cause of separation, is sufficiently gradual to delay separation. This usually calls for a pointed rear, like that of an airfoil. A typical streamlined shape is shown in Fig. 2-33. A perfectly streamlined body would have no wake and would have no form drag.

Stagnation Point. The streamlines in the fluid flowing past the bodies shown in Figs. 2-32 and 2-33 show that the fluid stream in the plane of the section is split by the body into two parts, one passing over the top of the body and the other over the bottom. Streamline AB divides the two parts and terminates at a definite point B at the nose of the body. This point is called a stagnation point. The velocity at a stagnation point is zero. If Eq. (2-38), the Bernoulli equation for incompressible fluids, is written for

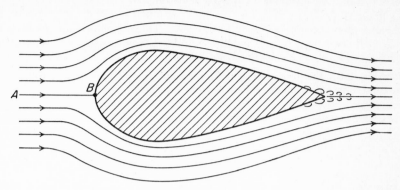

FIG. 2-33. Streamlined body: AB, streamline to stagnation point B.

streamline AB on the assumption that the flow is horizontal, the result is

$$\frac{p_0}{\rho} + \frac{u_0^2}{2g_c} = \frac{p_s}{\rho}$$

or
$$\frac{p_s - p_0}{\rho} = \frac{u_0^2}{2g_c} \qquad (2\text{-}86)$$

where p_s = pressure on body at stagnation point
p_0 = static pressure in undisturbed fluid
u_0 = velocity of the undisturbed fluid

The pressure increase $p_s - p_0$ for the streamline passing through a stagnation point is larger than that for any other streamline, because at that point the entire velocity head has been converted into pressure head. Friction along this streamline is negligible.

Equation (2-86) can also be used for compressible fluids when the Mach number is below about 0.2. A compressible fluid is compressed adiabatically from pressure p_0 to pressure p_s as it is stopped by the body, and at higher velocities the effect of this compression becomes appreciable. This situation is not considered in this book.

Example 2-13. A spherical storage tank 20 ft in diameter is to be mounted in the open several feet above the ground. It is desired to design the supports of the tank so that it cannot be moved by a 100-mph wind. If the temperature is 70°F and the pressure 1 normal atmosphere, what will be the total force on the tank?

Solution. The viscosity of the air is, from Appendix 7, 0.0185 centipoise. The density is

$$\rho = \frac{29 \times 492}{359(460 + 70)} = 0.075 \ \text{lb/ft}^3$$

The linear velocity of a 100-mph wind is

$$u_0 = \frac{5{,}280 \times 100}{3{,}600} = 146.6 \text{ ft/sec}$$

The Reynolds number is

$$N_{\text{Re}} = \frac{20 \times 146.6 \times 0.075}{0.0185 \times 6.72 \times 10^{-4}} = 1.77 \times 10^{\cdot}$$

From Fig. 2-31, the drag coefficient C_D for a sphere at $N_{\text{Re}} = 1.77 \times 10^7$ is 0.15. The total force acting on the sphere is, from Eq. (2-84),

$$F_D = \frac{C_D A_p u_0^2 \rho}{2g_c}$$

$$= \frac{0.15 \times 20^2 \times 0.7854 \times 0.075 \times 146.6^2}{2 \times 32.17} = 1{,}180 \text{ lb force}$$

Friction in Flow through Beds of Solids. In several technical processes, liquids or gases flow through beds of solid particles. In the unit operations, important examples of such flow processes are filtration and the flow of gas and liquid through packed towers. In filtration, the bed of solids consists of small particles that are being removed, by a filter cloth or fine screen, from the liquid. Packed towers are filled with solid particles usually in a size range between $\frac{1}{4}$ and 4 in., called tower packing. Typical tower packings are shown in Fig. 11-1. Liquid flows down through the packing and, by wetting the surface of the individual particles, offers a large contact area to a stream of gas flowing up through the interstices or voids in the bed. To estimate the pressure drop required to maintain flow of the fluids in a bed of solids, methods of correlating fluid friction in the bed with the geometry of the bed, the properties of the fluids, and the conditions of flow are desirable.

The resistance to the flow of a fluid through the voids in a bed of solids is the resultant of the total drag of all the particles in the bed. Depending on the Reynolds number, laminar flow, turbulent flow, boundary-layer separation, and wake formation occur. As in the drag of a single solid shape, there is no sharp transition between laminar flow and turbulent flow such as that occurring in flow through conduits.

Although attempts have been made to relate the pressure drop to the drag of the individual particles, the most successful methods of correlation are based on estimates of fluid friction in the fluid rather than of the drag on the solids. In these methods it is assumed that the total stream is divided into many small streams, which flow in parallel through the channels provided by the voids between the particles. Dimensional analysis predicts the usual relationship between the friction factor and the Reynolds number. This relationship, which has been found satisfactory for flow of a single fluid through beds of definite geometry, is

$$\frac{(p_a - p_b)g_c r_H}{L\rho \overline{V}^2} = f = \phi\left(\frac{r_H \overline{V} \rho}{\mu}\right) = \phi(N_{\text{Re}}) \qquad (2\text{-}87)$$

where $p_a - p_b$ = pressure loss in bed, lb force/ft^2

$\quad\quad g_c$ = Newton's-law conversion factor, 32.174 ft-lb/lb force-sec^2

$\quad\quad r_H$ = average hydraulic radius of channels, ft

$\quad\quad L$ = depth of bed, ft

$\quad\quad \overline{V}$ = average velocity of fluid through channels, ft/sec

$\quad\quad \rho$ = density of fluid, lb/ft^3

$\quad\quad \mu$ = viscosity of fluid, lb/ft-sec

$\quad\quad f$ = friction factor, dimensionless

The Reynolds number is $r_H \overline{V} \rho / \mu$, and the symbol $\phi(N_{\text{Re}})$ means, as usual, "function of the Reynolds number." The linear dimension used both in f and N_{Re} is the hydraulic radius r_H of the channels through the bed.

Further treatment of Eq. (2-87) is facilitated by the introduction of a geometrical characteristic of the bed called the porosity. The porosity ϵ is defined as the ratio of the volume of voids in the bed to the total volume of the bed. A porosity of 0.4, for example, means that for each cubic foot of bed volume the voids account for 0.4 ft^3 and the solid particles, 0.6 ft^3.

To calculate r_H, the average hydraulic radius in the bed, visualize a number of parallel channels, the total volume of which is the volume of the voids in the bed, and the total surface of contact between fluid and solid is the surface area of the solids. The hydraulic radius is then the ratio of the total volume of the voids to the total area of the solids. This definition is consistent with the definition of r_H for a conduit given by Eq. (2-58). Let N be the total number of particles in the bed, and let the surface and volume of a single particle be s_p and v_p, respectively. The total volume of the solids is $N v_p$, and the volume of the channels is $N v_p \epsilon / (1 - \epsilon)$. The total area of the particles is $N s_p$, and the hydraulic radius is

$$r_H = \frac{N v_p \epsilon / (1 - \epsilon)}{N s_p} = \frac{\epsilon v_p}{(1 - \epsilon) s_p} \tag{2-88}$$

It is convenient to use as linear velocity not $\overline{V}$, the actual velocity in the voids, but $\overline{V}_s$, the velocity based on the total cross-sectional area of the bed. For a packed tower, for example, $\overline{V}_s$ is based on the cross-sectional area of the empty tower before the packing is placed. These velocities are related by the equation

$$\overline{V} = \frac{\overline{V}_s}{\epsilon} \tag{2-89}$$

Substituting $\overline{V}$ from Eq. (2-89) and r_H from Eq. (2-88) into Eq. (2-87) gives

$$f = \frac{(p_a - p_b) g_c \epsilon^3 v_p}{L \rho \overline{V}_s^2 (1 - \epsilon) s_p} = \phi \left[\frac{\overline{V}_s \rho v_p}{\mu s_p (1 - \epsilon)} \right] = \phi(N_{\text{Re}}) \tag{2-90}$$ †

and

$$N_{\text{Re}} = \frac{\overline{V}_s \rho v_p}{\mu s_p (1 - \epsilon)} \tag{2-91}$$

† Equation (2-90) is known as the Kozeny-Carman equation.

The particles in a bed may be arranged at random, so there is no uniformity in the relation of one particle to the others. Random packing is the rule in filter cakes and in most packed towers, although certain types of packing are stacked in regular array. For random packing, the function $f = \phi(N_{\text{Re}})$ of Eq. (2-90) is, by experiment, the same for solid particles of all shapes and sizes provided the individual units do not contain holes or hollows. This relation is shown as the lower line in Fig. 2-34. It does not apply to the flow of gas through wetted packing.

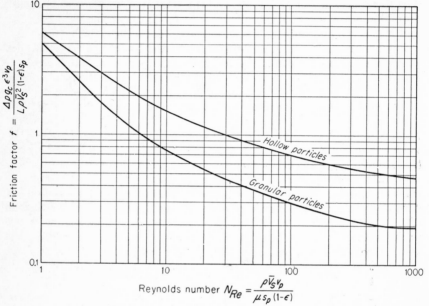

Fig. 2-34. Friction factor for beds of solids. (*Perry.*[12d])

For packing having holes or hollows, such as Berl saddles or rings, the upper line of Fig. 2-34 applies. Both lines in Fig. 2-34 are used in the same manner as the friction factor chart for flow through conduits.

To use Fig. 2-34 it is necessary to know the porosity of the bed. This is obtained either experimentally from measurement of the total bed volume and the number of units of packing contained therein or, for commercial packings, from data on porosities supplied by the manufacturer. The volume v_p and the area s_p of a unit of packing can be measured directly, or they are given by the manufacturer.

At low Reynolds numbers the relation between f and N_{Re} for both granular and hollow solids is found by experiment to be

$$f = \frac{5}{N_{\text{Re}}}$$

or
$$p_a - p_b = \frac{5L\overline{V}_s(1-\epsilon)^2 s_p^2 \mu}{v_p^2 g_c \epsilon^3} \qquad (2\text{-}92)$$

The significance of this relation is that at low Reynolds numbers the friction originates entirely in the viscous forces characteristic of laminar flow, and the inertia forces, characteristic of both turbulent flow and form drag are negligible. As the Reynolds number increases, turbulence and separation become important, and at very large Reynolds numbers form drag predominates. The friction factor then approaches asymptotically a constant magnitude, and viscous forces are unimportant.

For spherical particles, the geometric quotient v_p/s_p is $(\pi/6)D_p^3/\pi D_p^2$ $= D_p/6$, where D_p is the diameter of the sphere. For spherical particles, Eq. (2-90) becomes

$$\frac{(p_a - p_b)g_c\epsilon^3 D_p}{6L\rho \overline{V}_s^2(1 - \epsilon)} = \phi\left[\frac{\overline{V}_s\rho D_p}{6\mu(1 - \epsilon)}\right] = \phi(N_{Re}) = f \qquad (2\text{-}93)$$

If the particles are stacked in a regular arrangement, rather than at random, an individual relation of f vs. N_{Re} applies to that arrangement but will not necessarily apply to a second arrangement even if the shape, size, and porosity of the two arrangements are identical.[9] For beds containing oriented particles, the friction factor should be evaluated experimentally for each arrangement.

Example 2-14. Air at 70°F and 1 atm is to be forced through a bed of 1-in. spherical particles. The bed is 10 ft deep and 2 ft in diameter. The porosity of the bed is 0.38. The flow rate is 350 ft³/min. How much power is required to overcome friction in this bed?

Solution. The viscosity of the air is, from Appendix 7, 0.018 centipoise. The density of the air is

$$\rho = \frac{29 \times 492}{359(460 + 70)} = 0.075 \text{ lb/ft}^3$$

The velocity of the air, based on the total cross section of the bed, is

$$\overline{V}_s = \frac{350}{60 \times 2^2 \times 0.7854} = 1.86 \text{ ft/sec}$$

The diameter of the particle D_p is $\frac{1}{12}$ ft. By Eq. (2-93),

$$N_{Re} = \frac{1.86 \times 0.075 \times \frac{1}{12}}{6 \times 0.018 \times 6.72 \times 10^{-4}(1 - 0.38)} = 260$$

From Fig. 2-34, $f = 0.25$. Substituting in Eq. (2-93),

$$p_a - p_b = \frac{6 \times 0.25 \times 10 \times 0.075 \times 1.86^2(1 - 0.38)}{32.17 \times 0.38^3 \times \frac{1}{12}}$$

$$= 16.3 \text{ lb force/ft}^2$$

The power required is

$$\text{Power} = \frac{\text{lb force}}{\text{ft}^2} \frac{\text{ft}^3}{\text{sec}} \frac{1}{550} = \frac{\text{ft-lb force}}{\text{sec} \times 550}$$

$$P = \frac{16.3 \times 350}{60 \times 550} = 0.17 \text{ hp}$$

Example 2-15. A tower 4 ft in diameter is packed to a depth of 20 ft with 1-in. Berl saddles. Air at 1 atm and 70°F is to be forced through the bed at a rate of 2,000 ft^3/min. For this packing, $\epsilon = 0.69$, and $s_p/v_p = 76$ ft^{-1}. Calculate the pressure drop through the bed.

Solution. The velocity, based on the tower cross section, is

$$V_s = \frac{2,000}{60 \times 4^2 \times 0.7854} = 2.66 \text{ ft/sec}$$

The Reynolds number is, by Eq. (2-91),

$$N_{Re} = \frac{2.66 \times 0.075}{76 \times 0.018 \times 6.72 \times 10^{-4}(1 - 0.69)} = 700$$

The friction factor f is, from the upper curve of Fig. 2-34, 0.50, and the pressure drop is, by Eq. (2-90),

$$p_a - p_b = \frac{0.50 \times 20 \times 0.075 \times 2.66^2(1 - 0.69)76}{32.17 \times 0.69^3} = 12 \text{ lb force/ft}^2$$

MEASUREMENT OF FLOWING FLUIDS

To control industrial processes, it is desirable to know the amount of material entering and leaving the process. Because materials are transported in the form of fluids wherever possible, it is important to measure the rate at which a fluid is flowing through a pipe or other channel. Methods of measuring streams of flowing fluids may be classified as follows:

1. Direct weighing or measuring
2. Dynamic methods
   (a) Venturi meter
   (b) Orifice meter
   (c) Flow nozzle
   (d) Pitot tube
3. Area meters
   (a) Rotameter
   (b) Weirs
4. Current meters
   (a) Cup meters
   (b) Propeller meters
5. Positive-displacement meters
   (a) Disk meters
   (b) Piston meters
   (c) Rotary meters
   (d) Diaphragm meters
6. Dilution methods

The methods in the first class used for measuring liquids involve primarily mechanisms for weighing. Such mechanisms do not fall within the scope of this text. A gas cannot be weighed conveniently, but it can be measured by introducing it into a bell immersed in a liquid and determining the volume, temperature, and pressure. Such devices are simple and need no further discussion.

Venturi Meter. A venturi meter is shown in Fig. 2-35. It is constructed from a flanged inlet section A, consisting of a short cylindrical portion and a truncated cone; a flanged throat section B; and a flanged outlet section C, consisting of a long, truncated cone. In the upstream section, at the junction of the cylindrical and conical portions, an annular chamber D is provided, and a number of small holes E are drilled from the inside of the tube into the annular chamber. The annular ring and the small holes constitute a *piezometer ring*, which has the function of averaging the individual pressures transmitted through the several small holes. The

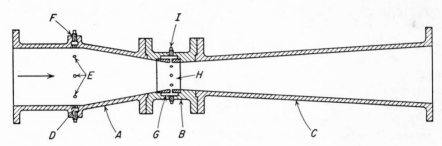

FIG. 2-35. Venturi meter: A, inlet section. B, throat section. C, outlet section. D, G, piezometer chambers. E, holes to piezometer chambers. F, upstream pressure tap. H, liner. I, downstream pressure tap. (*Builders-Providence Co., Inc., Providence, R.I.*)

average pressure is transmitted through the upstream pressure connection F. A second piezometer ring is formed in the throat section by an integral annular chamber G and a liner H. The liner is accurately bored and finished to a definite diameter, as the accuracy of the meter is reduced if the throat is not carefully machined to close tolerances. The throat pressure is transmitted through the pressure tap I. A manometer or other means for measuring pressure difference is connected between the taps F and I.

In the venturi meter, the velocity is increased, and the pressure decreased, in the upstream cone. The pressure drop in the upstream cone is utilized as shown below to measure the rate of flow through the instrument. The velocity is then decreased, and the original pressure largely recovered, in the downstream cone. To make the pressure recovery large, the angle of the downstream cone C is small, so boundary-layer separation is prevented and friction minimized. Since separation does not occur in a contracting cross section, the upstream cone can be made shorter than the downstream cone with but little friction, and space and material are thereby conserved. Although venturi meters can be applied to the measurement of gas, they are most commonly used for liquids, especially water. The following treatment is limited to incompressible fluids.

The basic equation for the venturi meter is obtained by writing the Bernoulli equation for incompressible fluids between the two pressure stations. Friction is neglected, the meter is assumed to be horizontal, and there is no pump. If $\overline{V}_a$ and $\overline{V}_b$ are the average upstream and downstream velocities, respectively, and ρ is the density of the fluid, Eq. (2-41) becomes

$$\frac{p_a}{\rho} + \frac{\overline{V}_a^2}{2\alpha_a g_c} = \frac{p_b}{\rho} + \frac{\overline{V}_b^2}{2\alpha_b g_c}$$

or

$$\frac{\overline{V}_b^2}{\alpha_b} - \frac{\overline{V}_a^2}{\alpha_a} = \frac{2g_c(p_a - p_b)}{\rho} \qquad (2\text{-}94)$$

The continuity relation [Eq. (2-15)] can be written, since the density is constant, as

$$\overline{V}_a = \left(\frac{D_b}{D_a}\right)^2 \overline{V}_b = \beta^2 \overline{V}_b \qquad (2\text{-}95)$$

where D_a = diameter of pipe
  D_b = diameter of throat of meter
  β = diameter ratio, D_b/D_a
If $\overline{V}_a$ is eliminated from Eqs. (2-94) and (2-95), the result is

$$\overline{V}_b = \frac{1}{\sqrt{1/\alpha_a - \beta^4/\alpha_b}} \sqrt{\frac{2g_c(p_a - p_b)}{\rho}} \qquad (2\text{-}96)$$

Venturi Coefficient. Equation (2-96) applies strictly to the frictionless flow of noncompressible fluids. To account for the small friction loss between locations a and b, Eq. (2-96) is corrected by introducing an empirical factor C_v and writing

$$\overline{V}_b = \frac{C_v}{\sqrt{1 - \beta^4}} \sqrt{\frac{2g_c(p_a - p_b)}{\rho}} \qquad (2\text{-}97)$$

The small effects of the kinetic-energy factors α_a and α_b are also taken into account in the definition of C_v. The coefficient C_v is determined experimentally. It is called the *venturi coefficient, velocity of approach not included.* The effect of the approach velocity $\overline{V}_a$ is accounted for by the term $1/\sqrt{1 - \beta^4}$. When D_b is less than $D_a/4$, the approach velocity and the term β can be neglected, since the resulting error is less than 0.2 per cent.

For well-designed venturi meters, the constant C_v is about 0.98 for pipe diameters of 2 to 8 in. and about 0.99 for larger sizes.[5]

Mass and Volumetric Flow Rates. The velocity through the venturi throat $\overline{V}_b$ usually is not the quantity desired. The flow rates of practical interest are the mass and volumetric flow rates through the meter. The mass flow rate is calculated by substituting $\overline{V}_b$ from Eq. (2-96) in Eq. (2-12)

$$w = \overline{V}_b S_b \rho = \frac{C_v S_b}{\sqrt{1 - \beta^4}} \sqrt{2g_c(p_a - p_b)\rho} \qquad (2\text{-}98)$$

where w = mass flow rate, lb/sec

S_b = area of throat, ft^2.

The volumetric flow rate is obtained by dividing the mass flow rate by the density, or

$$q = \frac{w}{\rho} = \frac{C_v S_b}{\sqrt{1 - \beta^4}} \sqrt{\frac{2g_c(p_a - p_b)}{\rho}} \qquad (2\text{-}99)$$

Pressure Recovery. If the flow through the venturi meter were frictionless, the pressure of the fluid leaving the meter would be exactly equal to that of the fluid entering the meter and the presence of the meter in the line would not cause a permanent loss in pressure. The pressure drop in the upstream cone $p_a - p_b$ would be completely recovered in the downstream cone. Friction cannot be completely eliminated, of course, and a permanent loss in pressure and a corresponding loss in power occur. Because of the small angle of divergence in the recovery cone, the permanent pressure loss from a venturi meter is relatively small. In a properly designed meter, the permanent loss is about 10 per cent of the venturi differential $p_a - p_b$, and approximately 90 per cent of the differential is recovered.

Example 2-16. A venturi meter is to be installed in a Schedule 40 4-in. line to measure the flow of water. The maximum flow rate is expected to be 325 gal/min at 60°F. The 50-in. manometer used to measure the differential pressure is to be filled with mercury, and water is to fill the leads above the mercury surfaces. The water temperature is to be 60°F throughout. What throat diameter should be specified for the venturi, to the nearest $\frac{1}{8}$ in., and what will be the power required to operate the meter at full load?

Solution. Equation (2-99) can be used to calculate the throat size. The quantities to be substituted in the equation are

$$q = \frac{325}{60 \times 7.48} = 0.725 \text{ ft}^3/\text{sec} \qquad \rho = 62.37 \text{ lb/ft}^3 \qquad \text{(Appendix 14)}$$

$$C_v = 0.98 \qquad g_c = 32.17 \text{ ft-lb/lb force-sec}^2$$

$$p_a - p_b = \tfrac{50}{12}(13.6 - 1.0)62.37 = 3{,}275 \text{ lb force/ft}^2$$

Substituting the above values in Eq. (2-99),

$$0.725 = \frac{0.98 S_b}{\sqrt{1 - \beta^4}} \sqrt{\frac{2 \times 32.17 \times 3{,}275}{62.37}}$$

from which

$$\frac{S_b}{\sqrt{1 - \beta^4}} = 0.01275 = \frac{0.7854 D_b^2}{\sqrt{1 - \beta^4}}$$

As a first approximation, call $\sqrt{1 - \beta^4} = 1.0$.   Then

$$D_b = 0.127 \text{ ft, or } 1.53 \text{ in.} \qquad \beta = \frac{1.53}{4.026} = 0.380$$

and $\qquad \sqrt{1 - \beta^4} = \sqrt{1 - 0.38^4} = 0.98$

The effect of this term is negligible in view of the desired precision of the final result.   To the nearest $\frac{1}{8}$ in., the throat diameter should be $1\frac{1}{2}$ in.

POWER LOSS.   The permanent loss in pressure is 10 per cent of the differential, or 327.5 lb force/ft². Since the maximum volumetric flow rate is $325/7.48 = 43.4$ ft³/min, the power required to operate the venturi at full flow is

$$P = \frac{43.4 \times 327.5}{33,000} = 0.43 \text{ hp}$$

Orifice Meter.   The venturi meter has certain practical disadvantages for ordinary plant practice.   It is expensive, it occupies considerable space, and its ratio of throat diameter to pipe diameter cannot be changed.   For a given meter and definite manometer system, the maximum measurable

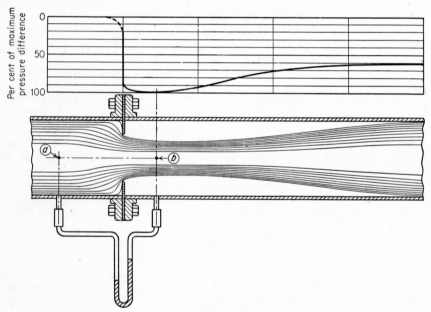

FIG. 2-36. Orifice meter.

flow rate is fixed, so if the flow range is changed, the throat diameter may be too large to give an accurate reading or too small to accommodate the new maximum flow rate.   The orifice meter meets these objections to the venturi but at the price of a larger power consumption.

A standard sharp-edged orifice is shown in Fig. 2-36. It consists of an accurately machined and drilled plate with the hole concentric with the pipe in which it is mounted, mounted between two flanges. The opening in the plate may be beveled on the downstream side. If the performance of an orifice meter is to be predicated accurately without calibration, certain definite construction and design standards must be met. These will be described later.

Pressure taps, one above and one below the orifice plate, are installed and are connected to a manometer or equivalent pressure-measuring device. The positions of the taps are arbitrary, and the coefficient of the meter will depend upon the position of the taps. Three of the recognized methods of placing the taps are shown in Table 2-2. The taps shown in Fig. 2-36 are vena-contracta taps.

TABLE 2-2. DATA ON ORIFICE TAPS

Type of tap	Distance of upstream tap from upstream face of orifice	Distance of downstream tap from downstream face
Flange.............	1 in.	1 in.
Vena contracta.....	1 pipe diameter (actual inside)	0.3 to 0.8 pipe diameter, depending on β
Pipe..............	$2\frac{1}{2}$ times nominal pipe diameter	8 times nominal pipe diameter

The principle of the orifice meter is identical with that of the venturi. The reduction of the cross section of the flowing stream in passing through the orifice increases the velocity head at the expense of the pressure head, and the reduction in pressure between the taps is measured by the manometer. Bernoulli's equation provides a basis for correlating the increase in velocity head with the decrease in pressure head.

One important complication appears in the orifice meter that is not found in the venturi. Because of the sharpness of the orifice, the fluid stream separates from the downstream side of the orifice plate and forms a free-flowing jet in the downstream fluid. A vena contracta forms, as shown in Fig. 2-36. The jet is not under the control of solid walls, as is the case in the venturi, and the area of the jet varies from that of the opening in the orifice to that of the vena contracta. The area at any given point, e.g., at the downstream tap, is not easily determinable, and the velocity of the jet at the downstream tap is not easily related to the diameter of the orifice. Orifice coefficients are more empirical than those for the venturi, and the quantitative treatment of the orifice meter is modified accordingly. Equation (2-94) is used as the basic relation for the orifice, by choosing stations a and b at the upstream and downstream taps, respectively.

To relate the velocity through the orifice plate to that at the position of the downstream tap define a coefficient of contraction

$$C_c = \frac{S_b}{S_o} \tag{2-100}$$

where S_b = cross-sectional area of jet at downstream tap
S_o = cross-sectional area of orifice

Since the diameter of the jet is less than that of the orifice, C_c is less than 1.0. Also, as before, use β as the ratio of the diameter of the orifice to that of the pipe. For an incompressible fluid, the continuity equation gives

$$\overline{V}_a D_a^2 = \overline{V}_o D_o^2 = \overline{V}_b D_b^2 \tag{2-101}$$

and also

$$\overline{V}_o S_o = \overline{V}_b S_b \tag{2-102}$$

Substituting $\overline{V}_a$ from Eq. (2-101) and $\overline{V}_b$ from Eq. (2-102) into Eq. (2-94) and introducing the coefficient C' to correct for friction and for α_a and α_b gives

$$\overline{V}_o = C_c C' \sqrt{\frac{1}{1 - C_c^2 \beta^4}} \sqrt{\frac{2g_c(p_a - p_b)}{\rho}} \tag{2-103}$$

Since the coefficients C_c and C' are not known, all terms containing them are combined in a single empirical orifice coefficient C, defined as

$$C = C_c C' \sqrt{\frac{1}{1 - C_c^2 \beta^4}} \tag{2-104}$$

The coefficient C is called the *orifice coefficient, velocity of approach included*. It is always determined experimentally.

Orifice Coefficients. Orifice meters are used to measure the flow of a wide variety of liquids and gases covering a large range of density and viscosity. Elaborate correlations of the effect of important variables on coefficient C have been worked out for the three types of taps described in Table 2-2.

For an incompressible fluid, a dimensional analysis shows that for geometrically similar orifices inserted in smooth pipes the coefficient C should depend only on a Reynolds number, the diameter ratio β, and the positions of the taps. Accurate data show that this prediction is essentially true for standard orifices, although there is a small specific effect of pipe size, probably from roughness and minor departures from strict geometric similarity among orifices of different sizes. In general, for incompressible fluids, for a given set of taps, C depends on N_{Re}, β, and D_a. In practice, C is separated into factors: (1) A basic coefficient C_o applies when the fluid is incompressible and the Reynolds number is very large. (2) A correction factor F_{Re}, depending only on N_{Re}, corrects for the viscosity of the fluid. (3) An expansion factor Y is introduced to extend the use of the orifice to gases. Then the coefficient C becomes

$$C = C_o F_{Re} Y \tag{2-105}$$

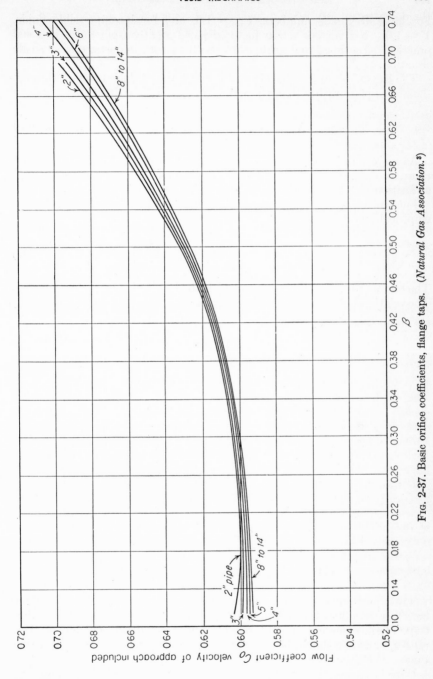

Fig. 2-37. Basic orifice coefficients, flange taps. (*Natural Gas Association.*[3])

For large Reynolds numbers, $F_{Re} = 1.0$, and for incompressible fluids, $Y = 1.0$. Neither correction factor departs greatly from unity in usual practical situations, and seldom are both factors important in a single situation.

The Reynolds number used in orifice calculations is based on flow through the orifice plate rather than through the pipe. It is defined as

$$N_{Re,o} = \frac{D_o G_o}{\mu} = \frac{D_o w}{S_o \mu} = \frac{4w}{\pi D_o \mu} \tag{2-106}$$

As an example of modern correlations of orifice coefficients, consider those for flange taps. Figure 2-37 shows that for these taps C_o is primarily a function of β but that it is also affected somewhat by pipe size.

For flange taps, the Reynolds-number factor F_{Re} is calculated from the empirical and dimensional equation

$$F_{Re} = 1 + \frac{\beta D_a'}{N_{Re,o}} \left(A + \frac{530}{\sqrt{D_a'}} \right) \tag{2-107}$$

The factor A is a function of β, and D_a' is the inside pipe diameter in inches. The relation between A and β is shown in Fig. 2-38.

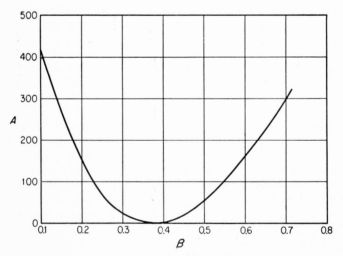

Fig. 2-38. Factor A, F_{Re} correction, flange taps.

The above equations and curves should not be used for Reynolds numbers less than 30,000 or for magnitudes of β less than 0.15 or greater than 0.70. Orifice meters are not reliable under conditions outside of these ranges, and they should be redesigned if necessary to enter the permissible ranges of $N_{Re,o}$ and β.

Correlations equivalent to those for flange taps have been prepared for vena-contracta taps and for pipe taps.[3]

For ideal gases, the expansion factor is given by the empirical equation

$$Y = 1 - \frac{0.41 + 0.35\beta^4}{\kappa} \frac{p_a - p_b}{p_a} \tag{2-108}$$

where κ is the specific-heat ratio for the gas.

The mass flow rate through the orifice is given by the following equation, which is applicable both to incompressible fluids and to ideal gases.

$$w = C_o F_{Re} Y S_o \sqrt{2g_c(p_a - p_b)\rho_a} \tag{2-109}$$

In gas flow, ρ_a is the density of the gas at the temperature and pressure of the gas flowing through the pipe upstream from the orifice. In liquid flow, ρ_a is the density of the liquid at the temperature of flow.

The volumetric flow rate of a gas is often referred to a standard or base temperature and pressure, and that of a hot liquid to a base temperature. The base density, denoted by ρ_B, is then different from the flowing density ρ_a. The volumetric flow rate, at base conditions, is

$$q_B = \frac{w}{\rho_B} = \frac{C_o F_{Re} Y S_o}{\rho_B} \sqrt{2g_c(p_a - p_b)\rho_a} \tag{2-110}$$

Equations (2-109) and (2-110) are final working equations for the precise calculation of the rate of flow of a given fluid through a given orifice when the differential $p_a - p_b$ is known. For liquids, $Y = 1.0$, and ρ_a and ρ_B are equal unless the flowing fluid is well above or well below the base temperature. For ideal gases, ρ_a may differ considerably from ρ_B, F_{Re} is usually unity, but Y may differ appreciably from 1.0. These equations do not apply without further correction to compressible fluids that do not obey the ideal-gas law. This correction is not within the scope of this book.

In calculating the flow rate through an orifice of known size from an observed reading of $p_a - p_b$, a preliminary magnitude for w is found by using Eq. (2-109) on the assumption that $F_{Re} = 1.00$. The Reynolds number is calculated from Eq. (2-106), and F_{Re} is determined from Fig. 2-38 and Eq. (2-107). Use of this quantity in Eq. (2-109) or (2-110) gives a final result for the flow rate.

Pressure Recovery. Because of the large friction losses from the eddies generated by the reexpanding jet below the vena contracta, the pressure recovery in an orifice meter is poor. The resulting power loss is one disadvantage of the orifice meter. The fraction of the orifice differential that is permanently lost depends on the value of β, and the relationship between the fractional loss and β is shown in Fig. 2-39. For a value of β of 0.5, the lost head is about 73 per cent of the orifice differential.

The pressure difference measured by pipe taps, where the downstream tap is eight pipe diameters below the orifice, is really a measurement of permanent loss, rather than of the orifice differential.

Approximate Orifice Equation. In the design of an orifice it is usually necessary to estimate the diameter of the orifice that gives a desired maximum reading for the largest flow rate. For this purpose, the accurate estimation of C is unnecessary, and the following simple equation based on C_v, the *orifice coefficient, velocity of approach not included*, is more convenient.

$$w = C_v S_o \sqrt{2g_c(p_a - p_b)\rho} \tag{2-111}$$

This equation is useful because C_v is nearly independent of β, and for ordinary values of $N_{Re,o}$, C_v can be taken as a constant equal to 0.61. Substi-

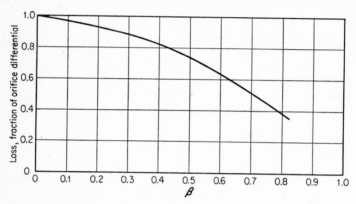

FIG. 2-39. Over-all pressure loss in orifice meters. ($ASME.$[2])

tuting the following equivalent value of S_o in Eq. (2-111),

$$S_o = \frac{D_a^2 S_o}{D_a^2} = \frac{D_a^2(\pi/4)D_o^2}{D_a^2} = \frac{\pi}{4}(D_a\beta)^2$$

and solving for β^2 gives the following equation, which can be used to calculate β directly.

$$\beta^2 = \frac{4w}{\pi D_a^2 C_v \sqrt{2g_c(p_a - p_b)\rho}} \tag{2-112}$$

D_a = diameter of pipe

Unless considerable precision is desired, Eq. (2-112) is adequate for orifice design. A check on the value of the Reynolds number should be made, however, since the coefficient 0.61 is not accurate when $N_{Re,o}$ is less than about 20,000.

Orifice-meter Standards. The coefficients and factors given above are accurate only if certain practical standards are followed in design and construction. Some of the more important of these are as follows.

The thickness of the plate must be at least $\frac{1}{16}$ in. For pipes larger than 4 in. ID, the thickness should be at least $\frac{1}{8}$ in. but not over $\frac{1}{4}$ in. unless the pipe is more than 16 in. in diameter, when the thickness can exceed $\frac{1}{4}$ in. The plate must be flat and perpendicular to the axis of the pipe. The upstream edge must be square and sharp, so it does not reflect light when viewed without magnification. The plate must be kept clean. The thickness at the orifice edge must not exceed the smallest of the following criteria: (1) one-thirtieth of the pipe diameter; (2) one-eighth of the orifice diameter; (3) one-fourth of the distance $(D_a - D_o)/2$. If the thickness of the plate exceeds the permissible thickness at the orifice edge, the downstream edge must be beveled at an angle of 45° or less to the face of the plate.

It is especially important that enough straight pipe be provided both above and below the orifice to ensure a flow pattern that is normal and undisturbed by fittings, valves, or other equipment. Otherwise the velocity distribution will not be normal, and the orifice coefficient will be affected in an unpredictable manner. Data are available for the minimum length, in pipe diameters, of straight pipe that should be provided upstream and downstream from the orifice to ensure normal velocity distribution.[13]

When the required length of straight pipe is not available upstream from the orifice, straightening vanes can be installed, and the minimum upstream length between the vanes and the orifice reduced to six diameters. A preferred form of vanes is provided by inserting a nest of small tubes in the pipe. The tubes must completely fill the cross section. The maximum diameter of a single tube is one-fourth of the pipe diameter, and the length of a tube is at least 10 times its diameter. Straightening vanes destroy transverse currents and thus aid in restoring normal velocity distribution.

Example 2-17. An orifice meter equipped with flange taps is to be installed to measure the flow rate of 27°API (at 60°F) topped crude to a cracking unit. The oil is flowing at 100°F through a Schedule 40 4-in. pipe. An adequate run of straight horizontal pipe is available for the installation. The expected maximum flow rate is 12,000 bbl/day (1 bbl = 42 U.S. gal), measured at 60°F. Mercury is to be used as a manometer fluid, and glycol (specific gravity = 1.11) is to be used in the leads as sealing liquid. The maximum reading of the meter is to be 30 in. The viscosity of the oil at 100°F is 46 SSU. The specific gravity (60°/60°) of the oil is 0.8927. The ratio of the density of the oil at 100°F to that at 60°F is 0.984.

Calculate: (a) the diameter of the orifice, approximate method, and (b) the power loss.

Solution. (a) Use Eq. (2-112). Assume $C_v = 0.61$. The values to be substituted are

$$\rho_B = 0.8927 \times 62.37 = 55.68 \text{ lb/ft}^3$$

$$\rho = 0.8927 \times 62.37 \times 0.984 = 54.79 \text{ lb/ft}^3$$

$$w = \frac{12,000 \times 42 \times 55.68}{24 \times 3,600 \times 7.48} = 43.42 \text{ lb/sec}$$

$$D_a = \frac{4.026}{12} = 0.3355 \text{ ft}$$

$$p_a - p_b = \frac{30}{12}(13.6 - 1.11)62.37 = 1,948 \text{ lb force/ft}^2$$

Then, from Eq. (2-112)

$$\beta^2 = \frac{4 \times 43.42}{\pi 0.3355^2 \times 0.61 \sqrt{2 \times 32.17 \times 1{,}948 \times 54.79}}$$

$$\beta^2 = 0.3073$$

from which $\beta = 0.554$, and $D_o = 0.554 \times 0.3355 = 0.186$ ft. The orifice diameter is $12 \times 0.186 = 2.23$ in. The viscosity is, from Eq. (2-29),

$$\mu = 0.8927 \times 0.984 \left(0.0022 \times 46 - \frac{1.80}{46}\right)$$

$$= 0.0545 \text{ poise, or } 0.0545 \times 0.0672 = 0.00367 \text{ lb/ft-sec}$$

The Reynolds number is, by Eq. (2-106),

$$N_{\text{Re},o} = \frac{4 \times 43.42}{\pi 0.186 \times 0.00367} = 81{,}000$$

The Reynolds number is sufficiently large to justify the value of 0.61 for C_v.

(b) Since $\beta = 0.554$, the fraction of the differential pressure that is permanently lost is, by Fig. 2-39, 68 per cent of the orifice differential. The maximum power consumption of the meter is

$$\frac{43.42 \times 1{,}948 \times 0.68}{0.984 \times 62.37 \times 0.8927 \times 550} = 1.9 \text{ hp}$$

Example 2-18. After the installation of the orifice meter of Example 2-17, the manometer reading at a definite constant flow rate is 1.75 in. Calculate the flow through the line in barrels per day measured at 60°F.

Solution. Assume, as a first approximation, that $F_{\text{Re}} = 1$. The quantities to be substituted into Eq. (2-109) are

$$C_o = 0.645 \text{ (Fig. 2-37)} \qquad S_o = 0.186^2 \frac{4}{\pi} = 0.0272 \text{ ft}^2$$

$$\rho_a = 54.79 \text{ lb/ft}^3$$

$$Y_a = 1 \qquad \text{(fluid is incompressible)}$$

$$p_a - p_b = \frac{1.75(13.6 - 1.11)62.37}{12} = 113.6 \text{ lb/ft}^2$$

Substituting into Eq. (2-109),

$$w = 0.645 \times 0.0272 \sqrt{2 \times 32.17 \times 113.6 \times 54.79} = 11.1 \text{ lb/sec}$$

The correction for $N_{\text{Re},o}$, from Eq. (2-106), is

$$N_{\text{Re},o} = \frac{4 \times 11.1}{\pi 0.186 \times 0.00367} = 20{,}700$$

From Fig. 2-38, when $\beta = 0.554$, $A = 110$, and, from Eq. (2-107),

$$F_{\text{Re}} = 1 + \frac{0.554 \times 4.026}{20,700}\left(110 + \frac{530}{\sqrt{4.026}}\right)$$

$$= 1 + 0.040 = 1.040$$

The corrected value of w is

$$w = 1.040 \times 11.1 = 11.54 \text{ lb/sec}$$

The flow at 60°F is

$$\frac{11.54 \times 7.48 \times 24 \times 3,600}{55.68 \times 42} = 3,190 \text{ bbl/day}$$

Example 2-19. Natural gas having a specific gravity relative to air of 0.60 and a viscosity of 0.011 centipoise is flowing through a standard 6-in. pipe in which is installed a standard sharp-edged orifice equipped with flange taps. The gas is at 100°F and 20 lb force/in.² abs at the upstream tap. The manometer reading is 46.3 in. of water at 60°F. The ratio of specific heats for natural gas is 1.30. The diameter of the orifice is 2.00 in.

Calculate the rate of flow of gas through the line in cubic feet per minute based on a pressure of 14.4 lb/in.² and a temperature of 60°F.

Solution. The quantities for substitution into Eq. (2-109) are

$$\text{Diameter of standard 6-in. pipe} = 6.065 \text{ in.}$$

$$D_a = \frac{6.065}{12} = 0.505 \text{ ft} \qquad D_o = 2.00 \text{ in.}$$

$$\beta = \frac{2.00}{6.065} = 0.330$$

From Fig. 2-37, $C_o = 0.602$.

$$S_o = \frac{\pi}{144} = 0.0218 \text{ ft}^2$$

$$\rho_a = \frac{29 \times 0.60 \times 20 \times 492}{359 \times 14.7 \times 560} = 0.0579 \text{ lb/ft}^3$$

$$\rho_B = \frac{29 \times 0.60 \times 492 \times 14.4}{359 \times 14.7 \times 520} = 0.0449 \text{ lb/ft}^3$$

$$p_a - p_b = \frac{46.3(62.37 - 0.06)}{12} = 240.4 \text{ lb/ft}^2$$

The calculation of Y is

$$\kappa = 1.30 \qquad \frac{p_a - p_b}{p_a} = \frac{240.4}{20 \times 144} = 0.0835$$

Substitute in Eq. (2-108).

$$Y_a = 1 - \left[\frac{0.41 + (0.35 \times 0.330^4)}{1.30}\right] 0.0835 = 1 - 0.027 = 0.973$$

Assume $F_{Re} = 1.00$.   Substituting into Eq. (2-109),

$$w = 0.0218 \times 0.602 \times 0.973\sqrt{64.35 \times 0.0579 \times 240.4} = 0.382 \text{ lb/sec}$$

Check on Reynolds number [Eq. (2-106)].

$$N_{Re,o} = \frac{4 \times 0.382 \times 12}{\pi 2.00 \times 0.011 \times 0.000672} = 395,000$$

At this large value of $N_{Re,o}$, the Reynolds-number correction is negligible, and the value of w calculated above is final.   The volumetric flow rate at 14.4 lb force/in.² and 60°F is, from Eq. (2-110),

$$q_B = \frac{0.382 \times 60}{0.0449} = 510 \text{ ft}^3/\text{min}$$

Flow Nozzles.   The flow nozzle is a device intermediate in character between the venturi meter and the orifice meter.   As shown in Fig. 2-40, a nozzle consists of a short, straight section following a well-rounded approach

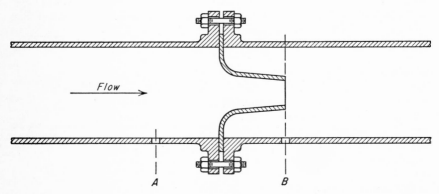

Flow

A   B

Fig. 2-40. Typical flow nozzle. A, upstream tap. B, downstream tap.

centered in a pipe by flanges.   The diverging section of the venturi is not used.   In the nozzle, the velocity is increased and the pressure decreased, as in the venturi, and the jet issues freely into the downstream pipe, as in the orifice.   During contraction, the stream is confined and controlled by definite solid boundaries, no vena contracta forms, and no uncertainty exists as to the velocity of the jet.   The upstream tap A is placed about one pipe diameter above the nozzle.   The downstream tap B is immediately behind the nozzle exit, so the pressure measured at the downstream station is essentially that at the nozzle exit.   The coefficient of a nozzle is approximately that of the venturi, and the over-all pressure loss is approximately that of the orifice.   The coefficient of a nozzle is usually supplied by the manufacturer.

Maximum Flow through a Nozzle.   If, at a constant upstream pressure, the downstream pressure on a nozzle of the type shown in Fig. 2-40 is reduced, the mass flow rate through the nozzle will increase to a definite maximum and will then remain constant if the downstream pressure is further reduced. The downstream pressure at which the maximum rate is attained is called the critical pressure, and the corresponding exit velocity is called the critical velocity. The critical velocity equals the acoustical velocity at the temperature and pressure of discharge. The same effect has been described on page 87 for flow through a long, straight pipe.

The phenomenon of critical velocity is not encountered in flow through a sharp-edged orifice, where the velocity in the unconfined jet can exceed the acoustical velocity. If the pressure downstream from a small orifice is greatly reduced, so that p_b is $p_a/10$ or less, the volumetric rate of flow through the orifice (based on upstream conditions) reaches a limiting value. Shock waves form in the jet below the orifice. Further reduction in the downstream pressure does not increase the volumetric rate of flow but alters the pattern of the jet and the shock waves. If a carefully made diverging or pressure-recovering diffuser, such as that of a venturi meter, is used downstream from the throat of a nozzle, velocities above that of sound can also be obtained.

Pitot Tube. The pitot tube is a device to measure the local velocity along a streamline. The principle of the device is shown in Fig. 2-41. The opening of the impact tube a is perpendicular to the flow direction. The opening of the static tube b is parallel to the direction of flow. The two tubes are connected to the legs of a manometer or equivalent device for measuring small pressure differences. The static tube measures the static pressure p since there is no velocity component perpendicular to its opening. The impact opening includes a stagnation point B. The streamline AB terminates

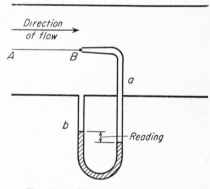

FIG. 2-41. Principle of pitot tube.

at the stagnation point B. The pressure p_s measured by the impact tube is given, for an incompressible fluid, by Eq. (2-86).

$$p_s = p + \frac{u^2 \rho}{2g_c}$$

The manometer measures $p_s - p$, and the manometer measurement is related to the velocity u at the impact opening by the equation

$$p_s - p = \frac{u^2 \rho}{2g_c}$$

$$u = \sqrt{\frac{2g_c(p_s - p)}{\rho}} \qquad (2\text{-}113)$$

OI

The form of pitot tube shown in Fig. 2-41 is not used in practice. Figure 2-42 shows a common type of standard tube. The instrument consists of two concentric tubes. The inner tube points directly upstream, and the total pressure p_s is transmitted through it. The small openings in the wall of the outer tube, immediately beyond the conical tip, transmit the static pressure p to the annular space between the tubes. The manometer or other pressure device is connected between the total and static pressure connections at the outer end of the tube and measures directly the pressure $p_s - p$.

The velocity measured by an ideal pitot tube would conform exactly to Eq. (2-113). Well-designed instruments, where the proportions of the tip and the number and position of the static-pressure taps meet certain standards, are in error by not more than 1 per cent of theory, but when precise measurements are to be made, the pitot tube should be calibrated and an appropriate correction factor applied. This factor is used as a coefficient before the radical in Eq. (2-113). It is nearly unity in well-designed pitot tubes.

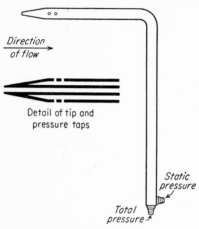

Direction of flow

Detail of tip and pressure taps

Static pressure

Total pressure

FIG. 2-42. Pitot tube.

It should be noted that, whereas the orifice and venturi meters measure the *average* velocity of the entire stream of fluid, the pitot tube measures the velocity at *one point* only. As discussed on page 54, this velocity varies over the cross section of the pipe. Consequently, to obtain the true average velocity over the cross section, one of two procedures is used. The tube may be accurately centered at the axis of the pipe and the average velocity calculated from the maximum velocity by means of Fig. 2-14. If this procedure is used, care must be taken to insert the pitot tube at least 100 pipe diameters from any disturbance in the flow, so that the velocity distribution will be normal. The other procedure is to take readings at a number of known locations in the cross section of the pipe and calculate the average velocity for the entire cross section by graphical integration.

The disadvantages of the pitot tube are: (1) it does not give the average velocity directly, and (2) its readings for gases are extremely small. When used for measuring low-pressure gases, some form of multiplying gauge, such as that shown in Fig. 2-3, must be used.

Example 2-20. Air at 200°F is forced through a long, circular flue 36 in. diameter. A pitot-tube reading is taken at the center of the flue at a sufficient distance from flow disturbances to ensure normal velocity distribution. The pitot reading is 0.54 in. of water, and the static pressure at the point of measurement is 15.25 in. of water. The coefficient of the pitot tube is 0.98.

Calculate the flow of air, in cubic feet per minute, measured at 60°F and normal barometric pressure.

Solution. The velocity at the center of the flue, which is that measured by the instrument, is calculated by Eq. (2-113), using the coefficient 0.98 to correct for imperfections in the flow pattern caused by the presence of the tube. The necessary quantities are as follows. The absolute pressure at the instrument is

$$p = 29.92 + \frac{15.25}{13.6} = 31.04 \text{ in. Hg}$$

The density of the air at flowing conditions is

$$\rho = \frac{29 \times 492 \times 31.04}{359(460 + 200)29.92} = 0.0625 \text{ lb/ft}^3$$

From the manometer reading,

$$p_s - p = \frac{0.54}{12} 62.37 = 2.81 \text{ lb force/ft}^2$$

By Eq. (2-113), the maximum velocity is

$$u_m = 0.98 \sqrt{2 \times 32.174 \frac{2.81}{0.0625}} = 52.7 \text{ ft/sec}$$

To obtain the average velocity from the maximum velocity, Fig. 2-14 is used. The Reynolds number, $N_{\text{Re,max}}$, is calculated as follows. From Appendix 7, the viscosity of air at 200°F is 0.022 centipoise.

$$N_{\text{Re,max}} = \frac{(36/12)52.7 \times 0.0625}{0.022 \times 0.000672} = 670,000$$

From Fig. 2-14,

$$\frac{\bar{V}}{u_m} = 0.82 \qquad \bar{V} = 0.82 \times 52.7 = 43.2 \text{ ft/sec}$$

The flow of air at 60°F and normal barometer is

$$q_B = 43.2 \left(\frac{36}{12}\right)^2 \frac{\pi}{4} \frac{520}{660} \frac{31.04}{29.92} 60 = 14,980 \text{ ft}^3/\text{min}$$

Flow Recorders. In plant practice, it is often essential to have a record of the instantaneous rate of flow through a flowmeter and of the average rate of flow over a considerable period. It is possible to attach to a venturi meter, orifice meter, or pitot tube a recording pressure gauge which shows the instantaneous values of the pressure drop, from which the instantaneous rates of flow may be computed. This is tedious, however; furthermore, it is unsound to compute the average rate of flow from the *average* recorded pressure drop. As shown by Eqs. (2-98), (2-109), (2-110), and (2-113), the velocity is proportional not to the pressure drop but to the square root of the pressure drop. If the pressure drop fluctuates through a considerable range, the square root of the pressure is not even approximately equal to the average of the square roots of the instantaneous pressure drops. Recording devices have been developed which directly record the square root of the pressure drop, and this can be used to give the true average flow rate and the total flow through the meter during a given period of time.

There are several ways of doing this, one of which is by using a formed bell or a formed tube, as shown in Fig. 2-43a and b. When a pressure differential is applied as indicated, the bell or float rises to an equilibrium position determined by its total weight on one hand and the buoyancy of

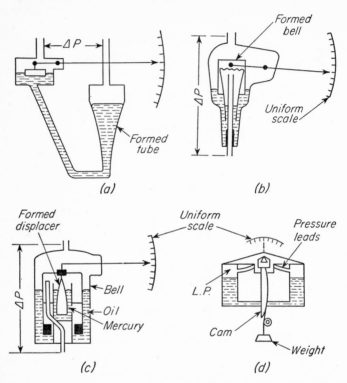

Fig. 2-43. Methods of obtaining manometer reading proportional to flow rate: (a) Formed-tube manometer. (b) Formed-bell manometer. (c) Formed-displacer manometer. (d) Slack-diaphragm flowmeter. (*Perry.*[12g])

the submerged parts plus the pressure difference on the other hand. Because of the curvature of the walls of the tube and the bell, the motion of the pointer or pen is proportional to the square root of the pressure difference. Other devices utilize a formed displacer connected to a bell and partially submerged in a pool of mercury (Fig. 2-43c) or, as in a ring-balance meter, a tilting manometer balanced on a rotating shaft or knife-edge bearings. The amount of tilt is made proportional to the flow rate, rather than to the pressure difference, by including a shaped cam bearing against a weighted tape, as shown in Fig. 2-43d.

Still another method is the electrical resistance flowmeter, the principle of which is illustrated in Fig. 2-44. One arm of the manometer is connected to chamber A, of large cross section; the mercury therefore rises and falls in chamber B in proportion to the pressure drop. In chamber B are resistance spools R_1, R_2, R_3, etc., each connected to contact rods C. A constant voltage is applied at E. As the mercury rises, it short-circuits the resistances one after the other and therefore changes the current in the circuit. The ends of the contact rods are spaced to form a square-root curve, so that the current is proportional to the square root of the difference be-

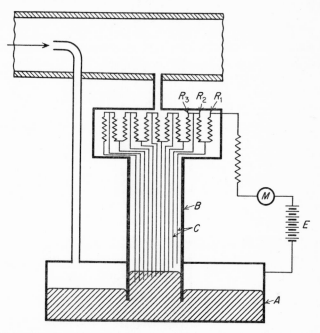

Fig. 2-44. Principle of resistance meter: A, mercury reservoir. B, metering chamber. C, contact rods. E, constant voltage supply. M, recording ammeter. R_1, R_2, R_3, resistance spools.

tween the levels in chambers A and B. With this instrument, a modified watthour meter may be used to integrate or totalize the flow.

Area Meters: Rotameters. In the orifice, nozzle, or venturi, the variation of flow rate through a constant area generates a variable pressure drop, which is related to the flow rate. Another class of meters, called area meters, consists of devices in which the pressure drop is constant, or nearly so, and the area through which the fluid flows varies with flow rate. The area is related, through proper calibration, to the flow rate.

The most important area meter is the rotameter, which is shown in Fig. 2-45. It consists essentially of a gradually tapered glass tube mounted vertically in a frame with the large end up. The fluid flows upward through the tapered tube and suspends freely a "float" (which actually does not float but is completely submerged in the fluid). The float is the indicating element, and the greater the flow rate, the higher the float rides in the tube. The entire fluid stream must flow through the annular space between the float and the tube wall.[15] The tube is marked in divisions, and the reading of the meter is obtained from the scale reading at the reading edge of the

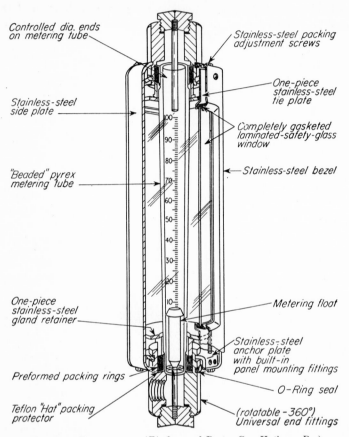

Controlled dia. ends on metering tube

Stainless-steel packing adjustment screws

Stainless-steel side plate

One-piece stainless-steel tie plate

Completely gasketed laminated-safety-glass window

"Beaded" pyrex metering tube

Stainless-steel bezel

One-piece stainless-steel gland retainer

Metering float

Stainless-steel anchor plate with built-in panel mounting fittings

Preformed packing rings

O-Ring seal

Teflon "Hat" packing protector

(rotatable - 360°) Universal end fittings

FIG. 2-45. Rotameter. (*Fischer and Porter Co., Hatboro, Pa.*)

float, which is taken at the largest cross section of the float. A calibration curve must be available to convert the observed scale reading to flow rate. Rotameters can be used for either liquid- or gas-flow measurement.

The rotameter tube is held between two stuffing boxes, which are carried by the inlet and outlet fittings. The fittings are tapped to standard pipe threads or in larger sizes are flanged. Rotameters are available over a size range of $\frac{1}{4}$ to 3 in. standard pipe size. As shown in Fig. 2-45 the rotameter tube may be placed between safety-glass windows supported by a metal case.

The bore of a glass rotameter tube is either an accurately formed plain conical taper or a taper with three beads, or flutes, parallel with the axis of the tube. The tube shown in Fig. 2-45 is a fluted tube. For opaque liquids, for high temperatures or pressures, or for other conditions where glass is impracticable, metal tubes are used. Metal tubes are plain tapered. Since, in a metal tube, the float is invisible, means must be provided for either indicating or transmitting the meter reading. This is accomplished by attaching a rod, called an extension, to the top or bottom of the float and using the extension as an armature. The extension is enclosed in a

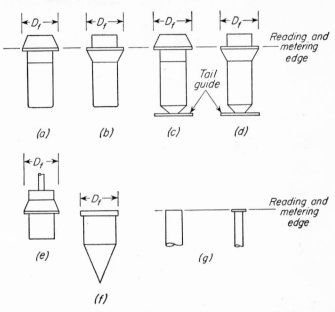

Fig. 2-46. Rotameter floats: (*a*), (*e*) Viscosity-compensating, guided by wire or extension. (*b*), (*f*) Streamlined, guided by wire or extension. (*c*) Viscosity-compensating, fluted tube. (*d*) Streamlined, fluted tube. (*g*) Indicating extension. (*a–d, Fischer and Porter Co., Hatboro, Pa.; e–g, Schutte and Koerting Co., Cornwells Heights, Pa.*)

fluid-tight tube mounted on one of the fittings. Since the inside of this tube communicates directly with the interior of the rotameter, no stuffing box for the extension is needed. The tube is surrounded by external induction coils. The length of the extension exposed to the coils varies with the position of the float. This in turn changes the induction of the coil, and the variation of the induction is measured electrically to operate a control valve or to give a reading on a recorder. Also, a magnetic follower, mounted outside the extension tube and adjacent to a vertical scale, can be used as a visual indicator for the top edge of the extension. By such modifications the rotameter has developed from a simple visual indicating instrument using only glass tubes into a versatile recording and controlling device. Rotameters are now used in many situations where orifice meters were formerly specified.

Floats may be constructed of metals of varying density from lead to aluminum or from glass or plastic. Stainless-steel floats are common. Float shapes and proportions are also varied for different applications. Several typical rotameter floats are shown in Fig. 2-46. Many other shapes have been used. For accuracy in reading, the float should be accurately centered in the tube. Float stability in a plain tapered tube is obtained by use of a guide wire supported by the end fittings, positioned in the center line of the tube, and passing through a small axial hole in the float. If the float carries an extension, the extension provides float stability without the use of the central wire. In fluted tubes, guide wires and extensions are unnecessary, but the float is equipped with a tail guide, as shown in Figs. 2-45 and 2-46c and d.

Two major types of float are used. The first type, examples of which are shown in Fig. 2-45b, f, and d, is streamlined to give a low pressure drop over the meter. The second type, examples of which are shown in Fig. 2-46a, e, and c, is so proportioned that the calibration of the meter is nearly constant over a wide range of fluid viscosities. This characteristic is obtained at the expense of an increase in pressure drop in comparison with that given by streamlined floats. Figure 2-46g shows how the top edge of the extension is used as the metering edge in extension meters.

Theory and Calibration of Rotameters. For a given flow rate, the equilibrium position of the float in a rotameter is established by a balance of three forces: (1) the weight of the float, (2) the buoyant force of the fluid on the float, and (3) the drag force on the float. Force (1) acts downward, and forces (2) and (3) upward. For equilibrium [8]

$$F_D g_c = v_f \rho_f g - v_f \rho g \tag{2-114}$$

where F_D = drag force, lb force
  g = acceleration of gravity, ft/sec^2
  g_c = Newton's-law conversion factor, 32.174 ft-lb/lb force-sec^2
  v_f = volume of float, ft^3
  ρ_f = density of float, lb/ft^3
  ρ = density of fluid, lb/ft^3

The quantity v_f can be replaced by m_f/ρ_f, where m_f is the mass of the float, and Eq. (2-114) becomes

$$F_D g_c = m_f g \left(1 - \frac{\rho}{\rho_f} \right) \tag{2-115}$$

For a given meter operating on a definite fluid, the right-hand side of Eq. (2-115) is constant and independent of the flow rate. Accordingly, F_D is also constant, and if the flow rate is changed, the effect of the change in flow rate must be countered by the effect of the change of the position of the float. If, for example, the flow rate is increased, the increase in F_D from the added flow rate is compensated for by the decrease in F_D from the reduction in velocity through the larger annular space at the new position of equilibrium.

The drag force can be investigated with the aid of dimensional analysis. Consider a float of definite shape having a diameter D_f and positioned at a level in the tube where the tube diameter is D_t. Let the mass flow rate of the fluid be w lb/sec, and let the viscosity of the fluid be μ lb/ft-sec. The drag force may then be assumed to depend on D_f, D_t, w, ρ, and μ. The effect on F_D of the angle of taper of the tube has been found by experiment to be negligible. Dimensional analysis, conducted in the usual manner, gives a result which may be written in the form

$$\frac{w}{D_f \sqrt{F_D g_c \rho}} = \chi \left(\frac{\sqrt{F_D g_c \rho}}{\mu}, \frac{D_t}{D_f} \right) \tag{2-116}$$

where χ means "function of." In this analysis, w, D_f, and μ were retained as primary quantities, so each appears only once in any of the dimensionless groups.

Substitution of $F_D g_c$ from Eq. (2-115) in Eq. (2-116) gives

$$\frac{w}{D_f \sqrt{m_f g \rho (1 - \rho/\rho_f)}} = \chi \left(\frac{\sqrt{m_f g \rho (1 - \rho/\rho_f)}}{\mu}, \frac{D_t}{D_f} \right) \tag{2-117}$$

Equation (2-117) suggests that, if $w/D_f \sqrt{m_f g \rho (1 - \rho/\rho_f)} = Y$ is plotted against $\sqrt{m_f g \rho (1 - \rho/\rho_f)}/\mu = X$, for constant magnitudes of the diameter ratio D_t/D_f, a family of curves should be obtained that depends only on the shape of the float. A set of curves of this type must be found experimentally for each float shape. A group of X vs. Y curves for float a of Fig. 2-46 is shown in Fig. 2-47. That this float gives a nearly constant calibration over a wide range of viscosities is shown by the constancy of the ordinates of the D_t/D_f lines at magnitudes of X greater than 10^4.

For use of Fig. 2-47, a relationship is necessary between D_t/D_f and R_r, the reading of the meter. Such relationships for various combinations of tubes and floats are supplied by the manufacturers of rotameters. For uniformly tapered tubes, of either glass or metal, this relationship is linear and is easily calculated from the dimensions of the tube and the diameter of the float. With the X vs. Y curves and the D_t/D_f vs. R_r line at hand, a curve of R_r vs. w can be prepared for any fluid flowing through a given meter. The magnitude of X is constant in any single situation, and a plot such as that of Fig. 2-47 yields corresponding numbers for D_t/D_f and Y, from which w can be calculated. The relationship between R_r and D_t/D_f is then used to relate w and R_r. Both gas and liquid flow are treated in the same manner. If q_B, the volumetric flow rate is wanted, it is obtained as usual from the relation $q_B = w/\rho_B$, where ρ_B is the density at base conditions.

By changing either the diameter or the mass of the float a single tube can be used for fluids of various densities or for different ranges of flow for a given fluid. If the shape of the float is not changed, the same X vs. Y curves can be used for all floats, tubes, and fluids within the accuracy required for ordinary practice.

The calibration of a rotameter, unlike that of the orifice meter, is not sensitive to the velocity distribution in the approaching stream, and neither long, straight approaches nor straightening vanes are necessary to use calibration curves like Fig. 2-47.

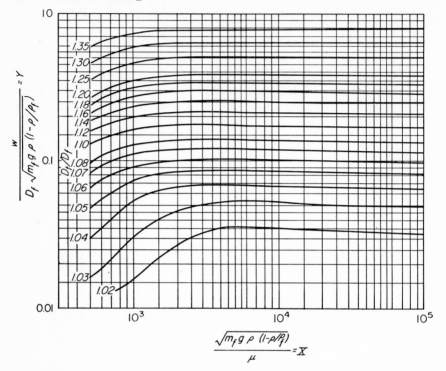

FIG. 2-47. Rotameter calibration curves. (*Based on Fischer and Porter Co., Catalogue 10-A-90, 1952, by permission.*)

Example 2-21. A 5-in. rotameter consists of a metal tube and an extension float, of the shape of float a of Fig. 2-46. The diameter of the tube at zero reading is 1.52 in., and that at the maximum reading of 130 mm is 1.92 in. The meter is to be used to measure the flow of oil having a specific gravity of 0.88 and a viscosity of 40 centipoises. A stainless-steel float, specific gravity = 8.02, mass = 550 g, and diameter = 1.505 in., is to be used. Plot a curve showing the flow rate in gallons per minute vs. reading in millimeters for this meter.

Solution. The quantity $\sqrt{m_f g \rho (1 - \rho/\rho_f)}$ is constant, and is equal to

$$\sqrt{\frac{550 \times 32.17 \times 0.88 \times 62.37(1 - 0.88/8.02)}{453.6}}, \text{ or } 43.66$$

The value of X to be used with Fig. 2-47 is, then, $43.66/(40 \times 6.72 \times 10^{-4})$, or 1,625, and Y is $w/(1.505/12)43.66$, or $0.1825w$. The volumetric flow rate is

$$q_B' = \frac{60w}{8.33 \times 0.88} = \frac{60Y}{8.33 \times 0.88 \times 0.1825} = 44.82Y \text{ gal/min}$$

Since the tube has a uniform taper, the D_t/D_f vs. R_r relationship is provided by a straight line passing through points (130, 1.92/1.505 = 1.275) and (0, 1.52/1.505 = 1.01). This line is line aa in Fig. 2-48.

The calculation of q'_B and R_r is shown in Table 2-3. Line bb in Fig. 2-48 is the desired plot of q'_B vs. R_r. It is seen that for a rotameter the relation between flow rate and reading is approximately linear, as compared with the calibration curve

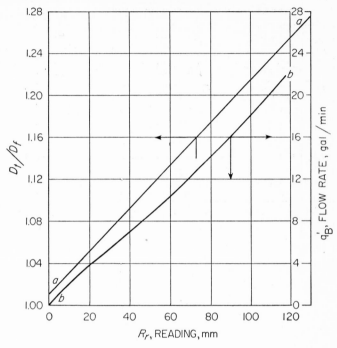

Fig. 2-48. Example 2-21.

for an orifice meter, where the flow rate is proportional to the square root of the reading.

TABLE 2-3. CALIBRATION OF ROTAMETER FOR EXAMPLE 2-21

$\dfrac{D_t}{D_f}$	Y	$q'_B,$ gal/min	$R_r,$ mm	$\dfrac{D_t}{D_f}$	Y	$q'_B,$ gal/min	$R_r,$ mm
1.275	.....	.....	130	1.08	0.133	5.95	34
1.25	0.49	21.9	117.5	1.07	0.116	5.2	29
1.20	0.37	16.6	93	1.06	0.098	4.4	24
1.18	0.33	14.8	83	1.05	0.0815	3.65	19
1.16	0.285	12.75	73.5	1.04	0.0625	2.8	13.5
1.14	0.245	10.95	63.5	1.03	0.0410	1.83	9.5
1.12	0.208	9.3	54	1.02	0.023	1.03	5
1.10	0.170	7.6	44				

Weirs. Weirs can be used only with liquids flowing in open channels, not in pipes or conduits that are completely full. They consist of a vertical partition or obstruction over which the liquid flows. Sometimes a notch is cut in the top or crest of the weir plate; more often the crest is horizontal. The depth of the liquid flowing over the weir is a direct function of the rate of flow. The flow rate may be measured by observing the height of the liquid surface above a fixed reference point.

In chemical engineering practice, weirs of many designs are occasionally used for measuring rates of flow, especially of slurries; but often they are employed for other purposes. They are widely used in connection with plate towers for absorption and distillation, in which the liquid on one plate overflows the weir into a downspout which leads it to the plate below. These weirs do not measure flow rate; instead they fix the depth of the liquid on the plate and ensure that the bubble caps are properly submerged. They must therefore be designed so that the liquid head above the weir crest never exceeds a reasonable small value. Two types of weirs are extensively used for this purpose: rectangular, or straight, weirs and circular weirs.

A rectangular weir in a plate tower is made from a vertical partition set across the plate perpendicular to the direction of liquid flow. The crest is horizontal. The length of the crest is related to the rate of flow by the equation

$$q = 3.33 b Z_{cr}^{1.5} \qquad (2\text{-}118)$$

where q = volumetric flow rate of liquid, ft^3/sec
b = length of weir crest, ft
Z_{cr} = height of liquid surface above weir crest, ft

Rectangular weirs are most used in large towers; in smaller towers, the weir is often the top of a circular downspout. A properly designed downspout does not run full. The liquid therefore flows inward over the weir crest, which here is the entire perimeter of the top of the spout. For this case, the recommended equation is

$$q = 3.0 b Z_{cr}^{1.4} \qquad (2\text{-}119)$$

A slotted weir is a combination of a circular weir and a special kind of rectangular weir. The liquid flows inward toward the center of a vertical pipe but not over the top of the pipe; instead it enters through a vertical slot of constant width cut through the pipe wall. Slotted weirs are sometimes used as reflux splitters on fractionating columns. In these devices, two slotted weirs are mounted in a single casing, into which flows the condensate from a total condenser. One weir returns liquid as reflux to the column; the other discharges it as overhead product. Since the widths of the slots in the two weir pipes bear a constant ratio to each other, the *ratio* of the flow rates of the two discharged streams remains constant regardless of the total flow rate of the condensate. This means that the column will operate with a constant reflux ratio despite momentary fluctuations or permanent changes in throughput. In some commercial reflux splitters the width of the weir slots can be changed to vary the reflux ratio by turning an external handwheel.

Notched weirs, such as V-notch and inverted-notch weirs, are valuable in special problems of measuring flow rate. V-notch weirs indicate very wide ranges of flow rate with the highest sensitivity at low rates of flow. An orifice meter, by contrast, has maximum sensitivity at high rates of flow. The inverted-notch weir may be designed so that the flow rate is directly proportional to the head, instead of to the square root of the head as in an orifice or venturi. This was once a considerable advantage in favor of weirs, but since the linear variation is also characteristic of area meters, in practice the inverted-notch weir has been completely displaced by the rotameter.

Current Meters. Like the flowmeters so far described, current meters are inferential meters; i.e., they measure velocity instead of the volume or weight of the fluid flowing, as do positive-displacement meters. Current meters are of two types: cup meters and propeller meters. Many different makes are on the market, but the principle of operation of all of them is the same. They all contain a rotating member turning with the least possible friction and so mounted that its speed of rotation is proportional to the velocity of the fluid.

Cup anemometers are most often used without a housing, for measuring wind velocity. By enclosing them in a casing and arranging the inlet pipe so that the fluid jet impinges directly on the inside of the cups, a cup meter can be used for measuring the flow rate of liquids. It is then called a turbine meter.

Propeller meters are used with both liquids and gases. Vane anemometers, which contain a light fan mounted in an open-ring housing through which the gas flows, are useful for measuring local air or gas velocities, especially in large ducts. In one type of propeller meter for liquids, there are two propellers on the same shaft, which is connected through gears to the indicator. One propeller is threaded to the right, the other to the left. The entering liquid stream is split so that part of it flows through a strainer past each propeller, meets the other part of the stream at the center of the instrument, and discharges through a common outlet. In another design, which can also be used for steam and gases, the main stream passes through an orifice, while a proportional part of the total flow bypasses the orifice through a shunt. The velocity in the shunt, which is directly related to the total flow rate, is measured by a small propeller driving a counter through a magnetic transmission.

Positive-displacement Flowmeters. A direct measure of the quantity of fluid flowing, not the velocity, is provided by positive-displacement flowmeters. They trap and discharge discrete quantities of the fluid and count the number of such increments that have passed through the meter. The main types of positive-displacement meters for liquids are nutating-disk meters; piston meters, either oscillating or reciprocating; and meters very much like the rotary pumps described in Chap. 3. For measuring gas flow rotary meters, wet-gas meters, and diaphragm, or dry-gas, meters are commonly used.

The principle of the *nutating-disk meter* is shown in Fig. 2-49. A circular disk is mounted on a spherical bearing in a chamber with a conical top and bottom. The disk is always tangent to the top cone at one point and to the

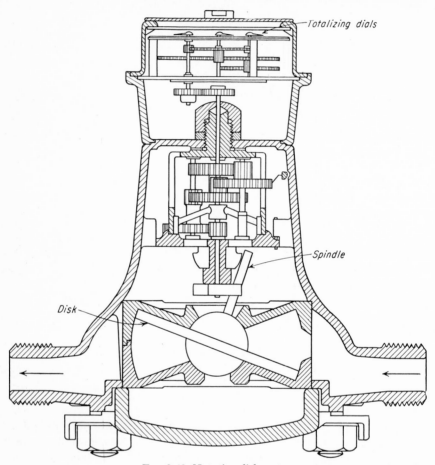

Fig. 2-49. Nutating-disk meter.

bottom cone at the point diametrically opposed. In the measuring chamber is a vertical partition that extends halfway across it and also projects into a slot in the disk. A short perpendicular spindle extends upward from the center of the disk. Liquid enters on one side of the partition, passes around through the chamber, and out on the other side of the partition. The disk does not turn; instead it wobbles, so that sometimes liquid enters above the disk and sometimes below it. In either event, the liquid must move the disk in order to pass, which makes the spindle move as though it were rotating around the surface of a cone with a vertical axis and its apex at the center of the disk. This motion is transmitted through a train of gears to the counting dials at the top of the meter.

Small *piston pumps* are often adapted to serve as metering instruments. These devices are driven mechanically instead of by fluid pressure, so that they serve both as pumps and as meters. Most rotary pumps can be used

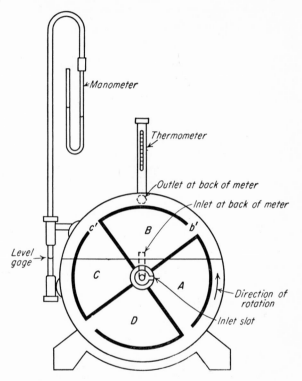

FIG. 2-50. Principle of wet-gas meter.

as meters in this way or by running them backward so that the fluid drives the pump instead of vice versa; gear pumps, two-lobe pumps, and other designs that do not require valves are most suitable. In fluid meters of this kind, the impeller is made light and adjusted to turn with a minimum of friction. The spindle, which is driven when the device is used for pumping, is attached to a totalizer or to an indicator of the instantaneous rate of flow. Several meters use rotating elements to measure gas flow. Each revolution of the sides is equivalent to a definite volume of gas. A *wet-gas meter* consists of a compartmented drum revolving in a cylindrical casing partly filled with liquid. It is illustrated in Fig. 2-50. Each compartment is filled in turn with gas from an inlet slot in a stationary axial feed pipe. The flow of gas into a submerged compartment causes it to rise, so that the drum rotates and brings another compartment below the liquid level.   In Fig. 2-50, gas is entering compartment A and is leaving compartments B and C through outlets b' and c'.   The sealing liquid is usually water but may be any moderately high-boiling liquid which does not absorb or react with the gas.   The meter must be precisely level and filled to the proper point at all times.   It measures the volume of *wet* gas, so that in precise work the measured volume must be corrected for its content of vapor from the sealing liquid.   Wet-gas meters are especially useful in laboratories but are also made in very large sizes for applications in gas-manufacturing plants and elsewhere.

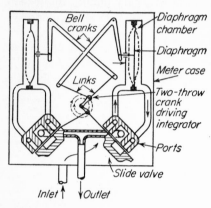

FIG. 2-51. Principle of dry-gas meter. (*By permission, from "Chemical Engineers' Handbook," 3d ed., by J. H. Perry.   Copyright, 1950.   McGraw-Hill Book Company, Inc.*)

Cycloidal, or two-lobe, blowers, illustrated in Fig. 3-25a, are adapted to the measurement of gas flow by running them backward and mounting the impeller so that it turns very easily.   Such meters handle up to 1 million standard cubic feet of gas per hour, introducing a pressure drop of only $\frac{1}{2}$ to 1 in. of water.   They can withstand internal pressure of at least 500 lb/in.^{2}, whereas the gas flowing through wet meters or dry-gas meters must be at or near atmospheric pressure.

Another device for measuring gas flow is the *dry-gas*, or *diaphragm, meter*. This contains no sealing liquid and directly measures the volume of dry gas passing through it.   The principle is shown in Fig. 2-51.[12e]   The casing contains two rigid chambers or pans, each divided into two compartments by a flexible diaphragm.   The diaphragms are connected through links to a counting device and to D slide valves, which direct the gas flow inside the meter.   Entering gas is admitted first to one side of one diaphragm and then to the other; it is then diverted into the other pan on a similar cycle.   In Fig. 2-51 gas is flowing into the right-hand chamber on the left side of the diaphragm and is pushing gas out through the outlet port and slide valve from the right side of the diaphragm.   The left-hand chamber is inactive in this part of the cycle; it will be activated by the crank when all the gas is discharged from the right side of the active chamber.

Dilution Methods.   In cases where no mechanical device is possible or convenient, a second fluid may be added at a known rate to the stream of fluid to be measured and the concentration of the added fluid determined by analysis after the two streams are well mixed.   Carbon dioxide or ammonia may be added to gas streams; salt brine of known concentration is used with large streams of water in open channels.   A way of measuring the flow rate of very large quantities of water in channels of known cross section is to add a quantity of solid salt to the stream and to note the time required for the salt to travel a known distance. The salt concentration at the downstream point is measured continuously, usually by a conductivity method; the flow rate is computed from the instant at which the conductivity downstream reaches a maximum.

Electrical energy instead of a second fluid may be added to the stream at a known rate by a resistance heater. The rate of flow of the fluid may be computed directly from its rise in temperature, provided its specific heat is known; or the meter may be calibrated in terms of the amount of current required to maintain a fixed temperature rise in the fluid.

SYMBOLS

A Area, ft^2; constant in Eqs. (2-29) and (2-107)

A_p Projected area, ft^2

B Constant in Eq. (2-29)

b Length of weir, ft

C Orifice coefficient; C_c, coefficient of contraction in orifice; C_o, basic orifice coefficient, velocity of approach included; C_v, venturi or orifice coefficient, velocity of approach not included; C', correction in orifice meter formula for α and friction

C_D Drag coefficient, dimensionless

D Diameter, ft; D_a, at upstream station; D_b, at downstream station; D_f, of rotameter float; D_o, of orifice; D_t, of rotameter tube; D_a', diameter of pipe above orifice, in.

E_k Kinetic energy, ft-lb force/lb

F Force, lb force; F_D, drag force; F_s, shear force

F_{Re} Reynolds-number factor, orifice meter

$\mathcal{F}$ Operator, meaning "function of"

G Mass velocity, w/S, lb/ft^2-sec

g Acceleration of gravity, ft/sec^2

g_c Newton's-law conversion factor, 32.174 ft-lb/lb force-sec^2

H_f Friction, ft-lb force/lb; H_{fc}, friction loss from sudden contraction; H_{fe}, friction loss from sudden enlargement; H_{fs}, skin friction

K_c Contraction-loss coefficient, dimensionless

K_e Enlargement-loss coefficient, dimensionless

k Roughness parameter, ft

L Length, ft; L_b, length of straight pipe

L_p Wetted perimeter, ft

M Molecular weight

m Mass, lb; m_f, mass of float in rotameter

N Number of particles in bed

n Exponent in Eq. (2-21)

P Power, hp

p Pressure, lb force/ft^2; p_A, pressure at acoustical velocity; p_a, at upstream station; p_b, at downstream station; p_0, static pressure in undisturbed stream; p_s, stagnation pressure; p', pressure, lb force/in.^{2}

q Volumetric flow rate, ft^3/sec; q_B, at base conditions; q_B', gal/min at base conditions

R_m Reading of manometer, ft; R_r, reading of rotameter

R_o Gas-law constant, 1,545 ft^3-lb force/(lb mole)(°R)

r_H Hydraulic radius, ft

S Cross section of conduit, ft^2; S_a, at upstream station; S_b, at downstream station; S_o, of orifice

s_p Area of one particle, ft^2

T Absolute temperature, °R; T_a, at upstream station; T_b, at downstream station; $\overline{T}$, average temperature, $(T_a + T_b)/2$

u Net local velocity, ft/sec; u_A, acoustical velocity; u_i, instantaneous local velocity in direction of flow; u_m, maximum local net velocity in closed channel; u_0, velocity of undisturbed stream approaching solid body; u', deviating velocity in direction of flow; u'', deviating velocity in direction perpendicular to flow

V Volume, ft^3

$\overline{V}$ Volumetric average velocity, q/S, ft/sec; $\overline{V}_a$, at upstream station; $\overline{V}_b$, at downstream station; $\overline{V}_o$, through orifice; $\overline{V}_s$, superficial velocity, based on total cross section of packed bed

v_f Volume of rotameter float, ft^3

v_p Volume of one particle, ft^3

X Abscissa (Fig. 2-47), $\sqrt{m_f g \rho (1 - \rho/\rho_f)}/\mu$, dimensionless

x_t Transition length, ft

$-W_s$ Shaft work done by pump, actual, ft-lb force

$-W_{sr}$ Shaft work done by pump, frictionless, ft-lb force

w Mass flow rate, lb/sec; w_A, maximum mass flow rate at acoustical velocity

Y Expansion factor, orifice meter, also ordinate (Fig. 2-47), $w/D_f\sqrt{m_f g \rho (1 - \rho/\rho_f)}$, dimensionless

y Distance from wall, ft

Z Height above datum plane, ft; Z_a, height of meniscus in manometer, height of upstream station; Z_b, height of meniscus in manometer, height of downstream station; Z_{cr}, height of liquid crest over weir; Z_m, height of pressure connections in manometer above measuring liquid; Z_x, thickness of boundary layer

Greek Letters

α Angle with horizontal; kinetic-energy correction factor; α_a, for upstream station, α_b, for downstream station

β Angle with vertical; ratio, diameter of orifice or venturi to diameter of pipe

γ Specific weight, $(g/g_c)\rho$, lb force/ft^3

ϵ Fraction voids, or porosity

η Efficiency of pump, over-all, W_{sr}/W_s

θ Time, sec

κ Specific heat at constant pressure/specific heat at constant volume

μ Viscosity, absolute, lb/ft-sec; μ_F, gravitational viscosity, μ/g_c, lb force-sec/ft^2; μ_s at conduit wall; μ_0, at $T = 492$°R

ν Kinematic viscosity, μ/ρ, ft^2/sec

ρ Density, lb/ft^3; ρ_A, of fluid A; ρ_B, of fluid B; density at base conditions; ρ_a, at upstream station; ρ_b, at downstream station; ρ_f, of rotameter float; $\bar{\rho}$, average density, $(\rho_a + \rho_b)/2$

τ Tractive force, or shear, F_s/A, lb force/ft^2; τ_0, minimum tractive force for plastic flow

$\daleth, \phi, \chi$ Operators, meaning "function of"

ψ Temperature-correction factor for skin friction, dimensionless

Dimensionless Groups

f Fanning friction factor, $H_{fs}Dg_c/2L\overline{V}^2$; f_a, at upstream station; f_b, at downstream station; $\bar{f}$, average friction factor, $(f_a + f_b)/2$

N_{Ma} Mach number, u/u_A

N_{Re} Reynolds number, $D\overline{V}\rho/\mu$, DG/μ, $D\overline{V}/\nu$; $N_{Re,o}$, Reynolds number at orifice

PROBLEMS

2-1. A simple U-tube manometer is installed across an orifice meter. The manometer is filled with mercury (specific gravity = 13.6), and the liquid above the mercury is carbon tetrachloride (specific gravity = 1.6). The manometer reads 8.30 in. What is the pressure difference over the manometer in inches of water?

2-2. A differential manometer of the type shown in Fig. 2-52 is sometimes used to measure small pressure differences. When the reading is zero, the levels in the two reservoirs are equal. Assume that fluid A is methane at atmospheric pressure and 60°F,

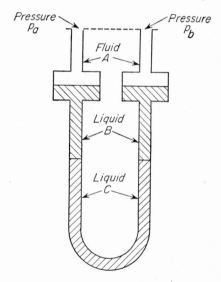

Fig. 2-52. Differential manometer.

that liquid B in the reservoirs is kerosene (specific gravity = 0.815), and that liquid C in the U tube is water. The inside diameters of the reservoirs and U tube are 2.0 and $\frac{1}{4}$ in., respectively. If the reading of the manometer is 5.72 in., what is the pressure difference over the instrument in inches of water: (*a*) when the change in levels in the reservoirs is neglected, (*b*) when the change in levels in the reservoirs is taken into account? What is the per cent error in the answer to part (*a*)?

2-3. Calculate the manometer reading in millimeters for each of the two manometers shown in Fig. 2-53. Explain the difference in readings if any. The following data apply in both cases. The pipe is 2-in., Schedule 40. The liquid flowing in the pipe is water at 80°F and at an average velocity of 10 ft/sec. The manometer fluid is mercury, and the leads above the mercury are filled with water. The distance L between the pressure taps is 20 ft.

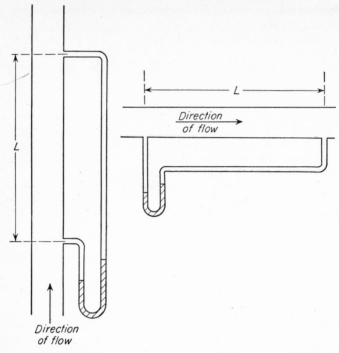

Fɪɢ. 2-53. Manometer and pipe arrangements for Prob. 2-3.

2-4. A noncompressible fluid is flowing through two horizontal pipes connected in series. The diameter of the upstream pipe is D_a, and that of the downstream pipe is D_b. Also $D_b > D_a$. At a given location in the upstream pipe, the pressure is p_a, and at another location in the downstream pipe, the pressure is p_b. Measurement of these pressures shows that $p_a < p_b$. Is this possible, or has there been an error in the pressure determinations?

2-5. Prove that the flow of a liquid in laminar flow between infinite parallel flat plates is given by the equation

$$p_a - p_b = \frac{12\mu \overline{V} L}{a^2 g_c}$$

where L = length of plate in direction of flow
$\quad\; a$ = distance between plates
Neglect end effects.

2-6. Calculate the horsepower required per foot of width of stream to force lubricating oil through the gap between two horizontal flat plates under the following conditions:

Distance between plates, $\frac{1}{4}$ in.
Flow rate of oil per foot of width, 135 gal/min
Viscosity of oil, 25 centipoises
Density of oil, 0.88 g/cm^3
Length of plates, 10 ft

Assume that the plates are very wide in comparison with the distance between them and that end effects are neglected.

2-7. Water at 60°F is pumped from a reservoir to the top of a mountain through a Schedule 120 6-in. pipe at an average velocity of 10 ft/sec. The pipe discharges into the atmosphere at a level 4,000 ft above the level in the reservoir. The pipeline itself is 5,000 ft long. If the over-all efficiency of the pump and the motor driving it is 70 per cent, and if the cost of electrical energy to the motor is $1\frac{1}{2}$ cents per kilowatthour, what is the hourly energy cost for pumping this water?

2-8. Crude oil having a specific gravity of 0.80 and a viscosity of 4 centipoises is draining by gravity from the bottom of a tank. The depth of liquid above the drawoff connection in the tank is 19 ft. The line from the drawoff is Schedule 40 4-in. pipe. Its length is 150 ft, and it contains one L and two gate valves. The oil discharges into the atmosphere 30 ft below the drawoff connection of the tank. What flow rate, in barrels per hour, may be expected through the line? 1 bbl = 42 gal.

2-9. Water is being heated in a horizontal shell-and-tube heater like that shown in Fig. 8-4. The heater has 500 tubes having an inside diameter of 0.620 in. and a length of 12 ft. The total flow of water into the heater is 2,400 gal/min. The inlet water temperature is 70°F, and the outlet temperature is 150°F. The temperature of the walls of the tubes is 230°F. The inlet and outlet connections are both 6 in. ID, and the area of each header is 2.5 times that of the total inside cross section of all the tubes. Estimate the pressure drop over this heater in inches of water.

2-10. The heater of Prob. 2-9 is used to heat air. The air enters at 20 in. water above atmospheric pressure and 70°F. It leaves at 200°F. The air flow through the unit is 3,500 ft³/min, measured at entrance conditions. The tube-wall temperature may be taken as 205°F. What is the pressure drop over the heater in inches of water?

2-11. Water at 68°F is to be pumped at a constant rate of 5 ft³/min to the top of an experimental absorber from a supply tank resting on the floor. The point of discharge is 15 ft above the floor, and the frictional losses in the Schedule 40 2-in. pipe are estimated to be 0.8 ft-lb force/lb. At what height in the supply tank must the water level be kept if the pump can develop a net power of only $\frac{1}{8}$ hp?

2-12. A centrifugal pump takes brine from the bottom of a supply tank and delivers it into the bottom of another tank. The brine level in the discharge tank is 200 ft above that in the supply tank. The line between the tanks is 700 ft of Schedule 40 6-in. pipe. The flow rate is 810 gal/min. In the line are two gate valves, four standard T's used as elbows, and four L's, with brine passing through the runs of the T's. What is the energy cost for running this pump for one 24-hr day? The specific gravity of brine is 1.18; the viscosity of brine is 1.2 centipoises; and the energy cost is $100 per horsepower-year on a basis of 300 days per year. The over-all efficiency of pump and motor is 60 per cent.

2-13. A fan draws air at rest and sends it through an 8- by 12-in. rectangular duct 180 ft long. The air enters at 60°F and 750 mm Hg abs pressure at a rate of 1,000 ft³/min. What is the theoretical power required?

2-14. Air at 25 lb force/in.² gauge and 60°F enters a horizontal 3-in. Schedule 40 pipe 3,000 ft long. The velocity at the entrance of the pipe is 50 ft/sec. Assuming isothermal flow, what is the pressure at the discharge end of the line?

2-15. Natural gas consisting essentially of methane is to be transported through a 20-in.-ID pipeline over flat terrain. Each pumping station increases the pressure to 100 lb force/in.² abs, and the pressure drops to 25 lb force/in.² abs at the inlet to the next pumping station 50 miles away. What is the gas flow rate in cubic feet per hour measured at 60°F and 30 in. Hg pressure?

2-16. A horizontal venturi meter having a throat diameter of 1.00 in. is located in a 3-in. Schedule 40 line. Water at 60°F is flowing through the line. A mercury manometer measures the differential over the instrument. The leads are filled with water. When the manometer reading is 15.2 in., what is the flow rate in gallons per minute? If 10 per cent of the differential is permanently lost, what is the power consumption of the meter?

2-17. Compressed air at 100 lb force/in.^{2} abs and 100°F flows through a 3-in. Schedule 40 pipe. A standard sharp-edged orifice equipped with flange taps and having a 1.00-in. hole is installed in the line. When the manometer reading is 14.1 in. Hg at 60°F, what is the flow rate of air in pounds per second?

2-18. Crude oil having a gravity of 16°API at 60°F is flowing at 150°F through a Schedule 40 6-in. pipe. The maximum flow rate through the line is 1,000 bbl/hr, measured at 60°F, where 1 bbl is 42 gal. It is desired to install a standard orifice in the line and to use flange taps. The maximum differential that can be read on the meter is 100 in. of water.

The kinematic viscosity of the oil is 6.0 centistokes at 150°F, and the density of the oil at 150°F is 96.5 per cent that at 60°F.

(a) Specify, to the nearest $\frac{1}{32}$ in., the size of the orifice. (b) Make a careful, freehand, dimensioned sketch showing the orifice, the position and sizes of the pressure taps, and other design details that must be used if the orifice is to be consistent with standard calibrations.

NOTE. The relationship between degrees API and specific gravity 60°F/60°F is

$$°API = \frac{141.5}{\text{sp. gr. } 60°F/60°F} - 131.5$$

2-19. Prepare an accurate calibration curve for the meter of Prob. 2-18 covering the range of flow rates for which it is applicable. Use the square root of the reading, in inches of water, for the abscissa and flow rate in barrels per hour at 60°F as the ordinate.

2-20. The rotameter of Example 2-21 is to be used to measure the flow rate of titanium tetrachloride, which has a specific gravity of 1.73 and a viscosity of 0.8 centipoise. Plot a calibration curve like that of Fig. 2-48 for the use of the rotameter with this substance.

2-21. A catalyst tower 50 ft high and 20 ft in diameter is packed with 1-in.-diameter spheres. Gas enters the top of the bed at a temperature of 500°F and leaves at the same temperature. The pressure at the bottom of the catalyst bed is 30 lb force/in.^{2} abs. The bed porosity is 0.40. If the gas has average properties similar to propane and the time of contact between the gas and the catalyst is 10 sec, what is the inlet pressure?

REFERENCES

1. Drew, T. B., E. C. Koo, and W. H. McAdams: *Trans. AIChE*, **28:** 56 (1932).
2. "Fluid Meters: Their Theory and Applications," 4th ed., p. 35, American Society of Mechanical Engineers, New York, 1937.
3. "Gas Measurement Committee Report," Natural Gas Association, New York, May 6, 1935.
4. Hunsaker, J. C., and B. G. Rightmire: "Engineering Applications of Fluid Mechanics," pp. 202–203, McGraw-Hill Book Company, Inc., New York, 1947.
5. Jorissen, A. L.: *Trans. ASME*, **74:** 905 (1952).
6. Kays, W. M.: *Trans. ASME*, **72:** 1067 (1950).
7. Lapple, C. E.: *Trans. AIChE*, **39:** 385 (1943).
8. Martin, J. J.: *Chem. Eng. Progr.*, **45:** 338 (1949).
9. Martin, J. J., W. L. McCabe, and C. C. Monrad: *Chem. Eng. Progr.*, **47:** 91 (1951).

10. McAdams, W. H.: "Heat Transmission," 2d ed., (a) p. 106, (b) pp. 109–111, (c) p. 119, (d) p. 121, McGraw-Hill Book Company, Inc., New York, 1942.
11. Moody, L. F.: *Trans. ASME*, **66**: 671 (1944).
12. Perry, J. H.: "Chemical Engineers' Handbook," 3d ed., (a) pp. 369–374, (b) p. 378, (c) pp. 380–381, (d) p. 394, (e) p. 411, (f) p. 1018, (g) p. 1286, McGraw-Hill Book Company, Inc., New York, 1950.
13. Rhodes, T. J.: "Industrial Instruments for Measurement and Control," p. 240, McGraw-Hill Book Company, Inc., New York, 1941.
14. Rothfus, R. R., and R. S. Prengle: *Ind. Eng. Chem.*, **44**: 1683 (1952).
15. Schoenborn, E. M., Jr., and A. P. Colburn: *Trans. AIChE*, **35**: 359 (1939).
16. Wattendorf, F. L., and A. M. Kuethe: *Physics*, **5**: 153 (1934).

Chapter 3

TRANSPORTATION OF FLUIDS

This chapter deals with the transportation of fluids, both liquids and gases. Solids are sometimes handled by similar methods by suspending them in a liquid to form a pumpable slurry or by conveying them in a high-velocity gas stream. It is cheaper to move fluids than solids, and materials are transported as fluids whenever possible. In the process industries, fluids are nearly always carried in closed channels, sometimes square or rectangular in cross section but much more often circular.

The first part of the chapter discusses the conduits in which fluids are transported—ducts, pipe, and tubing—and the accessories used for regulating flow. The second part describes the pumps, blowers, and compressors that impel the fluids through the pipe or tubing. The importance of pipe and fittings in a processing plant is often underestimated until it is recognized that in many plants the pipe costs as much as, or more than, the rest of the process equipment. The piping must be designed so that it does not introduce too large a pressure drop, does not become fouled, and does not leak. The satisfactory operation of many plants depends on the continued good performance of pumps and blowers. Their proper selection and installation, therefore, demand understanding and care.

PIPE, FITTINGS, AND VALVES

Pipe and Tubing. Fluids are transported most often in pipe or tubing which is circular in cross section and of widely varying sizes, wall thicknesses, and materials of construction. There is no clear-cut distinction between the terms pipe and tubing. Generally speaking, pipe is heavy-walled, relatively large in diameter, and supplied in moderate lengths of 20 to 40 ft; tubing is thin-walled and often comes in coils several hundred feet long. Metallic pipe can be threaded; tubing cannot. Pipe walls are usually slightly rough; tubing has very smooth walls. Lengths of pipe are joined by screwed, flanged, or welded fittings; pieces of tubing are connected by compression fittings, flare fittings, or soldered fittings. Finally, tubing is usually extruded or cold-drawn, while pipe is made by welding, casting, or piercing a billet in a piercing mill.

Materials of Construction. Pipe and tubing are made from many different materials, the list of which is constantly growing. Some are ductile,

136

some brittle, some plastic. Some are highly corrosion-resistant, while others are not; some have very high strength, while some are comparatively weak. The properties of the material dictate the methods of fabricating the pipe or tubing, the types of joints employed, and to a large extent the ways in which the pipe is used.

Most pipe, by far, is made of steel or iron. Common "black-iron pipe" in reality is made from low-carbon steel, welded or pierced, as described later. Wrought-iron pipe, which is somewhat more corrosion-resistant than steel pipe, can be had at a slight premium. Cast-iron pipe is widely used in underground lines under low pressure, especially for carrying water, but only sparingly in processing operations.

Pipe and tubing are also available in just about all metals that can be fabricated, from copper and brass to silver and tantalum. Their resistance to attack by corrosive fluids often justifies the high cost of the special metals and alloys. Increasingly, nonmetallic pipe is being used for carrying corrosive materials. Wood, clay, glass, asbestos-cement, and graphite make useful pipe for a variety of services. Most nonmetallic pipe is heavy-walled and relatively weak, and is used for low pressures only; but some, such as that made from plastic-bonded glass fiber, is light and very strong. Its use for fairly high-pressure service is rapidly growing.

High strength without sacrificing corrosion resistance can also be obtained with lined pipe. Metallic pipe, usually of steel, is lined with concrete, glass, rubber, or plastics, or is "clad" with a thin layer of another metal or alloy. Bimetallic tubing is used in heat exchangers where the corrosive nature of the fluid on one side of the tube wall differs greatly from that of the fluid on the other side. Table 3-1 lists materials of construction in which pipe and tubing are commonly available.

TABLE 3-1. MATERIALS OF CONSTRUCTION FOR PIPE AND TUBING

Metallic	Nonmetallic
Steel †	Wood (fir, maple, cypress)
Wrought iron †	Concrete
Cast iron	Portland cement and asbestos
High-silicon irons	Clay
Stainless steels †	Porcelain
Copper †	Silica
Brass †	Glass
Bronze †	Bonded glass fiber
Monel †	Graphite
Nickel †	Plastics (Haveg, Saran,† Polyethylene, etc.)
Aluminum †	Paper laminate
Magnesium †	
Tin	Lined pipe
Lead	Plastic-lined steel †
Silver	Concrete-lined steel †
Tantalum	Glass-lined steel †
	Rubber-lined steel †

† Available in IPS.

Methods of Fabrication. The chief methods of making tubular products are welding, piercing, casting, and extrusion. For short lengths forging and cupping are sometimes used; final sizing, especially of tubing, is often done by cold-drawing. Casting is done with brittle materials; the other operations require that the material be ductile or plastic.

Steel pipe is made by welding or by piercing. Pipe 2 in. in diameter or smaller is usually butt-welded; larger pipe is lap-welded. *Butt-welded pipe* is made from long, narrow strips of steel, called "skelp," which is heated in a furnace to 2600°F. The white-hot strips are then drawn through a welding "bell," which bends them into a circular shape and simultaneously welds the edges of the strips together. In *lap-welded pipe* the edges of the strip overlap after the strip is bent, giving a stronger joint than with butt welding. This is done by cutting the edges of the strip at an angle ("scarfing") before bending the skelp. Bending and welding cannot be done simultaneously with the larger pipes; instead the skelp is bent in a bell at about 1500°F and is then heated to 2600°F for welding. The welding is done by rolls, which press the scarfed edges together under high pressure. *Electric-welded pipe* is made by resistance-welding together the edges of bent steel strips; nearly all sizes of pipe are made by this method.

Seamless pipe is stronger than welded pipe, since it has a homogeneous wall. Its use is increasing as processing temperatures and pressures continue to rise. It is made by piercing a circular steel bar, known as a billet, in a piercing mill at a high temperature. The mill contains two rolls, which compress the billet radially, causing an opening to form in the center of the bar. The metal is sufficiently plastic at the elevated temperature to behave in this way. The size and position of the opening are controlled by a mandrel held between the rolls of the mill. The crude seamless pipe so formed is passed through dies to adjust the diameter and wall thickness to the desired tolerances.

Short pieces of seamless pipe are sometimes made by forging, or cupping, in which a central opening is hammered into a hot circular billet. Heavy-walled one-piece pressure vessels are also fabricated in this way. Tubing is usually made by cold-drawing a length of crude seamless pipe through a die surrounding a mandrel. Successive drawings change the diameter and accurately fix the dimensions. Cold-drawn tubing is known as mechanical tubing. With highly plastic materials seamless pipe and tubing can be made by extrusion, as macaroni is extruded. Lead pipe is also made in this way.

Casting is used with metals and other materials that are too brittle to roll or pierce. Molten metal is poured into a sand-lined mold and allowed to solidify. Cast-iron pipe, for example, is made in stationary molds, either horizontal or vertical, or in molds which rotate as the metal is poured in. This centrifugal casting puts the molten material under a centrifugal force of 40 or 50 times the force of gravity, giving a casting with uniform homogeneous wall that is stronger than can be made by the older methods. Flanges and other fittings are often cast into the pipe as it is made.

Sizes. Pipe and tubing are specified in terms of their diameter and their wall thickness. With steel pipe the standard nominal diameters range from $\frac{1}{8}$ to 30 in. For large pipe, more than 12 in. in diameter, the nominal diameters are the actual outside diameters; for small pipe the nominal diameter does not correspond to any actual dimension. The nominal value is close to the actual inside diameter for 3- to 12-in. pipe, but for very small pipe this is not true. Regardless of wall thickness, the outside diameter of all pipe of a given nominal size is the same, to ensure interchangeability of fittings. Standard dimensions of steel pipe are given in Appendix 4. Pipe of other materials is also made with the same outside diameters as steel pipe, to permit interchanging parts of a piping system. These standard sizes for steel pipe, therefore, are known as IPS (iron pipe size), or NPS (normal pipe size). Thus the designation "2-in. nickel IPS pipe" means nickel pipe having the same outside diameter as standard 2-in. steel pipe. Materials of construction in which IPS pipe is supplied are indicated in Table 3-1.

The wall thickness of pipe is indicated by the *schedule number*, given by the approximate equation:

$$\text{Schedule number} = \frac{1{,}000p'}{S} \tag{3-1}$$

where p' = internal working pressure, lb force/in.²

S = allowable stress, lb force/in.², for the alloy used

Ten schedule numbers, 10, 20, 30, 40, 60, 80, 100, 120, 140, and 160 are in use, but with pipe less than 8 in. in diameter only numbers 40, 80, 120, and 160 are common. For steel pipe the actual wall thicknesses corresponding to the various schedule numbers are given in Appendix 4; with other alloys the wall thickness may be greater or less than that of steel pipe, depending on the strength of the alloy. With steel at ordinary temperatures the allowable stress S is one-fourth the ultimate strength of the metal.

The size of tubing is indicated by the outside diameter. The nominal value is the actual outer diameter, to within very close tolerances. Wall thickness is ordinarily given by the BWG (Birmingham wire gauge) number, which ranges from 24 (very light) to 7 (very heavy). Sizes and wall thicknesses of heat-exchanger tubing are given in Appendix 5.

Selection of Pipe Size. The problem of selecting pipe of the proper size, like that of sizing any processing equipment, cannot usually be solved by precise methods. The requirements that a pipe or a machine has to meet may not be exactly known; more commonly the requirements, known accurately enough at first, may change markedly before the equipment has been in service very long. For a given service, both the initial cost and the cost of maintenance must be considered in selecting pipe size, for the added expense of keeping supplies of unusual sizes in stock may overshadow the small advantage gained by their use. For example, most plants do not carry $2\frac{1}{2}$- or $3\frac{1}{2}$-in. pipe in stock, even though these are standard sizes. In general it is best to specify the common sizes and to favor pipe that is a little too large.

The economic factors influencing the selection of pipe size have been considered by Sarchet and Colburn.[7] From their general equation comes the following simplified dimensional equation for turbulent flow in steel pipe [5a]

$$D'_e = \frac{0.098(w')^{0.45}}{\rho^{0.31}} \tag{3-2}$$

where D'_e = economic pipe diameter (actual inside diameter), in.

w' = mass rate of flow, lb/hr

ρ = fluid density, lb/ft^3

A nomograph of this equation is given by Perry.[5b]

The considerations leading to Eq. (3-2) do not include possible changes in demand or other effects, such as those of corrosion and fouling, which cannot be estimated precisely. Here the engineer must exercise judgment in selecting pipe of the proper size. In many instances a sound selection can be made by using rules of thumb based on experience. Table 3-2 shows the usual ranges of velocity for liquids and gases in processing equipment.

TABLE 3-2. FLUID VELOCITIES IN PIPE

Fluid	Type of flow	Velocity, ft/sec
Thin liquid..........	Gravity flow	0.5– 1
	Pump inlet	1 – 3
	Pump discharge	4 – 10
	Process line	4 – 8
Viscous liquid.......	Pump inlet	0.2– 0.5
	Pump discharge	0.5– 2
Steam...............	....	30 – 50
Air or gas...........	....	30 –100

These values are representative, not mandatory, for special conditions may dictate velocities considerably outside the indicated ranges. Low velocities should ordinarily be favored, especially when flow is by gravity from overhead tanks. This is another way of saying that the pipes should be large. Small pipe tends to clog with dirt and sediment, and since with gravity feed the head causing flow cannot be increased beyond that established by the arrangement of the equipment, the resulting flow rate may be too small. The carrying capacity of steel pipes is given in detail in Appendix 4.

Example 3-1. A condenser is to be supplied with 150 gal/min of cooling water at 80°F. What size of pipe should be used?

Solution. The density of water at 80°F (Appendix 14) is 62.1 lb/ft^3, or 8.30 lb/gal.

$$w' = 150 \times 60 \times 8.30 = 74,700 \text{ lb/hr}$$

$$\rho = 62.1 \text{ lb/ft}^3$$

From Eq. (3-2),

$$D'_e = \frac{0.098 \times 74,700^{0.45}}{62.1^{0.31}} = 4.26 \text{ in.}$$

The inside diameter of Schedule 40 4-in. steel pipe is 4.026 in. The next larger standard size is 5 in., and the next common standard size is 6 in. Therefore 4-in. pipe would be selected, even though it is slightly smaller than the computed size.

Joints and Fittings. The methods used to join pieces of pipe or tubing depend in part on the properties of the material but primarily on the thickness of the wall. Thick-walled tubular products are usually connected by screwed fittings, by flanges, or by welding. Pieces of thin-walled tubing are joined by soldering or by compression or flare fittings. Pipe made of brittle materials like glass, carbon, or cast iron is joined by flanges or bell-and-spigot joints.

Screwed Fittings. When screwed fittings are used, the ends of the pipe are threaded externally with a threading tool. The thread is tapered and the few threads farthest from the end of the pipe are imperfect, so that a tight joint is formed when the pipe is screwed into a fitting. Threading weakens the pipe wall, and the fittings, in general, are not so strong as the pipe itself; when screwed fittings are used, therefore, the schedule number computed from Eq. (3-1) should be doubled. Screwed fittings are standardized for pipe sizes up to 12-in., but because of the difficulty of threading and handling large pipe, they are rarely used in the field with pipe larger than 3-in. To lubricate and seal the threads, both in making the joint and to aid in breaking it later, pipe dope is used. A common type of dope is a suspension of graphite or other lubricant in an oil base. Dope is applied to the external threads before assembling a screwed joint.

Figure 3-1 shows common screwed fittings. Short pieces of pipe are known as *nipples*. A *close nipple* is threaded over its entire length; a *short nipple* has a narrow unthreaded central section on which a pipe wrench may be used without damaging the threads. *Couplings* are sleeves threaded internally, used to connect straight sections of pipe. The joint formed by a coupling cannot be opened without turning one of the pipes, which often involves dismantling a considerable amount of piping. To form a joint that can be opened easily a *union* is used. This consists of two mating pieces with ground metal faces pressed together tightly by a ring nut. By loosening the ring the halves of the union can be separated without disturbing the piping.

For turning corners *elbows*, or *L's*, either 45 or 90°, are used. A *street elbow* is threaded internally at one end and externally at the other. *T's* have three openings, ordinarily all the same size and all in the same plane; *crosses* have four. Special T's and crosses are made with openings of different sizes or with openings entering from the side.

Pipelines may be closed off completely by a *plug*, which screws into an elbow or coupling, or by a *cap*, which fits over the end of the pipe. Pipe sections of differing diameters are connected by *reducing couplings*, or *re-ducers*, and also by bushings. A *bushing* is a plug which screws into a fitting on the larger pipe and which is drilled and tapped to accommodate the smaller pipe. U bends and a variety of special fittings are also available.

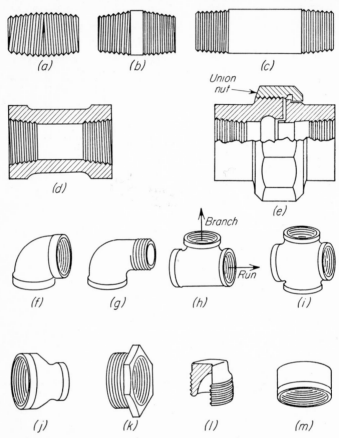

Fig. 3-1. Screwed pipe fittings: (*a*) Close nipple. (*b*) Short nipple. (*c*) Long nipple. (*d*) Coupling. (*e*) Union. (*f*) Elbow. (*g*) Street elbow. (*h*) T. (*i*) Cross. (*j*) Reducer. (*k*) Bushing. (*l*) Plug. (*m*) Cap.

For steel pipes the fittings are of malleable iron or gray cast iron. Similar fittings are also made in corrosion-resistant metals and alloys and in some plastics. Fittings having the same uses as those shown in Fig. 3-1 are made for welded and flanged piping. Couplings and unions, however, are not needed with welded or flanged piping.

Flanges. Lengths of pipe larger than about 2 in. are usually connected by flanges or by welding. Flanges are matching disks or rings of metal bolted together and compressing a gasket between their faces. The flanges themselves are attached to the pipe by screwing them on or by welding or brazing.   Figure 3-2 illustrates typical flanges for steel pipe and also the type used with glass pipe.

A flange with no opening, used to close a pipe, is called a blind flange or a blank flange.

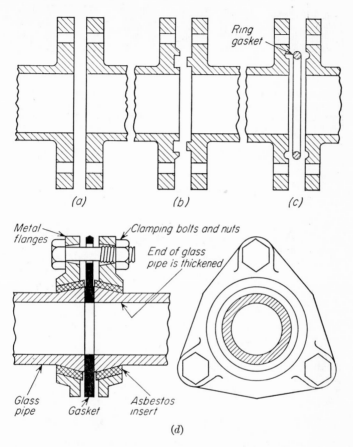

FIG. 3-2. Flanged joints for pipes: (*a*) Plain-faced.   (*b*) Tongue-and-groove joint.   (*c*) Ring gasket.   (*d*) Flanged joint for glass pipe.

The tightness and serviceability of a flanged joint depend to a large extent on the gasket, which must be properly made from the proper material. For low-pressure service soft gaskets of cork, asbestos, or rubber are used; higher pressures require hard gaskets, often of metal.   The trend in high-pressure processing is to eliminate gaskets and to depend on the contact of the flange faces, which must be ground to a high polish.

Welding, Brazing, and Soldering.   Welding has become the standard method of joining pieces of large steel pipe in process piping, especially for high-pressure service.   Typical fittings for welded pipe are shown in

Fig. 3-3. Welding makes stronger joints than do screwed fittings, and it does not weaken the pipe wall, so that lighter pipe can be used for a given pressure. Properly made welded joints are leakproof. The piping assemblies are lighter and cheaper than with other methods, and the disturbance to the flow pattern in the pipe is even less than with flanges. Almost the only disadvantage is that a welded joint cannot be opened without destroying it. Where the pipe must be opened, therefore, a flanged joint is included in the piping system.

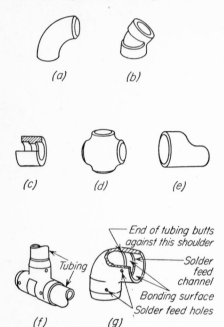

FIG. 3-3. Welding fittings: (a) Butt-welding 90° elbow. (b) Socket-welding 45° elbow. (c) Socket-welding coupling. (d) Butt-welding cross. (e) Eccentric butt-welding reducer. Solder-joint fittings. (f) T. (g) 90° elbow.

Sections of pipe and tubing that cannot be welded are often connected by soldering or brazing. Like welding, these methods give tight joints which do not weaken the pipe wall. The joints can usually be opened without destroying the pipe, however, by remelting the solder. Brazing is often used to hold flanges to pipe or tubing; high-melting solder (silver solder) is also used for this purpose. Soft, or low-melting, solder is used on tubing with the fittings of Fig. 3-3 f and g. Both the fitting and the end of the tube are heated; the tube is coated with solder, inserted in the fitting, and held in place until the solder solidifies. Soldered fittings are also known as "sweat fittings."

Compression and Flare Fittings. Tubing has walls that are too thin, usually, to be threaded. With big tubing, larger than $\frac{3}{4}$ in. in diameter, flanges and soldered fittings are used. Lengths of small tubing are most often connected by compression fittings or by flare fittings, both illustrated in Fig. 3-4. In a compression joint the end of the tubing is cut off square and inserted into a socket in the fitting, as shown in Fig. 3-4a. A ring which has previously been placed over the tubing is then compressed by a nut into the flared opening of the socket, forcing the ring into the outside of the tubing. Such fittings make tight joints, but it is best to use a new ring each time the joint is opened and remade.

For use with a flare fitting the end of the tubing is carefully flared with a special tool. It fits over the nose of the fitting, and is held there by the flare nut. Flared joints, properly made, will stand internal pressures of several hundred pounds per square inch and can be opened and recon-

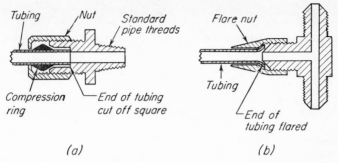

FIG. 3-4. Fittings for tubing: (a) Compression fitting. (b) Flare fitting.

nected several times without spoiling the joint. It requires more skill to make a good flared joint, however, than to use a compression fitting.

Bell-and-spigot Connections. Lengths of cast-iron pipe and pipe made of other brittle materials, like clay and porcelain, are often connected by bell-and-spigot joints. As shown in Fig. 3-5, one end of the pipe expands into a bell, forming a socket into which the plain end of the next length of pipe will fit. Usually the joint is made by stuffing a ring or two of oakum (rope of treated hemp fibers) into the bell around the plain pipe end, after which molten lead is poured in to fill the bell. The pipes do not have to be in perfect alignment, and special bell-and-spigot joints are made which withstand considerable vibration. This

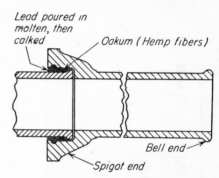

FIG. 3-5. Bell-and-spigot joint.

type of joint is most commonly used in low-pressure lines, nearly always below 50 lb/in.^{2} and often much lower.

Allowance for Expansion. Almost all pipe is subjected to varying temperatures, and in some high-temperature lines the temperature change is very large. Such changes cause the pipe to expand and contract. If the pipe is rigidly fixed to its supports, it may be torn loose, bend, or even break open. In large lines, therefore, fixed supports are not used; instead the pipe rests loosely on rollers or is hung from above by chains or rods. Provision is also made in all high-temperature lines for taking up expansion, so that the fittings and valves are not put under strain. This is done by bends or loops in the pipe, by packed expansion joints, by bellows, or "packless," joints, and sometimes by flexible metal hose.

Pipe Bends and Loops. Figure 3-6 shows the common ways of allowing for expansion. In assemblies of small pipe with numerous elbows and T's and only short runs of straight pipe, the temperature changes must be very large before the system can no longer absorb the expansion. In long,

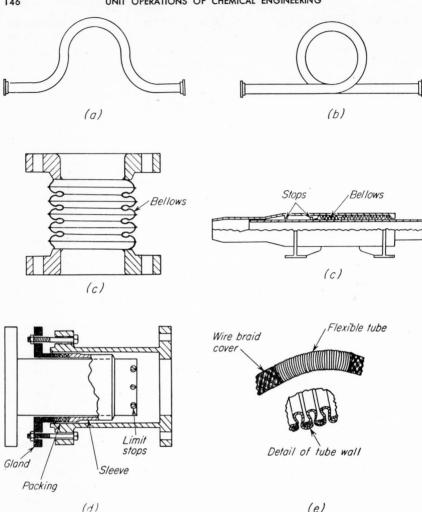

Fig. 3-6. Methods of allowing for pipe expansion: (*a*) Expansion U bend. (*b*) Circle bend. (*c*) Bellows expansion joints. (*d*) Slip joint. (*e*) Flexible metal hose.

straight runs of large pipe, however, as in outdoor steam service lines, frequent loops are provided, careful attention being paid to their placement and minimum radius.[3]

Packed Expansion Joints. A typical packed expansion joint, or slip joint, is shown in Fig. 3-6*d*. It contains a sliding joint sealed with packing in a conventional stuffing box of the type described later in this chapter. To allow for considerable expansion, double joints with two stuffing boxes are used. Packed joints can stand high internal pressures and permit large relative motions of the pieces of pipe, but they require maintenance to keep leakage to a minimum.

Packless Expansion Joints. Typical bellows, or packless, expansion joints are illustrated in Fig. 3-6c. The bellows expands or contracts as required by the motion of the pipe yet keeps the fluid inside the pipe at all times. These devices are now made for high-temperature and high-pressure service, and are generally more satisfactory than packed joints. Standard designs permit relative motion up to about 1 in.; for greater amounts of motion special designs, like the one on the right of Fig. 3-6c, are used.

Flexible Metal Hose. Metal tubing with a corrugated wall is highly flexible, but long pieces are weak. Flexible metal hose (Fig. 3-6e) has a corrugated (or more properly, a "convoluted") inner tube of brass, monel, Inconel, or stainless steel sheathed with a woven metal cover to give it strength. Short lengths are often used in piping systems to eliminate strain; longer pieces are useful in connecting process lines to vibrating or moving machinery and in places like drum-filling equipment where the line must frequently be moved.

Prevention of Leakage around Moving Parts. In many kinds of processing equipment it is necessary to have one part move in relation to another part without excessive leakage of a fluid around the moving member. This is true in packed expansion joints and in valves where the stem must enter the valve body and be free to turn without allowing the fluid in the valve to escape. It is also necessary where the shaft of a pump or compressor enters the casing, where an agitator shaft passes through the wall of a pressure vessel, and in other similar places. The devices used for this purpose are not pipe fittings, strictly speaking, but they are discussed here because familiarity with them is prerequisite to understanding the valves, pumps, and the other equipment covered in this chapter.

The common devices for reducing leakage while permitting relative motion are three: stuffing boxes, mechanical seals, and labyrinths. Note that none of these completely stop leakage, and if the pressure differential across the opening through which the moving part passes is large, the leakage may be considerable. There are ways, however, of minimizing leakage and of ensuring that only innocuous fluids escape from the equipment. The motion of the moving part may be reciprocating or rotational, or both together; it may be small and occasional, as in a packed expansion joint, or virtually continuous, as in a process pump.

Stuffing Boxes. A stuffing box can provide a seal around a rotating shaft and also around a shaft which moves axially. In this it differs from other types of seals, which are good only with rotating members. The "box" is a chamber cut into the stationary member surrounding the shaft or pipe, as shown in Fig. 3-7. Often a boss is provided on the casing or vessel wall to give a deeper chamber. The annular space between the shaft and the wall of the chamber is filled with *packing* consisting of a rope or rings of inert material, such as asbestos, containing a lubricant like graphite. The packing, when compressed tightly around the shaft, keeps the fluid from passing out through the stuffing box and yet permits the shaft to turn or move back and forth. The packing is compressed by a follower ring, or gland, pressed into the box by a flanged cap or packing nut. The shaft must have a smooth surface so that it does not wear away the packing; even so, the pressure of the packing considerably increases the force required to move the shaft. A stuffing box, even under ideal conditions, does not completely stop fluid from leaking out; in fact, when the box is operating properly there should be small leakage. Otherwise the wear on the packing and the power loss in the unlubricated stuffing box are excessive. When the liquid is toxic or corrosive, means must therefore be provided to prevent it from escaping from the equipment.

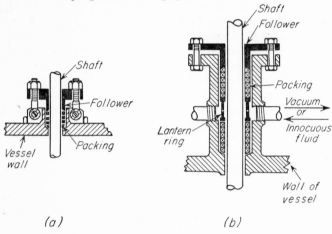

(a)   (b)

Fig. 3-7. Stuffing boxes: (a) Simple form. (b) Stuffing box with lantern gland.

This can be done by using a *lantern gland*, illustrated in Fig. 3-7b. This may be looked upon as two stuffing boxes on the same shaft, with two sets of packing separated by a lantern ring. The ring is H-shaped in cross section, with holes drilled through the bar of the H in the direction perpendicular to the axis of the shaft. The wall of the chamber of the stuffing box carries a pipe which takes fluid to or away from the lantern ring. By applying vacuum to this pipe, any dangerous fluid which leaks through one set of packing rings is removed to a safe place before it can get to the second set. Or by forcing a harmless fluid, usually water, under high pressure into the lantern gland, it is possible to ensure that no dangerous fluid leaks out the exposed end of the stuffing box.

For a stuffing box to give satisfactory service the packing must be selected with care. It must be inert to the fluid it seals against; it must be strong enough to resist abrasion by the moving shaft, even when pressed tightly against it; and yet it must be soft enough to yield to minor irregularities in the shaft surface. It is easier to seal around a rotating shaft than around a reciprocating shaft. Slow-speed shafts are less damaging to packing than are high-speed shafts, but where high speeds cannot be avoided, the stuffing box may be water-cooled.

Mechanical Seals. In a rotary, or mechanical, seal the sliding contact is between a ring of graphite and a polished metal face, usually of carbon steel. A typical seal is shown in Fig. 3-8. Fluid in the high-pressure zone is kept from leaking out around the shaft by the stationary graphite ring held by springs against the face of the rotating metal collar. A stationary ring of rubber or plastic is set in the space between the body of the seal and the chamber holding it around the shaft; this keeps fluid from leaking past

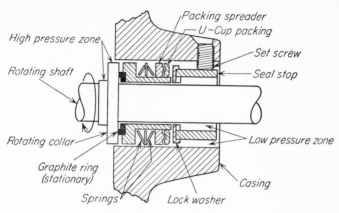

FIG. 3-8. Mechanical seal.

the nonrotating part of the seal and yet leaves the graphite ring free to move axially so that it can be pressed tightly against the collar. Rotary seals require less maintenance than do stuffing boxes and have come into wide use in equipment handling highly corrosive fluids.

Labyrinths. When shaft speeds and operating temperatures are too high for stuffing boxes or rotary seals, as in some turbines, recourse must be had to the labyrinths shown in Fig. 3-9. These devices do not stop leakage; instead they depend upon a small amount of flow to maintain the pressure difference between one part of the equipment and another. They are used only with gases. As the gas flows around the rotating shaft and through the labyrinth, it follows a tortuous path and undergoes repeated expansions and recompressions. As long as there is flow,

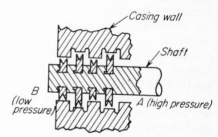

FIG. 3-9. Labyrinth. Clearances between shaft and wall are exaggerated.

therefore, there is a considerable pressure loss from point *A* to point *B*. If the flow stops, the pressure differential disappears. Thus labyrinths cannot be used where leakage must be very small, so that they find only limited application in chemical-processing equipment.

Valves. The parts of two typical valves are shown in detail in Fig. 3-10. Despite the wide variety in their design, all valves have a common primary purpose: to slow down or stop the flow of a fluid. Some valves work best in "on-or-off" service, i.e., fully open or fully closed. Others are designed to throttle, to reduce the pressure and flow rate of a fluid. Still others permit flow in one direction only or only under certain conditions of temperature and pressure. A steam trap, which is a special form of valve, allows some fluids to pass through while holding back another. Finally, through accessory devices, valves can be made to control the temperature, pressure, liquid level, or other properties of a fluid at points remote from the valve itself.

In all cases, however, the valve initially stops or controls flow. This is done by placing an obstruction in the path of the fluid, an obstruction

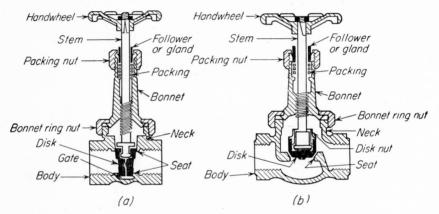

Fig. 3-10. (a) Gate valve. (b) Globe valve.

which can be moved about as desired inside the pipe with little or no leakage of the fluid from the pipe to the outside. Where the resistance to flow introduced by an open valve must be small, the obstruction and the opening which can be closed by it are large. For precise control of flow rate, usually obtained at the price of a large pressure drop, the cross-sectional area of the flow channel is greatly reduced, and a small obstruction is set into the small opening.

In most valves the obstruction is called the *disk*. Usually the disk is moved by an attached member, called a *stem*, which is inserted into the flow channel through a stuffing box. The stem is threaded so that turning it moves the disk into or out of the path of the fluid. When a valve is closed, the disk presses against the *seat*, forming a line of contact past which the fluid cannot seep. The seat is attached to, or is part of the *body* of, the valve. The neck of the body carries the valve *bonnet*, which contains the stuffing box and often the threads that mesh with the threaded part of the stem. At the top of the stem is the *handwheel*. The disk is usually removable for repair and replacement, and often the seat is removable as well.

Gate, Slide, and Butterfly Valves. A gate valve is shown in Fig. 3-10a; slide and butterfly valves are illustrated in Fig. 3-11. The simplest valves contain a flat plate, or disk, which can be set parallel with the direction of flow or turned 90° so that it impedes the flow. This is called a butterfly valve, most commonly used in large ducts for air or gas or in very large pipes carrying water. Another simple valve is a slide valve, consisting of a thin, flat plate inserted through a slot in the wall of the duct or pipe. The lips of the slot are covered with fabric or other soft material to form a partial seal. Slide valves are most often found at the bottom of storage bins

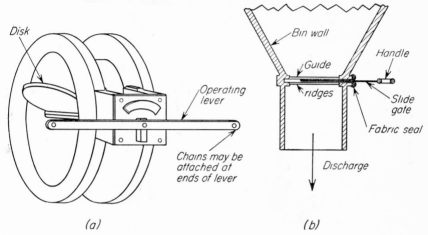

Fig. 3-11. (a) Butterfly valve. (b) Slide valve.

for granular solids, where they are used to control the flow of solids through the outlet.

In a gate valve the diameter of the opening through which the fluid passes is nearly the same as that of the pipe, and the direction of flow does not change. As a result, the valve introduces only a small pressure drop. The disk is tapered and fits into a tapered seat; when the valve is opened, the disk rises into the bonnet, completely out of the path of the fluid. In a *rising-stem valve* the disk and stem move together; in a *nonrising-stem valve* the disk moves on a threaded portion of the stem. In an *outside-screw-and-yoke (OSY) valve* the stem is moved by threads in the handwheel, which is itself kept from rising by a yoke attached to the bonnet. In this way the stem does not turn, minimizing wear on the stuffing box, and the threads are not submerged in, or exposed to, the fluid in the pipe. OSY valves are especially useful with highly corrosive fluids.

Gate valves work best when fully open or fully closed and function as "block" valves. When the valve is part-way open, the disk is subject to severe erosion by the fluid moving at high velocity through the restricted opening. It soon becomes so worn that flow can no longer be completely stopped. For controlling flow rates, therefore, some form of globe or needle valve should be used.

Globe and Needle Valves. In nearly all forms of globe valves (so called because in the earliest designs the valve body was spherical), the fluid passes through a restricted opening and changes direction several times in so doing. The pressure drop in this kind of valve, therefore, is large.

Various modifications of globe valves are shown in Figs. 3-10b and 3-12. In the commonest form the seat is a horizontal ring closed by a flat disk,

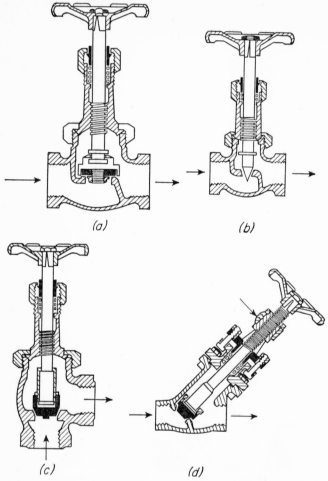

(a)   (b)

(c)   (d)

FIG. 3-12. Types of globe valves: (a) Disk globe valve. (b) Needle valve. (c) Angle valve. (d) Y valve.

as shown in Fig. 3-12a, or a plug with a slightly beveled edge as shown in Fig. 3-10b. When precise control of flow rate is required, the end of the stem carries a tapered point instead of a disk, making what is called a needle valve, as shown in Fig. 3-12b. The threads on the stem of a needle valve are fine, so that many turns of the handwheel are needed to effect a large change in the flow rate. This makes for precise control but also means that the valve cannot be quickly opened or quickly closed.

Still other modifications of globe valves include angle valves and Y valves. In an angle valve, as the name implies, the outlet is at 90° to the inlet, as shown in Fig. 3-12c. In a Y valve the seat and stem are at an angle of 45° to the direction of fluid flow. Such a valve is shown in Fig. 3-12d. In some designs the seat is held between separable flanged halves, which can be bolted together to make an angle valve if desired. A Y valve is a compromise between a gate valve, with its low pressure drop, and a globe valve, useful for controlling flow rates. Y valves do not clog easily and are good for controlling the flow rate of slurries.

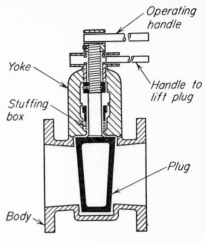

FIG. 3-13. Plug cock.

Plug Cocks. Figure 3-13 shows a typical plug cock, which is the metal counterpart of a glass laboratory stopcock. A tapered plug obstructs the flow when the cock is closed; turning the plug 90° opens the channel to almost unrestricted flow. Because of the large area of contact between the plug and the seat, the valve tends to stick. This is overcome by raising the plug slightly, by mechanical means or by forcing a lubricant under the plug, before the cock is opened. Plug cocks, like gate valves, offer little resistance to flow when fully open; they also are best for on-or-off service, not for throttling.

Check Valves. A check valve permits flow in one direction only. It is opened by the pressure of the fluid flowing in the proper direction; when the flow stops or tries to reverse, the valve automatically closes by gravity or by a spring pressing against the disk. Common types of check valves

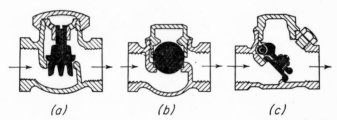

FIG. 3-14. Check valves: (a) Lift check. (b) Ball check. (c) Swing check.

shown in Fig. 3-14 are the *lift check, ball check,* and *swing check.* The principle of operation of these valves is obvious. A check valve operates satisfactorily if the disk and seat are in good repair and if nothing is trapped between disk and seat, but it should not be relied upon to guarantee that reverse flow will be stopped. Where reverse leakage cannot be tolerated, other means of stopping flow, such as auxiliary valves or a temporary blind flange, should be included in the line with the check valve.

Safety Devices. In an emergency it is sometimes necessary to open or close a valve in a hurry or to provide automatic relief of excessive pressure. The first is accomplished by a *quick-opening valve*, a gate valve fitted with a stem that can be pulled upward without turning it. An example is shown in Fig. 3-15b. Levers and a handle are provided so that the valve may be opened very rapidly. The second is done by *pressure-relief valves*, which are included on nearly all pressure vessels. One type is shown in Fig. 3-15a. In these valves the disk is held against the seat by a spring set

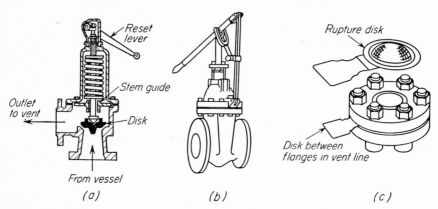

FIG. 3-15. Safety relief valves: (a) Safety valve. (b) Quick-opening valve. (c) Safety rupture disk.

to yield when the pressure in the vessel exceeds the safe working pressure by a definite amount, usually 25 per cent. Sometimes a relief valve recloses automatically after the excessive pressure is relieved; sometimes it stays open and must be reset by the operator.

When the pressure rise is very rapid, as in an explosion, a relief valve does not have time to open before the vessel shatters. Consequently other protective devices are used on pressure vessels in which there is danger of a sudden excessive rise in pressure. Such devices are called *frangible disks* or *safety disks*, an example of which is shown in Fig. 3-15c. They are disks of corrosion-resistant metal shallowly dished and held between two flanges in the vent line of the vessel. They are carefully made and guaranteed to rupture at a pressure within 5 per cent of the specified value. Disks are usually chosen which fail at a pressure $1\frac{1}{2}$ times the working pressure of the equipment. After failure, of course, they must be replaced.

Packless and Bellows Valves. In nearly all the valves described so far the stem passes through the wall of the pipe. Around the stem is a stuffing box. Thus there is a weak place in the system through which leakage can and does occur. Where there can be absolutely no leakage to the outside, a conventional stem and stuffing box cannot be used; instead the obstruction to fluid flow must be moved inside the valve body without providing an opening in the wall of the conduit.

With rubber tubing this is done simply enough by a pinch clamp. With metal pipe the same principle is employed in a packless valve, shown in Fig. 3-16b, in which a flexible diaphragm is pressed against a saddle-shaped seat to form the seal. The diaphragm, of rubber or thin metal, is moved by an oversize stem as shown in Fig. 3-16b. Packless valves are exten-

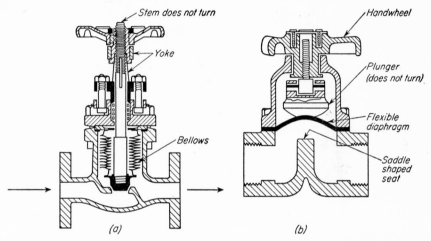

FIG. 3-16. Sealed valves: (a) Bellows valve. (b) Packless valve.

sively used in small gas lines and on cylinders for gas storage. They are also valuable for controlling the flow rate of slurries, since they do not foul easily. A bellows valve, an example of which is shown in Fig. 3-16a, is another way of making an absolutely tight system. It is especially useful for operations under very high vacuum.

A *pressure-control valve* is illustrated in Fig. 3-17. Such valves are often packless. They hold the downstream pressure constant despite changes in the upstream pressure. To achieve this the large pressure is made to operate against a very small area, with the resulting force opposed by the small pressure pushing against a much larger area. Through a balanced system the lower downstream pressure is held constant at the desired value. In the valve shown in Fig. 3-17, gas under high pressure enters through a very small opening in the nozzle at A. As it escapes into the valve body, it pushes the seat B downward a short distance. In so doing it deflects the diaphragm C, to which seat B is connected by a yoke. The low downstream pressure, acting against the entire surface of diaphragm C, tends to raise seat B and shut off the flow of high-pressure gas through the nozzle. Thus if the downstream pressure rises, seat B rises, and no more gas enters; if the downstream pressure falls, seat B is pushed away from the nozzle, and high-pressure gas enters until the correct downstream pressure is reestablished

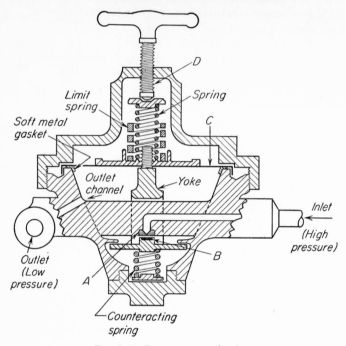

Fig. 3-17. Pressure-control valve.

The downstream pressure is set by turning stem D, which adjusts the position of the diaphragm and the tension in the springs.

Control Valves. As automatic control has become more and more extensively utilized in processing operations, control valves have concurrently gained in importance. Their design and selection are now an important and growing branch of technology, with a multitude of designs available for various functions. Just about all control valves are actuated by impulses from a sensing element, and adjust the flow rate of the fluid passing through the valve to keep conditions at the sensing element steady. The sensing element detects any deviations from the desired conditions and actuates the control valve accordingly. A closed control loop which acts in this way employs what is known as "feedback."

The two methods of actuating control valves are by compressed air and by a motor. A pneumatically operated control valve is shown in Fig. 3-18. Both methods have advantages and drawbacks, but the pneumatic method is more favored at present. The disk of an automatic control valve is specially shaped to give particular changes in flow rate in response to given changes in the position of the disk. V-port disks, parabolic plugs, and conical needles are most often used, frequently with two disks and two seats in the same valve to give a balanced load on the stem.

Steam Traps. A steam trap may be considered as a special kind of check valve which allows flow in one direction only and of certain fluids only. It allows hot water (steam condensate) and inert gases such as air to pass; it holds back the steam. Thus traps are used where steam is condensing under pressure in a system from which condensate and inerts must be removed. Examples are steam-heated household radiators, steam-jacketed process vessels, steam coils in evaporators and distilling columns, and so forth.

The three main types of steam traps are shown in Fig. 3-19. A *thermostatic*, or *expansion*, *trap* contains a bellows to which the valve disk is attached. When live steam or very hot water enters the trap, the bellows expands, shutting off the flow; when the condensate or inert gas in the trap cools sufficiently, the bellows contracts and the valve opens. Condensate and inerts are pushed out by the pressure of the steam in the main system, until steam or very hot condensate once more enters the trap, closing the valve. Thermostatic traps are slow-acting but very satisfactory for light duty.

The commonest form of trap for

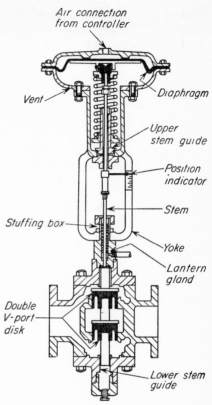

Fig. 3-18. Diaphragm-operated control valve.

general service is the *inverted-bucket trap* (Fig. 3-19b). In normal operation the chamber of this kind of trap is filled with water. The inlet is at the bottom. A small amount of vapor collects inside the inverted bucket, causing it to rise and through a series of links to close the trap outlet. Inert gas escapes through small holes in the floor of the bucket, as shown, and collects in the top of the trap. When the vapor in the bucket cools and condenses, the bucket loses its buoyancy. It sinks and opens the valve. The steam pressure then forces condensate out through the valve, entraining inert gases with it. As soon as live steam reenters the trap, it strikes the underside of the bucket, lifting it, and slams the valve shut. As in all steam traps, the action is intermittent. There is no flow of steam through the trap; flow occurs only when sufficient condensate and inerts have accumulated to cause the trap to open. For proper operation a trap of this kind should not be open more than about 60 per cent of the time.

For high-pressure service an *impulse trap* (Fig. 3-19c) is often used. The operation of this device involves a careful balance of differing pressures acting upon areas of different size. The chamber of the trap contains a tapered plug carrying a flange very slightly smaller in diameter than the chamber. The plug ordinarily rests in the seat, preventing loss of steam from the system. Very hot condensate in the space just below the flange "flashes" into vapor, which seeps past the flange into the space above the

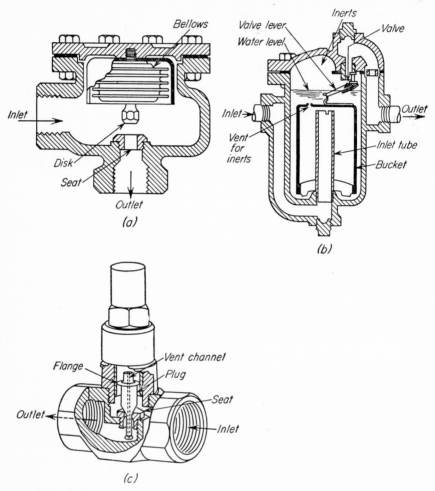

Fig. 3-19. Steam traps: (a) Thermostatic trap. (b) Inverted-bucket trap. (c) Impulse trap.

plug. The pressure here is lower than it is in the inlet line, but since the lower pressure acts on the entire area of the top of the plug while the inlet pressure pushes against the bottom of the flange only, the plug remains in the seat. When the condensate cools, it no longer vaporizes fast enough to maintain this balance. The higher pressure then lifts the plug. Condensate and inerts are swept through the large opening, until steam or very hot condensate which can vaporize rapidly once more enters the trap. The plug falls into the seat, and flow stops. A very small longitudinal channel through the plug keeps inert gas from accumulating in the trap. Impulse traps are fast-acting, suitable for high pressures, and easy to maintain.

Resistance of Valves and Fittings. When a fluid flows through a straight pipe, the resulting pressure drop can be calculated by the methods discussed in Chap. 2. Fittings and valves, however, introduce additional resistance which cannot be calculated directly but which must be found by actual measurement or from published data. As shown in Chap. 2, the resistance of fittings and valves is usually expressed in terms of the equivalent length, i.e., the length of straight pipe of the same nominal diameter which would introduce the same resistance as the fitting under the same conditions of flow. If the equivalent length is expressed in pipe diameters (the equivalent length, in inches, divided by the actual internal diameter of the pipe, in inches), the resistance of all similar fittings is almost independent of the size of the fitting. The equivalent resistance of various fittings and valves is given in Table 3-3.[5c] Actually these values are aver-

TABLE 3-3. EQUIVALENT RESISTANCE OF SCREWED FITTINGS AND VALVES †

Fittings	Equivalent resistance, pipe diameters
45° elbows	15
90° elbows (standard-radius)	32
90° elbows (long-radius)	20
90° square elbows	60
180° close return bends	75
T's (used as elbow, entering run)	60
T's (used as elbow, entering branch)	90
Couplings	Negligible
Unions	Negligible
Gate valves (open)	7
Globe valves (open)	300
Angle valves (open)	170

† From J. H. Perry (ed.), "Chemical Engineers' Handbook," 3d ed., p. 390, McGraw-Hill Book Company, Inc., New York, 1950.

ages for ordinary practice, and should not be considered exact. The resistance of valves, in particular, varies considerably from one manufacturer's design to another. Also, as shown by Fig. 3-20, the equivalent resistance is a function of the flow rate and at low Reynolds numbers may differ considerably from the published value.

In computing the total pressure drop in a piping system, the equivalent length of all the straight pipe plus the fittings is first computed, as shown in Chap. 2, page 74, after which the pressure drop may be found from Eq. (2-53).

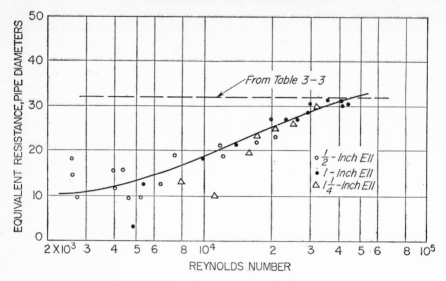

Fig. 3-20. Equivalent resistance of 90° elbows.

Example 3-2. Benzene at 40°C is flowing through a horizontal 1½-in. Schedule 40 steel pipe at the rate of 40 gal/min. The line contains six standard-radius 90° elbows, one T used as an elbow (with the liquid entering the run), a gate valve, and a globe valve. The length of straight pipe in the system is 69 ft. The density of benzene at 40°C is 53 lb/ft³; the viscosity is 0.5 centipoise. Compute the pressure loss in the pipe, in pounds force per square inch, when both valves are fully open.

Solution. Since there is no change either in elevation or in velocity, the pressure drop is entirely from friction. The equivalent resistance of the valves and fittings, from Table 3-3, is:

Fittings	Equivalent resistance, pipe diameters
Six elbows: 6 × 32.........	192
One T....................	60
One gate valve (open)......	7
One globe valve (open).....	300
Total.................	559

The internal diameter of 1½-in. Schedule 40 pipe (Appendix 4) is 1.610 in. The equivalent length of fittings and valves is $(1.610 \times 559)/12 = 75$ ft. The total equivalent length of pipe and fittings is $69 + 75 = 144$ ft.

The Reynolds number and friction factor are next computed.

$$D = \frac{1.610}{12} = 0.134 \text{ ft} \qquad \rho = 53 \text{ lb/ft}^3$$

$$\mu = 0.5 \times 6.72 \times 10^{-4} = 0.000336 \text{ lb/ft-sec}$$

The rate of flow is

$$\frac{40}{7.48 \times 60} = 0.0891 \text{ ft}^3/\text{sec}$$

The internal sectional area of $1\frac{1}{2}$-in. pipe is 0.01414 ft².

The average velocity is

$$\overline{V} = \frac{0.0891}{0.01414} = 6.31 \text{ ft/sec}$$

$$N_{\text{Re}} = \frac{D\overline{V}\rho}{\mu} = \frac{0.134 \times 6.31 \times 53}{0.000336} = 1.335 \times 10^5$$

$$\frac{k}{D} = \frac{0.00015}{0.134} = 0.0011$$

$$f = 0.0055 \text{ (Fig. 2-19)}$$

The friction, from Eq. (2-53), is

$$H_{fs} = 4f\frac{L}{D}\frac{\overline{V}^2}{2g_c}$$

$$= 4 \times 0.0055 \frac{144}{0.134}\frac{6.31^2}{2 \times 32.17} = 14.6 \text{ ft-lb force/lb}$$

$$p_a - p_b = H_{fs}\rho = \frac{14.6 \times 53}{144} = 5.37 \text{ lb force/in.}^2$$

Recommended Practice. In designing and installing a piping system many details must be given careful attention, for the successful operation of the entire plant may turn upon a seemingly insignificant feature of the piping arrangement. Some general principles are important enough to warrant mention here. In installing pipe, for example, the lines should be parallel and contain, as far as possible, right-angle bends. A maze of pipe running in all directions at odd angles is not good design. Provision should be made for opening the lines in order to change the piping or clean it out. This means that unions or flanged connections should be generously included. To facilitate cleaning, T's and crosses with their extra opening closed with plugs should be substituted for elbows in critical places. It then becomes easy to remove and clean out the line with a rod or brush. A process line should always be expected to become clogged, for nearly all process fluids contain some dirt that builds up in the line and eventually stops flow.

In gravity-flow systems the pipe should be oversize and contain as few bends as possible. Fouling of the lines is particularly troublesome where flow is by gravity, since the pressure head on the fluid cannot be increased to keep the flow rate up if the pipe becomes restricted.

Leakage through valves should also be expected. Where complete stoppage of flow is essential, therefore, where leakage past a valve would contaminate a valuable product or endanger the operators of the equipment, a valve or check valve is inadequate. In this situation a blind flange set between two ordinary flanges will stop all flow; or the line can be "broken" at a union or pair of flanges and the open ends capped or plugged.

In threading pipe to carry a screwed fitting the end of the pipe should be cut square, and the surface to be threaded should be free of wrench marks or other defects. Otherwise the threads will be imperfect and may not be pressure-tight. The threads should not be continued too far up the pipe. especially if a valve is to be placed on the threaded end, or the end of the pipe will enter too far into the valve body, damaging it or interfering with the motion of the disk. Pipe dope should be applied only to the male (outside) threads in a screwed joint; it should not be placed on the two or three threads near the end of the pipe, where there is danger of its getting into the pipe and contaminating the fluid.

Valves should be mounted vertically with their stems up, if possible. They should be accessible, well within convenient reach of the operator. They should be well supported without strain, with suitable allowance for thermal expansion of the adjacent pipe. Room should be allowed for fully opening the valve and for repacking the stuffing box.

Vent lines containing relief valves and safety disks must be large enough to accommodate very rapid flow without allowing the pressure in the vessel to build up to unsafe values; otherwise the safety device will be rendered ineffectual. Vent lines should have as few bends in them as possible and should discharge the contents of the vessel without hazard to the operators or others.

PUMPS, FANS, BLOWERS, AND COMPRESSORS

Fluids are moved through pipe, fittings, and equipment by pumps, fans, blowers, and compressors. The primary function of these devices, however, is not transportation; it is to add energy to the fluid. This energy increase may then be used to raise the pressure, elevation, or velocity. In the special case of liquid metals, energy may be added by magnetic pumps; air lifts, jet ejectors, and blow cases utilize pressure energy from a second fluid. Most common by far, however, is direct addition of mechanical energy by positive displacement or centrifugal action.

Operating Characteristics. The terms "pump," "fan," "blower," and "compressor" do not have precise meanings. "Air pump" and "vacuum pump" are used to designate machines for compressing a gas, but commonly a *pump* is a device which handles liquids. *Fans* discharge gases into open spaces or ducts at pressures up to 60 in. of water; *blowers* discharge at 1 to perhaps 15 lb force/in.^{2}; *compressors* discharge at 5 to 20,000 lb force/in.^{2} or higher.

In selecting a pump or blower the engineer is interested in its performance characteristics. The items of chief concern are the pressure change, the capacity, or throughput, the power required, and the efficiency. These are interrelated, as shown later, and vary in different ways, depending on the design of the machine. A given unit is designed to have a rated capacity with a specified gain in pressure; at these operating conditions the efficiency is usually a maximum. To the items listed, however, must be added reliability and ease of maintenance. In much process work the power and efficiency are relatively unimportant, for the amount of fluid pumped is small, and reliable performance is more important than saving a few kilowatts of power.

Suction Lift and Suction Head. Positive-displacement machines can draw gas or liquid to the inlet, sometimes creating considerable suction. Centrifugal pumps, on the other hand, cannot operate until they are filled with liquid, which must therefore be fed to the pump inlet under pressure from a storage tank or auxiliary priming device. Sometimes self-priming elements are included in the pump itself. Once primed, of course, a centrifugal pump or other pump can draw liquid up to an inlet considerably above the source of supply. The height to which the liquid is drawn is known as the *suction lift.* The theoretical maximum suction lift with cold water is a little over 33 ft at sea level; in practice actual lifts range from 15 to 28 ft. The theoretical minimum *absolute* pressure at the pump inlet is the vapor pressure of the liquid. Thus if the liquid is heated, the possible suction lift is diminished, until at the normal boiling point the theoretical lift is zero. In practice, when the vapor pressure of the liquid approaches or exceeds atmospheric pressure, it is necessary to have a positive pressure on the pump suction. Otherwise the hot liquid will vaporize in the pump, and flow will cease. The positive static pressure of the liquid on the pump suction minus the vapor pressure of the liquid is the *net positive suction head* (NPSH). The recommended minimum NPSH for liquids at or above their normal boiling point is 14 ft.[4]

Discharge Head. The discharge head is the pressure of the fluid leaving the pump or blower. It includes the static head and the dynamic, or velocity, head, but except with fans the dynamic head can ordinarily be neglected. The discharge head depends, of course, on the suction lift or the positive suction head; it is the *gain* in head, rather than the discharge head, which determines the power required, the capacity, and to a large extent the design of the machine. Positive-displacement pumps can raise the pressure of a liquid to a very high value in a single stage, but in centrifugal pumps and in all machines for handling gases the pressure gain in a single stage is relatively small. Here high discharge pressures can be obtained only through *multistaging,* in which fluid discharged from one stage is fed to the inlet of a second stage. Successive stages increase the pressure until the desired value is reached. Compressors are made with five or six stages in a single unit; some centrifugal pumps contain as many as 19 stages.

Positive-displacement Machines. In a positive-displacement pump, blower, or compressor a discrete quantity of fluid is trapped in a chamber, which is alternately filled from the inlet and emptied through the discharge. In reciprocating machines the chamber is stationary and contains a moving piston or plunger; in rotary machines the chamber moves from inlet to discharge. The chamber expands while it is being filled, then contracts to squeeze out the fluid against the discharge pressure. The capacity of these machines is almost independent of the discharge pressure.

Reciprocating Pumps. Typical reciprocating machines for liquids are piston pumps and plunger pumps, illustrated in Fig. 3-21. In a *piston pump* liquid is drawn through a check valve into the cylinder by a close-

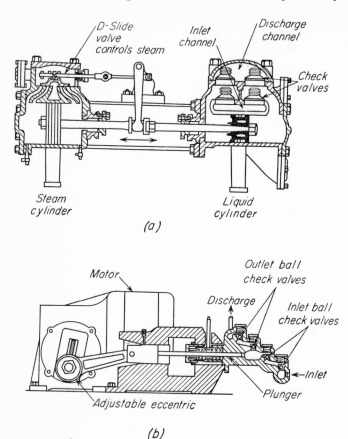

Fɪɢ. 3-21. Positive-displacement reciprocating pumps: (*a*) Piston pump. (*b*) Plunger pump.

fitting piston and is then forced out through a second check valve on the return stroke. The piston carries the packing or piston rings. Most industrial piston pumps are double-acting; i.e., liquid is admitted alternately on each side of the piston, so that one part of the cylinder is being filled while the other is being emptied. Often two or more cylinders are used in parallel with a common discharge manifold, and the relative positions of the pistons are adjusted to minimize fluctuations in the discharge rate. The piston shaft may be motor-driven but often is driven by a steam cylinder. Steam-driven pumps have the advantage of being completely explosion-proof; also, they can discharge against any pressure up to the limit set by the strength of the machine merely by increasing the steam pressure on the drive cylinder. The upper limit for commercial piston pumps is about 700 lb force/in.^{2}

For higher pressure *plunger pumps* are used. A heavy-walled cylinder of small diameter contains a close-fitting reciprocating plunger, which is really merely an extension of the driven shaft. The packing is stationary and is mounted in the cylinder. At the limit of its stroke the plunger fills virtually all the space in the cylinder. Plunger pumps are therefore always single-acting. They are usually motor-driven, and can discharge against pressures of 20,000 lb force/in.^{2} or more.

Volumetric Efficiency. The amount of fluid discharged by a positive-displacement machine operating at a given speed is almost, but not quite, constant. It drops a little as the discharge pressure rises, because of leakage. The ratio of the volume of fluid discharged to the volume theoretically displaced is the volumetric efficiency. For a reciprocating pump it varies from 94 to 100 per cent.

Power Required for Pumping. The power consumption of a pump depends on the total increase in head between the suction and discharge and on the efficiency of the pump, which is defined by Eq. (2-42). This can be demonstrated by use of the Bernoulli equation. Thus, choosing station a at the pump suction and station b at the pump discharge and writing Eq. (2-43) between these stations, the pump work, in feet times pounds force per pound, is

$$-W_s = \frac{[p_b/\rho + (g/g_c)Z_b + \overline{V}_b^2/2\alpha_b g_c] - [p_a/\rho + (g/g_c)Z_b + \overline{V}_a^2/2\alpha_a g_c]}{\eta}$$

$$(3\text{-}3)$$

The friction H_f is omitted from this equation because friction in the pump is accounted for by the efficiency η.

In pump practice it is customary to use, instead of pressure, a proportional quantity called head, measured in feet. Head h is defined by the equation

$$h = \frac{g_c}{g}\left(\frac{p}{\rho} + \frac{gZ}{g_c} + \frac{\overline{V}^2}{2\alpha g_c}\right) = \frac{g_c p}{g\rho} + Z + \frac{\overline{V}^2}{2\alpha g} \qquad (3\text{-}4)$$

Equation (3-3) may be written

$$-W_s\frac{g_c}{g} = \frac{(p_b g_c/g\rho + Z_b + \overline{V}_b^2/2\alpha_b g) - (p_a g_c/g\rho + Z_a + \overline{V}_a^2/2\alpha_a g)}{\eta}$$

$$= \frac{h_b - h_a}{\eta} = \frac{\Delta h}{\eta} \qquad (3\text{-}5)$$

The quantity $\Delta h = h_b - h_a$ is the developed head of the pump in feet.

Note that it takes into account pressure, velocity, and static height above datum. The pressure drop over the pump $p_b - p_a$ must be corrected for velocity and static effects to give an accurate picture of what the pump is doing.

If the mass flow rate is w lb/min, the brake horsepower P_B is

$$P_B = -\frac{wW_s}{33,000} = \frac{wg\,\Delta h}{33,000 g_c \eta} \tag{3-6}$$

When the volumetric flow rate is q' gal/min, Eq. (3-6) can be written in the practical dimensional form (by dropping the dimensional quantity g/g_c)

$$P_B = \frac{q'\rho\,\Delta h}{7.48 \times 33,000\eta} = \frac{q'\rho\,\Delta h}{246,800\eta} \tag{3-7}$$

Mechanical Efficiency. The over-all mechanical efficiency η is the ratio of the power theoretically required to the power actually delivered to the pump.† The efficiency of rotary and centrifugal pumps, fans, blowers, and compressors is discussed in subsequent sections of this chapter. The mechanical efficiency of small piston pumps, with a stroke length of 3 or 4 in., is 40 to 50 per cent; that of large pumps, with 30- to 60-in. strokes, is 70 to 90 per cent. As the discharge pressure rises, the efficiency of reciprocating pumps decreases.

Example 3-3. Compute the developed head of the pump and the brake horsepower required to pump the benzene through the piping system described in Example 3-2 if the pump is also to draw the benzene 10 ft upward from the supply through a $1\frac{1}{2}$-in. suction line and discharge it at the end of the piping system against a pressure of 200 lb force/in.² gauge. A piston pump with a mechanical efficiency of 60 per cent is to be used.

Solution. Figure 3-22 shows the pump and piping. The pump work $-W_s$ is found by writing the Bernoulli equation [Eq. (2-43)] between stations a, at the surface of the liquid in the supply tank, and b, at the outlet of the discharge line. By locating the datum plane at point a, noting that $\overline{V}_a = 0$, and calling atmospheric pressure zero, Eq. (2-43) gives

$$-\frac{W_s}{\eta} = \frac{gZ_b}{g_c} + \frac{p_b}{\rho} + \frac{\overline{V}_b^2}{2\alpha_b g_c} + H_f \tag{3-8}$$

Note that stations a and b chosen for this solution are not those chosen for Eq. (3-3). The latter are at the pump suction and discharge, shown at points a' and b' in Fig. 3-22.

The friction H_f is composed of three items: the contraction loss where the liquid enters the suction line, the friction in the suction line, and the friction in the discharge line. From Example 3-2 the velocity in the suction line is 6.31 ft/sec, the friction factor is 0.0055, and the pipe diameter is 0.134 ft. The friction in the

† This definition is consistent with the definition in terms of shaft work as given in Eq. (2-42).

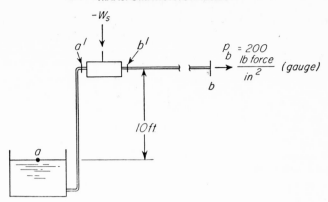

Fig. 3-22. Pump and pipe lines for Example 3-3.

suction line is

$$\frac{4 \times 0.0055(10/0.134)6.31^2}{2 \times 32.17} = 1.0 \text{ ft-lb force/lb}$$

The friction in the discharge line is, from Example 3-2, 14.6 ft-lb force/lb. The contraction loss is, from Eq. (2-64), taking K_c as 0.5 and α_b as 1.0,

$$\frac{0.5 \times 6.31^2}{2 \times 32.17} = 0.3 \text{ ft-lb force/lb}$$

The total friction is $H_f = 1.0 + 14.6 + 0.3 = 15.9$ ft-lb force/lb. The exit-velocity energy is

$$\frac{\overline{V}_b^2}{2g_c} = \frac{6.31^2}{2 \times 32.17} = 0.6 \text{ ft-lb force/lb}$$

The density of benzene is 53 lb/ft^3, and the pressure energy at station b, p_b/ρ, is $(200 \times 144)/53 = 543.4$ ft-lb force/lb. Substituting in Eq. (3-8),

$$-W_s\eta = \frac{g}{g_c} 10 + 0.6 + 543.4 + 15.9 = 569.9 \text{ ft-lb force/lb}$$

From Eq. (3-5) this is numerically equal to the developed head in feet, since g/g_c is numerically unity.

Since the flow rate is 40 gal/min, the brake horsepower is, by Eq. (3-7),

$$P_B = \frac{40 \times 53 \times 569.9}{246,800 \times 0.60} = 8.2 \text{ hp}$$

Diaphragm Pumps. A diaphragm pump is a reciprocating pump in which the "piston" is a flexible disk, movable at the center but fixed at the periphery. The pumping chamber expands and shrinks as the diaphragm is flexed, drawing in and discharging liquid through check valves. The diaphragm is made of thin metal, flexible plastic, or rubber. It may be connected at the axis to a reciprocating rod; it may be flexed by oil pressure from a piston-driven hydraulic cylinder; or it may be actuated by compressed air admitted and released on an automatic timed cycle. Air-operated diaphragm pumps are often used to feed slurries to the continuous filters described in Chap. 7.

Reciprocating Compressors. A typical compressor is shown in Fig. 3-23. Reciprocating machines for gases operate on the same principle as do those for liquids, with the important differences that leakage is harder to minimize and that the fluid is heated by the work of compression. Because of the heating, the gain in pressure in a single-stage compressor is limited, so that for high discharge pressures multistage compressors are required.

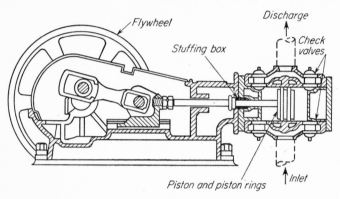

FIG. 3-23. Reciprocating compressor.

Reciprocating compressors are usually motor-driven and are nearly always double-acting.

The capacity of a compressor is measured in terms of standard cubic feet of gas. In industrial practice a *standard cubic foot* is based on a temperature of 60°F and an absolute pressure of 30 in. Hg. This corresponds to a molal volume of 378.7 ft^3/lb mole. Commercial compressors handle up to about 2,000 std ft^3/min. The capacity of a given machine depends on its volumetric efficiency η_v, which is the ratio of the volume of gas delivered to the volume swept out by the piston, and which falls as the discharge pressure rises according to the approximate equation [2]

$$\eta_v = 0.90 - 0.03 \frac{p_b}{p_a} \tag{3-9}$$

where p_a and p_b are the absolute pressures at suction and discharge, in consistent units.

In small compressors handling very small amounts of gas and in vacuum service, the rise in temperature of the gas on compression may be negligible. When there is no temperature change, the compression is said to be *isothermal*. More commonly, however, the gas is considerably hotter at the discharge than at the inlet. When there is no loss of heat to the surroundings, the compression is *adiabatic*. Formulas based on adiabatic compression closely approximate the performance of large compressors.

For frictionless adiabatic compression the shaft work required is given by

$$-W_{sv} = \frac{R_0 T_a}{n}\left[\left(\frac{p_b}{p_a}\right)^n - 1\right]$$

(3-10)

where $-W_{sv}$ = work, Btu/std ft^3

R_0 = gas constant, $1.99/378.7 = 0.00525$ Btu/(std ft^3)(°R)

T_a = absolute temperature at suction, °R

$n = (\kappa - 1)/\kappa$

κ = ratio of specific heats, C_p/C_v

p_a, p_b = absolute pressures at suction, discharge

The brake horsepower is

$$P_B = \frac{T_a}{520}\frac{0.0643 q_0}{n\eta}\left[\left(\frac{p_b}{p_a}\right)^n - 1\right]$$

(3-11)

where P_B = brake horsepower

q_0 = volume of gas compressed, std ft^3/min

η = mechanical efficiency

The power required to compress monatomic and diatomic gases, for which the ratio of specific heats is high, is somewhat greater than that required for gases like propane or butane with low ratios of specific heats. The power also depends on the inlet temperature, for a hot gas requires more work than a cold one, and on the mechanical efficiency, which is higher with heavy gases than with light ones. The efficiency varies with the *compression ratio*, which is the ratio of the absolute discharge pressure to the absolute inlet pressure. In reciprocating compressors the compression ratio is between 2.5 and 6 in each stage; the adiabatic efficiency is a maximum of 80 to 85 per cent at a compression ratio of about 4. In multistage compressors the compression ratio should be the same in each stage, since the power drawn by each stage is then the same. Thus in a four-stage compressor:

$$\frac{p_1}{p_a} = \frac{p_2}{p_1} = \frac{p_3}{p_2} = \frac{p_b}{p_3} = \sqrt[4]{\frac{p_b}{p_a}}$$

(3-12)

where p_a = absolute pressure at suction of stage 1

p_1, p_2, p_3, p_b = absolute pressure at discharge of stages 1, 2, 3, 4

Between the stages of multistage compressors are *intercoolers*, which are air-cooled or water-cooled exchangers to remove the heat of compression. Often an *aftercooler* or *aftercondenser* follows the last stage. The temperature of the gas leaving a compressor cylinder (before cooling) may be estimated from the expression for the adiabatic discharge temperature

$$T_b = T_a\left(\frac{p_b}{p_a}\right)^n$$

(3-13)

where T_a and T_b are the absolute temperatures at suction and discharge.

Vacuum blowers and compressors, often called *vacuum pumps*, remove gas from an evacuated space and discharge it—usually—against atmospheric pressure. As the absolute suction pressure is reduced, their volumetric efficiency drops, until it approaches zero at the lowest absolute suction pressure the machine can produce. The mechanical efficiency is often much lower than with compressors. The required displacement increases rapidly as the absolute suction pressure falls, so that to produce a very high vacuum a big machine is needed. The compression ratio in vacuum pumps is much larger than in compressors, ranging up to 100 or more, with a corresponding adiabatic discharge temperature of several hundred degrees Fahrenheit. In practice, however, the compression is nearly isothermal. The weight of gas compressed is so small in comparison with the weight of exposed metal in the machine that the discharge temperature is low even without intercoolers or aftercoolers.

Example 3-4. A reciprocating compressor is to compress 180 std ft³/min of methane from 14 to 900 lb force/in.² abs. The inlet temperature is 80°F. For the expected temperature range the average properties of methane are

$$C_p = 9.3 \text{ Btu/(lb mole)(°F)} \qquad \kappa = 1.31$$

(*a*) How many stages should be used? (*b*) How much power is needed if the mechanical efficiency is 80 per cent? (*c*) What is the approximate temperature at the discharge of the first stage? (*d*) If the temperature of the cooling water is to rise 20°F, how much water is needed in the intercoolers and aftercooler so that the compressed gas will leave each stage at 100°F?

Solution. (*a*) The number of stages must be found by trial to give a suitable compression ratio in each stage. Try two stages.

$$\frac{p_b}{p_1} = \sqrt{\frac{p_b}{p_a}} = \sqrt{\frac{900}{14}} = \sqrt{64.3} = 8.02$$

This compression ratio is too high, indicating that more stages are needed. Try three stages.

$$\frac{p_2}{p_1} = \sqrt[3]{\frac{p_b}{p_a}} = \sqrt[3]{\frac{900}{14}} = \sqrt[3]{64.3} = 4.01$$

This is satisfactory. Use a three-stage compressor.

(*b*) The power required per stage, from Eq. (3-11), is

$$n = \frac{\kappa - 1}{\kappa} = \frac{0.31}{1.31} = 0.236$$

$$P_B = \frac{80 + 460}{520} \frac{0.0643 \times 180}{0.236 \times 0.80} (4.01^{0.236} - 1)$$

$$= 1.038 \times 61.3(1.388 - 1) = 24.6 \text{ hp}$$

The total power for three stages is $3 \times 24.6 = 73.8$ hp.

(*c*) From Eq. (3-13),

$$T_b = (80 + 460)4.01^{0.236} = 540 \times 1.388 = 750°R = 290°F$$

(d) The adiabatic discharge temperature from the second and third stages is

$$T_b = T_2 = (100 + 460)4.01^{0.236} = 317°F$$

Since

$$1 \text{ lb mole} = 378.7 \text{ std ft}^3$$

the flow rate is

$$\frac{180 \times 60}{378.7} = 28.5 \text{ lb moles/hr}$$

The heat load is

First stage:	28.5(290 − 100)9.3 = 50,400 Btu/hr
Second stage:	28.5(317 − 100)9.3 = 57,500
Third stage:	Same as second stage = 57,500

Total = 165,400 Btu/hr

NOTE. 73.8 hp = 73.8 × 2,545 = 187,800 Btu/hr. The cooling water required is

First stage: $\dfrac{50,400}{20} = 2,520 \text{ lb/hr} = 5.04 \text{ gal/min}$

Second stage: $\dfrac{57,500}{20} = 2,875 \qquad = 5.75$

Third stage: $\dfrac{57,500}{20} = 2,875 \qquad = 5.75$

Total = 8,270 lb/hr = 16.54 gal/min

Rotary Pumps and Blowers.   In nearly all rotary machines a definite quantity of fluid is trapped in an expanding chamber near the inlet, moved to the outlet, and squeezed out into the discharge line.   Unlike reciprocating machines, they contain no check valves.   Close clearances between the moving and stationary parts minimize leakage and reverse flow; they also limit the operating speed to low values.   Rotary pumps work best on moderately viscous lubricating fluids, discharging them at pressures up to 3,000 lb force/in.² or more.   Rotary compressors can discharge at pressures up to about 80 lb force/in.²   Some rotary machines are effective for creating very high vacua.   The main types of pumps are gear pumps, lobe pumps, screw pumps, cam pumps, and vane pumps.   Devices similar to all these, except gear pumps and screw pumps, are used as blowers; in addition, liquid-piston blowers and compressors are used with gases only.

Gear Pumps.   Figure 3-24a shows a typical rotary gear pump.   Intermeshing gears turn fairly slowly with close clearance inside the casing. Liquid entering the bottom of the pump is caught in the spaces between the teeth and the casing and is carried around to the top of the chamber. It cannot return to the inlet, for the spaces between the teeth of each gear, in the center of the pump, are filled by the teeth of the other gear.   The liquid therefore must flow out the discharge. For light duty spur gears with straight teeth are used; for higher pressures the gear teeth are at an angle or shaped in a herringbone pattern. In some designs one gear drives the other, but commonly the gears are independently driven by internal or external timing gears.

In an internal-gear pump (Fig. 3-24b) a pinion or spur gear meshes with a ring gear carrying internal teeth. A crescent of solid metal fills the open space between the two gears. Liquid is carried from inlet to discharge by both gears, in the spaces between the gear teeth and the crescent. As in

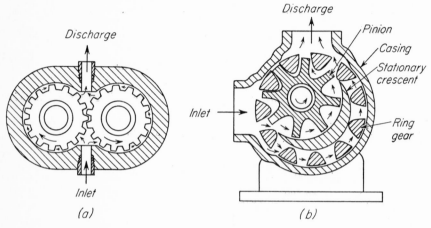

Fig. 3-24. Gear pumps: (a) Spur-gear pump. (b) Internal-gear pump.

other gear pumps, relief channels must sometimes be provided to avoid excessive build-up of pressure between the meshing teeth.

Lobe Pumps and Blowers. These machines operate in the same way as do gear pumps except that because of the special design of the "teeth" the minimum clearance is the same at all times. In lobe blowers the clearance is only a few thousandths of an inch. The relative position of the impellers is established by heavy external gears, to minimize backlash. A single-stage blower can discharge gas at 6 to 15 lb force/in.^{2}; in a two-stage blower the discharge pressure may be 30 lb force/in.^{2}

A two-lobe blower is shown in Fig. 3-25a; three-lobe machines are also common. Exactly similar devices are also used for pumping liquids. In lobe pumps the liquid is delivered to the discharge in a smaller number of larger increments than in a gear pump, so that the flow rate is not quite steady. In one design of blower (Fig. 3-25b) a helically threaded two-lobe rotor mates with a recessed, or "gate," rotor turning at half the speed. Gas enters at one end of the rotors on one side of the casing and discharges at 6 to 10 lb force/in.^{2} from the opposite side at the other end of the rotors. Flow is therefore axial (parallel to the axes of the impellers) as well as transverse.

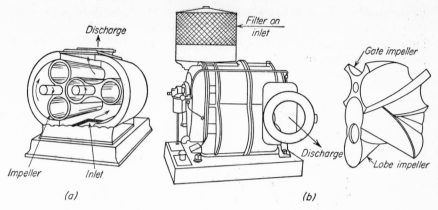

Discharge

Filter on inlet

Gate impeller

Discharge

Lobe impeller

Impeller Inlet

(a) (b)

Fig. 3-25. Lobe blowers: (a) Two-lobe blower. (b) Axial-flow positive-displacement blower. (*Read Standard Corp.*)

Screw Pumps. Intermeshing gears with a screw thread also give axial flow, but because of the nature of gears of this kind the seal between them cannot be highly effective. Screw pumps, therefore, cannot be used with gases, and do not work well with thin liquids because of excessive leakage. A screw pump for viscous liquids is illustrated in Fig. 3-26. Parallel shafts each carry two threaded sections, turning so as to move liquid from the ends of the casing to a central discharge. Both stuffing boxes are thus sub-

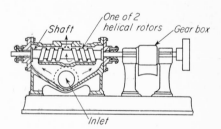

Shaft One of 2 helical rotors Gear box

Inlet

Fig. 3-26. Screw pump.

jected to the low suction pressure, not the high discharge pressure, and axial thrust on the shafts is eliminated.

Cam Pumps. In the pump shown in Fig. 3-27a the impeller itself does not rotate. Instead it oscillates axially on an eccentric mounted on a rotating shaft. Liquid entering the pump flows into the space on one side between the impeller and the casing. Simultaneously other liquid is leaving the pump on the other side of the divider vane. As the shaft turns, the impeller moves, causing liquid to flow around the inside of the casing to the discharge. Two impellers on the same shaft, opposed by 180°, are usually used. Such pumps handle clear liquids, even highly corrosive ones like nitric acid and hydrogen peroxide, against discharge pressures up to 100 lb force/in.[2] They cannot handle slurries since, like other rotary pumps, they depend for lubrication on the liquid being pumped. For the same reason they cannot be run dry without damage. Once full, however, these pumps are self-priming.

The action in a pump of this kind can be visualized by imagining that the space between the impeller and the casing is a stationary tube with flexible walls. The impeller acts like a movable clamp, pinching the walls of the tube together at one point and moving along the tube, driving the liquid before it. Some cam pumps operate in exactly this way, with a single or multiple cam rolling over a tube of rubber or flexible plastic.

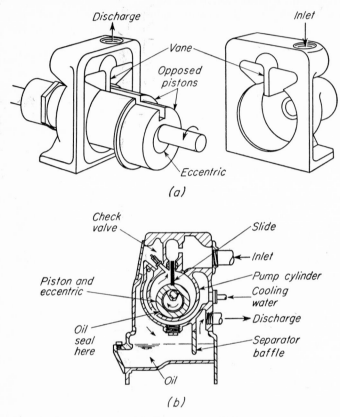

(a)

(b)

Fig. 3-27. Rotary pumps: (a) Cam pump. (b) Rotary vacuum pump.

Figure 3-27b shows a rotary vacuum pump which operates on the same principle as the cam pump first described. With gases a check valve on the discharge is needed to minimize reverse leakage. The pump cylinder is often water-cooled. Lubricating oil is drawn by the vacuum into the cylinder to form a seal between the impeller and the casing; it leaves the cylinder with the air and is separated from it at the base of the pump by the separator baffle. Such pumps can reduce the absolute pressure to 0.01 mm Hg, or 10 microns. For still lower pressures they are used in series with vapor-diffusion pumps.

Vane Pumps and Blowers. In a sliding-vane pump (Fig. 3-28a) flat vanes slide in and out of sockets cut in an eccentric rotor. The vanes are pressed against the bore of the casing by centrifugal force or, with very viscous liquids, by reciprocating push rods. Liquid is drawn into the expanding chamber between two vanes at the inlet, moved to the discharge, and forced out as the chamber contracts. In the pump shown in Fig. 3-28b the sliding vanes have been replaced by swinging vanes or buckets, which are pressed against the casing wall by centrifugal force. A swinging-vane pump operates in the same way as a sliding-vane pump. In both types a considerable amount of wear on the vanes does not affect the operation of the pump.

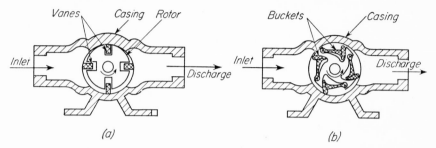

FIG. 3-28. Vane and bucket pumps: (a) Sliding-vane pump. (b) Swinging-bucket pump.

Eventually, of course, wear becomes so severe that the capacity of the pump drops sharply, and the vanes or buckets must be replaced.

Sliding vanes are used for thin volatile liquids, and—with push rods—for viscous liquids. Buckets are effective with nonvolatile liquids. Industrial vane pumps handle up to 1,500 gal/min at pressures up to 500 lb force/in.^{2} Sliding-vane machines also make effective vacuum pumps when oil is used to seal the line of contact between the vanes and the casing.

Liquid-piston Blowers. The Nash Hytor blower shown in Fig. 3-29 contains a vaned rotor turning in an elliptical casing. The space between the rotor and casing is partly filled with liquid, which is thrown outward

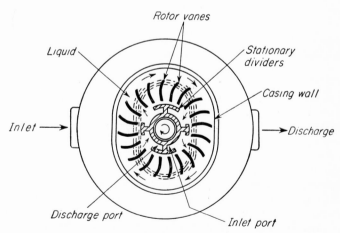

FIG. 3-29. Liquid-piston compressor. (*Nash Engineering Co.*)

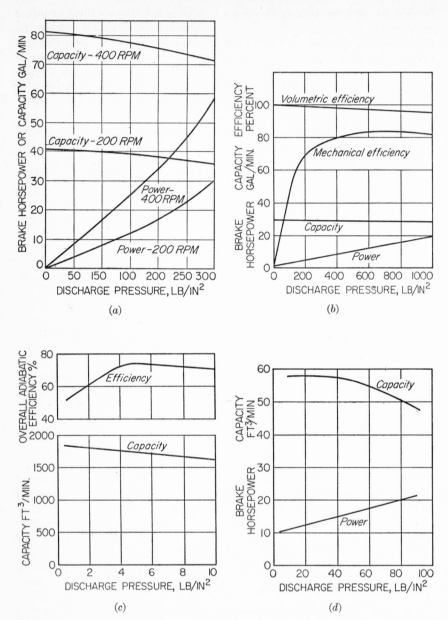

FIG. 3-30. Characteristic curves of positive-displacement pumps and blowers: (a) Spur-gear pump handling heavy oil. (b) Slide-vane pump. (c) Lobe blower of Fig. 3-25(a). (d) Liquid-piston compressor.

by centrifugal force as long as the rotor is turning. Thus a layer of liquid follows an elliptical path inside the casing, now coming inward toward the shaft, now moving outward and away. In so doing it alternately fills and empties the spaces between the rotor vanes. Twice in each revolution of the rotor an expanding chamber is formed, into which gas is drawn; twice the gas is forced out by the returning liquid. By setting two inlets and outlets at the proper points, gas can thus be compressed to about 80 lb force/in.^{2} The compressed gas is simultaneously cooled by the liquid, humidified or dehumidified as desired, and thoroughly cleaned. Unlike gas from a reciprocating compressor, it contains no oil. Liquid-piston blowers and compressors are therefore valuable where clean, dust-free gas is needed, as in making carbonated beverages. Wet chlorine and other gases can be simultaneously compressed and dried by using concentrated sulfuric acid as the sealing liquid.

Characteristics of Rotary Machines. Figure 3-30 shows how little the capacity of rotary pumps and blowers varies with the discharge pressure and what deviations from true "positive displacement" can be expected. In general the capacity drops from 5 to 15 per cent over the operating range of discharge pressures. The capacity is almost directly proportional to the shaft speed. The volumetric efficiency of rotary pumps is 95 to 100 per cent; the mechanical efficiency of both pumps and blowers is also high, ranging from 70 to 85 per cent.

Rotary pumps characteristically discharge relatively small amounts of clean liquids at steady rates against high pressures. They operate most efficiently on viscous liquids and fairly well on thin liquids. They are usually self-priming, once filled, but cannot be run dry. They can handle mixtures of gas and liquid but not slurries. Rotary blowers compress moderate amounts of gas to moderate pressures, again with only small variations in capacity with discharge head. As discussed later, rotary vacuum pumps can produce lower absolute pressures than reciprocating vacuum pumps or steam-jet ejectors.

Centrifugal Pumps and Blowers. The second broad class of machines for moving fluids consists of those which impart energy by centrifugal action. In most of them the fluid enters at or near the axis of a high-speed impeller and is thrown radially outward by vanes into a spiral casing or scroll. There the velocity head imparted by the vanes is converted into pressure head. There is a maximum discharge head which a machine of this kind can produce, even under ideal conditions; and losses in an actual machine reduce the developed head to considerably less than the theoretical maximum value. In some centrifugal pumps and in most centrifugal blowers the impeller is surrounded by stationary diffuser vanes, which guide the fluid to the discharge and minimize turbulence. In turbine pumps the liquid is moved around the periphery of the impeller by short radial vanes, instead of from the hub of the impeller outward. Axial-flow, or propeller, pumps discharge large volumes of liquid against low heads; axial-flow fans perform the same service with gases. In some pumps the flow is partly axial and partly radial; such a device is called a mixed-flow pump.

Axial-flow Pumps and Blowers. A typical propeller pump is shown in Fig. 3-31. Single machines of this kind can pump as much as a million gallons per minute against heads no greater than about 40 ft. They are most often used in irrigation work, for pumping out storm sewers, and in similar applications; in industrial processing they are used for circulating large volumes of liquid against low pressures. Axial-flow fans and blowers, shown in Fig. 3-32, are finding increasing application for moving industrial gases. They are somewhat easier to install than radial-flow fans, since they do not necessitate changing the direction of flow. They can be set right inside a gas-carrying duct. Stationary vanes on the motor swirl the gas before it strikes the rotating fan blades, in the direction opposite to the direction of rotation; in this way swirling of the gas leaving the fan is virtually eliminated, and the mechanical efficiency is increased to above that of radial-flow fans. Comparative performance curves are given in Fig. 3-49a and b. The power drawn by axial-flow fans may be computed from Eqs. (3-21) and (3-22), to be given later.

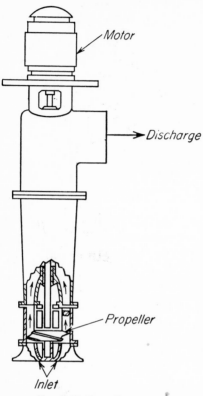

FIG. 3-31. Propeller pump.

Volute and Diffuser Pumps. Examples of common centrifugal pumps for liquids are shown in Fig. 3-33. In these pumps flow is primarily radial, away from an inlet at the axis of the shaft. Liquid enters on one side of the impeller in a single-suction pump, as shown in Fig. 3-33a, and from both sides in a double-suction pump, as in Fig. 3-33b. Double-suction pumps can be made stronger and better balanced than single-suction pumps. Pumps for corrosive chemicals are nearly always single-suction, however, since simplicity of design

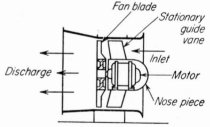

FIG. 3-32. Axial-flow fan.

and smaller requirements of corrosion-resistant metals favor them over the more massive double-suction units. The pumps shown in Fig. 3-33 are typical volute pumps; a diffuser pump with stationary vanes surround-

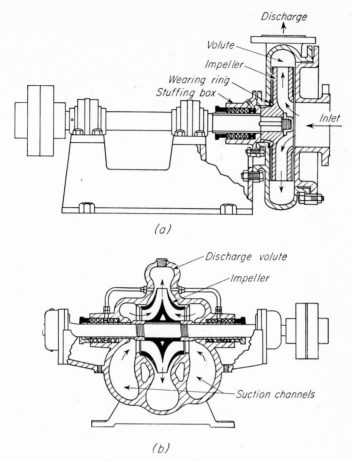

FIG. 3-33. Volute centrifugal pumps: (a) Single-suction. (b) Double-suction.

ing the impeller is illustrated in Fig. 3-34. Diffusers significantly increase the efficiency in large, high-speed pumps. In most process pumps, how-

ever, a high degree of reliability is much more important than high efficiency; in fact, many pumps are operated at slow speeds and far below their rated capacity. Mechanical efficiencies of 5 to 20 per cent are not uncommon in chemical-process pumps.[6]

Impellers for centrifugal pumps are "open," "semienclosed," or "enclosed," or "shrouded," as shown in Fig. 3-35. In very small pumps

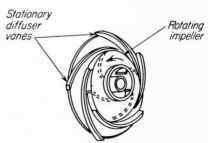

FIG. 3-34. Impeller and stationary vanes of diffuser pump.

and pumps handling scaling liquids or liquids carrying coarse solids the impellers are open or semienclosed; liquids carrying stringy solids may be pumped with an open two-bladed wide-sweep impeller. Otherwise, however, shrouded impellers are used because they give higher efficiencies and

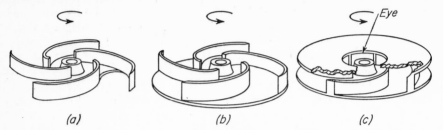

(a)   *(b)*   *(c)*

FIG. 3-35. Impellers for centrifugal pumps: (*a*) Open. (*b*) Semienclosed. (*c*) Shrouded.

because their performance is less affected by wear or erosion. In most shrouded impellers the vanes are curved backward from the direction of rotation.

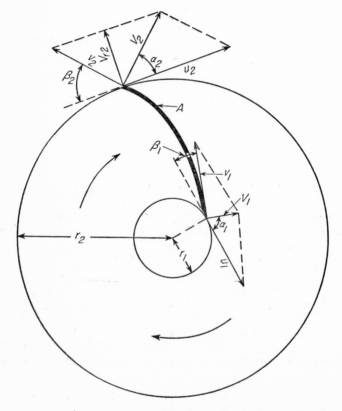

FIG. 3-36. Velocities at entrance and discharge of vanes in centrifugal pump.

Principle of Centrifugal Pump. In Fig. 3-36 let A be the section of a single vane of an impeller which is rotating in the direction shown by the arrow. The stream of liquid entering the suction end of the impeller at the tips of the vanes is moving along the impeller at a velocity of v_1 ft/sec, where the velocity is measured relative to the impeller. The velocity v_1 can be visualized as that apparent to an imaginary observer riding around on the vane. The entrance tip of the vane is itself moving at a velocity of u_1 ft/sec in a circular path around the center line of the pump shaft as a center. The velocity u_1 is measured relative to the ground. The vector v_1 makes an angle β_1 with the direction of vector u_1. The resultant velocity V_1 of the fluid at the entrance end of the vane is the vector sum of velocities v_1 and u_1 and is found by the usual parallelogram construction, as shown in Fig. 3-36. The vector V_1 makes an angle α_1 with vector u_1. The fluid leaving the impeller at the exit tip has a velocity along the impeller of v_2 ft/sec. The peripheral velocity at the exit end is u_2 ft/sec, and the resultant velocity is V_2 ft/sec. Angles α_2 and β_2 correspond to angles α_1 and β_1 at the suction. Velocities such as V_1 and V_2 are called absolute velocities, and are the velocities of the fluid that would be seen by an observer watching the action of the pump from the ground.

As shown in Fig. 3-37, the absolute-velocity vectors V_1 and V_2 may be resolved into components, one, V_r, parallel to the radius, and the other,

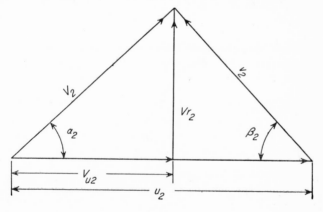

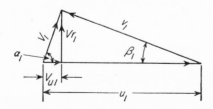

FIG. 3-37. Vector diagrams for centrifugal pump.

V_u, perpendicular to the radius. The velocity V_r is the radial component of the absolute velocity, and the velocity V_u is the peripheral component of the absolute velocity.

The radial components V_{r1} and V_{r2} are the velocities of the liquid entering and leaving the impeller, measured perpendicularly to the inlet and outlet peripheries, respectively. The product of the radial component of the absolute velocity by the corresponding peripheral area is the volumetric flow in or out of the impeller. By continuity, these flows are, of course, equal.

The peripheral area of the impeller is $2\pi b r_2$, where b is the width of the channels. If the vane thickness is neglected, this is also the cross-sectional area for flow out of the impeller. To account for the thickness of the vanes, the total peripheral area is corrected by a factor ϵ_2, which is the fraction of the perimeter not occupied by the cross sections of the vanes. This constant, which is fixed by the design of the impeller, is usually between 0.90 and 0.95. The theoretical volumetric flow rate q_r is, then, the product of the net peripheral area and the radial velocity, and

$$q_r = 2\pi\epsilon_2 b r_2 V_{r2} \tag{3-14}$$

Because of the mechanical work done on the fluid by the impeller, the velocity V_2 is much larger than the velocity V_1, and the velocity head of the fluid passing through the impeller is correspondingly increased. In the volute or diffuser, the velocity head of the fluid leaving the impeller is converted to pressure head in accordance with Bernoulli's equation, and the work done by the impeller finally appears as developed head of the pump.

Centrifugal-pump Theory. The basic equations for developed head and power consumption of a centrifugal pump can be derived from mechanical fundamentals for an ideal pump. The performance of an actual pump differs considerably from that of an ideal one, and actual pumps are studied and designed by applying experimentally determined corrections to the ideal case. More than is the case for any of the other equipment to be discussed in this book, the chemical engineer purchases pumps on the basis of the manufacturer's guaranteed characteristics. The basic theory may be found in standard works on pump design, and here the discussion will be limited to the deviations of actual performance from theoretical performance and the effect of any variables under the control of the user.

Head and Power Equations for the Ideal Pump. It may be shown [1] that in an ideal and frictionless pump, Δh_r, the head in feet developed (using the nomenclature of Fig. 3-37) is

$$\Delta h_r = \frac{u_2(u_2 - v_2 \cos \beta_2) - u_1(u_1 - v_1 \cos \beta_1)}{g} \tag{3-15}$$

The theoretical power P_r is, then,

$$P_r = \frac{q'\rho}{7.48 \times 33,000 g_c} [u_2(u_2 - v_2 \cos \beta_2) - u_1(u_1 - v_1 \cos \beta_1)] \quad (3\text{-}16)$$

where q' is the flow rate in gallons per minute.

Ordinarily the liquid enters the impeller along the shaft and changes direction through a right angle in a plane perpendicular to the impeller disk just before entering the vanes. This can be understood from Figs. 3-33 and 3-34. As a result of this, the inlet velocity vector u_1 is perpendicular to the plane of Fig. 3-36, and Eqs. (3-15) and (3-16) become, for such radial inlet pumps,

$$\Delta h_r = \frac{u_2(u_2 - v_2 \cos \beta_2)}{g} \quad (3\text{-}17)$$

$$P_r = \frac{q'\rho}{7.48 \times 33,000 g_c} [u_2(u_2 - v_2 \cos \beta_2)] \quad (3\text{-}18)$$

Actual Performance of Centrifugal Pumps. Both the developed head and the power consumption of an actual pump differ considerably from those predicted by the theoretical equations. The developed head is less than, and the power consumed is greater than, in an ideal pump. Head losses and power losses will be discussed separately.

Loss of Head: Circulation. The first difference between an actual and an ideal pump is the phenomenon of circulation that occurs in a real pump. In the theory of the ideal pump, it is assumed that the velocity of the liquid through the impeller is uniform in any cross section and the liquid leaves the impeller at the angle β_2 that actually exists at the impeller discharge. The fluid leaving an actual impeller does so at an angle consider-

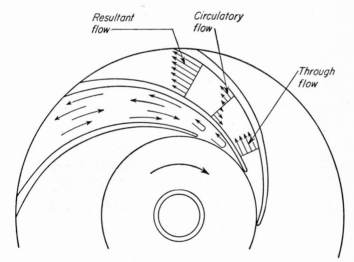

Fig. 3-38. Circulatory flow in pump impeller. (*By permission, from "Centrifugal Pumps and Blowers," by A. H. Church. Copyright, 1944. John Wiley & Sons, Inc.*)

ably less than the existing vane angle of the impeller. This is caused by a circulation flow in the impeller that occurs in addition to the through flow. The fluid passing through the impeller channel is being pushed by the vane opposite to the direction of rotation; and a pressure gradient exists across the stream, decreasing in the direction of rotation, so that the pressure head also decreases in the direction of rotation. By Bernoulli's principle, therefore, the velocity head must vary in the opposite sense and must be greater in the layers of fluid away from the vane applying the lateral force. The effect of the circulation thus induced is shown in Fig. 3-38.[1a] Because of circulation, the fluid leaves the impeller at an angle less than β_2 (the vane angle of the actual impeller). Hence the factor $v_2 \cos \beta_2$ is increased by circulatory flow, while the velocity u_2 is unchanged. From Eq. (3-17) it follows that the developed head is decreased. The circulatory flow does not cause a mechanical-energy loss, or friction, and the power consumption decreases in proportion to the loss in head. The developed head is less, however, than would be expected from the design of the pump.

The circulation effect is minimized by the use of long, narrow channels and the use of many vanes. Long, narrow channels may not be consistent with the design of the impeller, however, and the multiplicity of vanes adds to the cost of the impeller and to the friction loss, so the loss in head due to circulatory flow is accepted and a proper correction made by the designer.

Fluid Friction. The flow of liquid through the passages and channels of the pump is accompanied by the usual fluid friction, which is nearly proportional to the square of the velocity and hence to the square of the volumetric flow rate. It will also be large for rough surfaces and for small channels and ports. To minimize friction, smooth channels should be used, and expansion and contraction losses due to separation should be avoided by proper streamlining.

Shock Losses. A centrifugal pump is designed with a definite capacity, and a definite developed head at that capacity, in mind. These conditions establish the rated head and capacity of the pump, and are the design conditions. Although the pump is operated at constant speed, the flow rate can be varied over wide limits, both above and below the design condition, by simply varying the pressure increase caused by the pump. These design conditions fix the angle α_2 (Fig. 3-36) and therefore the direction of the stream of fluid leaving the impeller, as shown by vector V_2. These include the designer's corrections for circulation.

In a volute pump, the liquid leaving the impeller in the direction of the vector V_2 is suddenly introduced into a stream of liquid traveling circumferentially around the casing. This sudden change in direction causes turbulence, with consequent losses in both head and power. The direction and magnitude of the velocity u_2 are constant because the pump is operating at constant speed. Variations in discharge rate affect the angle α_2, which will be smaller than the design value at low flows and larger than the design value at high flows. Hence at low flow rates the vector V_2 is more nearly parallel to the velocity of the stream into which the liquid is discharged from the impeller, and at high flow rates the discrepancy is greater. Therefore losses due to turbulence are greater at high flows than at low flows.

In a diffuser pump (Fig. 3-34) the designer can make the slope of the entrance end of the vanes in the diffuser equal to his estimated value of α_2. Hence the liquid leaves the impeller and enters the diffuser practically without shock. If, however, the pump is operated at capacities above or below the design rating, the actual angle α_2 with which the liquid leaves

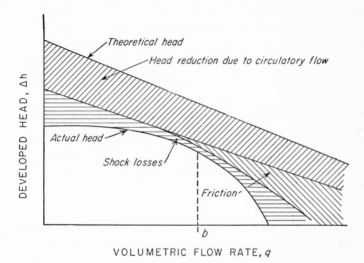

VOLUMETRIC FLOW RATE, q

FIG. 3-39. Theoretical head, actual head, and head losses in centrifugal pump. (*By permission, from "Centrifugal Pumps and Blowers," by A. H. Church. Copyright, 1944. John Wiley & Sons, Inc.*)

the diffuser will be less or more than the angle of the diffuser vanes. This causes a change in velocity, and hence in the shock turbulence losses, that becomes greater as the operating conditions deviate more and more from design conditions. The losses from this source should be negligible at design conditions. Point b in Figs. 3-39 and 3-40 represents the design conditions.

The net result of the three head losses—circulatory flow, fluid friction, and shock losses—is shown in Fig. 3-39.[1b]

The head that remains after the theoretical head is corrected for the losses is the actual developed head of the pump. It is denoted by Δh. The curve showing the actual developed head vs. the flow rate, at constant pump speed, is one of the characteristic curves of a centrifugal pump (Fig. 3-46).

Power Losses in Centrifugal Pumps. Circulatory flow does not cause a power loss, but fluid friction and shock losses do, as they are conversions of mechanical energy into heat. In each case, the power loss is proportional to the product of the lost head and the flow rate. Thus, power losses due to fluid friction are nearly proportional to the cube of the flow rate.

Three other sources of power loss are important in a centrifugal pump but do not cause a loss in head. They are leakage, disk friction, and mechanical friction.

A certain amount of unavoidable interior *leakage* occurs from the discharge of the impeller to the suction of the impeller. The pressure at the suction of the impeller is less than that at the discharge of the impeller, and the difference between these pressures is applied across the wearing ring. Although the clearance between the impeller hub and the wearing ring is small, it is not zero, and some leakage occurs. The effect of leakage is to reduce the volume of discharge per unit of power expended, and the extra work used to pump the recirculated liquid is converted into heat and lost.

Disk friction is the term used for the friction that occurs between the outer surfaces of the impeller (when a closed impeller is used) and the liquid in the space between the impeller and the casing and also for a considerably greater effect, the pumping action on the fluid that takes place in that space. The fluid in contact with the rotating impeller is picked up and thrown outward to the inside of the casing. The fluid must then flow back along the casing wall to the shaft, to be again picked up by the impeller and repumped. Power is required to maintain this secondary and useless pumping action.

In pumps the theoretical power P_r calculated from the actual discharge rate and the actual developed head is called the *fluid horsepower*. It is the power corresponding to $-W_{sr}$, the ideal work required per unit mass of liquid. The fluid horsepower is calculated from the equation

$$P_r = \frac{q'\rho g \, \Delta h}{2.47 \times 10^5 g_c} \qquad (3\text{-}19)$$

where q' is the actual flow rate in gallons per minute. The difference between the brake horsepower and the fluid horsepower is accounted for by the sum of the power losses described above. Figure 3-40[1c] shows how the brake horsepower is discounted by the various power losses to give the fluid horsepower.

The *efficiency* obviously varies with the flow rate and is a maximum at design conditions. The curve of efficiency vs. flow rate is another characteristic curve of a pump (Fig. 3-46).

Effect of Operating Variables on Performance of Centrifugal Pumps. The basic theory of the centrifugal pump leads to a prediction of the general effect of speed, vane angle, and flow rate on the developed head and power of a pump. The idealized results are modified by the realities of head and energy losses suffered by an actual pump, but the effects carry over in a large measure from the ideal to the actual case. In the following, an ideal radial-inlet pump is postulated.

The effect of changing the speed of a pump is to vary the velocity vector u_2 in proportion to the speed. If the flow rate is allowed to change in pro-

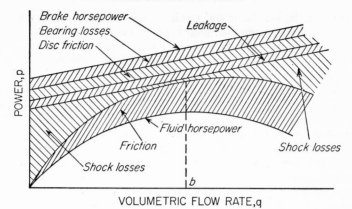

POWER, p

VOLUMETRIC FLOW RATE, q

Fig. 3-40. Fluid horsepower, brake horsepower, and power losses in centrifugal pump. *(By permission, from "Centrifugal Pumps and Blowers," by A. H. Church. Copyright, 1944. John Wiley & Sons, Inc.)*

portion to the speed, the velocity vector V_{r2} is also proportional to the speed, and all angles in the vector diagrams, including α_2, are unchanged. The developed head varies as the quantity $u_2(u_2 - v_2 \cos \beta_2)$, as shown by Eq. (3-17). Since both u_2 and $u_2 - v_2 \cos \beta_2$ are proportional to speed when angle β_2 is constant, the developed head is proportional to the square of the pump speed. The power, by Eq. (3-18), is proportional to the product of the head and the flow rate and therefore to the cube of the speed.

This relation between speed and power carries over quite accurately to an actual pump. Shock losses are determined largely by the angle α_2, and the effect of speed on power loss is largely confined to disk and fluid friction, both of which are relatively small over a range of flow rates. The efficiency at a given value of α_2 is nearly independent of speed.

The relation of theoretical head vs. theoretical flow rate at constant speed can be found from Eq. (3-17), noting that

$$v_2 = \frac{V_{r2}}{\sin \beta_2}$$

so

$$\Delta h_r = \frac{u_2(u_2 - V_{r2}/\tan \beta_2)}{g} \tag{3-20}$$

The vector V_{r2} is proportional to the flow rate. At constant speed u_2 is constant, and for a given pump β_2 is constant. The relationship of head vs. flow rate is therefore linear.

The slope of the line relating the theoretical flow rate to the theoretical head varies with the sign of $\tan \beta_2$ and therefore with the angle β_2. If β_2 is less than 90°, the line has a negative slope; if β_2 is 90°, the line is horizontal; if the angle is greater than 90°, the line has a positive slope. The first case corresponds to impeller blades convex in the direction of rotation, the second to straight blades, and the third to blades concave to the direction of motion. Losses, of course, considerably distort the curve of head vs. flow rate, but the general effect of the angle β_2 remains. In practice, it is desirable to have a characteristic of falling head with increase of flow rate, and the usual pump is constructed with a vane angle less than 90°.

Example 3-5. A single-suction diffuser pump operates at a speed of 1,750 rpm. The impeller diameter is 8 in., and the diameter at the inlet is 4 in. The width of the impeller at the inlet is 1.75 in. and at the outlet is 1.00 in. The inlet flow is radial, so α_1 is 90°. The inlet and outlet vane angles are 13 and 18°, respectively.

(a) Calculate the theoretical rated capacity of the pump in gallons per minute, and the developed head, in feet, at rated conditions, neglecting losses, circulatory flow, leakage, and vane thickness. (b) Plot the curve of theoretical head vs. flow rate from zero flow at maximum head to zero head at maximum flow. (c) Calculate values of the diffuser inlet angle α_2 for rated capacity, $\frac{1}{2}$ rated capacity, and $1\frac{1}{2}$ times rated capacity. What is the practical significance of the variation of α_2 with flow rate?

Solution. (a) The vector diagram for flow at the impeller inlet is shown in Fig. 3-41a. Since α_1 is 90°, the vector diagram is a right triangle, and $V_1 = V_{r1}$. The

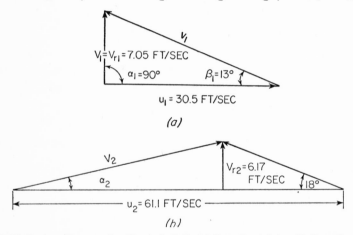

Fig. 3-41. Vector diagrams for Example 3-5: (a) Impeller inlet. (b) Impeller outlet.

inlet peripheral velocity u_1 is calculated from the speed of the pump and the inlet diameter.

$$u_1 = \frac{4}{12}\pi\frac{1,750}{60} = 30.5 \text{ ft/sec}$$

The radial component of the absolute velocity, which determines the flow rate through the pump, is, from Fig. 3-41a,

$$V_{r1} = u_1 \tan\beta_1 = 30.5 \tan 13° = 30.5 \times 0.2309 = 7.05 \text{ ft/sec}$$

The volumetric flow rate is the volume of water flowing each minute across a ring 4 in. in diameter and 1.75 in. wide at a velocity of 7.05 ft/sec.

$$q' = \frac{1.75}{12}\frac{4}{12}\pi 7.05 \times 60 \times 7.48 = 483 \text{ gal/min}$$

The vector diagram for flow at the impeller outlet is shown in Fig. 3-41b. The radial component of the absolute velocity at the outlet of the impeller is, by continuity,

$$V_{r2} = V_{r1} \frac{b_1}{b_2} \frac{D_1}{D_2}$$

where b_1 = width of the impeller at inlet
b_2 = width of the impeller at outlet
D_1 = inlet impeller diameter
D_2 = outlet impeller diameter

$$V_{r2} = 7.05 \frac{1.75}{1.00} \frac{4}{8} = 6.17 \text{ ft/sec}$$

The peripheral velocity at the outlet is

$$u_2 = \frac{8}{12} \pi \frac{1,750}{60} = 61.1 \text{ ft/sec}$$

Also, $\beta_2 = 18°$, and $\tan \beta_2 = 0.3249$. From Eq. (3-20), the developed head is

$$\Delta h_r = \frac{u_2(u_2 - V_{r2}/\tan \beta_2)}{32.17}$$

$$= \frac{61.1[61.1 - (6.17/0.3249)]}{32.17} = 80.0 \text{ ft}$$

(b) By Eq. (3-20), this relation is linear, and the line of q vs. Δh_r is straight. The developed head at zero flow is found from Eq. (3-20) by assuming $V_{r2} = 0$, so

$$\Delta h_r = \frac{u_2^2}{g_c} = \frac{61.1^2}{32.17} = 116.0 \text{ ft}$$

The maximum flow at zero head is found by solving Eq. (3-20) for V_{r2} on the assumption that $\Delta h_r = 0$, or

$$0 = \frac{61.1(61.1 - V_{r2}/\tan 18°)}{32.17}$$

from which

$$V_{r2} = 61.1 \times 0.3249 = 19.85 \text{ ft/sec}$$

The volumetric flow rate is

$$q' = 19.85 \frac{8}{12} \pi \frac{1.00}{12} 60 \times 7.48 = 1,555 \text{ gal/min}$$

The graph of the theoretical head vs. capacity for this case is shown in Fig. 3-42. It must be emphasized that this curve is especially unrealistic at the two extremes of zero head and zero flow. Actually, the head at zero flow (closed discharge) is approximately one-half of the theoretical value, or, in this case, about 58 to 60 ft. The actual flow rate at zero developed head will be but a fraction of the value calculated above, because friction and shock losses will convert the entire power input into heat as the flow rate increases beyond the rated capacity.

(c) In general, from Fig. 3-41b,

$$\tan \alpha_2 = \frac{V_{r2}}{u_2 - V_{r2}/\tan \beta_2}$$

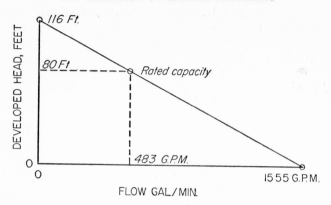

Fɪɢ. 3-42. Theoretical head-capacity relation for Example 3-5

At rated capacity $V_{r2} = 6.17$, and

$$\tan \alpha_2 = \frac{6.17}{61.1 - 6.17/\tan 18°} = 0.1466$$

$$\alpha_2 = 8°20'$$

At $\frac{1}{2}$ rated capacity $V_{r2} = 6.17 \times 0.5 = 3.085$, and

$$\tan \alpha_2 = \frac{3.085}{61.1 - 3.085/\tan 18°} = 0.05979$$

$$\alpha_2 = 3°25'$$

At $1\frac{1}{2}$ times rated capacity, $V_{r2} = 6.17 \times 1.5 = 9.255$, and

$$\tan \alpha_2 = \frac{9.255}{61.1 - 9.255/\tan 18°} = 0.2838$$

$$\alpha_2 = 15°51'$$

The actual diffuser angle should be fixed at the value of α_2 for rated conditions, or, in this case, at 8°20'.

Multistage Centrifugal Pumps. The maximum head that it is practicable to develop in a single impeller is limited by the peripheral speed reasonably attainable. If a head of more than 75 to 100 ft is needed, two or more impellers may be placed in series on a single shaft, and a multistage pump so obtained. Figure 3-43 shows a three-stage single-suction diffuser pump. The discharge from the first stage provides suction for the second stage, and the discharge from the second is the suction of the third. The capacities of all three stages are equal, and the capacity of the pump equals that of a single stage. The developed heads of all stages add to give a total head equal to three times that of a single stage. In the pump shown in Fig. 3-43, the first stage is shown as a section through both the impeller and the diffuser; the second stage shows the impeller in elevation and the diffuser in section; and the third stage shows the diffuser in elevation. The efficiency of a multistage pump is the efficiency (fractional) raised to a power equal to the number of stages. Thus, if the efficiency of each stage in the pump of Fig. 3-43 is 75 per cent, the efficiency of the pump is 0.75^3 = 0.42, or 42 per cent. Because of their low single-stage efficiencies, volute pumps are seldom multistaged. Multistage pumps are expensive, and are used only on clean, nonviscous, and noncorrosive liquids. Boiler feed pumps are an example of such service.

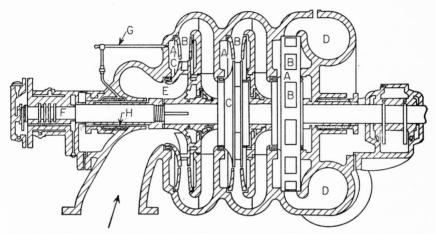

Fig. 3-43. Three-stage, single-suction, diffuser pump: A, diffuser rings. B, ports in diffuser rings. C, impeller. D, discharge volute. E, balance ports. F, thrust bearing. G, seal pipes. H, lantern.

Pump Priming and Self-priming Pumps. Equation (3-15) shows that the head developed by a centrifugal pump depends only on the velocities and the angles between their vectors. If these remain unchanged, the developed head is the same for fluids of all densities and is the same for both liquids and gases. The increase in pressure, however, is the product of the developed head and the fluid density. If the pump develops, say, a head of 100 ft, and if the casing is full of water, the increase in pressure is $(100 \times 62.3)/144 = 43$ lb force/in.^{2} If the casing is full of air at ordinary density, the increase in pressure is less than 0.1 lb force/in.^{2} A centrifugal pump that has its casing full of air can, then, neither draw liquid upward from an initially empty suction line nor force liquid along a full discharge line. When an operating pump has air in its casing, it is "airbound" and can accomplish nothing until the air has been replaced by a liquid. This can be done from an auxiliary priming tank connected to the suction line or by drawing liquid into the suction line by an independent source of vacuum. Sometimes a small positive-displacement pump is used for this purpose.

Several types of self-priming centrifugal pumps are made, which require only a small priming charge of liquid in the pump casing. This liquid remains in the pump at all times, even when the suction line is drained. Regardless of type, a self-priming centrifugal forms a froth of the priming liquid with the air or gas from the suction line, separates the froth into liquid and gas, discharges the gas, and reuses the liquid to make more froth until all the gas is removed from the suction line. The pump cannot

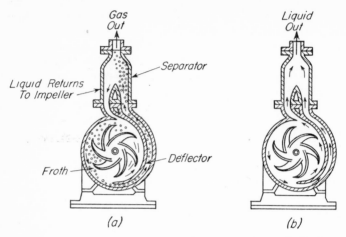

Fig. 3-44. Self-priming centrifugal pump: (a) During priming. (b) During pumping.

move gas alone, but it can move the froth if the froth has a reasonable density. As shown in Fig. 3-44, while the pump is priming, it draws gas into the casing and mixes it with the liquid. The impeller blades may be perforated in order to do this. A stationary vane cuts the froth away from the impeller, leading it to a separation chamber, where the gas is disengaged from the liquid. The gas is impelled into the discharge line; the liquid returns to the impeller. Once all the gas is out of the suction line, the entire casing becomes filled with liquid, and the pump then operates like any volute centrifugal.

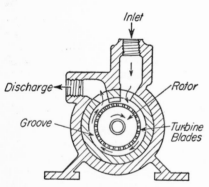

Fig. 3-45. Turbine pump.

Cavitation. In high-speed centrifugal pumps with a high suction lift a phenomenon known as cavitation may become troublesome. The sudden reduction in pressure at the inlet tips of the impeller vanes, resulting from the large increase in liquid velocity, causes tiny bubbles of dissolved gas to emerge from the liquid and collect on the vanes. As the liquid moves radially, its pressure increases rapidly—too rapidly for the bubbles to redissolve. They are therefore pressed with enormous force into the surfaces of the impeller, causing severe erosion.

Turbine Pumps. In present-day terminology a turbine pump often refers to a pump which operates on the impulse principle. Such a pump is illustrated in Fig. 3-45. Short radial vanes are cut out of the periphery of a thick disk-shaped rotor. Liquid enters at the edge of the impeller, is picked up by the vanes, and carried around to the outlet. In the casing surrounding the impeller is a deep groove leading from the inlet to the discharge. Recirculation is encouraged, not minimized, in this pump. As the liquid flows through the pump, it is thrown repeatedly into the groove and returns to the vanes for another impulse, all the while moving toward the discharge. The result is an approximation of multistaging, to give discharge heads up to 800 ft of liquid. A single-stage volute pump, by contrast, rarely discharges at heads greater than 300 ft of liquid.

Characteristics of Volute and Turbine Pumps. Figure 3-46 compares the performance characteristics of a turbine pump with those of a small single-

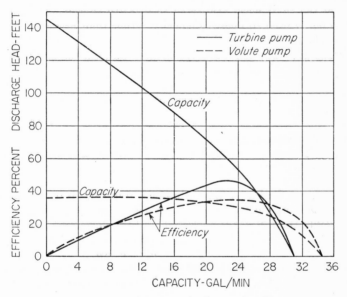

Fig. 3-46. Characteristic curves of volute and turbine pumps.

stage volute pump. The "shut-off head"—the head developed at zero flow with the discharge line closed—is much higher in the turbine pump, and the maximum efficiency of the turbine is considerably greater than that of the small volute pump. The volute pump, however, develops a nearly constant head over a wide range of discharge rates; the head developed by the turbine pump falls steadily as the throughput increases. Turbine pumps are limited to fairly small flows, below about 150 gal/min; volute pumps handle many thousands of gallons per minute. Large volute pumps have a much higher efficiency than indicated by Fig. 3-46; it may reach 85 per cent. At a specific maximum flow rate through each pump the developed head drops to zero. The pump cannot discharge more liquid than this even when the discharge line is wide open.

Centrifugal Fans. The operating principle of centrifugal fans is exactly the same as that of centrifugal pumps. Gas enters at the axis of the impeller and is thrown outward by the vanes into a scroll. The clearances are large, and the discharge heads are low. Because of the low density of the gas, fans rarely discharge at more than 60 in. H_2O, and often at 5 to 10 in. H_2O. Sometimes, as in ventilating fans, all the added energy is converted to velocity energy and almost none to static head. In any case the gain in velocity absorbs an appreciable fraction of the added energy, and must be considered in estimating efficiency and power requirements. The compressibility of the gas, however, may be neglected.

The static efficiency of a fan η_s is the fraction of the shaft-work input to the fan that is converted to pressure energy. It is of interest when the fan is used primarily to increase pressure. The dynamic efficiency η_d is the fraction of the shaft work converted to velocity energy. It is important when the fan is used to increase velocity rather than pressure. The total efficiency η is the fraction of the shaft work appearing as both pressure and velocity energy. The following values may be used for preliminary estimates of fan performance.

TABLE 3-4. EFFICIENCIES OF FANS

Type	Per cent
Static efficiency η_s...............................	60
Dynamic efficiency with zero pressure increase η_d.....	40
Total efficiency η................................	70

The power required by a fan is obtained from Eq. (3-6), with the aid of the appropriate fan efficiency. For example, if the static efficiency is used, the velocity in the exit gas is neglected, and the developed head is, by Eq. (3-5),

$$\Delta h = (p_b - p_a) \frac{g_c}{g\rho} = \frac{g_c \, \Delta p}{g\rho}$$

where

$$\Delta p = p_b - p_a$$

By Eq. (3-6),

$$P_B = \frac{wg \, \Delta p \, g_c}{33,000 g_c \eta_s g\rho} = \frac{w \, \Delta p}{33,000 \eta_s \rho}$$

If q is the volumetric rate of flow at density ρ in cubic feet per minute, and if $\Delta p'$ is the pressure increase in pounds force per square inch, $q = w/\rho$, and

$$P_B = \frac{q \, \Delta p}{33,000 \eta_s} = \frac{144 q \, \Delta p'}{33,000 \eta_s} = \frac{q \, \Delta p'}{229 \eta_s} \tag{3-21}$$

Also, when the pressure increase is negligible, the developed head is, from

Eq. (3-5) (neglecting the inlet velocity and calling $\alpha = 1.0$), $\Delta h = \overline{V}_b^2/2g$. Then, from Eq. (3-6),

$$P_B = \frac{wg\overline{V}_b^2/2g}{33,000g_c\eta_d} = \frac{w\overline{V}_b^2}{2 \times 32.17 \times 33,000\eta_d} = \frac{w\overline{V}_b^2}{2.125,000\eta_d} \quad (3\text{-}22)$$

In some fans the vanes extend from the hub of the impeller to its periphery; in others they are much shorter. Enclosed impellers with swept-back

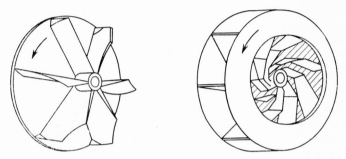

FIG. 3-47. Impellers for centrifugal fans.

blades are used to move clean gases and those with nearly straight blades for dust-laden air. Semienclosed or open impellers propel gas carrying long fibers, such as rags, shavings from woodworking machines, and similar materials. In enclosed impellers the "eye," or opening through which the fluid reaches the vanes, is much larger than that in pump impellers for liquids. Typically the diameter of the eye is 60 to 80 per cent of the diameter of the impeller. In many fans the vanes do not extend into the eye, but are confined to the enclosed space. Fan blades may be flat and radial, flat and inclined backward, curved forward or backward, or forward-curved with radial tips. Backward-curved blades give the highest total efficiency, radial blades a lower efficiency, and forward-curved blades the lowest efficiency. Typical impellers are shown in Fig. 3-47. Performance curves for a fan with forward-curving blades are given in Fig. 3-49b.

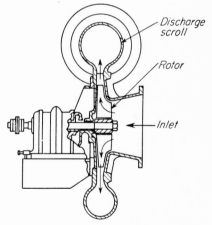

FIG. 3-48. Single-suction turboblower.

Turboblowers. A turboblower resembles a centrifugal pump in appearance, except that the casing is narrow and both the casing and volute are fairly large in diameter. Inside the casing a multibladed impeller turns with close clearance at high speed, commonly 3,600 rpm or higher. This speed must be great in order that the product of the head and the density will be appreciable. Gas enters at the impeller axis and is discharged into the scroll, as in a volute pump (see Fig. 3-48). Stationary diffusers are not

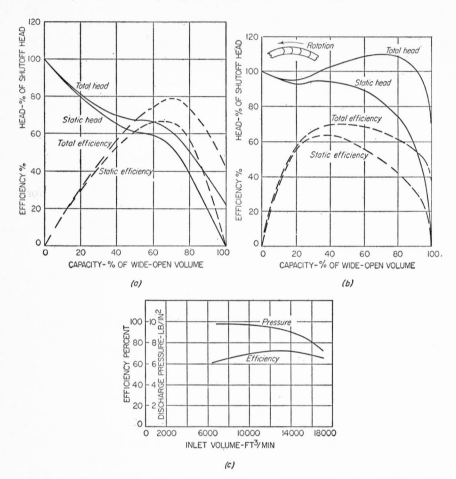

FIG. 3-49. Characteristic curves of fans and blowers: (*a*) Axial-flow fan. (*b*) Forward-curved fan. (*c*) Turbocompressor.

commonly used in pumps; in turboblowers they are essential. The discharge pressure from a single-stage turboblower is usually below 6 lb force/in.^{2} Multistage units are used for higher discharge pressures, up to several hundred pounds force per square inch. Such a multistage machine is called a *turbocompressor*.

Turboblowers and turbocompressors are small in size but discharge very large amounts of gas against considerable pressure. They have high efficiency, as shown in Fig. 3-49*c*, and require little maintenance. They have replaced reciprocating compressors in many operations where large volumes of compressed gas are needed.

In a *jet ejector* the fluid to be moved is entrained in a high-velocity stream of a second fluid. A common example is a laboratory water jet, in which the water stream entrains air from a suction flask or other vessel. The motive fluid and the fluid to be moved may be the same, as when compressed air is used to move air; but usually they are not. Industrially most use is made of steam-jet ejectors, which are valuable for drawing a fairly high vacuum. As shown in Fig. 3-50, steam at about 100 lb force/in.[2]

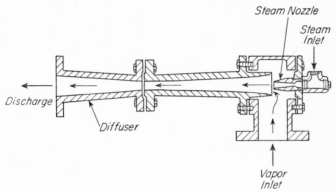

FIG. 3-50. Steam-jet ejector.

is admitted to a converging-diverging nozzle, from which it issues at supersonic velocity into a diffuser cone. The air or other gas to be moved is mixed with the steam in the first part of the diffuser, lowering the velocity to acoustic velocity or below; in the diverging section of the diffuser the kinetic energy of the mixed gases is converted to pressure energy so that the mixture can be discharged directly to the atmosphere. Often it is sent to a water-cooled condenser, particularly if more than one stage is used, for otherwise each stage would have to handle all the steam admitted to the preceding stages. As many as five stages are used in industrial processing.

Jet ejectors require very little attention and maintenance and are especially valuable with corrosive gases that would damage mechanical vacuum pumps. For difficult problems the nozzles and diffusers can be made of corrosion-resistant metal, graphite, or other inert material. Ejectors, particularly when multistage, use large quantities of steam and water. They are not used to produce absolute pressures below about 1 mm Hg.

Comparison of Devices for Moving Fluids. Positive-displacement machines, in general, handle smaller quantities of fluids at higher discharge pressures than do centrifugal machines. Positive-displacement pumps are not subject to air binding and are usually self-priming. In both positive-displacement pumps and blowers the discharge rate is nearly independent of the discharge pressure, so that these machines are extensively used for controlling and metering flow. Reciprocating devices require considerable maintenance but can produce the highest pressures. They deliver a pulsating stream. Rotary pumps work best on fairly viscous lubricating fluids, discharging a steady stream at moderate to high pressures. They cannot be used with slurries. Rotary blowers, except liquid-piston machines, discharge gas at a maximum pressure of 15 lb force/in.^{2} from a single stage. The discharge line of positive-displacement pumps cannot be closed without stalling or breaking the pump, so that a bypass line with a pressure-relief valve is required on the discharge.

Centrifugal machines, both pumps and blowers, deliver fluid at a uniform pressure without shocks or pulsations. They run at higher speeds than positive-displacement machines, and are directly connected to the motor drive instead of through a gear box. The discharge line can be completely closed without damage. Centrifugal pumps can handle a wide variety of corrosive liquids and slurries. Centrifugal blowers and compressors are much smaller, capacity for capacity, than reciprocating compressors and require less maintenance.

For producing vacuum, reciprocating machines are effective for absolute pressures down to 10 mm Hg. Industrial steam-jet ejectors pull down to 1 or 2 mm Hg abs and are valuable for handling corrosive gases. Rotary vacuum pumps can lower the absolute pressure to 0.01 mm Hg, and over a wide range of low pressures are cheaper to operate than multistage steam-jet ejectors.

SYMBOLS

b Width of impeller, ft; b_1, at inlet; b_2, at discharge

C_p Molal heat capacity at constant pressure, Btu/(lb mole)(°F)

C_v Molal heat capacity at constant volume, Btu/(lb mole)(°F)

D Diameter, ft; D_1, of impeller at inlet; D_2, of impeller at discharge; D'_e, economic pipe diameter, in.

g Acceleration of gravity, ft/sec^2

g_c Newton's-law conversion factor, 32.174 ft-lb/lb force-sec^2

H_f Friction, ft-lb force/lb; H_{fs}, skin friction

h Head, ft; h_a, at upstream station or pump inlet; h_b, at downstream station or pump discharge

K_c Coefficient, loss on sudden contraction of cross section

k Roughness parameter, ft

L Length, ft

n Exponent in Eq. (3-10); $n = (\kappa - 1)/\kappa$

P Power, hp; P_B, brake horsepower; P_r, theoretical horsepower of ideal pump

p Pressure, lb force/ft^2; p_a, at upstream station or pump suction; p_b, at downstream station or pump discharge; p_1, p_2, p_3, at discharge from stages 1, 2, and 3, respectively; p', internal working pressure, lb force/in.^{2}

q Volumetric flow rate, ft^3/min; q_r, theoretical value for ideal pump; q', flow rate, gal/min

R_0 Gas law constant, 0.00525 Btu/(std ft^3)(°R)

r Radius, ft; r_1, radius of impeller at suction; r_2, radius of impeller at discharge

S Allowable stress, lb force/in.^{2}

T Absolute temperature, °R; T_a, at suction of compressor; T_b, at discharge of compressor; T_1, T_2, discharge temperatures for stages 1 and 2

u Tangential velocity in pump impeller, ft/sec; u_1, at suction; u_2, at discharge

V Resultant velocity, absolute, in pump impeller, ft/sec; V_1, at suction; V_2, at discharge; V_{r1}, radial component of velocity V_1; V_{r2}, radial component of V_2; V_{u1}, tangential component of V_1; V_{u2}, tangential component of V_2

$\overline{V}$ Average velocity, ft/sec; $\overline{V}_a$, at upstream station; $\overline{V}_b$, at downstream station

v Average velocity of fluid relative to the impeller, ft/sec; v_1, at suction; v_2, at discharge

$-W_s$ Shaft work to pump, ft-lb force/lb; $-W_{sr}$, shaft work of ideal pump; $-W_{sv}$, shaft work of ideal compressor, ft-lb/std ft^3

w Mass flow rate, lb/min; w', flow rate, lb/hr

Z Height above datum plane, ft; Z_a, at upstream station; Z_b, at downstream station

Greek Letters

α Kinetic-energy factor; α_a, at upstream station; α_b, at downstream station; angle between absolute and tangential velocities in pump impeller; α_1, at suction; α_2, at discharge

β Vane angle in pump impeller; β_1, at suction; β_2, at discharge

Δh Head developed by pump, ft; Δh_r, theoretical head developed by ideal pump

Δp Increase in pressure over pump or blower, lb force/ft^2; $\Delta p'$, increase in pressure, lb force/in.^{2}

ϵ Fraction of impeller periphery not occupied by vanes; ϵ_1, at suction; ϵ_2, at discharge

η Over-all mechanical efficiency of pump, fan, or blower (η may include motor, also); η_d, dynamic efficiency of fan; η_s, static efficiency of fan; η_v, volumetric efficiency

κ Ratio, heat capacity at constant pressure to heat capacity at constant volume, C_p/C_v

μ Viscosity, lb/ft-sec

ρ Density, lb/ft^3

Dimensionless Groups

f Fanning friction factor, $2g_cH_{fs}D/4L\overline{V}^2$

N_{Re} Reynolds number, $D\overline{V}\rho/\mu$

PROBLEMS

3-1. A 3-in. steel pipe is to carry cold water at 800 lb force/in.^{2} gauge. The allowable stress in the steel is 13,000 lb force/in.^{2} What should be the schedule number of the pipe?

3-2. A steel pipe is to carry 250 gal/min of alcohol with a density of 49 lb/ft^3. What is the economical pipe diameter for this service?

3-3. The simplified equation [Eq. (3-2)] for economic pipe diameter contains no term for the viscosity of the liquid. Does this mean that the same size of pipe should be used for 50 gal/min of molasses (viscosity = 400 centipoises) as for the same amount of water (viscosity = 1 centipoise)? Explain.

3-4. Water at 40°C flows through a horizontal 2-in. Schedule 40 steel pipe at the rate of 72 gal/min. The line contains 216 ft of straight pipe, two 45° elbows, twelve 90° standard-radius elbows, an angle valve, and six globe valves. Compute the pressure drop in the line when all valves are fully open.

3-5. What is the theoretical suction lift to a pump handling water at 180°F? What maximum suction lift could actually be used?

3-6. Compute the developed head of the pump and the brake horsepower required to pump the water through the piping system described in Prob. 3-4 if the pump takes suction at atmospheric pressure and the water is discharged from the end of the system against a pressure of 35 lb force/in.^{2} gauge. The mechanical efficiency of the pump is 65 per cent.

3-7. Air entering at 70°F and atmospheric pressure is to be compressed to 4,000 lb force/in.^{2} gauge in a reciprocating compressor at the rate of 125 std ft^3/min. How many stages should be used? What is the theoretical shaft work per standard cubic foot for frictionless adiabatic compression? What is the brake horsepower if the efficiency of each stage is 85 per cent? For air $\kappa = 1.40$.

3-8. What is the discharge temperature of the air from the first stage in Prob. 3-7?

3-9. At rated capacity, assume that the following factors apply to the pump of Example 3-5.

Circulatory-flow Factor. Assume the actual value of V_{u2} is 75 per cent of the theoretical value.

Fluid Friction. Assume the actual developed head is 90 per cent of the theoretical developed head after correction for circulatory flow.

Leakage. Assume leakage is 2.5 per cent of the actual net flow through the pump.

Disk Friction. Assume disk friction is 3.5 hp.

Vane Thickness. Because of the thickness of the vanes, 92 per cent of the outlet peripheral area and 85 per cent of the inlet peripheral area of the impeller are available for water flow.

Mechanical-energy Loss. Assume mechanical-energy losses are 2.5 per cent of the brake horsepower.

Assume radial flow at the inlet of the impeller at design capacity.

Calculate: (*a*) the rated capacity, in gallons per minute; (*b*) the actual diffuser angle, corrected for circulatory flow; (*c*) the actual developed head at rated capacity; (*d*) the fluid horsepower at rated capacity; (*e*) the brake horsepower and over-all efficiency at rated capacity.

REFERENCES

1. Church, A. H.: "Centrifugal Pumps and Blowers," (*a*) p. 29, (*b*) p. 44, (*c*) p. 46, John Wiley & Sons, Inc., New York, 1944.
2. Clarke, L.: "Manual for Process Engineering Calculations," p. 368, McGraw-Hill Book Company, Inc., New York, 1947.
3. Hesse, H. C., and J. H. Rushton: "Process Equipment Design," pp. 225–238, D. Van Nostrand Company, Inc., New York, 1945.
4. Kristal, F. A., and F. A. Annett: "Pumps," 2d ed., McGraw-Hill Book Company, Inc., New York, 1953.
5. Perry, J. H.: "Chemical Engineers' Handbook," 3d ed., (*a*) p. 385, (*b*) p. 386, (*c*) p. 390, McGraw-Hill Book Company, Inc., New York, 1950.
6. Richardson, C. A.: *Chem. Eng. Progr.*, **1**(2): 17 (1947).
7. Sarchet, B. R., and A. P. Colburn: *Ind. Eng. Chem.*, **32**: 1249 (1940).

Chapter 4

SIZE REDUCTION

The term "size reduction" is applied to all the ways in which particles of solids are cut or broken into smaller pieces. Throughout the process industries solids are reduced by different methods for different purposes. Chunks of crude ore are crushed to workable size; synthetic chemicals are ground into powder; sheets of plastic are cut into tiny cubes or diamonds. Commercial products must often meet stringent specifications regarding the size and sometimes the shape of the particles they contain. Reducing the particle size also increases the reactivity of solids; it permits separation of unwanted ingredients by mechanical methods; it reduces the bulk of fibrous materials for easier handling.

Solids may be broken in eight or nine different ways, but only four of them are commonly used in size-reduction machines. They are: (1) compression, (2) impact, (3) attrition, or rubbing, and (4) cutting. A nutcracker, a sledge hammer, a file, and a pair of shears exemplify these four types of action. In general, compression is used for coarse reduction of hard solids, to give relatively few fines; impact gives coarse, medium, or fine products; attrition yields very fine products from soft, nonabrasive materials. Cutting gives a definite particle size and sometimes a definite shape, with few or no fines.

Size-reduction equipment is divided into crushers, grinders, ultrafine grinders, and cutting machines. *Crushers* do the heavy work of breaking large pieces of solid material into small lumps. A primary crusher operates on run-of-mine material, accepting anything that comes from the mine face and breaking it into 6- to 10-in. lumps. A secondary crusher reduces these lumps to particles perhaps $\frac{1}{4}$ in. in size. *Grinders* reduce crushed feed to powder. The product from an intermediate grinder might pass a 40-mesh screen; most of the product from a fine grinder would pass a 200-mesh screen. An *ultrafine grinder* accepts feed particles no larger than $\frac{1}{4}$ in.; the product size is typically 1 to 50 microns. *Cutters* give particles of definite size and shape, $\frac{1}{16}$ to $\frac{1}{2}$ in. in length.

The principal types of size-reduction machines are:

I. Crushers (coarse and fine)
   A. Jaw crushers
   B. Gyratory crushers
   C. Crushing rolls

II. Grinders (intermediate and fine)
    A. Hammer mills; impactors
    B. Rolling-compression mills
        1. Bowl mills
        2. Roller mills
    C. Attrition mills
    D. Revolving mills
        1. Rod mills
        2. Ball mills; pebble mills
        3. Tube mills; compartment mills
III. Ultrafine grinders
    A. Hammer mills with internal classification
    B. Fluid-energy mills
IV. Cutting machines
    A. Knife cutters; dicers; slitters

These machines do their work in distinctly different ways. Slow compression is the characteristic action of crushers. Grinders employ impact and attrition, sometimes combined with compression; ultrafine grinders operate principally by attrition. A cutting action is of course characteristic of cutters, dicers, and slitters.

PRINCIPLES OF COMMINUTION

Criteria for Comminution. Comminution is a generic term for size reduction; the machines listed above are types of comminuting equipment. An ideal crusher or grinder would: (1) have a large capacity, (2) require a small power input per unit of product, and (3) yield a product of the single size or the size distribution desired. The usual method of studying the performance of process equipment is to set up an ideal operation as a standard, compare the characteristics of the actual equipment with those of the ideal unit, and account for the difference between the two. When this method is applied to comminuting apparatus, the discrepancies between the ideal and the actual are considerable, and the gaps have not been completely accounted for, even theoretically. On the other hand, useful quantitative information is obtainable from the incomplete theory now at hand.

The capacities of comminuting machines are best discussed when the individual types of equipment are described. The fundamentals of product size and shape and of energy requirements are, however, common to most machines, and can be discussed more generally.

Characteristics of Comminuted Products. The objective of crushing and grinding is to produce small particles from larger ones. Smaller particles are desired either because of their large surface or because of their shape, size, and number. The energy efficiency of the operation is measured by the new surface created by the reduction in size. For these reasons, the geometric characteristics of particles, both alone and in mixtures, are important in evaluating the product from a crusher or grinder. The following sections deal with these matters.

Unlike an ideal crusher or grinder, an actual unit does not yield a uniform product, whether the feed is uniformly sized or not. The product always consists of a mixture of particles, ranging in size from a definite maximum to a submicroscopic minimum. Some machines, especially in the grinder class, are designed to control the magnitude of the largest particles in their products, but the fine sizes are not under control. In some types of grinders fines are minimized, but they are not eliminated. If the feed is homogeneous, both in the shapes of the particles and in chemical and physical structure, the shapes of the individual units in the product may be quite uniform; otherwise, the grains in the various sizes of a single product may vary considerably in proportions.

The smallest grain in a comminuted product may be comparable in size to that of a unit crystal,[6b] which is the smallest unit of the material that can exist as an independent crystal. This size is of the order of 10^{-3} micron, a micron being 10^{-3} mm, or 10^{-4} cm. If, for example, the largest particle in a product just passes a screen having 1-mm openings, the ratio of the diameters of the largest and smallest particles is of the order 10^{-1}: 10^{-7}, or 10^6. Because of this extreme variation in the sizes of the individual particles, relationships adequate for uniform sizes must be modified when applied to such mixtures. The term "average size," for example, is meaningless until the method of averaging is defined, and several different average sizes can be calculated.[4]

Unless they are smoothed by abrasion after crushing, comminuted particles resemble polyhedrons with nearly plane faces and sharp edges and corners. The number of major faces may vary, but is usually between four and seven. The particles may be compact, with length, breadth, and thickness nearly equal, or they may be platelike or needlelike. A compact grain with several nearly equal faces can be considered to be spherical, and the term "diameter" is generally used for particle size.

Particle Geometry for Uniform Sizes. Consider a single particle, and focus attention on its size, volume, and surface. To measure the size quantitatively, it is necessary to choose one major dimension as the defining length, and if other particles of the same shape are to be considered, the same choice must be made for all particles. For a cube or a sphere the simplest choice is the length of a side, or the diameter. For an irregular particle the choice of the defining dimension is arbitrary. Let the length of the defining dimension be D_p, and call this the diameter of the particle. The volume of the particle is proportional to D_p^3 and the surface to D_p^2. For example, the volume and surface of a cube are D_p^3 and $6D_p^2$; those of a sphere are $(\pi/6)D_p^3$ and πD_p^2. For both shapes the ratio of surface to volume is $6/D_p$.

The volume of a particle of any shape can be written as

$$v_p = aD_p^3 \tag{4-1}$$

and the surface as

$$s_p = 6bD_p^2 \tag{4-2}$$

where a and b are geometric constants which depend only on the shape of the particle. The ratio of surface to volume is, from Eqs. (4-1) and (4-2),

$$\frac{s_p}{v_p} = \frac{6(b/a)}{D_p} = \frac{6\lambda}{D_p} \tag{4-3}$$

where
$$\lambda = \frac{b}{a} \tag{4-4}$$

The shape factor λ is independent of particle size and is a function of shape only. It is unity for cubes and spheres. For particles of irregular shape it is greater than unity. For many products of comminution it is approximately 1.75.[6d]

In a sample of uniform particles of diameter D_p the total volume of the particles is m/ρ_p, where m and ρ_p are the total mass of the sample and the density of the particles, respectively. Since the volume of one particle is aD_p^3, N, the number of particles in the sample, is

$$N = \frac{m/\rho_p}{aD_p^3} \tag{4-5}$$

The total surface area of the particles is, from Eqs. (4-2), (4-4), and (4-5),

$$A = Ns_p = \frac{m/\rho_p}{aD_p^3} 6bD_p^2 = \frac{6\lambda m}{\rho_p D_p} \tag{4-6}$$

Mixed Particle Sizes and Screen Analysis. To apply Eqs. (4-1) to (4-6) to mixtures of particles having various sizes and densities, the mixture is sorted into fractions, each of constant density and approximately constant size. Each fraction can then be weighed, or the individual particles in it can be counted or measured by microscopic methods. The above equations can then be applied to each fraction and the results added. Methods for separating mixtures by size and density are discussed in Chap. 7. The simplest and most common method of separating mixtures by size alone is screening by testing sieves. The method is applicable to particles of uniform density and shape and for sizes ranging from approximately 3 in. (7,600 microns) to approximately 0.0015 in. (38 microns). The usual range of mesh sizes is between 1 and $\frac{1}{200}$ in.

Testing sieves are made of woven wire screens, the mesh and dimensions of which are carefully standardized. The openings are square. Each screen is identified in meshes per inch. The actual openings are smaller than those corresponding to the mesh numbers, however, because of the thickness of the wires. The characteristics of one common series, the Tyler standard screen series, are given in Appendix 18. This set of screens is based on the opening of the 200-mesh screen, which is established at 0.0074 cm. The area of the openings in any one screen in the series is exactly twice that of the openings in the next smallest screen. The ratio of the actual mesh dimension of any screen to that of the next smaller screen is, then, $\sqrt{2} = 1.41$. For closer sizing, intermediate screens are available, each of which has a mesh dimension $\sqrt[4]{2}$, or 1.189, times that of the next smaller standard screen. Ordinarily these intermediate screens are not used.

In practice, a set of standard screens is arranged serially in a stack, with the smallest mesh at the bottom and the largest at the top. An analysis is conducted by placing the sample on the top screen and shaking the stack mechanically for a definite time. The particles retained on each screen are removed and weighed, and the masses of the individual screen increments are converted to mass fractions or mass percentages of the total sample. Any particles that pass the finest screen are caught in a pan at the bottom of the stack.

The results of a screen analysis are tabulated to show the mass fraction of each screen increment as a function of the mesh size range of the increment. Since the particles on any one screen are passed by the screen immediately ahead of it, two numbers are needed to specify the size range of an increment, one for the screen through which the fraction passes and the other on which it is retained. Thus, the notation 14/20 means "through 14 mesh and on 20 mesh." An analysis tabulated in this manner is called a *differential analysis*. A typical differential analysis is shown in Table 4-1. The symbol $\Delta\phi_n$ is used for the mass fraction of the total sample that is retained by screen n, where the screens are numbered serially, starting at the top of the stack, so screen $n - 1$ is the screen immediately above screen n. The symbol D_{pn} is the particle diameter equal to the mesh opening of screen n.

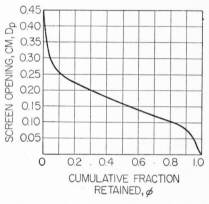

FIG. 4-1. Cumulative screen analysis.

The second type of screen analysis is the *cumulative analysis*. A cumulative analysis is obtained from a differential analysis by adding, cumulatively, the individual differential increments, starting with that retained on the largest mesh, and tabulating or plotting the cumulative sums against the mesh dimension of the retaining screen of the last to be added. If ϕ is defined by the equation

$$\phi = \Delta\phi_1 + \Delta\phi_2 + \cdots + \Delta\phi_n = \sum_1^n \Delta\phi \qquad (4\text{-}7)$$

the cumulative analysis is a relation between ϕ and D_p, where D_p is the mesh size of screen n. The quantity ϕ is the mass fraction of the sample that consists of particles larger than D_p. The value of ϕ for the entire sample is, of course, unity. The cumulative analysis corresponding to the differential analysis of Table 4-1 is shown in Table 4-2 and is plotted in Fig. 4-1.

TABLE 4-1. DIFFERENTIAL SCREEN ANALYSIS

Mesh	$\Delta\phi_n$	D_{pn}	Mesh	$\Delta\phi_n$	D_{pn}
4/6	0.0251	0.3327	35/48	0.0102	0.0295
6/8	0.1250	0.2362	48/65	0.0077	0.0208
8/10	0.3207	0.1651	65/100	0.0058	0.0147
10/14	0.2570	0.1168	100/150	0.0041	0.0104
14/20	0.1590	0.0833	150/200	0.0031	0.0074
20/28	0.0538	0.0589	Pan.......	0.0075	
28/35	0.0210	0.0417			

TABLE 4-2. CUMULATIVE SCREEN ANALYSIS

Mesh	D_p	ϕ	Mesh	D_p	ϕ
4	0.4699	0	35	0.0417	0.9616
6	0.3327	0.0251	48	0.0295	0.9718
8	0.2362	0.1501	65	0.0208	0.9795
10	0.1651	0.4708	100	0.0147	0.9853
14	0.1168	0.7278	150	0.0104	0.9894
20	0.0833	0.8868	200	0.0074	0.9925
28	0.0589	0.9406	Pan.......	......	1.0000

Calculations Based on Screen Analyses. Either the differential or the cumulative analysis can be used to calculate the surface area and particle population of a mixture. If the differential analysis is used, the assumption is made that all particles in a single fraction are equal in size and that the size is the arithmetical mean of the mesh dimensions of the two screens that define the fraction. Thus, the mesh dimensions of standard 10- and 14-mesh screens are 0.1651 and 0.1168 cm, respectively, and the 10/14 fraction is assumed to consist of uniform particles of diameter $(0.1651 + 0.1168)/2 = 0.1410$ cm. The symbol $\bar{D}$ is used for such arithmetic average diameters. If the cumulative analysis is used, the graph of ϕ vs. D_p is treated as a continuous function, and the methods of the calculus are utilized. In principle, the method based on the cumulative analysis is more precise than that based on the differential analysis, since when the cumulative analysis is used, the assumption that all particles in a single fraction are equal in size is not needed. The accuracy of a screen analysis is not good, however, and the precision of either method of calculation is better than that of the experimental data.

Specific Surface of Mixture. It is assumed that the particle density ρ_p and the shape factors a and b are known and that these quantities are independent of particle diameter. If the differential analysis is used, the surface of the particles in each fraction is calculated by Eq. (4-6), and the results for all fractions are added to give A_w, the total surface of one unit mass of sample.

$$A_w = \frac{6\lambda \, \Delta\phi_1}{\rho_p \overline{D}_1} + \frac{6\lambda \, \Delta\phi_2}{\rho_p \overline{D}_2} + \cdots + \frac{6\lambda \, \Delta\phi_n}{\rho_p \overline{D}_n}$$

$$= \frac{6\lambda}{\rho_p} \sum_{n=1}^{n_T} \frac{\Delta\phi_n}{\overline{D}_n} \tag{4-8}$$

where the subscripts refer to the individual screen increments, n_T is the number of screens, $\overline{D}_n$ is the arithmetic average of D_{pn} and $D_{p(n-1)}$, and the summation means the sum of all the $\Delta\phi_n/\overline{D}_n$ quantities of the individual fractions.

If the cumulative analysis is used, Eq. (4-6) is written differentially, and the total surface found by graphically integrating between the limits $\phi = 0$ and $\phi = 1.0$, or

$$A_w = \frac{6\lambda}{\rho_p} \int_0^{1.0} \frac{d\phi}{D_p} \tag{4-9}$$

The graphical integration is conducted in the usual manner by plotting $1/D_p$ as the ordinate vs. ϕ as the abscissa and measuring the area under the curve between $\phi = 0$ and $\phi = 1.0$ (see Chap. 1). The area per unit mass of sample is called the specific surface.

The specific surface A_w is related to an average particle size for the entire mixture. This average size is called the *volume-surface mean diameter* and is denoted by $\overline{D}_{vs}$. It is defined by the equation

$$\overline{D}_{vs} = \frac{6\lambda}{A_w \rho_p} \tag{4-10}$$

Number of Particles in Mixture. To calculate, from the differential analysis, the number of particles in a mixture, Eq. (4-5) is used to compute the number of particles in each fraction, and N_w, the total population in one mass unit of sample, is obtained by summation over all the fractions, or

$$N_w = \frac{\Delta\phi_1}{a\rho_p \overline{D}_1^3} + \frac{\Delta\phi_2}{a\rho_p \overline{D}_2^3} + \cdots + \frac{\Delta\phi_n}{a\rho_p \overline{D}_n^3}$$

$$= \frac{1}{a\rho_p} \sum_{n=1}^{n_T} \frac{\Delta\phi_n}{\overline{D}_n^3} \tag{4-11}$$

Using the cumulative analysis, Eq. (4-11) can be interpreted as follows:

$$N_w = \frac{1}{a\rho_p} \int_0^{1.0} \frac{d\phi}{D_p^3} \tag{4-12}$$

The integral is evaluated graphically.

Size Distribution of Fine Particles. Empirically, it has been found that for the fine sizes in a comminuted product the slope of the graph of ϕ vs. D_p is an exponential function of the particle diameter D_p,[6c] so

$$-\frac{d\phi}{dD_p} = BD_p^k \tag{4-13}$$

where B and k are constants. The minus sign accounts for the fact that as ϕ increases, D_p decreases. This equation can be used to extrapolate screen-analysis data to sizes too fine to sieve accurately.

Equation (4-13) can be integrated between the limits $\phi = \phi_1$ and $\phi = \phi_2$, and the corresponding limits $D_p = D_{p1}$ and $D_p = D_{p2}$. Integration and substitution of limits gives

$$\phi_2 - \phi_1 = \frac{B}{k+1} (D_{p1}^{k+1} - D_{p2}^{k+1}) \tag{4-14}$$

The constant k depends on the relative importance of the very fine sizes in the sample. It ranges from about -0.5 to about 0.1 for most comminuted products. The larger k is, the less important are the very small particles in the fraction between diameters D_{p1} and D_{p2}. If the product is overground, fine particles become predominant, and k is small. The constant B is a measure of the fraction of the entire product that falls between diameters D_{p1} and D_{p2}.

Constants B and k can be obtained from the differential screen analysis by the following method.[6c] It is assumed that the ratio of the screen opening of any screen in the series bears a constant ratio to that of the screen immediately below it. The Tyler screen series conforms to this requirement. If D_{pn} and $D_{p(n-1)}$ are the mesh sizes of the nth and $(n-1)$st screens, respectively, the mass fraction on the nth screen is $\phi_n - \phi_{n-1}$, and Eq. (4-14) can be written over screen n as follows

$$\phi_n - \phi_{n-1} = \Delta\phi_n = -\frac{B}{k+1} [D_{pn}^{k+1} - D_{p(n-1)}^{k+1}] \tag{4-15}$$

If the constant ratio between $D_{p(n-1)}$ and D_{pn} is r,

$$D_{p(n-1)} = rD_{pn} \tag{4-16}$$

where $r > 1$. Elimination of $D_{p(n-1)}$ from Eq. (4-15) by means of Eq.

(4-16) gives

$$\Delta\phi_n = \frac{B(r^{k+1} - 1)}{k + 1} D_{pn}^{k+1} = B'D_{pn}^{k+1} \qquad (4\text{-}17)$$

where
$$B' = \frac{B(r^{k+1} - 1)}{k + 1} \qquad (4\text{-}18)$$

The differential screen analysis provides the necessary relation between $\Delta\phi_n$ and D_{pn}. Equation (4-17) can be written logarithmically as

$$\log \Delta\phi_n = (k + 1) \log D_{pn} + \log B' \qquad (4\text{-}19)$$

The constants B' and k are evaluated by plotting $\Delta\phi_n$ vs. D_{pn} on logarithmic coordinates and drawing the best straight line through the points. The geometric slope of the line is $k + 1$. Once $k + 1$ is known, $\log B'$ can be found by use of Eq. (4-19) from the coordinates of any convenient point on the line. The constant B can then be calculated from B' by Eq. (4-18).

Equation (4-13) is used, in combination with Eqs. (4-9) and (4-12), for the calculation of specific surface and particle population. Elimination of $d\phi$ between Eqs. (4-9) and (4-13) and integration gives for A_w, the specific surface,

$$A_w = -\frac{6\lambda B}{\rho_p} \int_{D_{p_1}}^{D_{p_2}} D_p^{k-1} \, dD_p$$

$$= \frac{6B\lambda}{\rho_p k} (D_{p1}^k - D_{p2}^k) \qquad (4\text{-}20) \dagger$$

Elimination of $d\phi$ between Eqs. (4-12) and (4-13) and integration between limits gives for N_w, the number of particles per unit mass of mixture,

$$N_w = -\frac{B}{a\rho_p} \int_{D_{p_1}}^{D_{p_2}} \frac{dD_p}{D_p^{3-k}}$$

$$= \frac{B}{(2 - k)a\rho_p} \left(\frac{1}{D_{p2}^{2-k}} - \frac{1}{D_{p1}^{2-k}} \right) \qquad (4\text{-}22)$$

Example 4-1. The screen analysis shown in Tables 4-1 and 4-2 applies to a sample of crushed quartz. The specific gravity of the particles is 2.65, and the shape factors are $a = 2$ and $b = 3.5$. What is the specific surface, in square feet per pound?

Solution. A logarithmic plot of $\Delta\phi_n$ vs. D_n for the finer sizes is shown in Fig. 4-2. It is clear that, for diameters smaller than 0.0417 cm (35-mesh), the data

† Equation (4-20) is indeterminate if $k = 0$. In this case,

$$A_w = \frac{6 \times 2.303B\lambda}{\rho_p} \log \frac{D_{p1}}{D_{p2}} \qquad (4\text{-}21)$$

follow the empirical relationship of Eq. (4-19), so Eqs. (4-14) to (4-22) can be used for calculations in this size range. For all particles larger than 0.0417 cm, Eqs. (4-8) through (4-12) can be used. The constant λ is $3.5/2 = 1.75$.

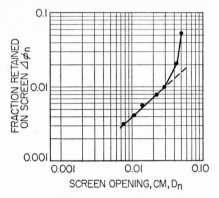

FIG. 4-2. Plot of log $\Delta\phi_n$ vs. log D_n for Example 4-1.

If the differential analysis is used directly for calculations in the size range between 0.4699 and 0.0417 cm, Eq. (4-8) can be written

$$A_w = \frac{6 \times 1.75}{2.65} \sum \frac{\Delta\phi_n}{\bar{D}_n} = 3.96 \sum \frac{\Delta\phi_n}{\bar{D}_n}$$

and Eq. (4-11) as

$$N_w = \frac{1}{2 \times 2.65} \sum \frac{\Delta\phi_n}{\bar{D}_n^3} = 0.189 \sum \frac{\Delta\phi_n}{\bar{D}_n^3}$$

To obtain $\bar{D}_n$ for a screen increment the arithmetic mean of the mesh dimensions for the defining screens for that increment is calculated from the mesh dimensions given in Appendix 18. Magnitudes of $1/\bar{D}_n$ and $1/\bar{D}_n^3$ for each increment are then calculated and weighted with respect to $\Delta\phi_n$, and the sums of $\Delta\phi_n/\bar{D}_n$ and $\Delta\phi_n/\bar{D}_n^3$ are obtained. These calculations are shown in Table 4-3.

TABLE 4-3. CALCULATION OF A_w AND N_w FOR EXAMPLE 4-1

Mesh	$\bar{D}_n$	$\Delta\phi_n$	$\dfrac{1}{\bar{D}_n}$	$\dfrac{1}{\bar{D}_n^3}$	$\dfrac{\Delta\phi_n}{\bar{D}_n}$	$\dfrac{\Delta\phi_n}{\bar{D}_n^3}$
4/6	0.4013	0.0251	2.49	15.5	0.063	0.4
6/8	0.2844	0.1250	3.52	43.5	0.439	5.4
8/10	0.2006	0.3207	4.98	124	1.599	39.7
10/14	0.1409	0.2570	7.1	358	1.824	92
14/20	0.1000	0.1590	10.0	1,000	1.590	159
20/28	0.0711	0.0538	14.1	2,800	0.757	150
28/35	0.0503	0.0210	19.9	7,860	0.417	165
Total....	......	......	.....	....	6.69	611

Then

$$A_w = 3.96 \times 6.69 = 26.5 \text{ cm}^2$$

$$N_w = 0.189 \times 611 = 115$$

If the cumulative analysis is used to calculate A_w and N_w, Eqs. (4-9) and (4-12) become

$$A_w = 3.96 \int_0^{0.9616} \frac{d\phi}{D_p}$$

and

$$N_w = 0.188 \int_0^{0.9616} \frac{d\phi}{D_p^3}$$

To integrate these equations graphically, plots of $1/D_p$ and $1/D_p^3$ are prepared, and the areas under the curves between $\phi = 0$ and $\phi = 0.9616$ are measured. The plots are shown in Figs. 4-3 and 4-4. In Fig. 4-4, changes in the scale of ordinates are made at $\phi = 0.7$ and $\phi = 0.925$ in order that reasonable precision can be obtained in all sections of the plot. The numerical values of the integrals are found to be 6.71 and 626, respectively. Then

$$A_w = 3.96 \times 6.71 = 26.6 \text{ cm}^2$$

and $N_w = 0.189 \times 626 = 118$

These results check closely with those found in calculations based directly on the differential analysis.

To obtain the particle area and population in the fraction passing the 35-mesh

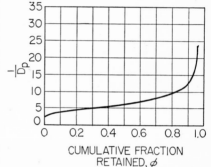

CUMULATIVE FRACTION RETAINED, ϕ

Fig. 4-3. Graphical integration, surface of particles for Example 4-1.

screen the constants k and B of Eq. (4-13) are needed. They are found from the line of Fig. 4-2. The geometric slope of the line is 0.886, which is also the value of $k + 1$, so k is -0.114. The coordinates of any point on the line can be used to evaluate B'. For example, when $\Delta\phi_n$ is 0.004, D_{np} is 0.01. Then, by Eq. (4-19),

$$\log 0.004 = 0.886 \log 0.01 + \log B'$$

from which $B' = 0.237$. For the Tyler screen series, $r = \sqrt{2} = 1.414$, and, from Eq. (4-18),

$$B = \frac{B'(k + 1)}{r^{k+1} - 1}$$

$$= \frac{0.237 \times 0.886}{1.414^{0.886} - 1} = 0.584$$

The largest particle in the pan fraction passes a mesh size of 0.0074 cm. If it is assumed that the relationship of ϕ with D_p in this fraction also conforms to Eq. (4-13), Eq. (4-14) can be used to estimate the diameter of the smallest particle in the pan. This gives

$$0.0075 = \frac{0.584}{0.886} (0.0074^{0.886} - D_{p2}^{0.886})$$

Solution of this equation gives $D_{p2} = 0.00072$ cm.

From Eq. (4-20) the area of the fraction having a size range of 0.0417 to 0.00072 cm is

$$A_w = \frac{6 \times 0.584}{2.65 \times -0.114} (0.0417^{-0.114} - 0.00072^{-0.114})$$

$$= 9.7 \text{ cm}^2$$

The total area of the entire sample is $26.6 + 9.7 = 36.3$ cm^2/g, or $(36.3 \times 453.6)/30.48^2 = 17.7$ ft^2/lb.

From Eq. (4-22),

$$N_w = \frac{0.584}{2.114 \times 2 \times 2.65} (0.00072^{-2.114} - 0.0417^{-2.114})$$

$$= 229,400$$

The total number of particles is then $118 + 229,500$ or about 229,600. In fine sizes, the number of particles per gram is very large. Calculation shows that approximately 227,900 of the total of 229,600 particles are in the pan fraction. The precision of this calculation of particle population is poor.

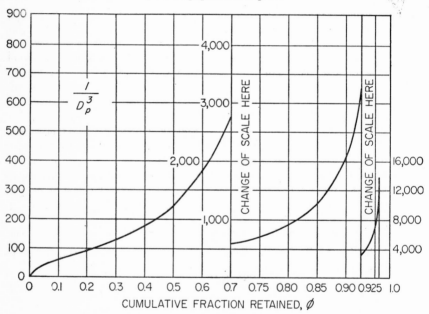

Fig. 4-4. Graphical integration, number of particles for Example 4-1.

Energy and Power Requirements in Comminution. The cost of power is a major expense in crushing and grinding, so the factors that control this cost are important. During size reduction, the particles of feed material are first distorted and strained. The work necessary to strain them is stored temporarily in the solid as mechanical energy of stress, just as mechanical energy can be stored in a coiled spring. As additional force is applied to the stressed particles, they are distorted beyond their ultimate strength, and they suddenly rupture into fragments. New surface is generated. Since a unit area of solid has a definite amount of surface energy, the creation of new surface requires work, which is supplied by the release of energy of stress when the particle breaks. By conservation of energy, all energy of stress in excess of the new surface energy created must appear as heat.

Crushing Efficiency. The ratio of the surface energy created by crushing to the energy absorbed by the solid is the crushing efficiency, and is denoted by η_c. If e_s is the surface energy per unit area, in feet times pounds force per square foot, and A_{wb} and A_{wa} are the areas of product and feed, respectively, in square feet per pound, the energy absorbed by the material W_n is, in feet times pounds force per pound,

$$W_n = \frac{e_s(A_{wb} - A_{wa})}{\eta_c} \tag{4-23}$$

The surface energy created by fracture is small in comparison with the total mechanical energy stored in the material at the time of rupture, and most of the latter is converted into heat. Crushing efficiencies are therefore low. They have been measured experimentally by estimating e_s from theories of the solid state, measuring W_n, A_{wb} and A_{wa}, and substituting into Eq. (4-23). The precision of the calculation is poor, primarily because of uncertainties in calculating e_s, but the results do show that crushing efficiencies range between 0.1 and 2 per cent.[7]

The energy absorbed by the solid W_n is less than that fed to the machine. Part of the energy input, W ft-lb force/lb, is used to overcome friction in the bearings and other moving parts, and the rest is available for crushing. The ratio of the energy absorbed to the energy input is η_m, the mechanical efficiency. Then, if W is the energy input,

$$W\eta_m = W_n = \frac{e_s(A_{wb} - A_{wa})}{\eta_c}$$

or

$$W = \frac{e_s(A_{wb} - A_{wa})}{\eta_m \eta_c} \tag{4-24}$$

If T is the feed rate, in tons per minute, the power required by the machine is, in horsepower,

$$P = \frac{2,000WT}{33,000} = \frac{2,000Te_s(A_{wb} - A_{wa})}{33,000\eta_c\eta_m} \tag{4-25}$$

Calculation of A_{wb} and A_{wa} from Eq. (4-10) and substitution in Eq. (4-25) gives

$$P = \frac{2{,}000 \times 6Te_s}{33{,}000\eta_c\eta_m\rho_p}\left(\frac{\lambda_b}{\overline{D}_{vsb}} - \frac{\lambda_a}{\overline{D}_{vsa}}\right)$$

$$= \frac{0.364Te_s}{\eta_c\eta_m\rho_p}\left(\frac{\lambda_b}{\overline{D}_{vsb}} - \frac{\lambda_a}{\overline{D}_{vsa}}\right) \tag{4-26}$$

where P = power, hp
 T = feed rate, tons/min
 e_s = specific surface energy, ft-lb force/ft^2
 η_m = mechanical efficiency
 η_c = crushing efficiency
 ρ_p = particle density, lb/ft^3
$\overline{D}_{vsa}, \overline{D}_{vsb}$ = volume-surface mean diameters of feed, product, ft
 λ_a, λ_b = shape factors of feed, product, dimensionless

Rittinger's Law. A crushing law proposed many years ago by Rittinger states that the work required in crushing is proportional to the new surface created. This law is equivalent to the statement that the crushing efficiency η_c is constant and, for a given machine and feed material, is independent of the sizes of feed and product.[6e] If the shape factors λ_a and λ_b are equal and the mechanical efficiency is constant, the various constants in Eq. (4-26) can be combined into a single constant K_r and Rittinger's law written as

$$\frac{P}{T} = K_r\left(\frac{1}{\overline{D}_{vsb}} - \frac{1}{\overline{D}_{vsa}}\right) \tag{4-27}$$

Rittinger's law has been shown to apply reasonably well under conditions where the energy input per unit mass of solid is not too great,[7] and it can be used as a first approximation for actual crushing processes where the constant K_r is determined experimentally in a test on a machine of the type to be used and with the material to be crushed.

Example 4-2. A certain crusher accepts a feed of rock having a volume-surface mean diameter of 0.75 in. and discharges a product of volume-surface mean diameter of 0.20 in. The power required to crush 12 tons/hr is 9.3 hp. What should be the power consumption if the capacity is reduced to 10 tons/hr and the volume-surface mean diameter to 0.15 in.? The mechanical efficiency remains unchanged.

Solution. Since the only variables are the feed rate and the diameters of feed and product, Eq. (4-27) is used, once for the original operation, and once for the new operation. It is unnecessary to convert units to those of Eq. (4-27), as conversion factors can be absorbed into constant K_r. The original operation is

$$\frac{9.3}{12} = K_r\left(\frac{1}{0.20} - \frac{1}{0.75}\right)$$

and the new operation is

$$\frac{P}{10} = K_r\left(\frac{1}{0.15} - \frac{1}{0.75}\right)$$

Dividing the second equation by the first gives

$$\frac{P}{9.3}\frac{12}{10} = \frac{1/0.15 - 1/0.75}{1/0.20 - 1/0.75}$$

from which

$$P = 11.4 \text{ hp}$$

Bond Crushing Law and Work Index. A recent method for estimating the power required for crushing and grinding is that proposed by Bond.[2,3] Based on semitheoretical reasoning, the Bond theory proposes that the work required to form particles of size D_p from very large feed is proportional to the square root of the surface-to-volume ratio of the product. Then

$$\frac{P}{T} = \frac{K_b}{\sqrt{D_p}} \tag{4-28}$$

where K_b is a constant which depends on the type of machine and on the material being crushed. This law calls for relatively less energy for the smaller product particles than does the Rittinger law. The Bond law is the more realistic in estimating the power requirements of commercial crushers and grinders.

To use Eq. (4-28), a work index W_i is defined as the gross energy in kilowatthours per ton of feed needed to reduce very large feed to such a size that 80 per cent of the product passes a 100-micron screen. This definition leads to a relation between K_b and W_i. If D_p is in feet, P in horsepower, and T in tons per minute,

$$\frac{60W_i}{0.746} = K_b \sqrt{\frac{30.48}{100 \times 10^{-4}}}$$

or

$$K_b = 1.46W_i \tag{4-29}$$

If 80 per cent of the feed passes a mesh size of D_{pa} ft and 80 per cent of the product a mesh of D_{pb} ft, it follows from Eqs. (4-28) and (4-29) that

$$\frac{P}{T} = 1.46W_i \left(\frac{1}{\sqrt{D_{pb}}} - \frac{1}{\sqrt{D_{pa}}} \right) \tag{4-30}$$

The work index includes the friction in the crusher, and the power given by Eq. (4-30) is gross power.

Table 4-4 gives typical work indexes for some common minerals. These data do not vary greatly among different machines of the same general type and apply to dry crushing or to wet grinding. For dry grinding, the power calculated from Eq. (4-30) is multiplied by $\frac{4}{3}$.

TABLE 4-4. WORK INDEXES FOR DRY CRUSHING † OR WET GRINDING ‡

Material	Sp. gr.	Work index W_i
Bauxite....................	2.20	8.78
Cement clinker.............	3.15	13.45
Cement raw material........	2.67	10.51
Clay......................	2.51	6.30
Coal......................	1.4	13.00
Coke.....................	1.31	15.13
Granite...................	2.66	15.13
Gravel...................	2.66	16.06
Gypsum rock..............	2.69	6.73
Iron ore (hematite)..........	3.53	12.84
Limestone.................	2.66	12.74
Phosphate rock............	2.74	9.92
Quartz....................	2.65	13.57
Shale.....................	2.63	15.87
Slate.....................	2.57	14.30
Trap rock.................	2.87	19.32

† For dry grinding, multiply by $\frac{4}{3}$.
‡ From Allis-Chalmers Co., Milwaukee, Wis., by permission.

Example 4-3. What is the power required to crush 100 tons/hr of limestone if 80 per cent of the feed passes a 2-in. screen and 80 per cent of the product a $\frac{1}{8}$-in. screen?

Solution. From Table 4-4, the work index for limestone is 12.74. Other quantities for substitution into Eq. (4-30) are

$$T = \frac{100}{60} = 1.67 \text{ tons/min}$$

$$D_{pb} = \frac{0.125}{12} = 0.0104 \text{ ft} \qquad D_{pa} = \frac{2}{12} = 0.167 \text{ ft}$$

The power required is

$$P = 1.67 \times 1.46 \times 12.74 \left(\frac{1}{\sqrt{0.0104}} - \frac{1}{\sqrt{0.167}}\right) = 228 \text{ hp}$$

SIZE-REDUCTION EQUIPMENT

The four major classes of size-reduction equipment are crushers, grinders, ultrafine grinders, and cutters. Machines that fall into these classes are tabulated on pages 201 and 202. As mentioned, the boundaries between the classes are indistinct, so that the same kind of machine may be found in more than one class. In general, however, a given type of mill is designed for a particular type of service or particular kind of feed. A size-reduction machine should therefore be selected with care.

Crushers. Crushers are slow-speed machines for coarse reduction of large quantities of solids. The main types are jaw crushers, gyratory crushers, smooth-roll crushers, and toothed-roll crushers. The first three operate by compression and can break large lumps of very hard materials, as in the primary and secondary reduction of rocks and ores. Toothed-roll crushers tear the feed apart as well as crushing it; they handle softer feeds like coal, bone, and soft shale.

Jaw Crushers. In a jaw crusher feed is admitted between two jaws, set to form a V open at the top. One jaw, the fixed, or anvil, jaw, is nearly

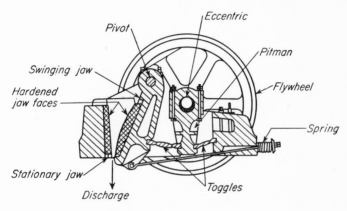

Fig. 4-5. Blake jaw crusher.

vertical and does not move; the other, the swinging jaw, reciprocates in a horizontal plane. It makes an angle of 20 to 30° with the anvil jaw. It is driven by an eccentric so that it applies great compressive force to lumps caught between the jaws. The jaw faces are flat or slightly bulged; they may carry shallow horizontal grooves. Large lumps caught between the upper parts of the jaws are broken, drop into the narrower space below, and are recrushed the next time the jaws close. After sufficient reduction they drop out the bottom of the machine. The jaws open and close 250 to 400 times per minute.

Types of jaw crushers differ in the way the swinging jaw moves. In a *Dodge crusher* the pivot point of the swinging jaw is at the bottom of the V, so that the greatest amount of motion is at the top of the jaws, and the size of discharge opening hardly changes. This type of crusher gives little oversize product and a great many fines. It also tends to "choke"; i.e., it becomes filled with partly crushed material that will not drop out the discharge. Because of this, Dodge crushers handle only moderate tonnages of dry free-flowing solids.

The most common type of jaw crusher is the *Blake crusher*, illustrated in Fig. 4-5. In this machine the eccentric drives a pitman connected to two toggles, one of which is pinned to the frame and the other to the swinging jaw. The pivot point is at the top of the movable jaw or above the top of the jaws on the center line of the jaw opening. The greatest amount of motion is at the bottom of the V, which means that there is little tendency for a crusher of this kind to choke. Blake crushers produce fewer fines than Dodge crushers and can handle stickier materials. They are made in much larger sizes than Dodge crushers. Some machines with a 72- by 96-in. feed opening can accept rocks 6 ft in diameter and crush 1,000 tons/hr to a maximum product size of 10 in. Smaller secondary crushers

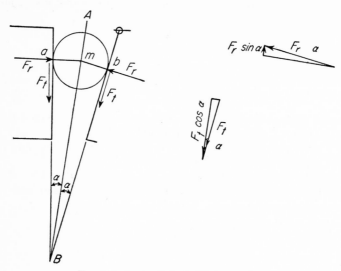

Fig. 4-6. Angle of nip in Blake crusher.

reduce the particle size of precrushed feed to $\frac{1}{4}$ to 2 in., at much lower rates of throughput.

Theory of Blake Crusher. Two points are of interest in considering the characteristics of the Blake crusher: (1) the *angle of nip*, which is the largest permissible angle between the faces of the crusher that will cause the particle to be gripped by the jaws and crushed; and (2) the pattern of forces that exist in the crusher.

For evaluating the angle of nip consider a spherical particle that has just been gripped between the jaws of a Blake crusher, as shown in Fig. 4-6. Contact between the particle and the fixed jaw exists at point a and between the particle and the movable jaw at point b. Let the angle between the jaws be 2α. The bisector of this angle is the line AB, which also passes through m, the center of the particle.

Neglecting the force of gravity, two kinds of force act on the particle. The first consists of the two equal forces F_r which act radially on the particle at points a and b, and the second consists of the two frictional forces F_t that act tangentially at the same points. A frictional force is related to a radial force by the coefficient of friction μ', so $F_t = \mu'F_r$. If the coefficient of friction is the same for both faces, the two frictional forces are also equal.

Each radial force has a component $F_r \sin \alpha$ acting upwardly in a direction parallel with line AB and tending to expel the particle from the crusher. These forces amount to $2F_r \sin \alpha$. Each frictional force also has a component $F_r\mu' \cos \alpha$ acting in a direction parallel with line AB but tending to pull the particle through the jaws. These forces amount to $2F_r\mu' \cos \alpha$. If the particle is to be crushed rather than expelled, the frictional force components must be equal to, or greater than, the radial components, or

$$2F_r\mu' \cos \alpha \geqq 2F_r \sin \alpha$$

and
$$\mu' \geqq \frac{\sin \alpha}{\cos \alpha} = \tan \alpha \qquad (4\text{-}31)$$

The limiting magnitude of angle 2α, where $\tan \alpha = \mu'$, is called the angle of nip. In ordinary designs, it lies between 20 and 30°.

Gravity also tends to pull the particle through the crusher, and its neglect in the analysis provides a factor of safety.

The *forces acting in a Blake crusher* are shown in Fig. 4-7. For simplicity, it is assumed that the angles between the pitman and its two

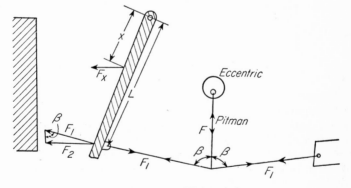

FIG. 4-7. Forces in Blake crusher.

toggles are equal, that the pitman and the fixed jaw are perpendicular, and that the effective crushing forces are the horizontal components exerted on the particle by the jaws. Let β be the angle between the pitman and the toggles and L be the distance between the pivot of the movable jaw and the toggle block on the jaw. The forces are as follows: F is the force on the pitman, F_1 is the force on the toggle, and F_2 is the horizontal force acting on the particle at the toggle block. By equilibrium of forces, F must equal the sum of the vertical components of the two forces F_1. One vertical component is $F_1 \cos \beta$, so

$$F = 2F_1 \cos \beta$$

and
$$F_1 = \frac{F}{2 \cos \beta} \tag{4-32}$$

The horizontal thrust F_2 at the toggle block is the horizontal component of the force F_1, or

$$F_2 = F_1 \sin \beta = \frac{F \sin \beta}{2 \cos \beta} = \frac{F \tan \beta}{2} \tag{4-33}$$

Let F_x be the horizontal thrust at the distance x from the pivot. Then, by moments,

$$xF_x = LF_2$$

and
$$F_x = \frac{LF_2}{x} = \frac{LF \tan \beta}{2x} \tag{4-34}$$

Angle β is nearly 90°, and $\tan \beta$ is therefore large. A small toggle force F is multiplied by the action of the toggle to give a much larger force F_2 for crushing. The amplitude of the motion of the movable jaw is, of course, much smaller than that of the pitman. Also, just as in a nutcracker, the nearer the point of thrust is to the pivot, the greater is force F_x, so the most powerful forces are applied nearest the pivot, where the largest particles are crushed.

Gyratory Crushers. A gyratory crusher may be looked upon as a jaw crusher with circular jaws, between which material is being crushed at some point at all times. A conical crushing head gyrates inside a funnel-shaped casing, open at the top. As shown in Fig. 4-8, the crushing head is carried on a heavy shaft pivoted at the top of the machine. An eccentric drives the bottom end of the shaft. At any point on the periphery of the casing, therefore, the bottom of the crushing head moves toward, and then away from, the stationary wall. Solids caught in the V-shaped space between the head and the casing are broken and rebroken until they pass out the bottom. The crushing head is free to rotate on the shaft and turns slowly because of friction with the material being crushed.

The speed of the crushing head is typically 125 to 425 gyrations per minute. Because some part of the crushing head is working at all times, the discharge from a gyratory is continuous instead of intermittent as in a jaw crusher. The load on the motor is nearly uniform; less maintenance is required than with a jaw crusher; and the power requirement per ton of material crushed is smaller. The biggest gyratories handle up to 3,500 tons/hr. The capacity of a gyratory crusher varies with the jaw setting, the impact strength of the feed, and the speed of gyration of the machine.

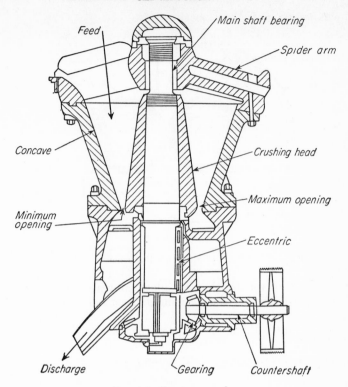

Feed

Main shaft bearing

Spider arm

Concave

Crushing head

Maximum opening

Minimum opening

Eccentric

Discharge

Gearing

Countershaft

FIG. 4-8. Gyratory crusher.

The capacity is almost independent of the compressive strength of the material being crushed.

Smooth-roll Crushers. Two heavy smooth-faced metal rolls turning on parallel horizontal axes are the working elements of the smooth-roll crusher illustrated in Fig. 4-9. Particles of feed caught between the rolls are broken in compression and drop out below. The rolls turn toward each other at the same speed. They have relatively narrow faces and are large in diameter so that they can "nip" moderately large lumps. Typical rolls are 24 in. in diameter with a 12-in. face to 78 in. in diameter with a 36-in. face. Roll speeds range from 50 to 300 rpm. Smooth-roll crushers are secondary crushers, with feeds $\frac{1}{2}$ to 3 in. in size and products $\frac{1}{2}$-in. to about 20-mesh.

The limiting size of particles that can be nipped by rolls of a given diameter is discussed below. The particle size of the product depends on the spacing between the rolls, as does the capacity of a given machine. Smooth-roll crushers give few fines and virtually no oversize. They operate most effectively when set to give a reduction ratio of 3 or 4 to 1; i.e., the maximum particle diameter of the product is one-third or one-fourth that of the feed. The forces exerted by the roll are very great, from 5,500 to 40,000

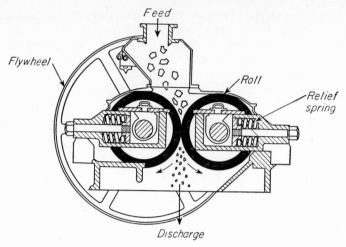

FIG. 4-9. Smooth-roll crusher.

lb force per inch of roll width.   To allow unbreakable material to pass through without damaging the machine at least one roll must be spring-mounted.

The angle of nip is the angle between the roll faces at the level where they will just take hold of a particle and draw it into the crushing zone. It is found as follows.   Figure 4-10 shows a pair of rolls and a spherical

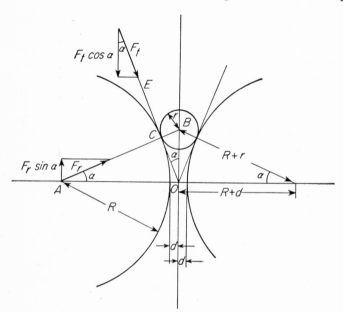

FIG. 4-10. Angle of nip in crushing rolls.

particle just being gripped between them. The radii of rolls and particle are R and r, respectively. The clearance between the rolls is $2d$. Line AB passes through the centers of the lefthand roll and the particle and through point C, the point of contact between roll and particle. Let α be the angle between line AB and the horizontal. Line OE is a tangent to the roll at point C, and this line makes the same angle α with the vertical.

Neglecting gravity, two forces act at point C: the tangential frictional force F_t, having a vertical component $F_t \cos \alpha$, and the radial force F_r, having a vertical component $F_r \sin \alpha$. Force F_t is related to force F_r through the coefficient of friction μ', so $F_t = \mu' F_r$. Force $F_r \sin \alpha$ tends to expel the particle from the rolls, and force $\mu' F_r \cos \alpha$ tends to pull it into the rolls to be crushed. If the particle is to be crushed,

$$F_r \mu' \cos \alpha \geqq F_r \sin \alpha$$

or
$$\mu' \geqq \tan \alpha \tag{4-35}$$

When $\mu' = \tan \alpha$, the angle α is half the angle of nip.

A simple relationship exists between the radius of the rolls, the size of the feed, and the gap between the rolls. Thus, from Fig. 4-10,

$$\cos \alpha = \frac{R + d}{R + r} \tag{4-36}$$

The largest particles in the product have a diameter $2d$, and Eq. (4-36) provides a relationship between the roll diameter and the size reduction that can be expected in the mill.

The *theoretical capacity* of a pair of crushing rolls is established by the volume of a continuous ribbon of material of width equal to the breadth of the roll faces, of thickness equal to the gap between the rolls, and of length equal to the peripheral speed of the roll surfaces. From this the theoretical capacity is

$$q = 60 \times 2d\pi nDb \tag{4-37}$$

where q = volumetric capacity, ft^3/hr
n = speed of rolls, rpm
D = diameter of rolls, ft
d = one-half roll spacing, ft
b = breadth of roll face, ft

The actual capacity of a pair of rolls is from one-third to one-tenth the theoretical quantity given by Eq. (4-37).

Example 4-4. A pair of rolls is to take a feed equivalent to spheres 1.5 in. in diameter and crush them to spheres having a diameter of 0.5 in. If the coefficient of friction is 0.29, what should be the diameter of the rolls?

Solution. Since $\tan \alpha = 0.29$, $\alpha = 16.2°$, and $\cos \alpha = 0.960$. Then, by Eq. (4-36),

$$0.960 = \frac{R + 0.50/2}{R + 1.50/2}$$

from which $R = 12$ in. The diameter of the rolls should be 24 in., or **2 ft**.

Toothed-roll Crushers. In many roll crushers the roll faces carry corrugations, breaker bars, or teeth. Such crushers may contain two rolls, as in smooth-roll crushers, or only one roll working against a stationary curved breaker plate. A single-roll toothed crusher is shown in Fig. 4-11. Machines known as disintegrators contain two corrugated rolls turning at different speeds, which tear the feed apart, or a small high-speed roll with transverse breaker bars on its face turning toward a large slow-speed smooth roll. Some crushing rolls for coarse feeds carry heavy pyramidal teeth. Other designs utilize a large number of thin toothed disks which

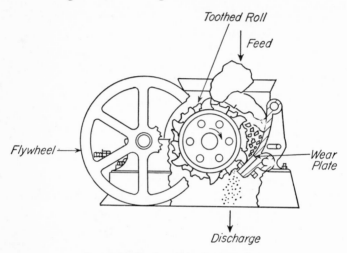

Fig. 4-11. Single-roll toothed crusher.

saw through slabs or sheets of material. Toothed-roll crushers are much more versatile than smooth-roll crushers, within the limitation that they cannot handle very hard solids. They operate by compression, impact, and shear, not by compression alone, as do smooth-roll machines. They are not limited by the problem of nip inherent with smooth rolls and can therefore reduce much larger particles. Some heavy-duty toothed double-roll crushers are used for the primary reduction of coal and similar materials. The particle size of the feed to these machines may be as great as 20 in.; their capacity ranges up to 500 tons/hr.

Grinders. The term "grinder" describes a variety of size-reduction machines for intermediate duty. The product from a crusher is often fed to a grinder, in which it is reduced to powder. The chief types of commercial grinders described in this section are hammer mills and impactors; rolling-compression machines; attrition mills; and revolving mills.

Hammer Mills and Impactors. These mills all contain a high-speed rotor turning inside a cylindrical casing. The shaft is usually horizontal. Feed is dropped into the top of the casing, broken, and falls out through a bottom opening. In a hammer mill the particles are broken by sets of swing hammers pinned to a rotor disk. A particle of feed entering the grinding zone cannot escape being struck by the hammers. It shatters into pieces, which fly against a stationary anvil plate inside the casing, and breaks into still smaller fragments. These in turn are rubbed into powder by the hammers and pushed through a grate or screen which covers the discharge opening.

Several rotor disks, 6 to 18 in. in diameter and each carrying four to eight swing hammers, are often mounted on the same shaft. The hammers may be straight bars of metal with plain or enlarged ends or with ends sharpened to a cutting edge. Intermediate hammer mills yield a product 1 in. to 20-mesh in particle size. In hammer mills for fine reduction the peripheral speed of the hammer tips may reach 22,000 ft/min; they reduce 0.1 to 15 tons/hr to sizes finer than 200-mesh. Hammer mills grind almost anything—tough fibrous solids like bark or leather, steel turnings, soft wet pastes, sticky clay, hard rock. For fine reduction they are limited to the softer materials.

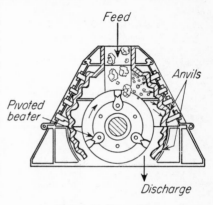

Fig. 4-12. Impactor.

The capacity and power requirement of a hammer mill vary greatly with the nature of the feed and cannot be estimated with confidence from theoretical considerations. They are best found from published information [9] or better from small-scale or full-scale tests of the mill with a sample of the actual material to be ground. Commercial mills typically reduce 100 to 400 lb of solid per hour per horsepower-hour of energy consumed.

An *impactor*, illustrated in Fig. 4-12, resembles a heavy-duty hammer mill except that it contains no grate or screen. Particles are broken by impact alone, without the rubbing action characteristic of a hammer mill. Impactors are often primary-reduction machines for rock and ore, processing up to 600 tons/hr. They give particles that are more nearly equidimensional (more "cubical") than the slab-shaped particles from a jaw crusher or gyratory crusher. The rotor in an impactor, as in many hammer mills, may be run in either direction to prolong the life of the hammers.

Rolling-compression Machines. In this kind of mill the solid particles are caught and crushed between a rolling member and the face of a ring or casing. The most common types are rolling-ring pulverizers, bowl mills, and roller mills. In the roller mill illustrated in Fig. 4-13, vertical cylindrical rollers press outward with great force against a stationary anvil ring or bull ring. They are driven at moderate speeds in a circular path. Plows lift the solid lumps from the floor of the mill and direct them between the ring and the rolls, where the reduction takes place. Product is swept out of the mill by a stream of air to a classifier separator, from which oversize particles are returned to the mill for further reduction. In a bowl

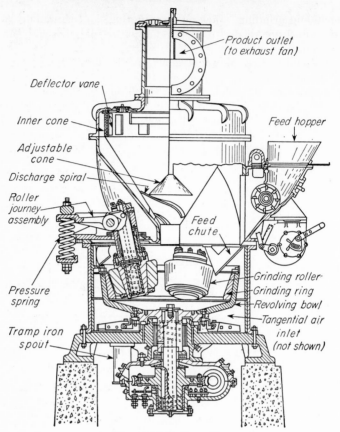

FIG. 4-13. Roller mill.

mill and some roller mills the bowl or ring is driven; the rollers rotate on stationary axes, which may be vertical or horizontal. Mills of this kind find most application in the reduction of limestone, cement clinker, and coal. They pulverize up to 50 tons/hr. When classification is used, the product may be as fine as 99 per cent through a 200-mesh screen.

Attrition Mills. In an attrition mill particles of soft solids are rubbed between the grooved flat faces of rotating circular disks. The axis of the disks is usually horizontal, sometimes vertical. In a single-runner mill one disk is stationary and one rotates; in a double-runner machine both disks are driven at high speed in opposite directions. Feed enters through an opening in the hub of one of the disks; it passes outward through the narrow gap between the disks and discharges from the periphery into a stationary casing. The width of the gap, within limits, is adjustable. At least one grinding plate is spring-mounted so that the disks can separate if unbreakable material gets into the mill. Mills with different patterns of grooves, corrugations, or teeth on the disks perform a variety of operations, including grinding, cracking, granulating, and shredding, and even some operations not related to size reduction at all, such as blending and feather curling.

A single-runner attrition mill is shown in Fig. 4-14. Single-runner mills contain disks of buhrstone or rock emery for reducing solids like clay and talc, or metal disks for solids like wood, starch, insecticide powders, and carnauba wax. Metal disks are usually of white iron, although for corrosive materials disks of stainless steel are sometimes necessary. Double-runner mills, in general, grind to finer products than do single-runner mills but process softer feeds. Air is often drawn through the mill to remove

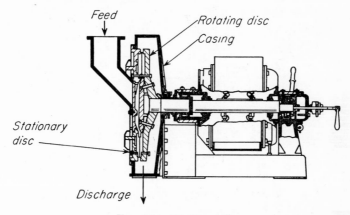

Fig. 4-14. Attrition mill.

the product and prevent choking. The disks may be cooled with water or refrigerated brine to take away the heat generated by the reduction operation. Cooling is essential with heat-sensitive solids like rubber, which would otherwise be destroyed.

The disks of a single-runner mill are 10 to 54 in. in diameter, turning at 350 to 700 rpm. Disks in double-runner mills turn faster, at 1,200 to 7,000 rpm. The feed is precrushed to a maximum particle size of about $\frac{1}{2}$ in. and must enter at a uniform controlled rate. Attrition mills grind from $\frac{1}{2}$ to 8 tons/hr to products which will pass a 200-mesh screen. The energy required depends strongly on the nature of the feed and the degree of reduction accomplished and is much higher than in the mills and crushers described so far. Typical values are between 10 and 100 hp-hr per ton of product.

Revolving Mills. A typical revolving mill is shown in Fig. 4-15. A cylindrical shell slowly turning about a horizontal axis and filled to about half its volume with a solid grinding medium forms a revolving mill. The shell is usually steel, lined with high-carbon steel plate, porcelain, silica rock, or rubber. The grinding medium is metal rods in a rod mill, lengths of chain or balls of metal, rubber, or wood in a ball mill, flint pebbles or porcelain or zircon spheres in a pebble mill. For intermediate and fine reduction of abrasive materials revolving mills are unequalled.

Unlike the mills previously discussed, all of which require continuous feed, revolving mills may be continuous or batch. In a batch machine a measured quantity of the solid to be ground is loaded into the mill through an opening in the shell. The opening is then closed and the mill turned on for several hours; it is then stopped and the product discharged. In a continuous mill the solid flows steadily through the revolving shell, entering at one end through a hollow trunnion and leaving at the other end through the trunnion or through peripheral openings in the shell.

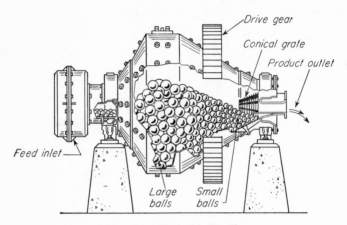

FIG. 4-15. Conical ball mill.

In all revolving mills, the grinding elements are carried up the side of the shell nearly to the top, from whence they fall on the particles underneath. The energy expended in lifting the grinding units is utilized in reducing the size of the particles. In some revolving mills, as in a *rod mill*, much of the reduction is done by rolling compression and by attrition as the rods slide downward and roll over one another. The grinding rods are usually steel, 1 to 5 in. in diameter, with several sizes present at all times in any given mill. The rods extend the full length of the mill. They are sometimes kept from twisting out of line by conical ends on the shell. Rod mills are intermediate grinders, reducing a $\frac{3}{4}$-in. feed to perhaps 10-mesh, often preparing the product from a crusher for final reduction in a ball mill. They yield a product with little oversize and a minimum of fines.

In a *ball mill* or *pebble mill* most of the reduction is done by impact as the balls or pebbles drop from near the top of the shell. In a large ball mill the shell might be 10 ft in diameter and 14 ft long. The balls are 1 to 5 in. in diameter; the pebbles in a pebble mill are 2 to 7 in. in size. A *tube mill* is a continuous mill with a long cylindrical shell, in which material is ground for two to five times as long as in the shorter ball mill.

Tube mills are excellent for grinding to very fine powders in a single pass where the amount of energy consumed is not of primary importance. Putting slotted transverse partitions in a tube mill converts it into a *compartment mill*. One compartment may contain large balls, another small balls, and a third pebbles. This segregation of the grinding media into elements of different size and weight aids considerably in avoiding wasted work, for the large, heavy balls break only the large particles, without interference by the fines. The small, light balls fall only on small particles, not on large lumps they cannot break.

Segregation of the grinding units in a single chamber is a characteristic of the *conical ball mill* illustrated in Fig. 4-15. Feed enters from the left through a 60° cone into the primary grinding zone, where the diameter of the shell is a maximum. Product leaves through the 30° cone to the right. A mill of this kind contains balls of different sizes, all of which wear and become smaller as the mill is operated. New large balls are added periodically. As the shell of such a mill rotates, the large balls move toward the point of maximum diameter, and the small balls migrate toward the discharge. The initial breaking of the feed particles, therefore, is done by the largest balls dropping the greatest distance; small particles are ground by small balls dropping a much smaller distance. The amount of energy expended is suited to the difficulty of the breaking operation, increasing the efficiency of the mill.

Action in Revolving Mills. The load of balls in a ball or tube mill should be such that, when the mill is stopped, the balls occupy somewhat more than one-half the volume of the mill. In operation, the balls follow a cyclical path. As shown in Fig. 4-16, they are picked up by the inside of the mill and carried nearly to the top, where they break contact with the wall and fall to the bottom to be picked up again and recycled. Centrifugal force keeps the balls in contact with the wall and with each other during the upward movement. While in contact with the wall, the balls do some grinding by slipping and rolling over each other, but most of the grinding occurs at the zone of impact, where the free-falling balls strike the bottom of the mill.

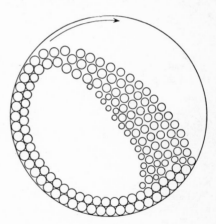

FIG. 4-16. Action in ball mill.

The faster the mill is rotated, the farther the balls are carried up the side of the mill, and the greater the power consumption. The added power is profitably utilized because the higher the balls are when they are released, the greater is the impact at the bottom, and the larger is the capacity of the mill. A typical example of the relationship between speed and power is shown in Fig. 4-17. If the balls are carried over, the mill is said to be centrifuging. The speed at which centrifuging occurs is called the critical

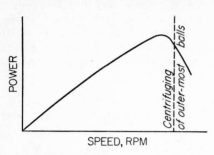

speed. Little or no grinding is done when a mill is centrifuging, and operating speeds must be less than the critical. An empirical rule for the best operating speed of a revolving mill is [3]

$$n = 57 - 40 \log D \quad (4\text{-}38)$$

where n = speed, rpm
D = inside diameter of mill, ft

SPEED, RPM

Fig. 4-17. Speed-power relationship in ball mill. (*Gaudin.*[6a])

The speed given by this equation is somewhat less than the critical.

The point at which the outermost balls lose contact with the wall of the mill depends on the balance between gravitational and centrifugal forces. This can be shown with the aid of Fig. 4-18. Consider the ball at point A on the periphery of the mill. Let the radii of the mill and of the ball be

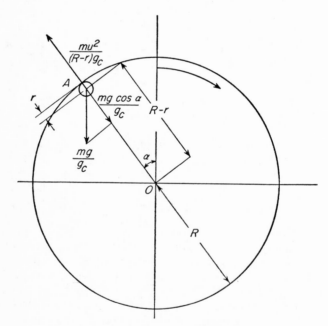

Fig. 4-18. Forces on ball in ball mill.

R and r, respectively. The center of the ball is, then, $R - r$ ft from the axis of the mill. Let the radius AO form the angle α with the vertical. Two forces act on the ball. The first is the force of gravity mg/g_c, where

m is the mass of the ball. The second is the centrifugal force $mu^2/(R - r)g_c$, where u is the peripheral speed of the center of the ball. The centripetal component of the force of gravity is (mg/g_c) cos α, and this force opposes the centrifugal force. As long as the centrifugal force exceeds the centripetal force, the particle will not break contact with the wall. As the angle α increases, however, the centripetal force increases, and unless the speed exceeds the critical, a point is reached where the opposing forces are equal and the particle is ready to fall away. The angle at which this occurs is found by equating the two forces, giving

$$m\frac{g}{g_c}\cos\alpha = \frac{mu^2}{(R - r)g_c}$$

$$\cos\alpha = \frac{u^2}{(R - r)g} \tag{4-39}$$

The speed u is related to the speed of rotation by the equation

$$u = 2\pi n(R - r) \tag{4-40}$$

and Eq. (4-39) can be written

$$\cos\alpha = \frac{4\pi^2 n^2(R - r)}{g} \tag{4-41}$$

At the critical speed, $\alpha = 0$, cos $\alpha = 1$, and n becomes the critical speed n_c. Then

$$n_c = \frac{1}{2\pi}\sqrt{\frac{g}{R - r}} \tag{4-42}$$

If the critical speed is in rpm, and R and r in feet, g is numerically 32.2×60^2, and the critical speed is given by the dimensional equation

$$n_c = 54.2\sqrt{\frac{1}{R - r}} \tag{4-43}$$

Capacity and Power Requirement of Revolving Mills. The maximum amount of energy that can be delivered to the solid being reduced may be computed from the mass of the grinding medium, the speed of rotation, and the maximum distance of fall. In an actual mill the useful energy is much smaller than this, and the total mechanical energy supplied to the mill is much greater. Energy is required to rotate the shell in its bearing supports. Much of the energy delivered to the grinding medium is wasted in overgrinding particles that are already fine enough and in lifting balls or pebbles that drop without doing much if any grinding. Good design, of course, minimizes the amount of wasted energy. Complete theoretical analysis of the many interrelated variables is virtually impossible, so the performance of revolving mills is judged by semiempirical comparisons.[8]

Rod mills yield 5 to 200 tons/hr of 10-mesh product; ball mills produce 1 to 50 tons/hr of powder of which perhaps 70 to 90 per cent would pass a 200-mesh screen. The total energy requirement for a typical rod mill grinding hard material is about 5 hp-hr/ton; for a ball mill it is about 20 hp-hr/ton.[9] Tube mills and compartment mills draw somewhat more power than this. As the product becomes finer, the capacity of a given mill diminishes, and the energy requirement increases.

Ultrafine Grinders. Many commercial powders must contain particles averaging 1 to 20 microns in size, with substantially all particles passing a standard 325-mesh screen which has openings 43 microns wide. Mills which reduce solids to such fine particles are called ultrafine grinders. Ultrafine grinding of dry powders is done by grinders, such as high-speed

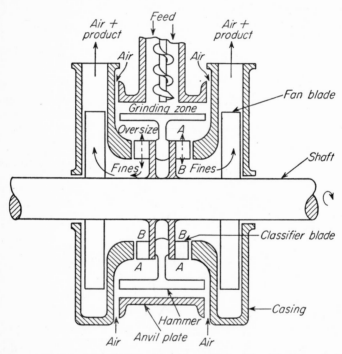

FIG. 4-19. Principle of Mikro-Atomizer. (*Pulverizing Machinery Co.*)

hammer mills, provided with internal or external classification and by fluid-energy or jet mills.†

A *hammer mill with internal classification* is the Mikro-Atomizer illustrated in Fig. 4-19. A set of swing hammers is held between two rotor disks, much as in a conventional hammer mill. In addition to the hammers the rotor shaft carries two fans, which draw air through the mill in the

† Details of the performance and characteristics of ultrafine grinders for dry solids are given by Berry.[1]

direction shown in the figure and discharge into ducts leading to collectors for the product.   On the rotor disks are short radial vanes for separating oversize particles from those of acceptable size.   In the grinding chamber the particles of solid are given a high rotational velocity.   Coarse particles are concentrated along the wall of the chamber because of centrifugal force acting on them.   The air stream carries finer particles inward from the grinding zone toward the shaft in the direction AB.   The separator vanes tend to throw particles outward in the direction BA.   Whether or not a given particle passes between the separator vanes and out to the discharge depends on which force—the drag exerted by the air or the centrifugal force exerted by the vanes—predominates.   Acceptably fine particles are carried through; particles that are too large are thrown back for further reduction in the grinding chamber.   The maximum particle size of the product is varied by changing the rotor speed or the size and number of the separator vanes. Mills of this kind reduce 1 or 2 tons/hr to an average particle size of 1 to 20 microns, with an energy requirement of about 50 hp-hr/ton.

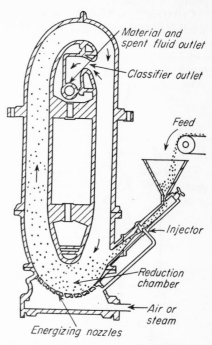

FIG. 4-20. Fluid-energy mill. (*Fluid Energy Processing and Equipment Co.*)

A typical *fluid-energy mill* is shown in Fig. 4-20.   In fluid-energy mills the solid particles are suspended in a gas stream and conveyed at high velocity in a circular or elliptical path.   Some reduction occurs when the particles strike or rub against the walls of the confining chamber, but most of the reduction is believed to be caused by interparticle attrition.   Internal classification keeps the larger particles in the mill until they are reduced to the desired size.

The suspending gas is usually compressed air or superheated steam, admitted at a pressure of 100 lb force/in.^{2} through energizing nozzles.   In the mill shown in the figure the grinding chamber is an oval loop of pipe 1 to 8 in. in diameter and 4 to 8 ft high.   Feed enters near the bottom of the loop through a venturi injector.   Classification of the ground particles takes place at the upper bend of the loop.   As the gas stream flows around this bend at high speed, the coarser particles are thrown outward against the outer wall while the fines congregate at the inner wall.   A discharge opening in the inner wall at this point leads to a cyclone separator and a bag collector for the product.   The classification is aided by the complex pattern of swirl generated in the gas stream at the bend in the loop of pipe.[1] Fluid-energy mills can accept feed particles as large as $\frac{1}{2}$ in. but are more effective when the feed particles are no larger than 100-mesh. They reduce up to 1 ton/hr of nonsticky solid to particles averaging $\frac{1}{2}$ to 10 microns in diameter, using 1 to 4 lb of steam or 6 to 9 lb of air per pound of product.

Cutting Machines. In some size-reduction problems the feed stocks are too tenacious or too resilient to be broken by compression, impact, or attrition. In other problems the feed must be reduced to particles of fixed dimensions. These requirements are met by devices which cut, chop, or tear the feed into a product with the desired characteristics. The saw-toothed crushers mentioned above do much of their work in this way. True cutting machines include rotary knife cutters and granulators. These devices find application in a variety of processes but are especially well adapted to size-reduction problems in the manufacture of rubber and plastics.

A *rotary knife cutter*, as shown in Fig. 4-21, contains a horizontal rotor turning at 200 to 900 rpm in a cylindrical chamber. On the rotor are two to twelve flying knives with edges of tempered steel or stellite passing with close clearance over one to seven stationary bed knives. Feed particles entering the chamber from above are cut several hundred times per minute and emerge at the bottom through a screen with $\frac{3}{16}$- to $\frac{5}{16}$-in. openings. Sometimes the flying knives are parallel with the bed knives; sometimes, depending on the properties of the feed, they cut at an angle. Rotary cutters and granulators are similar in design. A granulator yields more or less irregular pieces; a cutter may yield cubes, thin squares, or diamonds.

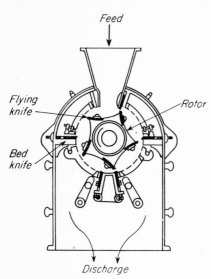

Feed

Flying knife

Rotor

Bed knife

Discharge

Fig. 4-21. Rotary knife cutter.

EQUIPMENT OPERATION

For the proper selection and economical operation of size-reduction machinery, attention must be given to many details of procedure and of auxiliary equipment. A crusher, grinder, or cutter cannot be expected to perform satisfactorily unless: (1) the feed is of suitable size and enters at a uniform rate; (2) the product is removed as soon as possible after the particles are of the desired size; (3) unbreakable material is kept out of the machine; and (4), in the reduction of low-melting or heat-sensitive products, the heat generated in the mill is removed. Heaters and coolers, metal separators, pumps and blowers, and constant-rate feeders are therefore important adjuncts to the size-reduction unit itself.

Open-circuit and Closed-circuit Operation. In many mills the feed is broken into particles of satisfactory size by passing it once through the mill. When no attempt is made to return oversize particles to the machine for further reduction, the mill is said to be operating in open circuit. This may require excessive amounts of power, for much energy is wasted in regrinding particles that are already fine enough. If a 50-mesh product is desired, it is obviously wasteful to continue grinding 100- or 200-mesh material. Thus it is often economical to remove partially ground material from the mill and pass it through a size-separation device. The undersize becomes the product and the oversize is returned to be reground. The separation device is sometimes inside the mill, as in ultrafine grinders; or, as is more common, it is outside the mill. Closed-circuit operation is the term applied to the action of a mill and separator connected so that oversize particles are returned to the mill.

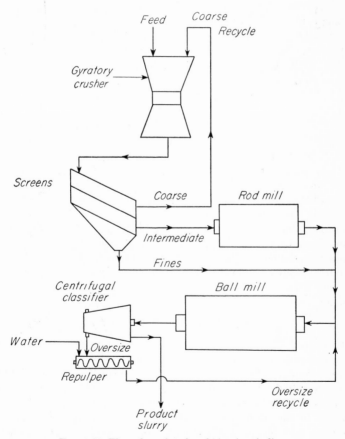

Fig. 4-22. Flow sheet for closed-circuit grinding.

For coarse particles the separation device is a screen or grizzly; for fine powders it is some form of classifier. A typical set of size-reduction machines and separators operating in closed circuit is diagramed in Fig. 4-22. The product from a gyratory crusher is screened into three fractions: fines, intermediate, and oversize. The oversize is sent back to the gyratory; the fines are fed directly to the final reduction unit, a ball mill. Intermediate particles are broken in a rod mill before they enter the ball mill. In the arrangement shown in the diagram the ball mill is grinding wet; i.e., water is pumped through the mill with the solid to carry the broken particles to a centrifugal classifier. The classifier throws down the oversize into a sludge, which is repulped with more water and returned to the mill. The undersize, or product, emerges from the classifier as a slurry containing particles of acceptable size. Although screens are simpler to operate than classifiers, they cannot economically make separations when the particles are smaller than about 150- to 200-mesh. It is the overgrinding of precisely these fine particles that results in excessive consumption of energy. Closed-circuit operation is therefore of most value in reduction to fine and ultrafine sizes, which demands that the separation be done by wet classifiers or air separators of the types described in Chap. 7. Energy must of course be supplied to drive the conveyors and separators in a closed-circuit system, but despite this, the reduction in total energy requirement over open-circuit grinding often reaches 25 per cent.

Feed Control. Of the operations auxiliary to the size reduction itself, control of the feed to the mill is the most important. The particles in the feed must be of appropriate size. Obviously they must not be so large that they cannot be broken by the mill; if too many of the particles are very fine, the effectiveness of many machines, especially intermediate crushers and grinders, is seriously reduced. With some solids precompression or chilling of the feed before it enters the mill greatly increases the ease with which it may be ground. In continuous mills the feed rate must be controlled within close limits to avoid choking or erratic variations in load and yet make full use of the capacity of the machine. In cutting sheet material into precise squares or thread into uniform lengths for flock, exact control of the feed rate is obviously essential.

Volumetric feeders, gravimetric feeders, and feeder rolls are commonly used for controlling the rate of flow of solids. Free-flowing powders and sheet stock lend themselves to such automatic control. Other solids, like wet, sticky clay, do not. The feed to a ball mill may be controlled by an "electric ear" actuated by the noise emanating from the grinding chamber. When the pulp level in the mill is too low, the falling balls strike the chamber floor and the other balls with a loud noise; when the level is too high, their fall is cushioned, and the sound they make is muffled. The controller admits just enough feed to keep the sound intensity at the proper value.

Mill Discharge. To avoid build-up in a continuous mill the rate of discharge must equal the rate of feed. Furthermore, the discharge rate must be such that the working parts of the mill can operate most effectively on the material to be reduced. In a jaw crusher, for example, particles may collect in the discharge opening and be crushed many times before they drop out. As mentioned before, this is wasteful of energy if many of the particles are crushed more than necessary. Operation of a crusher in this way is sometimes deliberate; it is known as *choke crushing*. Usually, however, the crusher is designed and operated so that the crushed particles readily drop out, perhaps carrying some oversize particles, which are separated and returned. This kind of operation is called *free-discharge crushing* or *free crushing*. Choke crushing is used only in unusual problems, for it requires large amounts of power and may lead to damaging the mill.

With fairly coarse comminuted products, as from a crusher, intermediate grinder, or cutter, the force of gravity is sufficient to give free discharge. The product usually drops out the bottom of the mill. In a revolving mill it escapes through openings in the chamber wall at one end of the cylinder ("peripheral discharge"); or it is lifted by scoops and dropped into a cone which directs it out through a hollow trunnion ("trunnion discharge"). A slotted grate or diaphragm keeps the grinding medium from leaving with the product. Figure 4-23 illustrates the two methods of dis-

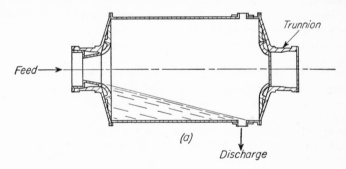

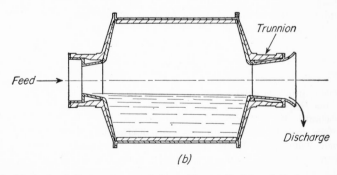

Fig. 4-23. Methods of discharging revolving mills: (*a*) Peripheral discharge. (*b*) Trunnion discharge.

charging revolving mills. Peripheral discharge is common in rod mills, trunnion discharge in ball mills and tube mills.

In discharging mills for fine and ultrafine grinding the force of gravity is replaced by the drag of a fluid carrier. The fluid may be a liquid or a a gas. Wet grinding with a liquid carrier is common in revolving mills. It gives more wear on the chamber walls and on the grinding medium than does dry grinding, but it saves energy, increases capacity, and simplifies handling and classification of the product. A sweep of air, steam, or inert gas removes the product from attrition mills, fluid-energy mills, and many hammer mills. The powder is taken out of the gas stream by cyclone separators or bag filters.

Removal or Supply of Heat. Only a very small fraction of the energy supplied to the solid in a grinding mill is used in creating new surface. The crushing efficiency η_c in Eqs. (4-23) and (4-24) is of the order of 0.1 to 2 per cent. The bulk of the energy, therefore, is converted to heat. The heat generated is sufficient, in fine and ultrafine grinding, to raise the temperature of the solid by many degrees. The solid may melt, decompose, or explode unless this heat is removed. For this reason cooling water or refrigerated brine is often circulated through coils or jackets in the mill. Sometimes the air blown through the mill is refrigerated, or solid carbon dioxide (dry ice) is admitted with the feed. Still more drastic temperature reduction is achieved with liquid nitrogen, to give grinding temperatures below $-100°F$. The purpose of such low temperatures is to alter the breaking characteristics of the solid, usually by making it more friable. In this way substances like lard and beeswax become hard enough to shatter in a hammer mill; tough plastics like polyethylene, which stall a mill at ordinary temperatures, become brittle enough to be ground without difficulty.[5]

Sometimes heat is deliberately added to the mill, usually by circulating hot air or flue gas through it. The mill can thus simultaneously grind and dry a moist solid. Particles of product lose their moisture to the hot gas during their brief passage through the discharge line to the collector. This "flash" drying is often complete within 1 to 5 sec, so that even heat-sensitive materials can be handled in this way. Part of the dried product is recirculated and mixed with the feed to make it more free-flowing and easier to admit at a controlled rate.

SYMBOLS

A Area, ft^2

A_w Specific surface of particles, ft^2/lb; A_{wa}, for feed; A_{wb}, for product

a Volume shape factor

B Constant in Eq. (4-13); B', constant, defined by Eq. (4-18)

b Surface shape factor; breadth of roll face in crushing rolls, ft

D Inside diameter of ball mill, ft

$\bar{D}$ Mean particle size, ft; $\bar{D}_1$, for fraction 1; $\bar{D}_2$, for fraction 2

$\bar{D}_n$ Arithmetic average of $\bar{D}_{pn}$ and $\bar{D}_{p(n-1)}$, ft

D_p Particle size, ft; D_{p1}, for largest particle in fraction; D_{p2}, for smallest particle in fraction

D_{pn} Mesh opening in screen n, ft
$\bar{D}_{vs}$ Mean volume-surface diameter, ft
d One-half distance between crushing rolls, ft
e_s Surface energy per unit area, ft-lb force/ft^2
F Force, lb force; F_r, radial force in rolls; F_t, tangential force in rolls; F_x, horizontal thrust in jaw crusher at distance x ft from pivot; F_1, force on toggle block of jaw crusher; F_2, horizontal thrust at toggle in jaw crusher
g Acceleration of gravity, ft/sec^2
g_c Newton's-law conversion factor, 32.174 ft-lb/lb force-sec^2
K_b Constant in Bond's law
K_r Constant in Rittinger's law
k Constant in Eq. (4-13)
L Distance between pivot and toggle block in jaw crusher, ft
m Mass, lb
N Particle population; N_w, total population per unit mass
n Speed, rpm; screen number, counting from large screen of series; n_c, critical speed in ball mill; n_T, total number of screens in series
P Power, hp
q Volumetric capacity of rolls, ft^3/hr
R Radius of rolls, ft
r Radius of feed particle, ft; radius of balls in ball mill, ft; ratio of $D_{p(n-1)}$ to D_{pn} in testing screens
s_p Area of particle, ft^2
T Feed rate to crusher, tons/min
u Peripheral speed of center of ball in ball mill, ft/sec
v_p Volume of particle, ft^3
W Energy input to crusher, ft-lb force/lb; W_n, energy absorbed by material during crushing
W_i Bond work index, kwhr/ton
x Variable distance from point of jaw crusher to pivot, ft

Greek Letters

α One-half angle of nip, also, angle between radius and vertical in ball mill
β Angle between pitman and toggle in jaw crusher
$\Delta\phi_n$ Mass fraction of total sample retained by screen n and passed by screen $n - 1$
η_c Crushing efficiency
η_m Mechanical efficiency of crusher
λ Shape factor b/a; λ_a, for feed; λ_b, for product, dimensionless
μ' Coefficient of friction
ρ_p Density of particle, lb/ft^3
Σ Operator, meaning "sum of"
ϕ Mass fraction of total sample cumulative to size D_{pn}; ϕ_1, for largest particle in fraction; ϕ_2, for smallest particle in fraction

PROBLEMS

4-1. In a mixture of particles of various sizes, a number of different average sizes may be defined. Some of these are: [4]

Arithmetic Mean Diameter. Defined by the equation

$$\bar{D}_N = \frac{\int_0^{N_w} D_p \, dN}{N_w} \tag{4-44}$$

Median Diameter. The diameter that divides the entire number of particles into two equal populations.

Volume-Surface Mean Diameter. Defined by Eq. (4-10).

Mass Mean Diameter. Defined by the equation

$$\bar{D}_w = \int_0^{1.0} D_p \, d\phi \tag{4-45}$$

Determine these various mean diameters for the sample of Example 4-1.

4-2. Trap rock is crushed in a gyratory crusher. The feed is nearly uniform 2-in. spheres. The differential screen analysis of the product is given in column 1 of Table 4-5. The power required to crush this material is 575 hp/ton. Of this 15 hp is needed

TABLE 4-5. DATA FOR PROB. 4-2

Mesh	Product	
	First grind (1)	Second grind (2)
4/6	3.1	
6/8	10.3	3.3
8/10	20.0	8.2
10/14	18.6	11.2
14/20	15.2	12.3
20/28	12.0	13.0
28/35	9.5	19.5
35/48	6.5	13.5
48/65	4.3	8.5
−65	0.5	
65/100	....	6.2
100/150	....	4.0
−150	....	0.3

to operate the empty mill. By reducing the clearance between the crushing head and the cone, the differential screen analysis of the product becomes that given in column 2 in Table 4-5. From Rittinger's law, what should the power be for the second operation?

4-3. Using the Bond method, estimate the power necessary in each of the operations in Prob. 4-2 per ton of rock.

4-4. In a certain jaw crusher it is necessary to apply a maximum force of 10 tons at the jaw at the point of the toggle block. The toggle block is 42 in. from the pivot. Angle β is 85° maximum. What is the force on the pitman when the moving jaw is closest to the fixed jaw? What is the force on the particle at a distance of 12 in. from the pivot?

4-5. What is a good operating speed, in rpm, of a ball mill 48 in. ID charged with 3-in. balls? What percentage is this speed of the critical speed?

4-6. The breadth of the rolls in Example 4-4 is 1 ft, the peripheral speed is 1,000 ft/min, and the actual capacity is 20 per cent of the theoretical. What is the actual volumetric capacity of the crusher, and what is the roll speed in rpm?

REFERENCES

1. Berry, C. E.: *Ind. Eng. Chem.*, **38**: 672 (1946).
2. Bond, F. C.: *Trans. Am. Inst. Mining. Met. Engrs.*, TP-3308B, and *Mining Eng.*, May, 1952.
3. Bond, F. C.: paper presented at Symposium on Mineral Dressing, Institution of Mining and Metallurgy, London, 1952.
4. DallaValle, J. M.: "Micromeritics," 2d ed., p. 47, Pitman Publishing Corporation, New York, 1948.
5. Foote, J. H.: *Chem. Eng. Progr.*, **49**: 68 (1953).
6. Gaudin, A. M.: "Principles of Mineral Dressing," (*a*) p. 107, (*b*) p. 126, (*c*) p. 129, (*d*) p. 132, (*e*) p. 136, McGraw-Hill Book Company, Inc., New York, 1939.
7. Kwong, J. N. S., J. T. Adams, J. F. Johnson, and E. L. Piret: *Chem. Eng. Progr.*, **45**: 508, 655, 708 (1949).
8. Piret, E. L.: *Chem. Eng. Progr.*, **49**:56 (1953)
9. Smith, J. C.: *Chem. Eng.*, **60**(8): 151 (1952).

Chapter 5

HANDLING OF SOLIDS

Solids, in general, are much more difficult to handle in processing operations than are liquids or gases. Solids exist in many forms—large, angular pieces; wide, continuous sheets; finely divided powders. They may be hot, abrasive, fragile, dusty, explosive, plastic, or sticky. The problems of storing solids, of moving them from place to place, and of feeding them at constant or controlled rates into a processing system are the subjects of this chapter.

Boxes, packages, and drums are handled as separate units or as separate lots. Bulk solids, on the other hand, are often handled as "fluids." Masses of granular particles will "flow," at least to some extent, and can be considered as fluids with somewhat unusual properties. True "fluidization" is achieved by suspending the particles in a gas or liquid; the suspensions can then be handled by modifications of the methods described in Chap. 3. Fluidization of solids and pneumatic transport are also discussed in this chapter.

PROPERTIES OF PARTICULATE MASSES

Understanding the problems of handling granular solids requires some knowledge of the properties of this kind of material. Masses of solid particles, especially when the particles are dry and not sticky, have many of the properties of a fluid. They exert pressure on the sides and walls of a container; they flow through openings or down a chute. They differ from liquids and gases in several ways, however, because the particles interlock under pressure and cannot slide over one another until the applied force reaches an appreciable magnitude. Unlike most fluids, granular solids and solid masses permanently resist distortion when subjected to at least some distorting force. When the force is large enough, failure occurs, and one layer of particles slides over another, but between the layers on each side of the failure there is appreciable friction. There is a close analogy between the flow of particulate solids and that of plastic non-Newtonian liquids.

Solid masses have the following distinctive properties:

1. The pressure is not the same in all directions. In general, a pressure applied in one direction creates some pressure in other directions, but it is

always smaller than the applied pressure. It is a minimum in the direction at right angles to the applied pressure.

2. A shear stress applied at the surface of a mass is transmitted throughout a static mass of particles unless failure occurs.

3. The density of the mass may vary, depending on the degree of packing of the grains. The density of a fluid is a unique function of temperature and pressure, as is that of each individual solid particle; but the bulk density of the mass is not. The bulk density is a minimum when the mass is "loose"; it rises to a maximum when the mass is packed by vibrating or tamping.

Depending on their flow properties, particulate solids are divided into two classes, cohesive and noncohesive. Noncohesive materials like grain, sand, and plastic chips readily flow out of a bin or silo. Cohesive solids, such as sticky, wet clay, are characterized by their reluctance to flow through openings.

Pressures in Masses of Particles. The minimum pressure in a solid mass is in the direction normal to that of the applied pressure. In a

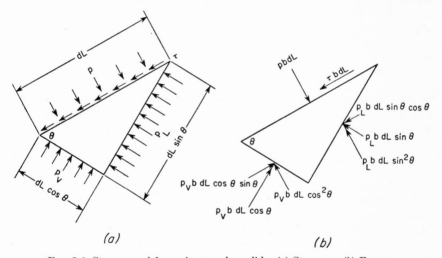

(a)   *(b)*

FIG. 5-1. Stresses and forces in granular solids: (a) Stresses. (b) Forces.

homogeneous mass the ratio of the normal pressure to the applied pressure is a constant K', which is characteristic of the material. K' depends on the shape and interlocking tendencies of the particles, on the stickiness of the grain surfaces, and on the degree of packing of the material. It is nearly independent of particle size until the grains become very small and the material is no longer free-flowing.

If the applied pressure is p_V and the normal pressure is p_L, the pressure at any intermediate angle p may be found as follows. A right-angled triangular differential section, of thickness b and hypotenuse dL, is shown in Fig. 5-1a. Pressure p_V acts on the base, p_L on the side, and p on the hypotenuse. The angle between base and hypotenuse is θ. At equilibrium the unequal pressures p_V and p_L cannot be balanced by a single pressure p; there must also be a shear stress τ. The forces resulting from these stresses are shown in Fig. 5-1b.

Equating the components of force at right angles to the hypotenuse gives

$$pb\ dL = p_L b\ dL \sin^2 \theta + p_V b\ dL \cos^2 \theta \qquad (5\text{-}1)$$

Dividing by $b\ dL$, and noting that $\sin^2 \theta = 1 - \cos^2 \theta$,

$$p = (p_V - p_L) \cos^2 \theta + p_L \qquad (5\text{-}2)$$

Similarly by equating forces parallel to the hypotenuse,

$$\tau = (p_V - p_L) \cos \theta \sin \theta \qquad (5\text{-}3)$$

When $\theta = 0°$, $p = p_V$; when $\theta = 90°$, $p = p_L$. In both these cases $\tau = 0$. When θ has an intermediate value, there is a shear stress at right angles to p. If corresponding values of p and τ are plotted for all possible values of θ, the resulting graph is a circle with a radius $(p_V - p_L)/2$ and its center on the horizontal axis at $p = (p_V + p_L)/2$. Such a graph is shown in Fig. 5-2; it is known as the *Mohr stress circle*.[16]

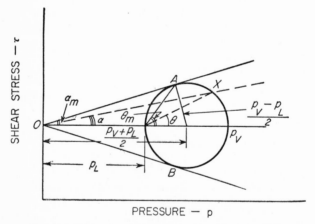

Fig. 5-2. Mohr stress diagram for noncohesive solids.

The ratio of τ to p for any value of θ is the tangent of an angle α formed by the p axis and the line OX through the origin and the point (p, τ). As θ increases from 0 to 90°, the ratio of τ to p rises to a maximum and then diminishes. It is a maximum when the line through the origin is tangent to the stress circle, as shown by the line OA in Fig. 5-2. Under these conditions α has a maximum value α_m. From Fig. 5-2 it may be seen that

$$\sin \alpha_m = \frac{(p_V - p_L)/2}{(p_V + p_L)/2} = \frac{p_V - p_L}{p_V + p_L} \qquad (5\text{-}4)$$

The lines OA and OB are tangent to all the stress circles for any value of p_V provided the material is noncohesive. They form the *Mohr rupture envelope*. With cohesive solids or solid masses the tangents forming the envelope do not pass through the origin but intercept the vertical axis at points above and below the horizontal axis.

The ratio of the normal pressure to the applied pressure p_L/p_V equals K'. Then

$$\sin \alpha_m = \frac{1 - K'}{1 + K'} \tag{5-5}$$

Also

$$K' = \frac{1 - \sin \alpha_m}{1 + \sin \alpha_m} \tag{5-6}$$

Angle of Internal Friction and Angle of Repose. The angle α_m is the "angle of internal friction" of the material. The tangent of this angle is the coefficient of friction between two layers of particles. When granular solids are piled up on a flat surface, the sides of the pile are at a definite reproducible angle with the horizontal. This angle, α_r, is the "angle of repose" of the material. Ideally, if the mass were truly homogeneous, α_r would equal α_m. In practice, the angle of repose is smaller than the angle of internal friction because the grains at the exposed surface are more loosely packed than those inside the mass and are often drier and less sticky. The angle of repose is low when the grains are smooth and rounded; it is high with very fine, angular, or sticky particles. Typical values of α_r are given in Table 5-2. K' approaches zero for a cohesive solid. For free-flowing granular materials K' is often between 0.35 and 0.6, which means that α_m is between 15 and 30°.

STORAGE OF SOLIDS

Bulk Storage. Pulverized solids like sulfur and coal are usually stored outdoors in large piles, unprotected from the weather. Where hundreds or thousands of tons of material are involved, this is the only economical method. The solids are removed from the pile by dragline or tractor shovel and are delivered to process or to a conveyor. Inventorying is done by estimating the volume of the pile through aerial or ground surveys and multiplying the volume by the bulk density of the material. A pile of finely divided solids is subject to erosion by wind and rain, so that this method of storage is ordinarily restricted to coarse water-insoluble materials. Substances like rock salt, however, are sometimes stored out of doors, usually in a shallow basin, which collects rain water and its load of dissolved material. Water may be deliberately circulated through and under such a pile of salt to form brine which is pumped to processing operations.

Bin Storage. Solids that are too valuable or too soluble to expose in outdoor piles are stored in bins, hoppers, or silos. These are cylindrical or rectangular vessels of concrete or metal. A silo is tall and relatively small in diameter; a bin is not so tall and usually fairly wide. A hopper is a small bin with a sloping bottom, for temporary storage before feeding solids to a process. All these containers are loaded from the top by some kind of elevator; discharging is ordinarily from the bottom. As discussed later, a major problem in bin design is to provide satisfactory discharge.

Capacity and Angle of Repose. When solids are dropped from a single point into a bin or hopper, a cone of solids is formed with its apex directly under the inlet. As the bin fills up, the surface of the solids retains its conical shape. The capacity of the bin is thus limited by the angle of repose of the material and the position of the inlet, because the apex of the cone can rise no higher than the bottom of the inlet pipe. Once the apex of the cone reaches the inlet, flow stops. As shown in Fig. 5-3, this means that the inlet should be above the center of the bin to give the maximum storage capacity. Offsetting the inlet, as in Fig. 5-3b, means that much of the capacity is not utilized. In wide bins several inlets should be provided, especially if the material has a large angle of repose.

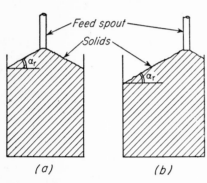

FIG. 5-3. Capacity of storage bins affected by location of feed spout: (*a*) Spout on center line. (*b*) Spout offset. (α_r, angle of repose.)

Pressures in Bins, Hoppers, and Silos. When granular solids are stored in a bin or hopper, the lateral pressure exerted on the walls at any point is less than predicted from the head of material above that point. Furthermore there usually is friction between the wall and the solid grains, and because of the interlocking of the particles, the effect of this friction is felt throughout the mass. The frictional force at the wall tends to offset the weight of the solid and reduces the pressure exerted by the mass on the floor of the container. In the extreme case this force causes the mass to arch or bridge so that it does not fall even when the material below is removed.

An expression for the pressure exerted by a granular solid on the floor of a circular bin with vertical walls is derived as follows. Figure 5-4 shows a differential horizontal layer of thickness dZ at a distance Z from the upper surface of the solids. The inside radius of the bin is R ft; the total height of solids is Z_T ft. At the level Z, assume that the differential layer is a piston pressing against the solid beneath and that this piston is acted upon by a concentrated vertical force F_V from above. The vertical pressure p_V at level Z therefore is

$$p_V = \frac{F_V}{\pi R^2} \tag{5-7}$$

From this

$$dF_V = \pi R^2 \, dp_V \tag{5-8}$$

The net increase in downward force caused by the differential layer is the force of gravity dF_g minus the frictional force dF_f. Thus

$$dF_V = dF_g - dF_f \tag{5-9}$$

The force of gravity on the layer is $(g/g_c)\pi\rho_b R^2\,dZ$, where ρ_b is the bulk density of the material. The frictional force is the product of the coeffi-

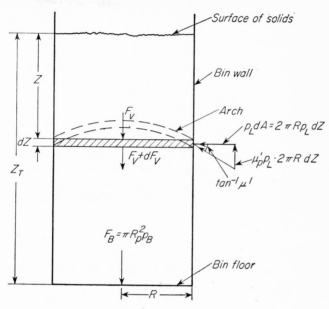

FIG. 5-4. Forces in circular bin with vertical sides.

cient of friction μ' at the bin wall and the lateral force F_L. The lateral force is, in turn, the product of the lateral pressure p_L and the area on which it acts, $2\pi R\,dZ$. Then

$$dF_V = \pi R^2\,dp_V = \pi R^2 \rho_b \frac{g}{g_c}\,dZ - \mu'(2\pi R p_L\,dZ) \tag{5-10}$$

Dividing through by πR, and noting that $p_L/p_V = K'$,

$$R\,dp_V = \left(R\rho_b\frac{g}{g_c} - 2\mu'\frac{p_L}{p_V}p_V\right)dZ$$

$$= \left(R\rho_b\frac{g}{g_c} - 2\mu'K'p_V\right)dZ \tag{5-11}$$

Let p_B equal the vertical pressure on the bin floor. Integrating Eq. (5-11) from top to bottom of the mass,

$$\int_0^{Z_T} dZ = \int_0^{p_B} \frac{R\,dp_V}{R\rho_b(g/g_c) - 2\mu'K'p_V}$$

$$Z_T = -\frac{R}{2\mu'K'}\left[\ln\left(R\rho_b\frac{g}{g_c} - 2\mu'K'p_V\right)\right]_0^{p_B} \tag{5-12}$$

Substituting the limits of integration and rearranging,

$$p_B = \frac{R\rho_b(g/g_c)}{2\mu'K'}\left(1 - e^{-2\mu'K'Z_T/R}\right) \tag{5-13}$$

Equation (5-13) is the Jannssen equation, which has been well substantiated by experiment.[2,4,13] A typical relation of base pressure to height is shown in Fig. 5-5. With many solids, when the height reaches about three

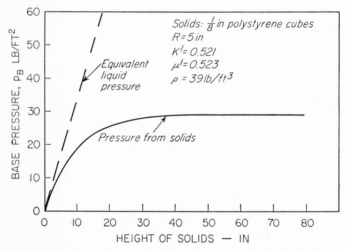

FIG. 5-5. Base pressure in circular bins. (*Rudd.*[13])

times the diameter of the bin, additional material has virtually no effect on the pressure at the base.

When the bin is other than circular in cross section, R is replaced by two times the hydraulic radius. The coefficient of friction is found experimentally by noting the angle at which the solids just begin to slide on an inclined surface. The coefficient μ' is the tangent of this angle. For granular materials on concrete or smooth metal surfaces, μ' ranges from 0.35 to 0.55.[4] When the sides of the bin or hopper are not vertical but slope at an angle less than about 5°, the pressure at the base is less than when the sides are vertical. When the angle is larger than about 10°, the base pressure is greater than with vertical sides.[13]

Example 5-1. A packed absorption tower 6 ft in diameter and 50 ft high is to be filled with crushed sized coke. Compute the vertical and lateral pressures at the base caused by the coke. Compare with the pressure that would be exerted by a liquid of the same density.

Solution. The bulk density (Table 5-2) is

$$\rho_b = 30 \text{ lb/ft}^3 \qquad \alpha_r \text{ (Table 5-2)} = 28°$$

$$\alpha_m \text{ (est.)} = 32° \qquad \sin \alpha_m = 0.5299$$

$$K' \text{ (Eq. 5-6)} = \frac{1 - 0.5299}{1 + 0.5299} = 0.307$$

$$R = 3 \text{ ft} \qquad Z_T = 50 \text{ ft} \qquad \mu' \text{ (est.)} = 0.5$$

The vertical pressure at the base (from Eq. 5-13) is

$$p_B = \frac{3 \times 30 \, (g/g_c)}{2 \times 0.5 \times 0.307} \, (1 - e^{-(2\times0.5\times0.307\times50)/3}) = 291 \text{ lb force/ft.}''$$

$$= 2.02 \text{ lb force/in.}^2$$

The lateral pressure at the base is

$$p_L = K'p_B = 0.307 \times 2.02 = 0.62 \text{ lb force/in.}^2$$

The equivalent liquid pressure is

$$p = \frac{\rho Z_t}{144} \frac{g}{g_c} = \frac{30 \times 50}{144} = 10.4 \text{ lb force/in.}^2$$

In estimating K' it is best to assume that α_m is fairly large, since this leads to conservative values of p_B.

Flow Out of Bins. Solids tend to flow out of any opening near the bottom of a bin, but are best discharged through an opening in the floor. As shown in Example 5-1, the pressure at a side outlet is smaller than the vertical pressure at the same level, so that the opening clogs more easily; furthermore, removal of solids from one side of a bin considerably increases the lateral pressure on the other side during the time the solids are flowing. A bottom outlet is less likely to clog and does not induce abnormally high pressures on the wall at any point.

Except in very small bins it is not feasible to open the entire bottom for discharge. Commonly a conical or pyramidal bottom leads to a fairly small circular outlet closed with a valve or to a rotary feeder of the type described later. The pressure at the bottom of the slope-sided section of the bin is appreciably less than that indicated by Eq. (5-13). In addition, the vertical pressure fluctuates as the material discharges and averages 5 to 10 per cent higher than when the mass is stationary.[4]

When the outlet at the bottom of a bin containing free-flowing solids is opened, the material immediately above the opening begins to flow. A central column of solids moves downward without disturbing the material at the sides. Eventually lateral flow begins, first from the topmost layer of solids. A conical depression is formed in the surface of the mass. The solids at the bin floor, at or near the walls, are the last to leave. The material slides laterally into the central column at an angle approximating the angle of internal friction of the solids. If additional material is added to the top of the bin at the same rate as material is flowing out the bottom, the solids near the bin walls remain stagnant and do not discharge no matter how long flow persists.

With cohesive solids it is often hard to start flow. Once flow does start, however, it again begins in the material directly above the discharge opening. Frequently the column of solids above the outlet moves out as a plug, leaving a "rat hole" with nearly vertical sides. Sticky solids and even some dry powders adhere strongly to vertical surfaces and have enough shear strength to support a plug of considerable diameter above an open discharge. Thus to get flow started and to keep the material moving, hammers and vibrators on the bin walls, internal plows near the bin floor, or jets of air in the discharge opening are often needed. A "poke hole" in the bin wall through which a rod may be inserted is sometimes used to initiate flow.

The discharge opening should be small enough to be readily closed when solids are flowing yet not so small that it will clog. It is best to make the opening large enough to pass the full desired flow when half open. It may then be opened further to clear a partial choke. If the opening is too large, however, the shutoff valve may be hard to close, and control of the flow rate will be poor.

Arching. The derivation of Eq. (5-13) is based on the assumption that at a bin wall there is a vertical force acting on the material even when the solid mass is stationary. The resultant of the vertical force and the lateral force is at an angle $\tan^{-1} \mu'$ with the horizontal, as shown in Fig. 5-4. This creates an arch, or dome, inside the bin, which in a loose granular solid rests on the material underneath it. In a cohesive solid the arch may be strong enough to support the overlying solid when the discharge is opened. During flow through the discharge the arches collapse and re-form and on occasion may interrupt the flow. Vertical sides encourage arching with cohesive solids.[5] Agitating the material while it is flowing keeps the arch from forming. Agitators or vibrators, however, should not be left in operation when the bin outlet is closed, for they pack the solid and encourage arching.

The sloping sides at the bottom of a bin or hopper should be at an angle a little greater than the angle of repose of the material. They are then self-cleaning, so that the bin may be completely emptied. Steeper slopes do not help and with cohesive solids may lead to arching. Hoppers should be drained completely fairly often, especially with solids that cake in storage or which vary from time to time in physical appearance.

Storage of Solid Units and Palletizing. Bags, drums, and other large solid units are stored, indoors or out, on floors of sufficient strength to support their weight. With most materials, therefore, storage is at ground level. Containers of raw materials or of product must be stored so as to simplify inventorying and withdrawal of material from storage. When many products are stored in one room, this is not a simple problem. In all storerooms adequate handling facilities should be provided; access aisles should be wide; lighting should be sufficiently bright.

Barrels, drums, and bags may be stacked by hand if they are light or by overhead cranes if they are heavy. Both these methods, however, involve repeated handling of single units. It is more efficient to handle several units at a time, in effect combining several smaller units into a large unit. This is done by placing several drums or bags on a small movable platform known as a *pallet* and moving the pallet and its load with a lift truck. Pallets are typically 4 to 6 ft square and about 6 in. thick. They may carry 4 large drums, 9 small drums, or up to 30 bags. One pallet load may be stacked on top of another load. The pallets increase the stability of the stacks, so the stacks may be higher than when pallets are not used. This leads to more effective use of storage area; it also means that more careful attention must be paid to the strength of the foundation and the floor.

CONVEYORS AND ELEVATORS

Bulk solids are moved from point to point in a plant by some form of *conveyor*, which carries the material or drags it along through a trough or conduit. Granular solids are sometimes suspended in a stream of liquid

TABLE 5-1. CONVEYING AND ELEVATING DEVICES

I. Conveyors that carry
    A. On upper surface or in pans
        1. Belt conveyors
        2. Slat, apron, and pan conveyors
        3. Pivoted-bucket conveyors
        4. Vibrating conveyors
        5. Gravity roller conveyors
    B. Inside closed tube
        1. Zipper conveyors
    C. By suspension from above
        1. Chain conveyors
        2. Overhead monorails
II. Conveyors that drag or push
    A. Drag and flight conveyors
    B. "En masse" conveyors
    C. Screw and ribbon conveyors
III. Conveyors depending on fluidization
    A. "Boiling-bed" type
        1. Airslide
    B. Pneumatic conveyors
        1. Vacuum system
        2. Pressure system
IV. Elevators
    A. Bucket elevators
        1. Centrifugal-discharge
        2. Positive-discharge
        3. Continuous-bucket type
    B. Pivoted-bucket conveyor-elevator
    C. Zipper conveyor-elevator
    D. Screw elevator
    E. "En masse" elevator
    F. Pneumatic conveyor-elevator

or gas, and moved as a "fluid." Conveyors that lift solids vertically are known as *elevators*. Large solid units are also handled on some kinds of conveyors but more commonly are moved by lift trucks or tractor-trailer combinations.

Equipment for the transportation of solids may be classified as shown in Table 5-1. This table includes devices which are driven mechanically, like belt conveyors, and devices which depend on gravity to move the solids, like gravity roller conveyors or Airslides. Units suspended from a monorail may be moved mechanically or pushed from place to place by the operator. Most of the conveyors listed can raise material to some extent as they move it along, but the term "elevator" is restricted to devices which lift solids vertically.

Conveyors: Carrying Type. Conveyors in industrial plants are nearly always permanent installations designed to move solids continuously or intermittently from one point to another. Some conveyors carry the solid; some drag or push it along. Single units like boxes, machine parts, and so forth are carried; bulk solids are either carried or dragged. Most carrying conveyors carry the load on their upper surface, although on some the load is suspended or is transported inside a closed tube.

Belt Conveyors. A belt conveyor, illustrated in Fig. 5-6, is the most common conveying device. An endless belt of rubber or canvas passes

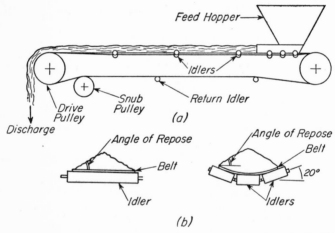

FIG. 5-6. Belt conveyor: (*a*) Conveyor layout. (*b*) Flat and troughed belts.

around two large pulleys, one of which is driven. Solids are dropped on the upper surface of the belt near the nondriven pulley and are carried to the other end of the belt and discharged. The driven pulley is at the discharge, or "head," end of the conveyor. The loaded belt is carried by a series of closely spaced idlers; the returning part of a long belt may be supported by widely spaced idlers. The belt is driven by a single pulley with small belts or light loads or by two pulleys in tandem with large belts carrying heavy loads. Floating pulleys, weighted or spring-mounted, automatically adjust for stretching of the belt. Conveyor belts range in size from a few inches wide and a foot or two long, in automatic packaging machines, to 60 in. wide and several miles long. They travel at speeds between 100 and 600 ft/min.

Flat belts transport boxes, bags, and other solid units. They are unloaded by allowing the solids to drop off over the head pulley, by plows set

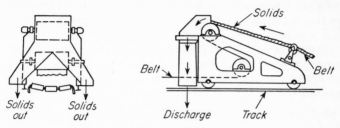

Fɪɢ. 5-7. Tripper for belt conveyor.

at an angle across the belt at an intermediate point, or by lifting off the units mechanically or by hand. Narrow flat belts are also used for small quantities of bulk solids. Belts wider than 14 in. are nearly always troughed, as shown in Fig. 5-6b, when handling bulk solids. A flat belt has limited capacity, especially when carrying solids with a low angle of repose. Troughing nearly doubles the carrying capacity of a belt. Troughed belts are discharged over the head pulley or at intermediate points by a "tripper," as shown in Fig. 5-7. The tripper contains two pulleys. As the

Tᴀʙʟᴇ 5-2. Pʀᴏᴘᴇʀᴛɪᴇs ᴏꜰ Tʏᴘɪᴄᴀʟ Mᴀᴛᴇʀɪᴀʟs Hᴀɴᴅʟᴇᴅ ʙʏ Tʀᴏᴜɢʜᴇᴅ Bᴇʟᴛ Cᴏɴᴠᴇʏᴏʀs †

Material	Density ρ_b, lb/ft^3	Angle of repose α_r, °	Max operating angle, °
Aluminum hydrate (ground)........	13.5	34	24
Aluminum sulfate (granular)........	54	32	15
Coke (crushed and sized)..........	30	28	13
Gypsum (ground).................	56	40	27
Iron ore (limonite)................	237	40	28
Lead sulfate (basic, pulverized)......	184	45	32
Phthalic anhydride (flaky)..........	42	24	10
Portland cement..................	95	39	28
Rock salt (crushed)...............	75	25	11
Soap chips.......................	10	30	18
Sodium bicarbonate...............	43	42	27
Wood chips......................	22	36	25

† Abstracted from H. L. Strube, *Chem. Eng.*, **61**(4): 203 (1954).

belt passes over the upper pulley, the solids drop off into a chute. Trippers are a permanent part of the conveyor system and are rarely justified economically except with large belts. They ride on rails parallel to the belt so that the discharge point can be moved. When trippers are used, the belt speed should be at least 300 ft/min so that the solids will discharge cleanly.

The angle of repose of the solids limits not only the capacity of the belt but also the maximum angle at which the belt may be inclined. This maximum safe angle is considerably less than the angle of repose and is usually between 10 and 30°. Table 5-2 lists typical values for representative granular solids.[15] Inclining a belt reduces its carrying capacity by 5 to 10 per cent. The capacities and speeds of belt conveyors are tabulated in Appendix 19.

Example 5-2. Crushed rock salt is to be conveyed at the rate of 50 tons/hr a horizontal distance of 600 ft and raised a vertical distance of 52 ft. An inclined belt conveyor has been proposed for this service. Determine whether one can be used; if so, specify the belt width required.

Solution. The angle of inclination is $\tan^{-1} (52/600) = 5°$. The maximum safe angle, from Table 5-2, is 11°. An inclined belt can therefore be used.

The bulk density of crushed rock salt, from Table 5-2, is 75 lb/ft^3. A horizontal troughed belt 16 in. wide traveling at 100 ft/min will carry 31 tons/hr (Appendix 19). Since the belt is inclined, its capacity would be reduced by 10 per cent. The speed required to carry 50 tons/hr is

$$\frac{50 \times 100}{31 \times 0.90} = 179 \text{ ft/min}$$

This is a little less than the normal operating speed of 200 ft/min. Therefore, a 16-in. belt operating at 200 ft/min would be satisfactory.

Pan, Apron, and Bucket Conveyors. For handling very hot solids or other difficult materials the belt of a belt conveyor may be replaced with metal pans or buckets. Dished pans fixed to an endless belt passing over large-diameter pulleys carry wet or sloppy solids, as in meat-rendering plants; pans and buckets held on double-strand steel roller chain convey hot solids and large, heavy lumps that would damage a belt. Pans that overlap give a continuous flat-bottomed moving trough known as an *apron conveyor.* Separate buckets may be suspended at the articulations of a double-strand chain in such a way that they always remain upright, regardless of the direction of travel of the chain. The buckets are emptied through inverting them by a movable tripper. Such a device, which can move solids horizontally, at any angle, or vertically, is known as a *pivoted-bucket carrier.*

In a somewhat similar design the buckets are not free to pivot. They elevate solids vertically in the same way as in a pivoted-bucket carrier but move material horizontally by dragging it over a stationary smooth flat surface. The buckets are loaded from a chute as they start a vertical run. They are emptied through an opening in the floor of the horizontal surface, for they are inclined steeply enough when traveling horizontally to empty completely as they pass over the opening. This kind of conveyor, operating partly by carrying and partly by dragging, is called a *gravity-discharge conveyor-elevator*.

Vibrating Conveyors. These conveyors were developed to move solids that must be completely enclosed or kept in an inert atmosphere. Their principle of operation is shown in Fig. 5-8. Solid particles rest on the floor of a trough beneath which is a vibrator, either mechanical or electrical. On the forward stroke of the vibrator the particles are impelled upward and somewhat forward; during their flight the trough moves backward.

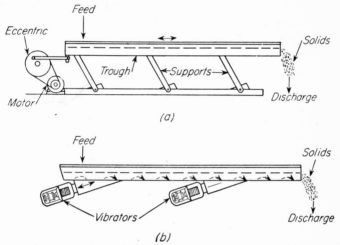

Fig. 5-8. Principles of vibrating conveyors: (*a*) Mechanical. (*b*) Electrical.

Particles therefore move along in a series of short hops. Electrical vibrators are set to work against a spring on the forward stroke and to be reinforced by the spring on the return, giving a relatively slow stroke forward and a very rapid one back. Mechanical vibrators are "tuned" to the natural frequency of vibration of the system to give considerable savings in power. A single vibrator is sufficient for very short runs; long conveyors require vibrators about every 10 ft. The trough may be open at the top, but usually it is completely closed, with flexible ducts attached to the inlet and discharge.

Gravity Flow through Chutes. Granular materials will flow down ducts, or chutes, which are inclined at an angle greater than the angle of repose of the solid. The rate of flow depends on the angle of inclination and on the amount of friction between the solid and the walls and floor of the chute. If the solid is likely to vary in stickiness or in moisture content, the angle of inclination should be considerably greater than the angle of repose of the dry solid. Steep angles, however, may lead to high velocities and excessive breakage of the particles. To retard the flow, transverse bars may be set in the floor of the chute, or the chute may be made to spiral around a central column.

Gravity Roller Conveyors. Boxes and other large solid units may "flow" by gravity from place to place on a series of rollers set in a stationary inclined track. Switches and guide bars are used to direct or retard the flow. In horizontal runs roller conveyors make the manual movement of heavy drums or other items much easier. They are especially useful where successive containers are to be weighed, for part of the track may be mounted on a scale platform.

Chain Conveyors and Monorails. Large pieces, such as metal sheets or plastic parts, which are to go through a number of processing operations are often suspended from an overhead chain. The chain carries the material through the various pieces of equipment, such as furnaces, coolers, spray booths, dip baths, and the like, without attention. Loading and unloading are ordinarily done by hand. Units that go through several successive operations but which must remain in each piece of equipment for a long time may be suspended from a monorail and pushed along by the operator. For example, the tray in a tray dryer may be carried on a suspended rack. The trays are loaded with moist solid at one point, perhaps at a filter press, and placed in the rack. The rack is pushed into the dryer, where it remains for 24 hr or so. It is then moved along the monorail to the discharge booth, where the trays of dry solid are dumped.

Zipper Conveyor. A zipper conveyor carries solids in a closed tube. A flat belt 4 in. wide carries flexible "wings" at its edges. The wings fold

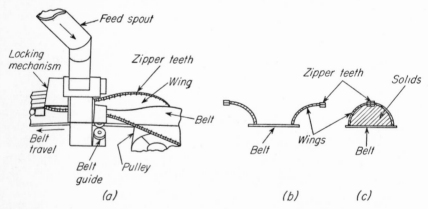

Fig. 5-9. Zipper conveyor: (*a*) Feeding and closing device. (*b*) Empty conveyor. (*c*) Loaded conveyor.

over and meet in the center to form a semicircular enclosure. The upper edges of the wings are fitted with large rubber "zipper" teeth, which interlock when closed. Solids are fed into the opened tube as shown in Fig. 5-9, after which the wings are "zipped" together. They are reopened at the discharge. A zipper conveyor can handle nearly any solid containing lumps no larger than $1\frac{1}{2}$ in. in diameter without breaking the particles or contaminating the material.

Conveyors: Drag or Push Type. Less power is needed to carry solids than to move them over a stationary surface, but conveyors which drag or push have advantages over carrying conveyors for some problems. Hot abrasive solids like clinker, ashes, or sintered ores and flaky materials like wood chips may be moved by a *flight conveyor* or a *drag conveyor*. In both of these a heavy-duty chain is pulled through a stationary trough. In a drag conveyor the chain is drawn along the trough floor, and the solids are pulled along by the chain links. In a flight conveyor the chain is above the floor and carries transverse plates, or flights, which scrape the walls and bottom of the trough. The edges of the flights may be thickened and made of hardened metal to minimize abrasion damage or may even be fitted with rollers. Flight and drag conveyors can operate at angles up to about 45°. They are less subject to damage than carrying conveyors and therefore solve some of the most difficult conveying problems. The increase in power required is more than offset by the reduction in maintenance.

"En Masse" Conveyors. These are modified flight conveyors with the chain and flights operating in a closed trough or tube. The solids completely fill the container and are moved as a unit without any relative motion between the particles. Hence the name "en masse." Some designs utilize solid flights set at frequent intervals on the chain. In the *skeleton-flight*, or *Redler, conveyor* shown in Fig. 5-10 the flight covers only a small part of the cross-sectional area of the trough, yet with many granular solids these flights move the

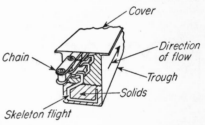

FIG. 5-10. Redler "en masse" conveyor.

mass as effectively as solid flights, with considerable saving in dead weight. The reason is that the solid particles interlock and cannot move relative to each other. Also the lateral pressure on the walls of the trough is much less than the pressure applied by the flight. A liquid would flow through the openings in the flight; the solid, if it moves at all, must move en masse.

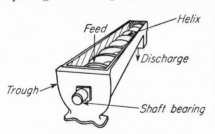

FIG. 5-11. Screw conveyor.

Redler conveyors are made with troughs as large as 24 in. wide and 18 in. deep. They handle solids in a closed system and, because they run full, can move several times as much as a belt conveyor of the same width. They can raise material at a steep angle or even vertically.

Screw Conveyors. In a screw conveyor solids are moved through a trough or tube by a metal helix, as shown in Fig. 5-11. As the screw turns, the particles are pushed along by the leading face of the helix at a level such that the lifting force is just balanced by the component of the force of gravity. Friction between the particles and the helix permits the particles to be lifted to this level; intergranular friction keeps the particles from flowing back, as would a liquid.

Screw conveyors have a number of desirable features:

1. They can be either closed or open.
2. They operate horizontally, at an angle, or vertically.
3. They can be loaded and discharged at several points.
4. They can convey both ways from a central loading point.
5. They can heat, cool, mix, or dry the solid as it passes through.

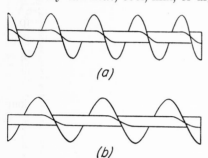

(a)

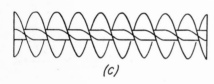

(b)

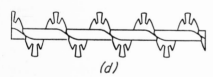

(c)

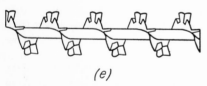

(d)

(e)

FIG. 5-12. Various designs of helix for screw conveyors: (*a*) Standard pitch, for inclines up to 20°. (*b*) Long pitch, for free-flowing material. (*c*) Double pitch, for even-flow feeders. (*d*) Cut, for mixing and retarding material. (*e*) Cut and folded, for mixing and retarding material. (*Strube*.[15])

6. They have no return leg and occupy little space.

Capacities of typical screw conveyors are listed in Appendix 20.

Ordinary screw conveyors range in diameter from 3 to 20 in. Figure 5-12 shows various designs of helix. The standard-pitch helix has a pitch equal to its maximum diameter, and is most common; but double-pitch, reducing-pitch, and other special helices are also available. A sectional-flight helix is made by welding sections of metal strip to a central pipe or shaft. In the helicoid type the metal strip is continuous. A helicoid helix is stronger than a sectional-flight helix and provides no crevices where material might lodge; it is also more expensive and harder to repair. For mixing purposes the flights of the helix may be cut, or inclined paddles may be used, or both. For moving gummy solids the center of the helix may be left out to give a *ribbon conveyor*.

Standard screw-conveyor flights are 8 to 12 ft long, with several sections joined to give the desired length. The stress developed in the shaft limits the length of a screw conveyor with a single drive to about 100 ft.

Elevators. Most conveyors can lift material at some angle of inclination, but not all of them can lift vertically. Those that can are indicated in Table 5-1. Most of these are no different when used as elevators than when conveying horizontally. A screw elevator is a little different from a screw conveyor. In an elevator a polished helicoid helix turns with close clearance inside a circular pipe which is filled with the solid. The maximum lift by a screw elevator is about 40 ft. Screw elevators have no return leg and are especially valuable where space is at a premium.

Bucket Elevators. A bucket elevator is used only for lifting solids vertically or at a very steep angle. The three main types are centrifugal-discharge, or spaced-bucket, elevators; continuous-bucket elevators; and positive-discharge elevators. These are shown in Fig. 5-13. In all types the buckets are suspended on a single-strand or double-strand chain which passes over upper and lower sprockets. In a centrifugal-discharge elevator the buckets scoop up their load as they pass through the bottom of the casing, or "boot." At the top, because of their high speed, they throw the solids outward into the discharge. This may lead to crystal breakage or erosion of the discharge chute and make it desirable to use the slower-speed continuous-bucket elevator. Here the buckets are more closely spaced and are fed from an inlet duct instead of from the boot. At the discharge each bucket as it turns over drops its contents on the bucket below, the sloping bottom of which guides the solids into the discharge.

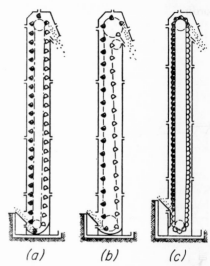

(a)   (b)   (c)

Fig. 5-13. Bucket elevators: (a) Spaced-bucket elevator. (b) Positive-discharge elevator. (c) Continuous-bucket elevator.

With cohesive solids the buckets in the usual elevators may not empty properly. In a positive-discharge elevator the chain passes over the upper sprocket and then past two snub sprockets, so that the widely spaced buckets are turned through more than 180°. Bucket speeds are lower than in a centrifugal-discharge elevator, so that the capacity of a positive-discharge elevator is small.

Centrifugal-discharge elevators are adequate for most granular solids. Buckets are typically 8 by 5 to 18 by 8 in. in size. Capacities range from 10 to 150 tons/hr.

SOLIDS FEEDERS

Solids are usually taken from storage in order to feed them into a processing system. Admission to the process may be continuous or intermittent; the control of the quantity fed may be very rough or highly precise. Feeding problems are aggravated by the uncooperative nature of most solids. Some solids are free-flowing, cool, and dry, but many of them are sticky, corrosive, erosive, hot, plastic, or pasty. Some, like portland cement, aerate, or "flood," when agitated, and leak with ease through ordinary sealing devices. Thus many types of solids feeders are needed to meet the varied problems.[11] Only a few of them, however, are described here.

Volumetric Feeders. The two main classes of solids feeders are volumetric, which deliver·controlled volumes, and gravimetric, which deliver

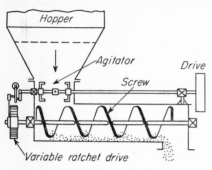

FIG. 5-14. Screw feeder.

by weight. Many volumetric feeders differ only in details from the conveyors already described. Belt conveyors, apron conveyors, vibrating conveyors, and screw conveyors deliver material at a steady rate when loaded uniformly from a bin or hopper. A screw-conveyor feeder is shown in Fig. 5-14. With uniform free-flowing granular solids this method is highly satisfactory. With solids that vary in bulk density or tend to "hang up" in the hopper, however, the rate of flow through the opening is far from constant, and a more positive feeding device, such as a star feeder, must be used.

A rotary-pocket, or "star," feeder is shown in Fig. 5-15. It resembles some of the positive-displacement rotary pumps described in Chap. 3. A rotor with peripheral pockets turns at constant speed between two hinged sectors which are pressed against the rotor by levers and springs. Solids from the hopper above the feeders flow into the pockets, are trapped, carried around to the bottom, and dropped out. Cast-iron balls inside the hollow rotor knock the solids out of the rotor pockets. Should a large hard lump enter the feeder, the hinged sector can swing out of the way to let it pass. The lower edge of each sector is cut at an angle so that discharge is nearly continuous. The delivery rate is controlled by the rotor speed.

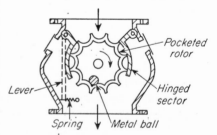

FIG. 5-15. Rotary-pocket (star) feeder.

Other similar feeders include vaned rotors of various designs, and drums with flexible rubber pockets actuated by a crank.[12b] The volume of the rubber pockets is a maximum at the inlet and diminishes almost to zero at the outlet. All star feeders, because of the seal between the rotor and the sectors, can discharge solids into a small vacuum or against gauge pressures up to about 15 lb force/in.[2] They can also handle solids which aerate.

Gravimetric Feeders: Weigh Belts. Although volumetric feeders sometimes give surprisingly constant rates of feed, they are rarely satisfactory where the weight rate of flow must be held constant within close limits. The discharge rate from a volumetric feeder depends on the bulk density of the material and its angle of repose, both of which may vary. The same solid, when ground to different degrees of fineness, when aerated or packed in different ways, or when containing different amounts of moisture, flows at widely different rates through a given opening. For precise control of the weight rate of flow the feeder must be gravimetric, as in the weigh belt shown in Fig. 5-16. This device contains a belt conveyor driven

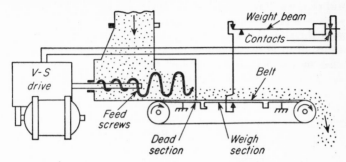

FIG. 5-16. Weigh belt with screw feeder.

at a constant speed. Part or all of the belt is connected to a scale beam so that the load is continuously weighed. If the loading per foot of belt and the belt speed are kept constant, the discharge rate from the belt must be constant. The weight of the belt is therefore made to control the rate at which solids are loaded onto the belt.

In the feeder illustrated the "live" central section of the belt is connected to one end of a weigh beam, the other end of which forms part of an electrical switch. The belt is loaded by a screw feeder driven by a motor and a variable-speed drive. Variations in the weight of the belt section cause the screw feeder to speed up or slow down to hold the belt loading very nearly constant. In operation the belt loading continually cycles above and below the control point, so that the discharge rate is not absolutely steady. In any 5-min period, however, these feeders deliver the same weight of solids to within 1 per cent or less.

Feeders for Semisolids. Materials with a consistency between that of liquid and that of a solid are called "semisolids." Thick suspensions such as gels, stiff foams, and wet filter cakes are examples. Semisolids are often sticky, plastic, and far from ideal in their rheological properties. In many ways they are the most difficult materials to handle in conveying and feeding machinery.

Plastic semisolids can sometimes be moved by screw and ribbon feeders; surprisingly viscous materials, if not abrasive, can be pumped by the spur-gear pumps described in Chap. 3. Through extrusion or pelleting, some of the stiffer semisolids, such as filter cakes, are often made easier to handle. In general the best way to handle a semisolid is to change its consistency, if at all possible. Semisolids may be heated or dried until they lose their stickiness, or they may be mixed with some dry material.

Going in the other direction, it may be possible to add a dispersing agent or to dilute them with a thin liquid to give a pumpable slurry or suspension. Lastly, a semisolid may be melted and pumped as a thin liquid, a technique that is applicable to nearly any low-melting solid.

FLUIDIZATION AND PNEUMATIC TRANSPORT

A liquid or a gas flowing at low velocities through a porous bed of solid particles, as in a packed tower, does not cause the particles to move. The fluid passes through the small, tortuous channels, losing pressure energy. The pressure drop in a stationary packed bed is large; it depends among other things on the porosity of the bed ϵ and on the superficial velocity of the fluid. The relationship is given by the Kozeny-Carman equation in Eq. (2-90) and Fig. 2-34. At higher fluid velocities, however, the particles no longer remain stationary but "fluidize" under the action of the liquid or gas.

Mechanism of Fluidization. Visualize a short, vertical tube partly filled with a granular material such as fine sand. Air is being admitted at

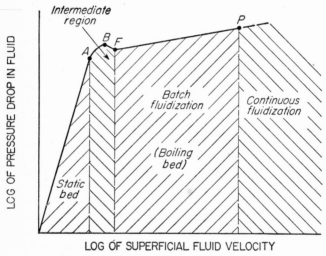

Fig. 5-17. Pressure drops in fluidized solids.

a very low rate to the bottom of the tube and flows upward through the sand without causing the grains to move. The air rate is now slowly and progressively increased. As the velocity rises, the pressure drop in the air passing through the bed increases, as is shown by the linear segment OA in Fig. 5-17. Eventually the pressure drop equals the force of gravity on the particles, and the grains begin to move. This is at point A on the graph. First the bed expands slightly with the grains still in contact. The porosity increases, and the pressure drop rises more slowly than before. When point B is reached, the bed is in the loosest possible condition with the grains still in contact. As the velocity is still further increased, the grains separate and true fluidization begins. The pressure drop sometimes diminishes a little from point B to point F. From point F onward the particles move more and more vigorously, swirling about and traveling in random directions. The contents of the tube strongly resemble a boiling liquid, and the term "boiling bed" has been given to solids fluidized in this way.

The linear velocity of the fluid between the particles is much higher than the velocity in the space above the bed. Consequently nearly all the particles drop out of the fluid above the bed. Even with vigorous fluidization only the smallest grains are entrained in the fluid and carried away. Suppose now, however, the fluid velocity is still further increased. The porosity of the bed rises; the bed of solids expands, and its density falls. Entrainment becomes appreciable, then severe, then complete. At point P all the particles have been entrained in the fluid, the porosity approaches unity, and the bed as such has ceased to exist. The phenomenon then becomes that of the simultaneous flow of two phases. From point F to point P and beyond the pressure drop rises with the fluid velocity but much more slowly than when the solid particles were stationary. Fluidization without entrainment of the solids is called *batch fluidization;* when entrainment is complete, the fluidization is said to be *continuous.*

Batch Fluidization and Boiling Beds. The precise behavior of a mass of fluidized solids depends on the particle size of the solid and on the nature of the fluid. When the fluid is a liquid, fluidization begins as a gentle rocking or oscillation of the solid particles. In the fully fluidized bed the particles move in random directions through all parts of the liquid. There are strong transient currents in the bed, with many particles traveling temporarily in the same direction, but in general the particles move as individuals. This mode of action is called *particulate fluidization.*[20]

When the fluid is a gas, the action in the bed is somewhat different and is strongly influenced by the particle size. Under conditions for good fluidization (with particles of the proper size and density) some of the gas travels through the bed between individual particles, but much of it travels through in "bubbles" or pockets containing almost no solids. At the bed surface the bubbles break, "splashing" individual particles or streamers into the space above. In the bed itself the particles move in distinct aggregates, which are lifted by the bubbles or which move aside to let the bubbles past. A fluidizing action of this kind is known as *aggregative fluidization.*[20]

If the solid particles are very small, with diameters less than about 10 microns, the interparticle attraction may be strong enough to give a kind of aggregative fluidization known as *cohesive fluidization.*[9] Particles in a bed of such material, when aerated, tend to "ball up" into spheres, which may be several millimeters in diameter. Thus with very small particles predictions of pressure drop and other characteristics may be greatly in error if based on the diameter of the ultimate particles.

Slugging is characteristic of large, heavy particles in tall, narrow vessels. The bubbles of gas tend to coalesce and grow as they rise through the fluidized bed. The rate of growth depends on the size and density of the particles: it is rapid when the particles are large and heavy, slow when they are small and light. If a vessel that is small in diameter contains a deep bed of solids, the bubbles may grow until they fill the entire cross section of the vessel. Successive bubbles then travel up the vessel, separated by slugs of solid particles. Operation is erratic and unstable.

Slugging may be avoided by the proper choice of particle size and by using shallow beds of solids. It has been pointed out that under slugging conditions the pattern of gas flow is similar to that observed when gas flows through a viscous liquid in a narrow tube. In a thin liquid the gas rises as many tiny bubbles; in a viscous liquid successive large bubbles fill the cross section of the tube. It has therefore been suggested that the critical ratio of height to diameter at which slugging begins is a function of the "viscosity" of the fluidized bed. This viscosity in turn has been related to the density and the size of the particles.[9] Note that the *ratio* of height to diameter is the controlling factor. The same solid which fluidizes well in a shallow bed may slug badly in a tall, narrow vessel.

Porosity and Maximum Bed Density. When a bed of fluidized solids expands, the porosity ϵ rises, approaching unity when entrainment becomes

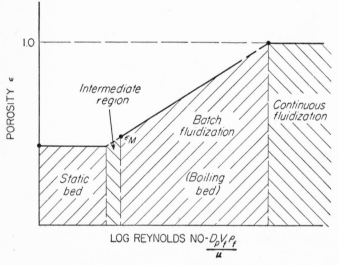

FIG. 5-18. Porosity of fluidized solids.

complete. From the beginning of fluidization to the point of complete entrainment the porosity rises linearly with the logarithm of the velocity and the Reynolds number based on the diameter of the particle and the superficial fluid velocity, as shown in Fig. 5-18. The velocity at the beginning of fluidization is known as the *critical velocity*. The velocity at which ϵ becomes unity agrees well with the terminal settling velocity of the particles, as discussed in Chap. 7, pages 357 to 361, provided the fluid is a liquid or the particles are coarse.[7] With fine particles fluidized by air, however, the graph of ϵ versus N_{Re} may extrapolate to values several times the computed terminal velocities.[8]

In particulate fluidization the Reynolds number of the particles may vary over a wide range. There is a close analogy between particulate fluidization and the settling of particles through a liquid. At any given value of porosity the density of all parts of the fluidized mass is the same. As the fluid velocity is raised, the bed thins out uniformly, and the density of the mass steadily diminishes.

In aggregative fluidization the picture is very different. The fluidized mass contains two "phases": a dense, continuous phase consisting of solid particles held slightly apart by slow streams of gas between them and a light, discontinuous phase consisting of bubbles of gas. Despite changes in the relative amounts of solid and gas, the properties of each phase remain almost constant. At the interface between the bubbles and the dense phase occasional particles may be violently thrown about, but in the bulk of the dense phase the passage of gas between the particles is slow and steady and almost always in laminar flow.[17] The density and porosity of the dense phase remain about constant; the over-all density and porosity of the bed are thus a function of the relative amounts of the two phases. Expressions for porosity based on analogies to the settling of particles in gases have not been derived.

The density of the dense phase is also the density of the fluidized bed when just enough gas is flowing through to separate the particles and there are no bubbles. This is called the *maximum bed density*. With large particles it approximates the density of the static bed; with small particles it is only 50 to 60 per cent of the density of the static bed. The maximum bed density corresponds to the minimum porosity for fluidization, data for which have been published.[6b] The following empirical equation has been proposed [9] for the maximum bed density of particles smaller than 500 microns and larger than 10 microns in diameter.

$$\rho_{MB} = 0.356\rho_s \, (\log D'_p - 1) \tag{5-14}$$

where D'_p = particle diameter, microns

ρ_{MB} = maximum bed density, lb/ft^3

ρ_s = ultimate density of the solid, lb/ft^3

For mixtures of particles the mass mean diameter $\bar{D}'_w$, also expressed in microns, is used in place of D'_p. $\bar{D}'_w$ is found from Eq. (4-45). The minimum porosity for fluidization ϵ_M is computed from the maximum bed density by the equation

$$\epsilon_M = \frac{\rho_s - \rho_{MB}}{\rho_s - \rho_a} \tag{5-15}$$

where ρ_a is the density of the gas.

When the gas velocity is greater than the critical value, the porosity can be approximated by assuming that all the additional gas, above that required to initiate fluidization, passes through the bed in bubbles containing no solids.

Bed Height. As the porosity rises, the volume of the fluidized bed increases. If the cross-sectional area of the vessel does not change with height, the porosity is a direct function of the height of the bed. Let Z_0 be the height the bed would have if the porosity were zero; i.e., if the solids existed as a single lump containing no voids. If Z is the height of the fluidized bed, the porosity is given by

$$\epsilon = \frac{Z - Z_0}{Z} = 1 - \frac{Z_0}{Z} \qquad (5\text{-}16)$$

Often the porosity at one condition, such as the minimum porosity for fluidization or the porosity of the static bed, is known. If the corresponding bed height is also known, the bed height for any new value of the porosity may be found from

$$Z_2 = Z_1 \frac{1 - \epsilon_1}{1 - \epsilon_2} \qquad (5\text{-}17)$$

where ϵ_1 and ϵ_2 are the porosities at heights Z_1 and Z_2, respectively.

Pressure Drop in Fluidized Bed. When fluidization just begins, the pressure drop through the bed is nearly equal to the force of gravity on the solids. It may be a little greater than this because of electrostatic or other effects,[8] but as a first approximation the pressure drop at fluidization may be found by equating the force it exerts on the solids to the force of gravity minus the buoyant force of the displaced fluid. If Z is the bed height, A its cross-sectional area, and ϵ_M the minimum porosity for fluidization,

$$-\Delta p\, A = \frac{g}{g_c} [\rho_s (1 - \epsilon_M) Z A - \rho_f (1 - \epsilon_M) Z A] \qquad (5\text{-}18)$$

Solving for the pressure drop per foot of height,

$$-\frac{\Delta p}{Z} = \frac{g}{g_c} (1 - \epsilon_M)(\rho_s - \rho_f) \qquad (5\text{-}19)$$

On the assumption that flow through a fluidized bed resembles that through a static bed of solids, the friction factors and pressure drops have been related to the modified Reynolds number as suggested by the Kozeny-Carman equation [Eq. (2-90)]. Data for particulate fluidization agree fairly well with those for flow through static beds; with aggregative fluidization the agreement is poor, especially at high Reynolds numbers.[10] Very small particles aggregate into balls of considerable size, so that friction factors computed on the basis of the true particle diameters are too low. With large particles and high Reynolds numbers the friction factors and pressure drops are several times those predicted from Eq. (2-90). The additional pressure drop over that in a static bed results from kinetic-energy losses. In a bed which is violently boiling, particles often collide with each other and with the vessel wall, losing kinetic energy of translation and rotation. This energy must come from pressure loss in the fluid.[17]

Modified friction factors have been correlated with the Reynolds number in a somewhat different way by Leva,[6a] as shown in Fig. 5-19. When

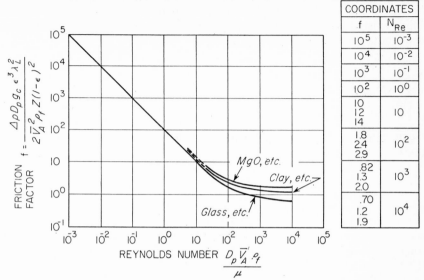

Fig. 5-19. Modified friction factors for fluidization of spherical particles. (*After Leva et al.*[6a])

the Reynolds number is below about 10, the pressure drop in both fixed and fluidized beds may be estimated from the equation

$$-\frac{\Delta p}{Z} = \frac{200\overline{V}_A\mu\lambda_L^2(1 - \epsilon)^2}{D_p^2 g_c \epsilon^3} \tag{5-20}$$

where $\overline{V}_A$ is the superficial gas velocity, in feet per second.

TABLE 5-3. SHAPE FACTORS FOR VARIOUS SOLIDS †

Material	Nature of particles	λ_L ‡
Average sand..........	....	1.33
Sharp sand...........	Jagged	1.50–1.54
Round sand..........	Rounded	1.16–1.20
Cork................	....	1.45
Pulverized coal.......	....	1.37
Natural coal dust......	Up to $\frac{3}{8}$ in.	1.54
Flue dust............	Fused, aggregates	1.82
	Fused, spherical	1.12
Mica................	Flakes	3.54

† From M. Leva, M. Grummer, M. Weintraub, and H. H. Storch, *Chem. Eng. Progr.*, **44:** 511 (1948); and R. D. Morse, *Ind. Eng. Chem.*, **41:** 1117 (1949).

‡ This shape factor is 0.205 $s_p/v_p^{\frac{2}{3}}$, where s_p and v_p are the surface and volume of one particle, respectively. This factor is unity for spheres.

This may be combined with Eq. (5-19) to give the critical velocity for fluidization

$$\bar{V}_M = \frac{0.005 g D_p^2 \epsilon_M^3 (\rho_s - \rho_f)}{\mu \lambda_L^2 (1 - \epsilon_M)} \tag{5-21}$$

A nomograph of Eq. (5-21) has been published [19] with suggested corrections for extending it into the transition range of Reynolds numbers between 10 and 200. The shape factor λ_L in Eqs. (5-20) and (5-21) is given for several typical solids in Table 5-3.

Example 5-3. A bed of sized pulverized coal 35-mesh in particle size is to be fluidized with a liquid petroleum fraction having a viscosity of 15 centipoises. The height of the static bed is 6 ft; the porosity of the static bed is 0.38. The density of the coal particles is 84 lb/ft^3; that of the liquid is 55 lb/ft^3.

Calculate the pressure drop required for fluidization.

Solution. The pressure drop is found from Eq. (5-19). Since $Z = 6$,

$$-\Delta p = 6 \frac{g}{g_c} (1 - 0.38)(84 - 55)$$

$$= 108 \text{ lb force/ft}^2, \text{ or } 0.75 \text{ lb force/in.}^2$$

Example 5-4. A bed containing 36 tons of 100-mesh sand is to be fluidized with air at 400°C and a pressure of 250 lb force/in.^{2} abs in a cylindrical vessel 10 ft in diameter. The ultimate density of the sand particles is 168 lb/ft^3. The viscosity of air at the operating conditions is 0.032 centipoise. Call g/g_c unity.

Calculate: (a) the maximum bed density and minimum porosity, (b) the minimum height of the fluidized bed, (c) the pressure drop in the bed, (d) the critical superficial air velocity.

Solution. The diameter of 100-mesh particles (Appendix 18) is 147 microns, or 4.82×10^{-4} ft. The density of air is

$$\rho_a = \frac{29}{359} \frac{273}{273 + 400} \frac{250}{14.7} = 0.558 \text{ lb/ft}^3$$

The viscosity of air is

$$\mu = 0.032 \times 0.000672 = 2.15 \times 10^{-5} \text{ lb/ft-sec}$$

(a) The maximum bed density, from Eq. (5-14), is

$$\rho_{MB} = 0.356 \times 168 \ (\log 147 - 1) = 69.8 \text{ lb/ft}^3$$

The minimum porosity, from Eq. (5-15), is

$$\epsilon_M = \frac{168 - 69.8}{168 - 0.558} = 0.586$$

(b) The minimum height of the fluidized bed is found as follows.

$$\text{Volume of solids} = \frac{36 \times 2,000}{168} = 428.6 \text{ ft}^3$$

$$\text{Volume of fluidized bed} = \frac{428.6}{1 - 0.586} = 1,035 \text{ ft}^3$$

The height of the fluidized bed is

$$Z = \frac{4 \times 1,035}{\pi 10^2} = 13.2 \text{ ft}$$

(c) The pressure drop when $Z = 13.2$, from Eq. (5-19), is

$$-\Delta p = 13.2 \times 1(1 - 0.586)(168 - 0.558)$$

$$= 915 \text{ lb force/ft}^2, \text{ or } 6.35 \text{ lb force/in.}^2$$

(d) The critical velocity is given by Eq. (5-21). The shape factor for average sand, from Table 5-3, is 1.33.

$$\overline{V}_M = \frac{0.005 \times 32.17(4.82 \times 10^{-4})^2 0.586^3(168 - 0.558)}{2.15 \times 10^{-5} \times 1.33^2(1 - 0.586)} = 0.080 \text{ ft/sec}$$

Now check to see that Eq. (5-21) applies by computing the particle Reynolds number.

$$N_{\text{Re}} = \frac{D_p \overline{V}_M \rho_a}{\mu} = \frac{4.82 \times 10^{-4} \times 0.080 \times 0.558}{2.15 \times 10^{-5}} = 1.00$$

Since N_{Re} is less than 10, Eq. (5-21) applies.

Uses of Fluidization. Extensive use of fluidization began in the enormous catalytic-cracking reactors in the petroleum industry. Fluidization now finds application in many catalytic processes and in other operations, such as the drying of crystals, as well.[14] The chief advantages of fluidization are that it ensures contact of the fluid with all parts of the solid particles; it prevents segregation of the solids by thoroughly agitating the bed; and it minimizes temperature variations even in a large reactor, again by virtue of the vigorous agitation.

When a fluid passes through a static bed of solids, some of the surface of the particles is screened from direct contact with the fluid. If the solid is a catalyst, some of its effectiveness is lost. In addition, if heat is generated in a static bed by an exothermic reaction, strong temperature gradients and often a localized zone of very high temperature may be formed. Such a zone is known as a "hot spot." Since the hot-spot temperature must not exceed a set value, which is usually close to the optimum temperature for the reaction, this means that much of the catalyst is operating at a temperature well below the optimum. In a fluidized bed, by contrast, the temperature throughout the mass rarely varies by more than 5 or 10°F. Because of contact of the fluid with all the catalyst surface and the elimination of hot spots, a pound of fluidized catalyst may convert from two to ten times as much as a pound of the same catalyst in a static condition. In drying crystalline solids, for the same reasons, the drying rate may be several times that in a static bed.

Offsetting these advantages are the increased power needed because of the higher pressure drop in a fluidized bed, the increased size of the vessel or reactor, and increased breakage of the solid particles. Facilities must usually be provided for recovering the fine solids carried out of the fluidized bed.

Airslide. Batch fluidization is also used in conveying powdered solids gently downward at a small angle. A metal trough inclined at 3 to 6° with the horizontal contains a flat floor of cloth or porous stone. Air admitted at a controlled rate through the floor fluidizes the solids and causes them to "flow" by gravity to the discharge end of the trough. Solids can thus be moved horizontally a considerable distance with only a small change in elevation. The only power required is that needed to compress the air. A device of this kind is known as an Airslide.

Continuous Fluidization and Pneumatic Transport. When the fluid velocity through a bed of solids becomes large enough, all the particles are entrained in the fluid and are carried along with it, to give what is called continuous fluidization. Its principal application is in transporting solids from point to point in a factory or plant. Sometimes the fluid is a liquid in which solids are suspended to form a pumpable slurry. More commonly, however, the suspending fluid is a gas, usually air, flowing at velocities between 50 and 100 ft/sec in pipes ranging from 2 to 16 in. in diameter.

Pneumatic conveyors utilizing air for transporting solids draw considerably more power than mechanical conveyors of the same capacity, but they have a number of advantages:

1. They are completely enclosed and therefore dustless.

2. They have no return leg or trough and occupy little space.

3. They may pick up material from different points or deliver it to any of several points.

4. They may carry solids long distances vertically or horizontally under or over roadways, railways, or canals, or through buildings, where mechanical conveyors could not go.

Vacuum and Pressure Systems. The air required for pneumatic transport is moved through the system by a rotary or centrifugal blower. An installation is termed a vacuum system or a pressure system, depending on whether the blower sucks or blows the air through the pipe. A vacuum system, illustrated in Fig. 5-20a, is best suited to pickup of material from different locations, with delivery to a single point. Air is drawn into the pipe through a nozzle on the end of a flexible hose, entraining solids as it enters. Figure 5-20b shows the details of a nozzle. Bulk solids from storage piles, boxcars, and barges are readily picked up by manipulation of the hose and nozzle. Air and solids are drawn through the pipe to a cyclone separator and a dust collector, which remove the solids from the air and drop them into a storage bin. The vacuum blower beyond the collectors handles only dust-free air.

A pressure system is used when solids are to be picked up at a single point and delivered to several points or where material is to be conveyed a very long way. A typical pressure system is sketched in Fig. 5-20c. The problem of getting the solids into the air stream is somewhat greater than in a vacuum system, and special injection and mixing devices are often necessary. A typical injector, the Fuller-Kinyon pump, is shown in Fig. 5-21.

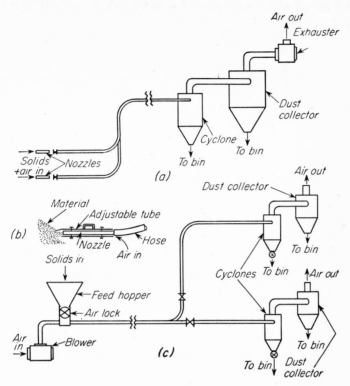

FIG. 5-20. Pneumatic conveying systems: (a) Vacuum system, multiple inlet.   (b) Nozzle detail.   (c) Pressure system, multiple discharge.

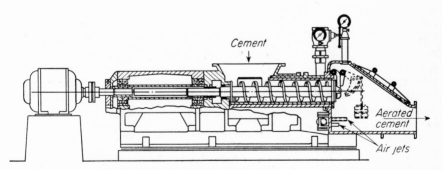

FIG. 5-21. Fuller-Kinyon pump.

Frictional Pressure Drop. In the turbulent flow of a dust-free fluid through a long, straight pipe the pressure drop due to friction is given by the Fanning equation [Eq. (2-53)]. The pressure drop per foot of pipe is a function of the pipe diameter and the superficial velocity, density, and viscosity of the fluid. Thus

$$\left(-\frac{\Delta p}{L}\right)_f = \psi_1(D, \overline{V}_A, \rho_f, \mu) \tag{5-22}$$

When solids are suspended in the fluid, additional variables are needed to define the system. The pressure drop per foot now also depends on the effective diameter of the solid particles D_p, the density of the particles ρ_s, and the mass ratio of the solids to the fluid r. Then

$$\left(-\frac{\Delta p}{L}\right)_s = \psi_2(D, \overline{V}_A, \rho_f, \mu, D_p, \rho_s, r, g) \tag{5-23}$$

In addition there may be a variation with the shape and surface roughness of the particles, but the effects are probably small.

The precise relationship of the pressure drop to these variables has not yet been established. Commercial systems are largely designed on the basis of experience and rules of thumb. Hudson[3] gives approximate methods and data for a number of solids. The underlying theory has been studied by Belden and Kassel,[1] Vogt and White,[18] and many others, but no completely satisfactory correlation has yet been proposed. The following method, however, will give approximately correct results for many materials.

In this method the frictional pressure drop is expressed as a relative pressure drop β, defined as the ratio of the pressure drop observed in the flow of the suspension to the pressure drop in the fluid alone flowing through the same pipe at the same velocity. Through dimensional analysis this ratio can be expressed as a function of various dimensionless quantities as follows:

$$\beta = \phi\left(\frac{D\overline{V}_A\rho_f}{\mu}, r, \frac{D_p}{D}, \frac{\rho_s}{\rho_f}, \sqrt{\frac{(\rho_s - \rho_f)\rho_f g D_p^3}{3\mu^2}}\right) \tag{5-24}$$

An empirical equation for the relative pressure drop is[18]

$$\beta - 1 = a\left(\frac{D}{D_p}\right)^2\left(\frac{\rho_f r}{\rho_s N_{\text{Re}}}\right)^k \tag{5-25}$$

Here a and k are functions of the group $\sqrt{(\rho_s - \rho_f)\rho_f g D_p^3/3\mu^2}$, as shown by Figs. 5-22 and 5-23.

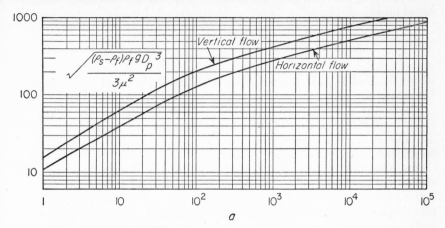

FIG. 5-22. $\sqrt{(\rho_s - \rho_f)\rho_f g D_p^3/3\mu^2}$ vs. a. (*Vogt and White.*[18] *Courtesy of Industrial and Engineering Chemistry.*)

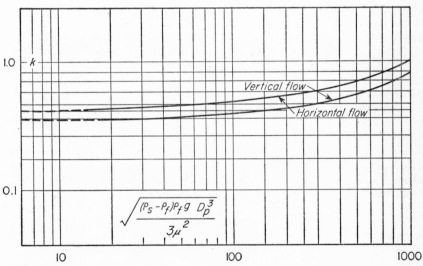

FIG. 5-23. k vs. $\sqrt{(\rho_s - \rho_f)\rho_f g D_p^3/3\mu^2}$. (*Vogt and White.*[18] *Courtesy of Industrial and Engineering Chemistry.*)

Total Pressure Drop. To lift and move the solid, energy is needed. For r lb of solid, this energy requirement, by a mechanical-energy balance based on Eq. (2-43), is

$$E_s = r \left[\frac{p_b - p_a}{\rho_s} + \frac{V_{sb}^2 - V_{sa}^2}{2g_c} + \frac{g}{g_c}(Z_b - Z_a) \right] \quad (5\text{-}26)$$

where V_{sa} and V_{sb} are the velocities of solids at inlet and outlet, in feet per second.

The energy E_s is supplied by the air. It is transmitted to the solid particles through the action of drag forces between the air and the solid. The energy E_s is a work term, and it must appear in the mechanical-energy balance for the air.

Assuming the pressure drop is a small fraction of the absolute pressure, the air can be considered to be an incompressible fluid of constant density $\bar{\rho}$, the average density of the air between the inlet and outlet. Neglecting the change in velocity head, assuming the kinetic energy factor to be unity, and allowing for E_s, the Bernoulli equation [Eq. (2-43)] becomes for 1 lb of air

$$\frac{p_b - p_a}{\bar{\rho}} + \frac{g}{g_c}(Z_b - Z_a) = -E_s - H_f \quad (5\text{-}27)$$

where H_f is the total friction in the stream. Eliminating E_s from Eqs. (5-26) and (5-27) and solving for $p_a - p_b$ gives

$$p_a - p_b = \frac{(g/g_c)(1 + r)(Z_b - Z_a) + r(V_{sb}^2 - V_{sa}^2)/2g_c + H_f}{1/\bar{\rho} + r/\rho_s} \quad (5\text{-}28)$$

By definition of β, the friction H_f is the product of β and the friction from the flow of air alone, or, using Eq. (2-53),

$$H_f = \frac{2\beta f L_b \bar{V}_{Am}^2}{D g_c} \quad (5\text{-}29)$$

where $\bar{V}_{Am}$ is the average superficial velocity of the air in the pipe, calculated from the density $\bar{\rho}$. Substituting H_f from Eq. (5-29) into Eq. (5-28) gives

$$p_a - p_b = \frac{g}{g_c} \left[\frac{(1 + r)(Z_b - Z_a) + (r/2g_c)(V_{sb}^2 - V_{sa}^2) + 2\beta f L_b \bar{V}_{Am}^2/gD}{1/\bar{\rho} + r/\rho_s} \right] \quad (5\text{-}30)$$

Slip Velocity. Equation (5-30) contains different terms for the velocity of the solids and the velocity of the gas. They are not equal. The solid moves along with the gas but at a slower rate. It is the drag of the gas past the particle that supplies the force for moving the solid. The difference between the velocity of the gas and that of the solid is the slip velocity.

In the vertical upward flow of a suspension through a pipe, the slip velocity is affected to a minor degree by the gas velocity and the solids feed rate.[8] Typical values are shown in Fig. 5-24. In horizontal flow the slip velocity is ordinarily smaller than when flow is vertical, and may be neglected in computing pressure drops and power requirements for pneumatic transport.

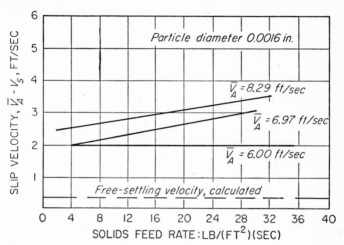

FIG. 5-24. Slip velocity vs. solids feed rate. (*Lewis, Gilliland, and Bauer.*[8] *Courtesy of Industrial and Engineering Chemistry.*)

Example 5-5. Air at 60°F is delivered to the entrance of a horizontal Schedule 40 3-in. steel pipe at a gauge pressure of 25 lb force/in.[2] The entering velocity is 50 ft/sec. Rounded sand 60-mesh in average particle size is dropped into the air just downstream from the entrance; the weight ratio of sand to air is 4:1. The density of the sand particles is 168 lb/ft³. If the pipe is 100 ft long, what is the pressure drop in the pipe?

Solution. Equation (5-30) is used. The friction factor and pressure drop for air alone are first calculated as in Example 2-12. The quantities needed are

$$D = \frac{3.068}{12} = 0.256 \text{ ft} \qquad L_b = 100 \text{ ft} \qquad \mu = 0.018 \times 6.72 \times 10^{-4} \text{ lb/ft-sec}$$

The density of air at the inlet is

$$\rho_a = \frac{29 \times 492(25 + 14.7)}{359(460 + 60)14.7} = 0.2064 \text{ lb/ft}^3$$

$$G = 50 \times 0.2064 = 10.32 \text{ lb/ft}^2\text{-sec}$$

$$N_{Re} = \frac{0.256 \times 10.32}{0.018 \times 6.72 \times 10^{-4}} = 2.2 \times 10^5$$

As in Example 2-12, $f = 0.0047$. Assume $\bar{\rho} = 0.205$ lb/ft³.

$$\bar{V}_{Am} = \frac{G}{\rho_m} = \frac{10.32}{0.205} = 50.3 \text{ ft/sec}$$

Compute the density at the outlet to check on $\bar{p}$. From Example 2-12,

$$p_b = 5{,}717 - \frac{100 \times 366}{600}$$

$$= 5.656 \text{ lb force/ft}^2$$

$$= \frac{5{,}656}{144} = 39.2 \text{ lb force/in.}^2$$

$$\rho_b = 0.2064 \frac{39.2}{25 + 14.7} = 0.204 \text{ lb/ft}^3$$

$$\bar{\rho} = \frac{0.2064 + 0.204}{2} = 0.205 \text{ lb/ft}^3$$

The multiplying factor β must now be calculated. The new quantities needed are as follows. For 60-mesh particles, from Appendix 18,

$$D_p = 0.246 \text{ mm} = \frac{0.246}{12 \times 25.4} = 8.07 \times 10^{-4} \text{ ft}$$

$$\sqrt{\frac{(\rho_s - \bar{p})\bar{\rho}gD_p^3}{3\mu^2}} = \sqrt{\frac{(168 - 0.2)0.205 \times 32.17 \times 8.07^3 \times 10^{-12}}{3 \times 0.018^2 \times 6.72^2 \times 10^{-8}}} = 36.4$$

From Figs. 5-22 and 5-23, for horizontal flow,

$$a = 8.0 \qquad k = 0.57$$

β is calculated from Eq. (5-25), with $r = 4$.

$$\beta - 1 = 8.0 \left(\frac{0.256}{8.07 \times 10^{-4}}\right)^2 \left(\frac{0.205 \times 4}{168 \times 2.2 \times 10^5}\right)^{0.57} = 34.7$$

$$\beta = 34.7 + 1 = 35.7$$

The initial velocity of the solids is zero; the final velocity is taken as equal to the air velocity, 50 ft/sec. In a horizontal pipe $Z_b = Z_a$. The total pressure drop, from Eq. (5-30), is

$$p_a - p_b = \frac{\dfrac{4}{64.4}(50^2 - 0) + \dfrac{2 \times 35.7 \times 0.0047 \times 100 \times 50.3^2}{32.2 \times 0.256}}{\dfrac{1}{0.205} + \dfrac{4}{168}}$$

$$= 2{,}130 \text{ lb force/ft}^2, \text{ or } 14.5 \text{ lb force/in.}^2$$

Strictly speaking, the values of $\bar{p}$ and $\overline{V}_{Am}$ should be recalculated and substituted in Eq. (5-30), but this refinement is not justified by the accuracy of the method.

Stable and Unstable Operation. If the superficial gas velocity through a pneumatic system is high, relatively large amounts of solids may be moved without difficulty. Commercial systems are often designed to have the solids occupy 3 to 12 per cent of the volume of the pipe, and air to occupy 88 to 97 per cent.[12a] This corresponds to values of r between 30 and 100 for solids with a density of 100 lb/ft^3. With high air velocities the flow is smooth and regular, with little variation in pressure.   The principal problem is that of erosion, especially at bends in the pipe.

If the air velocity is too low, however, or the load of solids too great, the flow changes from stable to unstable.   Solids collect at localized points; the gas flow slows down; the pressure builds up.   Finally the deposited solids are suddenly and violently blown out of the system.   The pressure

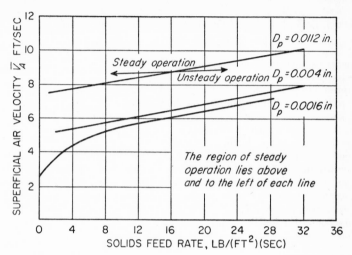

F_IG. 5-25. Stable and unstable flow, pneumatic conveying.   (*Lewis, Gilliland, and Bauer.*[8] *Courtesy of Industrial and Engineering Chemistry.*)

falls rapidly, then starts to rise once more as the cycle is repeated.   Limiting air velocities for stable operation in the transport of glass spheres are given in Fig. 5-25.   As would be expected, larger velocities are needed with larger particles to ensure stable operation.

SYMBOLS

a Constant in Eq. (5-25)

b Thickness, ft

D_p Particle diameter, ft; D'_p, particle diameter, microns; $\bar{D}'_w$, mass mean diameter, microns

E_s Energy needed to move solid, ft-lb force/lb solid

F Force, lb force; F_f, frictional force; F_g, force of gravity; F_L, lateral component; F_V, vertical component

G Mass velocity, lb/ft^2-sec

g Acceleration of gravity, ft/sec^2

g_c Newton's-law conversion factor, 32.174 ft-lb/lb force-sec^2

H_f Friction in air transport system, ft-lb force/lb air

K' Ratio, p_L/p_V

L Length, ft; L_b, length of channel, air transport

p Pressure, lb force/ft^2; p_a, p_b, inlet and outlet pressures; p_B, pressure on bin floor; p_L, p_V, directed pressures in stress analysis

R Radius, ft

r Mass ratio, solids to fluid

s_p Surface of particle, ft^2

$\overline{V}$ Average velocity, ft/sec; $\overline{V}_A$, superficial gas velocity; $\overline{V}_{Am}$, average superficial velocity; $\overline{V}_M$, critical velocity for fluidization; $\overline{V}_m$, average velocity of gas in pipe, based on density $\bar{\rho}$; V_{sa}, V_{sb}, velocity of solids at entrance and exit to channel, respectively

v_p Volume of particle, ft^3

Z Height, ft; Z_0, height of bed at zero porosity; Z_T, total height

Greek Letters

α Angle; α_r, angle of repose; α_m, angle of internal friction

β Relative pressure drop

Δp Pressure drop, $p_b - p_a$

ϵ Porosity; ϵ_M, minimum porosity for fluidization

θ Angle

λ_L Shape factor, Leva; $\lambda_L = 0.205 s_p/v_p^{\frac{2}{3}}$

μ Viscosity, lb/ft-sec or lb/ft-hr

μ' Coefficient of friction

ρ Density, lb/ft^3; ρ_a, density of gas; ρ_b, bulk density of solid; ρ_f, density of fluid; ρ_L, density of liquid; ρ_{MB}, maximum bed density; ρ_s, density of solid; $\bar{\rho}$, average air density

τ Shear stress, lb force/ft^2

ϕ Function in Eq. (5-24)

ψ_1 Function in Eq. (5-22); ψ_2, function in Eq. (5-23)

Dimensionless Groups

f Fanning friction factor, $H_{fs} g_c D/2 L_b \overline{V}_{Am}^2$

N_{Re} Reynolds number, $D_p \overline{V}_m \rho_a/\mu$

PROBLEMS

5-1. A circular silo 10 ft in diameter contains barley with a bulk density of 39 lb/ft^3. What are the vertical and lateral pressures at the base of the silo if the depth of the barley is 40 ft? If it is 80 ft? $K' = 0.40$; $\mu' = 0.45$.

5-2. An inclined troughed belt conveyor is to transport 10 tons/hr of ground gypsum a horizontal distance of 250 ft. What belt width should be specified? How far could the gypsum be elevated by this conveyor?

5-3. A tower 4 ft ID is packed with coke particles 1 in. in average diameter. The coke is fluidized by passing water at 30°C upward through the tower. If the height of the static bed is 10 ft, what is the pressure drop in the fluidized bed? The density of coke particles is 75 lb/ft^3, and the porosity of the static bed is 0.42. Assume fluidization is particulate.

5-4. A bed of 60-mesh alumina is to be fluidized with air at 400°C and 100 lb force/in.^{2} The ultimate density of the alumina particles is 230 lb/ft^3. The porosity of the static bed is 0.40. If the static bed is 12 ft deep and 9 ft in diameter, calculate: (a) the maximum bed density and minimum porosity, (b) the minimum height of the fluidized bed, (c) the pressure drop in the bed, (d) the critical air velocity.

5-5. Repeat Prob. 5-4 using hydrogen in place of air.

5-6. Pulverized coal, 35-mesh in particle size, is to be pneumatically conveyed 200 ft horizontally and then 50 ft vertically at the rate of 12,000 lb/hr in a pressure system. The carrying air enters at 80°F. If the pressure at the entrance to the cyclone separator is 5 lb force/in.^{2} gauge, what must the pressure be at the blower discharge? The density of the coal particles is 84 lb/ft^3. Assume $r = 10$ lb of coal per pound of air. The conduit is Schedule 40 3-in. steel pipe.

REFERENCES

1. Belden, D. H., and L. S. Kassel: *Ind. Eng. Chem.*, **41**: 1174 (1949).
2. Cain, W.: "Earth Pressure, Retaining Walls and Bins," p. 219, John Wiley & Sons, Inc., New York, 1916.
3. Hudson, W. G.: *Chem. Eng.*, **61**(4): 191 (1954).
4. Ketchum, M. W.: "Walls, Bins, and Grain Elevators," 3d ed., chap. 16, McGraw-Hill Book Company, Inc., New York, 1919.
5. Lee, C. A.: *Chem. Eng.*, **60**(5): 194 (1953).
6. Leva, M., M. Grummer, M. Weintraub, and H. H. Storch: *Chem. Eng. Progr.*, **44**: (a) 511, (b) 707 (1948).
7. Lewis, E. W., and E. W. Bowerman: *Chem. Eng. Progr.*, **48**: 603 (1952).
8. Lewis, W. K., E. R. Gilliland, and W. C. Bauer: *Ind. Eng. Chem.*, **41**: 1104 (1949).
9. Matheson, G. L., W. A. Herbst, and P. A. Holt II: *Ind. Eng. Chem.*, **41**: 1099 (1949).
10. Morse, R. D.: *Ind. Eng. Chem.*, **41**: 1117 (1949).
11. Olive, T. R.: *Chem. Eng.*, **59**(11): 163 (1952).
12. Riegel, E. R.: "Chemical Process Machinery," 2d ed., (a) p. 108, (b) p. 180, Reinhold Publishing Corporation, New York, 1953.
13. Rudd, J. K.: *Chem. Eng. News*, **32**(4): 344 (1954).
14. Sittig, M.: *Chem. Eng.*, **60**(5): 219 (1953).
15. Strube, H. L.: *Chem. Eng.*, **61**(4): 195 (1954).
16. Taylor, D. W.: "Fundamentals of Soil Mechanics," chap. 13, John Wiley & Sons, Inc., New York, 1948.
17. Toomey, R. D., and H. F. Johnstone: *Chem. Eng. Progr.*, **48**: 220 (1952).
18. Vogt, E. G., and R. R. White: *Ind. Eng. Chem.*, **40**: 1731 (1948).
19. Weintraub, M., and M. Leva: *Chem. Eng.*, **57**(1): 110 (1950).
20. Wilhelm, R. H., and M. Kwauk: *Chem. Eng. Progr.*, **44**: 201 (1948).

Chapter 6

MIXING

Mixing is used to prepare a uniform combination of two or more materials. The substances fed to a mixer may be solids, liquids, or gases. The combinations of materials considered in this chapter are: (1) two or more solids, (2) two or more liquids, (3) a liquid and a gas, and (4) a liquid and a solid. Mixing gases with gases is not considered.

Mixing may be done for various reasons. The most common objectives of mixing, and typical examples illustrating each, are shown in Table 6-1

The product from a mixer may be either homogeneous or heterogeneous. Homogeneous products are obtained in mixing miscible liquids and in dissolving soluble solids and gases. All other mixed products listed in Table 6-1 are heterogeneous and consist of mixtures of two or more phases. Except for mixtures of solids, the mixture consists of a liquid which encloses

TABLE 6-1. OBJECTIVES OF MIXING PROCESSES

Type	Examples
Dry blending of solids	Mixing pigments and toners for paints; mixing molding powders for plastics manufacture
Blending miscible liquids	Diluting concentrated solutions with solvent
Mixing immiscible liquids	Washing liquids with immiscible solvents; extracting liquids by liquids; manufacturing emulsions; chemically treating liquids
Suspending solids in liquids	Dissolving soluble solids in liquids; crystallizing solids from solutions; leaching and extracting solids by liquids; purifying liquids by solid adsorbents; promoting chemical reactions between liquids and solids; suspending solid catalysts in liquid reactants
Absorbing or dispersing gases in liquids	Absorbing soluble gas in liquids; preparing foams; oxidizing liquids by air or oxygen

particles of solids, drops of liquids, or bubbles of gas. The enclosed particles may be of colloidal dimensions, or they may be relatively coarse. The surrounding liquid is called the continuous phase, and the enclosed particles, drops, or bubbles are called the dispersed phase.

A heterogeneous product may either be permanently or temporarily dispersed. A concrete mix or an emulsion is expected to remain permanently dispersed, while a mixture of crystals and their mother liquor or a catalyst and its reaction mass is expected to settle rapidly when removed from the mixer, so the phases can be recovered individually. The problems encountered in such separations are discussed in Chap. 7.

The materials fed to mixers may be low- or moderate-viscosity liquids, highly viscous Newtonian or non-Newtonian liquids, plastic and deformable solids like rubber, or dry, free-flowing powders. The multiplicity of mixer types in commercial use is, in part, a result of the extreme range of viscosities and consistencies that must be treated by mixing machinery.

Basic Types of Mixers. Mixing machines may be grouped into three broad classes: impeller mixers for liquids; mixers for pastes and plastic solids; and mixers for dry powders. Impeller mixers utilize an impeller, such as a paddle, turbine, or propeller, to apply mechanical energy to the material. The performance of an impeller mixer depends on the creation of a current which penetrates to all points in the tank, and most of the mixing action occurs at a distance from the impeller. The tank acts as a container and as a baffle but otherwise does not take an essential part in the mixing process. Unless under pressure, it is usually constructed of light-gauge metal.

The second class of mixing machinery includes double-motion tank mixers, change cans, mixing rolls, pugmills, ribbon mixers, kneaders, Banbury mixers, and pan mixers. These are used when the material is too viscous or too plastic to flow readily to the suction side of an impeller and flow currents cannot be created. The material must all be brought to the agitator, or the agitator must visit all parts of the mix. The action in this machinery is well described as a "combination of low-speed shear, smearing, wiping, folding, stretching, and compressing." [2] The mechanical energy is applied by moving parts directly to the mass of material. In the closed types, such as Banbury mixers, the inner wall of the casing acts as part of the mixing means, and all mixing action occurs close to the moving parts. Clearances between mixing arms, rotors, and wall of casing are small. The forces generated in these mixers are large, the machinery must be ruggedly built, and the power consumption is high. The heat evolved per unit mass of material is sufficient to require cooling to prevent the temperature from reaching a level dangerous to the equipment or the material.

Mixers for dry powders constitute the third broad class of mixing machinery. This class includes some machines which are also used for heavy pastes and some machines which are restricted to free-flowing powders. Mixing is by slow-speed agitation of the mass with an impeller, by tumbling, or by centrifugal smearing and impact. These mixers are of fairly light construction, at least in comparison with mixers for tough plastic masses, and their power consumption per ton of material mixed is moderate.

The boundaries between the fields of application of the various types of mixers are not sharp. Impeller mixers are not often used when the viscosity is more than about 200,000 centipoises, especially if the liquid is not

Newtonian. Kneaders and mixer-extruders work on thick pastes and plastic masses; impact wheels are restricted to dry powders. Other mixers, however, can blend liquids, pastes, plastic solids, and powders. Typical ranges of application and power requirements are given in Table 6-2.

TABLE 6-2. APPLICATIONS AND POWER REQUIREMENTS OF MIXERS †

TYPE	POWER REQUIRED HP/1000 GALS	USUAL RANGE OF OPERATION			
		LIQUID VISCOSITY, CP (10 10^2 10^3 10^4 10^5)	PLASTIC MASSES (LIGHT MEDIUM HEAVY)	GRAINS AND POWDERS
PADDLES, GATES, ANCHORS	0.5 –10			
TURBINES	1–15			
PROPELLERS	1–5			
DISCS	10–50			
CONES	5–20			
CHANGE CANS	10–150			
KNEADERS	60–300			
DISPERSERS	100–1000			
MASTICATORS	500–8000			
CONTINUOUS KNEADERS	1–10 HP–HR/TON			
MIXER– EXTRUDERS	10–100 HP–HR/TON			
MIXING ROLLS	----			
MULLERS AND PANS	100–250			
PUG MILLS	2–20 HP–HR/TON			
RIBBON MIXERS	10–40			
TUMBLING MIXERS	5–40			
IMPACT WHEELS	1–2 HP–HR/TON			

† C. S. Quillen, *Chem. Eng.*, **61**(6): 178 (1954); J. C. Smith, *Chem. Inds.*, **64**: 399 (1949), **66**: 843 (1950).

This table is by no means an exhaustive list of mixers, for many other types are available. Extensive descriptions of commercial mixing machinery have been published.[7a, 9, 14]

IMPELLER MIXERS

In these machines mixing is done by a mechanically driven impeller, which creates a flow pattern in the liquid. The liquid circulates through the vessel, returning eventually to the impeller. Circulation by itself, however, is not the only important factor in mixing, for it is possible to have streams or currents travel side by side for some distance with little or no intermingling. Turbulence in the moving stream is the second important mixing factor. Turbulence provides the means of entraining material from the bulk of the tank contents and incorporating it in the flowing stream. As shown later, some applications call for large flows and relatively low turbulence, and others require high turbulence with relatively low flow.

Types of Impeller Mixers. The three main types of mixing impellers are paddles, turbines, and propellers. Each type contains many variations and subtypes for specific purposes, which will not be considered here. Two other impellers, disk agitators and cone agitators, find application in liquid mixing. They are useful in certain situations; but the first three types solve perhaps 95 per cent of all liquid-mixing problems.

Paddles. Various types of paddle agitators are shown in Fig. 6-1. For the simpler problems an effective agitator consists of a flat paddle turning on a vertical shaft. Two-bladed and four-bladed paddles are common.

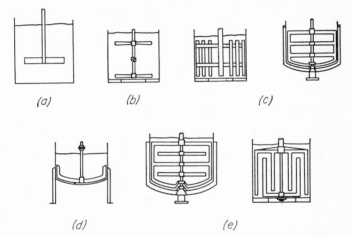

FIG. 6-1. Paddle agitators: (*a*) Flat paddles. (*b*) Pitched paddles. (*c*) Gate agitators. (*d*) Anchor agitator. (*e*) Double-acting agitators.

Sometimes the blades are pitched; more often they are vertical. Paddles turn at slow to moderate speeds in the center of a vessel; they push the liquid radially and tangentially with almost no vertical motion at the impeller unless the blades are pitched. The currents they generate travel outward to the vessel wall and then either upward or downward. In deep tanks several paddles are mounted one above the other on the same shaft.

Viscous liquids and thin pastes are agitated by multiple-blade paddles known as *gates*. In another design the blades conform to the shapes of a dished or hemispherical vessel so that they scrape the surface or pass over it with close clearance. A paddle of this kind is known as an *anchor agitator*. Anchors are useful for preventing deposits on a heat-transfer surface, as in a jacketed process vessel, but they are poor mixers. They nearly always operate in conjunction with a higher-speed paddle or other agitator, usually turning in the opposite direction. Multiple paddles turning between intermeshing stationary fingers, and counterrotating sets of multiple paddles, create the high localized shear needed in some dissolving operations, as in the manufacture of rubber cement.

Industrial paddle agitators turn at speeds between 20 and 150 rpm. The total length of a paddle impeller is typically 50 to 80 per cent of the inside diameter of the vessel. The width of the blade is one-sixth to one-tenth its length. At very slow speeds a paddle gives mild agitation in an unbaffled vessel; at higher speeds baffles become necessary. Otherwise the liquid is swirled around the vessel at high speed but with little mixing. A deep vortex around the shaft may be formed. The effect of baffles on the flow patterns in agitated vessels is discussed later in this section.

A single-paddle impeller is relatively gentle in its mixing action. It does not break fragile crystals or give the violent currents in the liquid that occur near turbines and propellers. Properly designed paddles can perform most of the services required of an agitator for liquids, including simple mixing, suspension and dissolution of solids, and dispersion of gases. They are especially useful when a process vessel must be emptied while its contents are being agitated, for slow-speed paddles do not create a deep vortex in shallow layers of liquid, as do impellers which turn at higher speeds.

Turbines. A turbine impeller submerged in a liquid operates like a centrifugal pump without a casing. Some of the many designs of turbine are shown in Fig. 6-2. Most of them resemble multibladed paddle agitators with short blades, turning at high speeds on a shaft mounted centrally in the vessel. The blades may be straight or curved, pitched or vertical. The impeller may be open, semienclosed, or shrouded. The diameter of the impeller is smaller than with paddles, ranging from 30 to 50 per cent of the diameter of the vessel.

Turbines are the most versatile of mixers for liquids. As shown in Table 6-2, they are effective over an enormous range of viscosities. Special turbines with curved blades have been used on liquids with kinematic viscosities of 700,000 centistokes. In thin liquids turbines generate strong currents which persist throughout the vessel, seeking out and destroying stagnant pockets. Near the impeller is a zone of rapid currents, high turbulence, and intense shear. The principal currents are radial and tangential. The tangential components induce vortexing and swirling, which must be stopped by baffles or by a diffuser ring if the impeller is to be most effective.

The semiopen turbine known as a *vaned disk*, shown in Fig. 6-2d, is used for dispersing or dissolving a gas in a liquid. Gas is admitted below the impeller, at its axis; the vanes throw the large gas bubbles outward and break them up into many tiny bubbles. This greatly increases the interfacial area between the gas and the liquid. Sometimes the gas is directed downward to the impeller, though this is somewhat less effective than admitting it from below; or a vortex may be formed deliberately to draw gas from the space above the liquid downward into the mixing zone.

A modified turbine known as a *radial impeller* is shown in Fig. 6-2e. It gives radial flow without baffles or a diffuser. The vertical mixing blades, carried at the ends of flat horizontal arms, are set to make an angle of 10

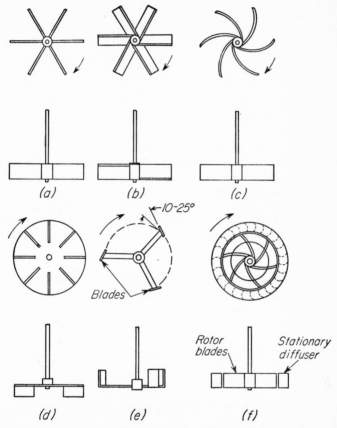

Fig. 6-2. Turbine impellers: (*a*) Straight blade. (*b*) 45° pitched blade. (*c*) Straight curved-blade. (*d*) Vaned disk. (*e*) Radial impeller. (*Struthers-Wells Co.*) (*f*) Shrouded curved-blade with diffuser ring.

to 25° with the tangent to their circle of rotation. As the impeller turns, the blades push the fluid outward, much in the manner of a snowplow, with little or no tangential velocity. Radial impellers give high shear and strong radial currents that are especially desirable in polymerization kettles for synthetic rubber.

Propellers. A propeller is a high-speed mixer for thin liquids. The ordinary marine propeller is representative of the group. Small ones turn at full motor speed, usually 1,750 rpm; larger ones turn at 400 to 800 rpm. The flow currents they generate are primarily longitudinal or axial, and continue through the liquid in a given direction until deflected by the floor or wall of the vessel. The highly turbulent swirling column of liquid leaving the impeller entrains stagnant liquid as it moves along, probably considerably more than would an equivalent column from a stationary

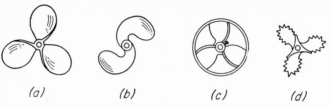

(a)   (b)   (c)   (d)

Fig. 6-3. Mixing propellers: (a) Standard three-blade. (b) Weedless. (c) Guarded. (d) Saw-toothed.

nozzle. The propeller blades vigorously cut or shear the liquid. Because of the persistence of the flow currents, propeller agitators are effective in very large vessels. Multiple side-entering propellers have been used to mix the contents of a 4,200,000-gal tank.

Propellers lose much of their effectiveness if mounted on a centrally located shaft, especially in an unbaffled vessel. Usually propeller agitators are used without baffles and are inclined or enter through the sides of the tank as shown in Figs. 6-6 and 6-7. Various propeller designs are illustrated in Fig. 6-3. Standard three-bladed marine propellers with a pitch equal to the diameter are most common; four-bladed, toothed, and other designs are employed for special purposes.

Propellers are much smaller in diameter than either paddles or turbines, rarely exceeding 18 in. in diameter regardless of the size of the vessel. In a deep tank two or more propellers may be mounted on the same shaft, usually directing the liquid in the same direction. Sometimes two propellers work in opposite directions, or in "push-pull," to create a zone of especially high turbulence between them.

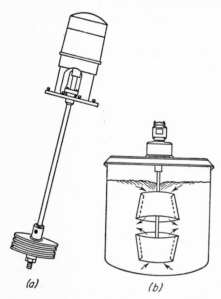

(a)   (b)

Fig. 6-4. Disk and cone agitators: (a) Disk. (H. E. Serner Co.) (b) Cone.

Disks and Cones. The agitators sketched in Fig. 6-4 operate by action of friction of the liquid with a flat or nearly flat surface. Disk impellers create high localized shear but relatively gentle flow currents; cones move the liquid steadily through a vessel without sudden changes in velocity. Some disk impellers are made up of several flat disks close together on a single shaft; others utilize single disks which are corrugated or which carry sharp peripheral teeth alternately bent upward and downward. This last type is effective in dispersing pigments and the like into heavy solvents or carriers. In the cone agitator shown in Fig. 6-4 liquid enters the impeller through the small open ends. Friction with the inside surface of the cones gives the liquid a gentle swirling motion and throws it outward by centrifugal action, causing it to flow to the point of maximum diameter. Here it escapes through openings in the impeller into the main body of liquid. Cones are useful in suspending heavy slurries of solids, such as cotton fibers, which must follow a smooth uninterrupted flow path.

Flow Mixers and Pump Mixers. Liquids can be mixed inside or outside a process vessel by devices known as flow mixers, which include pipe mixers, orifice mixers, and jet mixers. Passing liquids through a pipe in highly turbulent flow is often enough to blend them thoroughly, especially if the pipe contains several bends. Orifices and baffles may be included in the pipe to increase the turbulence. Jets of liquid or air properly directed into a vessel are sometimes adequate for mixing liquids with viscosities below about 1,000 centipoises.

A centrifugal pump is also an effective mixer for liquids. Considerable quantities of liquid are recirculated through a holdup tank and an oversize pump; smaller amounts of the liquids to be mixed are admitted to the pump inlet. The high turbulence in the volute and the discharge line gives rapid and effective blending.[8]

Principles of Impeller Mixers. In all impeller mixers, a current of fluid is maintained by a rotating impeller. To obtain adequate capacity, the volumetric flow rate must be sufficient to sweep out the entire volume of the mixer in a reasonable time. Also, the velocity of the stream leaving the impeller must be sufficient to carry the currents to the remotest places in the tank, or the material in these places will not be mixed. The stream leaving the impeller carries a definite amount of kinetic energy, and this energy is dissipated by shear friction as the current flows through the mass of liquid. If the kinetic energy is too low to carry the current to the far corners of the tank, this requirement will not be met. It is necessary, then, that the velocity of the fluid be sufficient to provide at least a minimum amount of kinetic energy.[6]

Quantitative data on the flow and velocity requirements of agitators for specific tasks are not available, but the importance of these factors is recognized,[3, 11] and they are taken into account by the designers of mixers.

The flow rate for a turbine or paddle agitator of definite size and shape mounted in a tank containing a specific liquid is proportional to the quantity nD_a^3, where n is the agitator speed, in rps, and D_a is the diameter of the impeller.[10] For propellers the flow rate is proportional to nD_a^2.[8]

Mechanism of Mixing. The basic mixing action in a flow mixer occurs where the high-velocity currents come into contact with the adjacent stationary or slow-moving liquid. The surface between a current and its surrounding liquid is a surface of discontinuity, and the shear at this surface rolls up eddies in the adjacent liquid that are incorporated, or entrained, by the current. Inside the current, turbulence exists, and entrained liquid is rapidly incorporated into the stream by the mechanism of eddy viscosity. The higher the velocity, the more intense the turbulence, and the more effective is the mixing.

Flow Patterns in Mixers.[1, 4, 12] The type of flow in a mixer depends on the type of impeller, the characteristics of the fluid, and the size and proportions of the tank, baffles, and agitator. The velocity of the fluid at any point in the tank has three components, and the over-all flow pattern in the tank depends on the variations in these three velocity components from point to point. The first velocity component is radial, and acts in a direction perpendicular to the shaft of the impeller. The second component is longitudinal, and acts in a direction parallel with the shaft. The third component is tangential, or rotational, and acts in a direction tangent to a circular path around the shaft. In the usual case of a vertical shaft, the radial and tangential components are in a horizontal plane, and the longitudinal component is vertical. The radial and longitudinal components are useful, and provide the flow necessary for the mixing action. When the shaft is vertical and centrally located in the tank, the tangential component is generally disadvantageous. The tangential flow follows a circular path around the shaft, creates a vortex at the surface of the liquid, as shown in Fig. 6-5, and tends to perpetuate, by a laminar-flow circulation, stratification at the various levels without accomplishing longitudinal flow between levels. If solid particles are present, circulatory currents tend to throw the particles to the outside by centrifugal force, from where they move downward and to the center of the tank at the bottom. Instead of mixing, its reverse, concentration, occurs. Since, in circulatory flow, the

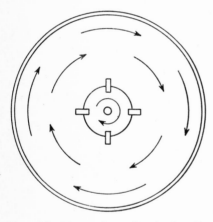

FIG. 6-5. Vortex formation in impeller mixer. (*After Lyons.*[4])

liquid flows with the direction of motion of the impeller blades, the relative velocity between the blades and the liquid is reduced, and the power that can be absorbed by the liquid is limited. If the speed is increased, the vortex deepens and reaches the suction of the impeller. When this happens, the power consumption drops sharply, and air is drawn into the charge, which is generally undesirable.

Circulatory flow and swirling can be prevented by either of two methods.[1] In small tanks, the impeller can be mounted off center as shown in Fig. 6-6. The shaft is moved away from the center line of the tank then tilted in a plane perpendicular to direction of the move. In larger tanks, the agitator may be mounted in the side of the tank, with the shaft in a horizontal plane, but at an angle with a radius, as shown in Fig. 6-7.

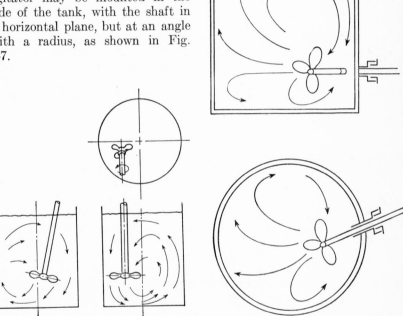

Fig. 6-6. Off-center impeller. (*Courtesy of AIChE.*)

Fig. 6-7. Side-entering propeller. (*After Bissell, Hesse, Everett, and Rushton.*[1])

For large tanks with vertical agitators, the preferable method of reducing swirling is to install *baffles*, which impede rotational flow without interfering with radial or longitudinal flow. A simple and effective baffling is attained by installing vertical strips perpendicular to the wall of the tank. Baffles of this type and the flow pattern resulting from their use are shown in Fig. 6-8. Except in very large tanks, four baffles are sufficient to prevent swirling and vortex formation. For turbines, the width of the baffle need be no more than one-twelfth the tank diameter; for propellers, no more than one-

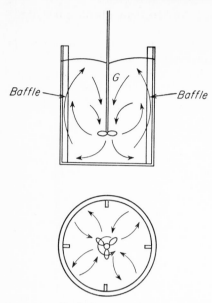

Fig. 6-8. Baffled tank. (*Courtesy of AIChE.*)

eighteenth the tank diameter.[1] With side-entering, inclined, or off-center propellers baffles are not needed.

The specific type of flow pattern in an impeller mixer depends on the type of impeller. Paddle mixers, which are the simplest and oldest impeller mixers, give good radial flow in the immediate plane of the blades but are poor in developing vertical currents. This is the main limitation of paddle mixers. The flow pattern in this type of mixer is shown in Fig. 6-9. Because of the poor vertical currents, paddle agitators are ineffective in suspending solids.

Propeller agitators drive the liquid straight down to the bottom of the tank, where the stream spreads radially in all directions toward the wall, flows upward along the wall, and returns to the suction of the propeller from the top. This pattern is shown in Fig. 6-8. Propellers are used when strong vertical currents are desired, e.g., when heavy solid particles are to be kept in suspension. They

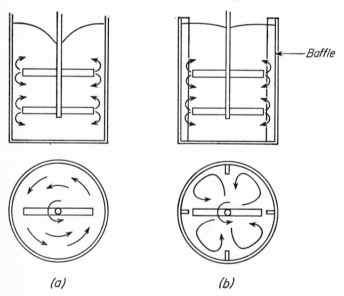

Fig. 6-9. Flow pattern in paddle agitator: (*a*) Without baffles. (*b*) With baffles. (*After Lyons.*[4])

are not ordinarily used for viscosities greater than about 5,000 centipoises.

Turbine impellers drive the liquid radially against the wall, where the stream divides, one portion flowing downward to the bottom and back to the center of the impeller from below, and the other flowing upward toward the surface and back to the impeller from above. As shown in Fig. 6-5, two separate circulation currents are generated. Turbines are especially effective in developing radial currents, but they also induce vertical flows, especially when baffled. They are excellent in mixing liquids having about the same specific gravity.

The most common shape of tank is the vertical cylinder, in which there are no distant corners difficult to reach by the currents. Tanks of other shapes are also used, however, and horizontal cylindrical tanks with horizontal agitators are common.

In a vertical cylindrical tank, the depth of the liquid should be equal to, or somewhat greater than, the diameter of the tank. If greater depth is desired, two or more impellers are mounted on the same shaft, and each impeller acts as a separate mixer. Two circulation currents are generated for each impeller, as shown in Fig. 6-10. The

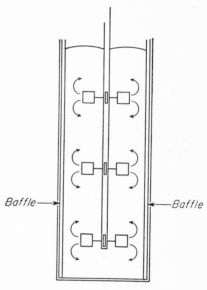

Fig. 6-10. Multiple turbines in tall tank.

bottom impeller, either of the turbine or the propeller type, is mounted about one impeller diameter above the bottom of the tank.

The return flow to an impeller of any type approaches the impeller from all directions, as it is not under the control of solid surfaces. The flow to and from a propeller, for example, is essentially similar to the flow of air to and from a fan operating in a room. In most applications of impeller mixers this is not a limitation, but when it is desired that the direction and velocity of flow to the suction of the impeller be controlled, *draft tubes* are used, as shown in Fig. 6-11. These devices may be useful when high shear in the impeller itself is desired, as in the manufacture of certain emulsions, or where solid particles that tend to float on the surface of the liquid in the tank are to be dispersed in the liquid. Draft tubes for propellers are mounted around the impeller, and those for turbines are mounted immediately above the impeller. This is shown in Fig. 6-11. Draft tubes add to the fluid friction in the system, and, for a given power input, they reduce the rate of flow, so they are not used unless they are required.

Shrouded impellers and *diffuser rings* can be used with turbines to stop swirling, in place of baffles, as shown in Fig. 6-2*f*. These devices are adapted from centrifugal-pump practice, but their main effect is to add friction and reduce circulation. Diffuser rings also tend to prevent the currents from reaching the far corners of the tank, and they are difficult to install and maintain. They are useful if intense shear and abnormal turbulence are desired at the impeller discharge.

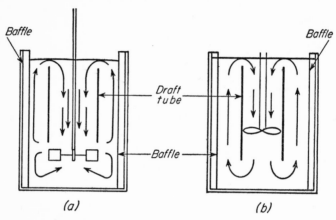

FIG. 6-11. Draft tubes, baffled tank: (*a*) Turbine. (*b*) Propeller. (*After Bissell, Hesse, Everett, and Rushton.*[1])

Power Consumption in Impeller Mixers. Quantitatively but little is known about the performance of mixers. Various methods of evaluating mixer performance are discussed in a later section. None of them are entirely satisfactory, but for the thin liquids handled by impeller mixers the power consumed is at least a crude measure of mixing effectiveness. The power consumed by an impeller can be studied by the same type of quantitative experiment, guided by dimensional analysis, that has been successfully applied to other problems in fluid mechanics. Thus, consider the flat-blade turbine agitator shown in Fig. 6-12. Assume that the tank is filled to a definite level (unagitated) with a liquid of known density and viscosity and that the impeller is driven at a speed of n revolutions per unit time. Assume also that all important measurements of the apparatus are known, including the diameter of the tank, the length and diameter of the impeller, the distance of the impeller from the bottom of the tank, the depth of the liquid, and the dimensions of the baffles if baffles are used. The number and arrangement of the baffles and the number of blades in the impeller are also assumed to be fixed. It is desired to find, by dimensional analysis, a relationship for the power required to maintain the speed of the impeller in terms of the above variables.

The power should be dependent on the flow pattern in the mixer and on the geometrical proportions of the equipment. The mechanism of flow is a complicated combination of laminar flow, turbulent flow, and boundary-layer separation. The controlling variables that must enter the analysis can be listed as follows: the important measurements of tank and impeller; the viscosity μ and the density ρ of the liquid; the speed n; and, because Newton's law applies, the dimensional constant g_c. Also, unless provision is made to eliminate swirling, a vortex will appear at the surface of the liquid. Some of the liquid must be lifted above the average, or unagitated, level of the liquid surface, and this lift must overcome the force

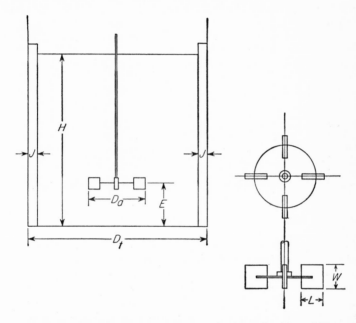

Fig. 6-12. Measurements of turbine mixer. (*Rushton, Costich, and Everett.*[12])

of gravity. Accordingly, the acceleration of gravity g must be considered as a factor in the analysis.

The various linear measurements can all be converted to dimensionless ratios, called shape factors, by dividing each of them by one of their number which is arbitrarily chosen as a basis. The diameter of the impeller D_a is a suitable choice for this base measurement, and the shape factors are calculated by dividing each of the remaining measurements by the magnitude of D_a. Let the shape factors, so defined, be denoted by $S_1, S_2, S_3,$ ..., S_n. The impeller diameter D_a is then also taken as the measure of the size of the equipment and used as a variable in the analysis, just as the diameter of the pipe was in the dimensional analysis of friction in pipes. Two mixers of the same geometrical proportions throughout but of different sizes will have identical shape factors but will differ in the magnitude of D_a. Devices meeting this requirement are said to be geometrically similar or to possess geometrical similarity.

Temporarily laying aside the shape factors, the power P is a function of the remaining variables, or

$$P = \psi\,(n,\,D_a,\,g_c,\,\mu,\,g,\,\rho\,) \tag{6-1}$$

Application of the method of dimensional analysis outlined in Chap. 1 gives the result [12]

$$\frac{Pg_c}{n^3 D_a^5 \rho} = \psi\left(\frac{nD_a^2\rho}{\mu},\frac{n^2 D_a}{g}\right) \tag{6-2}$$

By taking account of the shape factors, Eq. (6-2) can be written

$$\frac{Pg_c}{n^3 D_a^5 \rho} = \psi\left(\frac{nD_a^2\rho}{\mu},\frac{n^2 D_a}{g},S_1,S_2,\ldots,S_n\right) \tag{6-3}$$

The three dimensionless groups in Eq. (6-3) can be interpreted as follows. The product nD_a is proportional to the linear speed of the impeller tip. The group $nD_a^2\rho/\mu$ is proportional to $D_a u\rho/\mu$, where u is a linear velocity. The group, then, is a Reynolds number applying especially to an impeller rotating in a mass of fluid. The group $n^2 D_a/g$ is proportional to u^2/gD_a. This is called the Froude number and is denoted by N_{Fr}. It appears whenever wave motion on the surface of a liquid is important. It is a basic factor, for example, in evaluating the resistance to the motion of ships. The group $Pg_c/n^3 D_a^5\rho$ is called the power number, and is denoted by N_{po}. The power P is related to T, the torque on the impeller shaft, by the relation

$$P = \pi n T \tag{6-4}$$

The torque, in turn, is proportional to the product $F_D D_a$, where F_D is the drag force on the impeller. Since the geometrical proportions of the impeller have been fixed, A_p, the projected area of the impeller blades in the direction of motion of the blades, is proportional to D_a^2. The power number can then be written as

$$N_{po} \propto \frac{2nF_D D_a g_c}{(nD_a)^2 nD_a D_a^2\rho} = K\frac{2F_D g_c}{u^2 A_p\rho} = KC_D \tag{6-5}$$

where K is a constant and C_D is the drag coefficient, defined by Eq. (2-83). The power number is, then, equivalent to a drag coefficient. Physically, the power is applied to the liquid through the mechanism of the drag of the blades through the liquid.[3] Equation (6-3) can be written as

$$N_{po} = \psi(N_{Re},N_{Fr},S_1,S_2,\ldots,S_n) \tag{6-6}$$

The various shape factors in Eq. (6-6) depend on the type and arrangement of the equipment. A typical situation showing the necessary measurements and the corresponding shape factors is shown in Fig. 6-12. The measurements are tank diameter D_t, height of impeller above bottom of tank E, length of impeller blades L, width of impeller blades W, width of baffles J, and liquid depth H. Also, the number of baffles and of impeller blades must be specified. If a propeller is used, the pitch of the propeller is an important measurement. The shape factors for the mixer of Fig. 6-12 are $S_1 = D_t/D_a$, $S_2 = E/D_a$, $S_3 = L/D_a$, $S_4 = W/D_a$, $S_5 = J/D_a$, and $S_6 = H/D_a$.

The effect of the Froude number appears when there is vortex formation, and then only if the Reynolds number is greater than 300. For baffled tanks, or for side-entering propellers, or for Reynolds numbers less than 300, a vortex does not form, and the Froude number is not a factor. The Froude number is taken into account, when it is a factor, by the exponential equation

$$\frac{N_{po}}{N_{Fr}^m} = \psi(N_{Re},S_1,S_2,\ldots,S_n) = \phi \tag{6-7}$$

Quantity ϕ is called the power function.

The exponent m in Eq. (6-7) is, for a given set of shape factors, empirically related to the Reynolds number by the equation [12]

$$m = \frac{a - \log N_{Re}}{b} \tag{6-8}$$

where a and b are constants. The magnitudes of a and b for the curves of Figs. 6-13 and 6-14 are given in Table 6-3.

TABLE 6-3. CONSTANTS a AND b OF EQ. (6-8)

Fig.	Line	a	b
6-13	B	1.0	40.0
6-14	B	1.7	18.0
6-14	C	0	18.0
6-14	D	2.3	18.0

Equation (6-7) is applied by determining the quantity ϕ experimentally as a function of the Reynolds number for constant shape factors and plotting separate curves of ϕ vs. N_{Re} for each set of shape factors. Curves for a number of typical designs and types of impellers have been determined.[12]

A typical plot of ϕ vs. N_{Re} applying to tanks fitted with centrally located vertical flat-bladed turbines with six blades is shown in Fig. 6-13. The important shape factors are $S_1 = 3$, $S_2 = 1.0$, $S_3 = 0.25$, $S_6 = 1.0$. When baffled with four baffles, each of width one-tenth the tank diameter ($S_5 = 0.1$), curve A applies, and $\phi = N_{po}$. Without baffles, curve B applies, and the Froude number must be included in ϕ for all Reynolds numbers greater than 300.

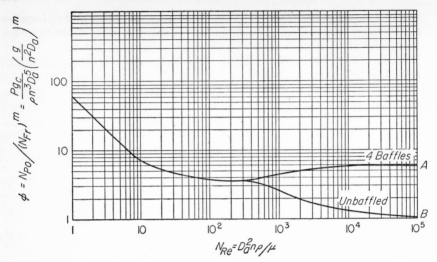

FIG. 6-13. Power function ϕ vs. N_{Re} for six-blade turbine. (*Rushton, Costich, and Everett.*[12] *Courtesy of AIChE.*)

Typical curves for three-bladed propeller mixers are shown in Fig. 6-14. For all curves, the propeller is one diameter from the bottom of the tank, and $S_2 = 1.0$. Curves A and B apply for blades having a pitch of 2.0 and a tank-to-propeller ratio $S_1 = 3.3$. Curve A applies to baffled tanks with four baffles where $S_5 = 0.1$. Curve B applies to unbaffled tanks. Curves C and D show the effect on N_{po} of changing the scale factor S_1,

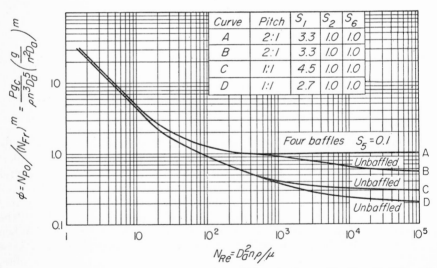

FIG. 6-14. Power function ϕ vs. N_{Re} for three-blade propellers. (*Rushton, Costich, and Everett.*[12] *Courtesy of AIChE.*)

the tank-to-propeller ratio, when the pitch is about 1.0. At low Reynolds numbers the effect of changing S_1 fades. Curve C applies where $S_1 = 4.5$, and curve D where $S_1 = 2.7$.

At low Reynolds numbers, the lines of N_{po} vs. N_{Re} for both baffled and unbaffled tanks coincide, and the slope of the line on logarithmic coordinates is -1. The flow becomes laminar in this range, density is no longer a factor, and Eq. (6-7) becomes

$$N_{po}N_{Re} = \frac{Pg_c}{n^2 D_a^3 \mu} = K_L = \psi_L(S_1,S_2,\ldots,S_n) \tag{6-9}$$

This relation can be used when N_{Re} is less than 10.

In baffled tanks at Reynolds numbers larger than about 10,000, the power function is independent of the Reynolds number, and viscosity is not a factor. In this range, the flow is fully turbulent, and Eq. (6-6) becomes

$$N_{po} = K_T = \psi_T(S_1,S_2,\ldots,S_n) \tag{6-10}$$

Magnitudes of the constants K_T and K_L for various types of impellers and tanks are shown in Table 6-4.

TABLE 6-4. VALUES OF CONSTANTS K_L AND K_T IN EQS. (6-9) AND (6-10) FOR BAFFLED TANKS HAVING FOUR BAFFLES AT TANK WALL, WITH WIDTH EQUAL TO 10 PER CENT OF THE TANK DIAMETER †

Type of impeller	K_L	K_T
Propeller (square pitch, three blades).......	41.0	0.32
Propeller (pitch of 2, three blades).........	43.5	1.00
Turbine (six flat blades).................	71.0	6.30
Turbine (six curved blades)..............	70.0	4.80
Turbine (six arrowhead blades)...........	71.0	4.00
Fan turbine (six blades).................	70.0	1.65
Flat paddle (two blades).................	36.5	1.70
Shrouded turbine (six curved blades).......	97.5	1.08
Shrouded turbine (with stator, no baffles)...	172.5	1.12

† From J. H. Rushton, *Ind. Eng. Chem.*, **44**: 2931 (1952).

The power delivered to the liquid is computed by combining Eq. (6-7) and the definition of N_{po} to give

$$P = \frac{\phi N_{Fr}^m n^3 D_a^5 \rho}{g_c} \tag{6-11}$$

When the Froude number is not a factor, the power is given by

$$P = \frac{\phi n^3 D_a^5 \rho}{g_c} \tag{6-12}$$

Example 6-1. A flat-blade turbine with six blades is installed centrally in a vertical tank. The tank is 6 ft in diameter; the turbine is 2 ft in diameter and is positioned 2 ft from the bottom of the tank. The tank is filled to a depth of 6 ft with a solution of 50 per cent caustic soda, at 150°F, which has a viscosity of 12 centipoises and a density of 93.5 lb/ft³. The turbine is operated at 90 rpm. The tank is unbaffled. What horsepower will be required to operate the mixer?

Solution. Curve *B* in Fig. 6-13 applies under the conditions of this problem. The Reynolds and Froude numbers are calculated. The quantities for substitution are, in consistent units,

$$D_a = 2 \text{ ft}$$

$$n = \tfrac{90}{60} = 1.5 \text{ rps}$$

$$\mu = 12 \times 6.72 \times 10^{-4} = 8.06 \times 10^{-3} \text{ lb/ft-se}$$

$$\rho = 93.5 \text{ lb/ft}^3$$

$$g = 32.17 \text{ ft/sec}^2$$

Then

$$N_{\text{Re}} = \frac{D_a^2 n \rho}{\mu} = \frac{2^2 \times 1.5 \times 93.5}{8.06 \times 10^{-3}} = 69{,}600$$

$$N_{\text{Fr}} = \frac{n^2 D_a}{g} = \frac{1.5^2 \times 2}{32.17} = 0.14$$

From Table 6-3, the constants *a* and *b* for substitution into Eq. (6-8) are *a* = 1.0 and *b* = 40.0. From Eq. (6-8),

$$m = \frac{1.0 - \log 69{,}500}{40.0} = -0.096$$

From curve *B* (Fig. 6-13), for $N_{\text{Re}} = 69{,}600$, $\phi = 1.07$, and from Eq. (6-11),

$$P = \frac{1.07 \times 0.14^{-0.096} \times 93.5 \times 1.5^3 \times 2^5}{32.17} = 406 \text{ ft-lb force/sec}$$

The power requirement is 406/550 = 0.74 hp.

Example 6-2. The tank of Example 6-1 is fitted with four baffles, each having a width of 7.5 in. What horsepower will be required to operate this baffled mixer?

Solution. Curve *A* of Fig. 6-13 now applies, and $\phi = N_{po}$. From curve *A*, at a Reynolds number of 69,500, $\phi = 6.0$. The Froude number is no longer a factor. Then, from Eq. (6-12),

$$P = \frac{6.0 \times 93.5 \times 1.5^3 \times 2^5}{32.17} = 1{,}880 \text{ ft-lb force/sec}$$

The power is 1,880/550 = 3.42 hp.

The increase in power resulting from the installation of the baffles is 360 per cent, but the effectiveness of the mixer is also greatly improved. It is important that, if baffles are installed in an unbaffled tank operating at high Reynolds numbers, the motor used to drive the mixer be sufficiently powerful. If, for example, a 1-hp motor were installed for operation under the conditions of Example 6-1, and if this motor were not replaced by a larger one when the baffles were installed, the 1-hp motor would probably burn out.

Example 6-3. The mixer of Example 6-1 is to be used to mix a rubber latex compound having a viscosity of 120,000 centipoises and a density of 70 lb/ft³. What horsepower will be required?

Solution. The Reynolds number is now

$$N_{Re} = \frac{2^2 \times 1.5 \times 70}{120,000 \times 6.72 \times 10^{-4}} = 5.2$$

This is well within the range of laminar flow, and $\phi = N_{po}$. For $N_{Re} = 5.2$, from Fig. 6-13, $N_{po} = 12.5$, and

$$P = \frac{12.5 \times 70 \times 1.5^3 \times 2^5}{32.17} = 2,940 \text{ ft-lb force/sec}$$

The power required is $2,940/550 = 5.35$ hp. This power requirement is independent of whether or not the tank is baffled. There is no reason for baffles in a mixer operated at low Reynolds numbers, as vortex formation does not occur under such conditions.

Note that a 10,000-fold increase in viscosity increases the power by only about 60 per cent over that required by the baffled tank operating on the low-viscosity liquid.

Scale-up of Agitator Design. A major problem in agitator design is to scale up from a laboratory or pilot-plant mixer to a full-scale agitator. Several methods have been proposed, all based on geometrical similarity between the laboratory and plant equipment. No satisfactory method is available for scaling up from an agitator of one design to one of another design. Given geometrical similarity, however, a conservative method of scale-up is to keep the power input per unit volume constant.[7b] This is done by the use of curves of ϕ vs. N_{Re}, as in Figs. 6-13 and 6-14. The optimum speed of the laboratory unit is first established. From the Reynolds number of the laboratory agitator the factor ϕ is read from the graph, and the speed required for the full-scale agitator to give the same power input per unit volume is computed by trial and error from the graph and the dimensions of the larger unit.

The optimum ratio of impeller diameter to tank diameter for a given power input is an important factor in scaling up impellers. Figure 6-15 shows how the rate of two

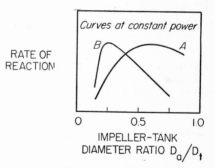

FIG. 6-15. Effect of impeller size on reaction rate at equal power input. (*Rushton*,[13] *courtesy of AIChE.*)

desired reactions varies with this ratio. Curve A applies to blending liquids and curve B to dispersing gas in a liquid. Both curves are based on the same power input. For blending, a large impeller gives the fastest mixing, and in dispersing, a smaller impeller is best. Since the power is the same in the two cases, the smaller impeller has the higher speed.

A small, high-speed impeller gives high turbulence but low flow. A large, slow impeller, on the other hand, gives large flow but low turbulence. In this way, the designer correlates speed and impeller diameter with the characteristics of the task to be performed.

MIXERS FOR PASTES AND PLASTIC MASSES

The mixing of heavy pastes, plastic solids, and rubber is much more of an art than a science. The properties of the materials to be mixed vary enormously from one problem to another. Even in a single material they may be widely different at various times during the mixing operation. A batch may start as a dry, free-flowing powder, become pasty on the addition of liquid, stiff and gummy as a reaction proceeds, and then perhaps dry, granular, and free-flowing once more. Indeterminate properties of the material such as "stiffness," "tackiness," and wettability are as significant in these mixing problems as viscosity and density. Mixers in this second class must, above all, be versatile. In a given problem the mixer chosen must handle the material when in its worst condition, and may not be so effective as other designs during other parts of the mixing cycle. As with other equipment, the choice of a mixer for heavy materials is often a compromise.

Mixing Action and Types of Mixers. All mixers in this class operate by direct action on the material adjacent to the mixing elements. They do not set up flow currents or mix at a distance, as do impeller mixers. Instead the agitator travels throughout the mass of material, or the material is brought to the agitator. Mixers described in this section are change-can mixers; kneaders, dispersers, and masticators; continuous kneaders; mixer-extruders; mixing rolls; mullers and pan mixers; and pugmills.

Change-can Mixers. These devices blend viscous liquids or light pastes, as in food processing or paint manufacture. A small removable can, 5 to 100 gal in size, holds the material to be mixed. In the "pony mixer" shown in Fig. 6-16a the agitator consists of several vertical blades or

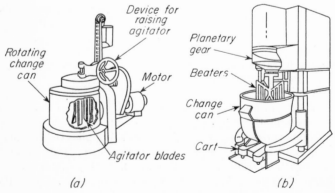

(a)   *(b)*

FIG. 6-16. Double-motion paste mixers: (a) Pony mixer. (b) Beater mixer.

fingers held on a rotating head and positioned near the wall of the can. The blades are slightly twisted. The agitator is mounted eccentrically with respect to the axis of the can. The can rests on a turntable driven in a direction opposite to that of the agitator, so that during operation all the liquid or paste in the can is brought to the blades to be mixed. When the mixing is complete, the agitator head is raised, lifting the blades out of the can; the blades are wiped clean; and the can is replaced with another containing a new batch.

In the beater mixer in Fig. 6-16b the can or vessel is stationary. The agitator has a planetary motion; as it rotates, it precesses, so that it repeatedly visits all parts of the vessel. Beaters are shaped to pass with close clearance over the side and bottom of the mixing vessel.

Kneaders, Dispersers, and Masticators. Kneading is a method of mixing used with deformable or plastic solids. It involves squashing the mass flat, folding it over on itself, and squashing it once more. Most kneading machines also tear the mass apart and shear it between a moving blade and a stationary surface. Considerable energy is required even with fairly thin materials, and as the mass becomes stiff and rubbery, the power requirements become very large.

A *two-arm kneader* handles suspensions, pastes, and light plastic masses. Typical applications are in the compounding of lacquer bases from pigments and carriers and in shredding cotton linters into acetic acid and acetic anhydride to form cellulose acetate. A *disperser* is heavier in construction and draws more power than a kneader; it works additives and coloring agents into stiff materials. A *masticator* is still heavier and draws even more power. It can disintegrate scrap rubber and compound the toughest plastic masses that can be worked at all. Masticators are often called "intensive mixers."

In all these machines the mixing is done by two heavy blades on parallel horizontal shafts turning in a short trough with a saddle-shaped bottom. The blades turn toward each other at the top, drawing the mass downward over the point of the saddle then shearing it between the blades and the wall of the trough. The circles of rotation of the blades are usually tangential, so that the blades may turn at different speeds in any desired ratio. The optimum ratio is about $1\frac{1}{2}:1$. In some machines the blades overlap and turn at the same speed or with a speed ratio of $2:1$. A small two-arm kneader with tangential blades is sketched in Fig. 6-17.

Designs of mixing blades for various purposes are shown in Fig. 6-18. The common sigma blade shown at the left is used for general-purpose kneading. Its edges may be serrated to give a shredding action. The double-naben, or fishtail, blade in the center is particularly effective with heavy plastic materials. The dispersion blade at the right develops the high shear forces needed to disperse powders or liquids into plastic or rubbery masses. Masticator blades are even heavier than those shown, sometimes being little larger in diameter than the shafts which drive them. Spiral, flattened, and elliptical designs of masticator blades are used.

Material to be kneaded or worked is dropped into the trough and mixed for 5 to 20 min or longer. Sometimes the mass is heated while in the machine, but more commonly it must be cooled to remove the heat generated by the mixing action. The trough is often unloaded by tilting it so

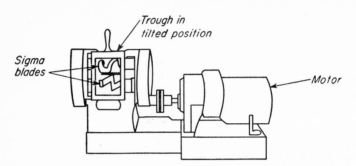

FIG. 6-17. Two-arm kneader.

that its contents spill out. In kneaders and some dispersers only one agitator blade is directly driven; the other is turned by timing gears. In masticators both shafts are directly driven, sometimes from both ends, so that the trough cannot be tilted and must be unloaded through an opening in the floor.

In many kneading machines the trough is open, but in some designs, known as "internal mixers," the mixing chamber is closed during the operating cycle. Thus, a cover, the underside of which conforms to the volume swept out by the blades, can be used on the kneader shown in Fig. 6-17. Such mixers do not tilt. This type is used for dissolving rubber and for making dispersions of rubber in liquids. The most common inter-

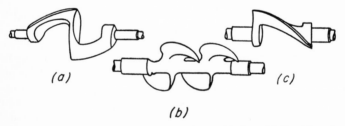

FIG. 6-18. Kneader and disperser blades: (a) Sigma blade. (b) Double-naben blade. (c) Disperser blade.

nal mixer is the *Banbury mixer*, shown in Fig. 6-19. This is a heavy-duty two-arm mixer in which the agitators are in the form of interrupted spirals. The shafts turn at 30 to 40 rpm. Solids are charged in from above and held in the trough during mixing by an air-operated piston under a pressure of 15 to as high as 150 lb force/in.^{2} Mixed material is discharged through a sliding door in the bottom of the trough. Banbury mixers compound rubber and plastic solids, masticate crude rubber, devulcanize rubber scrap, and make water dispersions and rubber solutions. They also accomplish the same tasks as kneaders but in a shorter time and with smaller batches. The heat generated in the material is removed by cooling

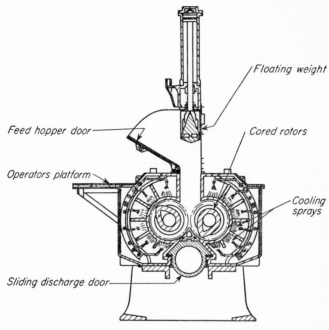

Fig. 6-19. Banbury internal mixer. (*Farrel-Birmingham Mfg. Co.*)

water sprayed on the walls of the mixing chamber and circulated through the hollow agitator shafts.

Continuous Kneaders. The machines just described operate batchwise on relatively small amounts of material. The more difficult the material is to mix, the smaller the batch size must be. Many industrial processes are continuous, with steady uniform flow into and out of units of equipment; into such processes batch equipment is not readily incorporated. Continuous kneading machines have been developed which can handle light and medium stiff pastes. A typical design, the Ko-Kneader, is shown in Fig. 6-20. A single horizontal shaft, slowly turning in a mixing chamber, carries rows of teeth arranged in a spiral pattern to move the material through the chamber. The teeth on the rotor pass with close clearance between stationary teeth set in the wall of the casing. The shaft not only turns; it also reciprocates in the axial direction. Material between the meshing teeth is therefore smeared in an axial or longitudinal direction as well as being subjected to radial shear. Solids enter the machine near the driven end of the rotor, as shown, and discharge through an opening surrounding the shaft bearing in the opposite end of the mixing chamber. The chamber is an open trough with light solids, a closed cylinder with plastic masses. These machines can mix several tons per hour of heavy, stiff, or gummy materials.

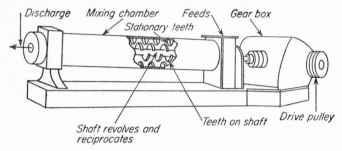

FIG. 6-20. Ko-Kneader mixer. (*Baker-Perkins, Inc.*)

Mixer-Extruders. If the discharge opening of the Ko-Kneader is restricted by covering it with an extrusion die, the pitched blades of the rotor build up considerable pressure in the material. The mix is cut and folded while in the mixing chamber and subjected to additional shear as it flows through the die. Other mixer-extruders function in the same way. They contain one or two horizontal shafts, rotating but not reciprocating, carrying a helix or blades set in a helical pattern. Pressure is built up by reducing the pitch of the helix near the discharge, reducing the diameter of the mixing chamber, or both. In the Roto-feed mixer shown in Fig. 6-21

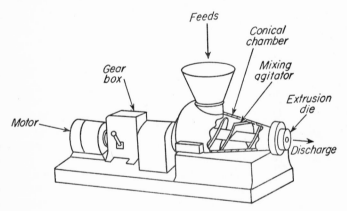

FIG. 6-21. Roto-feed mixer. (*Marco Co., Inc.*)

a helical agitator turns in a conical chamber. Material is fed into the large end of the cone and discharges through a die head at the small end. Mixing in the chamber is enhanced by a blade on the agitator, which folds the material back on itself as it flows through. The design of the agitator is varied to suit the requirements of the problem. Mixer-extruders continuously mix, compound, and work thermoplastics, doughs, clays, and other hard-to-mix materials.

Mixing Rolls. Another way of subjecting pastes and deformable solids to intense shear is to pass them between smooth metal rolls turning at different speeds. By repeated pass-es between such mixing rolls solid additives can be thoroughly dis-persed into pasty or stiff plastic materials. Thin fluid pastes in the paint industry are passed through continuous mills containing three to five horizontal rolls, set one above the other in a vertical stack. The paste passes from the slower rolls to successively faster rolls. Rubber products and some plastics are com-

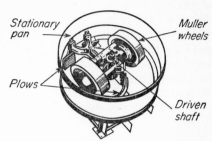

Fig. 6-22. Muller mixer.

pounded on batch roll mills with two rolls set in the same horizontal plane. Solids are picked up on the faster roll, cut at an angle by the operator, and folded back into the "bite" between the rolls. Additives are sprinkled on the material as it is being worked. Batch roll mills require long mixing times and careful attention by the operator, and in general are less fa-vored for most services than internal mixers.

Muller Mixers. A muller gives a distinctly different mixing action from that of other machines. Mulling is a smearing or rubbing action similar to that in a mortar and pestle. In large-scale processing this action is given by the wide, heavy wheels of the mixer shown in Fig. 6-22. In this particular design of muller the pan is stationary and the central vertical shaft is driven, causing the muller wheels to roll in a circular path over a layer of solids on the pan floor. The rubbing action results from the slip of the wheels on the solids. Plows guide the solids under the muller wheels, or to an opening in the pan floor at the end of the cycle when the mixer is being discharged. In another design the axis of the wheels is held stationary and the pan is rotated; in still another the wheels are not centered in the pan but are off-set, and both the pan and the wheels are driven. Mixing plows may be substituted for the muller wheels to give what is called a *pan mixer*. Mull-ers are good mixers for batches of heavy solids and pastes; they are especially effective in uniformly coating the particles of a granular solid with a small amount of liquid.

Fig. 6-23. Pugmill.

Pugmills. In a pugmill, shown in Fig. 6-23, the mixing is done by blades or knives set in a helical pattern on a horizontal shaft turning in an open trough or closed cylinder. Solids continuously enter one end of the mixing chamber and discharge from the other. While in the chamber, they are cut, mixed, and moved forward to be acted upon by each succeeding blade. Single-shaft mills utilize an enclosed mixing chamber; open-trough double-shaft mills are used where more rapid or more thorough mixing is required. The chamber of most enclosed mills is cylindrical, but in some it is polygonal in cross section to prevent sticky solids from being carried around with the shaft. Pugmills blend and homogenize clays, break up agglomerates in plastic solids, and mix liquids with solids to form thick, heavy slurries. Sometimes they operate under vacuum to deair clay or other materials. They are built with jackets for heating or cooling.

Power Requirements. Large amounts of mechanical energy are needed to mix heavy plastic masses. The material must be sheared into elements which are moved relative to one another, folded over, recombined, and redivided. In continuous mixers the material must also be moved through the machine. Only part of the energy supplied to the mixer is directly useful for mixing, and in many machines the useful part is small. Probably mixers which work intensively on small quantities of material, dividing it into very small elements, make more effective use of energy than those which work more slowly on large quantities. Machines which weigh little per pound of material processed waste less energy than heavier machines. Other things being equal, the shorter the mixing time required to bring the material to the desired degree of uniformity, the larger the useful fraction of the energy supplied will be. Regardless of the design of the machine, however, the power needed to drive a mixer for pastes and deformable solids is many times greater than that needed by a mixer for liquids, as shown by Table 6-2. The energy supplied appears as heat, which must ordinarily be removed to avoid damaging the machine or the material. In a large internal mixer this means that a considerable volume of coolant must be circulated through the machine. This is illustrated by the following example.

Example 6-4. A large Banbury mixer masticates 1,500 lb of scrap rubber with a density of 70 lb/ft³. The power load is 7,000 hp per 1,000 gal of rubber. How much cooling water, in gallons per minute, is needed to remove the heat generated in the mixer if the temperature of the water is not to rise more than 20°F?

Solution. Assume that all the energy delivered to the mass appears as heat.

$$\text{Volume of batch} = \frac{1,500 \times 7.48}{70} = 160 \text{ gal}$$

$$\text{Power load} = \frac{7,000 \times 160}{1,000} = 1,120 \text{ hp}$$

$$1 \text{ hp} = 2545 \text{ Btu/hr}$$

The heat to be removed is

$$1,120 \times 2545 = 2,850,000 \text{ Btu/hr}$$

The specific heat of water is 1.0 Btu/(lb)(°F), and the cooling water required is

$$\frac{2,850,000}{20 \times 1.0} = 142,500 \text{ lb/hr} = \frac{142,500}{60 \times 8.33} = 285 \text{ gal/min}$$

MIXERS FOR DRY POWDERS

Many of the machines described in the last section can blend solids when they are dry and free-flowing as well as when they are damp, pasty, rubbery, or plastic. Mullers, pan mixers, and pugmills are examples. Such versatile machines are needed when the properties of the material change markedly during the mixing operation. In general, however, these devices are less effective on dry powders than on other materials and are heavier and more powerful than necessary for free-flowing particulate solids.

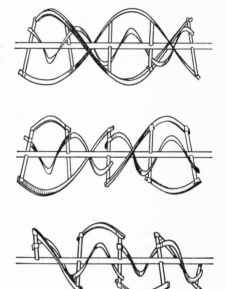

The lighter machines discussed here handle dry powders and—sometimes—thin pastes. They mix by mechanical shuffling, as in ribbon blenders; by repeatedly lifting and dropping the material and rolling it over, as in tumbling mixers and vertical screw mixers; or by smearing it out in a thin layer over a rotating disk or impact wheel.

Ribbon Blenders. A ribbon blender consists of a horizontal trough containing a central shaft and a helical ribbon agitator. Typical designs of mixing ribbons are shown in Fig. 6-24. Two counteracting ribbons are mounted on the same shaft, one moving the solid slowly in one direction, the other moving it quickly in the other. The ribbons may be continuous or interrupted. Mixing results from the

FIG. 6-24. Continuous and interrupted agitator ribbons.

"turbulence" induced by the counteracting agitators, not from mere motion of the solids through the trough. Some ribbon blenders operate batchwise, with solids charged in and mixed until satisfactory; others mix continuously, with solids fed in one end of the trough and discharged from the other. The trough is open or lightly covered for light duty, closed and heavy-walled for operation under pressure or vacuum. Ribbon blenders are effective mixers for thin pastes and for powders that do not flow readily. Some batch units are very large, holding up to 9,000 gal of material. The power they require, as shown in Table 6-2, is moderate.

Tumbling Mixers. Many materials are mixed by tumbling them in a partly filled container rotating about a horizontal axis. The ball mills described in Chap. 4 are often used as mixers. Designs of tumbling mills which do not contain internal grinding elements are illustrated in Fig. 6-25. The mixer shown at (a) resembles a ball mill without the balls. It

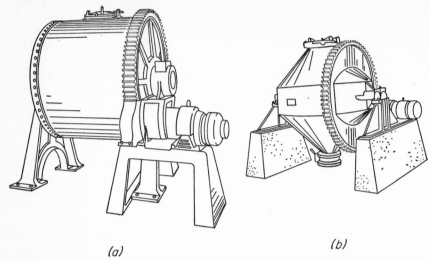

(a)

(b)

FIG. 6-25. Tumbling mixers: (a) Tumbling barrel. (b) Double-cone mixer.

mixes suspensions of dense solids in liquids and heavy dry powders, as in the manufacture of glass. The other designs handle lighter dry solids only. The double-cone mixer shown at (b) is a popular mixer for free-flowing dry powders. A batch is charged into the body of the machine from above until it is 50 to 60 per cent full. The ends of the container are closed and the solids tumbled for 5 to 20 min. The machine is stopped; mixed material is dropped out the bottom of the container into a conveyor or bin. Tumbling mixers are made in a wide range of sizes and materials of construction. They draw a little less power, ordinarily, than do ribbon blenders.

Impact Wheels. Fine, light powders such as insecticides may be blended continuously by spreading them out in a thin layer under centrifugal action. A premix of the several dry ingredients is fed continuously near the center of a high-speed spinning disk 10 to 27 in. in diameter, which throws it outward into a stationary casing. The intense shearing forces acting on the powders during

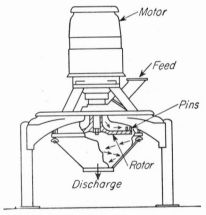

FIG. 6-26. Entoleter mixer. (*Safety Car Heating and Lighting Co.*)

their travel over the disk surface thoroughly blend the various materials. The disk in some machines is vertical; in others it is horizontal. In the device shown in Fig. 6-26 the premix is dropped onto a double rotor carrying short vertical pins near its periphery to increase the mixing effectiveness. A 14-in. disk turns at 1,750 rpm for easy problems and 3,500 rpm for materials that are hard to mix. Sometimes several passes through the same machine, or through machines in series, are necessary. For good results the premix fed to an impact wheel must be fairly uniform, for there is almost no holdup of material in the mixer and no chance for recombining material that has passed through with that which is entering. Impact wheels blend 1 to 25 tons/hr of light free-flowing powders.

MIXER OPERATION

Criteria of Mixer Effectiveness: Mixing Index. The performance of an industrial mixer is judged by the time required, the power load, and the properties of the product. Both the requirements of the mixing device and the properties desired in the mixed material vary widely from one problem to another. Sometimes a very high degree of uniformity is required; sometimes a rapid mixing action; sometimes a minimum amount of power. Because of the lack of quantitative measures of mixing performance, highly arbitrary criteria of performance have been the rule, especially with mixers for pastes, plastic solids, and dry powders.

With thin liquids the amount of power consumed by the impeller per unit volume of liquid has been used as a measure of mixing effectiveness, based on the reasoning that increased amounts of power mean a higher degree of turbulence and a higher degree of turbulence means better mixing. Studies have shown this to be at least roughly true. In a given mixer the amount of power consumed can be directly related to the rate of solution of a gas or the rate of certain reactions, such as oxidations, that depend on the intimacy of contact of one phase with another. In a rough qualitative way it may be said that $\frac{1}{2}$ to 1 hp per 1,000 gal of thin liquid gives "mild" agitation, 2 to 3 hp per 1,000 gal gives "vigorous" agitation, and 4 to 6 hp per 1,000 gal gives "intense" agitation. These figures refer to the power that is actually delivered to the liquid and do not include power used in driving gear-reduction units or in turning the agitator shaft in bearings and stuffing boxes.

It should be remembered also that there is not necessarily any direct relation between power consumed and amount or degree of mixing. The energy delivered by any mixing device to the mixer contents is used partly for blending and partly for merely lifting the material or moving it about. Under some conditions with a thin liquid, as when it is swirled about in an unbaffled vessel, individual particles may follow parallel circular paths almost indefinitely and mix very slowly or not at all. Almost none of the energy supplied is used for mixing. The same liquid in a baffled vessel, however, may be effectively mixed by the same agitator. In the baffled vessel most of the power is utilized for mixing, and much less of it is wasted.

The power drawn by a given impeller in a given vessel and a given liquid is a fairly good measure of mixing effectiveness, but when comparing one mixer or one system with another, the power consumption is a poor criterion. In general the "best" mixer is the one which does the desired job with the smallest amount of power.

With mixers for pastes, plastic solids, and powders even such approximate methods of estimating performance are not available. The power load has not been related to the properties of the mixed material or to the shape and speed of the mixing elements. Changes in design have been made almost entirely on the basis of empirical tests.

The only true measure of *mixing*, as opposed to mere movement of the material in the vessel, is the degree of uniformity of the product. The factors influencing the rate of attainment of uniformity have been studied experimentally by adding a small amount of tracer material to the batch and taking numerous sets of small samples for analysis after various increments of mixing time. The amount of agreement, or lack of agreement, among the analyses can be expressed by statistical methods as a "mixing index." [5] The index is a measure of the homogeneity of the mix. Mixing in reality is the creation of a state of disorder, not one of order; it is the dispersion of one material into another in a completely random, disordered way. With coarse, particulate solids, therefore, complete homogeneity cannot be reached. Small samples will always differ somewhat in analysis no matter how long mixing proceeds.[8] With liquid solutions or very small particles the mass may approach true homogeneity with long mixing times.

Batch and Continuous Mixing. Mixing is more often a batchwise operation than a continuous one. Simple blending of thin liquids and dry powders is fairly easy to make continuous, but mixing viscous liquids, pastes, and plastic solids is not. When the materials fed to a mixer vary from time to time in physical or chemical properties or change markedly in consistency during mixing, batchwise operation is often the only answer. Each batch can be processed for a different amount of time or under different conditions at the discretion of the operator until it is satisfactory.

Continuous operation is much less flexible. All material passing through the mixer receives the same treatment regardless of its properties. Off-quality material may escape detection in the mixer and on discharge spoil a large amount of otherwise acceptable product. A continuous mixer is usually designed for a specific problem, and is not applicable to other problems. The rates of feed to a continuous mixer must be continuously and precisely controlled, whereas in a batch mixer the ingredients need be weighed only once. Often more of one ingredient may be added to a batch if the need arises.

On the other hand, continuous equipment is much smaller than batch equipment and usually makes more effective use of power. Sometimes the intensive working given a small quantity of material in a continuous mixer results in a distinctly different or better product than does the gentler, slower action of a batch mixer. Since continuous machines are small, they may economically be made of expensive corrosion-resistant materials throughout. They can be more easily designed to avoid pockets or crevices in which material may lodge and escape being mixed, as sometimes occurs in batch mixers.

Small batch mixers operating in parallel on staggered cycles may be adapted for use in continuous processing where a continuous mixer cannot be employed. The batch mixers successively discharge small amounts of mixed material into a small bin or storage vessel, from which flow may then be continuous. They may, however, require considerable attention during operation. This compromise of batch with continuous processing gives what has been called "plug" or "slugwise" continuous flow.[8]

SYMBOLS

A_p Projected area of impeller in direction of motion, ft^2
a Constant in Eq. (6-8)
b Constant in Eq. (6-8)
D Diameter, ft; D_a, of agitator or impeller; D_t, diameter of tank
E Height of impeller above tank bottom, ft
F_D Drag force, lb force
g Acceleration of gravity, ft/sec^2
g_c Newton's-law conversion factor, 32.174 ft-lb/lb force-sec^2
H Depth of liquid in tank, ft
J Width of baffle, ft
K Constant in Eq. (6-5); K_L, constant in Eq. (6-9); K_T, constant in Eq. (6-10)
L Length of impeller blade, ft
m Exponent in Eq. (6-7)
n Agitator speed, rps
P Power, ft-lb force/sec
T Torque, ft-lb force
u Linear velocity, ft/sec
W Width of impeller, ft

Greek Letters

μ Viscosity, lb/ft-sec
ρ Density, lb/ft^3
ϕ Power function, N_{po}/N_{Fr}^m
ψ Function; ψ_L, function in Eq. (6-9); ψ_T, function in Eq. (6-10)

Dimensionless Groups

C_D Drag coefficient, $2F_D g_c/\rho u^2 A_p$
N_{Fr} Froude number, $n^2 D_a/g$
N_{po} Power number, $Pg_c/n^3 D_a^5 \rho$
N_{Re} Reynolds number, $nD_a^2 \rho/\mu$
S Shape factor; $S_1 = D_t/D_a$; $S_2 = E/D_a$; $S_3 = L/D_a$; $S_4 = W/D_a$; $S_5 = J/D_a$; $S_6 = H/D_a$

PROBLEMS

6-1. A tank 4 ft in diameter and 6 ft high is filled to a depth of 4 ft with a latex having a viscosity of 1,000 centipoises and a density of 47 lb/ft^3. The tank is not baffled. A three-blade 12-in.-diameter propeller is installed in the tank 1 ft from the bottom. The pitch is 1:1 (pitch equals diameter). The motor available develops 10 hp. Is the motor adequate to drive this agitator at a speed of 1,000 rpm?

6-2. What is the maximum speed at which the agitator of the tank described in Prob. 6-1 may be driven if the liquid is replaced by one having a viscosity of 100 centipoises and the same density?

6-3. What power is required for the mixing operation of Prob. 6-1 if a propeller 12 in. in diameter with a 2:1 pitch (pitch twice the diameter) is used and if four baffles, each 5 in. wide, are installed?

REFERENCES

1. Bissell, E. S., H. C. Hesse, H. J. Everett, and J. H. Rushton: *Chem. Eng. Progr.*, **43**: 649 (1947).
2. Bullock, H. L.: *Chem. Eng. Progr.*, **47**: 397 (1951).
3. Hixson, A. W., and S. L. Baum: *Ind. Eng. Chem.*, **34**: 194 (1942).
4. Lyons, E. J.: *Chem. Eng. Progr.*, **44**: 341 (1948).
5. Michaels, A. S., and V. Puzinauskas: *Chem. Eng. Progr.*, **50**: 604 (1954).
6. Miller, F. D., and J. H. Rushton: *Ind. Eng. Chem.*, **36**: 499 (1944).
7. Perry, J. H.: "Chemical Engineers' Handbook," 3d ed., (*a*) pp. 1202–1214, (*b*) pp. 1228–1229, McGraw-Hill Book Company, Inc., New York, 1950.
8. Quillen, C. S.: *Chem. Eng.*, **61**(6): 178 (1954).
9. Riegel, E. R.: "Chemical Process Machinery," 2d ed., chaps. 10, 11, Reinhold Publishing Corporation, New York, 1953.
10. Rushton, J. H.: *Chem. Eng. Progr.*, **47**: 485 (1951).
11. Rushton, J. H.: *Ind. Eng. Chem.*, **44**: 2931 (1952).
12. Rushton, J. H., E. W. Costich, and H. J. Everett: *Chem. Eng. Progr.*, **46**: 395, 467 (1950).
13. Rushton, J. H.: *Chem. Eng. Progr.*, **50**: 587 (1954).
14. Smith, J. C.: *Chem. Ind.*, **64**: 399 (1949); **66**: 843 (1950).

Chapter 7

MECHANICAL SEPARATIONS

Frequently it is necessary to separate the components of a mixture into individual fractions. The fractions may differ from each other in particle size, in phase, or in chemical composition. Thus, a crude product may be purified by removing from it contaminating impurities; two or more products in a mixture may be separated into the individual pure products; the stream discharged from a process step may consist of a mixture of product and unconverted raw material, which must be separated and the unchanged raw material recycled to the reaction zone for further processing; or a valuable substance, such as a metallic ore, dispersed in a mass of inert material must be liberated for recovery and the inert material discarded. A number of methods have been invented for accomplishing such separations, and several unit operations are devoted to them. In practice, many separation problems are encountered, and the engineer must choose the method best fitted to the problem.

Procedures for separating the components of mixtures fall into two classes. The first class includes techniques, called mechanical separations, that are useful for separating particles of solid or drops of liquid. The second class includes methods that depend on phase changes, such as vaporization, solution, precipitation, or condensation. The present chapter considers the mechanical operations that constitute the first class.

Mechanical separations are applicable to heterogeneous mixtures, not to homogeneous solutions. Colloids, which are an intermediate class of mixtures, are not usually treated by the methods discussed in this chapter, which deals primarily with particles larger than $\frac{1}{10}$ micron. The techniques are based on physical differences among the particles such as size, shape, density, wettability, or electric and magnetic characteristics. They are applicable to separating liquids from liquids, solids from gases, liquids from gases, solids from solids, and solids from liquids. The general methods are the use of a sieve, septum, or membrane, such as a screen or a filter, which retains one component and allows the other to pass; the utilization of the velocity of particles or drops of different sizes or densities through liquids or gases; the use of centrifugal force to replace the force of gravity in utilizing density differences or to develop pressure for filtration through a septum; and the exploitation of differences in the electrical or magnetic properties of the substances.

313

SCREENING

Screening is a method of separating particles according to size alone. In industrial screening the solids are dropped on, or thrown against, a screening surface. The undersize, or "fines," pass through the screen openings; oversize, or "tails," do not. A single screen can make but a single separation into two fractions. These are called unsized fractions, because although either the upper or lower limit of the particle sizes they contain is known, the other limit is unknown. Material passed through a series of screens of different sizes is separated into sized fractions, i.e., fractions in which both the maximum and minimum particle sizes are known. Screening is occasionally done wet but much more commonly dry. The discussion in this section is limited to the screening of dry particulate solids.

Industrial screens are made from metal bars, perforated or slotted metal plates, woven wire cloth, or fabric, such as silk bolting cloth. Metals used include steel, stainless steel, bronze, copper, nickel, and monel. The mesh size of woven screens ranges from 4-in. to 400-mesh, but screens finer than 100- or 150-mesh are rarely used. With very fine particles other methods of separation are usually more economical. Many varieties and types of screens are available for different purposes.[19a] A few representative types are described below.

Types of Screens. In most screens the particles drop through the screen opening by gravity. In a few designs, which will not be discussed here,

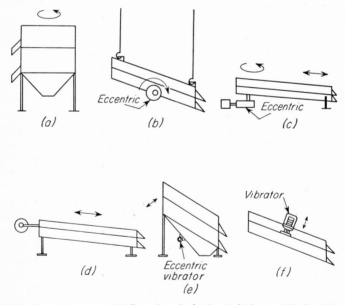

Fig. 7-1. Motions of screens: (a) Gyrations in horizontal plane. (b) Gyrations in vertical plane. (c) Gyrations at one end, shaking at other. (d) Shaking. (e) Mechanically vibrated. (f) Electrically vibrated.

they are pushed through by a brush or by centrifugal force. Coarse particles drop quickly and easily through large openings in a stationary surface; with finer particles the screening surface must be agitated in some way. Common ways are by revolving a cylindrical screen about a horizontal axis; or, with flat screens, by shaking, gyrating, or vibrating them mechanically or electrically. Typical screen motions are illustrated in Fig. 7-1.

Stationary Screens and Grizzlies. A grizzly is a grid of parallel metal bars set in an inclined stationary frame. The slope and the path of the material are usually parallel to the length of the bars. Very coarse feed, as from a primary crusher, falls on the upper end of the grid. Large chunks roll

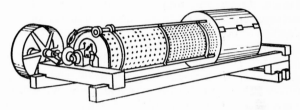

FIG. 7-2. Revolving screen.

and slide to the tails discharge; small lumps fall through to a separate collector. In cross section the top of each bar is wider than the bottom, so that the bars can be made fairly deep for strength without being choked by lumps passing part way through. The spacing between the bars is 2 to 8 in.

Stationary inclined woven-metal screens operate in the same way, separating particles $\frac{1}{2}$ to 4 in. in size. They are effective only with very coarse free-flowing solids containing few fine particles.

Revolving Screens. Another design of screen for coarse materials is the revolving screen, or trommel, shown in Fig. 7-2. The screening surface is a metal cylinder perforated with sets of holes of various sizes. The cylinder revolves about a slightly inclined longitudinal axis. Feed enters the cylinder at the upper end; oversize leaves from the lower end. During their travel through the machine the solids pass first over a set of small holes, perhaps $\frac{1}{4}$ in. in diameter, then over sets of progressively larger holes. The smallest particles drop through first, then successively larger particles in separate sized fractions.

Gyrating Screens. In most other screens which produce sized fractions the coarse material is removed first and the fines last. This is illustrated by the gyrating flat screens shown in Figs. 7-3 and 7-4. These machines contain several "decks" of screens, one above the other, held in a box or casing. The coarsest screen is at the top and the finest at the bottom, with suitable discharge ducts to permit removal of the several fractions. The mixture of particles is dropped on the top screen. Screens and casing are gyrated to sift the particles through the screen openings.

In the design shown in Fig. 7-3 the casing is inclined at an angle between 16 and 30° with the horizontal. The gyrations are in a vertical plane about a horizontal axis. They are caused by an eccentric shaft set in the floor of the casing halfway between the feed point and the discharge. The screens are rectangular and fairly long, typically $1\frac{1}{2}$ by 4 ft to 5 by 14 ft. The speed of gyration and the amplitude of throw are adjustable, as is the angle of tilt. One particular combination of speed and throw usually gives the maximum yield of desired product from a given feed. The rate of gyration is 600 to 1,800 rpm; the motor size is 1 to 3 hp. The angle of tilt, as discussed later, greatly influences the capacity of the screen. It is best to

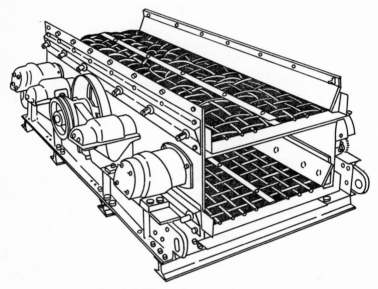

Fig. 7-3. Heavy-duty gyrating screen.

use the steepest angle possible. Very steep angles, however, can be used only with coarse products; good separation into fine fractions usually requires an angle of not more than 20° with the horizontal.

The screens shown in Fig. 7-4 are gyrated in a horizontal plane. The design in Fig. 7-4a utilizes square screens, perhaps 2 ft on a side, stacked three to six high and driven by an eccentric at the bottom. The throw circle is $1\frac{1}{2}$ to $2\frac{1}{2}$ in. in diameter. The rate of gyration is about 300 rpm. In this machine the screens are often all of the same mesh size to ensure complete removal of oversize particles. By using graded screens, however, as many as eight separations (into two unsized and six sized fractions) can be made in a single unit.

The design in Fig. 7-4b contains rectangular slightly inclined screens which are gyrated at the feed end. The discharge end reciprocates but does not gyrate. This combination of motions stratifies the feed, so that fine particles travel downward to the screen surface, where they are pushed through by the larger particles on top. Often the screening surface is double, as shown in the figure. Between the two screens are rubber balls held in separate compartments. As the screen operates, the balls strike the screen surface and free the openings of any material that tends to plug them. Dry, hard, rounded or cubical grains ordinarily pass without trouble through screens, even fine screens; but elongated, sticky, flaky, or soft

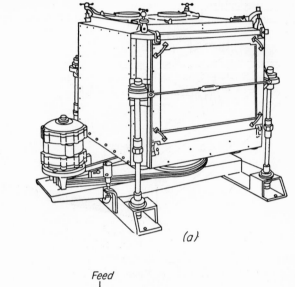

(a)

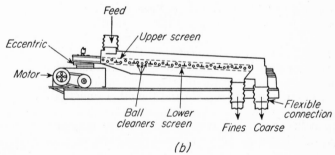

(b)

FIG. 7-4. Horizontally gyrated screens: (a) Gyratory sifter. (b) Ro-ball screen.

particles do not. Under the screening action such particles may become wedged into the openings and prevent other particles from passing through. A screen plugged with solid particles is said to be *blinded*.

Vibrating Screens. Screens which are rapidly vibrated with small amplitude are less likely to blind than are gyrating screens. The vibrations may be generated mechanically or electrically. Mechanical vibrations are usually transmitted from high-speed eccentrics to the casing of the unit and from there to steeply inclined screens. Electrical vibrations from heavy-duty solenoids are transmitted to the casing or directly to the screens.

Figure 7-5 shows a directly vibrated unit. Ordinarily no more than three decks are used in vibrating screens. Between 1,800 and 3,600 vibrations per minute are usual. A screen 12 in. wide and 24 in. long draws about $\frac{1}{3}$ hp; a 48- by 120-in. screen draws 4 hp.

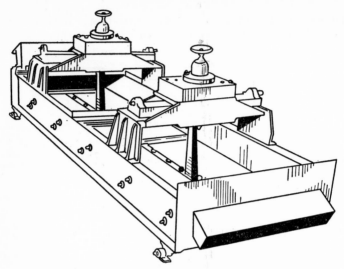

Fig. 7-5. Electrical-vibration screen. (*W. S. Tyler Co.*)

Comparison of Ideal and Actual Screens. The objective of a screen is to accept a feed containing a mixture of particles of various sizes and separate it into two fractions, an underflow that is passed through the screen and an overflow that is rejected by the screen. Either one, or both, of these streams may be a "product," and in the following discussion no distinction is made between the overflow and underflow streams from the standpoint of one's being desirable and the other undesirable.

An ideal screen would sharply separate the feed mixture in such a way that the smallest particle in the overflow would be just larger than the largest particle in the underflow. Such an ideal separation defines a cut diameter D_{pc} which marks the point of separation between the fractions. By diameter is meant a typical particle dimension, chosen as described on page 203. Although the choice of defining length is arbitrary, it is convenient to so choose the diameter that the cut diameter D_{pc} is nearly equal to the mesh opening of the screen. If the particles are spheres, the diameter of the particle is the logical choice. If a cube, the length of a side is satisfactory; if the particles are rectangular parallelepipeds or some other irregular shape, the second largest dimension is recommended as the defining dimension.[3b]

The performance of an ideal screen in terms of the screen analysis of the feed is shown in Fig. 7-6a. The cut point is point c in the curve. Fraction A consists of all particles larger than D_{pc}, and fraction B consists of all particles smaller than D_{pc}. Materials A and B are the overflow and the underflow, respectively. Screen analyses of the ideal fractions A and B are plotted in Fig. 7-6b. The first point on the curve for B has the same abscissa as the last point on the curve for A, and there is no overlap of these curves.

Actual screens do not give a sharp separation. Rather, the screen analyses of the overflow and underflow are like those shown in Fig. 7-6c.

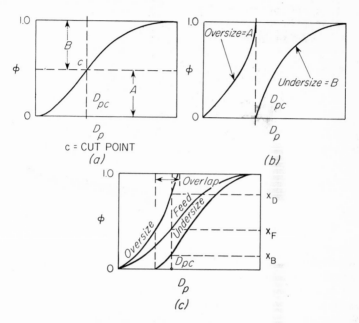

FIG. 7-6. Ideal vs. actual screening: (a) Ideal screening. (b) Screen analyses of products from ideal screening. (c) Actual screening.

The overflow has an appreciable content of particles smaller than the desired cut diameter, and the underflow has particles larger than the cut diameter. The two curves overlap, as shown in Fig. 7-6c. Also, the mass of the two leaving streams will not equal the individual masses of A and B unless it happens that the oversize material in the underflow is equal to the undersize in the overflow.

The closest separations are obtained with spherical particles on standard testing screens, but even here there is an overlap between the smallest particles in the overflow and the largest ones in the underflow. The overlap is especially pronounced when the particles are needlelike or fibrous or where the particles tend to aggregate into clusters that act as large particles. Some long, thin particles may strike the screen surface endwise and pass through easily, while other particles of the same size and shape may strike the screen sidewise and be retained. Commercial screens usually give poorer separations than do testing screens of the same mesh opening operating on the same mixture.

Material Balances over Screen. Simple material balances, of the type described in Chap. 1, can be written over a screen. The equations are useful in calculating the ratios of feed, oversize, and underflow from the screen analyses of the three streams and knowledge of the desired cut diameter. Let

F = mass of feed stream, lb/hr
D = mass of overflow stream, lb/hr
B = mass of underflow stream, lb/hr
x_F = mass fraction of material A in feed
x_D = mass fraction of material A in overflow
x_B = mass fraction of material A in underflow

The mass fractions of material A are shown in Fig. 7-6c. The mass fractions of material B in the feed, overflow, and underflow are $1 - x_F$, $1 - x_D$, and $1 - x_B$.

Since the total material fed to the screen must leave it either as underflow or as overflow,

$$F = D + B \tag{7-1}$$

The material A in the feed must also leave in these two streams, and

$$Fx_F = Dx_D + Bx_B \tag{7-2}$$

Elimination of B from Eqs. (7-1) and (7-2) gives

$$\frac{D}{F} = \frac{x_F - x_B}{x_D - x_B} \tag{7-3}$$

Elimination of D gives

$$\frac{B}{F} = \frac{x_D - x_F}{x_D - x_B} \tag{7-4}$$

Screen Effectiveness. The effectiveness of a screen is a measure of the success of the screen in closely separating materials A and B. If the screen functioned perfectly, all of material A would be in the overflow and all of material B would be in the underflow. Actually, the fraction of the entering A that is in the overflow is Dx_D/Fx_F, and the fraction of the entering B that is in the underflow is $B(1 - x_B)/F(1 - x_F)$. The first fraction is a measure of the effectiveness of the screen in keeping material A out of the undersize, and the second a measure of the effectiveness in keeping material B out of the oversize. A combined effectiveness can be defined as the product of the two individual ratios,[3a] and if this product is denoted by E,

$$E = \frac{DBx_D(1 - x_B)}{F^2 x_F(1 - x_F)}$$

Substituting D/F and B/F from Eqs. (7-3) and (7-4) into this equation gives

$$E = \frac{(x_F - x_B)(x_D - x_F)x_D(1 - x_B)}{(x_D - x_B)^2(1 - x_F)x_F} \tag{7-5}$$

Example 7-1. The quartz mixture having the screen analysis shown in Tables 4-1 and 4-2 is screened through a 10-mesh-per-inch screen constructed of 0.04-in. wire. The cumulative screen analyses of overflow and underflow are given in Table 7-1.

TABLE 7-1. SCREEN ANALYSES FOR EXAMPLE 7-1

Mesh	D_p, cm	ϕ	
		Overflow	Underflow
4	0.4699	0	
6	0.3327	0.071	
8	0.2362	0.43	0
10	0.1651	0.85	0.195
14	0.1168	0.97	0.58
20	0.0833	0.99	0.83
28	0.0589	1.00	0.91
35	0.0417	.....	0.94
65	0.0208	.....	0.96
Pan........	......	.....	1.00

Calculate the mass ratios of overflow and underflow to feed and the effectiveness of the screen.

Solution. The cumulative analyses of feed, overflow, and product are plotted in Fig. 7-7. The cut-point diameter is the mesh size of the screen, which is $(0.10 - 0.04) \times 2.54 = 0.152$ cm. From Fig. 7-7, for $D_{pc} = 0.152$ cm

$$x_F = 0.540 \qquad x_D = 0.895 \qquad x_B = 0.275$$

From Eq. (7-3), the ratio of overflow to feed is

$$\frac{D}{B} = \frac{0.540 - 0.275}{0.895 - 0.275} = \frac{0.265}{0.620} = 0.427$$

The ratio of underflow to feed is

$$\frac{B}{F} = 1 - \frac{D}{F} = 1 - 0.427 = 0.573$$

The over-all effectiveness is, from Eq. (7-5),

$$E = \frac{(0.540 - 0.275)(0.895 - 0.540)(1 - 0.275)0.895}{(0.895 - 0.275)^2 0.460 \times 0.540} = 0.64$$

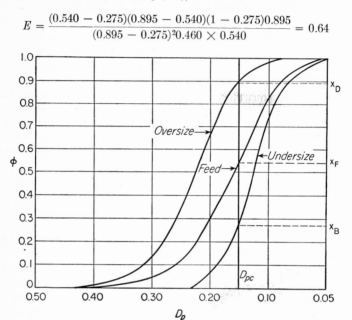

Fig. 7-7. Analyses for Example 7-1.

Capacity and Effectiveness of Screens. In addition to effectiveness, capacity is important in industrial screening. The capacity of a screen is measured by the mass of the material fed per unit time to unit area of the screen. Capacities are usually expressed in tons per hour per square foot.

Capacity and effectiveness are opposing factors. To obtain maximum effectiveness the capacity must be small, and large capacity is obtainable only at the expense of a reduction in effectiveness. In practice, a reasonable balance between capacity and effectiveness is desired. Although accurate relationships are not available for estimating these operating characteristics of screens, certain fundamentals apply, which can be used as guides in understanding the basic factors in screen operation.

The capacity of a screen is controlled simply by varying the rate of feed to the unit. The effectiveness obtained for a given capacity depends on the nature of the screening operation. The over-all chance of passage of a given undersize particle is a function of the number of times the particle strikes the screen surface and the probability of passage during a single contact. If the screen is overloaded, the number of contacts is small, and the chance of passage on contact is reduced by the inteference of the other particles. The improvement of effectiveness attained at the expense of reduced capacity is a result of more contacts per particle and better chances for passage on contact.

Ideally, a particle would have the greatest chance of passing through the screen if it struck the surface perpendicularly, if it were so oriented that its minimum dimensions were parallel with the screen surface, if it were unimpeded by any other particles, and if it did not stick to, or wedge into, the screen surface. None of these conditions apply to actual screening, but this ideal situation can be used as a basis for estimating the effect of mesh size and wire dimensions on the performance of screens.

If the width of the wire in a screen were negligible in comparison with the size of the screen openings, the wires would not interfere with the passing of the particles, and practically the entire screen surface would be active. The probability of passage of a particle that strikes the screen would then be nearly unity. In an actual screen the diameter of the wire, or the fraction of the surface that is not in openings, is appreciable, and the solid meshes strongly affect the performance of the screen, especially in retarding the passage of particles nearly as large as the screen openings.

Effect of Mesh Size on Capacity of Screens. The probability of passage of a particle through a screen depends on the fraction of the total surface represented by openings, on the ratio of the diameter of the particle to the width of an opening in the screen, and on the number of contacts between the particle and the screen surface. When these factors are all constant, the average number of particles passing through a single screen opening in unit time is a constant, independent of the size of the screen opening. If the size of the largest particle that can just pass through a screen opening is taken equal to the width of a screen opening, both dimensions may be represented by D_{pc}. For a series of screens of different mesh sizes, the number of openings per unit screen area is proportional to $1/D_{pc}^2$. The mass of one particle is proportional to D_{pc}^3. The capacity of the screen, in mass per unit time, is, then, proportional to $(1/D_{pc}^2)D_{pc}^3 = D_{pc}$. Then the capacity of a screen, in mass per unit time, divided by the mesh size should be constant for any specified conditions of operation. This is a well-known practical rule of thumb.[9a]

Capacities of Actual Screens. Although the preceding analysis is useful in analysing the fundamentals of screen operation, in practice, a number of complicating factors appear that cannot be treated theoretically. Some of these disturbing factors are the interference of the bed of particles with the motion of any one; blinding; cohesion of particles to each other; the adhesion of particles to the screen surface; and the oblique direction of approach of the particles to the surface. When large and small particles are present, the large particles tend to segregate in a layer next to the screen and so prevent the smaller particles from reaching the screen. All of these factors tend to reduce capacity and lower effectiveness. Moisture content of the feed is especially important. Either dry particles or particles moving in a stream of water screen more readily than do damp particles, which are prone to stick to the screen surface and to each other and to screen slowly and with difficulty.

Approximate capacities for screens operating on dry ores are shown in Table 7-2, where the capacities are given in tons/(hr)(ft^2)(mm mesh size).

TABLE 7-2. APPROXIMATE SCREEN CAPACITIES †

Type of screen	Capacity range
Grizzlies	1 – 5
Trommels	0.3– 2
Shaking screens	2 – 8
Vibrating screens	5 –20

† From A. M. Gaudin, "Principles of Mineral Dressing," p. 162, McGraw-Hill Book Company, Inc., New York, 1939.

The capacities in Table 7-2 will be considerably modified by variations in density and moisture content. Capacities of specific screens, operating on definite materials, should be obtained from the manufacturer of the equipment.

Particle-size Limitation of Screens. Screening becomes progressively more difficult as the sizes of the particles are reduced below approximately 100-mesh. The reasons for the poor performance of screens on very fine particles are low capacity; blinding; cohesion and adhesion of particles; low fraction opening to provide sufficient strength for the screen surface; and the small pull of gravity on small particles.

FILTRATION

Filtration is the removal of solid particles from a fluid by passing the fluid through a filtering medium, or septum, on which the solids are deposited. Industrial filtrations range from simple straining to highly complex separations. The fluid may be a liquid or a gas; the solid particles may be coarse or fine, rigid or plastic, round or elongated, separate individuals or aggregates. The feed suspension may carry a heavy load of solids or almost none. It may be very hot or very cold, or under vacuum or high pressure. Further complexities are introduced by the relative value of the two phases. Sometimes the fluid is the valuable phase; sometimes the solid; sometimes both. In some problems the separation of the phases must be virtually complete; in others only a partial separation is desired. A multitude of filters has therefore been developed to meet the different problems.[17, 19d] Some of the commoner types are discussed in the following pages.

Equipment for Liquid-Solid Filtrations

Liquid-solid filters may be divided into four groups, depending on the service they perform: strainers, clarifiers, cake filters, and filter thickeners. A strainer is usually little more than a metal screen set across a flow channel to remove dirt or rust from a flowing liquid. An example of the use of a simple strainer is in a line leading to a steam trap, to clean the steam and condensate of trash which would interfere with the operation of the trap. When the screen becomes plugged, it is easily replaced. Clarifiers also remove small quantities of solids, usually to produce sparkling clear liquids, as printing inks or beverages. The removed solids are most often discarded. The filter medium in a clarifier is a septum of cloth or paper or a cartridge of metal disks. Cake filters separate large amounts of solids from a liquid as a cake of crystals or sludge. Often they include provisions for washing the solids and for removing as much residual liquid from the solids as possible before discharge. A filter-thickener gives partial separation of a thin slurry, discharging some clear liquid and a thickened but still flowable suspension of solids.

Liquid flows through a filter medium by virtue of a pressure differential across the medium. Filters are also classified, therefore, into those which operate with a pressure above atmospheric on the upstream side of the filter medium and those which operate with atmospheric pressure on the upstream side and a vacuum on the downstream side. Pressures above atmospheric may be developed by the force of gravity acting on a column of liquid, by a pump, or by centrifugal force. Centrifugal filters are discussed

TABLE 7-3. TYPES OF FILTRATION EQUIPMENT

Method of operation	Type	Typical applications
Pressure filters:		
Discontinuous...	Plate-and-frame press	Filtration and washing of pigments and dyes; clarification of viscose; dewaxing of petroleum
	Shell-and-leaf filter	High-pressure filtration and washing of fine solids
	Cartridge filter	Clarification of process liquids, cooling water, etc.
Continuous......	Rotary-drum filter	Pressure filtration of sludges; precoat clarification of liquids
	Filter-thickener	Partial filtration of thin slurries; displacement washing
Vacuum filters:		
Discontinuous...	Nutsch	Semiworks or pilot-plant filtrations
	Leaf filter	Filtration and washing of pigments
	Filter-thickener	Partial filtration of thin slurries
Continuous......	Rotary-drum filter:	
	Bottom feed	Filtration and washing of fine crystals, sludges
	Top feed	Filtration and drying of coarse crystals
	Horizontal, or table, filter	Filtration and washing of free-draining solids

in a later section of this chapter. In a gravity filter the filter medium can be no finer than a coarse screen or a bed of coarse particles like sand. Gravity filters are therefore restricted in their industrial applications to the draining of liquor from very coarse crystals and to the clarification of water.

Most industrial filters are either pressure filters or vacuum filters. They are also either continuous or discontinuous, depending on whether the discharge of filtered solids is steady or intermittent. During much of the operating cycle of a discontinuous filter the flow of liquid through the device is continuous, but it must be interrupted periodically to permit discharging the accumulated solids. In a continuous filter the discharge of both solids and liquid is uninterrupted as long as the equipment is in operation. Typical filters and the kinds of services they perform are listed in Table 7-3.

Discontinuous Pressure Filters. Pressure filters can apply a large pressure differential across the septum to give economically rapid filtration of viscous liquids or fine solids. The most common types of pressure filters are filter presses, shell-and-leaf filters, and cartridge filters. Because of the difficulty of discharging solids against pressures above atmospheric, pressure filters are usually discontinuous. Continuous pressure cake filters find only limited applications in processing operations.

Filter Press. A filter press contains a set of plates designed to provide a series of chambers, or compartments, in which solids may collect. The plates are covered with a filter medium such as canvas. Slurry is admitted to each compartment under pressure; liquor passes through the canvas and out a discharge pipe, leaving a wet cake of solids behind.

The plates of a filter press may be square or circular, vertical or horizontal. In some designs the compartments for solids are formed by recesses in the faces of the plates. More commonly, however, they are formed as in the *plate-and-frame press* shown in Fig. 7-8. In this design, square plates 6 to 56 in. on a side alternate with open frames. The plates are $\frac{1}{4}$ to 2 in. thick, the frames $\frac{1}{4}$ to 8 in. thick. Details of the plates and frames are shown in Fig. 7-8b. Plates and frames sit vertically in a metal rack, with cloth covering the face of each plate, and are squeezed tightly together by a hand-operated screw or a hydraulic ram. Slurry enters at one end of the assembly of plates and frames, as shown. It passes through a channel running lengthwise through one corner of the assembly. Auxiliary channels carry slurry from the main inlet channel into each frame. Here the solids are deposited on the cloth-covered faces of the plates. Liquor passes through the cloth, down grooves or corrugations in the plate faces, and out of the press. In an *open-discharge press* each plate carries a liquor outlet leading from the corner of the plate to an open trough, or launder. A cock on each outlet permits any plate to be shut off if necessary. In a *closed-discharge press* liquor leaves through a longitudinal channel similar to that which brings in the slurry but in the opposite corner of the assembly of plates and frames. A closed-discharge press is needed where the liquid must not be exposed to the air. It gives a completely closed system. Open-discharge presses have the advantage of permitting the operation of each individual plate to be observed; if the discharge from any plate is cloudy, that plate can be shut off without interrupting the filtration cycle.

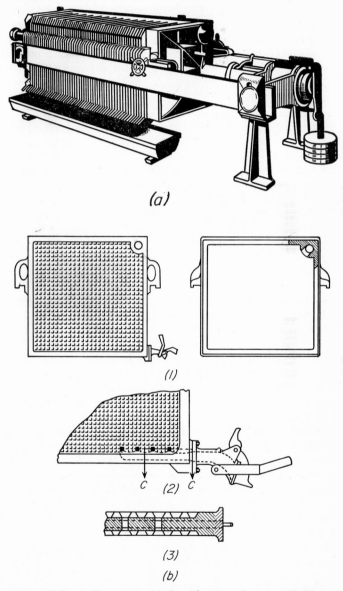

(a)

(1)

(2)

(3)

(b)

F IG. 7-8. Plate-and-frame filter press: (a) Outside view of press. (b) Plate and frame of open-delivery, nonwashing press: (1) Elevation. (2) Details of open-delivery ports. (3) Section along line CC.

In operating a plate-and-frame press, the plates are first "clothed" with suitable lengths of the filter medium. Holes are cut in each cloth to match the slurry and discharge channels, and wash channels if there are any. Plates and frames are set in the rack and pushed together by hand, then squeezed tightly by the ram. The cloth acts as a gasket, sealing the machined faces of the plates and frames. In large presses overhead cranes and monorails are used to move the heavy plates. After assembly of the press, slurry is admitted from a pump or blow case under a pressure of 40 to 150 lb force/in.² Filtration is continued until liquor no longer flows out

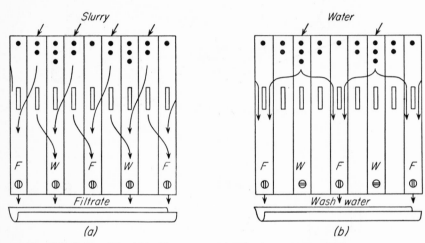

Fig. 7-9. Paths of liquid in filter press: (a) Filtering. (b) Washing. (Riegel.[19])

the discharge or the filtration pressure suddenly rises. These occur when the frames are full of solid and no more slurry can enter. The press is then said to be jammed. Wash liquid may then be admitted to remove soluble impurities from the solids, after which the cake may be "blown" with steam or air to displace as much residual liquid as possible. The press is then opened, and the cake of solids scraped off the filter medium and dropped to a conveyor or storage bin.

If the solids are to be effectively washed, the wash liquid must pass through as many of the channels in the cake as possible. This is achieved by having every other plate of a special design known as a washing plate. A longitudinal channel carries wash liquid into the corner of the press diagonally across from the slurry channel. Auxiliary channels carry the wash into each washing plate behind the filter medium. The plates alternating with the washing plates are called filter plates; to them wash liquid is not admitted. During washing the liquid passes through the layer of cloth covering the washing plates, through the cake, and through the septum of the filter plates to the discharge. The flow path through open-discharge plates is shown in Fig. 7-9. The cocks on the washing plates are open during the filtering part of the cycle and shut during the washing part. In a closed-delivery press, two filtrate channels are supplied. One is cored into the washing plates and the other into the nonwashing plates. During washing the discharge from the washing plates is closed. Some manufacturers cast little buttons on the outside of their plates and frames for convenience in giving directions to the operators. In the usual system a filter plate is a one-button plate and a wash plate is a three-button plate. The frames carry two buttons.

Thorough washing in a filter press may take several hours, for the wash liquid tends to follow the easiest paths and to bypass tightly packed parts

Fig. 7-10. Sweetland press.

of the cake. If the cake is less dense in some parts than in others, as is usually the case, much of the wash liquid will be ineffective. If washing must be exceedingly good, it may be best to reslurry a partly washed cake with a large volume of wash liquid and refilter it or to use a shell-and-leaf filter, which permits more effective washing than does a plate-and-frame press.

Shell-and-leaf Filters. For filtering under higher pressures than are possible in a plate-and-frame press, to economize on labor, or where more effective washing of the cake is necessary, a shell-and-leaf filter may be used. The *Sweetland press*, shown in Fig. 7-10, contains closely spaced circular leaves in a horizontally split casing with a hinged bottom. The leaves are made from metal hoops and wire screen covered with a canvas septum. In operation the casing is bolted shut and slurry admitted under pressure. Liquor passes through the septum into the inside of the leaves, from which it discharges into a manifold at the back of the press. Cake builds up on the vertical faces of the leaves. When filtration is complete, excess slurry is drained from the casing and replaced with wash liquid, which passes through the solids to the manifold. At the end of the cycle the casing is drained once more; the counterweighted bottom of the shell is swung downward to open the press; and the cake is scraped off the filter leaves, blown off with compressed air, or sluiced off with jets of water. The shell is then closed and the cycle repeated.

In some leaf filters the hollow shaft which carries the leaves rotates slowly inside a heavy casing. Slurry is admitted to the casing under pressure. Solids are deposited on the filter cloth, washed, then sluiced off the leaves with strong jets of water and pumped as a sludge from an outlet in the bottom of the casing. The cycles of filtration, washing, and cake removal may be repeated without opening the press. The rotation of the leaves keeps the slurry agitated so the solids do not settle, and results in a more uniform cake than in a press with stationary leaves. A rotating-leaf filter is most applicable when a small amount of solids having little or no value is to be removed from the liquid, as in the filtration of sugar liquors.

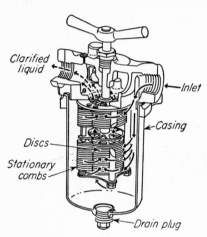

FIG. 7-11. Cartridge filter.

Cartridge Filters. A typical cartridge filter, used principally for removing small amounts of solids from process fluids, is the filter illustrated in Fig. 7-11. The filtering cartridge is a series of thin metal disks 3 to 10 in. in diameter set in a vertical stack with very narrow uniform spaces between them. The disks are carried on a vertical hollow shaft, and fit into a closed cylindrical casing. Liquid is admitted to the casing under pressure. It flows inward between the disks to openings in the central shaft and out through the top of the casing. Solids are trapped between the disks and remain in the filter. Since most of the solids are removed at the periphery of the disks, this kind of device is known as an *edge filter.* Periodically the accumulated solids must be dislodged from the cartridge. In the design shown in Fig. 7-11 this is done by giving the cartridge a half turn. The stationary teeth of a comb cleaner pass between the disks, causing the solids to drop to the bottom of the casing, from which they may be removed at fairly long intervals.

Continuous Pressure Filters. Batch filters require considerable operating labor, which in large-scale processing may be prohibitively expensive. The continuous vacuum filters described below were developed to reduce the labor required for filtration. Sometimes, however, vacuum filtration is not feasible or not economical, as when the solids are very fine and filter very slowly, or when the liquid has a high vapor pressure, has a viscosity greater than 100 centipoises, or is a saturated solution which will crystallize if cooled at all. With slow-filtering slurries the pressure differential across the septum must be greater than can be obtained in a vacuum filter;

with liquids that vaporize or crystallize at reduced pressure the pressure on the downstream side of the septum cannot be less than atmospheric. Thus the continuous rotary-drum filters described later are occasionally adapted for operation under positive pressures up to 40 lb force/in.[2]; however, the mechanical problems of discharging the solids from these filters, their expense and complexity, and their small size limit their application to unusual problems. Where vacuum filtration cannot be used, other means of separation, such as continuous centrifugal filters, should also be considered.

Pressure Filter-Thickener. The purpose of a filter-thickener is to remove some of the liquid from a thin slurry to produce a thick suspension. The

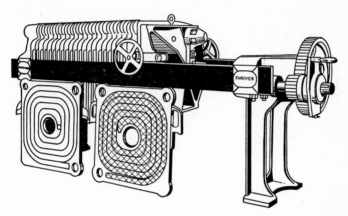

FIG. 7-12. Pressure filter-thickener. (*T. Shriver and Co., Inc.*)

design shown in Fig. 7-12 continuously discharges clear liquor and a stream of thickened slurry. In appearance this device resembles a filter press; it contains no frames, however, and the plates are modified as shown in the detail drawing. Successive plates carry matching channels which form, when the press is assembled, a long winding path for the slurry. The suspension flows continuously all the way through the press, first in the channel between the faces of one pair of plates, through an opening, then back between the faces of the next pair. The sides of the channels are covered with filter medium held between the plates. As slurry passes through the channel under pressure, some of the liquid flows through the filter medium into grooves in the plate faces and from there to a clear-liquor discharge manifold. The thickened slurry is kept moving fast enough so that the channel does not clog. The number of plates is chosen so that the pressure drop through the press does not exceed 80 lb force/in.[2] Under these conditions it is usually possible to about double the concentration of the inlet slurry, as for example from 5 per cent solids at the inlet to 10 per cent solids at the outlet. When a greater degree of concentration than this is needed, the partly thickened slurry from one press is repumped to a second press for additional thickening.

Discontinuous Vacuum Filters. Pressure filters are usually discontinuous; vacuum filters are usually continuous. A discontinuous vacuum filter, however, is sometimes a useful tool. A vacuum *nutsch* is little more than a large Büchner funnel, 3 to 10 ft in diameter and forming a layer of solids 4 to 12 in. thick. Because of its simplicity, a nutsch can readily be made of corrosion-resistant material, and is valuable where experimental batches of a variety of corrosive materials are to be filtered. Nutsches are not recommended for production operations because of the high labor cost involved in digging out the solid cake.

Continuous Vacuum Filters. In all continuous vacuum filters liquor is sucked through a moving septum to deposit a cake of solids. The cake

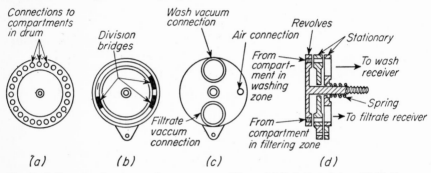

FIG. 7-13. Rotary valve for continuous vacuum filter. (*a*) Rotating plate. (*b*) Stationary plate, for timing control. (*c*) Stationary plate, with outside connections. (*d*) Cross-section of assembly.

is moved out of the filtering zone, washed, sucked dry, and dislodged from the septum, which then reenters the slurry to pick up another load of solids. Some part of the septum is in the filtering zone at all times, part is in the washing zone, and part is being relieved of its load of solids, so that the discharge of both solids and liquids from the filter is uninterrupted. The pressure differential across the septum in a continuous vacuum filter is not high, ordinarily between 10 and 20 in. Hg. Various designs of filter differ in the method of admitting slurry, the shape of the filter surface, and the way in which solids are discharged. They all, however, apply vacuum from a stationary source to the moving parts of the unit, usually through a rotary valve of the type described below.

Rotary Valve. A typical rotary valve is shown in Fig. 7-13. It consists of three circular metal plates, two stationary and one rotating. To the rotating plate are attached pipes leading to various sectors or compartments in the filter drum; to the stationary cover plate are attached the vacuum lines and sometimes a compressed-air line. The stationary intermediate plate consists of two concentric rings with an annular space between them. The openings in the cover plate and those in the rotating plate match with the annular space between the rings; the annulus thus forms a passage through which air or liquid may flow from the rotating parts of the filter to the stationary cover plate. Metal "bridges" divide the annulus into three or more zones, one corresponding to the filtration zone of the filter, one or more to the washing and drying zones, and one to the discharge zone. In the valve illustrated in Fig. 7-13 there are but three zones, with vacuum applied to two of them and compressed air to the third. Any given opening in the rotating plate passes first through the bottom, or filtering, zone, then through the drying zone, and then into the short third zone, where the vacuum is cut off and a small amount of air introduced. The sliding seal between the rotating parts of the valve is formed by the wide machined faces of the plates, which are pressed tightly together by springs. Under the low pressure differential between the atmosphere and the inside of the valve there is virtually no leakage between the plate faces.

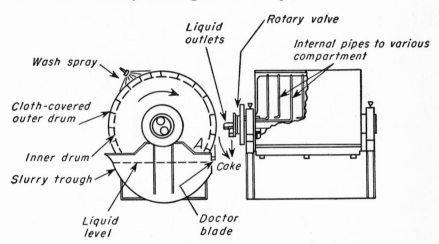

FIG. 7-14. Continuous vacuum rotary filter.

Rotary-drum Filter. The most common type of continuous vacuum filter is the rotary-drum filter illustrated in Fig. 7-14. A horizontal drum with a slotted face turns at 0.1 to 2 rpm in an agitated slurry trough. A filter medium, such as canvas, covers the face of the drum, which is partly submerged in the liquid. Under the slotted cylindrical face of the main drum is a second, smaller drum with a solid surface. Between the two drums are radial partitions dividing the annular space into separate compartments, each connected by an internal pipe to one hole in the rotating plate of the rotary valve. Vacuum and air are alternately applied to each compartment as the drum rotates. A strip of filter cloth covers the exposed face of each compartment to form a succession of "panels."

Consider now the panel shown at *A* in Fig. 7-14. It is just about to enter the slurry in the trough. As it dips under the surface of the liquid, vacuum is applied through the rotary valve. A layer of solids builds up on the face of the panel as liquid is drawn through the cloth into the compartment, through the internal pipe, through the valve, and into a collecting tank. As the panel leaves the slurry, the corresponding opening in the rotating valve plate passes over the first bridge in the rotary valve, leaving the filtering zone and entering the washing and drying zone. Here vacuum is applied to the panel from a separate system, sucking wash liquid and air through the cake of solids. As shown in the flow sheet of Fig. 7-15, wash liquid is drawn through the filter into a separate collecting tank. After

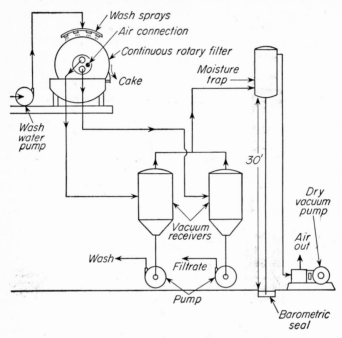

Fig. 7-15. Flow sheet for continuous vacuum filtration.

the cake of solids on the face of the panel has been sucked as dry as possible, the panel leaves the drying zone, vacuum is cut off, and the cake is removed by scraping it off with a horizontal knife known as a doctor blade. A little air is blown in under the cake to belly out the cloth. This cracks the cake away from the cloth and makes it unnecessary for the knife to scrape the drum face itself. Once the cake is dislodged, the panel reenters the slurry and the cycle is repeated. The operation of any given panel, therefore, is cyclic, but since some panels are in each part of the cycle at all times, the operation of the filter as a whole is continuous.

Many variations of the rotary-drum filter are commercially available. In some designs there are no compartments in the drum; vacuum is applied to the entire inner surface of the filter medium. Filtrate and wash liquid are removed together through a dip pipe; solids are discharged by air blown through the cloth from a stationary shoe inside the drum, bellying out the filter cloth and cracking off the cake. In other models the cake is lifted from the filter surface by a set of closely spaced parallel strings. Wash liquid may be sprayed directly on the cake surface, or, with cakes that crack when air is drawn through them, it may be sprayed on a cloth blanket that travels with the cake through the washing zone and is tightly pressed against its outer surface.

The amount of submergence of the drum is also variable. Most bottom-feed filters operate with about 30 per cent of their filter area submerged in the slurry. When high filtering capacity and no washing are desired, a high-submergence filter, with 60 to 70 per cent of its filter area submerged, may be used. The capacity of any rotary filter depends strongly on the characteristics of the feed slurry and particularly on the thickness of the cake which may be deposited in practical operation. The cakes formed on industrial rotary vacuum filters are $\frac{1}{8}$ to about $1\frac{1}{2}$ in. thick. Standard drum sizes range from 1 ft in diameter with a 1-ft face to 10 ft in diameter with a 14-ft face.

Continuous or semicontinuous clarifying devices are exemplified by a *precoat filter*, which is a rotary-drum filter modified for filtering small amounts of fine or gelatinous solids that ordinarily plug a filter cloth. In the operation of this machine a layer of a porous filter aid, such as diatomaceous earth, is first deposited on the filter medium. Process liquid is then sucked through the layer of filter aid, depositing a very thin layer of solids. This layer and a little of the filter aid is then scraped off the drum by a slowly advancing knife, which continually exposes a fresh surface of porous material for the subsequent liquor to pass through. A precoat filter may also operate under pressure. In the pressure type the discharged solids and filter aid collect in the housing, to be removed periodically at atmospheric pressure while the drum is being recoated with filter aid. Precoat filters can be used only where the solids are to be discarded or where their admixture with large amounts of filter aid introduces no serious problem. The usual submergence of a precoat-filter drum is 50 per cent.

Top-feed Filter. A top-feed filter is similar in operation to the rotary filter just described except that the slurry trough is above the filter drum instead of below it. Top-feed filters are used when the solids are too coarse and free-draining to be retained on the vertical face of a filter drum as it rises out of a slurry. Instead the feed is admitted to a distributor trough set across the top of the filter from which slurry flows on to the top of the drum. Here it drops off or is scraped off with a doctor blade. Wash water may be sprayed on the cake during its travel; in some top-feed filters the lower part of the drum is enclosed so that hot air may be sucked through the cake to dry it. Top-feed filters are fairly small, with drums 3 to 6 ft in diameter. The cake of solids, however, is 1 to 3 in. thick, and the drum speed is high, so that even though they make use of only about half of their filtering surface, top-feed filters have high capacities.

Horizontal Filter. When free-draining solids are to be thoroughly washed, especially when for any reason the filtrate piping must be easily accessible, a horizontal, or table, filter may be used. This is a horizontal ring-shaped trough with a flat cloth-covered false floor. The trough turns slowly about a central axis. The space beneath the false floor is divided into compartments; a large rotary valve applies vacuum to each compartment and then releases it at appropriate intervals. Filtrate and any wash liquid which may be sprayed on the cake during its travel are sucked through the cloth, into the compartments beneath the false floor, and out through the rotary valve. At the discharge point the vacuum is released and the solids removed radially outward by a helical conveyor-scraper. Each section of a table filter may be considered as a vacuum nutsch, carried successively through filtering, washing, and drying zones to the discharge. On freedraining solids table filters have very high capacity.

Filter Media. The septum in any filter must meet the following requirements:

1. It must retain the solids to be filtered, giving a reasonably clear filtrate.

2. It must not plug or blind.

3. It must be resistant chemically and strong enough physically to withstand the process conditions.

4. It must permit the cake formed to discharge cleanly and completely.

5. It must not be prohibitively expensive.

In industrial filtration the most common filter medium is canvas cloth, either duck or twill weave. Many different weights and patterns of weave are available for different services. Corrosive liquids require the use of other filter media, such as woolen cloth, metal cloth of monel or stainless steel, glass cloth, or paper. Synthetic fabrics like nylon, Vinyon N, Saran, Orlon, Dynel, and Acrilan are also highly resistant chemically.

In a cloth of a given mesh size, smooth synthetic or metal fibers are less effective in removing very fine particles than are the more ragged natural fibers. Ordinarily, however, this is a disadvantage only at the start of a filtration, because except with hard, coarse particles containing no fines the actual filtering medium is not the septum but the first layer of deposited solids. Filtrate may first come through cloudy, then grow clear. Cloudy filtrate is returned to the slurry tank for refiltration.

Filter Aids. Slimy or very fine solids which form a dense, impermeable cake quickly plug any filter medium that is fine enough to retain them. Practical filtration of such materials requires that the porosity of the cake be increased to permit passage of the liquor at a reasonable rate. This is done by adding a filter aid, such as diatomaceous earth, asbestos, purified wood cellulose, or other inert porous solid, to the slurry before filtration. The filter aid may subsequently be separated from the filter cake by dissolving away the solids or by burning out the filter aid. If the solids have no value, they and the filter aid are discarded together.

Another way of using a filter aid is by precoating, i.e., depositing a layer of it on the filter medium before filtration. In batch filters the precoat layer is usually thin; in a continuous precoat filter the layer of precoat is thick, and the top of the layer is continually scraped off by an advancing knife to expose a fresh filtering surface. Precoats prevent gelatinous solids from plugging the filter medium and give a clearer filtrate. The precoat is really a part of the filter medium, rather than of the cake.

Filter Auxiliaries. Any filtration operation requires several pieces of equipment in addition to the filter. The slurry must be stored in an agitated vessel and fed at an appropriate rate to the filter. Sources of compressed air and vacuum may be needed. Means must be provided for supplying wash liquid under pressure, for removing the filtrate, and for collecting, transporting, and possibly reslurrying the filtered solids. A typical arrangement of equipment for a continuous rotary-drum vacuum filter is shown in Fig. 7-15. The pieces of auxiliary equipment are usually of standard design. Pumps, compressors, agitators, and conveyors are of the types described in other chapters for handling liquids, gases, slurries, and semisolids such as filter cakes. The power consumed by the auxiliaries may greatly exceed that consumed by the filter itself, as shown by the following typical distribution: [19e] filter drive, 4 per cent of the total power required for filtration; vacuum pump, 67 per cent; filtrate pump, 14 per cent; wash-water pump, 11 per cent; and air compressor, 4 per cent.

Principles of Filtration

From the standpoint of fluid mechanics a filter is a flow device. By means of a pressure difference applied between the slurry inlet and the filtrate outlet, filtrate is forced through the equipment. During filtration the solids in the slurry are retained in the unit, and form a bed of particles through which filtrate must flow. The filtrate passes through three kinds of resistance in series. These are: (1) the resistances of the channels conducting the slurry to the upstream face of the cake and the filtrate away from the filter medium, (2) the resistance of the cake, and (3) the resistance associated with the filter medium.

During the washing of the cake in a pressure-leaf filter or on a rotary filter wash water flows through the same resistances that were encountered by the filtrate. In a plate-and-frame press the wash water follows a different path with respect to the cake than did the filtrate, but again the three kinds of resistance are found.

Distribution of Over-all Pressure Drop. Since the flow is in series, the total pressure drop over the filter can be equated to the sum of the individual pressure drops. In a well-designed filter the resistances of the inlet and outlet connections are small, and can be neglected in comparison with those over the cake and filter medium. In actual operation the resistance associated with the filter medium is greater than that offered by a clean filter medium to flow of clear filtrate. During the first moments of the filtration solid particles become embedded in the meshes of the filter medium and so develop abnormal resistance to the subsequent flow. The entire resistance built up in the filter medium, including that from the embedded particles, is called the filter-medium resistance. It is important during the early stages of the filtration. The resistance offered by all solids not associated with the filter medium is called the cake resistance. The cake resistance is zero at the beginning of the filtration, and because of the continuous deposition of solids on the medium, this resistance increases steadily with time of filtration. During washing all resistances, including that of the cake, are constant, and that of the filter medium is usually negligible.

Since the resistance of the channels can be neglected, the over-all pressure drop at any time is the sum of the pressure drops over medium and cake. If p_a is the inlet pressure, p_b the outlet pressure, and p' the pressure at the boundary between cake and medium,

$$-\Delta p = p_a - p_b = (p_a - p') + (p' - p_b) = -\Delta p_c - \Delta p_m \quad (7\text{-}6)$$

where $-\Delta p$ = over-all pressure drop
$\quad -\Delta p_c$ = pressure drop over cake
$\quad -\Delta p_m$ = pressure drop over medium

The operator Δ is defined as the outlet condition minus the inlet condition. Therefore Δp equals $p_b - p_a$ and is inherently negative.

Types of Filtration. For a given slurry in a specific filter the major variable under the control of the operator is the over-all pressure drop; e.g., when the outlet pressure is constant, the pressure drop is controlled by varying the inlet pressure. If the pressure drop is constant, the flow rate is a maximum at the start of the filtration and decreases continuously to the end. This method of operation is called *constant-pressure* filtration. If the pressure drop is varied, usually it is kept small at the start of the filtration and then increased, either continuously or in steps, and is a maximum at the end of the filtration. One method is to maintain the flow rate constant by continuously increasing the inlet pressure. This method is called *constant-rate* filtration. Various combinations of these basic methods are used. A common one is to use constant rate until the inlet pressure reaches a specified maximum and then to continue at constant pressure until the end of the filtration. This minimizes loss of solids through the septum when the resistance is low and avoids packing solids into the septum. In washing both the flow rate and pressure drop are usually constant.

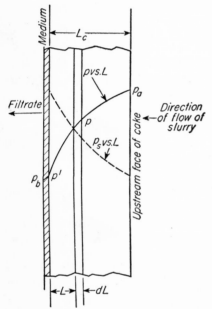

FIG. 7-16. Section through filter medium and cake, showing pressure gradients: p, fluid pressure. p_s, stress pressure. L, thickness of cake.

Pressure Drop through Filter Cake. Figure 7-16 shows diagrammatically a section through a filter cake and filter medium at a definite time θ sec from the start of the flow of filtrate. At this time the thickness of the cake, measured from the filter medium, is L_c ft. The filter area, measured perpendicularly to the direction of flow, is A ft^2. Consider the thin layer of cake of thickness dL ft lying in the cake at a distance L ft from the medium. Let the pressure at this point be p lb force/ft^2. This layer consists of a thin bed of solid particles through which the filtrate is flowing. In a filter bed the velocity is sufficiently low to ensure laminar flow. Accordingly, as a starting point for treating the pressure drop through the cake, the Kozeny-Carman relation [Eq. (2-90)] can be used in the following form

$$\frac{dp}{dL} = \frac{k\mu u(1 - \epsilon)^2(s_p/v_p)^2}{g_c\epsilon^3} \tag{7-7}$$

where dp = pressure drop through layer of thickness dL, lb force/ft^2
  μ = viscosity of filtrate, lb/ft-sec
  u = linear velocity of filtrate, based on filter area, ft/sec
  s_p = surface of single particle, ft^2
  v_p = volume of single particle, ft^3
  ϵ = porosity of cake
  k = constant
  g_c = Newton's-law conversion factor, 32.174 ft-lb/lb force-sec^2
For random packed particles of definite size and shape, $k = 5$.

The linear velocity u is given by the equation

$$u = \frac{dV/d\theta}{A} \tag{7-8}$$

where V is the volume, in cubic feet, of filtrate collected from the start of the filtration to time θ. Since the filtrate must pass through the entire cake, V/A is the same for all layers, and u is independent of L.

The volume of solids in the layer is $A(1 - \epsilon)\,dL$ ft^3, and if ρ_p is the density of the particles, the mass dm of solids in the layer is, in pounds,

$$dm = \rho_p(1 - \epsilon)A\,dL \tag{7-9}$$

Elimination of dL from Eqs. (7-7) and (7-9) gives

$$dp = \frac{k\mu u(s_p/v_p)^2(1 - \epsilon)}{g_c\rho_p A\,\epsilon^3}\,dm \tag{7-10}$$

Compressible and Incompressible Sludges. In some special situations, e.g., in the filtration of sludges composed of rigid, uniform particles under low pressure drops, all factors on the right-hand side of Eq. (7-10) except m are independent of L, and the equation is integrable directly, over the thickness of the cake. If m_c is the total mass of solids in the cake, the result is

$$-\int_{p_a}^{p'} dp = \frac{k\mu u(s_p/v_p)^2(1 - \epsilon)}{g_c\rho_p A\,\epsilon^3}\int_0^{m_c} dm$$

$$p_a - p' = \frac{k\mu u(s_p/v_p)^2(1 - \epsilon)m_c}{g_c\rho_p A\,\epsilon^3} = -\Delta p_c \tag{7-11}$$

Sludges of this type are called incompressible.

Most sludges encountered industrially do not have the simple structure of a bed of individual rigid particles. The usual slurry is a mixture of agglomerates, or flocs, consisting of loose assemblies of very small particles, and the resistance characteristics of the cake depend upon the properties of the flocs rather than on the geometry of the individual particles.[10] The flocs are deposited from the slurry on the upstream face of the cake, and form a complicated network of channels to which the Kozeny-Carman relationship does not precisely apply. The resistance of such a sludge is sensitive to the method used in preparing the slurry and to the age and temperature of the material. Also, the flocs are distorted and broken down by the forces existing in the cake, and the factors ϵ, k, and s_p/v_p vary from layer to layer. Sludges the resistance of which vary in this manner are called compressible sludges. Although Eq. (7-10) does not apply accurately to compressible sludges, it is a useful guide for developing empirical relationships applicable to such materials. Some sludges are more compressible than others, and a valid treatment should provide for a quantitative measure of compressibility.

Mechanics of Filter Cakes. The variation in resistance of a cake from layer to layer is the result of mechanical effects in the cake. The fluid pressure in a filter cake is a maximum at the upstream face and a minimum at the filter medium. It might seem, therefore, that the porosity of the cake should be a minimum at the upstream face and a maximum at the filter medium. Actually the reverse is true. A compressible cake is relatively open and porous at the upstream face and compressed at the filter medium. The reason for this is as follows. Consider a particle in the bed at distance L from the medium. The stream of filtrate flowing past the particle imparts a drag force on the particle tending to move it in the direction of the medium. This drag is resisted by an equal and opposite force from the particles immediately ahead of the particle under consideration. The particle is also acted upon by forces from those immediately upstream from itself. Since each layer of solids, starting from the upstream face of the cake, transmits its drag force to the particles ahead of itself, the drag forces act cumulatively through the bed, and each layer is acted on by a force equal to the sum of the drags of all layers between itself and the upstream face of the cake.

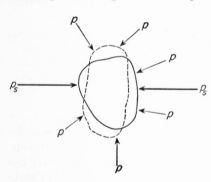

Fig. 7-17. Distortion of particle by stress pressure.

Each layer transmits this cumulative drag, augmented by its own drag, to the next layer. These cumulative forces can be converted to pressures by dividing them by A. Such pressures are called stress pressures. As shown in Fig. 7-16, the stress pressure is zero at the upstream face of the cake, where the fluid pressure is p_a, and is a maximum at the filter medium, where the fluid pressure is p'. As the fluid pressure decreases, the stress pressure increases. The fluid pressure at a given point acts equally in all directions, but the stress pressure acts in a direction parallel with that of flow and tends to squeeze the particles and flatten them.

Figure 7-17 shows how an undistorted particle, represented by the solid boundary, is distorted by stress pressure to the shape shown by the dotted boundary. The effect is equivalent to squeezing a particle of putty between the fingers while it is immersed under a static head of water.

The cumulative drag at the layer L ft from the filter medium equals $p_a - p$, the fluid pressure drop in the portion of the cake from the upstream face to the plane at L. It follows that k, s_p/v_p, and ϵ each depends on $p_a - p$ only. It is convenient to lump all of these factors as they appear in Eq. (7-10) into a single quantity α_L, called the local specific cake resistance. Mathematically

$$\alpha_L = \frac{k(s_p/v_p)^2(1 - \epsilon)}{\epsilon^3 \rho_p} \tag{7-12}$$

This coefficient depends only on $p_a - p$.

Equation (7-10) may now be written

$$dp = \frac{\mu u \alpha_L}{g_c A} \, dm \tag{7-13}$$

Since $d(p_a - p) = -dp$, $dp/\alpha_L = -d(p_a - p)/\alpha_L$, and Eq. (7-13) can be written

$$-\frac{d(p_a - p)}{\alpha_L} = \frac{\mu u}{g_c A} \, dm \tag{7-14}$$

Assume, now, that the local coefficient α_L is known as a function of $p_a - p$, through an equation or an experimental curve. Equation (7-14) may then be integrated, either analytically or graphically. The limits are when $m = 0$, $p_a - p = 0$, and when $m = m_c$, $p_a - p = p_a - p'$, where m_c is the total mass of solids in the cake. Then

$$-\int_{p_a-p'}^{0} \frac{d(p_a - p)}{\alpha_L} = \int_{0}^{p_a-p'} \frac{d(p_a - p)}{\alpha_L} = \frac{\mu u}{g_c A} \int_{0}^{m_c} dm = \frac{\mu u m_c}{g_c A} \tag{7-15}$$

In Fig. 7-18 is shown a characteristic graphical integration of the integral of Eq. (7-15). The reciprocal of α_L, $1/\alpha_L$, is plotted against $p_a - p$, and the area under the curve is measured between the axis of ordinates and the ordinate $p_a - p'$. For any functional relationship between α_L and $p_a - p'$ the integral depends only on $p_a - p'$, the pressure drop through the cake.

A coefficient α may now be defined by the equation

$$\int_{0}^{p_a-p'} \frac{d(p_a - p')}{\alpha_L} = \frac{p_a - p'}{\alpha} = \frac{-\Delta p_c}{\alpha} \tag{7-16}$$

Coefficient α is the reciprocal of the average ordinate under the curve of Fig. 7-18. Equation (7-15) may be written

$$\frac{p_a - p'}{\alpha} = \frac{-\Delta p_c}{\alpha} = \frac{\mu u m_c}{g_c A} \tag{7-17}$$

Equation (7-17) is the basic equation for pressure drop through a filter cake.

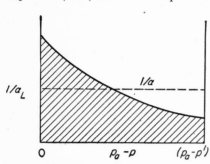

FIG. 7-18. Integration of Eq. (7-16).

The coefficient α is the average specific cake resistance. Physically, it is the pressure drop required to give unit velocity of filtrate flow when the viscosity is unity and the cake contains one unit mass of solid per unit filter area. This factor is an average quantity for the entire cake, and it must be measured experimentally for each sludge. For a given sludge, its value depends only on the pressure drop through the cake. The dimensions of α are, from Eq. (7-17) $\overline{L}\overline{M}^{-1}$, and it is measured in feet per pound.

For incompressible sludges α_L is independent of $p_a - p$, and $\alpha_L = \alpha$.

Filter-medium Resistance. The filter-medium resistance, denoted by R_m, can be defined, by analogy with Eq. (7-17), by the equation

$$\frac{p' - p_b}{R_m} = \frac{-\Delta p_m}{R_m} = \frac{\mu u}{g_c} \tag{7-18}$$

The dimension of R_m is $\overline{L}^{-1}$, and the units are expressed in reciprocal feet.

The factors controlling the magnitude of R_m have been studied. It is known that pressure drop, and perhaps flow rate, affect it, and that an old, used filter medium has a much larger resistance than a clean, new one. Since the medium resistance usually is important only during the early stages of the filtration, it is satisfactory to assume that R_m is a constant during any given filtration and to determine its magnitude empirically from experimental data. The filter-medium resistance varies somewhat from experiment to experiment even for the same sludge and filter. When R_m is treated as an empirical constant, it also includes any resistance to flow that may exist in the leads to and from the filter.

From Eqs. (7-17) and (7-18),

$$-\Delta p = -\Delta p_c - \Delta p_m = \frac{\mu u}{g_c}\left(\frac{m_c \alpha}{A} + R_m\right) \tag{7-19}$$

Strictly, the cake resistance α is a function of Δp_c rather than of Δp. During the important stage of the filtration, when the cake is of appreciable thickness, Δp_m is small in comparison with Δp_c, and the effect on the magnitude of α of carrying the integration of Eq. (7-16) over a range

$-\Delta p$ instead of $-\Delta p_c$ can be safely ignored. In Eq. (7-19), then, α is taken as a function of Δp.

In using Eq. (7-19) it is convenient to replace u, the linear velocity of the filtrate, and m_c, the total mass of solid in the cake, by functions of V, the total volume of filtrate collected to time θ. Equation (7-8) relates u and V, and a material balance relates m_c and V. If c is the mass of the particles deposited in the filter per unit volume of filtrate,† in pounds per cubic foot, the mass of solids in the filter at time θ is Vc, and

$$m_c = Vc \qquad (7\text{-}21)$$

Substituting u from Eq. (7-8) and m_c from Eq. (7-21) in Eq. (7-19) gives

$$\frac{d\theta}{dV} = \frac{\mu}{Ag_c(-\Delta p)}\left(\frac{\alpha c V}{A} + R_m\right) \qquad (7\text{-}22)$$

Constant-pressure Filtration. When Δp is constant, the only variables in Eq. (7-22) are V and θ. The equation may be integrated as follows

$$\int_0^\theta d\theta = \frac{\mu}{Ag_c(-\Delta p)}\left(\frac{c\alpha}{A}\int_0^V V\,dV + R_m\int_0^V dV\right)$$

$$\theta = \frac{\mu}{g_c(-\Delta p)}\left[\frac{c\alpha}{2}\left(\frac{V}{A}\right)^2 + R_m\frac{V}{A}\right] \qquad (7\text{-}23)$$

where V is the total volume of filtrate, in cubic feet, collected to time θ sec, assuming that time is counted from the instant the first drop of filtrate is obtained, so that when $V = 0$, $\theta = 0$.

Equation (7-23) is that of a parabola with its vertex displaced from the origin of coordinates ($\theta = 0$, $V = 0$). A plot of V vs. θ for a constant-pressure filtration is, therefore, a section of this parabola.

To evaluate the constants α and R_m for a definite pressure drop, data of V vs. θ from an experimental run at that pressure drop are needed. The

† The concentration of solid in the slurry fed to the filter is slightly less than c, since the wet cake includes sufficient liquid to fill its pores, and V, the actual volume of filtrate, is slightly less than the total liquid in the original slurry. Correction for this retention of liquid in the cake may be made by material balances, if desired. Thus, let w be the ratio of the mass of the wet cake, including the filtrate retained in its voids, to the mass of the dry cake obtained by washing the cake free of soluble material and drying. Also, let ρ be the density of the filtrate. Then if c_s is the concentration of solids in the slurry in pounds per cubic foot of liquid fed to the filter, material balances give

$$c = \frac{c_s}{1 - (w-1)c_s/\rho} \qquad (7\text{-}20)$$

treatment of such data is facilitated by using Eq. (7-22) in the form

$$\frac{d\theta}{dV} = K_p V + B \tag{7-24}$$

where
$$K_p = \frac{c\alpha\mu}{A^2(-\Delta p)g_c} \tag{7-25}$$

and
$$B = \frac{R_m\mu}{A(-\Delta p)g_c} \tag{7-26}$$

Assume that a number of observations of V vs. θ have been made. Then for any two successive observations, the quantity $\Delta\theta/\Delta V$ can be calculated, where $\Delta\theta$ is the time between observations and ΔV is the increment of filtrate collected over time period $\Delta\theta$. Since, by Eq. (7-24), $d\theta/dV$ is linear with V, a value of $\Delta\theta/\Delta V$ is the true slope of the θ vs. V line at a point $(V_1 + V_2)/2$, or halfway between the observed values of V that define ΔV. Then, the arithmetic mean of each two successive observations of V can be plotted against $\Delta\theta/\Delta V$ for the same pair of readings. The best straight line is drawn through these points, and the slope of the line is K_p, and the V-axis intercept is B. This construction is shown for several constant-pressure runs in Fig. 7-19. Since the various simplifying assumptions made during the derivation of Eq. (7-21) are most questionable during the early stages of the filtration, the points taken during these stages may not fall accurately on the graph, and these points should be given little weight in plotting the line. When K_p and B are known, α and R_m are then calculated from Eqs. (7-25) and (7-26).

Empirical Equations for Cake Resistance. By conducting constant-pressure experiments at various pressure drops, the variation of α with Δp may be found. If α is independent of Δp, the sludge is incompressible. Ordinarily K_p increases with $-\Delta p$, as most sludges are at least to some extent compressible. For highly compressible sludges, K_p increases rapidly with $-\Delta p$.

Empirical equations may be fitted to observed data for Δp vs. α. Two such equations are common. They are

$$\alpha = \alpha_0(-\Delta p)^s \tag{7-27}$$

and
$$\alpha = \alpha_0'[1 + \beta(-\Delta p)^{s'}] \tag{7-28}$$

where α_0, α_0', β, s, and s' are empirical constants. Equation (7-27) is somewhat more restricted than Eq. (7-28), but it is simpler to use, and only two constant-pressure runs are needed to evaluate the constants α_0 and s. This equation is obviously wrong at low pressure drops, as it indicates a zero resistance at zero Δp. Three experiments are needed to evaluate the constants in Eq. (7-28), but it can be used over a wider range of pressure drops, from zero up, than Eq. (7-27).

In Eq. (7-27), s is a quantitative measure of compressibility. It is zero for incompressible sludges and is positive for compressible ones. Constant s

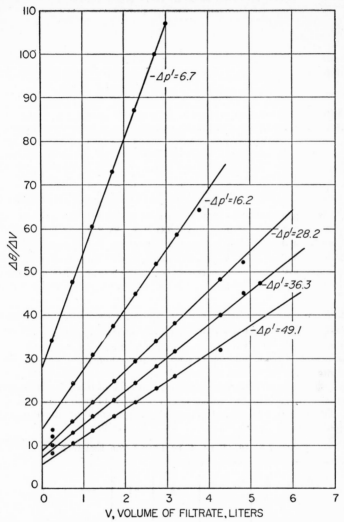

Fig. 7-19. Plot of $\Delta\theta/\Delta V$ vs. $\overline{V}$ for Example 7-2.

usually falls between 0.1 and 1.0, the large values applying to the more compressible sludges. This constant is commonly called the compressibility coefficient.

Neither Eq. (7-27) nor (7-28) should be used in a pressure range much different from that covered by the experiments conducted to evaluate its constants.

Example 7-2. Laboratory filtrations conducted at constant pressure drop on a slurry of $CaCO_3$ in H_2O gave the data shown in Table 7-4.[20] The filter area was

TABLE 7-4. VOLUME-TIME DATA FOR EXAMPLE 7-2

Test number............	I	II	III	IV	V
Pressure drop $-\Delta p$, lb force/in.^{2}............	6.7	16.2	28.2	36.3	49.1
Filtrate volume V, liters	\multicolumn{5}{c}{Time, sec}				

Filtrate volume V, liters	I	II	III	IV	V
0.5	17.3	6.8	6.3	5.0	4.4
1.0	41.3	19.0	14.0	11.5	9.5
1.5	72.0	34.6	24.2	19.8	16.3
2.0	108.3	53.4	37.0	30.1	24.6
2.5	152.1	76.0	51.7	42.5	34.7
3.0	201.7	102.0	69.0	56.8	46.1
3.5	.....	131.2	88.8	73.0	59.0
4.0	.....	163.0	110.0	91.2	73.6
4.5	.....	.....	134.0	111.0	89.4
5.0	.....	.....	160.0	133.0	107.3
5.5	.....	.....	.....	156.8	
6.0	.....	.....	.....	182.5	

440 cm^2, the mass of solid per unit volume of filtrate was 23.5 g/liter, and the temperature was 25°C. Evaluate, in foot, pound force, pound, second, units the quantities α and R_m as a function of pressure drop, and fit an empirical equation to the results for α.

Solution. The first step is to prepare plots, for each of the five constant-pressure experiments, of $\Delta\theta/\Delta V$ vs. $\overline{V}$, which is $(V_1 + V_2)/2$ of each increment of filtrate volume. The data and calculations for the first experiment are given in Table 7-5, and the plots for all experiments are shown in Fig. 7-19.

TABLE 7-5. $\Delta\theta/\Delta V$ vs. $\overline{V}$ IN TEST I FOR EXAMPLE 7-2

Filtrate volume V, liters	Time θ, sec	$\Delta\theta$	ΔV	$\dfrac{\Delta\theta}{\Delta V}$	$\overline{V}$
0	0				
0.5	17.3	17.3	0.5	34.6	0.25
1.0	41.3	24.0	0.5	48.0	0.75
1.5	72.0	30.7	0.5	61.4	1.25
2.0	108.3	36.3	0.5	72.6	1.75
2.5	152.1	43.8	0.5	87.6	2.25
3.0	201.7	49.6	0.5	99.2	2.75

The slope of each line of Fig. 7-19 is K_p, in seconds per liter per liter. To convert to seconds per cubic foot per cubic foot the conversion factor is $28.31^2 = 801$. The intercept of each line on the axis of ordinates is B, in seconds per liter. The conversion factor to convert this to seconds per cubic foot is 28.31. The slopes and intercepts, in the observed and converted units, are given in Table 7-6.

TABLE 7-6. VALUES OF K_p, B, R_m, AND α FOR EXAMPLE 7-2

Test	Pressure drop $-\Delta p$		Slope K_p		Intercept B		R_m, ft^{-1} $\times 10^{10}$	α, ft/lb $\times 10^{11}$
	Lb force/in.^{2}	Lb force/ft^2	Sec/liter^2	Sec/ft^6	Sec/liter	Sec/ft^3		
I	6.7	965	25.8	20,700	28.5	807	1.99	1.65
II	16.2	2,330	13.9	11,150	13.2	374	2.23	2.15
III	28.2	4,060	9.3	7,450	8.5	241	2.50	2.50
IV	36.3	5,230	7.7	6,170	7.0	198	2.65	2.67
V	49.1	7,070	6.4	5,130	5.5	156	2.82	3.00

The viscosity of water is, from Appendix 14, 0.886 centipoise, or $0.886 \times 6.72 \times 10^{-4} = 5.95 \times 10^{-4}$ lb/ft-sec. The filter area is $440/30.48^2 = 0.474$ ft^2. The concentration c is $(23.5 \times 28.31)/454 = 1.47$ lb/ft^3.

From the values of K_p and B in Table 7-6, corresponding quantities for α and R_m are found from Eqs. (7-25) and (7-26). Thus

$$\alpha = \frac{A^2(-\Delta p)g_c K_p}{c\mu} = \frac{0.474^2 \times 32.17(-\Delta p)K_p}{5.95 \times 10^{-4} \times 1.47} = 8.28 \times 10^3(-\Delta p)K_p$$

$$R_m = \frac{A(-\Delta p)g_c B}{\mu} = \frac{0.474 \times 32.17(-\Delta p)B}{5.95 \times 10^{-4}} = 2.56 \times 10^4(-\Delta p)B$$

Table 7-6 shows the values of α and R_m for each test. Figure 7-20 is a plot of R_m vs. $-\Delta p$.

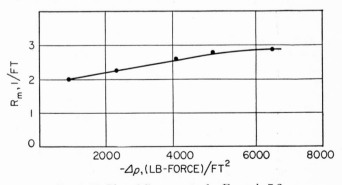

FIG. 7-20. Plot of R_m vs. $-\Delta p$ for Example 7-2.

Figure 7-21 is a logarithmic plot of α vs. $-\Delta p$. The points fall on a straight line, so Eq. (7-27) is suitable as an equation for α as a function of $-\Delta p$. The slope of the line is 0.30, and this is the value of s for the sludge. The sludge is moderately compressible.

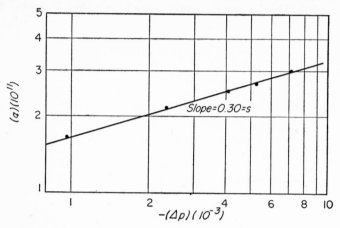

FIG. 7-21. Plot of log α vs. log $(-\Delta p)$ for Example 7-2.

Constant α_0 can be calculated by reading the coordinates of any convenient point on the line of Fig. 7-21 and calculating α_0 by Eq. (7-27). For example, when $-\Delta p = 1,000$, $\alpha = 1.65 \times 10^{11}$, and

$$\alpha_0 = \frac{1.65 \times 10^{11}}{1,000^{0.30}} = 2.08 \times 10^{10}$$

Equation (7-27) becomes for this sludge

$$\alpha = 2.08 \times 10^{10}(-\Delta p)^{0.30}$$

Constant-rate Filtration. At constant rate the linear velocity u is constant, and

$$u = \frac{dV/d\theta}{A} = \frac{V}{A\theta} \tag{7-29}$$

Equation (7-17) can be written, after substituting m_c from Eq. (7-21) and u from Eq. (7-29), as

$$\frac{-\Delta p_c}{\alpha} = \frac{\mu c}{\theta g_c}\left(\frac{V}{A}\right)^2 \tag{7-30}$$

The specific cake resistance α is retained on the left-hand side of Eq. (7-30) because it is a function of Δp for compressible sludges.†

† The concentration c may also vary somewhat with pressure drop. In operation, c_s rather than c is constant, and, by Eq. (7-20), since w changes with pressure, c also changes when $(w - 1)(c_s/\rho)$ is appreciable in comparison with unity. Any such variation in c with Δp can be ignored, in view of the other approximations made in the general theory of filtration.

If α is known as a function of Δp_c, and if $-\Delta p_m$, the pressure drop through the filter medium can be estimated, Eq. (7-30) may be used directly to relate the over-all pressure drop to time when the rate of flow of filtrate is constant. A more direct use of this equation may be made, however, if Eq. (7-27) is accepted to relate α and Δp_c.[12] If α from Eq. (7-27) is substituted in Eq. (7-30), and if $-(\Delta p - \Delta p_m)$ is substituted for $-\Delta p_c$, the result is

$$(-\Delta p_c)^{1-s} = \frac{\alpha_0 \mu c \theta}{g_c} \left(\frac{V}{A\theta} \right)^2 = [-(\Delta p - \Delta p_m)]^{1-s} \qquad (7\text{-}31)$$

Again, the simplest method of correcting the over-all pressure drop for the pressure drop through the filter medium is to assume the filter medium resistance is constant during a given constant-rate filtration. Then, by Eq. (7-18), $-\Delta p_m$ is also constant in Eq. (7-31).

The constants $-\Delta p_m$, α_0, and s can be evaluated by the following method from measurements of θ vs. $-\Delta p$. Since the only variables in Eq. (7-31) are $-\Delta p$ and θ, the equation can be written

$$[-(\Delta p - \Delta p_m)]^{1-s} = K_r \theta \qquad (7\text{-}32)$$

where K_r is defined by the equation

$$K_r = \frac{\mu u^2 c \alpha_0}{g_c} \qquad (7\text{-}33)$$

Taking logarithms of both sides of Eq. (7-32) gives

$$\log \theta = (1 - s) \log [-(\Delta p - \Delta p_m)] - \log K_r \qquad (7\text{-}34)$$

If $-(\Delta p - \Delta p_m)$ is plotted as the abscissa against θ as the ordinate, a straight line should be obtained, the slope of which is $1 - s$. Since the actual data are the over-all pressure drop $-\Delta p$ and the time θ, the pressure drop $-\Delta p_m$ is not known and must be estimated. A tentative magnitude for $-\Delta p_m$ can be found by plotting θ vs. $-\Delta p$ on rectangular coordinates, passing a smooth curve through the points, and extrapolating the curve to the pressure axis, where $\theta = 0$. A tentative result for $-\Delta p_m$ is thus found, which can be used for preparing a plot of $-(\Delta p - \Delta p_m)$ vs. θ on logarithmic coordinates. If the line so obtained is straight, the tentative value of $-\Delta p_m$ can be taken as final. If the line is curved, additional approximations for $-\Delta p_m$ can be made until a straight line is achieved.

When a straight line has been obtained on the plot of $-(\Delta p - \Delta p_m)$ vs. θ, the constant s is obtained from the slope of the line. The factor K_r is calculated, by means of Eq. (7-32), from the coordinates of any convenient point on the line, and α_0 is then evaluated by means of Eq. (7-33).

Example 7-3. Table 7-7 shows data obtained in a constant-rate filtration of a sludge consisting of $MgCO_3$ and H_2O. The rate was 0.100 lb/ft²-sec, the viscosity of the filtrate was 0.92 centipoise, and the concentration of solids in the slurry

TABLE 7-7. PRESSURE DROP–TIME DATA IN CONSTANT-RATE FILTRATION
FOR EXAMPLE 7-3 †

$-\Delta p$, lb force/in.²	θ, sec	$-\Delta p$, lb force/in.²	θ, sec
4.4	10	11.8	70
5.0	20	13.5	80
6.4	30	15.2	90
7.5	40	17.6	100
8.7	50	20.0	110
10.2	60		

† From O. D. Hughes, R. W. Ver Hoeve, and C. D. Luke, paper given at meeting of AIChE, Columbus, Ohio, December, 1950.

was 1.08 lb/ft³ filtrate. Evaluate, in foot, pound force, second units the constants R_m, s, and α_0 for this sludge.

Solution. To obtain a preliminary result for $-\Delta p_m$, $-\Delta p$ is plotted against θ on rectangular coordinates, as shown in Fig. 7-22. Extrapolation to the pressure-drop

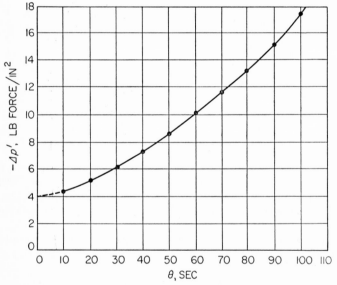

FIG. 7-22. Plot of $-\Delta p$ vs. θ for Example 7-3.

axis gives $-\Delta p_m = 4$ lb force/in.² Figure 7-23 is a logarithmic plot of $-(\Delta p - 4)$ vs. θ. A satisfactory straight line is established by the points on Fig. 7-23 and the preliminary evaluation of $-\Delta p_m$ can be accepted as final. From Eq. (7-18),

$$R_m = \frac{-\Delta p_m g_c}{\mu u} = \frac{4 \times 144 \times 32.17}{0.92 \times 6.72 \times 10^{-4} \times 0.100} = 3 \times 10^8 \text{ ft}^{-1}$$

The slope of the line of Fig. 7-23, equal to $1 - s$, is 0.642, and $s = 0.358$.

From Fig. 7-23, when $-(\Delta p - \Delta p_m) = 10$ lb force/in.^{2} (1,440 lb force/ft^2), $\theta = 81$ sec. From Eq. (7-32),

$$K_r = \left(\frac{1,440}{81}\right)^{0.642}$$

$$= 1.32$$

From Eq. (7-33),

$$\alpha_0 = \frac{K_r g_c}{\mu u^2 c} = \frac{1.32 \times 32.17}{0.92 \times 6.72 \times 10^{-4} \times 0.100^2 \times 1.08} = 6.35 \times 10^6$$

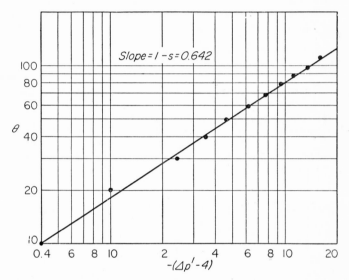

Fig. 7-23. Plot of log $-(\Delta p - \Delta p_m)$ vs. log θ for Example 7-3.

Washing Filter Cakes. To wash soluble material that may be retained by the filter cake after a filtration, a solvent miscible with the filtrate may be used as a wash. Water is the most common wash liquid. The rate of flow of the wash liquid and the volume of liquid needed to reduce the solute content of the take to a desired degree are important in the design and operation of a filter. Although the following general principles apply to the problem, these questions cannot be completely answered without experiment.[7]

The volume of wash liquid required is related to the concentration-time history of the wash liquid leaving the filter. During washing the concentration-time relation is of the type shown in Fig. 7-24. The first portion of the recovered liquid is represented by segment ab in Fig. 7-24. The effluent consists essentially of the filtrate that was left on the filter, which is swept out by the first wash liquid without appreciable dilution. This stage of washing is called *displacement wash*, and is the ideal method of washing a cake. Under favorable conditions where the particle size of the cake is small, as much as 90 per cent of the solute in the cake can be recovered during this state. The volume of wash liquid needed for a displacement wash is equal to the volume of filtrate left in the cake, or $\bar{\epsilon}AL$, where L is the cake thickness and $\bar{\epsilon}$ is the average porosity of the cake. The second stage of washing, shown by the segment bc in Fig. 7-24, is characterized by a rapid drop in concentration of the effluent. The volume of wash liquid used in this stage is also of the order of magnitude of that used in the first stage. The third stage is shown by segment cd. The concentration of solute in the effluent is low, and the remaining solute is slowly leached from

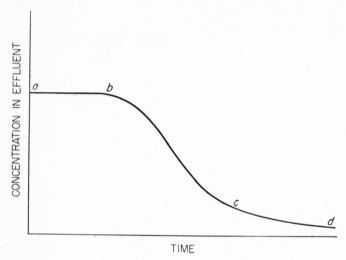

FIG. 7-24. Washing filter cake.

the cake. If sufficient wash liquid is used, the residual solute in the cake can be reduced to any desired point, but the washing should be stopped when the value of the unrecovered solute is less than the cost of recovering that solute.

In pressure-leaf filters and rotary vacuum filters, the wash liquid follows the same path as that of the filtrate. The rate of flow of the wash liquid is, in principle, equal to that of the last of the filtrate, provided the pressure drop remains unchanged in passing from filtration to washing. If the viscosities of filtrate and wash liquid differ, correction for this difference can be made. This rule is only approximate, however, as during washing the liquid does not actually follow exactly the path of the filtrate because of channeling, formations of cracks in the cake, and short-circuiting.

In a plate-and-frame press using three-button plates, the actual conditions are even more divergent from the ideal than for other types of filter. The surface available for the wash liquid is only one-half that used in the filtration, and the wash water must pass through the entire cake thickness rather than one-half of it. The theoretical rate of flow is, then, one-fourth that of the final flow rate of the filtrate. In filter presses, also, the effects of channeling, poor distribution of wash liquid, and short-circuiting are more pronounced, and more wash liquid is required to remove a definite fraction of solute than in the other types.

Filtration of Solids from Gases by Cloth Collectors

Filtration through a cloth filter medium is used to remove solid particles or dusts from air or gas. The particles, which may be smaller than the meshes in the cloth, are first caught in the channels of the cloth itself and form the true filter medium on which a layer of solids subsequently forms in the same manner as does the cake in the filtration of a slurry. These filters are also called bag filters. The filter on an ordinary vacuum cleaner is a small bag filter. Cloth collectors are effective in removing particles of diameters between about 0.1 and 100 microns.[15a]

A typical industrial bag filter is shown in Fig. 7-25. Tall cylindrical bags of cotton cloth, 6 in. in diameter and about 9 ft high, are held vertically in a large metal box. The floor of the box is perforated with holes about 6 in. in diameter. Over each perforation is a collar to which the bottom of a bag is clamped. The upper ends of the bags are sewn shut and attached to a motor-driven shaker bar. Beneath the box floor is a series of hoppers with discharge gates for the collected dust. When the filter is operating, dust-laden air enters a large inlet, impinging on a wearing

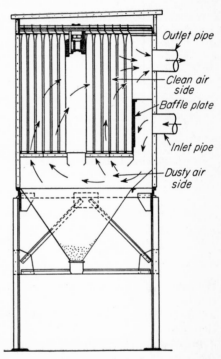

FIG. 7-25. Bag filter.

plate; travels beneath the box floor and over the hoppers; and flows up and through the bags into the box. An outlet duct carries clean air to a fan. Dust is deposited in a thin layer on the inside of the bags. Periodically the bags are vigorously shaken upward and downward by the motor, causing the dust to drop out into the hoppers. The air flow may be shut off during the shaking period; in some designs the cleaning is assisted by a small flow of air in the reverse direction.

Typical bag filters contain 800 to 16,000 ft^2 of filtering surface. If the concentration of dust in the air is low, up to $3\frac{1}{2}$ ft^3/min may be passed through each square foot of filter surface. With high dust concentrations the throughput should not exceed 2 ft^3/ft^2-min.[19b]

Pressure Drop in Cloth Collectors. Equation (7-23) can be used as the basis for calculating the pressure drop through a bag filter. The specific resistance α is measured experimentally, and the "cake" is assumed to be incompressible. For a given kind of solid particle, α depends primarily on

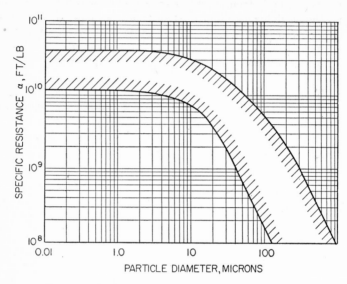

FIG. 7-26. Specific resistance of dust layers. (*Lapple.*[15e])

particle size.[15e] An approximate relation between α and particle size is shown in Fig. 7-26, in which the diameter of the particles is given in microns.

SEPARATIONS BASED ON THE MOTION OF PARTICLES THROUGH FLUIDS

Many methods for mechanical separation are based on the movement of solid particles or liquid drops through a fluid. The fluid may be gas or liquid, and it may be flowing or at rest. In some situations the objective of the process is to remove particles from a stream of fluid in order to eliminate contaminants from the fluid or to recover the particles. Examples are the elimination of dust and fumes from air or flue gas, the removal of solids from liquid wastes to allow discharge into public drainage systems, and the recovery of acid mists from the waste gas of an acid plant. In other problems, particles are deliberately suspended in fluids to obtain separations of the particles into fractions differing in size or density. The fluid is then recovered, sometimes for reuse, from the fractionated particles. Examples are the use of the stream of air in a roller mill, as shown in Fig. 4-13, and ore dressing, in which the valuable constituents in ores are recovered from the waste material called gangue.

Two simple methods of utilizing the movements of particles through fluids are shown in Fig. 7-27. A settling chamber is shown in Fig. 7-27a. A stream of dust-laden air enters the chamber and flows horizontally at a low velocity through the unit. The dust particles settle under the influence of gravity to the bottom of the chamber, from which they can be removed, and the clean air leaves at the exit from the chamber. Solid particles can be removed from a slurry in the same way by allowing them to settle in an open tank, drawing off the clear liquid from the top of the tank, and recovering the solid as a thick sludge from the bottom.

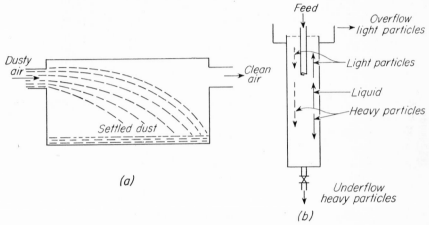

Fig. 7-27. Simple gravity separators: (a) Dust-settling chamber. (b) Hydraulic separator. Solid vectors: velocity relative to ground. Dotted vectors: velocity relative to liquid.

Figure 7-27b shows a method of separating two kinds of particles having the same size but different densities. The heavier particles settle through a liquid at a greater speed than the lighter ones. This fact can be utilized by passing the liquid upward in a vertical tank at a velocity intermediate between the settling velocities of the two materials. The lighter substance is swept out of the tank with the liquid, and the heavier particles collect at the bottom of the apparatus. The lighter particles can then be recovered by settling from the overflow, and the heavier ones recovered from the bottom of the separator. Methods of this kind have been refined and perfected mechanically to give the more elaborate methods described later in this section.

Principles of Particle Mechanics

All methods based on the movement of particles through fluids require that a density difference exist between the particles and the fluid. Also, an external force is needed to impart motion to the particle relative to the fluid. The external force is usually gravity, but when gravity is not sufficiently strong, centrifugal force, which can be many times that of gravity, is used. If the densities of particle and fluid are equal, the buoyant force from the immersion of the particle in the fluid will counterbalance an external force, however large, and the particle will not move through the fluid. The greater the density difference, the more effective the process.

Three forces act on a particle moving through a fluid: (1) the external force, gravitational or centrifugal; (2) the buoyant force, which acts parallel with the external force, but in the opposite direction; and (3) the drag force, which appears whenever there is relative motion between the particle and the fluid. The drag force acts to oppose the motion and acts parallel with the direction of movement but in the opposite direction.

In the general case, the direction of movement of the particle relative to the fluid may not be parallel with the direction of the external and buoyant forces, and the drag force then makes an angle with the other two. In this situation, which is called two-dimensional motion, the drag must be resolved into components, which complicates the treatment of particle mechanics. Equations are available for two-dimensional motion,[16] but only the one-dimensional case, where the lines of action of all forces acting on the particle are collinear, will be considered in this book.

Equations for One-dimensional Motion of Particle through Fluid.
Consider a particle of mass m lb moving through a fluid under the action of an external force F_e lb force. Let the velocity of the particle relative to the fluid be u ft/sec. Let the buoyant force on the particle be F_b lb force, and the drag be F_D lb force. Then, the resultant force on the particle is $F_e - F_b - F_D$, the acceleration of the particle is $du/d\theta$ ft/sec^2, and, by Eq. (1-10),

$$\frac{m}{g_c}\frac{du}{d\theta} = F_e - F_b - F_D \tag{7-35}$$

The external force can be expressed as a product of the mass and the acceleration a_e of the particle from this force, and

$$F_e = \frac{ma_e}{g_c} \tag{7-36}$$

The buoyant force is, by Archimedes' principle, the product of the mass of the fluid displaced by the particle and the acceleration from the external force. The volume of the particle is m/ρ_p ft^3, where ρ_p is the density of the particle, and the particle displaces this same volume of fluid. The mass of fluid displaced is $(m/\rho_p)\rho$ lb, where ρ is the density of the fluid. The buoyant force is, then,

$$F_b = \frac{m\rho a_e}{\rho_p g_c} \tag{7-37}$$

The drag force is, from Eq. (2-84),

$$F_D = \frac{C_D u^2 \rho A_p}{2g_c} \tag{7-38}$$

where C_D is the dimensionless drag coefficient and A_p is the projected area, in square feet, of the particle measured in a plane perpendicular to the direction of motion of the particle.

Substituting the forces from Eqs. (7-36) through (7-38) into Eq. (7-35) gives

$$\frac{du}{d\theta} = a_e - \frac{\rho a_e}{\rho_p} - \frac{C_D u^2 \rho A_p}{2m}$$

$$= a_e \frac{\rho_p - \rho}{\rho_p} - \frac{C_D u^2 \rho A_p}{2m} \qquad (7\text{-}39)$$

Motion from Gravitational Force. If the external force is gravity, a_e is g, the acceleration of gravity in feet per second per second, and Eq. (7-39) becomes

$$\frac{du}{d\theta} = g \frac{\rho_p - \rho}{\rho_p} - \frac{C_D u^2 \rho A_p}{2m} \qquad (7\text{-}40)$$

Motion in a Centrifugal Field. A centrifugal force appears whenever the direction of movement of a particle is changed. From elementary physics, the acceleration from a centrifugal force from circular motion is

$$a_e = r\omega^2 \qquad (7\text{-}41)$$

where r = radius of path of particle, ft
    ω = angular velocity, radians/sec
Substituting into Eq. (7-39) gives

$$\frac{du}{d\theta} = r\omega^2 \frac{\rho_p - \rho}{\rho_p} - \frac{C_D u^2 \rho A_p}{2m} \qquad (7\text{-}42)$$

In this equation, u is the velocity of the particle relative to the fluid, and is directed outwardly along a radius.

Terminal Velocity. In gravitational settling, g is constant. Also, the drag always increases with velocity. Equation (7-40) shows that the acceleration decreases with time and approaches zero. The particle quickly reaches a constant velocity, which is the maximum attainable under the circumstances, and which is called the terminal velocity. The equation for the terminal velocity u_t is found, for gravitational settling, by taking $du/d\theta = 0$. Then, from Eq. (7-40),

$$u_t = \sqrt{\frac{2g(\rho_p - \rho)m}{A_p \rho_p C_D \rho}} \qquad (7\text{-}43)$$

In motion from a centrifugal force, the velocity depends on the radius, and the acceleration is not constant if the particle is in motion with respect to the fluid. In many practical uses of centrifugal force, however, $du/d\theta$ is small in comparison with the other two terms in Eq. (7-40), and if $du/d\theta$

is neglected, a "terminal velocity" at any given radius can be defined by the equation

$$u_t = \omega \sqrt{\frac{2rm(\rho_p - \rho)}{\rho_p \rho C_D A_p}} \tag{7-44}$$

Drag Coefficients. The quantitative use of Eqs. (7-39) to (7-44) requires that numerical values be available for the drag coefficient C_D. Figure 2-31, which shows the drag coefficient as a function of Reynolds number, provides such a relationship. A portion of the curve of C_D vs. N_{Re} for spheres is reproduced in Fig. 7-28. The drag curve shown in Fig.

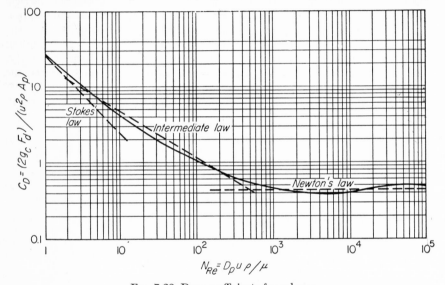

Fig. 7-28. Drag coefficients for spheres.

7-28 applies, however, only under restricted conditions. The particle must be a solid sphere, it must be free to move without being influenced by other particles or by the wall or bottom of the vessel, it must be moving at a constant velocity, it must not be too small, and the fluid through which it is moving must be still.

Variations in particle shape can be accounted for by obtaining separate curves of C_D vs. N_{Re} for each shape, as shown in Fig. 2-31 for cylinders and plates. Certain correlations are available that can be used for particles of known geometry, but in the following treatment spheres will be assumed, as the same principles apply to any shape.[11,18]

When the particle is at sufficient distance from the boundaries of the container and from other particles, so that its fall is not affected by them, the process is called *free settling*. If the motion of the particle is impeded by other particles, which will happen when the particles are near each other even though they may not actually be colliding, the process is called *hin-*

dered settling. The drag coefficient in hindered settling is greater than in free settling.

If the particle is accelerating, the drag is influenced by the changes in velocity gradients near the surface of the particle.[13] This causes a drag greater than that shown in Fig. 7-28 for the same Reynolds number. Also, if the particles are liquid drops, circulatory currents are generated in them, and rapid oscillations or changes in drop shape may occur. Additional drag is necessary to supply the energy required to maintain these motions in the drop itself.

If the particles are very small, Brownian movement appears. This is a random motion imparted to the particle by collisions between the particle and the molecules of the surrounding fluid. This effect becomes appreciable at a particle size of about 2 to 3 microns and predominates over the force of gravity with a particle size of 0.1 micron or less. The random movement of the particle tends to suppress the effect of the force of gravity, so settling does not occur. Application of centrifugal force reduces the relative effect of Brownian movement.

Motion of Spherical Particles. If the particles are spheres of diameter of D_p ft,

$$m = \frac{\pi D_p^3 \rho_p}{6} \tag{7-45}$$

and

$$A_p = \frac{\pi D_p^2}{4} \tag{7-46}$$

Substitution of m and A from Eqs. (7-45) and (7-46) into Eq. (7-39) gives

$$\frac{du}{d\theta} = a_e \frac{\rho_p - \rho}{\rho_p} - \frac{3 C_D u^2 \rho}{4 \rho_p D_p} \tag{7-47}$$

At the terminal velocity, $du/d\theta = 0$, and

$$a_e(\rho_p - \rho) = \frac{3 C_D u_t^2 \rho}{4 D_p} \tag{7-48}$$

Approximate Equations for Drag Coefficients of Spheres. Although the relationship of C_D vs. N_{Re} in Fig. 7-31 is a continuous curve, it can, for use in calculations, be replaced by three straight lines without serious loss in accuracy.[15a] These lines, each of which covers a definite range of Reynolds numbers, are shown as dotted lines in Fig. 7-28. The equations for the lines and the ranges of the Reynolds number over which each applies are, for $N_{\text{Re}} < 1.9$,

$$C_D = \frac{24}{N_{\text{Re}}} \tag{7-49}$$

and

$$F_D = \frac{3\pi\mu u_t D_p}{g_c} \tag{7-50}$$

This is the Stokes' law range. For $1.9 < N_{\text{Re}} < 500$, they are

$$C_D = \frac{18.5}{N_{\text{Re}}^{0.6}} \tag{7-51}$$

and
$$F_D = \frac{2.31\pi(u_t D_p)^{1.4}\mu^{0.6}\rho^{0.4}}{g_c} \tag{7-52}$$

This is the intermediate range. For $500 < N_{\text{Re}} < 200{,}000$, they are

$$C_D = 0.44 \tag{7-53}$$

and
$$F_D = \frac{0.055\pi(u_t D_p)^2\rho}{g_c} \tag{7-54}$$

This is Newton's law range.

Equations (7-49), (7-51), and (7-53) can be written in the single general form

$$C_D = \frac{b_1}{N_{\text{Re}}^n} \tag{7-55}$$

Then Eqs. (7-50), (7-52), and (7-54) can be written as

$$F_D = \frac{\mu^n b_1 \pi (D_p u_t)^{2-n}\rho^{1-n}}{8g_c} \tag{7-56}$$

where b_1 and n are constants, summarized in Table 7-8.

TABLE 7-8. CONSTANTS IN EQUATIONS FOR DRAG COEFFICIENTS

Range	b_1	n
Stokes' law..........	24	1
Intermediate range...	18.5	0.6
Newton's law........	0.44	0

A general equation for the terminal velocity of spheres is obtained by substituting C_D from Eq. (7-55) into Eq. (7-48) and solving for u_t.

$$a_e(\rho_p - \rho) = \frac{3u_t^2 b_1 \rho \mu^n}{4D_p(D_p u_t \rho)^n}$$

and
$$u_t = \left[\frac{4a_e D_p^{1+n}(\rho_p - \rho)}{3b_1\mu^n\rho^{1-n}}\right]^{1/(2-n)} \tag{7-57}$$

For gravity settling, $a_e = g$, and for centrifugal settling, $a_e = \omega^2 r$. The special cases of Eq. (7-57) are, for Stokes' law range,

$$u_t = \frac{a_e D_p^2 (\rho_p - \rho)}{18\mu} \tag{7-58}$$

for intermediate range,

$$u_t = \frac{0.153 a_e^{0.71} D_p^{1.14} (\rho_p - \rho)^{0.71}}{\rho^{0.29} \mu^{0.43}} \tag{7-59}$$

and for Newton's law range,

$$u_t = 1.74 \sqrt{\frac{a_e D_p (\rho_p - \rho)}{\rho}} \tag{7-60}$$

If the terminal velocity of a particle of known diameter is desired, the Reynolds number is unknown, and a choice of equation cannot be made. To identify the range in which the motion of the particle lies, the following equation provides a criterion.

$$K = D_p \left[\frac{a_e \rho (\rho_p - \rho)}{\mu^2} \right]^{\frac{1}{3}} \tag{7-61}$$

If the size of the particle is known, K can be calculated from Eq. (7-61). If it is less than 3.3, Stokes' law applies. If K lies between 3.3 and 44.0, the intermediate law should be used, and if it is between 44.0 and 2,360, Newton's law must be chosen.

Example 7-4. Drops of oil 15 microns in diameter are to be settled from their mixture with air. The specific gravity of the oil is 0.90, and the air is at 70°F and 1 atm. A settling time of 1 min is available. How high should the chamber be to allow settling of these particles?

Solution. The effects of flow within the drops and of the initial period of acceleration are neglected. Also, the density of the air is so small in comparison with that of the drops that ρ_p can be used in place of $\rho_p - \rho$. The quantities for use in Eq. (7-61) are

$$D_p = \frac{15 \times 10^{-4}}{30.48} = 4.92 \times 10^{-5} \text{ ft}$$

The density of air at 70°F and 1 atm is 0.075 lb/ft³. From Appendix 7, the viscosity of the air is 0.018 centipoise, and

$$\mu = 0.018 \times 6.72 \times 10^{-4} = 1.21 \times 10^{-5} \text{ lb/ft-sec}$$

Also $\rho_p = 0.90 \times 62.37 = 56.1 \text{ lb/ft}^3$

From Eq. (7-61),

$$K = 4.92 \times 10^{-5} \left(\frac{32.17 \times 56.1 \times 0.075}{1.21^2 \times 10^{-10}} \right)^{\frac{1}{3}} = 0.479$$

The motion of the drops is well within the Stokes' law range, and Eq. (7-58) can be used to calculate the terminal velocity.

$$u_t = \frac{32.17 \times 4.92^2 \times 10^{-10} \times 56.1}{18 \times 1.21 \times 10^{-5}} = 0.020 \text{ ft/sec}$$

In 1 min the particles will settle $0.02 \times 60 = 1.2$ ft, so the depth of the chamber should not be greater than this.

Hindered Settling. In hindered settling the particles are sufficiently close together to cause the velocity gradients surrounding each particle to be affected by the presence of the neighboring units. Also, the particles, in settling, displace liquid, the upward velocity of which is appreciable. The velocity of the liquid is, then, greater with respect to the particle than with respect to the apparatus. The effective density of the fluid can be taken as that of the slurry itself and can be calculated from the composition of the slurry and the densities of the particles and of the fluid.

For the hindered settling of spherical particles the drag force given by Eq. (7-56) must be modified by a correction factor, usually divided into the viscosity term, that depends upon the fractional volume of the slurry occupied by the liquid. This is equivalent to the porosity of the aggregation of particles, and is denoted by ϵ. Then, mathematically,

$$F_D = \frac{\pi b_1 (D_p u_{tr})^{2-n} \rho_m^{1-n} [\mu/\psi(\epsilon)]^n}{8 g_c} \tag{7-62}$$

where u_{tr} is the terminal velocity of the particle *relative to the fluid*, $\psi(\epsilon)$ is a function of the porosity ϵ, and ρ_m is the density of the slurry. The terminal velocity with respect to the apparatus u_t is less than that with respect to the fluid u_{tr}, and the relationship between the velocities is

$$u_t = \epsilon u_{tr} \tag{7-63}$$

The drag is equal to the difference between the external and buoyant forces, so, from Eqs. (7-36), (7-37), (7-45), (7-62), and (7-63),

$$\frac{\pi a_e D_p^3 (\rho_p - \rho_m)}{6 g_c} = \frac{\pi b_1 (D_p u_t)^{2-n} \mu^n \rho_m^{1-n}}{8 \epsilon^{2-n} g_c [\psi(\epsilon)]^n}$$

Solution of this equation for u_t gives

$$u_t = \left\{ \frac{4 a_e D_p^{1+n} \epsilon^{2-n} [\psi(\epsilon)]^n (\rho_p - \rho_m)}{3 b_1 \rho_m^{1-n} \mu^n} \right\}^{1/(2-n)} \tag{7-64}$$

A general relation between ϵ and $\psi(\epsilon)$ over the entire range of Reynolds numbers has not been determined. A relationship for the settling of spheres in the Stokes' law range has been found and is shown graphically in Fig. 7-29. This relationship does not apply without further correction to sharp-edged particles.

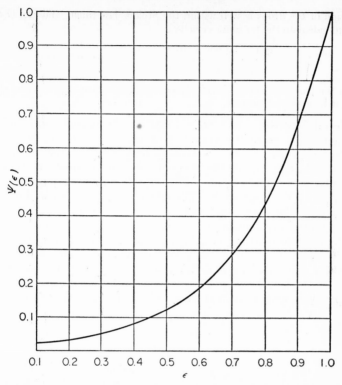

FIG. 7-29. Correction factor $\psi(\epsilon)$, hindered settling of spheres. (*Steinour.*[24])

Assuming that the Stokes' law range terminates at a Reynolds number of 1.9, based on velocity u_{tr}, that of the particle relative to the fluid, the criterion for hindered settling in the Stokes' law range is

$$K = D_p \left\{ \frac{a_e \rho_m (\rho_p - \rho_m)[\psi(\epsilon)]^2}{\mu^2} \right\}^{\frac{1}{3}} \tag{7-65}$$

If $K < 3.3$, the settling is in the Stokes' law range.

Example 7-5. Particles of sphalerite (specific gravity = 4.0) are settling under the force of gravity through a slurry consisting of 25 per cent by volume of quartz particles (specific gravity = 2.65) and water. The diameter of the sphalerite particles is 0.006 in. The volumetric ratio of sphalerite to slurry is 0.25. The temperature is 50°F. What is the terminal velocity of the sphalerite?

Solution. The density of the settling medium is that of the slurry. This can be calculated from the composition. The specific gravity of the slurry is $0.25 \times 2.65 + 0.75 = 1.41$, and the specific-gravity difference between particles and medium is $4.00 - 1.41 = 2.59$. The density difference $\rho_p - \rho_m$ is $62.37 \times 2.59 = 161.6$ lb/ft³. The density of the medium ρ_m is $62.37 \times 1.41 = 87.9$ lb/ft³. The viscosity of water at 50°F is, from Appendix 14, 1.310 centipoises. The porosity is $\epsilon =$

$1 - 0.25 = 0.75$. From Fig. 7-29, for $\epsilon = 0.75$, $\psi(\epsilon) = 0.35$. The criterion K is

$$K = \frac{0.006}{12} \left[\frac{32.17 \times 161.6 \times 87.9 \times 0.35^2}{(6.72 \times 1.31 \times 10^{-4})^2} \right]^{\frac{1}{3}} = 2.1$$

The settling occurs in the Stokes' law range.

The terminal velocity can be calculated from Eq. (7-64), taking $n = 1$ and $b_1 = 24$.

$$u_t = \frac{4 \times 32.17(0.006/12)^2 0.75 \times 0.35 \times 161.6}{3 \times 24 \times 1.31 \times 6.72 \times 10^{-4}} = 0.022 \text{ ft/sec}$$

Applications of Mechanics of Particles

If a particle starts at rest with respect to the fluid in which it is immersed and is then moved through the fluid by an external force, its motion can be divided into two stages. The first stage is a short period of acceleration, during which the velocity increases from zero to the terminal velocity. The second stage is the period during which the particle is at its terminal velocity.

Since the period of initial acceleration is short, usually of the order of tenths of a second or less, initial-acceleration effects are short-range. Terminal velocities, on the other hand, can be maintained as long as the particle is under treatment in the equipment. Equations such as (7-40), (7-42), and (7-47) apply during the acceleration period, and equations such as (7-64) and (7-65) during the terminal-velocity period.

Separations Based on Particle Size. Many different devices are used to separate particles into fractions based on particle size. Under some circumstances the objective of the process is to remove as many of the particles as possible. The particles usually consist of the same material and have the same density, although there may be differences in particle shape. Preferably, free settling or settling under slightly hindered conditions is used to obtain maximum capacity and settling rate. Devices for separating particles according to size rather than kind are called sizing classifiers.

Removal of Solids or Liquids from Gases. Both gravitational and centrifugal methods are used to remove particles or drops from gas streams. It may be desired either to remove substantially all the particles or to remove only those particles larger than a definite maximum size and to allow the fine particles to leave the separator with the gas for subsequent recovery in a separate unit. The use of air separators in close-circuit grinding, as described in Chap. 4, is an important example of partial removal of particles. In these operations, usually free settling in the Stokes' law range prevails.

Gravity Settling Chambers. For removing particles of dust larger than about 325-mesh, gravity settling chambers are useful. Such a device is a large box at one end of which dust-laden air enters and from the other end of which clarified air leaves. In the absence of air currents, particles settle toward the floor at a speed equal to their terminal velocities. If the air remains in the chamber a sufficient length of time, the particles reach the floor of the chamber, from which they can subsequently be removed. To prevent the air stream from lifting the particles from the floor and reentraining them, the air velocity should not be greater than about 10 ft/sec.[15b]

Example 7-6. It is desired to remove dust particles 50 microns in diameter from 8,000 ft³/min of air, using a settling chamber for the purpose. The air contains 0.5 grain of dust per cubic foot, and the temperature and pressure are 70°F and 1 atm. The particle density is 150 lb/ft³. What minimum dimensions of the chamber are consistent with these conditions?

Solution. To establish the range in which the particles are settling, Eq. (7-61) is used. The quantities needed are

$$D_p = \frac{50 \times 10^{-4}}{30.48} = 1.64 \times 10^{-4} \text{ ft}$$

From Example 7-4,

$$\mu = 1.21 \times 10^{-5} \text{ lb/ft-sec}$$

$$\rho = 0.075 \text{ lb/ft}^3 \qquad \rho_p = 150 \text{ lb/ft}^3$$

Then

$$K = \frac{1.64}{10^4} \left(\frac{32.17 \times 0.075 \times 150}{1.21^2 \times 10^{-10}} \right)^{\frac{1}{3}} = 2.22$$

Stokes' law must be used. The terminal velocity is, from Eq. (7-60),

$$u_t = \frac{32.17 \times 1.64^2 \times 10^{-8} \times 150}{18 \times 1.21 \times 10^{-5}} = 0.60 \text{ ft/sec}$$

If V_c is the volume of the chamber, the time of residence for the air is $60 V_c/8,000 = V_c/133$ sec. The maximum distance that a particle can settle during this time is $V_c u_t/133$ ft. If Z_c is the height of the chamber, for complete removal of the particles, those starting from the roof of the chamber must be given sufficient time to fall this same distance, or

$$Z_c = \frac{u_t V_c}{133} = \frac{0.60 V_c}{133} = \frac{V_c}{222} \qquad \text{ft}$$

Since V_c/Z_c is the floor area of the chamber, this area must be at least 222 ft².

The minimum cross-sectional area of the chamber is dictated by the maximum permissible velocity of the air. If this is 10 ft/sec, the cross-sectional area of the chamber is $8,000/(60 \times 10) = 13.3$ ft².

Since the floor area is 222 ft² and the cross-sectional area is 13.3 ft², the ratio of length to height is $222/13.3 = 16.7$. This is the only proportion of the chamber that is fixed by the conditions of the design. Any single dimension can be chosen at will, and the other two will follow. For example, if the height is taken as 3 ft, the length is 50 ft, and the breadth is $13.3/3 = 4.4$ ft. The volume of the chamber is $3 \times 50 \times 4.4 = 660$ ft³. The smaller the height, the shorter and wider the chamber and the smaller the volume. If the height is too small, however, it is difficult to remove the solids from the floor of the chamber.

The size of the chamber is independent of the dust load of the entering gas, but the heavier the load, the more difficult is the cleaning and the practical operation of the chamber. In this example, the dust load per unit ground area is, since 1 **lb** is 7,000 grains,

$$\frac{0.5 \times 60 \times 8,000}{7,000 \times 222} = 0.155 \text{ lb/ft}^2\text{-hr}$$

Cyclones. Settling chambers have low capacity for small particles and occupy considerable space. To remove smaller particles in compact equipment, the cyclone separator, which uses centrifugal force to amplify the settling rate, is often used. A typical cyclone is shown in Fig. 7-30. It consists of a vertical cylinder with a conical bottom, a tangential inlet near the top, and an outlet for dust at the bottom of the cone. The inlet is usually rectangular. The outlet pipe is extended into the cylinder to prevent short-circuiting of air from inlet to outlet.

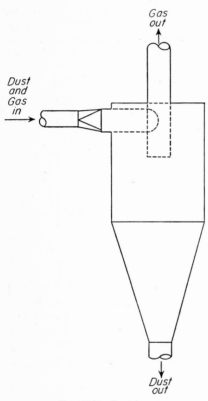

The incoming dust-laden air receives a rotating motion on entrance to the cylinder. The vortex so formed develops centrifugal force, which acts to throw the particles radially toward the wall. Basically, a cyclone is a settling device in which a strong centrifugal force, acting radially, is used in place of relatively weak gravitational force acting vertically. The centrifugal force in a cyclone is from 5 times gravity in large, low-velocity units to 2,500 times gravity in small, high-pressure units.[15c]

FIG. 7-30. Cyclone.

The path of the air in a cyclone follows a downward vortex, or spiral, adjacent to the wall and reaching to the bottom of the cone. The air stream then moves upward in a tighter spiral, concentric with the first, and leaves through the outlet pipe, still whirling. Both spirals rotate in the same direction.

In a cyclone, dust particles quickly reach the terminal velocities corresponding to their sizes and their radial position in the cyclone, in accordance with Eq. (7-58). The radial acceleration in a cyclone depends on the radius of the path being followed by the air, and is given by the following empirical equation.[21]

$$a_e = \omega^2 r = \frac{b_2}{r^n} \tag{7-66}$$

where b_2 and n are constants and the exponent n lies between 2.0 and 2.4. Combining Eqs. (7-58) and (7-66) gives, for the terminal velocity of a particle of diameter D_p rotating around the axis of the cyclone at a distance r ft from the center,

$$u_t = \frac{b_2 D_p^2 (\rho_p - \rho)}{18 \mu r^n} \tag{7-67}$$

For a given particle size, the terminal velocity is a maximum in the inner vortex, where r is small, and the finest particles separated from the air are eliminated in the inner vortex. These pass through the outer vortex to the wall of the cyclone and drop through the dust outlet. Smaller particles,

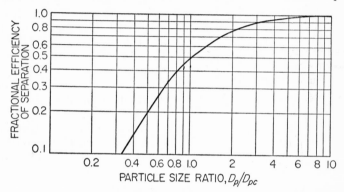

FIG. 7-31. Efficiency vs. particle size for cyclones.[15c]

which do not have time to reach the wall, are retained by the air and accompany it to the outlet, while larger ones are eliminated early in the travel of the air through the unit. Although the chance of a particle for separation decreases with the square of the particle diameter, the fate of a particle depends also on its position in the cross section of the entering stream and on its history in the cyclone, so the separation according to size is not sharp.

A definite diameter, called the cut diameter, can be defined as that diameter for which one-half the inlet particles, by mass, are separated and the other half retained by the air. Also, the efficiency of separation for a definite particle size can be defined as the mass fraction of the particles of that size that is separated and collected by the unit. A typical relation between efficiency and particle diameter for a definite cyclone is shown in Fig. 7-31. The abscissa is the ratio of the diameter of the particle to the cut diameter, and the ordinate is the efficiency. For example, only 80 per cent of the particles twice the size of the cut diameter are removed, and of the particles one-half the cut diameter, 20 per cent are separated.

Cyclones are also extensively used for separating solids from liquids, especially for purposes of classification.

Removal of Liquid from Vapor: Entrainment Separators. The vapors from evaporators, boilers, and stills carry with them, or "entrain," small drops of liquid. The drops are formed in the bursting of the bubbles generated in forming vapor from boiling liquid. It is often important to eliminate these drops. Although many methods are used, cyclones are effective for this purpose. The drops, after their removal from the vapor by centrifugal action, form a liquid that can be drained from the bottom of the separator. The size of entrained drops is relatively great, and the practical elimination of them does not require large centrifugal forces. Care must be taken, however, to prevent reentrainment of the liquid by the vapor stream as it passes over the recovered liquid. The velocity of the vapor must be sufficiently small to prevent this. Mass velocities of about 10 lb/ft^2-sec in the inlet should not be exceeded.

Mechanical Air Classifiers. Mechanical air classifiers are used in closed-circuit grinding, as described in Chap. 4. Some of them are integral with the size-reduction machine; others are separate units. They separate the fine product particles from the larger, partially ground material and return the latter to the mill for further grinding. In these devices, air classification and centrifugal and gravitational settling are all active. The centrifugal force is generated either by cyclones or internal fans. The centrifugal force opposes the drag of an air stream on the particles. The drag of the air pulls the acceptably fine particles through the classifier to a collector; with larger particles the centrifugal force predominates and throws them down into a separate collector or returns them to the grinding mill. By an appropriate balance of opposing forces precise splits can be made even with particles as fine as 1 to 10 microns in diameter.

In the simplest air classifiers a horizontal bladed disk, known as a "whizzer," revolves in a vertical cylindrical casing with a conical bottom. An external fan draws air and particles of solid into the separator, past the whizzer blades, and out a duct in the roof of the casing. The blades "precipitate" the larger particles; i.e., they throw them outward to the wall of the casing, where the air no longer acts on them. They fall to the bottom of the separator and back into the mill for regrinding. Fine particles passing between the whizzer blades are deflected by the centrifugal force but not enough to move them out of the air stream. They pass out the top of the casing to a collector. The critical size of particles which just escape being thrown back depends on the speed, number, and design of the whizzer blades. Simple whizzer classifiers do not make highly precise splits, but they can handle sticky solids which clog the close-set stationary vanes of more elaborate separators.

Highly precise air classifiers are exemplified by the double-cone separator illustrated in Fig. 7-32. The casing is a cone-bottomed cylinder containing a smaller similarly shaped inner chamber, which in turn surrounds a bladed distributor plate turning at 245 to 750 rpm. Above the distributor plate, and turning with it, is a circulating fan. Solids fall from a hopper at the top onto the distributor plate, which throws them outward. Fine particles are picked up by the air stream and carried over into the outer cone, where they flow down the wall to the discharge outlet at the bottom. The air returns to the inner cone through baffled openings, about halfway down.

Large particles are thrown outward by the distributor plate and its rejector blades with such high velocity that they escape being picked up by the air. They impinge on the inner wall of the inner cone and fall to a separate outlet.

The fineness of the particles carried over into the outer cone is controlled by adjustable plates, or "valves," which fix the size of the opening at the top of the inner cone, and by the size, speed, and the number of rejector blades. Casing diameters range from 3 to 18 ft; motor sizes from 2 to

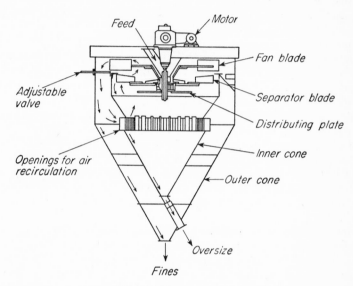

FIG. 7-32. Double-cone air separator.

100 hp. The biggest classifiers handle up to 200,000 lb of solids per hour. Typical specifications for the particle size of fine products from a double-cone separator are 90 per cent through a 200-mesh screen (74 microns) to 99.99 per cent through 325-mesh (43 microns).

Separation of Coarse Particles from Liquids. A common problem in both chemical and metallurgical practice is that of separating relatively coarse particles, which are called sands, from a slurry of fine particles, which are called slimes. The most common method is to use continuous settling equipment called classifiers. Sufficient time is provided to allow the sands to settle to the bottom of the device; the slimes leave in the effluent liquid. Either mechanical or gravitational means can be used to remove the sands continuously from the bottom of the classifier.

Classifying Equipment. Types of classifiers for coarse particles are shown in Fig. 7-33. In both these devices the settling vessel is an inclined trough with a liquid overflow at the lower end. Slurry is continuously fed to the trough at an intermediate point. The flow rate and slurry concentration are adjusted so that the fines do not have time to settle but are carried out with the liquid leaving the classifier. Larger particles sink to the floor of the trough, from which they must be removed. The two classifiers illustrated differ in the means by which they do this.

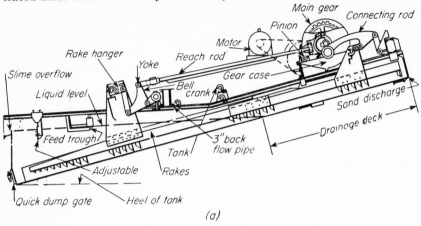

(a)

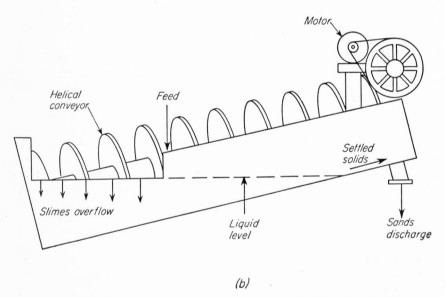

(b)

FIG. 7-33. Wet-settling classifiers: (a) Rake classifier. (b) Cross-flow classifier.

In the rake classifier in Fig. 7-33a heavy particles are moved up the sloping floor of the trough by a series of mechanically operated rakes. The rakes slowly scrape upward for a foot or so along the trough floor; they are then lifted away from the floor and quickly returned to their original position for another stroke. The rakes not only move the coarse solids out of the pool of liquid and up the trough to the discharge, they also agitate the liquid and resuspend fine particles that may have settled prematurely or been trapped by the coarse material. Additional liquid may be added near the fines discharge to dilute the slurry and permit the operation to approach free settling.

The crossflow classifier in Fig. 7-33*b* moves the solids out of the liquid by a helical conveyor. The trough is semicylindrical, set at an angle of about 12° with the horizontal. Feed slurry is admitted below the liquid level. The applications of crossflow classifiers are much the same as those of rake classifiers.

Both rake and crossflow classifiers work well with coarse particles where exact splits are not required. Typical applications are in connection with ball or rod mills for reduction to particle sizes between 8- and 20-mesh. These classifiers have high capacities. They lift coarse solids for return to the mill, so that auxiliary conveyors and elevators are not needed. For close separations with finer particles, however, other types of classifier must be used. One such device is the centrifugal classifier described in a later section; its action bears a strong resemblance to that of a crossflow classifier, but the settling is greatly accelerated by the substitution of centrifugal force for the force of gravity.

Removal of Fine Solids from Liquids; Sedimentation and Thickening. To remove relatively coarse sands, which have reasonable settling velocities, gravity classification under free or hindered settling is satisfactory. To remove fine particles of diameters of a few microns or less settling velocities are too low, and for practicable operation the particles must be agglomerated, or flocculated, into large particles that do possess a reasonable settling speed.

Flocculation. Many slimes consist of particles that carry electrical charges, which may be either positive or negative, and such particles, because of the mutual repulsion of like charges, tend to remain dispersed. If an electrolyte is added, the ions formed in solution neutralize the charges on the particles. The neutralized particles may then agglomerate to form flocs, each containing many particles. When the original particles are negatively charged, the cation of the electrolyte is effective, and if the charge is positive, the anion is active. In either situation, the greater the valence of the ion, the more effective is the ion as a flocculation agent. Other methods of flocculation include the use of surface-active agents and the addition of materials, such as glue, lime, alumina, or sodium silicate, that drag down the slime particles with them. In a properly treated slime the flocs are visible to the naked eye.

Flocculated particles possess two important settling characteristics. The first characteristic is the complicated structure of the flocs. The aggregates are loose, the bond between the particles in them is weak, and they retain a considerable amount of water in their structures, which accompanies the flocs when they settle. Although initially flocs settle in either free or hindered settling and the usual equations apply in principle, it is not practical to use the laws of settling quantitatively, because the diameter and shape of a floc are not readily definable.

The second characteristic of a flocculated pulp is the complexity of its settling mechanism. The history of the settling of a typical flocculated sludge is as follows.[5] Figure 7-34a shows a sludge uniformly distributed in the liquid and ready to settle. If there are no sands in the mixture, the first appearance of solids is the deposit, on the bottom of the settler, of flocs originating in the lower portion of the mixture. As shown in Fig. 7-34b, these solids, which consist of flocs resting lightly on one another, form a layer, called zone D. Above zone D forms another layer, called

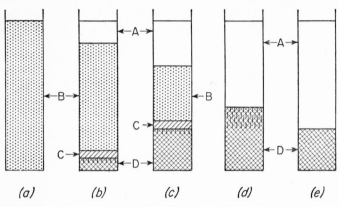

(a)   (b)   (c)   (d)   (e)

Fig. 7-34. Batch sedimentation.

zone C, which is a transition layer, the solid content of which varies from that in the original pulp to that in zone D. Above zone C is zone B, which consists of a homogeneous suspension of the same concentration as that of the original pulp. Above zone B is zone A, which, if the particles have been fully flocculated, is a clear liquid. In well-flocculated pulps, the boundary between zones A and B is sharp. If unagglomerated particles remain, zone A is turbid, and the boundary between zones A and B is hazy.

As settling continues, the depths of zones D and A increase, that of zone C remains constant, and that of zone B decreases. This is shown in Fig. 7-34c. After further settling, zones B and C disappear, and all the solids are in zone D. This is shown in Fig. 7-34d. Then a new effect, called compression, begins. The moment when compression first is evident is called the critical point. In compression, a portion of the liquid, which has accompanied the flocs into the compression zone D, is expelled when the weight of the deposit breaks down the structure of the flocs. During compression, some of the liquid in the flocs spurts out of zone D like small geysers, and the thickness of this zone decreases. Finally, when the weight of the solid reaches mechanical equilibrium with the compressive strength of the flocs, the settling process stops, as shown in Fig. 7-34e. At this time, the sludge has reached its ultimate height. The entire process shown in Fig. 7-34 is called sedimentation.

Rate of Sedimentation. A typical plot of height of sludge (the boundary between zones *A* and *B*) vs. time is shown in Fig. 7-35. During the early stage of settling the velocity is constant, as shown by the first portion of the curve. As solid accumulates in zone *D*, the rate of settling decreases and steadily drops until the ultimate height is reached. The critical point is reached at point *C* in Fig. 7-35.

Sludges vary greatly in their settling rates and in the relative heights of various zones during settling. Experimental study of each individual sludge is necessary to appraise accurately its settling characteristics.

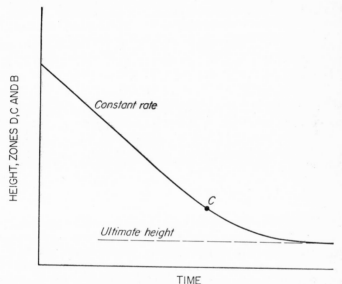

Fig. 7-35. Rate of sedimentation.

Equipment for Sedimentation: Thickeners. Industrially, the above process is conducted on a large scale in equipment called thickeners. These include settling tanks, settling cones, and mechanically agitated thickeners. For relatively fast-settling particles a batch *settling tank* may be adequate. This is a fairly tall tank provided with several side outlets, one above the other. With all side outlets closed the tank is filled with slurry, which is then allowed to settle. After several minutes or hours the topmost outlet is opened and clear liquor withdrawn. Then the next outlet is opened, and the next, until the liquor no longer runs clear. The thickened sludge is dropped through a large bottom outlet. A *settling cone* provides a means of making simple decantation continuous. Feed is continuously admitted to the large upper end of a conical tank, usually a few inches below the liquid surface. Clarified liquor is withdrawn continuously through an overflow pipe near the top of the cone or spills over the top of the vessel into a peripheral trough. Solids settle into the small end of the cone. When the density of the sludge reaches a set value, an internal float rises, opening a discharge valve in the sludge outlet. Sludge flows from the bottom of the cone. The thin suspension above the thickened sludge moves downward as the sludge flows out, lowering the density of the material in the lower part of the cone. The float then sinks, closing the discharge valve until the sludge density rises once again.

Settling tanks and cones will not work with sludges that do not flow readily through a discharge opening. For many duties a *mechanically agitated thickener* like the one shown in Fig. 7-36 must be employed. This is a large fairly shallow tank with slow-moving radial rakes driven from a central shaft. Its bottom may be flat or a shallow cone. Dilute feed

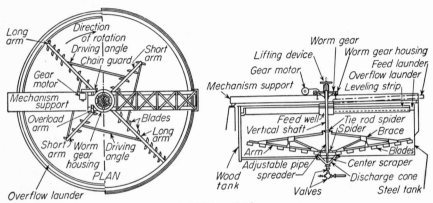

FIG. 7-36. Dorr thickener.

slurry flows from an inclined trough into the center of the thickener. Liquor moves radially at a constantly decreasing velocity, allowing the solids to settle to the bottom of the tank. Clear liquor spills over the edge of the tank into a launder. The rake arms gently agitate the sludge and move it to the center of the tank, where it flows through a large opening to the inlet of a sludge pump. In some designs of thickener the rake arms are pivoted so that they can ride over an obstruction, such as a hard lump of mud, on the tank floor.

Mechanically agitated thickeners are usually large, typically 30 to 300 ft in diameter and 8 to 12 ft deep. In a large thickener the rakes may make 1 revolution every 30 min. These thickeners are especially valuable when large volumes of dilute slurry must be thickened, as in cement manufacture or the production of magnesium from sea water. They are also used extensively in sewage treatment and in water purification.

The volume of clear liquor produced in a unit time by a continuous thickener depends primarily on the cross-sectional area available for settling and in industrial separators is almost independent of the liquid depth. Higher capacities per square foot of floor area are therefore obtained by using a *multiple-tray thickener*, with several shallow settling zones, one above the other, in a cylindrical tank. Rake or scraper agitators move the settled sludge downward from one tray to the next. Multistage countercurrent displacement washing is possible in these devices.[19c] They are considerably smaller in diameter, however, than single-stage thickeners.

A given particle in a given fluid settles under gravitational force at a fixed maximum rate. To increase the rate of sedimentation the effective

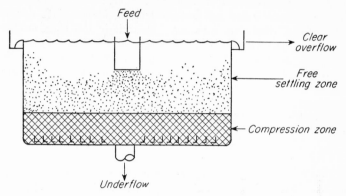

FIG. 7-37. Zones in continuous thickener.

diameter of the particle may be increased by flocculation. Another way is to replace the force of gravity by a much stronger centrifugal force. Centrifugal settling devices have to some extent displaced gravity thickeners in production operations, because of their increased effectiveness with fine particles and their much smaller size for a given capacity. When very large volumes of dilute liquor must be thickened, however, the power required to accelerate the slurry to high velocity in a centrifuge becomes prohibitively expensive, and a gravity thickener, despite its size, is the more economical device.

Sedimentation Zones in Continuous Thickeners. In a continuous thickener, equipped with rakes to remove the underflow, the feed pulp is admitted at the center line of the unit at a depth of 2 or 3 ft

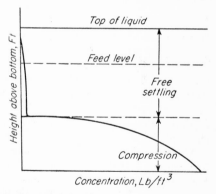

FIG. 7-38. Solid concentrations in continuous thickener. (*Comings.*[6])

below the surface of the liquid. On entrance, the slurry spreads radially through the cross section of the thickener, and the liquid then flows upward to be withdrawn at the overflow launder, and the solids settle toward the bottom. As shown in Fig. 7-37, two zones are established. The upper zone is free from particles in the upper levels and increases slightly in solid content below the entrance of the feed. This is shown by the relationship of solids concentration vs. height in Fig. 7-38. Particles settle in this zone by free settling. Below the dilute zone is a compression zone, in which the concentration of solids increases rapidly with distance from the boundary between the zones. This zone corresponds to the compression zone D in a batch thickener. Zones B and C of the batch process are not found in a continuous thickener. The rakes, which operate in the bottom of the compression zone, tend to break the floc structure and to compact the underflow to a solid content greater than that in the zone D of a batch thickener.

In practice, a clear overflow can be obtained if the upward velocity of the liquid in the dilute zone is less than the minimum terminal velocity of the solid at all points in the zone. The velocity of the liquid is proportional to the overflow rate, and if the overflow rate is too high, a cloudy overflow is obtained. If this happens, the equipment is being overloaded. The solid content and the degree of thickening in the underflow depend on the time of retention of the material in the compression zone, which is, in turn, proportional to the depth of this zone. Since the depth of the zone can be increased only at the expense of a deeper and more costly tank, the economic limit is reached at an underflow concentration somewhat less than the ultimate obtainable with a long process time.

SEPARATIONS BASED ON DIFFERENCE IN DENSITY

In the methods considered to this point solid particles either have been eliminated completely from fluids or have been segregated into fractions depending on particle size. Another set of operations has the objective of segregating solid particles into fractions differing in composition, such as the recovery of valuable mineral from its mixture with worthless gangue. These are often called sorting methods. Those that depend on differences in density are the subject of this section. They include the gravity separation of immiscible liquids and the hydraulic separation of solid particles. Devices for this latter are called sorting classifiers.

Liquid-Liquid Separations. Immiscible liquids may be separated in a *cone-bottom settling tank*, which operates in exactly the same way as a laboratory separatory funnel. The tank is filled with the mixture of liquids and allowed to stand. The liquid separates into layers. The lower layer is drained off through a bottom valved outlet and a sight glass until the interface between the liquid layers appears in the glass. The valve is shut and the liquid allowed to settle a little longer to yield additional heavy liquid. The cut is then made and the light liquid sent through appropriate valves to a separate receiver.

The interface between the liquid layers is usually much less clear-cut in an industrial separator operating on process liquids than it is in the laboratory with clean materials. Often dirt and other solid impurities collect at the interface, making it thick and rather indefinite. The purpose of the conical bottom of the separator is to make the interface more definite as it approaches the outlet and thus facilitate making the cut.

A conical bottom is not needed in the *continuous decanter* shown in

Fig. 7-39. The vessel may be of any reasonable shape. It is provided with an overflow for the light liquid at the side and a leg carrying heavy liquid up from the bottom of the tank to a second, somewhat lower, overflow point. The overflow lines must be vented to the air or to the open space above the liquid in the vessel; they must also be large enough to im-

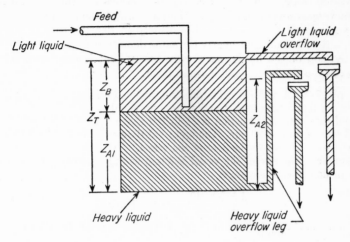

FIG. 7-39. Continuous gravity decanter for immiscible liquids.

pose a negligible resistance to flow. Feed enters the body of the separator about halfway down.

The position of the interface between the liquid layers is fixed by the relative densities of the liquids and the positions of the overflows. Note that the discharge lines are unrestricted: the rate of flow through each one automatically adjusts to equal the rate of flow of the corresponding phase entering the separator, so that the position of the interface does not change. If for a time only light liquid is fed to the unit, the discharge of heavy liquid stops. The equilibrium level of the interface is found by balancing the head of heavy liquid in the overflow leg against the head of heavy liquid plus light liquid in the body of the vessel. If ρ_A and ρ_B are the densities of the heavy and light liquids, respectively, it may be seen from Fig. 7-39 that

$$Z_B \rho_B + Z_{A1} \rho_A = Z_{A2} \rho_A \qquad (7\text{-}68)$$

where Z_B = thickness of layer of light liquid in vessel
$\quad Z_{A1}$ = thickness of layer of heavy liquid in vessel
$\quad Z_{A2}$ = height of liquid in heavy-liquid overflow leg
Solving Eq. (7-68) for Z_{A1},

$$Z_{A1} = Z_{A2} - Z_B \frac{\rho_B}{\rho_A} = Z_{A2} - (Z_T - Z_{A1}) \frac{\rho_B}{\rho_A} \qquad (7\text{-}69)$$

where the total depth of liquid in the vessel is $Z_T = Z_B + Z_{A1}$. From this

$$Z_{A1} = \frac{Z_{A2} - Z_T(\rho_B/\rho_A)}{1 - \rho_B/\rho_A} \qquad (7\text{-}70)$$

Equation (7-70) shows that as ρ_A approaches ρ_B, the position of the interface becomes very sensitive to changes in Z_{A2}, the height of the heavy-liquid leg. With liquids that differ widely in density this height is not critical, but with liquids of nearly the same density it must be set with care. Often the top of the leg is made movable so that it may be adjusted in service to give the best possible separation.

Example 7-7. A vertical cylindrical continuous decanter is to separate 1,500 bbl/day of a liquid petroleum fraction from an equal volume of wash acid. The density of the petroleum fraction is 54 lb/ft³; that of the acid is 72 lb/ft³. The required settling time is 20 min. Compute: (a) the size of the vessel, (b) the height of the acid overflow above the floor of the vessel.

Solution. (a) The size of the vessel is found as follows. Since 1 bbl = 42 gal, the rate of flow of each stream is

$$\frac{1{,}500 \times 42}{24 \times 60} = 43.8 \text{ gal/min}$$

The total liquid holdup is

$$2 \times 43.8 \times 20 = 1{,}752 \text{ gal}$$

The vessel size should be about $1.1 \times 1{,}752$ or 2,000 gal.

The height of the tank should approximately equal its diameter. A tank 7 ft in diameter and 7 ft deep would be satisfactory.

(b) The volume of the liquid in the separator is 1,752 gal. In a 7-ft vessel the depth of the liquid Z_T is

$$Z_T = \frac{1{,}752 \times 4}{7.48\pi 7^2} = 6.10 \text{ ft}$$

Set the interface halfway between the vessel floor and the liquid surface. Then $Z_{A1} = 3.05$ ft. Solving Eq. (7-70) for Z_{A2}, the height of the heavy-liquid overflow,

$$Z_{A2} = Z_{A1} + (Z_T - Z_{A1})\frac{\rho_B}{\rho_A} \qquad (7\text{-}71)$$

$$= 3.05 + (6.1 - 3.05)\tfrac{54}{72} = 5.34 \text{ ft}$$

Hydraulic Separation. Two principal methods are used in hydraulic separation, sink and float and differential settling. Sorting classifiers use one or the other of these methods.

A sink-and-float method uses a liquid sorting medium, the density of which is intermediate between that of the light material and that of the heavy. Then the heavy particles settle through the medium, and the lighter ones float, and a separation is thus obtained. This method has the advantage that, in principle, the separation depends only on the difference in the densities of the two substances and is independent of the particle size. These methods are also called *heavy-fluid separations*. The process shown in Fig. 7-27b is an idealized example of sink and float.

Differential-settling methods utilize the difference in terminal velocities that can exist between substances of different density. The density of the medium is less than that of either substance. The disadvantage of the method is that, since the mixture of materials to be separated covers a range of particle sizes, the larger, light particles settle at the same rate as the smaller, heavy ones, and a mixed fraction is obtained.

Sink-and-float Methods. Heavy-fluid processes are used to treat relatively coarse particles, usually greater than 10-mesh. The first problem in the use of sink and float is the choice of a liquid medium of the proper gravity to allow the light material to float and the heavy to sink. True liquids can be used, but since the specific gravity of the medium must be in the range 1.3 to 3.5 or greater, there are but few liquids that are sufficiently heavy, cheap, nontoxic, and noncorrosive to be practicable. Halogenated hydrocarbons are used for the purpose. Calcium chloride solutions are used for cleaning coal. A more common choice of medium is a pseudo liquid consisting of a suspension in water of fine particles of a heavy mineral. Magnetite (specific gravity = 5.17), ferrosilicon (specific gravity = 6.3 to 7.0), and galena (specific gravity = 7.5) are used. The ratio of mineral to water can be varied to give a wide range of medium densities.

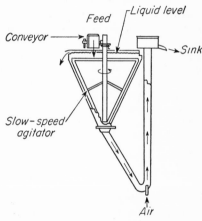

Fig. 7-40. Sink-and-float separatory cone.

Provision must be made for feeding the mixture to be separated, for removing overflow and underflow, and for recovering the separating fluid, which may be expensive relative to the value of the materials being treated. In the process particles fall or rise at their terminal velocities, and hindered settling is used. Cleaning coal and concentrating ores are the common applications of sink and float. Under proper conditions, clean separations between materials differing in specific gravity by only 0.1 have been claimed.[25a]

Typical equipment for sink-and-float separations is shown in Fig. 7-40. Fines and slimes that interfere with the separation are removed by a preliminary sizing screen. The oversize from this screen is fed to a *separatory cone*, which is kept full of the liquid separating medium. Light particles float across the surface and over a weir to a drainage screen; heavy particles sink to the J-shaped outlet line at the bottom of the cone. There a stream of air elevates them to a separate drainage screen. Most of the liquid that is carried over with the particles is removed on the drainage screens and pumped back to the separatory cone; the rest is washed off by water sprays on sets of washing screens. Finished "sink" and finished "float" discharge from the washing screens. The washings are thickened and purified before returning to the cone. With separating fluids containing magnetite or ferrosilicon this is done by magnetic separators, which recover the magnetic substance from the nonmagnetic fines.

Differential-settling Methods. In differential settling, both light and heavy materials settle through the same medium. This method brings in the concept of equal-settling particles.

Equal-settling Particles. Consider particles of two materials A and B settling through a medium of density ρ_m. Let material A be the heavier; e.g., component A might be galena (specific gravity = 7.5) and component B quartz (specific gravity = 2.65). The terminal velocity of a particle of size D_p and of density ρ_p settling under gravity through a medium of density ρ_m is given by Eq. (7-64), using $a_e = g$. This equation can be written, for a galena particle of density ρ_{pA} and diameter D_{pA}, as

$$u_{tA} = \left\{ \frac{4gD_{pA}^{1+n}\epsilon^{2-n}[\psi(\epsilon)]^n(\rho_{pA} - \rho_m)}{3b\rho_m^{1-n}\mu^n} \right\}^{1/(2-n)} \tag{7-72}$$

The terminal velocity of a quartz particle of density ρ_{pB} and diameter D_{pB} is

$$u_{tB} = \left\{ \frac{4gD_{pB}^{1+n}\epsilon^{2-n}[\psi(\epsilon)]^n(\rho_{pB} - \rho_m)}{3b\rho_m^{1-n}\mu^n} \right\}^{1/(2-n)} \tag{7-73}$$

Although $\rho_{pA} > \rho_{pB}$ and a galena particle of definite size settles faster than a quartz particle of the same size, a larger quartz particle can have the same velocity as a smaller galena one. A relationship between the diameters of such equal-settling particles is found by assuming $u_{tA} = u_{tB}$, equating the right-hand sides of Eqs. (7-72) and (7-73), canceling common factors, and obtaining

$$(\rho_{pA} - \rho_m)D_{pA}^{1+n} = (\rho_{pB} - \rho_m)D_{pB}^{1+n}$$

or
$$\frac{D_{pA}}{D_{pB}} = \left(\frac{\rho_{pB} - \rho_m}{\rho_{pA} - \rho_m} \right)^{1/(1+n)} \tag{7-74}$$

Particles of substances A and B whose diameters conform to Eq. (7-74) are called *equal-settling particles.*

If the particles settle in the Stokes' law range, $n = 1$, and Eq. (7-74) becomes

$$\frac{D_{pA}}{D_{pB}} = \sqrt{\frac{\rho_{pB} - \rho_m}{\rho_{pA} - \rho_m}} \tag{7-75}$$

The significance, in a separation process, of the equal-settling ratio of diameters is shown by Fig. 7-41, in which curves of u_t vs. D_p are plotted for components A and B, for Stokes' law settling. Assume that the diameter range of both substances lies between points D_{p1} and D_{p4} on the size axis. Then, all particles of the light component B having diameters between D_{p1} and D_{p2} will settle more slowly than any particle of the heavy substance A, and can be obtained as a pure fraction. Likewise, all particles of heavy material A having diameters between D_{p3} and D_{p4} settle faster than any particle of substance B, and can also be obtained as a pure fraction. But any light particle having a diameter between D_{p2} and D_{p4} settles at the same speed as a particle of component A in a size range from D_{p1} to D_{p3}, and all particles in these size ranges form a mixed fraction.

Inspection of Eq. (7-74) shows that the sharpness of separation is improved if the density of the medium is increased. Since, in hindered set-

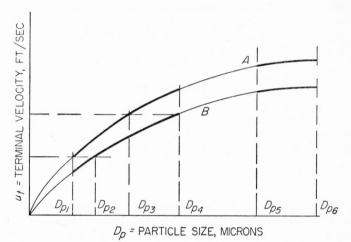

Fig. 7-41. Equal-settling velocities.

tling, the apparent density of the medium ρ_m is that of the entire suspension and is greater than ρ, the density of the liquid itself, hindered settling is more effective in separation than is free settling. The rate of hindered settling is less, however, than that of free settling. It is also clear from Fig. 7-41 that the intermediate fraction can be reduced or eliminated by closer sizing of the feed. For example, if the size range of the feed is from D_{p5} to D_{p6} in Fig. 7-41, complete separation is possible.

Although close sizing and increased medium density help to reduce the size of the intermediate fraction, practical limits prevent these methods from being completely effective. Various methods have been worked out to break up equal-settling mixtures. Of these the most important is flotation.

Flotation

Hydraulic separations based on differential-settling velocities have two limitations. They cannot treat fine particles, because fines do not have a practicable settling rate, and they require that the substances to be separated differ appreciably in density. Flotation is free from these limitations. It is a process in which the apparent density of one constituent is reduced by the adhesion of bubbles of air to finely ground particles of that constituent. The buoyancy of the aerated particles is such that they float and are thereby separated by gravity from the particles of the remaining constituents, which do not attract air, and which remain suspended in the water phase. Since for effective aeration the particles should be small and the original density of the floated material is unimportant, flotation can be applied where hydraulic methods fail.

Originally, flotation was used to separate sulfide ores of copper, lead, and zinc from their gangues. So successful and versatile has flotation become that it has supplanted the older hydraulic methods in a number of separation problems. Flotation is used for concentrating nonsulfide ores, for cleaning coal, for separating salts from their mother liquors, and for recovering elements such as sulfur and graphite.

The preferred method of removing the floated material is to form a froth, or foam, in the slurry during aeration. The froth collects the aerated particles and floats with them on the top of the liquid, from which it can be removed. This process is called froth flotation. It is conducted continuously in equipment called flotation cells.

The success of flotation depends on so controlling conditions in the slurry that air is selectively retained by one constituent and rejected by the others. To attain this objective, the pulp must be treated by the addition of small amounts of chemicals which render one constituent floatable. Thus, a complete flotation process is conducted in several steps: (1) the feed is ground, usually to a size between about 60- and 200-mesh; (2) a slurry containing from 10 to 35 per cent solids in water is prepared; (3) the necessary chemicals are added and sufficient agitation and time provided to distribute the chemicals on the surface of the grains to be floated; (4) the treated slurry is aerated in a flotation cell by agitation in the presence of a stream of air or by blowing air in fine streams through the pulp; and (5) the aerated particles in the foam are withdrawn from the top of the cell as an overflow and the remaining solids and water taken from the bottom as an underflow.

No attempt is made to recover the chemicals, as they are in too small amount. The froth may be broken by passing it over vacuum filters or through thickeners, or it may be sent directly to the next step in the process.

Flotation Equipment. The solids to be separated are usually ground wet to the proper size in a ball mill. Flotation reagents are mixed with the pulp from the ball mill in an agitated vessel called a *conditioner*, which contains a high-speed propeller or turbine agitator and a draft tube. Reagents are admitted to the conditioner at carefully controlled rates. Pulp enters the top of the draft tube; treated slurry spills out an overflow in the side of the tank. Inasmuch as 1 lb of reagent is ordinarily used with 1 to 20 *tons* of solid to be treated, mixing must be exceedingly thorough. Sometimes reagents are added to the ball mill during grinding, so that some conditioning is done before the material enters the agitated vessel. Mechanically agitated flotation cells also help disperse the reagents over the surface of the solid particles after they leave the conditioner.

Following the conditioner the treated suspension passes through one or more *flotation cells*, in which it is mixed with air to form a froth. The two

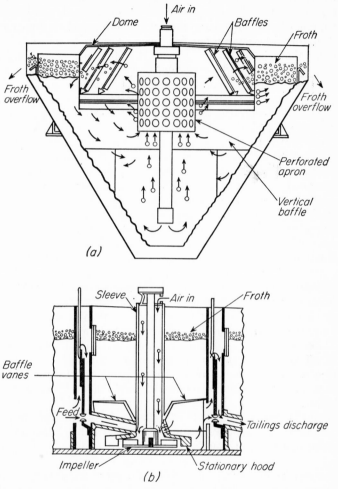

FIG. 7-42. Flotation cells: (a) Cross section of air-lift cell. (*Bethlehem Steel Co.*) (b) Mechanical cell. (*Denver Equipment Co.*)

main types of cell, the air-lift cell and the mechanically agitated cell, are shown in Fig. 7-42. In both machines the slurry is held in a compartmented vessel or trough 4 to 6 ft wide at the top and several times as long. Each compartment forms a single cell. Air enters a central pipe leading nearly to the bottom of each compartment and flows upward through the body of the liquid. In the *air-lift cell* it carries some of the pulp upward through a perforated apron to agitate the slurry and form a froth of small bubbles. The froth is deflected by a cover plate, or "dome," against sloping baffles and from there to overflow lips on the sides of the trough. Feed enters the compartment at one end of the trough and passes from cell to cell. Nonfloating particles, or "tailings," sink to the bottom and are removed through outlets in the floor of the trough. In a *mechanically agitated cell* air is drawn downward through a sleeve and dispersed into the liquid by a turbine impeller turning inside a stator. Feed enters each

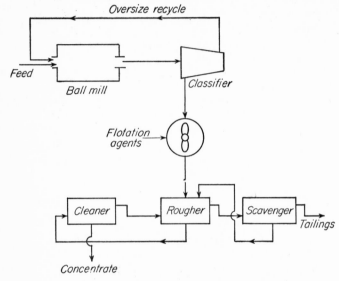

FIG. 7-43. Flotation flow sheet.

compartment over an adjustable weir; tailings leave the compartment over a similar weir. Froth overflows from the side of the trough into a froth launder. In some designs its discharge is aided by a slowly rotating froth skimmer.

In most processes a single flotation separation is not adequate, and the material must be passed through a series of cells, as shown in Fig. 7-43. It first enters a rougher, a cell which makes a crude separation. The tailings from the rougher go to a scavenger, which removes as much floatable material as possible and discharges "final tailings." The overflow is returned to the rougher. Overflow from the rougher is further treated in a cleaner, which overflows "final concentrate" containing virtually no tailings. Tailings from the cleaner are also recycled to the rougher. Rougher, scavenger, and cleaner cells are often merely different compartments in a single trough, not completely separate machines. Multiple scavengers or cleaners are necessary for difficult separations. Separate sets of equipment, including separate conditioners and flotation cells, are operated in series to make successive separations of different materials from a single ore.[3c]

Principles of Flotation. On aeration, particles float if the tendency of air to adhere to the particles is stronger than that of water to wet them. Otherwise, when air comes into contact with the solid, the water at the surface of the solid is not displaced, and adhesion of bubbles to the solid does not occur. Fundamentally, the adhesion or nonadhesion of air depends on the mechanics of surfaces of contact between liquid, solid, and gas phases.

Figure 7-44 shows a cross section through the interfaces of three phases in contact with each other. Line *ao* represents the interface between air and solid, line *bo* that between air and water, and line *co* that between water and solid. For each interface between two phases there exists a definite interfacial tension, measured in force per unit length, which is a measure of the mechanical energy needed to create a unit area of the interface. This is the same type of energy that must be used to create new solid-air interfaces in crushing and grinding. The usual surface tension of a liquid is simply the interfacial tension be-

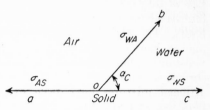

Fig. 7-44. Interfacial tensions.

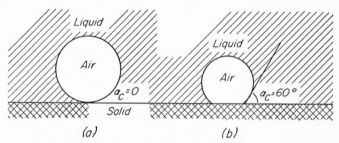

Fig. 7-45. Contact angles for air bubbles: (*a*) Contact angle = 0, no adhesion. (*b*) Contact angle = 60°, adhesion.

tween the liquid and air. Let the interfacial tension between solid and air be σ_{AS}, that between air and water be σ_{WA}, and that between water and solid be σ_{WS}. Also, let the angle that forms between the air-water boundary and the water-solid boundary be α_c. This is called the *angle of contact*. For equilibrium of forces, σ_{AS} must be balanced by the sum of σ_{WS} and the component of σ_{WA} in the direction of σ_{AS}, or

$$\sigma_{AS} = \sigma_{WS} + \sigma_{WA} \cos \alpha_c \qquad (7-76)$$

If $\sigma_{AS} \leq \sigma_{WS}$, the contact angle α_c either is zero or cannot exist, a liquid film will separate the air from the solid, and attachment of air to the solid cannot take place. If $\sigma_{AS} > \sigma_{WS}$, the contact angle is greater than zero, and the air adheres to the solid. A bubble with zero contact angle, which has no tendency to adhere to the solid, is shown in Fig. 7-45*a*. A bubble having a contact angle of 60° is shown in Fig. 7-45*b*. This bubble is well adhered to the solid. The larger the contact angle, the greater the tendency of air to adhere. For most flotation agents, an angle of 50 to 75° is obtained, which ensures that the particle can be floated.

Interfacial tensions are sensitive to the physical and chemical nature of the interface. Small quantities of certain chemicals adsorbed on the surface of solids, and forming thin layers thereon, can modify drastically the interfacial tensions of the solid and change the contact angle within wide limits. The effect is highly selective. A material effective on one material may be inert on another. Successful flotation depends on this selectivity. Substances that are effective in changing interfacial tensions are called surface-active agents. Large numbers of pure chemicals and natural or artificial mixtures have been evaluated as surface-active materials in flotation, and many useful agents have been found. Since the action of the materials is restricted to the surface of the solid and the thickness of the film is a matter of 1 to 100 molecules, only a few tenths of a pound of chemical per ton of solid is needed.

Chemicals Used in Flotation. Several chemicals are used in each flotation process. Each material added has a definite function, and no one chemical can meet all requirements. Flotation agents are classified into groups. The more important classes of addition agents are as follows.[9b, 25b]

Collectors form the main film on the particles and reduce the interfacial tension σ_{WS} or increase tension σ_{AS}, so a positive contact angle is formed for the particles to be floated. Air then adheres to these particles. Typical examples of collectors are xanthates and dithiophosphates, which are effective in floating sulfides of lead and copper, and oleic and palmitic acids, or their sodium salts, which are used for nonsulfide ores.

Collectors used alone are not effective. *Conditioners*, or *promoters*, are needed, either to improve the action of the collector or to suppress the adhesion of air to other constituents which are not to be floated. If the conditioner improves adhesion of air to the solid, it is called an *activator*. If it depresses the action of the collector, it is called a *depressor*. A typical activator is copper sulfate, used in conjunction with xanthates in floating sphalerite (ZnS). Cyanides and sulfides are effective depressors. Close control of pH by the addition of calcium hydroxide, sodium carbonate, or sodium hydroxide is usually necessary to obtain the full effect of collectors and activators.

Frothers are used to promote bubble formation and to prevent bubble collapse before the foam has been removed from the cell. Frothers modify the interfacial tension σ_{WA}. Pine oil and higher alcohols are effective frothers.

Other agents may be used to break up agglomerates or to increase selectivity in difficult separations.

Impingement Methods

For removing dust or fog from gas streams, a variety of methods is used which depend, entirely or partially, on impingement of the particles against solid surfaces placed in the flowing stream. In these methods, the particles, because of their inertia, are expected to cross the streamlines of the fluid and strike and adhere to the solid, from which they can subsequently be removed. The principle of impingement separation is shown in Fig. 7-46. The solid lines are the streamlines passing around a sphere, and the dotted lines show the paths followed by the particles. Particles initially moving along the streamlines between A and B strike the solid, and can be

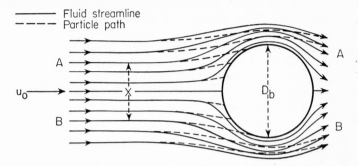

Fig. 7-46. Principle of impingement. (*Lapple.*[15b])

removed if they adhere to the wall and are not reentrained. Particles initially following streamlines outside lines A and B do not strike the solid and cannot be removed from the gas stream by impingement. The *target efficiency* η_t is defined as the fraction of the particles in the gas stream that strike the solid. For particles that would settle through still fluid in the Stokes' law range, the target efficiencies for ribbons, spheres, and cylinders are shown in Fig. 7-47. The abscissa is the dimensionless group, called

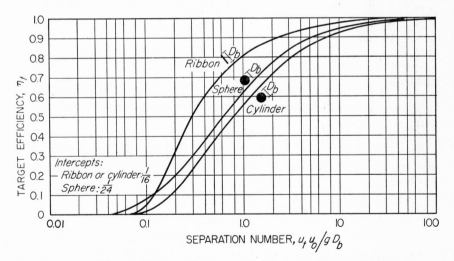

Fig. 7-47. Target efficiency of spheres, cylinders, and ribbons. (*Lapple.*[15b])

the *separation number,* $u_t u_0/g D_b$, where u_t is the terminal velocity of the particle in still fluid, in feet per second; u_0 is the velocity of the fluid approaching the solid, in feet per second; g is the acceleration of gravity, in feet per second per second; and D_b is the width of the ribbon or the diameter of the sphere or cylinder, in feet.

Since, in settling in the Stokes' law range, the terminal velocity u_t is proportional to D_b^2, the separation number is proportional to D_b. The smaller the particle, the lower the target efficiency. For each solid shape, a minimum separation number exists, below which all particles follow streamlines around the solid, impingement does not occur, and the target efficiency is zero. The target efficiency approaches unity for separation numbers of 100 or more.

Actual equipment using impingement is built in a wide variety of forms.[15d] The simplest device, used especially in treating acid mists, consists of a box or tower filled with an inert broken solid, such as coke, through which the gas passes with a low velocity. Tappers are frequently used to clear solids from the impingement surfaces. In bag filters, the initial formation of the cake in the filter cloth or other medium is by impingement. Once the first layer of solid is formed, subsequent deposition of solid is by filtration.

Scrubbers, in which a stream of liquid is passed over the impingement surfaces to wash away the precipitated particles; spray chambers, in which drops of liquid are used as the impingement agent; combination spray chambers and cyclones; and packed towers of the type described in Chap. 10—these constitute a second class of impingement separators.[15f]

A third kind of equipment used to remove dust from air includes a variety of impingement filters.[15g] Glass fibers, wood shavings, metal screens, or corrugated paper can be used as filter media. In one group, the filter medium is covered with a viscous oil, to retain dust and eliminate reentrainment. Metallic filters can be cleaned and reused, but nonmetallic units are usually discarded without cleaning. The ordinary household air filter is an example of viscous-fluid throwaway filters. Dry filters, which have smaller channels than the viscous type, are used for industrial applications.

ELECTRICAL AND MAGNETIC METHODS

Two principles are used to separate particles electrically. One method uses differences in the magnetic properties of the materials; the other is based on the ability of particles to take on electrostatic charges, which render them attractive to a charged electrode carrying a charge opposite in sign to that on the particles. Electrostatic methods can be divided into precipitation methods, in which all particles are removed from a gas stream, and separation methods, in which a mixture of different substances is sorted in accordance with temporary differences in the electrostatic charge on the substances.

Magnetic Separators. Magnetic separators are used most commonly to separate iron, steel, or magnetic iron oxide from materials that have a low magnetic attractability. The methods have been refined to such a degree that they can also be used to separate almost any two materials that differ appreciably in magnetic characteristics. Magnetic devices are used to remove tramp iron from nonmagnetic materials being fed to a crusher or other mechanical equipment, to eliminate small amounts of magnetic impurities from substances, to reclaim iron or steel from wastes, or to sort materials in accordance with their magnetic attractability.

Table 7-9 shows the relative magnetic attractability of some common substances. It is feasible to separate, by magnetic methods, a material having a relative attractability of about 0.4 from a nonmagnetic substance.

TABLE 7-9. RELATIVE MAGNETIC ATTRACTABILITY OF MINERALS †

Substance	Relative attractability
Iron (standard)	100
Magnetite (Fe_3O_4)	40
Hematite (Fe_2O_3)	1.3
Quartz (SiO_2)	0.37
Pyrite (FeS_2)	0.23
Gypsum ($CaSO_4 \cdot 2H_2O$)	0.12
Galena (PbS)	0.04

† From R. E. Kirk and D. F. Othmer, "Encyclopedia of Chemical Technology," vol. 8, p. 625, Interscience Publishers, Inc., New York, 1952.

Since magnetic separators have a wide variety of functions, several devices are available. A simple method of using a magnet to remove tramp iron from the feed to a crusher is to place the magnet close to the stream of contaminated material as it passes along a conveyor. Iron or steel particles are collected on the face of the magnet, from which they can be removed periodically by moving the magnet over a bin and cutting off the current to the magnet.

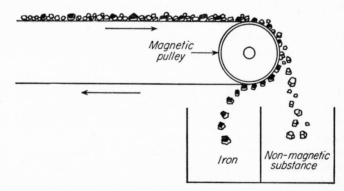

Magnetic pulley

Iron

Non-magnetic substance

FIG. 7-48. Magnetic pulley. (*By permission, from "Encyclopedia of Chemical Technology," vol. 8, by R. E. Kirk and D. F. Othmer. Copyright, 1952. Interscience Publishers, Inc.*)

Magnetic pulleys are used to remove magnetic substances from a stream of material. A magnetic pulley contains permanent magnets or electromagnets mounted between the hub and the periphery of a pulley with their poles pointed towards the periphery. A magnetic field is thus created around the pulley. A simple application of a magnetic pulley is shown in Fig. 7-48. In this figure, a belt conveyor carrying the mixture to be treated passes around a magnetic pulley. Nonmagnetic material falls from the

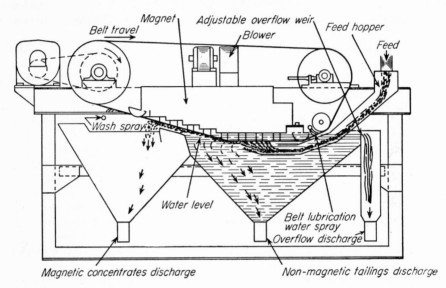

Fig. 7-49. Wet magnetic separator. (*By permission, from "Encyclopedia of Chemical Technology," vol. 8, by R. E. Kirk and D. F. Othmer. Copyright, 1952. Interscience Publishers, Inc.*)

pulley by gravity, just as from an ordinary belt-conveyor discharge. Magnetic particles remain in contact with the belt because of their attraction toward the pulley, and are forced off only when the belt removes them from the magnetic field. A baffle directs the magnetic particles to one hopper and the nonmagnetic ones to another.

Figure 7-49 shows a magnetic separator for sorting magnetic and nonmagnetic materials in a slurry. The equipment consists essentially of a tank, to which the slurry is fed and in which the nonmagnetic material is collected; a hopper, in which the magnetic substance is received; a set of electromagnets; and a continuous belt and drive.

The feed slurry enters through a chute leading to the top of the tank. The poles of the electromagnet point downward, and several poles dip below the surface of the liquid. The belt, moving in a direction away from the feed, is located above the feed chute. Magnetic particles are held against the bottom of the belt by attraction of the magnet, and the nonmagnetic substance settles by gravity to the bottom of the tank, from which it can be removed as a concentrated slurry. Excess water in the feed overflows from the tank.

Additional magnets are so mounted that their poles are on an incline, above the surface of the water. These magnets hold the magnetic particles against the bottom of the belt until the material is over the hopper. Further travel of the belt removes the particles from the magnetic field and drops them into the hopper. Water sprays are used to wash the belt as it leaves the magnets and to lubricate the back of the belt as it approaches the magnets. A blower forces air around the poles of the magnets to prevent condensation.

An important application of the separator shown in Fig. 7-49 is the recovery of the ferrosilicon or the magnetic iron oxide from the sorting fluid used in heavy-medium separation. A recovery of 99.9 per cent is obtainable in this separator.

Many other types of magnetic separators are available.[14, 25c]

Electrostatic Methods. Any particle charged with static electricity is attracted by an electrode that carries a charge opposite in sign to that on the particle. Thus, a particle carrying a negative charge is attracted by a positively charged electrode, and a particle carrying a positive charge is attracted by a negatively charged electrode. All methods of electrostatic separation depend on this fact. The force of attraction depends on the magnitudes of the charges on particles and electrode, on the distances between particle and electrode, and on the fluid in which they are immersed. Commercial separations are conducted in air. The force of attraction acts on solids and liquids alike and is independent of the material of which the particles and electrode are made.

To operate an electrostatic separator, some or all of the particles must be charged with static electricity. Three methods are most important commercially.[8] In the first method, called the frictional method, dissimilar materials are rubbed together. One substance thereby is charged with positive electricity, and the other with an equal charge of negative electricity. The familiar static developed by shuffling across a rug in a dry atmosphere is an example of frictional electricity. Frictional charging is effective in charging particles that are poor conductors of electricity but is not effective in charging conductive particles. In the second method, called the conductive method, the particles are placed between two charged electrodes differing in sign and then allowed to touch one of the electrodes. The conductive method is effective in charging conducting particles but is ineffective for nonconductors. In the third method, the ionized-gas method, the particles are carried by a gas through which a stream of charged particles, called ions, is flowing. This stream is sometimes called an ion spray. Solid or liquid particles suspended in the gas collect ions and become charged, whether the particles are conductive or not. If, however, the particles are grounded while being sprayed, only nonconductive particles are charged.

In all electrostatic units high potentials, of the order of magnitude of several thousand volts, are used in order that the charges be sufficiently

large to give practical forces of attraction. Since electrostatic charges
leak away rapidly in moist air or in the presence of moisture, the equipment and air must be dry. Air conditioning is often required for proper
operation of electrostatic separators.
Skillful electrical engineering is required to provide proper insulation
and to achieve safe operation.

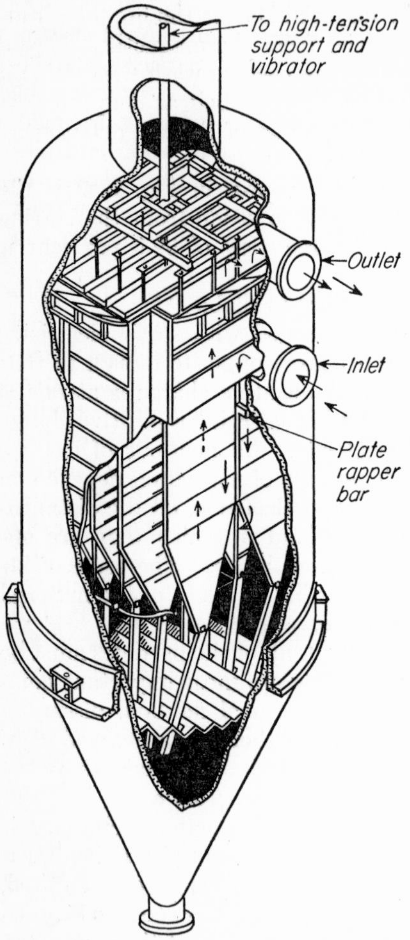

Electrostatic Precipitators. An important application of the electrostatic method is the removal of dust
or mist from a stream of gas. Devices used for this purpose are called
precipitators and are used to remove
all particles from the gas rather
than to sort particles into fractions. Since they were introduced
into American practice by F. G.
Cottrell, precipitators are often
called Cottrells. The movement of
the charged particles in a precipitator is independent of settling
velocity and is not impeded by
Brownian movement, so these units
are especially effective in treating
fine particles. They remove particles as small as 0.01 micron, a size
approaching that of large molecules,
which include all dusts and fumes
encountered industrially.

*Principle of Electrostatic Precipi-
tators.* The ionized-gas method is
used to charge the particles in a
precipitator. When a high, unidirectional electric potential is applied over an air gap between two
electrodes, a unidirectional † current
is generated, which is called a corona
discharge. This is a partial breakdown in the insulating power of the
air. If a potential much greater
than that giving the corona is applied across the gap, a complete

Fig. 7-50. Plate-type electrostatic precipitator. (*By permission, from "Encyclopedia
of Chemical Technology,"* vol. 8, *by R. E.
Kirk and D. F. Othmer. Copyright, 1952.
Interscience Publishers, Inc.*)

breakdown and sparking occur. The current from a corona is much less
than that from a spark.

The current in a corona is carried by ionized gas molecules, which provide

---

† A unidirectional current is one that flows only in one direction, but not necessarily
steadily.

the ionic spray for charging the particles in the gas stream. The charged particles are then attracted by the electrode carrying a charge opposite in sign to that on the particles. On contact with this electrode, the charges on the particles are discharged, and the particles coalesce and are removed in bulk from the unit.

Equipment for Electrostatic Precipitation. The essential steps in electrostatic precipitation are ionizing the gas and charging the particles, attracting the charged particles to the precipitating electrodes, and removing the precipitated substance. Two types of equipment are used. In a single-stage unit, which is the type most used industrially, ionization and precipitation occur in the same apparatus and utilize the same electrodes. In a two-stage unit, which ordinarily is used to remove dust and pollen from air in air-conditioning practice, separate units are used for ionization and precipitation.

The details of design of a single-stage precipitator vary with the application and the scale of operation. Units are usually specifically designed by experts for each installation. A typical example of a plate-type single-stage precipitator is shown in Fig. 7-50. The dust-laden inlet gas enters toward the top of the unit, passes downward along one bank of electrodes, reverses, flows upward along a second bank of electrodes, and leaves, free from dust, through the outlet at the top of the case. The precipitated material collects in the bottom of the shell. To facilitate removal of material from the discharge electrodes, mechanical tappers, which periodically vibrate the electrodes, are used. The temperature in an electrostatic precipitator is sufficiently high to render humidity in the air unimportant.

Electrostatic Separators. The principles of operation of two typical separation devices using electrostatic methods are shown in Fig. 7-51. The objective of these units is to separate two materials differing appreciably in electrical conductivity.

In Fig. 7-51a, the conductive particles, represented by the open circles, are charged positively by conduction from the positively charged roll B.

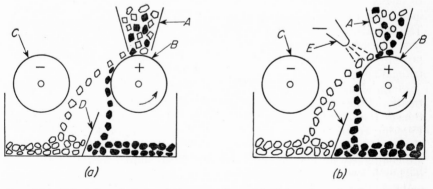

(a) (b)

FIG. 7-51. Principles of electrostatic separation: (a) Using charged roll. (b) Using charged roll and ionic spray. (*Courtesy of Industrial and Engineering Chemistry.*)

The nonconducting particles, represented by the solid circles, are unaffected by roll B, but they may be charged negatively by friction in rubbing against the other particles or the wall of the hopper A. The positively charged conductive particles are attracted by the negatively charged electrode C, and the negatively charged nonconductive particles tend to adhere to the positively charged electrode B. These forces divert the particles from their normal paths of fall and cause them to land on opposite sides of the baffle D.

In Fig. 7-51b, the separating action is enhanced by the use of an ionic spray from the negatively charged electrode E, which gives a corona discharge toward the positive electrode B. The ions are effective in increasing the negative charge on the nonconducting particles and increase their adhesion to electrode B. Ionic spraying does not affect the conducting particles, because they are in contact with grounded roll B during ionization.

Other types of electrostatic separators are available.[4,8] They are used for cleaning foods and for specialized mineral and ore separations that cannot be easily conducted by more conventional methods.

CENTRIFUGAL SEPARATIONS

Many industrial separators utilize centrifugal force in place of the force of gravity. The cyclones described earlier in this chapter remove solid particles or liquid drops from gases by centrifugal action; air classifiers balance the drag of an air stream against a centrifugal force. These devices are not considered as centrifuges, however. Industrial centrifuges nearly always operate on liquids, not on gases. In this section the centrifugal separations of two immiscible liquids and of solids from liquids are discussed.

Centrifugal separators are divided into two broad classes: those which settle and those which filter. In the first class, centrifugal force is utilized to increase the settling rate over that obtainable by gravity settling by increasing the apparent difference between the densities of the phases. The operation of centrifugal settlers is analogous to that of gravity decanters, clarifiers, thickeners, and classifiers. In filtering centrifugals the pressure needed to force the liquid through a septum is generated by centrifugal action.

Material may leave a centrifugal separator continuously or intermittently. Clarifying centrifuges are usually classed as continuous, even though the discharge of the small amount of solids may be intermittent. Filtering centrifugals are considered to be discontinuous unless the discharge of solids is continuous. Some discontinuous filtering centrifugals are completely automatic and operate on such short cycles that they are readily incorporated in an otherwise continuous process. Many designs and modifications of centrifugal separators are commercially available.[1,22,23] Representative types are listed in Table 7-10.

TABLE 7-10. TYPES OF CENTRIFUGES

Class	Type	Liquid discharge	Solids discharge
Settling centrifuges:			
Decanters and clar-ifiers	Tubular †	Continuous ‡	Intermittent
	Disk †	Continuous ‡	Intermittent
	Nozzle-discharge disk †	Continuous	Continuous
	Suspended solid-bowl centrifugal	Continuous ‡	Intermittent
Sludge separators...	Suspended solid-bowl centrifugal	Intermittent	Intermittent
	Solid-bowl helical-conveyor centrifuge	Continuous	Continuous
Classifiers..........	Tubular	Continuous ‡	Intermittent
	Solid-bowl conveyor centrifuge	Continuous	Continuous
Filtering centrifugals:			
Suspended.........	Top-suspended	Intermittent	Intermittent
Automatic.........	Short-cycle batch	Intermittent ¶	Intermittent ¶
Conveyor..........	Slotted-bowl helical	Continuous	Continuous
	Reciprocating	Continuous	Continuous

† May clarify a single liquid or separate two immiscible liquids from each other and from a solid phase.

‡ Must be interrupted during discharge of solids.

¶ May be used in continuous process because of very short cycles.

Settling Centrifuges

Settling centrifuges include decanters for separating immiscible liquids; clarifiers for removing traces of solids from a liquid, often with the simultaneous separation of two liquid phases; sludge separators for separating a slurry into a clear liquid and a very thick sludge; and classifiers for sorting solid particles according to size or specific gravity. In all settling machines the material to be separated is spun in a revolving container or bowl with a solid wall containing no slots or perforations. Liquids flow across and over, not through, the settled solids. Solids travel away from the center of rotation to the wall of the bowl, from which they are removed mechanically or by hand.

Centrifugal Decanters. Immiscible liquids are separated industrially in centrifugal decanters. Their operation is similar to that of the gravity decanters described on page 377 except that the separating force is much larger than that of gravity and it acts in the direction away from the axis of rotation instead of downward toward the earth's surface. The main types of centrifugal decanters are tubular centrifuges and disk centrifuges.

These machines are small and turn at high speeds, developing very large forces for separation.

Principle of Centrifugal Decanters. Figure 7-52 shows diagrammatically the layers of liquid in the bowl of a liquid-liquid centrifuge. The bowl is

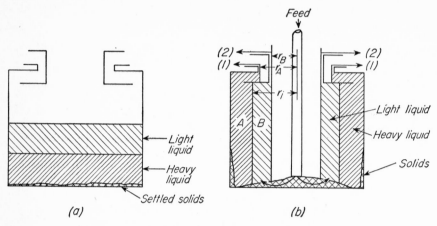

FIG. 7-52. Centrifugal separation of immiscible liquids: (*a*) Bowl at rest. (*b*) Bowl rotating. Zone *A*, separation of light liquid from heavy. Zone *B*, separation of heavy liquid from light. (1) Heavy-liquid drawoff. (2) Light-liquid drawoff.

cylindrical and revolves about a vertical axis. In Fig. 7-52*a* the bowl is at rest. The heavier liquid forms a layer on the floor of the bowl beneath a layer of the lighter liquid. Any solid particles which are present sink to the bottom of the bowl provided they are more dense than either liquid.

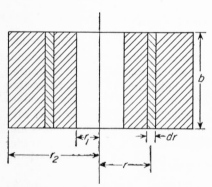

FIG. 7-53. Single liquid in rotary-centrifuge bowl.

In Fig. 7-52*b* the bowl is revolving. The heavier liquid now forms a layer, denoted as zone *A* in Fig. 7-52*b*, next to the inside wall of the bowl. A layer of lighter liquid, denoted as zone *B*, forms inside the first. A cylindrical interface of radius r_i separates the two layers, and since the force of gravity can be neglected in comparison with that of centrifugal action, the interface is vertical. It is called the *neutral zone*.

The feed mixture is introduced at a radius near that of the neutral zone. The density of the heavy liquid is ρ_A; that of the light liquid is ρ_B. The light liquid flows over the discharge lip at a radius r_B, and the heavier liquid over a discharge lip at radius r_A. Assuming that both liquids rotate with the bowl and that friction is negligible, the pressure drop in the light liquid between r_i and r_B equals that in the heavier liquid between r_i and r_A.

The pressure drop over any ring of rotating liquid is calculated as follows. Consider the ring of liquid shown in Fig. 7-53 and the volume element of thickness dr at a radius r. The breadth of the ring is b ft. The centrifugal force on the volume element is

$$dF_e = \frac{\omega^2 r \, dm}{g_c}$$

where dm is the mass of liquid in the element. If ρ is the density of the liquid,

$$dm = 2\pi\rho rb \, dr$$

Eliminating dm gives

$$dF_e = \frac{2\pi\rho b\omega^2 r^2 \, dr}{g_c}$$

The change in pressure over the element is

$$dp = \frac{dF_e}{2\pi rb} = \frac{\omega^2 \rho r \, dr}{g_c}$$

The pressure drop over the entire ring is

$$p_2 - p_1 = \int_{r_1}^{r_2} \frac{\omega^2 \rho r \, dr}{g_c}$$

and, integrating,

$$p_2 - p_1 = \frac{\omega^2 \rho (r_2^2 - r_1^2)}{2g_c} \tag{7-77}$$

For the liquids shown in Fig. 7-52

$$p_i - p_B = \frac{\omega^2 \rho_B (r_i^2 - r_B^2)}{2g_c}$$

and

$$p_i - p_A = \frac{\omega^2 \rho_A (r_i^2 - r_A^2)}{2g_c}$$

Equating these pressure drops and simplifying,

$$\rho_B(r_i^2 - r_B^2) = \rho_A(r_i^2 - r_A^2)$$

Solving for r_i gives

$$r_i = \sqrt{\frac{r_A^2 - (\rho_B/\rho_A)r_B^2}{1 - \rho_B/\rho_A}} \tag{7-78}$$

Equation (7-78) is analogous to Eq. (7-70) for a gravity settling tank. It shows that r_i, the radius of the neutral zone, is sensitive to the density ratio, especially when the ratio is nearly unity.[2] If the densities of the fluids are too nearly alike, the neutral zone may be unstable even if the speed of rotation is sufficient to separate the liquids quickly. The difference between ρ_A and ρ_B should not be less than approximately 3 per cent for operation to be stable.

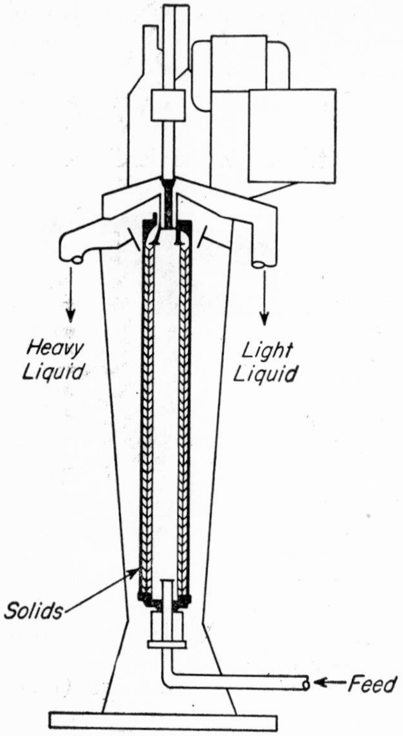

Equation (7-78) also shows that, if r_B is held constant and r_A, the radius of the discharge lip for the heavier liquid, increased, the neutral zone is shifted toward the wall of the bowl. If r_A is decreased, the zone is shifted toward the axis. An increase in r_B, at constant r_A, also shifts the neutral zone toward the axis, and a decrease in r_B causes a shift toward the wall. The position of the neutral zone is important practically. In zone A, the lighter liquid is being removed from a mass of heavier liquid, and in zone B, heavy liquid is being stripped from a mass of light liquid. If one of the processes is more difficult than the other, more time should be provided for the more difficult step. For example, if the separation in zone B is more difficult than that in zone A, zone B should be large and zone A small. This is accomplished by moving the neutral zone toward the wall, by increasing r_A or decreasing r_B. To obtain a larger time factor in zone A, the opposite adjustments would be made. Many centrifugal separators are so constructed that either r_A or r_B can be varied to control the position of the neutral axis.

Fig. 7-54. Tubular centrifuge.

Tubular Centrifuge. A tubular liquid-liquid centrifuge is shown in Fig. 7-54. The bowl is tall and narrow, 4 to 6 in. in diameter, and turns in a stationary casing at very high speed. Feed enters from a stationary nozzle inserted through an opening in the bottom of the bowl. It separates into two concentric layers of liquid inside the bowl. The inner, or lighter, layer spills over a weir at the top of the bowl; it is thrown outward into a stationary discharge cover and from there to a spout. Heavy liquid flows over another weir into a separate cover and discharge spout. The weir over which the heavy liquid flows is removable and may be replaced with another having an opening of a different size. Since this weir is ring-shaped, it is known as a ring dam.

The discharge of the two liquids over appropriate weirs is exactly like that from a gravity decanter, as shown by Fig. 7-55. In Fig. 7-55b the discharge end of a centrifuge bowl lying on its side is sketched to show its similarity to the gravity decanter in Fig. 7-55a. In practice the bowl operates vertically, but since the centrifugal force is many times that of gravity, the position of the bowl makes no difference. Changing the ring dam is analogous to changing the height of the heavy-liquid overflow from a settling tank. Since the centrifuge bowl is small, and since the liquids to be separated often differ only slightly in density, the size of the opening in the

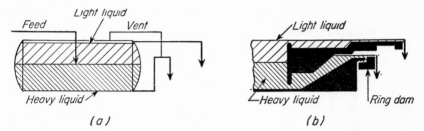

Fig. 7-55. Control of interface position in liquid-liquid separator: (a) Gravity decanter. (b) Discharge end of tubular centrifuge.

ring dam is critical. A very small change in the diameter of this opening results in a large percentage change in the position of the liquid-liquid interface, as shown by the following example.

Example 7-8. A tubular centrifuge is separating two liquids, one having a density of 62.0 lb/ft³, the other having a density of 63.9 lb/ft³. The inner radius of the centrifuge bowl is 2.05 in. The thickness of each liquid layer in the bowl is $\frac{3}{4}$ in. If the height of the heavy-liquid overflow is lowered by 0.005 in., how far will the liquid-liquid interface move?

Solution. Before the overflow is lowered, the radii indicated in Fig. 7-52 are

$$r_B = 2.05 - (2 \times 0.75) = 0.55 \text{ in.}$$

$$r_i = 2.05 - 0.75 = 1.30 \text{ in.}$$

Solving Eq. (7-78) for r_A,

$$r_A = \sqrt{r_i^2 - \frac{\rho_B}{\rho_A}(r_i^2 - r_B^2)}$$

$$\frac{\rho_B}{\rho_A} = \frac{62.0}{63.9} = 0.9702$$

$$r_A = \sqrt{1.30^2 - 0.9702(1.30^2 - 0.55^2)} = 0.5865 \text{ in.}$$

After the overflow is lowered, r_B does not change, and $r_A = 0.5865 + 0.005 = 0.5915$ in. From Eq. (7-78), r_i becomes

$$r_i = \sqrt{\frac{0.5915^2 - (0.9702 \times 0.55^2)}{1 - 0.9702}} = 1.373 \text{ in.}$$

The distance the interface moves is 1.373 − 1.300 = 0.073 in.

Note that in this particular case the interface moves 15 times as far as does the height of the overflow dam. Note also that the position of the liquid-liquid interface is independent of the speed of rotation of the bowl.

Disk Centrifuge. For some liquid-liquid separations the disk-type centrifuge illustrated in Fig. 7-56 is highly effective. A short wide bowl 8 to 20 in. in diameter turns on a vertical axis. The bowl has a flat bottom and a conical top. Feed enters from above through a stationary pipe set into the neck of the bowl. Two liquid layers are formed as in a tubular centri-

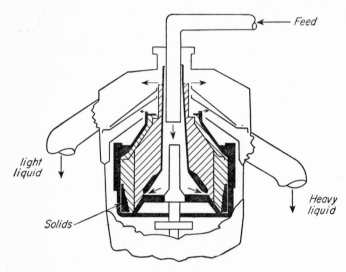

FIG. 7-56. Disk centrifuge.

fuge; they flow over adjustable dams into separate discharge spouts. Inside the bowl and rotating with it are closely spaced "disks," which are actually cones of sheet metal set one above the other. Matching holes in the disks about halfway between the axis and the wall of the bowl form channels through which the liquids pass. In operation feed liquid enters the bowl at the bottom, flows into the channels, and upward past the disks. Heavier liquid is thrown outward, displacing lighter liquid toward the center of the bowl. In its travel the heavy liquid very soon strikes the underside of a disk and flows beneath it to the periphery of the bowl without encountering any more light liquid. Light liquid similarly flows inward and upward over the upper surfaces of the disks. Since the disks are closely spaced, the distance a drop of either liquid must travel to escape from the other phase is short, much shorter than in the comparatively thick liquid layers in a tubular centrifuge. In addition, in a disk machine there is considerable shearing at the liquid-liquid interface, as one phase flows in one direction and the other phase in the opposite direction. This shearing helps break certain types of emulsions. Although disk centrifuges do not develop so high a centrifugal force as do tubular machines, they are more effective for certain duties. In particular they are valuable where the purpose of the centrifuging is not complete separation but the concentration of one fluid phase as in the separation of cream from milk and the concentration of rubber latex.

Centrifugal Clarification. If the liquid fed to a disk or tubular centrifuge contains dirt or other heavy solid particles, the solids accumulate inside the bowl and must periodically be discharged. This is accomplished

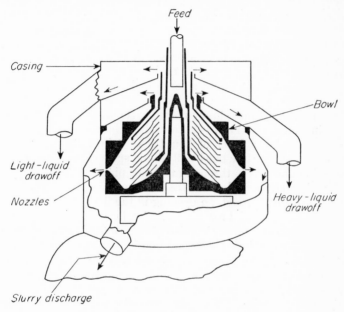

FIG. 7-57. Nozzle-discharge centrifuge.

by stopping the machine, removing and opening the bowl, and scraping out its load of solids. This requires considerable time, even for an experienced operator, and since the bowl holds only 5 to 40 lb of solids, manual cleaning becomes uneconomical if the solids are more than a few per cent of the feed.

Tubular and disk centrifuges are used to advantage for removing traces of solids from lubricating oil, process liquids, ink, and beverages that must be perfectly clean. They can take out gelatinous or slimy solids that would quickly plug a filter. Usually they clarify a single liquid in a bowl provided with but a single liquid overflow; they also, however, may throw down solids while simultaneously separating two liquid phases.

Nozzle-discharge Centrifuge. When the feed liquid contains more than a few per cent of solids means must be provided for discharging the solids automatically. One way of doing this is shown in Fig. 7-57. This separator is a modified disk-type centrifuge with a double conical bowl. In the periphery of the bowl at its maximum diameter is a set of small holes, or nozzles, perhaps $\frac{1}{8}$ in. in diameter. The central part of the bowl operates in the same way as the usual disk centrifuge, overflowing either one or two streams of clarified liquid. Solids are thrown to the periphery of the bowl and escape continuously through the nozzles, together with considerable liquid. In some designs part of the slurry discharge from the nozzles is recycled through the bowl to increase its concentration of solids; wash liquid may also be introduced into the bowl for displacement washing. In still other designs the nozzles are closed most of the time by plugs or valves which open periodically to discharge a moderately concentrated slurry.

Suspended Solid-bowl Centrifuge. A suspended solid-bowl centrifuge is useful when small amounts of solids must be removed from large volumes of liquid provided the solids settle rapidly under the modest centrifugal force. These machines contain a large cylindrical bowl, 24 to 72 in. in diameter, attached to the free lower end of a vertical shaft. In external appearance these machines closely resemble the suspended basket centrifugal shown in Fig. 7-59. Feed is admitted to the rotating bowl through a large nozzle, forming a layer of liquid 3 to 6 in. thick inside the bowl. Solids settle out against the bowl wall; liquid spills over the top of the bowl into the casing and leaves through an outlet pipe. When sufficient solids have accumulated, feed is shut off, the bowl nearly stopped, and the sludge cut out and dropped through an opening in the floor of the bowl.

The bowl of an industrial machine may hold up to 8 ft^3 of settled sludge. If the feed contains no more than 1 or 2 per cent of solids, therefore, the flow of liquid may continue for several hours without interruption. Eventually the feed must be shut off while the bowl is being unloaded, but since this rarely requires more than 2 or 3 min, the operation of the machine is virtually continuous.

Principle of Centrifugal Clarification. In a clarifying centrifuge, a particle of given size is removed from the liquid if sufficient time is available for the particle to reach the wall of the separator bowl. If it is assumed that the particle is at all times moving radially at its terminal velocity, the diameter of the smallest particle that should just be removed can be calculated.

Consider the volume of liquid in a centrifuge bowl shown in Fig. 7-53, and examine the cylindrical volume element of thickness dr ft and radius r ft. Let the height of the cylinder be b ft, and assume the volumetric flow rate is q ft^3/sec. The volume of the element is $2\pi br\ dr$ ft^3, and the settling time provided by the element is, in seconds,

$$d\theta = \frac{2\pi br\ dr}{q}$$

If the particle settles in the Stokes' law range, the terminal velocity at radius r is, by Eq. (7-58),

$$u_t = \frac{\omega^2 r(\rho_p - \rho)\ D_p^2}{18\mu}$$

The distance traversed by the particle in $d\theta$ sec is

$$dy = u_t \, d\theta = \frac{\pi b \omega^2 (\rho_p - \rho) D_p^2 r^2 \, dr}{9 \mu q}$$

The distance traveled by the particle during its entire residence time is, in feet,

$$y = \frac{\pi b \omega^2 (\rho_p - \rho) D_p^2}{9 \mu q} \int_{r_-}^{r_2} r^2 \, dr$$

$$= \frac{\pi b \omega^2 (\rho_p - \rho) D_p^2 (r_2^3 - r_1^3)}{27 \mu q} \tag{7-79}$$

where r_2 = inside radius of bowl

r_1 = radius of inside surface of liquid

A cut point can be defined[2] as the diameter of that particle which just covers one-half the distance between r_1 and r_2. If D_{pc} is the cut diameter, a particle of this size moves a distance $y = (r_2 - r_1)/2$ ft during the settling time allowed, and Eq. (7-79) can be written

$$q = \frac{2 \pi \omega^2 (\rho_p - \rho) D_{pc}^2 (r_2^2 + r_2 r_1 + r_1^2) b}{27 \mu} \tag{7-80}$$

Most particles whose diameters are larger than D_{pc} will be eliminated by the centrifuge, and most particles having smaller diameters will remain in the liquid.

Sludge Separators. In a nozzle-discharge centrifuge the solids leave the bowl from below the liquid surface, and therefore carry with them considerable quantities of liquid. For separating a feed slurry into a clear liquid fraction and a heavy "dry" sludge, the settled solids must be moved mechanically out of the liquid and given a chance to drain while still under centrifugal force. This is done in continuous sludge separators, a typical example of which is illustrated in Fig. 7-58. In this helical-conveyor centrifuge a conical bowl lying on its side rotates about a horizontal axis. Feed enters through a stationary axial pipe, spraying outward into a pond of liquid. Liquid overflows through ports in the plate covering the large end of the cone. The position of the ports fixes the depth of liquid in the bowl. Solids settle through the liquid to the inner surface of the bowl; a helical conveyor turning slightly slower than the bowl moves the solids out of the pond and up the "beach" to discharge openings in the small end of the cone. Wash liquid may be sprayed on the solids as they travel up the beach, to remove soluble impurities. The wash flows into the pond and discharges with the liquor. Drained sludge and clarified liquor are thrown out from the bowl into different parts of the casing, from which they leave through suitable openings.

Helical-conveyor centrifuges are made with maximum bowl diameters from 4 to 54 in. They separate large amounts of material. An 18-in.

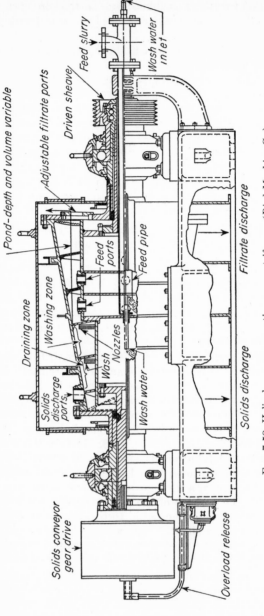

Fig. 7-58. Helical-conveyor continuous centrifuge. (*Bird Machine Co.*)

machine, for example, might handle 1 ton of solids per hour; a 54-in. machine, 50 tons/hr. With thick feed slurries the capacity of a given machine is limited by the allowable torque on the conveyor. With dilute slurries the liquid-handling capacity of the bowl and overflow ports limits the throughput.

Practical operation of a sludge separator, of course, requires that the solids be heavier than the liquid and not be resuspended by the action of the conveyor. The liquid effluent from these machines is usually not completely free from solids, and may require subsequent clarification. Within these restrictions sludge separators solve a wide variety of problems. They separate fine particles from liquids, dewater and wash free-draining crystals, and are often used as classifiers, as discussed below.

Centrifugal Classifiers. During the passage of the liquid through a centrifuge bowl the heavier, larger solid particles are thrown out of the liquid. Finer, lighter particles may not settle in the time available and be carried out with the liquid effluent. As in a gravity hydraulic classifier, solid particles can be sorted according to size, shape, or specific gravity. Because the separating force is much larger than that of gravity, precise splits can be made with very small particles.

For the removal of dirt and agglomerates from colloidal or near-colloidal suspensions, a tubular centrifuge may be operated in series with a colloid mill. The suspension is passed through the mill and then through the centrifuge, where a force as large as 13,000 times that of gravity throws down oversize particles. The flow rate through the bowl is kept high enough to that the extremely fine solid particles in the suspension do not settle out in the centrifuge. The suspension may be passed through the mill and classifier several times to ensure complete cleaning. The settled solids are removed at intervals from the centrifuge bowl by hand.

Continuous helical-conveyor centrifuges are used as classifiers in closed circuit with wet-grinding ball mills. Their operation is like that of the sludge separators except that the bowl speed and slurry feed rate are adjusted and controlled so that acceptably fine particles leave with the liquid in the form of a dilute slurry. Oversize particles are conveyed to the discharge as a sludge, repulped, and returned to the mill for further grinding. Since long settling times and a fairly wet sludge are desirable, the pond of liquid in this type of classifier nearly fills the rotating bowl.

Note the similarity in operating principle of these machines and that of the crossflow classifier shown in Fig. 7-33b. In the centrifugal classifier the separating force is of the order of 600 times the force of gravity, permitting sharp separations of particles 1 micron or less in diameter. Much coarser particles than this, however, are also classified in centrifugal machines.

Filtering Centrifugals

Solids which form a porous cake may be separated from liquids in a filtering centrifugal. Slurry is fed to a rotating basket having a slotted or perforated wall. Covering the wall is a filter medium such as canvas or metal cloth. Pressure resulting from the centrifugal action forces the liquor through the filter medium, leaving the solids behind. If the feed to the basket is then shut off and the cake of solids spun for a short time, much of the residual liquid in the cake drains off the particles, leaving the solids much "drier" than those from a filter press or even from a vacuum filter. Typically if a cake from a filter press contains 7 per cent of residual liquid, a centrifuged cake of the same solid might contain 3 per cent; if the filter yielded a cake containing 30 per cent liquid, a centrifuge might yield one containing 15 per cent. This reduction in residual moisture is the chief advantage of centrifugal filters over other types of filtration equipment. When the filtered material must subsequently be dried by thermal means, considerable savings may result from the use of a centrifuge. In general it is less costly to remove moisture from a solid by mechanical means than by thermal vaporization.

The main types of filtering centrifugals are suspended batch machines, which are discontinuous in their operation; automatic short-cycle batch machines; and continuous conveyor centrifuges (see Table 7-10). In suspended centrifugals the filter media are canvas or other fabric, woven metal cloth, or—very occasionally—paper. In automatic machines fine metal screens are used; in conveyor centrifuges the filter medium may be a metal screen or the slotted wall of the basket itself.

Suspended Batch Centrifugals. The most common batch centrifugal in industrial processing is the top-suspended centrifugal shown in Fig. 7-59. The perforated baskets range from 30 to 48 in. in diameter and from 18 to 30 in. deep. They turn at speeds between 600 and 1,800 rpm. The basket, as in the suspended solid-bowl clarifying centrifuge described earlier, is held at the lower end of a free-swinging vertical shaft driven from above. A filter medium lines the perforated wall of the basket. Feed slurry enters the rotating basket through an inlet pipe or chute. Liquor drains through the filter medium into the casing and out a discharge pipe; the solids form a cake 2 to 6 in. thick inside the basket. Wash liquid may be sprayed through the solids to remove soluble material. The cake is then spun as dry as possible, sometimes at a higher speed than during the loading and washing steps. The motor is shut off and the basket nearly stopped by means of a brake. With the basket slowly turning, at perhaps 30 to 50 rpm, the solids are discharged by cutting them out with an unloader knife, which peels the cake off the filter medium and drops it through an opening in the basket floor. The filter medium is rinsed clean, the motor turned on, and the cycle repeated.

Top-suspended centrifugals are used most extensively in sugar refining, where they operate on short cycles of 2 to 3 min per load and produce up to 5 tons/hr of crystals per machine. Automatic controls are often provided for some or all of the steps in the cycle. In most processes where large tonnages of crystals are separated, however, other automatic centrifugals or continuous conveyor centrifuges are used. Suspended centrifugals, except in sugar manufacture, find application where smaller amounts of solids are to be dewatered. Most suspended centrifugals operate on cycles

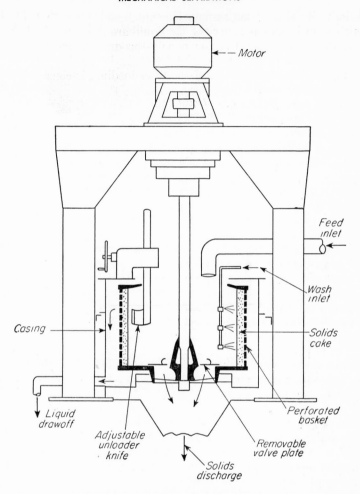

Fig. 7-59. Top-suspended basket centrifugal.

of 10 to 30 min per load, discharging 700 to 4,000 lb of solids per hour.[23]

Principles of Centrifugal Filtration. The basic theory of constant-pressure filtration can be modified to apply to filtration in a centrifuge. The treatment applies after the cake has been deposited and during flow of clear filtrate or fresh wash water through the cake. Figure 7-60 shows such a cake. In this figure,

r_1 = radius of inner surface of liquid, ft
r_i = radius of inner face of cake, ft
r_2 = inside radius of basket, ft

The following simplifying assumptions are made. The effects of gravity and of changes in kinetic energy of the liquid are neglected, and the pressure drop from centrifugal action equals the drag of the liquid flowing through the cake; the cake is completely filled with liquid; the flow of the liquid is laminar; the resistance of the filter medium is constant; and the

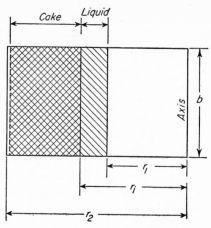

Fig. 7-60. Centrifugal filter.

cake is nearly incompressible, so an average specific resistance can be used as a constant.

The equation obtained by integrating Eq. (7-13) in the light of the assumptions can be written [10]

$$q = \frac{\rho\omega^2(r_2^2 - r_1^2)}{2\mu(\alpha m_c/\bar{A}_L\bar{A}_a + R_m/A_2)} \tag{7-81}$$

where q = volumetric rate of flow of filtrate, ft^3/sec
   ρ = density of liquid, lb/ft^3
   μ = viscosity of liquid, lb/ft-sec
   m_c = mass of solid in cake, lb
   R_m = resistance of filter medium, lb/ft
   α = specific cake resistance, ft/lb
   A_2 = area of filter medium (inside area of centrifuge basket), ft^2
   ω = angular velocity, radians/sec
   $\bar{A}_a$ = arithmetic mean cake area, ft^2
   $\bar{A}_L$ = logarithmic mean cake area, ft^2
The average areas $\bar{A}_a$ and $\bar{A}_L$ are defined by the equations

$$\bar{A}_a = (r_i + r_2)\pi b \tag{7-82}$$

$$\bar{A}_L = \frac{2\pi b(r_2 - r_i)}{\ln (r_2/r_i)} \tag{7-83}$$

where b is the height of the basket in feet. Note that Eq. (7-81) applies to a cake of definite mass and is *not* an integrated equation over an entire filtration starting with an empty centrifugal.

Automatic Batch Centrifugals. A completely automatic centrifugal is illustrated in Fig. 7-61. In this machine the basket rotates at constant speed about a horizontal axis. Feed slurry, wash liquid, and screen rinse are successively sprayed into the basket at appropriate intervals for controlled lengths of time. The basket is unloaded while turning at full speed by a heavy knife which rises periodically and cuts the solids out with

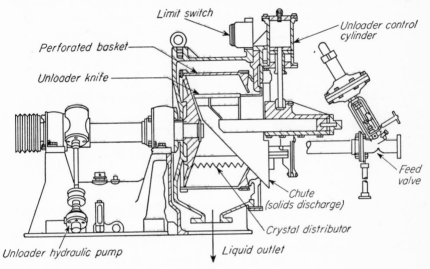

FIG. 7-61. Automatic batch centrifugal. (*Sharples Corp.*)

considerable force through a discharge chute. Cycle timers and solenoid-operated valves control the various parts of the operation: feeding, washing, spinning, rinsing, and unloading. Any part of the cycle may be lengthened or shortened as desired.

The basket in these machines is usually 20 or 27 in. in diameter. Despite their small size, automatic centrifugals have high productive capacity with free-draining crystals. Usually they are not used when the feed contains many particles finer than 150-mesh. With coarse crystals the total operating cycle ranges from 35 to 90 sec, so that although each basket load is small—30 to 100 lb—the hourly throughput is large. Because of the short cycle and the small amount of holdup required for feed slurry, filtrate, and discharged solids, automatic centrifugals are easily incorporated into continuous manufacturing processes. The small batches of solid can be effectively washed with small amounts of wash liquid, and—as in any batch machine—the amount of washing can be temporarily increased to clean up off-quality material should it become necessary. Automatic centrifugals cannot handle slow-draining solids, which would give uneconomically long cycles, or solids which do not discharge cleanly through the chute. There is also considerable breakage or degradation of the crystals by the unloader knife.

Continuous Filtering Centrifugals. Liquids and crystalline solids may be separated in a machine known as a *slotted-bowl conveyor centrifuge*, which is similar in many respects to the conical solid-bowl conveyor centrifuge described earlier. The bowl rotates about a horizontal axis and contains a helical conveyor, which moves solids to discharge outlets at one

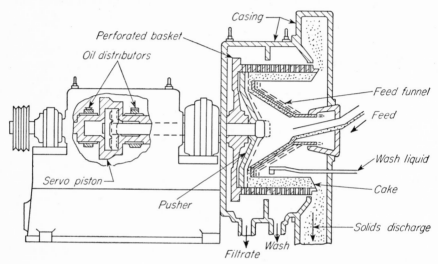

FIG. 7-62. Reciprocating conveyor centrifuge. (*Baker-Perkins, Inc.*)

end of the machine. The bowl is cylindrical, however, not conical, and has a slotted wall through which liquid escapes. Wash liquid may be sprayed over the cake as it travels along. The particles of solid must be fairly large, usually coarser than 100-mesh, in order not to pass through the slots in the wall of the bowl. These machines dewater several tons per hour of coarse crystalline solids like sodium chloride and ammonium sulfate.

Another continuous separator for coarse crystals is the *reciprocating conveyor centrifuge* shown in Fig. 7-62. A rotating basket with a slotted wall is fed through a revolving feed funnel. The purpose of the funnel is to accelerate the feed slurry gently and smoothly. Feed enters the small end of the funnel from a stationary pipe at the axis of rotation of the basket. It travels toward the large end of the funnel, gaining speed as it goes, and when it spills off the funnel onto the wall of the basket, it is moving in the same direction as the wall and at very nearly the same speed. Liquor flows through the basket wall, which may be covered with a woven metal cloth. A layer of crystals 1 to 3 in. thick is formed. This layer is moved over the filtering surface by a reciprocating pusher. Each stroke of the pusher moves the crystals a few inches toward the lip of the basket; on the return stroke a space is opened on the filtering surface in which more cake can be deposited. When the crystals reach the lip of the basket, they fly outward into a large casing and drop into a collector chute. Filtrate and any wash liquid that is sprayed on the crystals during their travel leave the casing through separate outlets. The gentle acceleration of the feed slurry and deceleration of the discharged solids minimize breakage of the crystals. Reciprocating centrifuges are made with baskets ranging in diameter from 12 to 48 in. They dewater and wash 0.3 to 25 tons/hr of solids containing no more than 10 per cent by weight of material finer than 100-mesh.

SYMBOLS

A Area, ft^2; filter area; A_p, projected area perpendicular to direction of motion of fluid or solid; A_1, area of inner surface of material in centrifuge; A_2, area of outer surface of material in centrifuge; $\bar{A}_a$, arithmetic mean of A_1 and A_2; $\bar{A}_L$, logarithmic mean of A_1 and A_2

a Acceleration, ft/sec^2; a_e, acceleration of particle from external forces

B Constant in constant-pressure filtration equation, defined by Eq. (7-28); also, underflow from screen, lb/hr

b Breadth of material in centrifuge, ft

b_1 Constant in Eq. (7-55)

b_2 Constant in Eq. (7-66)

c Mass of solid deposited in filter per volume of filtrate, lb/ft^3

c_s Concentration of solids in slurry to filter, lb/ft^3

D Overflow from screen, lb/hr

D_b Width of ribbon, or diameter of sphere or cylinder, impingement equipment, ft

D_p Particle size, ft; D_{pA}, diameter of heavy particle; D_{pB}, diameter of light particle; D_{pc}, cut diameter

E Screen effectiveness, dimensionless

F Feed to screen, lb/hr; also, force, lb force; F_b, buoyant force; F_D, drag force; F_e, external force

g Acceleration of gravity, ft/sec^2

g_c Newton's-law conversion factor, 32.174 ft-lb/lb force-sec^2

K Criterion for settling [Eq. (7-61)], dimensionless

K_p Constant in constant-pressure filtration equation, defined by Eq. (7-27)

K_r Constant in constant-rate filtration equation, defined by Eq. (7-33)

k Constant in Kozeny-Carman filtration equation, dimensionless

L Distance in cake measured from filter medium, ft; L_c, cake thickness

m Mass, lb; m_c, mass of solid in filter cake

n Exponent

p Pressure, lb force/ft^2; pressure in cake at distance L ft from filter medium; p_A, at discharge of heavy liquid in centrifuge; p_B, at discharge of light liquid from centrifuge; p_a, at inlet to filter; p_b, at discharge from filter; p_i, at neutral zone in centrifuge; p', at boundary between cake and medium in filter

q Volumetric flow rate, ft^3/sec

R_m Filter-medium resistance, ft^{-1}

r Radius, ft; r_A, at discharge of heavy liquid from centrifuge; r_B, at discharge of light liquid from centrifuge; r_i, in centrifuge, radius of neutral zone or of interface between cake and liquid layer; r_1, inner radius of material in centrifuge; r_2, outer radius of material in centrifuge

s Compressibility coefficient, dimensionless; exponent in Eq. (7-27)

s_p Surface area of single particle, ft^2

s' Exponent in Eq. (7-28)

u Linear velocity, ft/sec; of liquid through filter; u_0, of undisturbed fluid approaching solid; u_t, terminal velocity of settling particle; u_{tA}, terminal velocity of heavy particle; u_{tB}, terminal velocity of light particle; u_{tr}, terminal velocity relative to the fluid

V Volume, ft^3 or liters; of filtrate collected to time θ; V_c, of settling chamber; $\overline{V}$, average filtrate (cumulative) in time increment $\Delta\theta$

v_p Volume of single particle, ft^3

w Ratio, mass of wet filter cake to mass of same cake after drying

x Mass fraction of cut in mixture of particles; x_B, in underflow from screen; x_D, in overflow from screen; x_F, in feed to screen

y Distance, variable, traveled by particle, ft

Z Height, ft; Z_{A1}, of heavy liquid in decanter; Z_{A2}, of heavy liquid in overflow leg of decanter; Z_B, of layer of light liquid in decanter; Z_c, height of settling chamber; Z_T, total height of both light and heavy liquids in decanter, $Z_T = Z_{A1} + Z_B$

Greek Letters

α Specific cake resistance, ft/lb; α_c, contact angle; α_L, local specific cake resistance; α_0, constant in Eq. (7-27); α_0', constant in Eq. (7-28)

β Constant in Eq. (7-28)

$-\Delta p$ Over-all pressure drop through filter, lb force/ft^2, $p_b - p_a$; $-\Delta p_c$, pressure drop through cake, $p' - p_a$; $-\Delta p_m$, pressure drop through filter medium, $p' - p_b$

ΔV Volume increment of filtrate, ft^3

$\Delta\theta$ Time increment, sec

ϵ Porosity or volume fraction voids in bed of solids, or volume fraction of liquid in slurry, dimensionless; $\bar{\epsilon}$, average porosity of filter cake

η_t Target efficiency, impingement equipment

θ Time, sec

μ Viscosity, lb/ft-sec

ρ Density, lb/ft^3; of fluid; ρ_A, of heavy fluid; ρ_B, of light fluid; ρ_m, of slurry; ρ_p, of particle; ρ_{pA}, of heavy particle; ρ_{pB}, of light particle

σ Interfacial tension, lb force/ft; σ_{AS}, solid to air; σ_{WA}, water to air; σ_{WS}, water to solid

ϕ Mass fraction in screen analysis, cumulative to particle size D_p

Ψ Function [Eq. (7-62)]

ψ Function [Eq. (7-72)]

ω Angular velocity, radians/sec

Dimensionless Groups

C_D Drag coefficient, $2g_cF_D/u^2\rho A_p$

N_{Re} Reynolds number, $D_p u \rho/\mu$

PROBLEMS

7-1. It is desired to separate a mixture of crystals into three fractions, a coarse fraction retained on an 8-mesh screen, a middle fraction passing an 8-mesh but retained on a 14-mesh screen, and a fine fraction passing a 14-mesh. Two screens in series are used, an 8-mesh and a 14-mesh, conforming to the Tyler standard. Screen analyses of feed, coarse, medium, and fine fractions are given in Table 7-11. Assuming the analyses are

TABLE 7-11. SCREEN ANALYSES FOR PROB. 7-1

Screen	Feed	Coarse fraction	Middle fraction	Fine fraction
3/4	3.5	14.0		
4/6	15.0	50.0	4.2	
6/8	27.5	24.0	35.8	
8/10	23.5	8.0	30.8	20.0
10/14	16.0	4.0	18.3	26.7
14/20	9.1	.....	10.2	20.2
20/28	3.4	.....	0.7	19.6
28/35	1.3	.....	.....	8.9
35/48	0.7	.....	.....	4.6
Total......	100.0	100.0	100.0	100.0

accurate, what do they show as to the ratio by weight of each of the three fractions actually obtained? What is the effectiveness of each screen?

7-2. The screens used in Prob. 7-1 are shaking screens with a capacity of 4 tons/ (hr)(ft^2)(mm mesh size). How many square feet of screen are needed for each of the screens in Prob. 7-1 if the feed to the first screen is 100 tons/hr?

7-3. The data in Table 7-12 were taken in a constant-pressure filtration of a slurry of CaCO$_3$ in H$_2$O.[†] The filter was a 6-in. filter press with an area of 1.0 ft^2. The mass

TABLE 7-12. DATA FROM CONSTANT-PRESSURE FILTRATION

5-lb force/in.^{2} pressure drop		15-lb force/in.^{2} pressure drop		30-lb force/in.^{2} pressure drop		50-lb force/in.^{2} pressure drop	
Filtrate, lb	Time, sec	Filtrate, lb	Time, sec	Filtrate, lb	Time, sec	Filtrate, lb	Time, sec
0	0	0	0	0	0	0	0
2	24	5	50	5	26	5	19
4	71	10	181	10	98	10	68
6	146	15	385	15	211	15	142
8	244	20	660	20	361	20	241
10	372	25	1,009	25	555	25	368
12	524	30	1,443	30	788	30	524
14	690	35	2,117	35	1,083	35	702
16	888						
18	1,188						

† E. L. McMillen and H. A. Webber, *Trans. AIChE,* **34:** 213 (1938).

fraction of solids in the feed to the press was 0.139; the mass ratio of wet cake to dry cake was 1.59 in the experiment when the pressure was 5 lb force/in.^{2} and 1.47 in the other three; and the dry cake density was 63.5 in the run at 5 lb force/in.^{2}, 73.0 in the runs at 15 and 30 lb force/in.^{2}, and 73.5 in the run at 50 lb force/in.^{2} Calculate the values of β, R_m, and cake thickness for each of the experiments.

7-4. The sludge of Prob. 7-3 is to be filtered in a large press having a total area of 10 ft^2 and operated at a constant pressure drop of 25 lb force/in.^{2} The frames are 1.5 in. thick. Assume the filter-medium resistance in the large press is the same as that in the laboratory filter. Calculate the filtration time required and the volume of filtrate obtained in one cycle.

7-5. Assuming the actual rate of washing is 80 per cent of the theoretical rate, how long will it take to wash the cake in the press of Prob. 7-4 with a volume of wash water equal to that of the filtrate?

7-6. The following relation between α and Δp for Superlight CaCO$_3$ has been determined.[10]

$$\alpha = 8.8 \times 10^{10}[1 + 3.36 \times 10^{-4}(-\Delta p)^{0.86}]$$

where $-\Delta p$ is in pounds force per square foot. This relation is followed over a pressure range from 0 to 1,000 lb force/in.^{2}

A slurry of this material giving 3.0 lb of cake solid per cubic foot of filtrate is to be filtered at a constant pressure drop of 80 lb force/in.^{2} and a temperature of 70°F. Experiments on this sludge and the filter cloth to be used gave a value of $R_m = 1.5 \times 10^{10}$ ft^{-1}. A pressure filter of the Sweetland type is to be used. How many square feet of filter surface are needed to give 1,500 gal of filtrate in a 1-hr filtration?

7-7. The filter of Prob. 7-6 is washed at 70°F and 80 lb force/in.^{2} with a volume of wash water equal to one-third that of the filtrate. The washing rate is 80 per cent of the theoretical value. How long should it take to wash the cake?

7-8. The filter of Prob. 7-6 is operated at a constant rate of 0.5 gal/ft^2-min from the start of the run until the pressure drop reaches 80 lb force/in.^{2} and then at constant pressure drop of 80 lb force/in.^{2} until a total of 1,500 gal of filtrate is obtained. The operating temperature is 70°F. What is the total filtration time required?

7-9. An aqueous slurry of calcium carbonate is to be settled in a continuous sedimentation unit. The particle size is uniform at 50 microns, and the particles may be considered spheres. The temperature is 75°F. The specific gravity of the particles is 2.93. At what consistency (fraction of solids by volume) will the settling capacity, in pounds per square foot of settling area, be a maximum? What is the capacity under these conditions?

7-10. How long will it take for the following spherical particles to settle, under free-settling conditions, through 5 ft of water at 70°F?

Substance	Sp. gr.	Diameter, in.
Galena..........	7.5	0.01
		0.001
Quartz..........	2.65	0.01
		0.001
Coal.............	1.3	0.25
Steel............	7.7	1.0

7-11. Galena spheres 0.001 in. in diameter are rotated in a water suspension by a centrifuge at 70°F. The speed of the centrifuge is 600 rpm. The inside diameter of the rotating liquid is 6 in., and the outside diameter is 12 in. Assuming the particles are at all times at the terminal velocities corresponding to their locations, what time is required to completely separate the particles from the liquid? The initial distribution of the particles in the suspension is uniform, and the settling is under free-settling conditions.

7-12. Air carrying particles of diameter 50 microns and density 60 lb/ft³ enters a cyclone at a linear velocity of 50 ft/sec. The temperature is 70°F. The diameter of the cyclone is 24 in., and the diameter of the outlet is 8 in. The diameter corresponding to the center line of the inlet connection is 20 in. Assuming that the inlet velocity equals that at the diameter of 20 in. and that the exponent n of Eq. (7-67) is 2.0, what is the terminal velocity of the particles at the radius of the outlet connection?

7-13. Quartz and galena are to be separated by differential hydraulic classification. The mixture has a size range of 65- to 200-mesh. What fractions may be expected if free-settling conditions are used?

This mixture is subjected to hindered settling in a sorting bed of specific gravity 2.0 having equal volumes of quartz and galena. What fractions may be expected? Compare quantitatively the velocities of the particles under hindered settling with those under free settling.

7-14. A sink-and-float process is used to separate jasper (quartz) from hematite by means of a ferrosilicon medium. The specific gravities of hematite and ferrosilicon are 5.1 and 6.7, respectively. At what consistency, in volume per cent solids, should the medium be maintained?

7-15. The dust-laden air of Prob. 7-12 is passed through an impingement separator at a linear velocity of 20 ft/sec. The separator consists essentially of ribbons 1 in. wide. What is the maximum fraction of the particles that can be removed in this manner?

7-16. The bowl of the tubular centrifuge in Example 7-8 has a depth of 12 in. The volumetric ratio of light to heavy liquid is 1:1. For satisfactory separation it is necessary that the heavy liquid remain in the centrifuge bowl at least 5.5 sec and the light liquid 4.0 sec. What is the capacity of the centrifuge in gallons per minute of total feed both before and after the change in the position of the heavy-liquid overflow?

7-17. What is the capacity in gallons per minute of a clarifying centrifuge operating under these conditions?

Diameter of bowl, 24 in.
Thickness of liquid layer, 3 in.
Depth of bowl, 16 in.
Speed, 1,000 rpm
Specific gravity of liquid, 1.3
Specific gravity of solid, 1.6
Viscosity of liquid, 3 centipoises
Cut size of particles, 30 microns

7-18. A batch centrifugal filter having a bowl diameter of 30 in. and a bowl height of 18 in. is used to filter a suspension having the following properties:

Liquid, water
Temperature, 25°C
Concentration of solid in feed, 60 g/liter
Porosity of cake, 0.835
Density of dry solid in cake, 125 lb/ft³

Final thickness of cake, 6 in.

Speed of centrifuge, 2,000 rpm

Specific cake resistance, 9.5×10^{10} ft/lb

Filter medium resistance, 2.6×10^{10} ft^{-1}

The final cake is washed with water under such conditions that the radius of the inner surface of the liquid is 8 in. Assuming that the rate of flow of wash water equals the final rate of flow of filtrate, what is the rate of washing in gallons per minute?

REFERENCES

1. Ambler, C. M.: *Chem. Eng. Progr.*, **44**: 405 (1948).
2. Ambler, C. M.: *Chem. Eng. Progr.*, **48**: 150 (1952).
3. Brown, G. G., and associates: "Unit Operations," (*a*) p. 15, (*b*) p. 22, (*c*) p. 106, John Wiley & Sons, Inc., New York, 1950.
4. Bullock, H. L.: *Ind. Eng. Chem.*, **33**: 1119 (1941).
5. Coe, H. S., and G. H. Clevenger: *Trans. AIMME*, **55**: 356 (1916).
6. Comings, E. W.: *Ind. Eng. Chem.*, **32**: 663 (1940).
7. Crozier, H. E., and L. E. Brownell: *Ind. Eng. Chem.*, **44**: 631 (1952).
8. Fraas, F., and O. C. Ralston: *Ind. Eng. Chem.*, **32**: 600 (1940).
9. Gaudin, A. M.: "Principles of Mineral Dressing," (*a*) p. 144, (*b*) pp. 367–398, McGraw-Hill Book Company, Inc., New York, 1939.
10. Grace, H. P.: *Chem. Eng. Progr.*, **49**: 303, 367, 427 (1953).
11. Heiss, J. F., and J. Coull: *Chem. Eng. Progr.*, **48**: 133 (1952).
12. Hughes, O. D., R. W. Ver Hoeve, and C. D. Luke: paper given at meeting of AIChE, Columbus, Ohio, December, 1950.
13. Hughes, R. R., and E. W. Gilliland: *Chem. Eng. Progr.*, **48**: 497 (1952).
14. Kirk, R. E., and D. F. Othmer: "Encyclopedia of Chemical Technology," vol. 8, pp. 624–638, Interscience Publishers, Inc., New York, 1952.
15. Lapple, C. E., in J. H. Perry (ed.): "Chemical Engineers' Handbook," 3d ed., (*a*) p. 1019, (*b*) p. 1022, (*c*) pp. 1023–1028, (*d*) pp. 1023, 1028–1030, (*e*) p. 1031, (*f*) pp. 1034–1039, (*g*) pp. 1045–1050, McGraw-Hill Book Company, Inc., New York, 1950.
16. Lapple, C. E., and C. B. Shepherd: *Ind. Eng. Chem.*, **32**: 605 (1940).
17. Miller, S. A.: *Chem. Inds.*, **66**: 38 (1950).
18. Pettyjohn, E. S., and E. B. Christiansen: *Chem. Eng. Progr.*, **44**: 157 (1948).
19. Riegel, E. R.: "Chemical Process Machinery," 2d ed., (*a*) pp. 44–63, (*b*) p. 234, (*c*) pp. 315–318, (*d*) pp. 324–359, (*e*) p. 356, Reinhold Publishing Corporation, New York, 1953.
20. Ruth, B. F.: personal communication.
21. Shepherd, C. B., and C. E. Lapple: *Ind. Eng. Chem.*, **31**: 972 (1939); **32**: 1246 (1940).
22. Smith, J. C.: *Chem. Inds.*, **65**: 519 (1949).
23. Smith, J. C.: *Chem. Eng.*, **59**(4): 140 (1952).
24. Steinour, H. H.: *Ind. Eng. Chem.*, **36**: 618 (1944).
25. Taggart, A. F.: "Handbook of Mineral Dressing: Ores and Industrial Minerals," (*a*) p. **11**-123, (*b*) pp. **12**-03–**12**-18, (*c*) pp. **13**-15–**13**-39, John Wiley & Sons, Inc., New York, 1945.

Chapter 8

FLOW OF HEAT

Practically all the operations that are carried out by the chemical engineer involve the production or absorption of energy in the form of heat. The laws governing the transfer of heat and the types of apparatus that have for their main object the control of heat flow are therefore of great importance. This chapter will consider: (1) the basic mechanisms of heat flow, (2) the fundamental quantitative methods of calculation with especial reference to chemical engineering, and (3) the application of these principles to the design of heating and cooling equipment.

Classification of Heat-flow Processes. Heat may flow by one or more of three basic mechanisms.

Conduction. When heat flows through a body by the transference of the momentum of individual molecules without mixing, it is said to flow by conduction, e.g., the flow of heat through the brick wall of a furnace or the metal shell of a boiler.

Convection. When heat flows by mixing or turbulence, the mechanism is known as convection. Convection occurs only in fluids. In practice, heat flow by conduction through fluids is restricted to certain cases of laminar flow or heat flow through thin films. Heat flow by convection only, unaccompanied by conduction, is never encountered, because of the laminar layer through which heat must pass by conduction. In the usual case of industrial heat transfer in fluids, conduction and convection occur together, and it is neither simple nor useful to attempt to separate them. In this text, the transfer of heat through a fluid is considered from the point of view of combined conduction and convection.

NATURAL AND FORCED CONVECTION. In many instances of heat transfer it is necessary to distinguish between natural and forced convection. If the mixing action is due to a current that is artificially generated by a pump, agitator, or other mechanical device, the action is called forced convection. If the mixing originates in currents set up when a mass of fluid, otherwise stationary, is heated, the effect is called natural or free convection. The two kinds of mixing are independent, and in certain cases, both natural and forced convection are simultaneously active.

Radiation. Radiation is a term given to the transfer of energy through space by means of electromagnetic waves. If radiation is passing through empty space, it is not transformed to heat or any other form of energy, nor is it diverted from its path. If, however, matter appears in its path, the radiation will be transmitted, reflected, or absorbed. It is only the absorbed energy that appears as heat, and this transformation is quantitative. For example, fused quartz transmits practically all of the radiation that strikes it; a polished opaque surface or mirror will reflect most of the radiation impinging on it; a black or matte surface will absorb most of the radiation received by it and will transform such absorbed energy quantitatively into heat.

Most gases are nearly transparent to radiation, and it is quite common to find that heat is flowing through a fluid both by radiation and by conduction-convection. Examples are the loss of heat from a radiator or unlagged steam pipe to the ambient air of the room and heat transfer in furnaces and other high-temperature gas-heating equipment. The two mechanisms are mutually independent and occur in parallel, so one type of heat flow can be controlled or varied independently of the other. Conduction-convection and radiation can be studied separately and their separate effects added together in cases where both are important. In very general terms, radiation becomes important at high temperatures and is independent of the circumstances of the flow of the fluid. Conduction-convection is sensitive to flow conditions and is relatively unaffected by temperature level.

CONDUCTION

Conduction is most easily understood by the study of conduction through solids, because in this case convection is not present. The basic law of heat transfer by conduction can be written in the form of the rate equation [Eq. (1-11)].

$$\text{Rate} = \frac{\text{driving force}}{\text{resistance}} \tag{8-1}$$

The driving force is the temperature drop across the solid, since it is apparent that heat can flow only when there is an inequality of temperature.

Fourier's Law. The resistance term in Eq. (8-1) is defined by means of Fourier's law. Consider a flat wall of thickness x and an area A. Let the temperature drop across the wall be Δt. Fourier's law states that the rate of heat flow through the wall is proportional to the temperature drop and to the area, and inversely proportional to the thickness of the wall. If Δt does not vary with time, the rate of heat flow is constant, and the process is one of *steady state*. If Q is the heat flow in the time θ, then Q/θ is the rate of heat flow, and Fourier's law can be written in mathematical form as

$$\frac{Q}{\theta} = \frac{kA\,\Delta t}{x} = q = \frac{\Delta t}{R} \tag{8-2}$$

where k is a proportionality constant and q replaces Q/θ. By comparing Eqs. (8-2) and (8-1), remembering that Δt is the driving force, it is seen that the resistance R is x/kA.

Thermal Conductivity. The proportionality constant k in Eq. (8-2) is a physical characteristic or property of the material from which the layer is made. It is called the thermal conductivity. In engineering units, Q is measured in Btu, θ in hours, A in square feet, t in degrees Fahrenheit, and x in feet. The units of k then become $Btu/(ft^2)(hr)(°F/ft)$ or, alternately, $Btu\text{-}ft/(ft^2)(hr)(°F)$. It must be clearly understood that area A is that of the plane *perpendicular* to the direction of the flow of heat and that x is the length of the path of the heat flow and is measured *parallel* to the direction of heat flow.

The numerical value of k depends upon the material of which the body is made and upon its temperature. The variation with temperature is not great, and for small values of Δt, k can be considered constant. In general, the variation of thermal conductivity is linear with temperature; i.e.,

$$k = a + bt \tag{8-3}$$

where a and b are constants and t is the temperature. For larger values of Δt, and where k changes significantly with temperature, it is mathematically correct to use an average value $\bar{k}$ for k, which can be found either by averaging the individual values of k for the two end temperatures or by calculating the arithmetical average temperature and using the value of k at that temperature.

Data showing typical values of thermal conductivity are shown in Appendixes 9, 10, and 11. The thermal conductivities of liquids and gases are small in comparison with those of most solids. For example, at 212°F the thermal conductivity of silver is 238 $Btu\text{-}ft/(ft^2)(hr)(°F)$, that of building brick is about 0.4, that of water is about 0.35, and that of air is about 0.017.

Accurate data on thermal conductivities are few. This is because thermal conductivity is difficult to measure accurately and because this property is very sensitive to small changes in chemical composition and to variations in physical structure. It is difficult to measure the value of k for a fluid because of the difficulty of eliminating convection currents.

Example 8-1. A layer of pulverized cork 6 in. thick is used as a layer of insulation in a flat wall. The temperature of the cold side of the cork is 40°F, and that of the warm side is 180°F. The thermal conductivity of the cork at 32°F is 0.021 $Btu\text{-}ft/(ft^2)(hr)(°F)$, and that at 200°F is 0.032. The area of the wall is 25 ft^2. What is the heat flow through the wall, in Btu per hour?

Solution. The arithmetic average temperature in the cork layer is $(40 + 180)/2 = 110°F$. By linear interpolation, the thermal conductivity at 110°F is

$$\bar{k} = 0.021 + \frac{(110 - 32)(0.032 - 0.021)}{200 - 32}$$

$$= 0.021 + 0.005 = 0.026 \ Btu\text{-}ft/(ft^2)(hr)(°F)$$

Also, $\quad A = 25 \ ft^2 \qquad \Delta t = 180 - 40 = 140°F \qquad x = \tfrac{6}{12} = 0.5 \ ft$

Substituting into Eq. (8-2) gives

$$= \frac{0.026 \times 25 \times 140}{0.5} = 182 \ Btu/hr$$

Compound Resistances in Series. Consider a flat wall constructed of a series of layers, as in Fig. 8-1. Let the thickness of the layers be x_A,

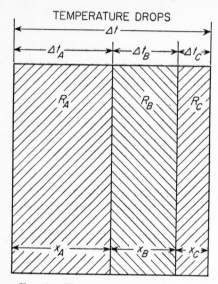

x_B, and x_C, and the conductivities of the materials of which the layers are made be k_A, k_B, and k_C, respectively. Furthermore, let the area of the compound wall, at right angles to the plane of the illustration, be A. Let Δt_A be the temperature drop across the first layer, Δt_B that across the second, and Δt_C that across the third. Let Δt be the temperature drop in all three layers. Therefore

$$\Delta t = \Delta t_A + \Delta t_B + \Delta t_C \quad (8\text{-}4)$$

It is desired, first, to derive a formula giving the rate of heat transfer through this series of resistances and, second, to determine what expression must be used for the over-all resistance if Δt is used as the over-all driving force.

FIG. 8-1. Thermal resistances in series.

Equation (8-4) can be written for each of the layers in the following form.

$$\Delta t_A = q_A \frac{x_A}{k_A A} \qquad \Delta t_B = q_B \frac{x_B}{k_B A} \qquad \Delta t_C = q_C \frac{x_C}{k_C A} \quad (8\text{-}5)$$

If Eqs. (8-5) are added together, Eq. (8-6) results.

$$\Delta t_A + \Delta t_B + \Delta t_C = \frac{q_A x_A}{A k_A} + \frac{q_B x_B}{A k_B} + \frac{q_C x_C}{A k_C} = \Delta t \quad (8\text{-}6)$$

Since all the heat that passes through the first resistance must pass through the second and in turn pass through the third, q_A, q_B, and q_C must be equal and can all be represented by q. Using this fact and solving for q,

$$q = \frac{\Delta t}{x_A/k_A A + x_B/k_B A + x_C/k_C A} = \frac{\Delta t}{R_A + R_B + R_C} \quad (8\text{-}7)$$

where R_A, R_B, and R_C are the resistances as defined on page 418. It is not necessary to memorize Eq. (8-7), because it is written in the form

$$\text{Rate} = \frac{\text{driving force}}{\text{resistance}}$$

and the over-all resistance is equal to the sum of the individual resistances, just as is the case in the flow of electric current through a series of resistances.

Example 8-2. A flat furnace wall is constructed of a 4.5-in. layer of Sil-o-cel brick, with a thermal conductivity of 0.08, backed by a 9-in. layer of common brick, of conductivity 0.8. The temperature of the inner face of the wall is 1400°F, and that of the outer face is 170°F. Calculate the heat loss through this wall in Btu per square foot per hour.

Solution. Thermal resistance has been defined as x/kA. Considering 1 ft^2 of wall ($A = 1$), the thermal resistances are for Sil-o-cel brick

$$R_A = \frac{4.5/12}{0.08 \times 1} = 4.687$$

and for common brick

$$R_B = \frac{9/12}{0.8 \times 1} = 0.938$$

Since the total resistance is the sum of the individual resistances,

$$R = 4.687 + 0.938 = 5.625$$

$$\text{Rate of heat flow} = \frac{\text{temperature drop}}{\text{resistance}}$$

Hence,
$$\frac{q}{A} = \frac{1400 - 170}{5.625} = 219 \text{ Btu/ft}^2\text{-hr}$$

It is often useful to recall the analogies between the flow of heat and the steady flow of electricity through a conductor. The flow of heat is covered by the expression

$$\text{Rate} = \frac{\text{temperature drop}}{\text{resistance}}$$

In the flow of electricity the potential factor is the electromotive force, and the rate of flow is coulombs per second, or amperes. The rate equation for electrical flow is

$$\text{Amperes} = \frac{\text{volts}}{\text{ohms}}$$

By comparing this equation with Fourier's law it is seen that rate of flow in Btu per hour is analogous to amperes, temperature drop to voltage, and thermal resistance to electrical resistance. The various units for the electrical circuit have been given names such as amperes, volts, and ohms, while the corresponding units for heat flow have never been given names.

The rate of flow of heat through several resistances in series has been shown to be analogous to the current flowing through several electrical resistances in series. In an electric circuit the potential drop over any one of several resistances is to the total potential drop in the circuit as the individual resistances are to the total resistance. In the same way the potential drops in a thermal circuit, which are the temperature differences, are to the total temperature drop as the individual thermal resistances are to the total thermal resistance. This may be expressed mathematically as

$$\frac{\Delta t}{R} = \frac{\Delta t_A}{R_A} = \frac{\Delta t_B}{R_B} = \frac{\Delta t_C}{R_C} \tag{8-8}$$

Example 8-3. In Example 8-2, what is the temperature of the interface between the refractory brick and the common brick?

Solution. The temperature drop in any one of a series of resistances is to the individual resistance as the total drop is to the total resistance, or

$$\frac{\Delta t_A}{4.687} = \frac{1400 - 170}{5.625}$$

from which,

$$\Delta t_A = 1025°F$$

The temperature of the interface is $1400 - 1025 = 375°F$.

Heat Flow through a Cylinder. Consider the hollow cylinder represented by Fig. 8-2. The inside radius of the cylinder is r_i, the outside radius is r_o, and the length of the cylinder is L. The thermal conductivity of the material of which the cylinder is made is k. The temperature of the inside surface is t_i, and that of the outside is t_o. It will be assumed that t_o is larger than t_i and hence that heat is flowing from the outside to the inside. It is desired to calculate the rate of heat flow for this case.

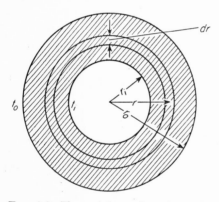

Consider a very thin cylinder, concentric with the main cylinder, of radius r, where r is between r_i and r_o. The thickness of the wall of this cylinder is dr; and if dr is small enough with respect to r so that the lines of heat flow may be considered parallel, Eq. (8-2) can be applied and written in the form

Fig. 8-2. Flow of heat through thick-walled cylinder.

$$q = k \frac{dt}{dr} 2\pi r L \tag{8-9}$$

since the area perpendicular to the heat flow is equal to $2\pi r L$ and the x of Eq. (8-2) is equal to dr. In order to integrate Eq. (8-9) it is necessary only to separate the variables t and r as follows.

$$\frac{dr}{r} = \frac{2\pi L k}{q} dt \tag{8-10}$$

Since all quantities except dr and dt are constant, Eq. (8-10) can be integrated as follows.

$$\int_{r_i}^{r_o} \frac{dr}{r} = \frac{2\pi Lk}{q} \int_{t_i}^{t_o} dt$$

$$\ln r_o - \ln r_i = \frac{2\pi Lk}{q}(t_o - t_i)$$

$$q = \frac{k(2\pi L)(t_o - t_i)}{\ln (r_o/r_i)} \tag{8-11}$$

Eq. (8-11) can be used to calculate the flow of heat through a thick-walled cylinder. It can be put in a more convenient form by expressing the rate of flow of heat as

$$q = \frac{k\bar{A}_L(t_o - t_i)}{r_o - r_i} \tag{8-12}$$

This is of the same general form as the equation for heat flow through a flat wall [Eq. (8-2)] with the exception of $\bar{A}_L$, which must be so chosen

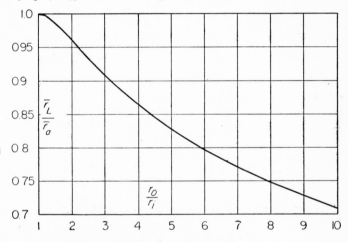

FIG. 8-3. Relation between logarithmic and arithmetic means.

that the equation is correct. The term $\bar{A}_L$ can be determined by equating the right-hand sides of Eqs. (8-11) and (8-12) and solving for $\bar{A}_L$.

$$\bar{A}_L = \frac{2\pi L(r_o - r_i)}{\ln (r_o/r_i)} \tag{8-13}$$

Note from Eq. (8-13) that $\bar{A}_L$ is the area of a cylinder of length L and radius $\bar{r}_L$, where

$$\bar{r}_L = \frac{r_o - r_i}{\ln (r_o/r_i)} = \frac{r_o - r_i}{2.303 \log (r_o/r_i)} \tag{8-14}$$

The form of the right-hand side of Eq. (8-14) is important enough to repay memorizing. It is known as the *logarithmic mean*, and in the particular case of Eq. (8-14), $\bar{r}_L$ is called the *logarithmic mean radius*. It is the radius which, when applied to the integrated equation for a flat wall, will give the correct rate of heat flow through a thick-walled cylinder.

The logarithmic mean is less convenient than the arithmetic mean, and the latter can be used without appreciable error for thin-walled tubes where r_o/r_i is nearly 1. The ratio of the logarithmic mean $\bar{r}_L$ to the arithmetic mean $\bar{r}_a$ is a function of r_o/r_i, as shown in Fig. 8-3. Thus, when $r_o/r_i = 2$, the logarithmic mean is $0.96\bar{r}_a$, and the error in the use of the arithmetic mean is 4 per cent. The error is 1 per cent when $r_o/r_i = 1.4$.

Example 8-4. A tube, 2.5 in. OD, is lagged with a 2-in. layer of asbestos (conductivity 0.12), which is followed with a 1.5-in. layer of cork (conductivity 0.03). If the temperature of the outer surface of the pipe is 290°F and the temperature of the outer surface of the cork is 90°F, calculate the heat loss in Btu per hour per foot of pipe.

Solution. These layers are too thick to use the arithmetic mean radius, and the logarithmic mean should be used. For asbestos

$$\bar{r}_L = \frac{3.25 - 1.25}{2.303 \log (3.25/1.25)} = 2.09 \text{ in.}$$

and for cork

$$\bar{r}_L = \frac{4.75 - 3.25}{2.303 \log (4.75/3.25)} = 3.95 \text{ in.}$$

Calling asbestos substance A and cork substance B,

$$R_A = \frac{x_A}{k_A A_A} = \frac{2/12}{0.12 \times 2\pi(2.09/12)L} = \frac{1.269}{L}$$

$$R_B = \frac{x_B}{k_B A_B} = \frac{1.5/12}{0.03 \times 2\pi(3.95/12)L} = \frac{2.015}{L}$$

$$q/L = \frac{290 - 90}{1.269 + 2.015} = \frac{200}{3.284} = 60.9 \text{ Btu/ft-hr}$$

HEAT FLOW IN FLUIDS BY CONDUCTION AND CONVECTION

The flow of heat from a fluid through a solid wall to a cooler fluid is often encountered in chemical engineering practice. The heat transferred may be latent heat accompanying phase changes such as condensation or vaporization, or it may be sensible heat coming from increasing or decreasing the temperature of a fluid without phase change. Typical examples are reducing the temperature of a fluid by transfer of sensible heat to a cooler fluid, the temperature of which is increased thereby; condensing steam by cooling water; and vaporizing water from a solution at a given pressure by condensing steam at a higher pressure. All such cases require that heat be transferred by conduction and convection.

Typical Heat-exchange Equipment. To establish a basis for specific discussion of heat transfer to and from flowing fluids, consider the simple tubular heater of Fig. 8-4. The function of the heater is to increase the temperature of a fluid by transferring to it the latent heat of condensation of vapor. The unit can equally well be called a condenser if the emphasis is placed on condensing the vapor and heating the fluid is considered incidental.

The heater consists essentially of a bundle of parallel tubes A, the ends of which are expanded into tube sheets B_1 and B_2. The tube bundle is inside a cylindrical shell C and is provided with two channels D_1 and D_2,

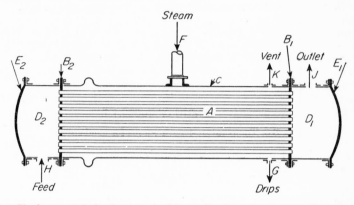

FIG. 8-4. Single-pass tubular heater: A, tubes. B_1, B_2, tube sheets. C, shell. D_1, D_2, channels. E_1, E_2, channel covers. F, steam inlet. G, condensate outlet. H, liquor inlet. J, liquor outlet. K, noncondensed-gas vent.

one at each end, and two channel covers E_1 and E_2. Steam or other vapor is introduced through nozzle F into the shell-side space surrounding the tubes, condensate is withdrawn through drip connection G, and any noncondensable gas that might enter with the inlet vapor is removed through vent K. The fluid to be heated is pumped through connection H into channel D_2. It flows through the tubes into the other channel D_1, and is discharged through connection J. The two fluids are physically separated but are in thermal contact with the thin metal tube walls separating them. Heat flows through the tube walls from the condensing vapor to the cooler fluid in the tubes.

If the vapor entering the heater is not superheated, and if the condensate is not subcooled below its boiling temperature, the temperature throughout the shell side of the heater is constant. This is because the temperature is that of condensing vapor at the pressure of the shell-side space, and the pressure in that space is constant. The temperature of the fluid in the tubes increases continuously as the fluid flows through the tubes.

The temperatures of the condensing vapor and of the liquid are plotted against the tube length in Fig. 8-5. The horizontal line represents the temperature of the condensing vapor, and the curved line below it the rising temperature of the tube-side fluid. In Fig. 8-5, the inlet and outlet fluid temperatures are t_{ca} and t_{cb}, respectively, and the constant temperature of the vapor is t_h. At a length L from the entrance end of the tubes, the fluid temperature is t_c, and the local difference between the tempera-

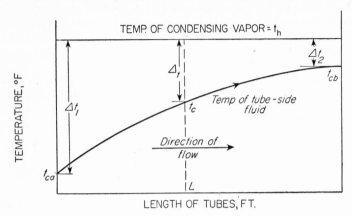

FIG. 8-5. Temperature-length curves for condenser.

tures of vapor and fluid is $t_h - t_c$. This temperature difference is called a point temperature difference, and is denoted by Δt. The point temperature difference at the inlet of the tubes is $t_h - t_{ca}$, denoted by Δt_1, and that at the exit end is $t_h - t_{cb}$, denoted by Δt_2. The terminal point temperature differences Δt_1 and Δt_2 are called the *approaches*.

The change in temperature of the fluid, $t_{cb} - t_{ca}$, is called the temperature range or, simply, the *range*. In a condenser or heater there is but one range, that of the cold fluid being heated.

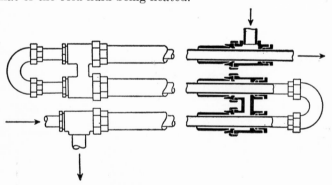

FIG. 8-6. Double-pipe heat exchanger.

A second example of simple heat-transfer equipment is the double-pipe exchanger shown in Fig. 8-6. It is assembled of standard iron pipe and standardized return bends and return heads, the latter equipped with stuffing boxes. One fluid flows through the inside pipe and the second fluid through the annular space between the outside and the inside pipe. The function of a heat exchanger is to increase the temperature of a cooler fluid and decrease that of a hotter fluid. In a typical exchanger, the inner pipe may be $1\frac{1}{4}$ in. and the outer pipe $2\frac{1}{2}$ in., both IPS. Such an exchanger may consist of several passes arranged in a vertical stack. Double-pipe exchangers are useful when not more than 100 to 150 ft^2 of surface are required. For larger capacities, more elaborate shell-and-tube exchangers, containing hundreds of square feet of area, and described on pages 499 to 505, are used.

Countercurrent and Parallel-current Flows. The two fluids enter at different ends of the exchanger shown in Fig. 8-6 and pass in opposite directions through the unit. This type of flow is that commonly used and is

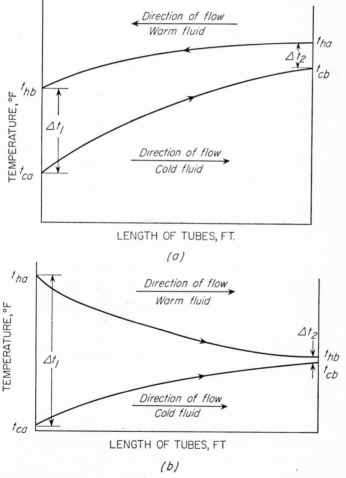

Fig. 8-7. Temperature in countercurrent and parallel flow: (a) Countercurrent flow. (b) Parallel-current flow.

called countercurrent flow. The temperature-length curves for this case are shown in Fig. 8-7a. The four terminal temperatures are denoted as follows:

Temperature of entering hot fluid, t_{ha}
Temperature of leaving hot fluid, t_{hb}
Temperature of entering cold fluid, t_{ca}
Temperature of leaving cold fluid, t_{cb}

The approaches are

$$t_{ha} - t_{cb} = \Delta t_2 \quad \text{and} \quad t_{hb} - t_{ca} = \Delta t_1 \tag{8-15}$$

The warm-fluid and cold-fluid ranges are $t_{ha} - t_{hb}$ and $t_{cb} - t_{ca}$, respectively.

If the two fluids enter at the same end of the exchanger and flow in the same direction to the other end, the flow is called parallel-current flow. The temperature-length curves for parallel flow are shown in Fig. 8-7b. Again, the subscript a refers to the entering fluids and subscript b to the leaving fluids. The approaches are $\Delta t_1 = t_{ha} - t_{ca}$ and $\Delta t_2 = t_{hb} - t_{cb}$.

Parallel-current flow is rarely used in a single-pass exchanger such as that shown in Fig. 8-6, because, as inspection of Fig. 8-7a and b will show, it is not possible with this method of flow to bring the exit temperature of one fluid nearly to the entrance temperature of the other, and the heat that can be transferred is less than that possible in countercurrent flow. In the multipass exchangers, described on pages 501 to 505, parallel flow is used in some passes, largely for mechanical reasons, and the capacity and approaches obtainable are thereby affected. Parallel flow is used in special situations where it is necessary to limit the maximum temperature of the cooler fluid or where it is important to change the temperature of at least one fluid rapidly.

Enthalpy Balances in Heat-exchange Equipment. In heat exchangers there is no shaft work, and mechanical-potential and mechanical-kinetic energies are small in comparison with the other terms in the enthalpy-balance equation. Equation (1-23) may therefore be used. It is convenient to apply this equation on an hour basis. Thus, for one stream through the exchanger

$$w(i_b - i_a) = q \tag{8-16}$$

where w = flow rate of stream, lb/hr
q = heat flow into stream, Btu/hr
i_a, i_b = enthalpies of stream at entrance and exit, respectively, Btu/lb
Equation (8-16) may be written for each stream flowing through the exchanger.

A further simplification in the use of the heat flow rate q is justified. One of the two fluid streams, that outside the tubes, can gain or lose heat by transfer with the ambient air if the fluid is colder or hotter than the ambient. Heat flow to or from the ambient is not desired in practice, and it is usually reduced to a small magnitude by suitable lagging. It is customary to neglect it in comparison with the heat transfer through the walls of the tubes from the warm fluid to the cold fluid, and q is interpreted accordingly.

Accepting the above assumptions, Eq. (8-16) may be written for the warm fluid as

$$w_h(i_{hb} - i_{ha}) = q_h \qquad (8\text{-}17)$$

and for the cold fluid as

$$w_c(i_{cb} - i_{ca}) = q_c \qquad (8\text{-}18)$$

where w_h = mass flow rate of warm fluid, lb/hr

w_c = mass flow rate of cold fluid, lb/hr

i_{hb} = enthalpy of leaving warm fluid, Btu/lb

i_{ha} = enthalpy of entering warm fluid, Btu/lb

i_{cb} = enthalpy of leaving cold fluid, Btu/lb

i_{ca} = enthalpy of entering cold fluid, Btu/lb

q_h = heat added to warm fluid, Btu/hr (The sign of q_h is obviously negative, since the warm fluid loses, rather than gains, heat.)

q_c = heat added to the cold fluid, Btu/hr (The sign of q_c is positive.)

The heat lost by the warm fluid is gained by the cold fluid, and

$$q_c = -q_h$$

Therefore, from Eqs. (8-17) and (8-18),

$$w_h(i_{ha} - i_{hb}) = w_c(i_{cb} - i_{ca}) = q \qquad (8\text{-}19)$$

Equation (8-19) is called the over-all enthalpy balance.

If constant specific heats are assumed, the over-all enthalpy balance for a heat exchanger becomes

$$w_h c_{ph}(t_{ha} - t_{hb}) = w_c c_{pc}(t_{cb} - t_{ca}) = q \qquad (8\text{-}20)$$

and for a condenser, from Eq. (1-18),

$$w_h \lambda = w_c c_{pc}(t_{cb} - t_{ca}) = q \qquad (8\text{-}21)$$

where w_h = rate of condensation of vapor, lb/hr

λ = latent heat of vaporization of vapor, Btu/lb

c_{pc} = specific heat of cold fluid, Btu/(lb)(°F)

c_{ph} = specific heat of warm fluid, Btu/(lb)(°F)

Equation (8-21) is based on the assumption that the vapor enters the condenser as saturated vapor (no superheat) and that the condensate leaves at condensing temperature without being further cooled. If either of these sensible heat effects is important, it must be accounted for by an added term in the left-hand side of Eq. (8-21). For example, if the condensate leaves at a temperature t_{hb} which is less than t_h, the condensing temperature of the vapor, Eq. (8-21) must be written

$$w_h[\lambda + c_{ph}(t_{ha} - t_{hb})] = w_c c_{pc}(t_{cb} - t_{ca}) \qquad (8\text{-}22)$$

where c_{ph} is now the specific heat of the condensate.

RATE OF HEAT TRANSFER

The enthalpy relationships [Eqs. (8-17) to (8-22)] relate the rate of heat transfer q to the rates of flow of the fluids and to the changes in enthalpy and temperature of the fluids. They do not help in the solution of such important engineering problems as the amount of heat a given heat exchanger can transfer or the size of heat-transfer unit needed to transfer a required amount of heat. The methods used to attack these problems are the subject of the next sections.

Heat-transfer Areas. Heat-transfer calculations are based on the area of the heating surface and are expressed in Btu per hour per square foot of surface through which the heat flows. Except in certain specialized equipment, heating surfaces are constructed from tubes or pipe. Heat-transfer rates may be based either on the inside area or on the outside area of the tubes. Although the choice is arbitrary, it must be clearly stated, because the numerical magnitudes of the heat-transfer rates per unit area will not be equal for the two choices.

Average Temperature of Fluid Stream. When a fluid is being heated or cooled, the temperature will vary throughout the cross section of the stream. If the fluid is being heated, the temperature of the fluid is a maximum at the wall of the heating surface and decreases toward the center of the stream. If the fluid is being cooled, the temperature is a minimum at the wall and increases toward the center. Because of these temperature gradients throughout the cross section of the stream, it is necessary, for definiteness, to state what is meant by the temperature of the stream. It is agreed that it is the temperature that would be attained if the entire fluid stream flowing across the section in question were withdrawn and mixed adiabatically to a uniform temperature. The temperature so defined is called the average stream temperature. The temperatures used in Eqs. (8-20) to (8-22) conform to this definition, and those plotted in Fig. 8-7 are all average fluid temperatures also.

Over-all Heat-transfer Coefficient. The rate of heat transfer per unit area is referred to as the *heat flux density*. It is usually expressed in Btu per square foot per hour. By Eq. (1-11), the flux must be proportional to a driving force. In heat flow, the driving force is taken as $t_h - t_c$, where t_h is the average temperature of the hot fluid and t_c is that of the cold fluid. The quantity $t_h - t_c$ is the *over-all local temperature difference*. It is denoted by Δt. It is clear from Fig. 8-7 that Δt can vary considerably from point to point along the tube, and, therefore, since q/A is proportional to Δt, the flux density also varies with tube length. It is necessary to start with a differential equation, by focusing attention on a differential area dA through which a differential heat flow dq occurs under the driving force of a local value of Δt. The local flux density is then dq/dA, and is related to the local value of Δt by the equation

$$\frac{dq}{dA} = U \, \Delta t = U(t_h - t_c) \tag{8-23}$$

The quantity U, defined by Eq. (8-23) as a proportionality factor between dq/dA and Δt, is called the *local over-all heat-transfer coefficient*.

To complete the definition of U in a given case, it is necessary to specify the area. If A is taken as the outside tube area A_o, U becomes a coefficient based on that area, and is written U_o. Likewise, if the inside area A_i is chosen, the coefficient is also based on that area, and is denoted by U_i. Since Δt and dq are independent of the choice of area, it follows that

$$\frac{U_o}{U_i} = \frac{dA_i}{dA_o} = \frac{D_i}{D_o} \tag{8-24}$$

where D_i and D_o are the inside and outside tube diameters, respectively.

Integration over Total Heating Surface. To apply Eq. (8-23) to the entire area of a heat exchanger, the equation must be integrated. In general

$$\int_{\Delta t_1}^{\Delta t_2} \frac{dq}{U\,\Delta t} = \int_0^{A_T} dA = A_T \tag{8-25}$$

where A_T = total area of heating surface

Δt_2, Δt_1 = approaches

In the general case, U, q, and Δt vary along the heat-exchange surface, and to integrate Eq. (8-25) all must be expressed in terms of a single variable, which usually is the temperature of one of the streams. In an important special case where certain assumptions can be accepted, a simple result is obtained. The assumptions are: (1) the over-all coefficient U is constant, (2) the specific heats c_{pc} and c_{ph} are constant, (3) heat exchange with the ambient is negligible, (4) the flow is steady and either parallel or countercurrent, as shown in Fig. 8-6. The result cannot be used unchanged for multipass exchangers.

The most questionable of these assumptions is that of a constant over-all coefficient. The coefficient does in fact vary with the temperatures of the fluids, but its change with temperature is gradual, so when the temperature ranges are moderate, the assumption of constant U is not seriously in error.

By assumptions (2) and (4), $c_{pc}w_c$ and $c_{ph}w_h$ are constant, and if t_c and t_h are plotted against q, as shown in Fig. 8-8, straight lines are obtained. Since t_c and t_h vary linearly with q, Δt does likewise, and $d(\Delta t)/dq$, the slope of the graph of Δt vs. q, is constant. Therefore

$$\frac{d(\Delta t)}{dq} = \frac{\Delta t_2 - \Delta t_1}{q} \tag{8-26}$$

where q is the total heat transferred. Elimination of dq from Eqs. (8-23) and (8-26) gives

$$\frac{d(\Delta t)}{U\,\Delta t\,dA} = \frac{\Delta t_2 - \Delta t_1}{q} \tag{8-27}$$

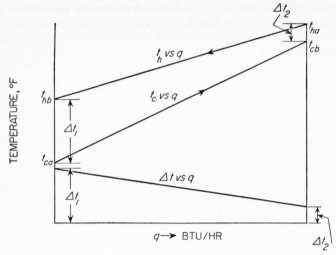

FIG. 8-8. Temperature vs. heat flow in countercurrent flow.

The variables Δt and A can be separated, and if U is constant, the equation can be integrated over the limits A_T and 0 for A and Δt_2 and Δt_1 for Δt.

$$\int_{\Delta t_1}^{\Delta t_2} \frac{d(\Delta t)}{\Delta t} = \frac{U(\Delta t_2 - \Delta t_1)}{q} \int_0^{A_T} dA$$

or

$$\ln \frac{\Delta t_2}{\Delta t_1} = \frac{U(\Delta t_2 - \Delta t_1)}{q} A_T \qquad (8\text{-}28)$$

Equation (8-28) can be written

$$q = UA_T \frac{\Delta t_2 - \Delta t_1}{\ln (\Delta t_2/\Delta t_1)} = UA_T \, \overline{\Delta t_L} \qquad (8\text{-}29)$$

where

$$\overline{\Delta t_L} = \frac{\Delta t_1 - \Delta t_2}{\ln (\Delta t_1/\Delta t_2)} = \frac{\Delta t_1 - \Delta t_2}{2.303 \log (\Delta t_1/\Delta t_2)} \qquad (8\text{-}30)$$

Equation (8-30) defines the *logarithmic mean temperature drop*. It is of the same form as that of Eq. (8-14) for the logarithmic mean radius of a thick-walled tube. When Δt_1 and Δt_2 are nearly equal, their arithmetic average may be used for $\overline{\Delta t_L}$ within the same limits of accuracy given for Eq. (8-14), as shown in Fig. 8-3.

If one of the fluids is at constant temperature, as in a heater or condenser, no difference exists among countercurrent flow, parallel flow, or multipass flow, and Eq. (8-30) applies to all of them.

Variable Over-all Coefficient. The use of Eq. (8-30), which depends upon the assumption of constancy of U, leads to appreciable error if U

varies considerably from one end of the exchanger to the other. A more accurate calculation can be made by using Eq. (8-31), which is based on the assumption that U varies linearly with temperature drop over the entire heating surface.[7]

$$q = A_T \frac{U_2\,\Delta t_1 - U_1\,\Delta t_2}{\ln\,(U_2\,\Delta t_1/U_1\,\Delta t_2)} \tag{8-31}$$

where U_1, U_2 = local over-all coefficients at ends of exchanger, Btu/$(\text{ft}^2)(\text{hr})(°F)$

Δt_1, Δt_2 = temperature approaches at corresponding ends of exchanges, °F

Equation (8-31) calls for use of a logarithmic mean value of the $U\,\Delta t$ cross product, where the over-all coefficient at one end of the exchanger is multiplied by the temperature approach at the other. The derivation of this equation requires that assumptions (2), (3), and (4) above be accepted.

In countercurrent flow, Δt_1, the cold-end approach, may be less than Δt_2, the warm-end approach. In this case, for convenience and to eliminate negative numbers and logarithms, the subscripts in Eqs. (8-30) and (8-31) can be interchanged.

Individual Heat-transfer Coefficients

The over-all coefficient depends upon so many variables that it is necessary to break it into its parts. The reason for this becomes apparent if a typical case is examined. Consider, for this purpose, the local over-all coefficient at a specific point in the double-pipe exchanger shown in Fig. 8-6. For definiteness, assume that the warm fluid is flowing through the inside pipe and that the cold fluid is flowing through the annular space. Assume also that the Reynolds numbers of the two fluids are sufficiently large to ensure turbulent flow and that both surfaces of the inside tube are clear of dirt or scale. If, now, a plot is prepared, as shown in Fig. 8-9, with temperature as the ordinate and distance perpendicular to the wall as the abscissa, several important facts become evident. In the figure, the metal wall of the tube separates the warm fluid on the right from the cold fluid on the left. The change in temperature with distance is shown by the broken line $t_a t_b t_{wh} t_{wc} t_e t_g$. The temperature gradient is thus divided into three separate parts, one through each of the two fluids and the other through the metal wall. The over-all effect, therefore, should be studied in terms of these individual parts.

It has been shown in Chap. 2 that, in turbulent flow, three zones exist, even in a single fluid, so the study of one fluid is, itself, complicated. In each fluid shown in Fig. 8-9 there is a thin laminar sublayer at the wall, a turbulent core occupying most of the cross section of the stream, and a buffer zone between them. The velocity gradients have been described in Chap. 2. The velocity gradient is large near the wall, small in the turbulent core, and in rapid change in the buffer zone. It has been found that the temperature gradient in a fluid being heated or cooled when flowing in turbulent flow follows much the same course. The temperature gradient is large at the wall and through the laminar layer, small in the turbulent core, and in rapid change in the buffer zone. Basically, the reason for this is that heat must flow through the laminar layer by conduction, which

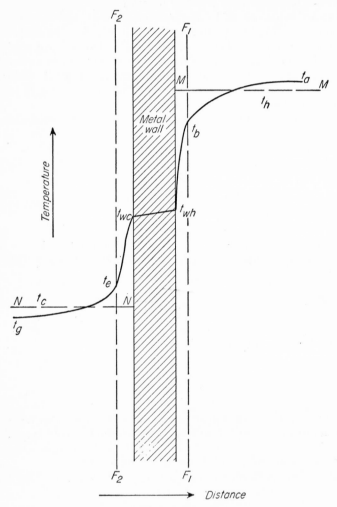

FIG. 8-9. Temperature gradients in forced convection.

calls for a steep temperature gradient in most fluids because of the low thermal conductivity, whereas the rapidly moving eddies in the core are effective in equalizing the temperature in the turbulent zone. In Fig. 8-9 the dotted lines F_1F_1 and F_2F_2 represent the boundaries of the laminar layers.

The average temperature of the warm stream is somewhat less than the maximum temperature t_a and is represented by the horizontal line MM, which is drawn at temperature t_h. Likewise, line NN, drawn at temperature t_c, represents the average temperature of the cold fluid.

The over-all resistance to the flow of heat from the warm fluid to the cold fluid is a result of three separate resistances operating in series. Two resistances are those offered by the individual fluids, and the third is that of the solid wall. In general, also, as shown in Fig. 8-9, the wall resistance is small in comparison with that of the fluids. The over-all coefficient is best studied by analyzing it in terms of the separate resistances and treating each separately. The separate resistances can then be combined to form the over-all coefficient. This approach requires the use of individual heat-transfer coefficients.

The individual, or surface, heat-transfer coefficient, denoted by h, is defined generally by the equation

$$h = \frac{dq/dA}{t - t_w} \tag{8-32}$$

where dq/dA = local heat-flux density, based on area in contact with fluid, $Btu/(ft^2)(hr)$
t = local average temperature of fluid, °F
t_w = temperature of wall in contact with fluid, °F

Equation (8-32), when applied to the two fluids of Fig. 8-9, becomes, for the warm side (inside of tube),

$$h_i = \frac{dq/dA_i}{t_h - t_{wh}} \tag{8-33}$$

and for the cold side (outside of tube),

$$h_o = \frac{dq/dA_o}{t_{wc} - t_c} \tag{8-34}$$

where A_i and A_o are the inside and outside areas of the tube, respectively.

Calculation of Over-all Coefficients from Individual Coefficients. The over-all coefficient is constructed from the individual coefficients and the resistance of the tube wall in the following manner.

The rate of heat transfer through the tube wall is given by the differential form of Eq. (8-12),

$$\frac{dq}{d\bar{A}_L} = \frac{k_m(t_{wh} - t_{wc})}{x_w} \tag{8-35}$$

where $t_{wh} - t_{wc}$ = temperature difference through tube wall, °F
k_m = thermal conductivity of wall, $Btu\text{-}ft/(ft^2)(hr)(°F)$
x_w = tube-wall thickness, ft
$dq/d\bar{A}_L$ = local heat-flux density, based on logarithmic mean of inside and outside areas of tube, $Btu/(ft^2)(hr)$

If Eqs. (8-33) to (8-35) are solved for the temperature differences and the temperature differences added, the result is

$$(t_h - t_{wh}) + (t_{wh} - t_{wc}) + (t_{wc} - t_c)$$

$$= t_h - t_c = \Delta t$$

$$= dq \left(\frac{1}{dA_i \, h_i} + \frac{x_w}{d\bar{A}_L \, k_m} + \frac{1}{dA_o \, h_o} \right) \qquad (8\text{-}36)$$

Assume that the heat-transfer rate is arbitrarily based on the outside area. If Eq. (8-36) is solved for dq, and if both sides of the resulting equation are divided by dA_o, the result is

$$\frac{dq}{dA_o} = \frac{t_h - t_c}{\dfrac{dA_o}{dA_i \, h_i} + \dfrac{x_w}{k_m} \dfrac{dA_o}{d\bar{A}_L} + \dfrac{1}{h_o}} \qquad (8\text{-}37)$$

Now

$$\frac{dA_o}{dA_i} = \frac{D_o}{D_i} \quad \text{and} \quad \frac{dA_o}{d\bar{A}_L} = \frac{D_o}{\bar{D}_L}$$

where D_o, D_i, and $\bar{D}_L$ are the outside, inside, and logarithmic mean diameters of the tube, respectively. Therefore

$$\frac{dq}{dA_o} = \frac{t_h - t_c}{\dfrac{D_o}{D_i h_i} + \dfrac{x_w}{k_m} \dfrac{D_o}{\bar{D}_L} + \dfrac{1}{h_o}} \qquad (8\text{-}38)$$

Comparing Eq. (8-23) with Eq. (8-38) shows that

$$U_o = \frac{1}{\dfrac{D_o}{D_i h_i} + \dfrac{x_w}{k_m} \dfrac{D_o}{\bar{D}_L} + \dfrac{1}{h_o}} \qquad (8\text{-}39)$$

If the inside area A_i is chosen as the base area, division of Eq. (8-36) by dA_i gives for the over-all coefficient

$$U_i = \frac{1}{\dfrac{1}{h_i} + \dfrac{D_i}{\bar{D}_L} \dfrac{x_w}{k_m} + \dfrac{D_i}{D_o h_o}} \qquad (8\text{-}40)$$

Resistance Form of Over-all Coefficient. A comparison of Eqs. (8-7) and (8-38) suggests that the reciprocal of an over-all coefficient can be considered to be an over-all resistance, composed of three resistances in series. The total, or over-all, resistance is given by the equation

$$\frac{1}{U_o} = \frac{D_o}{D_i h_i} + \frac{x_w}{k_m} \frac{D_o}{\bar{D}_L} + \frac{1}{h_o} \qquad (8\text{-}41)$$

The individual terms on the right-hand side of Eq. (8-41) represent the individual resistances of the two fluids and of the metal wall. The over-all temperature drop is proportional to $1/U$, and the temperature drops in the two fluids and the wall are proportional to the individual resistances, or, for the case of Eq. (8-41),

$$\frac{\Delta t}{1/U_o} = \frac{\Delta t_i}{D_o/D_i h_i} = \frac{\Delta t_w}{(x_w/k_m)(D_o/\overline{D}_L)} = \frac{\Delta t_o}{1/h_o} \tag{8-42}$$

where Δt = over-all temperature drop
Δt_i = temperature drop through inside fluid
Δt_w = temperature drop through metal wall
Δt_o = temperature drop through outside fluid

Fouling Factors. In actual service, heat-transfer surfaces do not remain clean. Scale, dirt, and other solid deposits form on one or both sides of the tubes, provide additional resistances to heat flow, and reduce the over-all coefficient. The effect of such deposits is taken into account by adding a term $1/dA\, h_d$ to the bracketed term of Eq. (8-36) for each scale deposit. Thus, assuming that scale deposits on both the inside and the outside surface of the tubes, Eq. (8-36) becomes, after correction for the effects of scale,

$$\Delta t = dq \left(\frac{1}{dA_i\, h_{di}} + \frac{1}{dA_i\, h_i} + \frac{x_w}{d\overline{A}_L\, k_m} + \frac{1}{dA_o\, h_o} + \frac{1}{dA_o\, h_{do}} \right) \tag{8-43}$$

where h_{di} and h_{do} are the fouling factors for the scale deposits on the inside and outside tube surfaces, respectively. The following equations for the over-all coefficients based on outside and inside areas, respectively, follow from Eq. (8-43).

$$U_o = \frac{1}{\dfrac{D_o}{D_i h_{di}} + \dfrac{D_o}{D_i h_i} + \dfrac{x_w}{k_m}\dfrac{D_o}{\overline{D}_L} + \dfrac{1}{h_o} + \dfrac{1}{h_{do}}} \tag{8-44}$$

and

$$U_i = \frac{1}{\dfrac{1}{h_{di}} + \dfrac{1}{h_i} + \dfrac{x_w}{k_m}\dfrac{D_i}{\overline{D}_L} + \dfrac{D_i}{D_o h_o} + \dfrac{D_i}{D_o h_{do}}} \tag{8-45}$$

The actual thicknesses of the deposits are neglected in Eqs. (8-44) and (8-45).

Numerical values of fouling factors corresponding to "satisfactory performance in normal operation, with reasonable service time between cleanings" [29d, 33] are recommended. They cover a range of approximately 100 to 2000 Btu/(ft^2)(hr)(°F). The smaller value applies to polluted water and the larger to sea water or organic vapors. Fouling factors for ordinary industrial liquids fall in the range 300 to 1000. Factors for scale deposited from solutions possessing inverted solubility curves, such as those described in Chap. 9, are not included.

Example 8-5. Methyl alcohol flowing in the inner pipe of a double-pipe exchanger is cooled with water flowing in the jacket. The inner pipe is made from Schedule 40 1-in. steel pipe. The thermal conductivity of steel is 26 Btu-ft/ $(ft^2)(hr)(°F)$. The film coefficients and fouling factors, in $Btu/(ft^2)(hr)(°F)$, are:

Alcohol film coefficient:	$h_i = 180$
Water film coefficient:	$h_o = 300$
Inside fouling factor:	$h_{di} = 1000$
Outside fouling factor:	$h_{do} = 500$

What is the over-all coefficient, based on the outside area of the inner pipe?

Solution. The diameters and wall thickness of Schedule 40 1-in. pipe, from Appendix 4, are

$$D_i = \frac{1.049}{12} = 0.0874 \text{ ft}$$

$$D_o = \frac{1.315}{12} = 0.1096 \text{ ft}$$

$$x_w = \frac{0.133}{12} = 0.0111 \text{ ft}$$

The logarithmic mean diameter D_L is calculated as in Eq. (8-14).

$$\bar{D}_L = \frac{D_o - D_i}{2.303 \log (D_o/D_i)} = \frac{0.1096 - 0.0874}{2.303 \log (0.1096/0.0874)} = 0.0983 \text{ ft}$$

The over-all coefficient is found from Eq. (8-44).

$$U_o = \frac{1}{\dfrac{0.1096}{0.0874 \times 1000} + \dfrac{0.1096}{0.0874 \times 180} + \dfrac{0.0111 \times 0.1096}{26 \times 0.0983} + \dfrac{1}{300} + \dfrac{1}{500}}$$

$$= 71.3 \text{ Btu}/(ft^2)(hr)(°F)$$

Special Cases of the Over-all Coefficient. Although the choice of area to be used as the basis of an over-all coefficient is arbitrary, sometimes one particular area is more convenient than others. Suppose, for example, that one individual coefficient, h_i, is large numerically in comparison with the other, h_o, and that fouling effects are negligible. Also, the term representing the resistance of the metal wall is usually small in comparison with $1/h_o$, the ratios D_o/D_i and $D_o/\bar{D}_L$ have so little significance that they can be disregarded, and Eq. (8-39) can be replaced by the simpler form

$$U_o = \frac{1}{1/h_o + x_w/k_m + 1/h_i} \tag{8-46}$$

In such a case it is advantageous to base the over-all coefficient on that area that corresponds to the largest resistance, or the lowest value of h.

For thin-walled tubes of large diameter, flat plates, or any other case where a negligible error is caused by using a common area for A_i, $\bar{A}_L$, and A_o, Eq. (8-46) can be used for the over-all coefficient, and U_i and U_o are identical.

Sometimes one coefficient, e.g., h_o, is so very small in comparison with both x_w/k and the other coefficient h_i that the term $1/h_o$ is very large compared with the other terms in the resistance sum. When this is true, it is sufficiently accurate to equate the over-all coefficient to the small individual coefficient, or, in this case, $h_o = U_o$.

Classification of Individual Heat-transfer Coefficients. The problem of predicting the rate of heat flow from one fluid to another through a retaining wall reduces to the problem of predicting the numerical values of the individual coefficients of the fluids concerned in the over-all process. A wide variety of individual cases is met in practice, and each type of phenomenon must be considered separately. The following classification will be followed in this text.

1. Heat flow to or from fluids inside tubes, without phase change.
2. Heat flow to or from fluids outside tubes, without phase change.
3. Heat flow from condensing fluids.
4. Heat flow to boiling liquids.

Magnitude of Heat-transfer Coefficients. The ranges of values covered by the coefficient h vary greatly, depending upon the character of the process.[24a] Some typical ranges are shown in Table 8-1.

TABLE 8-1. MAGNITUDES OF HEAT-TRANSFER COEFFICIENTS †

Type of process	Range of values of h, Btu/(ft²)(hr)(°F)	
Steam (dropwise condensation)	5,000	–20,000
Steam (film-type condensation)	1,000	– 3,000
Boiling water	300	– 9,000
Condensing organic vapors	200	– 400
Water (heating or cooling)	50	– 3,000
Oils (heating or cooling)	10	– 300
Steam (superheating)	5	– 20
Air (heating or cooling)	0.2–	10

† From W. H. McAdams, "Heat Transmission," 3d ed., p. 5, McGraw-Hill Book Company, Inc., New York, 1954. By permission of the author.

Heating and Cooling Fluids inside Tubes

In engineering practice, the range of variables encountered in heat transfer to fluids inside pipes, tubes, or other enclosed channels is so great that several separate cases must be recognized and studied individually.

A fundamental division of the subject appears when heat transfer to a single phase, either gas or liquid, is compared with heat transfer to a two-phase stream of liquid and vapor or of solid particles and a fluid. Heat flow to two-phase streams is of great technical importance in such equipment as steam boilers or petroleum-cracking units, but this problem is beyond the scope of this book, and attention will be restricted to heating or cooling single-phase fluids flowing under forced convection through closed channels.

To discuss properly heat transfer in single-phase flow it is necessary to differentiate among turbulent flow, laminar flow, and flow in the transition zone (the so-called "dip" region) between them; to distinguish between heating and cooling; to recognize the effect of natural convection when that effect is appreciable; to account for the effect of roughness; to cover a wide range of viscosities and other fluid properties; to recognize the effects of very high velocities; and to account for stream cross sections other than circular. Also, the relation between local values of the coefficient h_i and the average value of h_i over the entire tube length must be clarified.

As a starting point, consider heat flow to or from a fluid, either liquid or gas, in turbulent flow through a long, straight, smooth pipe. Assume that temperature variations throughout the cross section of the stream are sufficiently small so that values of the various fluid properties, such as viscosity, are essentially constant and single values can be used for them. It is desired to obtain a method of predicting, from a knowledge of the conditions of flow and the characteristics of the fluid, the value of h_i at a given point along the tube.

Although considerable mathematical attention has been given to this problem and some useful equations have been derived relating the heat-transfer coefficient to the friction factor, the most fruitful approach has been that of dimensional analysis.

The variables to be considered in the analysis can be chosen by examining the mechanism of heat flow from the metal wall to the stream of fluid. The heat first flows by true conduction through the laminar sublayer immediately adjacent to the wall, and is then picked up by the eddies which are present in the buffer layer and which constitute the turbulent core that makes up the bulk of the flowing stream. The transfer of heat by conduction through the laminar layer depends upon the thickness of the layer and the thermal conductivity k of the fluid. From fluid mechanics, the thickness of the layer is known to depend upon the variables that constitute the Reynolds number: the mass velocity G, the diameter of the pipe D, and the viscosity of the fluid μ. These variables, along with k, must be chosen for the analysis. The ability of an eddy of a given size to carry away heat from the laminar layer should be proportional to the specific heat of the fluid c_p, and the magnitude and distribution of the eddies is a function of the Reynolds-number variables that have already been chosen. On the other hand, if the velocity distribution is fully developed at the point under study, the length of the tube should not enter the problem.

The basic assumption for the dimensional analysis is, then,

$$h_i = \Psi_i'(D,G,\mu,k,c_p) \tag{8-47}$$

The result of the dimensional analysis is

$$\frac{h_i D}{k} = \Psi_i \left(\frac{DG}{\mu}, \frac{c_p \mu}{k} \right) \tag{8-48}$$

The Reynolds number again appears, which should not be unexpected, and two other dimensionless groups are found. The first, $h_i D/k$, is called the Nusselt number, and is denoted by N_{Nu}. The second, $c_p \mu/k$, is called the Prandtl number, and is denoted by N_{Pr}. These groups are named after noted investigators in the fields of heat transfer and fluid mechanics.

The Prandtl number assembles into one group the three important properties of the fluid. Its numerical value depends upon the fluid and on temperature and pressure. The Prandtl numbers of both gases and liquids are nearly independent of pressure except at high pressure; they are nearly independent of temperature in the case of gases but decrease with increasing temperature for liquids. Near the critical point, the Prandtl numbers of both liquids and gases change rapidly with both temperature and pressure. Prandtl numbers range from approximately 0.02 for molten metals to over 100 for some liquids. The value for diatomic gases is 0.74, and for triatomic gases, it is 0.80. For water, the Prandtl number is about 1 at 160°F and about 5 at 45°F. Values of this function for a number of fluids are given in Appendixes 15 and 16.

The form of the function of Eq. (8-48) has been the subject of a large amount of research, and many experimental data have been used to derive an empirical equation by means of which the Nusselt number can be calculated from the other two groups. The correlations show that a simple exponential equation is satisfactory, and

$$N_{Nu} = b N_{Re}^n N_{Pr}^m \tag{8-49}$$

where b, m, and n are constants. The best data are consistent with the following values for these constants.

$$b = 0.023 \qquad n = 0.80 \qquad m = \tfrac{1}{3}$$

Equation (8-49) then becomes

$$\frac{h_i D}{k} = 0.023 \left(\frac{DG}{\mu} \right)^{0.80} \left(\frac{c_p \mu}{k} \right)^{\frac{1}{3}} \tag{8-50}$$

Equation (8-50) should not be used for Reynolds numbers below 10,000 or for molten metals, which possess abnormally low Prandtl numbers. The success of this equation in correlating data over a range of Reynolds numbers from 10,000 to about 400,000 and of Prandtl numbers from 0.7 to 120 demonstrates the soundness of the choice of variables used in the dimensional analysis.

Average Value of h_i in Turbulent Flow. The temperature of the fluid changes from one end of the tube to the other. Since the fluid properties μ, k, and c_p are all functions of temperature, the local value of h_i varies from point to point along the tube.

For gases, the temperature effect on h_i is small. The Prandtl number is nearly independent of temperature, and since the mass velocity G and the diameter D are also constant, h_i is proportional to the factor $k/\mu^{0.8}$. For gases, both k and μ increase slowly with temperature, and so h_i changes but little unless the temperature range is large. For example, for air, h_i increases only about 6 per cent if the air temperature changes from 100 to 200°F.

For liquids, the temperature effect is much greater than for gases. Equation (8-50) can be condensed to read

$$h_i = 0.023 \frac{G^{0.8} k^{\frac{2}{3}} c_p^{\frac{1}{3}}}{D^{0.2} \mu^{0.47}} \tag{8-51}$$

Inspection of Eq. (8-51) shows that, for liquids, the effect of a temperature rise is to increase h_i. Both c_p and k increase slowly, and μ decreases rapidly, with increase in temperature, and the increase in h_i with temperature is due principally to the effect of temperature on viscosity. For water, for example, h_i increases about 50 per cent over a temperature range from 100 to 200°F.

In practice, unless the variation in h_i over the length of the tube is more than about 2:1, an average value of h_i is calculated and used as a constant in calculating the over-all coefficient U. This procedure neglects the variation of U over the tube length and allows the use of the logarithmic mean temperature drop in calculating the area of the heating surface. The average value of h_i is computed by evaluating the fluid properties c_p, k, and μ at the average fluid temperature, which is defined as the arithmetic mean between the inlet and outlet temperatures. The value of h_i calculated from Eq. (8-50), using these property values, is called the average coefficient. For example, assume that the fluid enters at 100°F and leaves at 200°F. The average fluid temperature is $(100 + 200)/2 = 150°F$, and the values of the properties used to calculate the average value of h_i are those at 150°F.

For larger changes in h_i, two procedures can be used: (1) The values of h_i at the inlet and outlet can be calculated, corresponding values of U_1 and U_2 found, and the procedure called for by Eq. (8-31) used. (2) For even larger variations in h_i, and therefore in U, the tube can be divided into sections and an average U used for each section. Then the lengths of the individual sections can be added to account for the total length of the tube.

Effect of Tube Length. At the entrance to the tube h_i is greater than that calculated from Eq. (8-50) because of the effect of the formation of the velocity distribution in the first portion of the tube. For a long tube having a ratio of length to diameter of 60 or more the entrance effect is negligible. For short tubes the coefficient calculated from Eq. (8-53) should be multiplied [24g] by the factor $1 + (D/L)^{0.7}$. In heat transfer to liquid metals, or to any fluid in laminar flow, the effect of length is greater than that given by this correction.

Heating and Cooling: Viscosity Correction. Equation (8-50) does not differentiate between heating and cooling and predicts the same value of h_i for both situations if the mass velocity, the tube diameter, and the fluid properties are unchanged. This assumption is justified when the variation in viscosity throughout the cross section of the stream is small. It can be accepted for gases at moderate temperature drops and for liquids of low viscosity at small temperature drops. In general, however, the

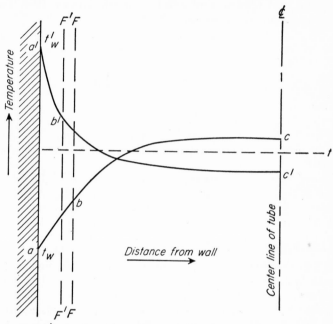

FIG. 8-10. Temperature distribution, heating and cooling of fluid: $t'w$, wall temperature, fluid heated. t_w, wall temperature, fluid cooled. t, average fluid temperature. $F'F'$, boundary of laminar layer, liquid heated. FF, boundary of laminar layer, liquid cooled.

variation in viscosity across the stream is too large to ignore, and, for liquids, the value of h_i is greater when the liquid is being heated than when it is being cooled. For gases, the effect, though small, is reversed, and the coefficient under heating conditions is slightly less than under cooling conditions.

Qualitatively, the difference between heating and cooling may be understood by the aid of Fig. 8-10, in which the local temperature of the fluid is plotted against the distance from the wall of the tube. Curve $a'b'c'$ shows the temperature distribution if the fluid is being heated by the tube wall and curve abc the distribution if the fluid is being cooled. The average fluid temperature t and, therefore, the fluid properties c_p, μ, and k are the same in both cases. Inspection of Fig. 8-10 shows that the average temperature of the laminar film is greater than t when the fluid is being heated and less than t when the fluid is being cooled. If the fluid is a liquid, the viscosity in the laminar film is less in the heating case than in the cooling case. It can be expected that the thickness of the laminar film during heating, the boundary of which is represented by line $F'F'$, will be smaller than that during cooling, represented by the line FF, and the coefficient h_i greater for heating than for cooling.

If the fluid is a gas, the viscosity in the laminar film is greater for heating than for cooling, the film thickness is greater for heating, and the coefficient is smaller.

Quantitatively, the difference between heating and cooling can be taken into account by multiplying the right-hand side of Eq. (8-50) by the empirical, dimensionless viscosity-correction factor [30] ϕ_v

$$\phi_v = \left(\frac{\mu}{\mu_w}\right)^{0.14} \tag{8-52}$$

where μ is the viscosity of the fluid at mean fluid temperature and μ_w is the viscosity of the fluid at the wall temperature. The viscosity μ is the same as that used in the Reynolds and Prandtl numbers in Eq. (8-50), and is found in the same manner as described above. The viscosity μ_w is that corresponding to the temperature of the wall in contact with the fluid. In Fig. 8-10 the temperature of the wall is shown as t_{wh} for heating and as t_{wc} for cooling. For liquids, $\mu_{wh} < \mu$ for heating, and $\mu_{wc} > \mu$ for cooling. In heating a liquid, therefore, $\phi_v > 1.0$ for heating, and $\phi_v < 1.0$ for cooling. In heating and cooling gases, the inequalities are reversed.

Equation (8-50) can be written to include the factor ϕ_v.

$$\frac{h_i D}{k} = 0.023 \left(\frac{DG}{\mu}\right)^{0.8} \left(\frac{c_p \mu}{k}\right)^{\frac{1}{3}} \left(\frac{\mu}{\mu_w}\right)^{0.14} \tag{8-53}$$

Equation (8-50) can be considered a special case of Eq. (8-53), applicable where ϕ_v is nearly unity.

Estimation of Wall Temperature t_w. To evaluate μ_w, the viscosity of the fluid at the wall, t_w must be found. The estimation of this temperature requires a trial-and-error calculation based on the resistance equation [Eq. (8-42)]. If the individual resistances can be estimated, the total temperature drop Δt can be split into the individual temperature drops by the use of this equation and an approximate value for the wall temperature found. To determine t_w in this way the wall resistance (x_w/k_m) $(D_o/\overline{D}_L)$ can be neglected, and Eq. (8-42) used as follows:

From the first two members of Eqs. (8-42)

$$\Delta t_i = \frac{D_o/D_i h_i}{1/U_o} \Delta t$$

Substituting $1/U_o$ from Eq. (8-41) gives

$$\Delta t_i = \frac{1/h_i}{1/h_i + D_i/D_o h_o} \Delta t \tag{8-54}$$

Use of Eq. (8-54) requires preliminary estimates of the coefficients h_i and h_o. To estimate h_i Eq. (8-50) can be used. The calculation of h_o will be described later. The wall temperature t_w is then obtained from the following equations. For heating

$$t_w = t + \Delta t_i \tag{8-55a}$$

For cooling

$$t_w = t - \Delta t_i \tag{8-55b}$$

where t is the average fluid temperature.

If the first approximation is not so accurate as desired, a second calculation of t_w based on the results of the first can be made. Unless the factor ϕ_v is quite different from unity, however, the second approximation is unnecessary.

Alternate Form of Eq. (8-53) and Colburn j_H' Factor. Equation (8-53) can be written in an equivalent form. The alternate equation, which has certain advantages for some purposes, is obtained by multiplying both sides of Eq. (8-53) by $(1/N_{\mathrm{Re}})(1/N_{\mathrm{Pr}})$ and then transferring the Prandtl number and viscosity-ratio terms to the left-hand side.

$$\frac{h_i D}{k} \frac{k}{c_p \mu} \frac{\mu}{DG} = 0.023 \left(\frac{DG}{\mu}\right)^{0.8} \left(\frac{c_p \mu}{k}\right)^{\frac{1}{3}} \left(\frac{\mu}{\mu_w}\right)^{0.14} \frac{k}{c_p \mu} \frac{\mu}{DG}$$

From this,

$$\frac{h_i}{c_p G} \left(\frac{c_p \mu}{k}\right)^{\frac{2}{3}} \left(\frac{\mu_w}{\mu}\right)^{0.14} = 0.023 \left(\frac{DG}{\mu}\right)^{-0.2} \tag{8-56}$$

The left-hand side of Eq. (8-56) is used to define the j_H' *factor.*[6]

$$j_H' = \frac{h_i}{c_p G} \left(\frac{c_p \mu}{k}\right)^{\frac{2}{3}} \left(\frac{\mu_w}{\mu}\right)^{0.14} \tag{8-57}$$

Equation (8-56) is a special case of a more general statement that the j_H' factor is a function of the Reynolds number. The j_H' factor occupies the same place in heat-transfer considerations that the friction factor has in friction correlations. Many heat-transfer correlations are found by plotting experimental data as j_H' against N_{Re}.

Colburn has shown that j_H' is a fairly satisfactory measure of skin friction and that, for smooth pipes and turbulent flow,

$$\tfrac{1}{2} f = j_H' \tag{8-58}$$

approximately. Equation (8-58) is an *analogy equation,* as it presents an analogy between heat transfer and fluid friction. Several other analogy equations have been published.[24f]

Cross Sections Other than Circular. To use Eqs. (8-53) or (8-56) for cross sections other than circular it is only necessary to replace the diameter D in both Reynolds and Nusselt numbers by the equivalent diameter D_e, defined as four times the hydraulic radius r_H. The method is the same as that used in calculating friction loss, as described on page 71.

Example 8-6. Benzene is cooled from 130 to 90°F in the inner pipe of a double-pipe exchanger. Cooling water flows countercurrently to the benzene, entering the jacket at 60°F and leaving at 80°F. The exchanger consists of an inner pipe of $\frac{7}{8}$-in. 16 BWG copper tubing jacketed with Schedule 40 $1\frac{1}{2}$-in. steel pipe. The linear velocity of the benzene is 5 ft/sec; that of the water is 4 ft/sec. Neglecting the resistances of the wall and scale films, compute the film coefficients of the benzene and water and the over-all coefficient based on the outside area of the inner pipe.

Solution. The average temperature of the benzene is $(130 + 90)/2 = 110°F$; that of the water is $(60 + 80)/2 = 70°F$. The physical properties at these temperatures are:

Property	Value at average fluid temp.	
	Benzene	Water
Density ρ, lb/ft^3	53.1 [29a]	62.3 (Appendix 14)
Viscosity μ, lb/ft-hr	1.16 (Appendix 8)	$2.42 \times 0.982 = 2.34$ (Appendix 14)
Thermal conductivity k, Btu-ft/(ft^2)(hr)(°F)	0.089 (Appendix 11)	0.346 (Appendix 14)
Specific heat c_p, Btu/(lb)(°F)	0.435 (Appendix 13)	1.000

The diameters of the inner tube are

$$D_{it} = \frac{0.745}{12} = 0.0621 \text{ ft}$$

$$D_{ot} = \frac{0.875}{12} = 0.0729 \text{ ft}$$

The inside diameter of the jacket is, from Appendix 4,

$$D_{ij} = \frac{1.610}{12} = 0.1342 \text{ ft}$$

The equivalent diameter of the annular jacket space is found as follows. The cross-sectional area is $\pi/4 \,(0.1342^2 - 0.0729^2)$. The wetted perimeter is $\pi(0.1342 + 0.0729)$. The hydraulic radius is

$$r_H = \frac{(\pi/4)(0.1342^2 - 0.0729^2)}{\pi(0.1342 + 0.0729)}$$

$$= \tfrac{1}{4}(0.1342 - 0.0729) = \tfrac{1}{4} \times 0.0613 \text{ ft}$$

The equivalent diameter is

$$D_e = 4 \times \tfrac{1}{4} \times 0.0613 = 0.0613 \text{ ft}$$

The Reynolds number and Prandtl number of each stream are next computed. For benzene they are

$$N_{Re} = \frac{D_{it}\bar{V}\rho}{\mu} = \frac{0.0621 \times 5 \times 3,600 \times 53.1}{1.16} = 5.12 \times 10^4$$

$$N_{Pr} = \frac{c_p\mu}{k} = \frac{0.435 \times 1.16}{0.089} = 5.67$$

and for water

$$N_{Re} = \frac{D_e\bar{V}\rho}{\mu} = \frac{0.0613 \times 4 \times 3,600 \times 62.3}{2.34} = 2.35 \times 10^4$$

$$N_{Pr} = \frac{1.00 \times 2.34}{0.346} = 6.76$$

Preliminary estimates of the coefficients are obtained from Eq. (8-50).

Benzene:
$$h_i = \frac{0.089 \times 0.023(5.15 \times 10^4)^{0.8}5.6^{\frac{1}{3}}}{0.0621}$$

$$= 346 \text{ Btu/(ft}^2)(\text{hr})(°F)$$

Water:
$$h_o = \frac{0.346 \times 0.023(2.35 \times 10^4)^{0.8}6.76^{\frac{1}{3}}}{0.0613}$$

$$= 771 \text{ Btu/(ft}^2)(\text{hr})(°F)$$

The temperature drop over the benzene resistance, from Eq. (8-54), is

$$\Delta t_i = \frac{1/346}{1/346 + 0.0621/(0.0729 \times 771)}(110 - 70) = 29°F$$

$$t_w = 110 - 29 = 81°F$$

The viscosities of the liquids at t_w are now found.

Benzene: $\mu_w = 1.45 \text{ lb/ft-hr}$

Water: $\mu_w = 0.852 \times 2.42 = 2.06 \text{ lb/ft-hr}$

The viscosity-correction factors ϕ_v, from Eq. (8-52), are

Benzene:
$$\phi_v = \left(\frac{1.16}{1.45}\right)^{0.14} = 0.969$$

Water:
$$\phi_v = \left(\frac{2.34}{2.06}\right)^{0.14} = 1.018$$

The corrected coefficients, from Eq. (8-53), are

Benzene: $h_i = 346 \times 0.969 = 335 \text{ Btu/(ft}^2)(\text{hr})(°F)$

Water: $h_o = 771 \times 1.018 = 785 \text{ Btu/(ft}^2)(\text{hr})(°F)$

The temperature drops over the benzene resistance and the wall temperature become

$$\Delta t_i = \frac{1/335}{1/335 + 0.0621/(0.0729 \times 785)}(110 - 70) = 29.3°F$$

$$t_w = 110 - 29.3 = 80.7°F$$

This is so close to the wall temperature calculated previously that a second approximation is unnecessary.

The over-all coefficient is found from Eq. (8-41), neglecting the resistance of the tube wall.

$$\frac{1}{U_o} = \frac{0.0729}{0.0621 \times 335} + \frac{1}{785} = 0.00478$$

$$U_o = \frac{1}{0.00478} = 209 \text{ Btu/(ft}^2\text{)(hr)(°F)}$$

Effect of Roughness. The heat-transfer coefficient in turbulent flow is, for equal Reynolds numbers, somewhat greater for a rough tube than for a smooth one. The effect of roughness on heat transfer is much less than on fluid friction, and economically it is usually more important to use a smooth tube for minimum friction loss than to rely on roughness to yield a larger heat-transfer coefficient. The effect of roughness on h_i is neglected in practical calculations.

Heat Transfer at High Velocities.[24o] When a compressible fluid flows through a tube at high velocities, temperature gradients, caused by friction, appear even when there is no heat transfer through the wall, and $q = 0$. The friction, which is a maximum at the wall, raises the temperature of the fluid at the wall above the average fluid temperature. The temperature difference between wall and fluid causes a flow of heat from wall to fluid, and a steady state is reached when the rate of heat developed by friction at the wall equals the rate of heat transfer back into the fluid stream. The constant wall temperature thus attained is called the *adiabatic wall temperature*. Further treatment of this subject is beyond the scope of this text. The effect becomes appreciable for Mach numbers above approximately 0.4, and appropriate equations must be used in this range of velocities instead of Eqs. (8-53) and (8-56).

Heat Transfer to Molten Metals. Molten metals are being used for transferring heat at temperatures that are not reachable by other fluids. Liquid mercury, sodium, potassium, and "NaK," a mixture of sodium and potassium, can be so used. Mercury vapor is also often used as a carrier of latent heat. Temperatures close to the "metallurgical limit" of about 1500°F are obtainable by sensible heat transfer to and from these materials. Molten metals possess moderate specific heats, high thermal conductivities, and low viscosities. Their Prandtl numbers are therefore low. This property for NaK is, for example, 0.019 at 212°F and 0.048 at 660°F.

Equations (8-53) and (8-56) are not applicable at low Prandtl numbers, because the mechanism of heat transfer differs from that at the usual values of N_{Pr} covered by these equations. For Prandtl numbers greater than about 1 the heat is transferred in the laminar layer entirely by conduction, in the turbulent core almost entirely by eddies, and in the buffer zone by a combination of conduction and eddy action. In heat transfer at low Prandtl numbers, on the other hand, conduction is important, not only in the laminar and buffer zones but also in the turbulent core. Diagrammatically, the effect of N_{Pr} on the temperature distribution is shown in

Fig. 8-11, which gives, for equal Reynolds numbers and equal center-line temperatures, the variation of temperature with radius from the center line of the tube to the wall. Curves are drawn for Prandtl numbers of less than 1.0, 1.0, and greater than 1.0. For Prandtl numbers of 1 or more, the temperature-distribution curves in the turbulent core are identical for all values of N_{Pr}, but the curves diverge in the buffer zone. For low Prandtl numbers, however, the temperature-distribution curves coincide only near the center line of the tube and diverge in the remainder of the turbulent core. It is this difference in the temperature distribution that leads to the inapplicability of Eqs. (8-53) and (8-56).

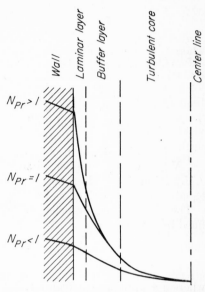

The following equations [21] have been found for the calculation of sensible heat to and from molten metals.

$$N_{Nu} = 7 + 0.025 N_{Pe}^{0.8} \qquad (8-59)$$

where N_{Pe} is the Peclet number $D\overline{V}\rho c_p/k$, or $N_{Re}N_{Pr}$. For annular spaces Eq. (8-59) is modified to

$$N_{Nu} = 4.9 + 0.0175 N_{Pe}^{0.8} \qquad (8-60)$$

Fig. 8-11. Effect of Prandtl number on temperature distribution. (*Courtesy of AIChE.*)

The equivalent diameter D_e is used in Eq. (8-60).

Heat Transfer to Fluids in Laminar Flow inside Tubes. The mechanism of heat flow to fluids in laminar flow differs from that of heat flow to fluids in turbulent flow. In laminar flow eddies do not form, and the heat must flow largely by the mechanism of conduction. As shown in Chap. 2, the velocity distribution in a pipe, after the flow regime is fully developed, across the flowing stream in laminar flow is parabolic. This is true only if the flow is isothermal, so the viscosity does not change from point to point in the cross section. Theoretical equations have been derived for heat flow occurring under these idealized isothermal conditions. For actual fluids that are being either heated or cooled the viscosity does vary across the stream and, also, in practical equipment, the full development of the parabolic velocity distribution does not always occur. The effect of the viscosity gradients is to cause secondary flows either from the wall toward the center of the stream or from the center toward the wall of the tube. These crossflows carry heat. Experimental data agree with the theoretical equation only at very small temperature drops and fully developed laminar flow. The most important applications of heat flow to fluids in laminar flow are those in which the fluid is a viscous liquid, such as heavy petroleum fractions, and for these the viscosity effects are especially large.

The theoretical analysis does give, however, one important result, namely, the ratio of the temperature range to the inlet temperature drop depends only on the dimensionless group called the *Graetz number*. This statement can be written

$$\frac{t_b - t_a}{t_w - t_a} = \chi' \left(\frac{wc_p}{kL}\right) = \chi'(N_{Gz}) \tag{8-61}$$

where t_a = average fluid temperature at tube inlet, °F
    t_b = average fluid temperature at tube outlet, °F
    t_w = temperature of wall in contact with fluid, °F. (This temperature is, by assumption, constant.)
    w = mass flow rate, lb/hr
    c_p = specific heat of fluid, at constant pressure, Btu/(lb)(°F)
    k = thermal conductivity of fluid, Btu-ft/(ft²)(hr)(°F)
    L = heated length of tube, ft
    N_{Gz} = Graetz number, dimensionless, defined as wc_p/kL

Equation (8-61) does not contain the heat-transfer coefficient h_i. An equation in h_i is derived as follows.

The average heat-transfer coefficient, over the entire tube, can be defined by the equation

$$h_i = \frac{q}{A_T \, \overline{\Delta t_i}} \tag{8-62}$$

where q = rate of heat transfer, Btu/hr
    A_T = area of heating surface, ft²
    $\overline{\Delta t_i}$ = average temperature drop over fluid, from tube wall to average fluid temperature, °F

From an enthalpy balance,

$$q = wc_p(t_b - t_a) \tag{8-63}$$

Also
$$A_T = \pi DL \tag{8-64}$$

where D is the inside diameter of the tube, in feet. Elimination of q and A_T from Eqs. (8-62) to (8-64) gives

$$h_i = \frac{wc_p(t_b - t_a)}{\pi DL \, \overline{\Delta t_i}} \tag{8-65}$$

The left-hand side of Eq. (8-65) can be expressed in the form of a Nusselt number by multiplying through by D/k.

$$N_{Nu} = \frac{h_i D}{k} = \frac{1}{\pi} \frac{wc_p}{kL} \frac{t_b - t_a}{\overline{\Delta t_i}} = \frac{N_{Gz}(t_b - t_a)}{\pi \, \overline{\Delta t_i}} \tag{8-66}$$

In turbulent flow, the coefficient is nearly independent of the tube length, and the logarithmic mean temperature difference is the logical form for practical use. In laminar flow, however, the average coefficient over the tube depends upon the heated length, and, in fact, $h_i \propto 1/L$. The choice of the logarithmic temperature is no longer logical, and the simpler arithmetic mean is more convenient. Thus, let $\overline{\Delta t_i}$ be

$$\overline{\Delta t_i} = \frac{(t_w - t_a) + (t_w - t_b)}{2} = \overline{\Delta t_a} \tag{8-67}$$

where $\overline{\Delta t_a}$ is the arithmetic mean temperature drop over the inside resistance, in degrees Fahrenheit.

Substitution of $\overline{\Delta t_a}$ for $\overline{\Delta t_i}$ in Eq. (8-66) gives

$$\frac{h_{ia}D}{k} = \frac{2N_{\text{Gz}}}{\pi} \frac{t_b - t_a}{(t_w - t_a) + (t_w - t_b)} \tag{8-68}$$

The subscript a on h_{ia} calls attention to the fact that the coefficient is based on the arithmetic average temperature drop.

If Eq. (8-61) is solved for t_b and the resulting value of t_b substituted into the denominator of Eq. (8-68), the result is

$$\frac{h_{ia}D}{k} = \frac{2N_{\text{Gz}}}{\pi} \frac{t_b - t_a}{(t_w - t_a)[2 - \chi'(N_{\text{Gz}})]}$$

Then, substituting for $(t_b - t_a)/(t_w - t_a)$ from Eq. (8-61),

$$\frac{h_{ia}D}{k} = \frac{2N_{\text{Gz}}}{\pi} \frac{\chi'(N_{\text{Gz}})}{2 - \chi'(N_{\text{Gz}})} = \chi(N_{\text{Gz}}) \tag{8-69}$$

where $\chi(N_{\text{Gz}})$ is another function of the Graetz number.

Experimental data are satisfactorily correlated, for fluids of moderate viscosity, by the simple equation [24h]

$$\chi(N_{\text{Gz}}) = 2N_{\text{Gz}}^{\frac{1}{3}} \tag{8-70}$$

and

$$\frac{h_{ia}D}{k} = N_{\text{Nu},a} = 2N_{\text{Gz}}^{\frac{1}{3}} = 2\left(\frac{wc_p}{kL}\right)^{\frac{1}{3}} \tag{8-71}$$

For viscous liquids and large temperature drops, Eq. (8-71) is modified by the same viscosity factor, $\phi_v = (\mu/\mu_w)^{0.14}$, used in the turbulent-flow correlation, so

$$\frac{h_{ia}D}{k} = 2\left(\frac{wc_p}{kL}\right)^{\frac{1}{3}}\left(\frac{\mu}{\mu_w}\right)^{0.14} \tag{8-72}$$

or, more concisely,

$$N_{\text{Nu},a} = 2N_{\text{Gz}}^{\frac{1}{3}}\phi_v \tag{8-73}$$

Equations (8-71) and (8-73) are subject to an important limitation, which arises from the fact that h_{ia} is based on the arithmetic mean temperature drop. If the tube is sufficiently long, the exit fluid temperature t_b approaches the temperature of the wall t_w. An asymptote is reached, then, where $t_b - t_a = t_w - t_a$ and $t_b - t_w = 0$. Equation (8-68) becomes, for asymptotic conditions,

$$\frac{h_{ia}D}{k} = \frac{2wc_p}{\pi kL} = \frac{2}{\pi}N_{Gz} \tag{8-74}$$

The asymptotic condition is closely approached for values of N_{Gz} of 10 to 13, and Eq. (8-74) should be used rather than Eq. (8-72) or (8-73) for values of N_{Gz} less than this. If a heat exchanger is operating under conditions such that the asymptotic range is entered, part of the heating surface is inactive, and such surface is wasted.

Heat Transfer in Transition Region between Laminar and Turbulent Flow. Equations (8-53) and (8-56) apply only for Reynolds numbers greater than 10,000, and Eqs. (8-72) and (8-73) apply only for Reynolds numbers less than 2,100. The range of Reynolds numbers between 2,100 and 10,000 is called the transition region, and no simple equation applies in this region. A graphical method is therefore used. The method is based on plotting Eqs. (8-56) and (8-72) on a common plot of j''_H vs. N_{Re}, which shows graphically the j''_H factor as a function of N_{Re} over the entire range of Reynolds numbers, from the laminar region through the transition region and into the turbulent region. To accomplish this, it is necessary to convert Eq. (8-72) to the j''_H form, which can be done in the following manner.

The Reynolds number may be written in the form

$$N_{Re} = \frac{D\overline{V}\rho}{\mu} = \frac{Dw\rho}{\rho(\pi/4)D^2\mu} = \frac{4w}{\pi\mu D} \tag{8-75}$$

Thus the term wc_p/kL can be written

$$N_{Gz} = \frac{wc_p}{kL} = \frac{4w}{\pi\mu D}\frac{\pi}{4}\frac{c_p\mu}{k}\frac{D}{L} = \frac{\pi}{4}\frac{D}{L}N_{Re}N_{Pr} \tag{8-76}$$

Substitution of wc_p/kL from Eq. (8-76) into Eq. (8-72) and multiplication by $(1/N_{Re})(1/N_{Pr})$ gives

$$\frac{h_{ia}D}{k}\frac{\mu}{GD}\frac{k}{c_p\mu} = 2\left(\frac{\pi}{4}\frac{D}{L}N_{Re}N_{Pr}\right)^{\frac{1}{3}}\frac{1}{N_{Pr}N_{Re}}\left(\frac{\mu}{\mu_w}\right)^{0.14}$$

and, simplifying,

$$\frac{h_{ia}}{Gc_p} = 1.85\left(\frac{D}{L}\right)^{\frac{1}{3}}N_{Re}^{-\frac{2}{3}}N_{Pr}^{-\frac{2}{3}}\left(\frac{\mu}{\mu_w}\right)^{0.14}$$

or

$$j''_H = \frac{h_{ia}}{Gc_p}N_{Pr}^{\frac{2}{3}}\left(\frac{\mu_w}{\mu}\right)^{0.14} = 1.85\left(\frac{D}{L}\right)^{\frac{1}{3}}N_{Re}^{-\frac{2}{3}} \tag{8-77}$$

From Eq. (8-77) it can be seen that, for each value of the length-diameter ratio L/D, a logarithmic plot of j''_H vs. N_{Re} gives a straight line with a slope of $-\frac{2}{3}$. The straight lines on the left-hand portion of Fig. 8-12 are plots of this equation for various values of L/D. The lines all terminate at a Reynolds number of 2,100.

Equation (8-56), when plotted on the same coordinates, gives a straight line with a slope of -0.20 for Reynolds numbers above 10,000. This line is drawn in the right-hand region of Fig. 8-12.

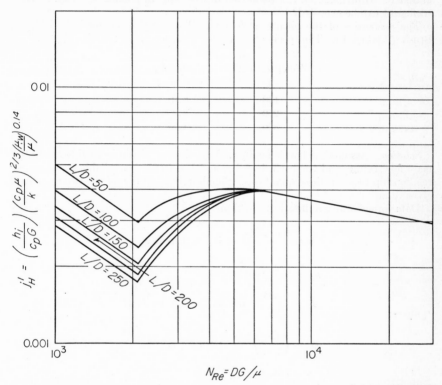

FIG. 8-12. Heat transfer in transition range. (*By permission of author and publishers, from "Heat Transmission," 3d ed., by W. H. McAdams. Copyright by author, 1954. McGraw-Hill Book Company, Inc.*)

The curved lines between Reynolds numbers of 2,100 and 10,000 represent the relationships of j'_H vs. N_{Re} for the transition region. The effect of L/D is pronounced at the lower Reynolds numbers in this region and fades out as a Reynolds number of 10,000 is approached.

Figure 8-12 is a summary chart which can be used for the entire range of Reynolds numbers from 1,000 to 100,000. Beyond its lower and upper limits, Eqs. (8-77) and (8-56), respectively, can be used.

In using Fig. 8-12 it must be understood that, for Reynolds numbers below 2,100, the coefficient h_{ia} is the average value of the coefficient over the entire length of the pipe based on the arithmetic mean temperature drop. For values of N_{Re} above 2,100, the coefficient h_i is either the local value of h_i or the average value of h_i over the tube length based on the logarithmic mean temperature drop when conditions are such that the use of this is correct.

Example 8-7. A light motor oil of the characteristics given below is to be heated from 150 to 250°F in a $\frac{1}{4}$-in. Schedule 40 pipe 15 ft long. The pipe wall is at 350°F. How much oil can be heated in this pipe, in pounds per hour? What coefficient can be expected?

The properties of the oil are as follows. The thermal conductivity is 0.082 Btu-ft/(ft²)(hr)(°F). The specific heat is 0.48 Btu/(lb)(°F). The viscosity is:

Temperature, °F	Viscosity, centipoises
150	6.0
250	3.3
350	1.37

Solution. Assume the flow is laminar. Equation (8-72) can be used to relate the mass flow rate w and the coefficient h_{ia}, and Eq. (8-65) is available for a second relationship between these quantities. Simultaneous solution of the two equations gives values of w and h_{ia}.

Data for substitution into Eq. (8-72) are

$$\mu = \frac{6.0 + 3.3}{2} = 4.65 \text{ centipoises}$$

$$\mu_w = 1.37 \text{ centipoises}$$

$$D = \frac{0.364}{12} = 0.0303 \qquad \text{(Appendix 4)}$$

$$\phi_v = \left(\frac{4.65}{1.37}\right)^{0.14} = 1.187$$

$$k = 0.082 \qquad c_p = 0.48$$

From Eq. (8-72),

$$\frac{0.0303 h_{ia}}{0.082} = 2 \times 1.187 \left(\frac{0.48 w}{0.082 \times 15}\right)$$

From this, $h_{ia} = 4.69 w^{\frac{1}{3}}$.

Data for substitution into Eq. (8-65) are

$$\overline{\Delta t_a} = \frac{(350 - 150) + (350 - 250)}{2} = 150°F$$

$$L = 15 \qquad D = 0.0303$$

$$t_b - t_a = 250 - 150 = 100°F$$

From Eq. (8-65),

$$h_{ia} = \frac{0.48 \times 100 w}{\pi 0.0303 \times 15 \times 150} = 0.224 w$$

Then
$$4.69w^{\frac{1}{3}} = 0.224w$$

$$w = \left(\frac{4.69}{0.224}\right)^{\frac{3}{2}} = 95.9 \text{ lb/hr}$$

and
$$h_{ia} = 0.224 \times 95.9 = 21.5 \text{ Btu/(ft}^2)(\text{hr})(°\text{F})$$

To check the assumption of laminar flow, the maximum Reynolds number, which exists at the outlet end of the pipe, should be calculated.

$$N_{Re} = \frac{4 \times 95.9}{\pi 2.42 \times 3.3 \times 0.0303} = 500$$

This is well within the laminar range.

Effect of Natural Convection. At low velocities, at low viscosities, in large pipes, and with large temperature drops, natural convection may occur in pipes to such an extent that the above equations give coefficients that are too low. This effect is found almost entirely in laminar flow, as the higher velocities characteristic of flow in the transition and turbulent regions overcome the relatively gentle currents of natural convection.

The mechanism of natural convection is described on pages 466 and 468. The magnitude of the natural-convection effect is measured, as might be expected, by still another dimensionless group. The natural-convection group is called the *Grashof number*, which is defined by the equation

$$N_{Gr} = \frac{D^3 g \rho^2 \beta \, \Delta t_i}{\mu^2} \tag{8-78}$$

where D = diameter of the tube, ft
g = acceleration due to gravity, $32.17 \times 3{,}600^2 = 4.17 \times 10^8 \text{ ft/hr}^2$
β = coefficient of volume expansion of fluid, which is the fractional increase in volume, at constant pressure, $1/°\text{F}$ (see page 469)
Δt_i = temperature drop between fluid and wall, °F
μ = viscosity of fluid, lb/ft-hr

The effect of natural convection on the heat transfer to fluids in laminar flow through horizontal tubes can be accounted for by multiplying the coefficient h_{ia}, computed from Eq. (8-77) or Fig. 8-12, by the factor [19]

$$\phi_n = \frac{2.25(1 + 0.010N_{Gr}^{\frac{1}{3}})}{\log N_{Re}} \tag{8-79}$$

Natural-convection effects occur also in vertical tubes. The equations for this are too complicated to include here.[22]

Heating and Cooling of Fluids outside Tubes

The important engineering cases of heat flow, in the absence of change of phase, between tubes and fluids surrounding them are heat flow under forced convection with the fluid flowing in a direction perpendicular to the axis of the tube and heat flow under natural convection. Heat flow to fluids in forced convection parallel to outside surfaces has been discussed in the last section.

Forced Convection. The mechanism of heat flow in forced convection outside tubes differs from that of flow inside tubes, because of differences in fluid-flow mechanism. As has been shown on pages 49 and 50, no form drag exists inside tubes except perhaps for a short distance at the entrance end, and all friction is skin friction. Because of the lack of form friction, there is no variation in the local heat transfer at different points in a given circumference, and a close analogy exists between friction and heat transfer. An increase in heat transfer is obtainable at the expense of added friction simply by increasing the fluid velocity. Also, a sharp distinction exists

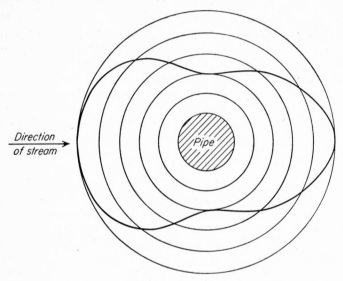

Fig. 8-13. Heat flow around cylinder, air flow normal to axis. (*By permission of author and publishers, from "Heat Transmission," 3d ed., by W. H. McAdams. Copyright by author, 1954. McGraw-Hill Book Company, Inc.*)

between laminar and turbulent flow, which calls for different treatment of heat-transfer relations for the two flow regimes.

On the other hand, as has been shown on pages 91 and 92, in flow of fluids across a cylindrical shape, boundary-layer separation occurs, and a wake develops that causes form friction. No sharp distinction is found between laminar and turbulent flow, and a common correlation can be used for both low and high Reynolds numbers. Also, the local value of the heat flux varies from point to point around a circumference. In Fig. 8-13 the local value of the heat-transfer rate per unit area is plotted radially for all points around the circumference of the tube. The rate is a maximum at the front and back of the tube and a minimum at the sides. The ratio of the maximum rate to the minimum rate is 2.5. In practice, the variation in flux, and in local coefficient h, is of no importance, and average values based on the entire circumference are used.

Radiation may be important in heat transfer to outside tube surfaces.

Inside tubes, the surface cannot "see" surfaces other than the inside wall of the same tube, and heat flow by radiation does not occur. Outside tube surfaces, however, are necessarily "in sight of" external surfaces, if not nearby, at least at a distance, and the surrounding surfaces may be appreciably hotter or cooler than the tube wall. Heat flow by radiation, especially when the fluid is a gas, is appreciable in comparison with heat flow by conduction and convection. The total heat flow is then a sum of two independent flows, one by radiation and the other by conduction and convection. The relations given in the remainder of this section have to do with conduction and convection only. Radiation, as such and in combination with conduction and convection, is discussed on pages 507 to 521.

Dimensionless Equation. The variables affecting the coefficient of heat transfer to a fluid in forced convection outside a tube, by the same reasoning applied in studying the case of heat flow to fluids inside tubes, are D_o, the outside diameter of the tube; c_p, μ, and k, the specific heat at constant pressure, the viscosity, and the thermal conductivity, respectively, of the fluid; and G, the mass velocity of the fluid approaching the tube. Dimensional analysis gives, then, an equation of the type of Eq. (8-48).

$$\frac{h_o D_o}{k} = \Psi_o \left(\frac{D_o G}{\mu}, \frac{c_p \mu}{k} \right) \tag{8-80}$$

Here, however, the similarity between the two types of process—the flow of heat to fluids inside of tubes and the flow of heat to fluids outside of tubes—ends, and the functional relationships in the two cases differ.

Heat Transfer to and from Fluids Flowing Normally to a Single Tube or Cylinder. For air, where the Prandtl number is nearly independent of temperature, Eq. (8-80) becomes simply

$$\frac{h_o D_o}{k_f} = \Psi_o \left(\frac{D_o G}{\mu_f} \right)$$

Experimental data are correlated by the curve of Fig. 8-14. The effect of radiation is not included in this curve, and radiation must be calculated separately. The range of variables covered by this correlation is very great, as can be seen from Table 8-2.[24j]

TABLE 8-2. RANGE OF VARIABLES FOR FIG. 8-14 †

Variable	Range
Diameter of cylinder, in.	0.0004–5.9
Air velocity, ft/sec	0–62
Air temperature, °F	60–500
Cylinder temperature, °F	70–1,840
Pressure, atm abs	0.4–4
$D_o G / \mu_f$, consistent units	0.2–235,000
$h_o D_o / k_f$, consistent units	0.5–500

† From W. H. McAdams, "Heat Transmission," 3d ed., p. 258, McGraw-Hill Book Company, Inc., New York, 1954. By permission of the author.

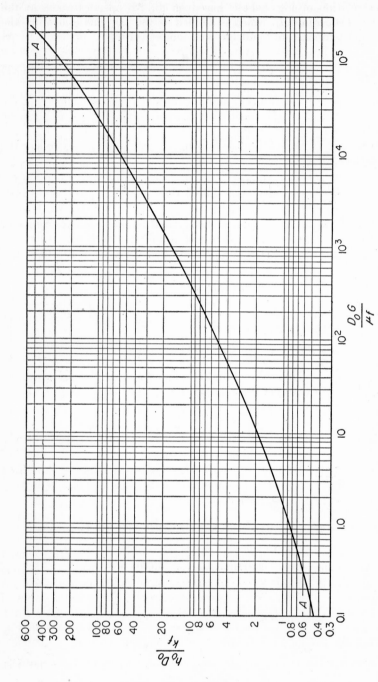

FIG. 8-14. Heating and cooling air flowing normal to single cylinder.

The correlation of Fig. 8-14 is based on fluid properties taken at the *average film temperature* t_f, which is defined as the arithmetic mean of the average fluid temperature t and the wall temperature † t_w, or

$$t_f = \frac{t + t_w}{2} \tag{8-81}$$

The subscript f on the terms k_f and μ_f calls attention to the fact that the correlation is based on the mean film temperature. Figure 8-14 can be used for both heating and cooling.

For heating and cooling liquids flowing normal to single cylinders the following equation is used.[24l]

$$\frac{h_o D_o}{k_f} \left(\frac{c_p \mu_f}{k_f}\right)^{-0.3} = 0.35 + 0.56 \left(\frac{D_o G}{\mu_f}\right)^{0.52} \tag{8-82}$$

Heat Transfer to and from Fluids Flowing Perpendicularly to Tube Banks.[24m] Banks of tubes, across which fluids flow in a direction perpendicular to the tubes, are often used to heat gases, especially air, and liquids. Two tube arrangements are common. In the first, shown in Fig. 8-15a, the centers

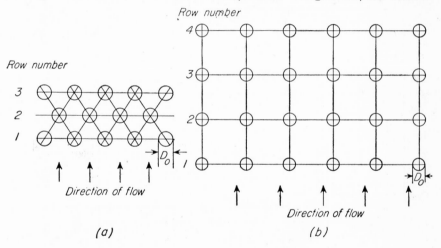

FIG. 8-15. Tube-bank arrangements: (a) Triangular pitch. (b) Square pitch.

of the tubes are set at the corners of equilateral triangles. The arrangement is called triangular pitch. In the second, shown in Fig. 8-15b, the centers of the tubes form squares. The arrangement is called square pitch. A bank consists of a number of rows of tubes, where a row is the series of tubes in a single plane perpendicular to the direction of flow of the fluid. The rows are numbered from front to back. The bank shown in Fig. 8-15a has three rows, and that in Fig. 8-15b has four rows.

† The definition and method of calculating the wall temperature t_w are given on page 444.

The value of h for all tubes in a single row is the same, but h varies from row to row. The deeper the row, the larger is h, because of the greater turbulence of the fluid in contact with the deeper rows. The effect of row number becomes small after the first three or four rows. In practice, an average value of h, denoted by $\bar{h}$, is used and applied to the entire bank.

Data for heating air are correlated by the following equation.

$$\frac{\bar{h}D_o}{k_f} = b \left(\frac{D_o G_{\max}}{\mu_f} \right)^n \tag{8-83}$$

where $G_{\max}$ is the mass velocity through the minimum cross section available for the flow of air. This may be either the area perpendicular to the direction of flow or the area between the tubes on a diagonal. The constants b and n depend on the tube arrangement, on the diameter and clearance of the tubes, and on the number of rows of tubes in the bank. Tables are available for these constants.[24n] As another example, Eq. (8-84) is used in the following dimensional form for heating gases flowing in a direction normal to tubes on triangular centers.

$$\bar{h} = \frac{0.133 c_p G_{\max}^{0.6}}{D_o^{0.4}} \tag{8-84}$$

where pound, foot, hour, Btu units are used. For fewer than 10 rows, $\bar{h}$ from Eq. (8-84) is multiplied by the factor given in Table 8-3.[24n]

TABLE 8-3. RATIO OF $\bar{h}$ FOR N-ROW BANK TO $\bar{h}$ FOR 10-ROW BANK †

Number of rows.........	1	2	3	4	5	6	7	8	9	10
Triangular pitch........	0.68	0.75	0.83	0.89	0.92	0.95	0.97	0.98	0.99	1.0
Square pitch............	0.64	0.80	0.87	0.90	0.92	0.94	0.96	0.98	0.99	1.0

† From W. H. McAdams, "Heat Transmission," 3d ed., p. 274, McGraw-Hill Book Company, Inc., New York, 1954.

A special type of flow across tube banks is encountered in heat exchangers in which the shell side is baffled. This situation is discussed on pages 506 and 507.

Extended Surfaces

Heat-exchange problems are encountered where one of the two fluid streams has a much lower heat-transfer coefficient than the other. A typical case is heating a fixed gas, such as air, by means of condensing steam. The coefficient for the air stream is of order of magnitude of 10, and that for the condensing steam is of order of magnitude of 1,000 to 2,000. The over-all coefficient is essentially equal to the individual coefficient for the air, the capacity of a unit area of heating surface will be low, and many feet of tube will be required to provide reasonable capacity. Other variations of the same problem are found in heating or cooling viscous liquids or in treating a stream of fluid at low flow rate, because of the low rate of heat transfer in laminar flow.

To conserve space and to reduce the cost of the equipment in these cases, certain types of heat-exchange surfaces, called extended surfaces, have been developed, in which the outside area of the tube is multiplied, or extended, by fins, pegs, disks, and other appendages and the outside area in contact with the fluid thereby made much larger than the inside area. The fluid stream having the lower coefficient is brought into contact with the extended surface, and flows outside the tubes, while the other fluid, having the high coefficient, flows through the tubes. The quantitative effect of extending the outside surface can be seen from the over-all coefficient, written in the following form, in which the resistance of the tube wall is neglected,

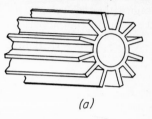

(a)

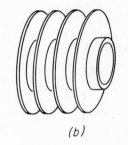

(b)

$$U_i = \frac{1}{1/h_i + A_i/A_o h_o} \qquad (8\text{-}85)$$

Equation (8-85) shows that, if h_o is small and h_i large, the value of U_i will be small; but if the area A_o is made much larger than A_i, the resistance $A_i/A_o h_o$ becomes small, and U_i increases just as if h_o were increased, with a corresponding increase in capacity per foot length of tube or per square foot of inside area.

Types of Extended Surface. Three common types of extended surfaces are available, examples of which are shown in Fig. 8-16. Longitudinal fins are used when the direction of flow of the fluid is parallel to the axis of the tube, e.g., as in the hairpin heat exchanger shown in Fig. 8-17. Transverse fins are used when the direction of flow of the fluid is across tubes, e.g., as in the air heater shown in Fig. 8-18. Spikes, pins, studs, or spines

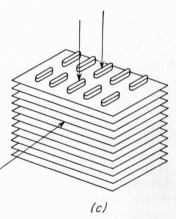

(c)

FIG. 8-16. Types of extended surface: (a) Longitudinal. (b) Transverse fins. (c) Flattened tubes, continuous fins.

are also used to extend surfaces, and tubes carrying these can be used for either direction of flow. In all types, it is important that the fins be in tight contact with the tube, both for structural reasons and to ensure good thermal contact between the base of the fin and the wall.

FIG. 8-17. Hairpin heat exchanger. (*Brown Fintube Co.*)

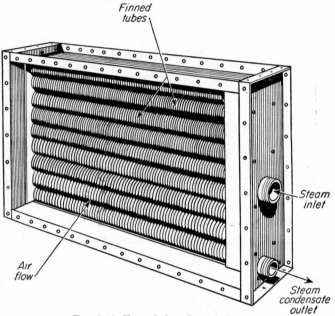

FIG. 8-18. Extended-surface air heater.

Fin Efficiency. The outside area of a finned tube consists of two parts, the area of the fins, and the area of the bare tube not covered by the bases of the fins. A unit area of fin surface is not so efficient as a unit area of bare tube surface because of the added resistance to the heat flow by conduction through the fin to the tube. Thus, consider a single longitudinal fin attached to a tube, as shown in Fig. 8-19, and assume that the heat is flowing to the tube from the fluid surrounding the fin. Let the temperature of the fluid be t and that of the bare portion of the tube t_w. The temperature at the base of the fin will also be t_w. The temperature drop available for heat transfer to the bare tube is $t - t_w$, or Δt_o. Consider the heat transferred to the fin at the tip, the point farthest away from the tube wall. This heat, in order to reach the wall of the tube, must flow, by conduction, through the entire length of the fin, from tip to base. Other increments of heat, entering the fin at points intermediate between tip and base, also must flow through a part of the fin length. A temperature gradient will be necessary, therefore, from the tip of the fin to the base, and the tip will be warmer than the base. If t_F is the temperature of the fin at a distance x from the base, the temperature drop available for heat transfer from fluid to fin at that point will be $t - t_F$. Since $t_F > t_w$, $t - t_F < t - t_w = \Delta t_o$, and the efficiency of any unit area away from the fin base is less than that of a unit area of bare tube. The difference between $t - t_F$ and Δt_o is zero at the base of the fin and is a maximum at the tip of the fin. Let the average value of $t - t_F$, based on the entire fin area, be denoted by $\overline{\Delta t_F}$. The efficiency of the fin is defined as the ratio of $\overline{\Delta t_F}$ to Δt_o, and is denoted by η_F.

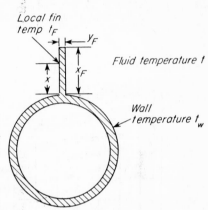

$$\eta_F = \frac{\overline{\Delta t_F}}{\Delta t_o} \qquad (8\text{-}86)$$

The efficiency can, of course, be expressed on a percentage basis. An efficiency of unity (or of 100 per cent) means that a unit area of fin is as effective as a unit area of bare tube, as far as temperature drop is concerned. Any actual fin will have an efficiency less than 100 per cent.

FIG. 8-19. Tube and single longitudinal fin.

Calculations for Extended-surface Exchangers. Consider, as a basis, 1 ft of tube. Let A_F be the area of the fins and A_b the area of the bare tube. Let h_o be the heat-transfer coefficient of the fluid surrounding the fins and tube. Assume that h_o is the same for both fins and tube. An over-all coefficient, based on the inside area A_i, can be written

$$U_i = \cfrac{1}{\cfrac{A_i}{h_o(\eta_F A_F + A_b)} + \cfrac{x_w D_i}{k_m \overline{D}_L} + \cfrac{1}{h_i}} \qquad (8\text{-}87)$$

where h_i = heat-transfer coefficient inside tube, Btu/(ft²)(hr)(°F)
k_m = thermal conductivity of tube wall, Btu-ft/(ft²)(hr)(°F)
x_w = thickness of tube wall, ft
D_i = inside tube diameter, ft
$\overline{D}_L$ = logarithmic mean tube diameter, ft

To use Eq. (8-87) it is necessary to know the values of the fin efficiency η_F and of the individual coefficients h_i and h_o. The coefficient h_i is calculated by the usual methods. The calculation of the coefficient h_o will be discussed later.

The fin efficiency η_F can be calculated mathematically, on the basis of certain reasonable assumptions, for fins of various types.[18f,g] For example, the efficiency of longitudinal fins is given in Fig. 8-20, in which η_F is plotted as a function of the quantity $a_F x_F$, where x_F is the length of the fin from base to tip and a_F is defined by the equation

$$a_F = \sqrt{\frac{h_o L_p / S}{k_m}} \qquad (8\text{-}88)$$

where h_o = coefficient outside tube, Btu/(ft^2)(hr)(°F)
  k_m = thermal conductivity of metal in fin, Btu-ft/(ft^2)(hr)(°F)
  L_p = perimeter of fin, ft
  S = cross-sectional area of fin, ft^2

The product $a_F x_F$ is dimensionless.

Fin efficiencies for other types of extended surface are available.[18h] Figure 8-20 shows that the fin efficiency is nearly unity when the ratio of h_o to k is small. Extended surfaces are neither efficient nor necessary if the coefficient h_o is large. Also, fins increase the pressure drop.

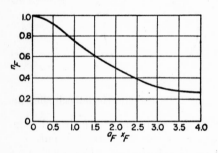

The coefficient h_o cannot be accurately found by the use of the equations normally used for calculating the heat-transfer coefficients for bare tubes. The fins change the flow characteristics of the fluid, and the coefficient for an extended surface differs from that for a smooth tube. Individual coefficients for extended surfaces must be determined experimentally and correlated for each type of surface, and such correlations are supplied by the manufacturer of the tubes. A

Fig. 8-20. Fin efficiency, longitudinal fins.

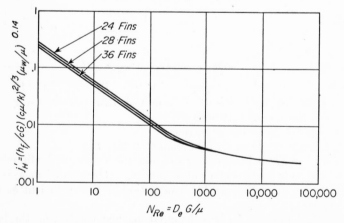

Fig. 8-21. Heat-transfer coefficients, longitudinal finned tubes. 1½-in. IPS with ½- by 0.035-in. fins in 3-in. IPS shell. (*Brown Fintube Co.*)

typical correlation for longitudinal finned tubes is shown in Fig. 8-21. The quantity D_e is the equivalent diameter, defined as usual as four times the hydraulic radius, which is, in turn, the cross section of the fin-side space divided by the total wetted perimeter of fins and inside tube calculated as in Example 8-8.

Example 8-8. Air is heated in the shell of an extended-surface exchanger similar to that shown in Fig. 8-17. The inner pipe is $1\frac{1}{2}$-in. IPS Schedule 40 steel pipe carrying 28 longitudinal fins $\frac{1}{2}$ in. high and 0.035 in. thick. The shell is 3-in. Schedule 40 steel pipe. The exposed outside area of the inner pipe (not covered by the fins) is 0.416 ft² per lineal foot; the total surface area of the fins and pipe is 2.830 ft²/ft. Steam condensing at 250°F inside the inner pipe has a film coefficient of 1500 Btu/(ft²)(hr)(°F). The thermal conductivity of steel is (from Appendix 9) 26 Btu-ft/(ft²)(hr)(°F). The wall thickness of the inner pipe is 0.145 in. If the mass velocity of the air is 5,000 lb/(hr)(ft²) and the average air temperature is 130°F, what is the over-all heat-transfer coefficient based on the inside area of the inner pipe? Neglect fouling factors.

Solution. The film coefficient h_o of the air is found from Fig. 8-21. To use this correlation the Reynolds number of the air must first be calculated, as follows.

The viscosity of air at 130°F is 0.046 lb/ft-hr (Appendix 7). The equivalent diameter of the shell space is

$$\text{ID of shell (Appendix 4)} = \frac{3.068}{12} = 0.2557 \text{ ft}$$

$$\text{OD of inner pipe (Appendix 4)} = \frac{1.900}{12} = 0.1583 \text{ ft}$$

The cross-sectional area of the shell space is

$$\frac{\pi(0.2557^2 - 0.1583^2)}{4} - \frac{28 \times 0.5 \times 0.035}{144} = 0.0282 \text{ ft}^2$$

The perimeter of the air space is

$$\pi 0.2557 + 2.830 = 3.633 \text{ ft}$$

The hydraulic radius is

$$r_H = \frac{0.0282}{3.633} = 0.00776 \text{ ft}$$

The equivalent diameter is

$$D_e = 4 \times 0.00776 = 0.0310 \text{ ft}$$

The Reynolds number of air is therefore

$$N_{\text{Re}} = \frac{0.0310 \times 5,000}{0.046} = 3.37 \times 10^3$$

From Fig. 8-21, the heat-transfer factor is

$$j_H'' = \frac{h_o}{c_p G} \left(\frac{c_p \mu}{k}\right)^{\frac{2}{3}} \left(\frac{\mu}{\mu_w}\right)^{-0.14} = 0.0031$$

The quantities needed to solve for h_o are

$$c_p = 0.25 \text{ Btu/(lb)(°F)} \quad \text{(Appendix 12)}$$

$$k = 0.0162 \text{ Btu-ft/(ft}^2\text{)(hr)(°F)} \quad \text{(Appendix 10)}$$

In computing μ_w the resistance of the wall and the steam film are considered negligible, so $t_w = 250°F$, and $\mu_w = 0.0528$ lb/ft-hr.

$$\phi_v = \left(\frac{\mu}{\mu_w}\right)^{0.14} = \left(\frac{0.046}{0.0528}\right)^{0.14} = 0.981$$

$$N_{Pr} = \frac{c_p\mu}{k} = \frac{0.25 \times 0.046}{0.0162} = 0.710$$

$$h_o = \frac{0.0031 \times 0.25 \times 5,000 \times 0.981}{0.710^{\frac{2}{3}}} = 4.78 \text{ Btu/(ft}^2)(\text{hr})(°F)$$

For rectangular fins, disregarding the contribution of the ends of the fins to the perimeter, $L_p = 2L$, and $S = Ly_F$, where y_F is the fin thickness and L is the longitudinal length of the fin. Then, from Eq. (8-88),

$$a_F x_F = x_F \sqrt{\frac{h_o(2L/Ly_F)}{k_m}} = x_F \sqrt{\frac{2h_o}{k_m y_F}}$$

$$= \frac{0.5}{12} \sqrt{\frac{2 \times 4.78}{26(0.035/12)}} = 0.467$$

From Fig. 8-20, $\eta_F = 0.933$.

The over-all coefficient is found from Eq. (8-87). The additional quantities needed are

$$D_i \text{ (Appendix 4)} = \frac{1.610}{12} = 0.1342 \text{ ft}$$

$$D_L = \frac{0.1583 - 0.1342}{2.303 \log (0.1583/0.1342)} = 0.1454 \text{ ft}$$

$$A_i = \pi 0.1342 \times 1.0 = 0.422 \text{ ft}^2/\text{lineal ft}$$

$$A_F + A_b = 2.830 \text{ ft}^2/\text{lineal ft}$$

$$A_F = 2.830 - 0.416 = 2.414 \text{ ft}^2/\text{lineal ft}$$

$$L = \frac{0.145}{12} = 0.0121 \text{ ft}$$

$$U_i = \frac{1}{\dfrac{0.422}{4.78(0.933 \times 2.414 + 0.416)} + \dfrac{0.0121 \times 0.1342}{26 \times 0.1454} + \dfrac{1}{1500}}$$

$$= 29.3 \text{ Btu/(ft}^2)(\text{hr})(°F)$$

Note that the over-all coefficient, when based on the small inside area of the inner pipe, may be much larger than the air-film coefficient based on the area of the extended surface.

Natural Convection

As an example of natural convection, consider a hot, vertical plate in contact with the air in a room. The temperature of the air in contact with the plate will be that of the surface of the plate, and a temperature gradient will exist from the plate out into the room. At the bottom of the plate, the temperature gradient is steep, as shown by the full line marked "Z = 1 cm" in Fig. 8-22. At distances above the bottom of the plate, the gradient becomes less steep, as shown by the full curves marked "Z = 24 cm" of Fig. 8-22. At a height of about 2 ft from the bottom of the plate, the temperature-distance curves approach an asymptotic condition and do not change with further increase in height.

The density of the heated air immediately adjacent to the plate is less than that of the unheated air at a distance from the plate, and the buoyancy

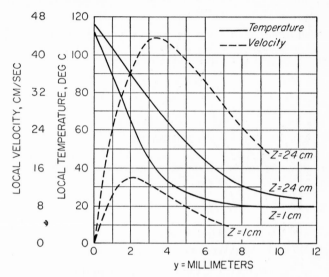

FIG. 8-22. Velocity and temperature gradients, natural convection from heated vertical plate. (*By permission of author and publishers, from "Heat Transmission," 3d ed., by W. H. McAdams. Copyright by author, 1954. McGraw-Hill Book Company, Inc.*)

of the hot air causes an unbalance between the vertical layers of air of differing density. As a result of the unbalanced forces, a circulation is generated, by which hot air near the plate rises and cold air flows toward the plate from the room to replenish the rising air stream. A velocity gradient near the plate is formed. Since the velocities of the air in contact with the plate and that out in the room are both zero, the velocity is a maximum at a definite distance from the wall. The velocity reaches its maximum a few millimeters from the surface of the plate. The dotted curves in Fig. 8-22 show the velocity gradients for heights of 1 and 24 cm above the bottom of the plate. For tall plates, an asymptotic condition is approached.

The temperature difference between the surface of the plate and the air in the room at a distance from the plate causes a transfer of heat by conduction into the current of gas next to the wall, and the stream carries the heat away by convection in a direction parallel to the plate.

The natural convection currents surrounding a hot, horizontal pipe are more complicated than those adjacent to a vertical heated plate, but the mechanism of the process is similar. The layers of air immediately next to the bottom and sides of the pipe are heated, and tend to rise. The rising layers of hot air, one on each side of the pipe, separate from the pipe at points short of the top center of the pipe and form two independent rising

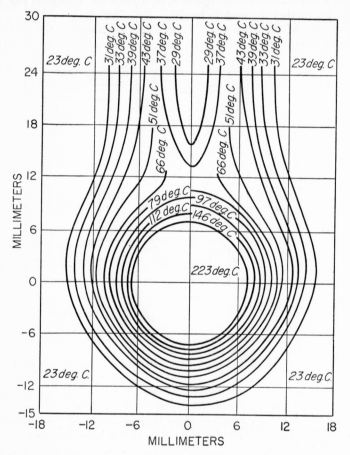

FIG. 8-23. Temperature field around heated horizontal cylinders. (*By permission of author and publishers, from "Heat Transmission," 2d ed., by W. H. McAdams. Copyright by author, 1942. McGraw-Hill Book Company, Inc.*)

currents with a zone of relatively stagnant and unheated air between them. Temperature explorations around a heated pipe, shown in Fig. 8-23, illustrate this effect.

Natural convection in liquids follows the same pattern, because liquids are also less dense hot than cold. The buoyancy of heated liquid layers near a hot surface generates convection currents just as in gases.

On the assumption that h depends upon pipe diameter, specific heat, thermal conductivity, viscosity, coefficient of thermal expansion, the acceleration of gravity, and temperature difference, dimensional analysis gives

$$\frac{hD_o}{k} = \Phi\left(\frac{c\mu}{k}, \frac{D_o^3 \rho g}{\mu^2}, \beta \, \Delta t\right) \tag{8-89}$$

Since the effect of β is through buoyancy in a gravitational field, the product $g\beta \, \Delta t$ acts as a single factor and the last two groups fuse into the Grashof number N_{Gr}.

For single horizontal cylinders, the heat-transfer coefficient can be correlated by an equation containing three dimensionless groups, the Nusselt number, the Prandtl number, and the Grashof number, or, specifically,

$$\frac{\bar{h}D_o}{k_f} = \Phi\left(\frac{c_p\mu_f}{k_f}, \frac{D_o^3 \rho_f^2 \beta g \, \Delta t_o}{\mu_f^2}\right) \tag{8-90}$$

where $\bar{h}$ = average heat-transfer coefficient, based on entire pipe surface, $Btu/(ft^2)(hr)(°F)$

D_o = outside pipe diameter, ft

k_f = thermal conductivity of fluid, $Btu\text{-}ft/(ft^2)(hr)(°F)$

c_p = specific heat, at constant pressure, of fluid, $Btu/(lb)(°F)$

ρ_f = density of fluid, lb/ft^3

β = coefficient of thermal expansion of fluid, $1/°F$

g = acceleration of gravity, ft/hr^2

Δt_o = average difference in temperature between outside of pipe and fluid distant from wall, °F

The fluid properties μ_f, ρ_f, and k_f are evaluated at the mean film temperature. Radiation is not accounted for in this equation.

The coefficient of thermal expansion β is a property of the fluid, defined as the fractional increase in volume, at constant pressure, of the fluid, per degree Fahrenheit, or mathematically,

$$\beta = \frac{(\partial v/\partial t)_p}{v} \tag{8-91}$$

where v = specific volume of fluid, ft^3/lb for liquids and ft^3/lb mole for gases

$(\partial v/\partial t)_p$ = rate of change of specific volume with temperature, at constant pressure, $ft^3/(lb)(°F)$ for liquids and $ft^3/(lb\ mole)(°F)$ for gases

For liquids, β can be considered constant over a definite temperature range and Eq. (8-91) written as

$$\beta = \frac{\Delta v/\Delta t}{\bar{v}}$$

where $\bar{v}$ is the average specific volume in cubic feet per pound. In terms of density,

$$\beta = \frac{1/\rho_2 - 1/\rho_1}{(t_2 - t_1)(1/\rho_1 + 1/\rho_2)/2} = \frac{\rho_1 - \rho_2}{\bar{\rho}_a(t_2 - t_1)} \tag{8-92}$$

where $\bar{\rho}_a = (\rho_1 + \rho_2)/2$

ρ_1 = density of fluid at temperature t_1

ρ_2 = density of fluid at temperature t_2

For an ideal gas, since $v = R_o T/p$,

$$\left(\frac{\partial v}{\partial T}\right)_p = \left(\frac{\partial v}{\partial t}\right)_p = \frac{R_o}{p}$$

and
$$\beta = \frac{R_o/p}{R_o T/p} = \frac{1}{T} \qquad (8\text{-}93)$$

The coefficient of thermal expansion of an ideal gas equals the reciprocal of the absolute temperature.

In Fig. 8-24 is shown a relationship, based on Eq. (8-90), which satisfactorily correlates experimental data for heat transfer from a single hori-

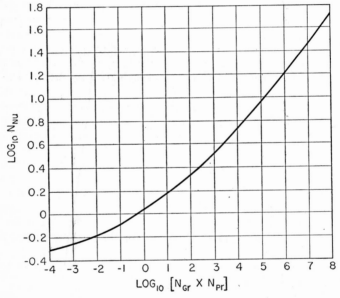

FIG. 8-24. Heat transfer between single horizontal cylinders and fluids in natural convection.

zontal cylinder to liquids or gases. The range of variables covered by the single line of Fig. 8-24 is very great.

For magnitudes of log $N_{Gr}N_{Pr}$ of 4 or more, the line of Fig. 8-24 follows closely the empirical equation [24e]

$$N_{Nu} = 0.53(N_{Gr}N_{Pr})_f^{0.25} \qquad (8\text{-}94)$$

Natural Convection to Air from Vertical Shapes and Horizontal Planes.

Equations for heat transfer in natural convection between fluids and solids of definite geometric shape are of the form

$$\frac{hL}{k_f} = b \left[\frac{L^3 \rho_f^2 g \beta_f \, \Delta t}{\mu_f^2} \left(\frac{c_p \mu}{k} \right)_f \right]^n \tag{8-95}$$

where b and n are constants and L is the height of a vertical surface or the length of a horizontal square surface, in feet.[24d] Properties are taken at the mean film temperature. Equation (8-95) can be written

$$N_{\mathrm{Nu},f} = b(N_{\mathrm{Gr}} N_{\mathrm{Pr}})_f^n$$

Values of the constants b and n for various conditions are given in Ref. 24d.

Heat Transfer to Jackets and Coils.

In the formulas given above, the fluid is either flowing at a known velocity through a definite channel or is moving in natural convection across a simple shape, such as a horizontal or vertical tube or plate. Many practical heat-transfer problems are encountered that cannot be solved by the use of these equations. For example, consider a tank of liquid heated with a steam coil or steam jacket. The velocity of the liquid as it circulates through the tank is due only to natural convection, which in turn depends on the dimensions and proportions of the tank, the shape and area of the coil, the viscosity of the liquid, and other factors. No general equation is available for calculation of the coefficient for this process, although some data on over-all coefficients are known.[29e] Where the expense is justified, experimental data can be determined and correlated by the same methods that were used to find the correlations now available.

In an agitated vessel, the agitator develops forced-convection currents. In Chap. 6 it was shown that the dimensionless group $D_a^2 n \rho / \mu$ is a Reynolds number useful in correlating data on the power consumption in a stirred tank. This same group has been found to be satisfactory as a correlating variable for heat transfer to jackets or coils in an agitated tank.[5] The following equations are typical of those that have been offered for this purpose.

For heating or cooling liquids in a cylindrical tank equipped with a single heating or cooling coil,[10]

$$\frac{hD_t}{k} = 1.01 \left(\frac{D_a^2 n \rho}{\mu} \right)^{0.62} \left(\frac{c_p \mu}{k} \right)^{\frac{1}{3}} \left(\frac{\mu}{\mu_w} \right)^{0.14} \tag{8-96}$$

For heating and cooling liquids in a cylindrical vessel equipped with a heating or cooling jacket,

$$\frac{hD_t}{k} = 0.40 \left(\frac{D_a^2 n \rho}{\mu} \right)^{\frac{2}{3}} \left(\frac{c_p \mu}{k} \right)^{\frac{1}{3}} \left(\frac{\mu}{\mu_w} \right)^{0.14} \tag{8-97}$$

where h = heat-transfer coefficient, heating surface to liquid,
$\qquad$ Btu/(ft^2)(hr)(°F)
D_a = diameter of agitator, ft
D_t = tank diameter, ft
$\quad k$ = thermal conductivity of liquid, Btu-ft/(ft^2)(hr)(°F)
$\quad n$ = agitator speed, rph
$\quad \rho$ = density of liquid, lb/ft^3
$\quad \mu$ = viscosity of liquid at bulk-liquid temperature, lb/ft-hr
$\quad \mu_w$ = viscosity of liquid at temperature of wall of coil or jacket,
$\qquad$ lb/ft-hr
$\quad c_p$ = specific heat of liquid, Btu/(lb)(°F)

Equations of this type are not applicable to other situations differing from those for which the equations were derived. In the absence of specific data and equations, the engineer is faced, as usual, with the necessity of basing his answer to the problem on a combination of experience and judgment.

Heat Transfer from Condensing Vapors

The condensation of vapors on the surfaces of tubes cooler than the condensing temperature of the vapor is important when vapors such as those of water, hydrocarbons, and other volatile substances are processed. Some examples will be met later in this text, in discussing the unit operations of evaporation, distillation, and drying.

The condensing vapor may consist of a single substance, a mixture of condensable and noncondensable substances, or a mixture of two or more condensable vapors. Friction losses in a condenser are normally small, so that condensation is essentially a constant-pressure process. The condensing temperature of a single pure substance depends only on the pressure, and therefore the process of condensation of a pure substance is isothermal. Also, the condensate is a pure liquid. Mixed vapors, condensing at constant pressure, condense over a temperature range and yield a condensate of variable composition until the entire vapor stream is condensed, when the composition of the condensate equals that of the original uncondensed vapor.† A common example of the condensation of one constituent from its mixture with a second noncondensable substance is the condensation of water from a mixture of steam and air.

The condensation of mixed vapors is complicated and beyond the scope of this text.[18e, 24q] The following discussion is directed to the heat transfer from a single volatile substance condensing on a cold tube.

Dropwise and Film-type Condensation. A vapor may condense on a cold surface in one of two ways, which are well described by the terms dropwise and film-type. In film condensation, which is more common than dropwise condensation, the liquid condensate forms a film, or continuous layer, of liquid that flows over the surface of the tube under the action of gravity. It is the layer of liquid interposed between the vapor

---

† Exceptions to this statement are found in the condensation of azeotropic mixtures, which are discussed on pp. 668 to 670.

and the wall of the tube that provides the resistance to the flow of heat and, therefore, controls the heat-transfer coefficient. In dropwise condensation, the condensate forms as fine drops, like those often seen on the outside of a cold-water pitcher in a humid room. The fine drops tend to coalesce into rivulets, which flow down the tube under the action of gravity, sweep away condensate, and clear the surface for more droplets. During dropwise condensation, large areas of the cold tube are bare and are directly exposed to the vapor. Because of the absence of the liquid film, the resistance to heat flow at these bare areas is very low and the heat-transfer coefficient correspondingly high. The average coefficient for dropwise condensation may be from five to eight times that for film-type condensation. On long tubes, condensation on some of the surface may be film condensation and the remainder dropwise condensation.

Dropwise condensation is observed only with steam. Its appearance depends upon the wetting or nonwetting of the surface by liquid water. Fundamentally, the phenomenon lies in the field of surface chemistry. A summary of the results of considerable experimental work on the dropwise condensation of steam is given in the following paragraphs.[11]

1. Film-type condensation occurs if both the steam and the tube are clean, in the presence or absence of air, on rough or on polished surfaces.

2. Dropwise condensation is obtainable only when the cooling surface is contaminated. It is more easily maintained on a smooth contaminated surface than on a rough contaminated surface.

3. The quantity of contaminant or promoter required to cause dropwise condensation is minute, and apparently only a monomolecular film is necessary.

4. Effective drop promoters are strongly adsorbed by the surface, and substances that merely prevent wetting are ineffective. Some promoters are especially effective on certain metals, e.g., mercaptans on copper alloys; other promoters, such as oleic acid, are quite generally effective. Some metals, e.g., steel and aluminum, are difficult to treat to give dropwise condensation.

5. The average coefficient obtainable in pure dropwise condensation is as high as 14,000 $Btu/(ft^2)(hr)(°F)$.

Although attempts are sometimes made to realize practical benefits from these large coefficients by artificially inducing dropwise condensation, this type of condensation is so unstable and the difficulty of maintaining it so great that the method is not common. Also the resistance of the layer of steam condensate even in film-type condensation is ordinarily small in comparison with the resistance inside the condenser tube, and the increase in the over-all coefficient is relatively small when dropwise condensation is achieved. For normal design, therefore, film-type condensation is assumed.

Coefficients for Film-type Condensation. The basic equations for the rate of heat transfer in film-type condensation were first derived by Nusselt.[18c,24p, 27] The Nusselt equations are based on the assumption that the only resistance to the flow of heat is that offered by the layer of condensate

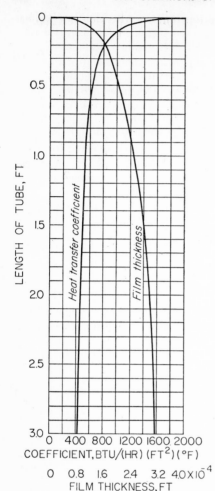

flowing downward in laminar flow under the action of gravity. It is also assumed that the velocity of the liquid at the wall is zero, that the velocity of the liquid at the outside of the film is not influenced by the velocity of the vapor, and that the temperatures of the wall and the vapor are constant. Superheat in the vapor is neglected, the condensate is assumed to leave the tube at the condensing temperature, and the physical properties of the liquid are taken at mean film temperature.

Vertical Tubes. In film-type condensation, the Nusselt theory shows that the condensate film starts to form at the top of the tube and that the thickness of the film increases rapidly in the first few inches at the top and then more and more slowly in the remaining tube length. The *local* heat-transfer coefficient changes inversely with the film thickness, and is large where the film is thin and small where the film is thick. The variations of the film thickness and of the local coefficient, both with length of tube, are shown, for a typical case, in Fig. 8-25.

The average heat-transfer coefficient, based on the entire tube, is, from the Nusselt theory,

Fig. 8-25. Film thickness and local coefficients, descending film of condensate. (*By permission, from "Process Heat Transfer," by D. Q. Kern. Copyright, 1950. McGraw-Hill Book Company, Inc.*)

$$\bar{h}\left(\frac{\mu_f^2}{k_f^3 \rho_f^2 g}\right)^{\frac{1}{3}} = 1.47 \left(\frac{4\Gamma}{\mu_f}\right)^{-\frac{1}{3}} \tag{8-98}$$

where $\bar{h}$ = average coefficient, based on entire tube, Btu/(ft^2)(hr)(°F)
k_f = thermal conductivity of condensate, at mean film temperature, Btu-ft/(ft^2)(hr)(°F)
μ_f = viscosity of condensate, at mean film temperature, lb/ft-hr
ρ_f = density of condensate, at mean film temperature, lb/ft^3
g = acceleration of gravity, 4.17 × 10^8 ft/hr^2
Γ = condensate loading per unit tube perimeter, at bottom of tube, lb/ft-hr

The mean film temperature is evaluated from the following relation.[24p]

$$t_f = t_h - \frac{3(t_h - t_w)}{4} \qquad (8\text{-}99)$$

where t_f = mean film temperature for evaluation of k_f, μ_f, and ρ_f
  t_h = temperature of condensing vapor
  t_w = temperature of outside surface of tube wall
The loading per unit perimeter is defined by the equation

$$\Gamma = \frac{w}{\pi D_o} \qquad (8\text{-}100)$$

where w = condensation rate, for entire tube, lb/hr
  D_o = outside tube diameter, ft
The term $4\Gamma/\mu_f$ is a Reynolds number for the flow of a liquid film down a vertical surface. This can be shown with the aid of the basic equation for a Reynolds number,

$$N_{\text{Re}} = \frac{4r_H G}{\mu} \qquad (8\text{-}101)$$

Thus, let S_c be the cross-sectional area of the annular ring of condensate flowing down the tube. Then, from the definitions of hydraulic radius r_H and mass velocity G,

$$r_H = \frac{S_c}{\pi D_o} \qquad (8\text{-}102)$$

$$G = \frac{w}{S_c} \qquad (8\text{-}103)$$

Substituting Eqs. (8-102) and (8-103) into Eq. (8-101) and noting the definition of Γ,

$$N_{\text{Re}} = \frac{4S_c}{\pi D_o} \frac{w}{S_c \mu} = \frac{4\Gamma}{\mu} \qquad (8\text{-}104)$$

Equation (8-98) is often used in an equivalent form, in which the term Γ has been eliminated by means of the basic definition of the coefficient $\bar{h}$. This definition is

$$\bar{h} = \frac{q}{A_o(t_h - t_w)} = \frac{q}{A_o \, \Delta t_o} \qquad (8\text{-}105)$$

where q = total heat transferred, Btu/hr
  A_o = outside tube area, ft^2
  Δt_o = difference between condensing temperature and outside temperature of tube wall, °F
Since only latent heat of condensation is transferred,

$$q = w\lambda \qquad (8\text{-}106)$$

where λ is the latent heat of condensation in Btu per pound.

The heat-transfer area is

$$A_o = \pi D_o L \tag{8-107}$$

where L is the heated length of the tube in feet.

From Eqs. (8-105) to (8-107),

$$\bar{h} = \frac{w\lambda}{\pi \, \Delta t_o \, D_o L} = \frac{\Gamma\lambda}{L \, \Delta t_o} \tag{8-108}$$

Eliminating Γ from Eq. (8-98) by means of Eq. (8-108) and solving for $\bar{h}$ gives

$$\bar{h} = 0.943 \left(\frac{k_f^3 \rho_f^2 g \lambda}{\Delta t_o \, L \mu_f} \right)^{\frac{1}{4}} \tag{8-109}$$

Horizontal Tubes. Corresponding to Eqs. (8-98) and (8-109) for vertical tubes, the following equations apply to horizontal tubes,

$$\bar{h} \left(\frac{\mu_f^2}{k_f^3 \rho_f^2 g} \right)^{\frac{1}{3}} = 1.51 \left(\frac{4\Gamma'}{\mu_f} \right)^{-\frac{1}{3}} \tag{8-110}$$

and

$$\bar{h} = 0.725 \left(\frac{k_f^3 \rho_f^2 g \lambda}{\Delta t_o \, D_o \mu_f} \right)^{\frac{1}{4}} \tag{8-111}$$

where Γ' is the condensate loading per unit length of tube, in pounds per foot per hour, and all other symbols have the usual meaning.

The close resemblance between Eqs. (8-98) and (8-110) is obvious. The equations can be made identical by using the coefficient 1.5 for each. The quantity $4\Gamma'/\mu_f$ in Eq. (8-110) is not, however, a Reynolds number but is twice the Reynolds number for condensate flow around a horizontal tube.

Practical Use of Nusselt Equations. In the absence of high vapor velocities, experimental data check Eqs. (8-110) and (8-111) well, and these equations can be used as they stand for calculating heat-transfer coefficients for film-type condensation on a single horizontal tube. Also, Eq. (8-110) can be used for film-type condensation on a vertical stack of horizontal tubes, where the condensate falls cumulatively from tube to tube and the total condensate from the entire stack finally drops from the bottom tube. It is necessary only to define Γ as the average loading per tube, based on the total flow from the bottom tube.

$$\Gamma = \frac{w}{LN} \tag{8-112}$$

where w = total condensate for entire stack, lb/hr

L = length of one tube, ft

N = number of tubes in stack

Practically, because of the fact that some condensate splashes away from each individual tube instead of dripping completely to the tube below, it is more accurate to use a value of Γ calculated from the equation [18d]

$$\Gamma = \frac{w}{LN^{\frac{2}{3}}} \tag{8-113}$$

Equation (8-111) becomes

$$\bar{h} = 0.725 \left(\frac{k_f^3 \rho_f^2 g \lambda}{N^{\frac{2}{3}} \, \Delta t_o \, D_o \mu_f} \right)^{\frac{1}{4}} \tag{8-114}$$

Equations (8-98) and (8-109), for vertical tubes, were derived on the assumption that the condensate flow was laminar. This limits their use to cases where $4\Gamma/\mu_f$ is less than 2,100. For long tubes, the condensate film becomes sufficiently thick, and its velocity sufficiently large, to cause turbulence to occur in the lower portions of the tube. Also, even when the flow remains laminar throughout, coefficients measured experimentally are about 20 per cent larger than those calculated from the equations. This is attributed to the effect of ripples on the surface of the falling film. For practical calculations the coefficients in Eqs. (8-98) and (8-109) should be taken as 1.76 and 1.13, respectively. When the velocity of the vapor phase in the condenser is appreciable, the drag of the vapor induces turbulence in the condensate layer. The coefficient of the condensing film is then considerably larger than predicted by Eqs. (8-98) and (8-109), and should be computed by the method of Carpenter and Colburn.[3] In the presence of a high vapor velocity the flow of the condensate layer changes from streamline to turbulent at a Reynolds number, $4\Gamma/\mu$, of about 240.

In Fig. 8-26 is plotted $\bar{h}(\mu_f^2/k_f^3\rho_f^2 g)^{\frac{1}{3}}$ vs. $4\Gamma/\mu_f$. Line AA is the theoretical relation for both horizontal and vertical tubes for magnitudes of $4\Gamma/\mu_f$

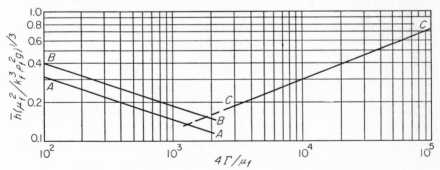

FIG. 8-26. Film condensation on vertical tubes or plates.

less than 2,100. This line can be used directly for horizontal tubes. Line BB, lying above line AA, is used for vertical tubes, and includes the effect of ripples.

For magnitudes of $4\Gamma/\mu_f$ greater than 2,100, the coefficient $\bar{h}$ increases with increase in Reynolds number. Line CC in Fig. 8-26 can be used for calculation of values of $\bar{h}$ when the value of $4\Gamma/\mu_f$ exceeds 2,100. This value is not reached in condensation on horizontal pipes, and a line for turbulent flow is not needed.

The quantity $(k_f^3 \rho_f^2 g / \mu_f^2)^{\frac{1}{3}}$ is, for a given substance over a moderate pressure range, a function of temperature. Use of Fig. 8-26 is facilitated if this quantity, which can be denoted by ψ_f, is calculated and plotted as a function of temperature for a given vapor. If the individual factors of ψ_f are expressed in Btu, hours, feet, pounds, and degrees Fahrenheit, the dimensions of ψ_f are $Btu/(ft^2)(hr)(°F)$. In Appendix 14 is given the magnitude of ψ_f for water as a function of temperature. Corresponding tables can be prepared for other vapors when desired.

Example 8-9. A shell-and-tube condenser with vertical $\frac{3}{4}$-in. 16 BWG copper tubes has chlorobenzene condensing at atmospheric pressure in the shell. The tubes are 5 ft long. Cooling water at an average temperature of 175°F is flowing in the tubes. If the water-side coefficient is 400 Btu/(ft²)(hr)(°F), (a) what is the coefficient of the condensing chlorobenzene; (b) what would the coefficient be in a horizontal condenser with the same number of tubes if the average number of tubes in a vertical stack is six? Neglect fouling factors and tube-wall resistance.

Solution. (a) Equation (8-109) applies, but the properties of the condensate must be evaluated at the film temperature t_f, which is given by Eq. (8-99). In computing t_f the wall temperature t_w must be estimated from $\bar{h}$, the condensate-film coefficient. A trial-and-error solution is therefore necessary.

The quantities in Eq. (8-109) that may be specified directly are

$$\lambda = 139.7 \text{ Btu/lb } ^{29b}$$

$$g = 4.17 \times 10^8 \text{ ft/hr}^2$$

$$L = 5 \text{ ft}$$

The condensing temperature t_h is 267°F.

The wall temperature t_w must be between 175 and 267°F. The resistance of a condensing film of organic liquid is usually greater than the thermal resistance of flowing water; consequently t_w is probably closer to 175°F than it is to 267°F. As a first approximation, t_w will be taken as 205°F.

The temperature difference Δt_o is $267 - 205 = 62°F$.

The average film temperature, from Eq. (8-99), is

$$t_f = 267 - \tfrac{3}{4}(267 - 205) = 220°F$$

The density and thermal conductivity of liquids vary so little with temperature that they may be assumed constant at the following values.

$$k_f = 0.083 \text{ Btu-ft}/(ft^2)(hr)(°F) \qquad \text{(Appendix 11)}$$

$$\rho_f = 65.4 \text{ lb/ft}^3$$

The viscosity of the film is

$$\mu_f = 0.30 \times 2.42 = 0.726 \text{ lb/ft-hr} \qquad \text{(Appendix 8)}$$

The first estimate of $\bar{h}$ is then

$$\bar{h} = 1.13 \left(\frac{0.083^3 \times 65.4^2 \times 4.17 \times 10^8 \times 139.7}{62 \times 5 \times 0.726} \right)^{\frac{1}{4}}$$

$$= 179 \text{ Btu}/(ft^2)(hr)(°F)$$

The corrected wall temperature is found from Eq. (8-54). The outside diameter of the tubes D_o is $0.75/12 = 0.0625$ ft. The inside diameter D_i is $0.0625 - (2 \times 0.065)/12 = 0.0517$ ft.

The temperature drop over the water resistance is, from Eq. (8-54),

$$\Delta t_i = \frac{1/400}{1/400 + 0.0517/(0.0625 \times 179)} (267 - 175) = 32°F$$

The wall temperature is

$$t_w = 175 + 32 = 207°F$$

This is sufficiently close to the estimated value, 205°F, to make further calculation unnecessary. The coefficient $\bar{h}$ is 179 Btu/(ft²)(hr)(°F).

(b) For a horizontal condenser, Eq. (8-114) is used. The coefficient of the chlorobenzene is probably greater than in part (a), so the wall temperature t_w is now estimated to be 215°F. The new quantities needed are

$$N = 6 \qquad \Delta t_o = 267 - 215 = 52°F \qquad D_o = 0.0625 \text{ ft}$$

$$t_f = 267 - \tfrac{3}{4}(267 - 215) = 228°F$$

$$\mu_f = 0.280 \times 2.42 = 0.68 \text{ lb/ft-hr} \qquad \text{(Appendix 14)}$$

$$\bar{h} = 0.725 \left(\frac{0.083^3 \times 65.4^2 \times 4.17 \times 10^8 \times 139.7}{6^{\frac{3}{3}} \times 52 \times 0.0625 \times 0.68} \right)^{\frac{1}{4}}$$

$$= 271 \text{ Btu/(ft²)(hr)(°F)}$$

$$\Delta t_i = \frac{1/400}{1/400 + 0.0517/(0.0625 \times 271)} 92 = 41°F$$

$$t_w = 175 + 41 = 216°F$$

This checks the assumed value, 215°F, and no further calculation is needed. The coefficient $\bar{h}$ is 271 Btu/(ft²)(hr)(°F).

In general, the coefficient of a film condensing on a horizontal tube is considerably larger than that on a vertical tube under otherwise similar conditions unless the tubes are very short or there are very many horizontal tubes in each stack. Vertical tubes are preferred when the condensate must be appreciably subcooled below its condensation temperature. Mixtures of vapors and noncondensing gases are usually cooled and condensed *inside* vertical tubes, so that the inert gas is continually swept away from the heat-transfer surface by the incoming stream.

Condensation of Superheated Vapors. If the vapor entering a condenser is superheated, both the sensible heat of superheat and the latent heat of condensation must be transferred through the cooling surface. For steam, because of the low specific heat of the superheated vapor and the large latent heat of condensation, the heat of superheat is usually small in comparison with the latent heat. For example, 100 degrees of superheat represents only 50 Btu/lb, as compared with approximately 1000 Btu/lb for latent heat. In the condensation of organic vapors, such as petroleum fractions, the superheat may be appreciable in comparison with the latent heat. When the heat of superheat is important, either it can be calculated from the degrees of superheat and the specific heat of the vapor and added to the latent heat, or if tables of thermal properties are available, the total heat transferred per pound of vapor can be calculated by subtracting the enthalpy of the condensate from that of the superheated vapor.

The effect of superheat on the rate of heat transfer depends upon whether the temperature of the surface of the tube is higher or lower than the condensation temperature of the vapor. If the temperature of the tube is lower than the temperature of condensation, the tube is wet with condensate, just as in the condensation of saturated vapor, and the temperature of the outside boundary of the condensate layer equals the saturation temperature of the vapor at the pressure existing in the equipment. The situation is complicated by the presence of a thermal resistance between the bulk of the superheated vapor and the outside of the condensate film and by the existence of a temperature drop, which is equal to the degrees of superheat in the vapor, across that resistance. Practically, however, the net effect of these complications is small, and it is satisfactory to assume that the entire heat load, which consists of the heats of superheat and of condensation, is transferred through the condensate film; that the temperature drop is that across the condensate film; and that the coefficient is the average coefficient for condensing vapor as read from Fig. 8-26. The procedure is summarized by the equation

$$q = \bar{h}A(t_h - t_w) \tag{8-115}$$

where q = total heat transferred, including latent heat and superheat, Btu/hr

A = area of heat-transfer surface in contact with vapor, ft^2

$\bar{h}$ = coefficient of heat transfer, from Fig. 8-26, $Btu/(ft^2)(hr)(°F)$

t_h = saturation temperature of vapor, °F

t_w = temperature of tube wall, °F

When the vapor is highly superheated and the exit temperature of the cooling fluid is close to that of condensation, the temperature of the tube wall may be greater than the saturation temperature of the vapor, condensation cannot occur, and the tube wall will be dry. The tube wall remains dry until the superheat has been reduced to a point where the tube wall becomes cooler than the condensing temperature of the vapor, and condensation takes place. The equipment can be considered as two sections, one a desuperheater and the other a condenser. In calculations, the two sections must be considered separately. The desuperheater is essentially a gas cooler. The logarithmic mean temperature applies, and the heat-transfer coefficient is that for cooling a fixed gas. The condensing section is treated by the methods described in the previous paragraphs.

Because of the low individual coefficient on the gas side, the over-all coefficient in the desuperheater section is small, and the area of the heating surface in that section is large in comparison with the amount of heat removed. This situation should be avoided in practice. Superheat can be eliminated more economically by injection of liquid directly into the superheated vapor, the desuperheating section thereby eliminated, and the high coefficients from condensing vapors so obtained.

Effect of Noncondensable Gases on Rate of Condensation. The presence of even small amounts of noncondensing gas in a condensing vapor seriously reduces the rate of condensation. A layer of gas accumulates on the film of condensate and establishes an added thermal resistance which

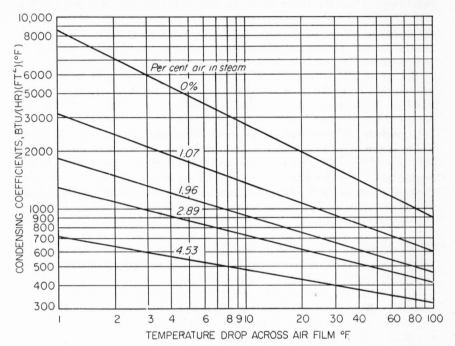

FIG. 8-27. Effect of air on condensation of steam. *(By permission, from "Process Heat Transfer," by D. Q. Kern. Copyright, 1950. McGraw-Hill Book Company, Inc.)*

is not present during the condensation of a pure vapor. Also, the vapor of the condensing substance must flow by diffusion through the layer of gas to reach the film of condensate, and the resistance offered by the gas to diffusion further impedes the process of condensation. Rigorous methods of solving the general problem are based on equating the heat flow to the condensate surface at any point to the heat flow away from the surface. This involves trial-and-error solutions for the point temperatures of the condensate surface, and from these an estimation of the point values of the heat flux $U \Delta t$. The values of $1/(U \Delta t)$ for each point are plotted against the heat transferred to that point and the area of the condenser surface found by graphical integration.[8,31]

An important example of condensation in the presence of small amounts of fixed gas is met in the condensation of steam containing air.[28] Figure

8-27 shows how, in a typical case, the coefficient for condensing steam is reduced by small concentrations of air.

Heat Transfer to Boiling Liquids

Heat transfer to a boiling liquid is a necessary step in the unit operations of evaporation and distillation and also in other kinds of general processing, such as steam generation, petroleum processing, and control of the temperatures of chemical reactions. The boiling liquid may be contained in a vessel equipped with a heating surface fabricated from horizontal or vertical plates or tubes, which supplies the heat necessary to boil the liquid. Or the liquid may flow through heated tubes, under either natural or forced convection, and the heat transferred to the fluid through the walls of the tubes. The most important application of boiling in tubes is the evaporation of water from solutions. Evaporation is described in detail in Chap. 9, and the present section is limited to a discussion of boiling by means of hot immersed surfaces.

In practice, either one of two types of boiling may be used. In the first, which is the more common, the temperature of the mass of the liquid is the same as the boiling point of the liquid under the pressure existing in the equipment. Bubbles of vapor are generated at the heating surface, rise through the mass of liquid, and disengage from the surface of the liquid. Vapor accumulates in a vapor space over the liquid; a vapor outlet from the vapor space removes the vapor as fast as it is formed. This type of boiling can be described as the *boiling of saturated liquid* since the vapor leaves the liquid in equilibrium with the liquid at its boiling temperature. In the second type of boiling, which is used in special problems, such as the liquid cooling of airplane engines, the temperature of the mass of the liquid is below that of its boiling point, but the temperature of the heating surface is above the boiling point of the liquid. Bubbles form on the heating surface, but on release from the surface are absorbed by the mass of the liquid. This type of boiling is called *surface boiling*. The mass of the liquid is at a temperature below the boiling temperature corresponding to the pressure of the liquid. In surface boiling, the stream of boiling liquid is confined by the channel through which the fluid stream is flowing, and vapor separation does not occur, so no vapor space is necessary. The two methods of boiling will be discussed separately.

Boiling of Saturated Liquid. Consider a horizontal tube immersed in a vessel containing a boiling liquid. Assume that q/A, the heat-flux density in Btu per square foot per hour, and Δt, the difference between the temperature of the tube wall and that of the boiling liquid, are measured. Start with a very low temperature drop Δt. Increase the temperature drop by steps, measuring q/A and Δt at each step, until very large values of Δt are reached. A plot of q/A vs. Δt on logarithmic coordinates will give a curve of the type shown in Fig. 8-28. This curve can be divided into four segments. In the first segment, at low temperature drops, the line AB is straight, and log (q/A) changes linearly with log Δt. The slope of the line is 1.25. The equivalent mathematical statement is

$$\frac{q}{A} = b \, \Delta t^{1.25} \qquad (8\text{-}116)$$

where b is a constant. The second segment, line BC, is also straight, but
its slope is greater than that of line AB. The slope of the line BC depends
upon the specific experiment; it usually lies between 3 and 4. The second
segment terminates at a definite point of maximum flux, which is point C
in Fig. 8-28. The temperature drop corresponding to point C is called the

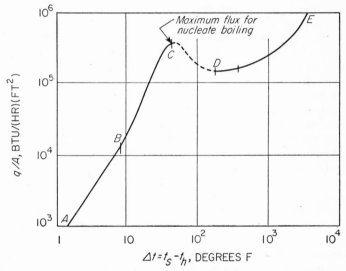

Fig. 8-28. Flux density vs. temperature drop, boiling of water at 212°F: AB, natural
convection. BC, nucleate boiling. CD, partial film boiling. DEF, film boiling. (Mc-
Adams, Addoms, Rinaldo, and Day.[25] Courtesy of AIChE.)

critical temperature drop, and the flux at point C is the peak flux. In the
third segment, line CD in Fig. 8-28, the flux decreases as the temperature
drop rises and reaches a minimum at point D. Point D is called the Leidenfrost point. In the last segment, line DE, the flux again increases with Δt
and, at large temperature drops, surpasses the previous maximum reached
at point C.

Because, by definition, $h = (q/A)/\Delta t$, the plot of Fig. 8-28 is readily
convertible into a plot of h vs. Δt. This curve is shown in Fig. 8-29. A
maximum coefficient, corresponding to point C of Fig. 8-28, and a minimum coefficient, corresponding to point D of Fig. 8-28, are evident in
Fig. 8-29. The coefficient is proportional to $\Delta t^{0.25}$ in the first segment of
the line in Fig. 8-29 and to between Δt^2 and Δt^3 in the second segment.

Each of the four segments of the graph in Fig. 8-28 corresponds to a
definite mechanism of boiling. In the first section, at low temperature
drops, the mechanism is that of heat transfer to a liquid in natural con-

vection, and the variation of h with Δt agrees with that given by Eq. (8-95). Bubbles form on the surface of the heater, are released from it, rise to the surface of the liquid, and are disengaged into the vapor space; but they are too few to disturb appreciably the normal currents of free convection.

At larger temperature drops, lying between 9 and 45°F in the case shown in Fig. 8-29, the rate of bubble production is large enough so that the stream of bubbles moving up through the liquid increases the velocity of the circulation currents in the mass of liquid, and the coefficient of heat transfer becomes greater than that in undisturbed natural convection. As Δt is increased, the rate of bubble formation increases, and the coefficient increases rapidly.

The action occurring at temperature drops below the critical temperature drop is called *nucleate boiling*, in reference to the formation of tiny

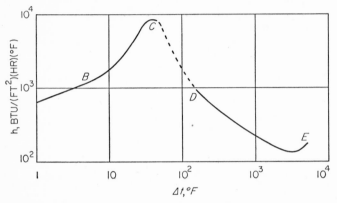

FIG. 8-29. Heat-transfer coefficients vs. Δt, boiling of water at 1 atm.

bubbles, or vaporization nuclei, on the heating surface. During nucleate boiling, the bubbles occupy but a small portion of the tube surface at a time, and most of the surface is in direct contact with liquid. As the flux increases, however, more and more of the heating surface is covered by vapor, which, because of its poor thermal conductivity, tends to blanket the surface. The insulating effect of the vapor becomes so large at the critical temperature drop that the flux diminishes as the temperature drop is further increased. The vapor blanketing that occurs at this point is sometimes called "vapor binding." At the Leidenfrost point vapor binding becomes complete, and the entire heat current must flow by conduction and radiation through a fixed gas. This action is called *film boiling*. The mechanism at temperature drops between the critical temperature drop and the Leidenfrost point is a combination of nucleate and film boiling, and is called "metastable film boiling." Film boiling can be seen when a drop of water is placed on a hot metal plate.

Clearly, film boiling is not desired in commercial equipment, because the heat-transfer rate is low in comparison with the temperature drop, and temperature drop is not utilized efficiently. Heat-transfer apparatus should be so designed and operated that the temperature drop in the film of boiling liquid is smaller than the critical temperature drop.

The effectiveness of nucleate boiling depends primarily on the ease with which bubbles form and free themselves from the heating surface. The layer of liquid next to the hot surface is superheated by contact with the wall of the heater. The superheated liquid tends to form vapors spontaneously and so relieve the superheat. It is the tendency of superheated liquid to "flash" into vapor that provides the impetus for the boiling process. Physically, the flash can occur only by forming vapor-liquid interfaces in the form of small bubbles. It is not easy to form a small bubble, however, in a superheated liquid, because, at a given temperature, the vapor pressure in a very small bubble is less than that in a large bubble or that from a plane liquid surface. A very small bubble can exist in equilibrium with superheated liquid, and the smaller the bubble, the greater the equilibrium superheat and the smaller the tendency to flash. By taking elaborate precautions to eliminate all gas and other impurities from the liquid and to prevent shock, it is possible to superheat water by several degrees Fahrenheit without formation of bubbles.

A second difficulty appears if the bubble does not readily leave the surface once it is formed. The important factor in controlling the rate of bubble detachment is the interfacial tension between the liquid and the heating surface. If the interfacial tension is large, the bubble tends to spread along the surface and blanket the heat-transfer area, as shown in Fig. 8-30c, rather than leaving the surface to make room for other bubbles. If the interfacial tension is low, the bubble will pinch off easily, in the manner shown in Fig. 8-30a. An example of intermediate interfacial tension is shown in Fig. 8-30b.

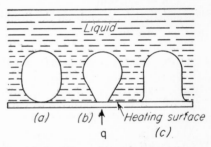

Fig. 8-30. Effect of interfacial tension on bubble formation. (*Jakob and Fritz.*[17])

The coefficient obtained during nucleate boiling is sensitive to a number of variables, including the nature of the liquid, the type and condition of the heating surface, the composition and purity of the liquid, the presence or absence of agitation, and the temperature or pressure. Minor changes in some variables cause major changes in the coefficient. The reproducibility of check experiments is poor, and general correlations in this field are nearly unknown.

Qualitatively, the effects of some variables can be predicted from a consideration of the mechanisms of boiling. A roughened surface provides centers for nucleation that are not present on a polished surface. Thus roughened surfaces usually give larger coefficients than smooth surfaces. This effect, however, is often due to the fact that the total surface of a rough tube is larger than that of a smooth surface of the same projected area. A very thin layer of scale may increase the coefficient of the boiling liquid, but even a thin scale will reduce the over-all coefficient by adding a resistance that reduces the over-all coefficient more than the improved boiling liquid coefficient increases it. Gas or air adsorbed on the surface of the heater or contaminants on the surface often facilitate boiling by either the formation or the disengaging of bubbles. A freshly cleaned surface may give a higher or lower coefficient than the same surface after it has been stabilized by a previous period of operation. This effect is associated with a change in the condition of the heating surface. Agitation increases the coefficient by increasing the velocity of the liquid across the surface, which helps to sweep away bubbles.

The temperature of a boiling liquid can, of course, be changed by changing the pressure. The change in coefficient with change in temperature can be calculated approximately by the equation [9]

$$\log \frac{h_a}{h} = b(t_a - t) \tag{8-117}$$

where t_a = boiling temperature of liquid at 1 atm
h_a = coefficient at temperature t_a
h = coefficient at temperature t
b = constant

The constant b depends on the liquid and on the kind and condition of the heating surface.

Maximum Flux and Critical Temperature Drop. The critical temperature drop for water is between 40 and 50°F. That for an organic liquid is higher, in a range of 60 to 120°F. The critical temperature drop is relatively insensitive to the temperature of boiling and the corresponding pressure. It is less sensitive to variations in the type and condition of the heating surface than is the peak flux.

The critical temperature drop can be exceeded in actual equipment unless precautions are taken to keep the actual temperature drop below the critical. This is especially true for water, because of its low critical temperature drop. If the source of heat is another fluid, such as condensing steam or hot liquid, the only penalty for exceeding the critical temperature drop is a decrease in flux to a level between that at the peak and that at the Leidenfrost point. If the heat is supplied by an electric heater, exceeding the critical temperature drop usually will burn out the heater, as the boiling liquid cannot absorb heat fast enough at a large temperature drop, and the heater immediately becomes very hot.

The peak flux at the critical temperature drop is large. For water it is in the range 115,000 to 400,000 Btu/(ft²)(hr) depending on the purity of the water, the pressure, and the type and condition of the heating surface. For organic liquids, the peak flux lies in the range 40,000 to 130,000 Btu/(ft²)(hr). These limits apply to boiling under atmospheric pressure. They can be exceeded by increase in pressure, as shown by Eq. (8-117).

Surface Boiling. Surface boiling may be demonstrated by pumping a gas-free liquid upward through a vertical annular space consisting of a transparent outer tube and an internal heating element and observing the effect on the liquid of a gradual increase in heat flux and temperature of the heating element. It is observed that, when the temperature of the element exceeds a definite magnitude, which depends on the conditions of the experiment, bubbles form, just as in nucleate boiling, and then condense in the adjacent cooler liquid. For a definite velocity and pressure, the flux density q/A and the temperature drop from heating element to liquid Δt can be measured and a logarithmic plot of flux density vs. temperature drop prepared. Figure 8-31 shows the results of such an experiment conducted on degassed distilled water flowing at a velocity of 4 ft/sec under an absolute pressure of 60 lb force/in.² through the annular space between a jacket 0.77 in. ID and an electric heating element 0.55 in. in diameter.[26]

The line in Fig. 8-31 consists of two distinct segments connected by a short transition curve. The lower branch, line AB, covering temperature drops below 80°F is straight and has a slope of 1.0. The flux in this range is proportional to Δt, and the coefficient h is independent of Δt. The heat-transfer coefficients in this range agree well with those predicted from the usual equations for heat transfer to nonboiling liquids in turbulent flow, such as Eqs. (8-53) and (8-56). As the temperature drop exceeds 90°F, the curve

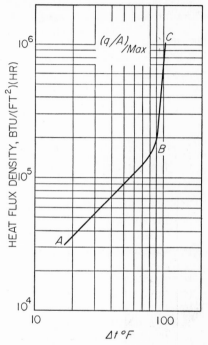

FIG. 8-31. Example of surface boiling. (*McAdams, Addoms, Rinaldo, and Day.*[25] *Courtesy of AIChE.*)

bends sharply upward, where surface boiling sets in, and section BC is nearly straight. The slope is very great, and a small increase in temperature drop corresponds to a large change in flux. It is this increase in capacity accompanying the appearance of surface boiling that makes this type of operation important for heat-transfer equipment that must pack great capacity into small space. When the temperature drop corresponding to point C was reached, the heater became blanketed with vapor and burned out.

The break point in the curve of flux vs. temperature drop for surface boiling does not occur at the boiling temperature of the liquid at the pressure in the equipment. Surface boiling does not take place until the liquid is appreciably superheated. The temperature at which surface boiling appeared in the experiment represented by Fig. 8-31 was about 40°F above the boiling temperature corresponding to the pressure in the equipment.

The most significant characteristic of surface boiling is the very high heat flux that can be obtained by this technique, in comparison with the flux obtained in heat transfer to fluids without change in phase. The maximum flux density obtained at the burn-out point shown in Fig. 8-31 is over 1,000,000 Btu/(ft^2)(hr). A flux density as high as 11,400,000 Btu/(ft^2)(hr) has been reported.[24r]

Surface boiling is used for specialized situations. Additional data covering the effects of important variables on the rate of heat transfer by this method are available in the references given above.

UNSTEADY-STATE HEAT TRANSFER

The foregoing discussion has been based throughout on the assumption of steady-state operation; i.e., that at any point in the system conditions do not change with time. In many chemical engineering operations, however, conditions do change with time. In a batch process a charge of material may be slowly heated from room temperature to some elevated temperature and later cooled to its original temperature. In a cyclic process such as catalytic cracking, a given particle of material is alternately heated and cooled. Even in truly continuous processes there are periods of unsteady-state operation when the equipment is being started or shut down. It often becomes necessary, therefore, to predict the length of time required to heat or cool a quantity of material a specified amount in a given piece of equipment or to design heat-exchange equipment to transfer the desired amount of heat in a specified time.

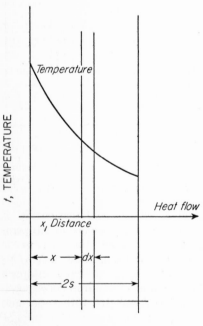

FIG. 8-32. Unsteady-state conduction in solid slab.

Unsteady-state Conduction in Solids. A full treatment of unsteady-state heat conduction lies in the field of applied mathematics and is outside the scope of this text. The basic assumption of such treatments is that the Fourier law [Eq. (8-2)] still applies when the temperature gradient $\partial t/\partial x$ varies with both time and place.

As a simple example of unsteady-state conduction consider a flat wall as shown in Fig. 8-32. Focus attention on the thin slice of solid of thickness dx located a distance x from the hot side of the slab. The temperature gradient at x is, at a definite instant of time, $\partial t/\partial x$, and the heat input in time $d\theta$ at x is $-kA(\partial t/\partial x)\,d\theta$, where A is the area of the slab perpendicular to the flow of heat and k is the thermal conductivity of the solid. The gradient at distance $x + dx$ is slightly greater than that at x, and may be represented as

$$\frac{\partial t}{\partial x} + \frac{\partial}{\partial x}\left(\frac{\partial t}{\partial x}\right) dx$$

The heat flow out of the slab at $x + dx$ is, then,

$$-kA\left[\frac{\partial t}{\partial x} + \frac{\partial}{\partial x}\left(\frac{\partial t}{\partial x}\right) dx\right] d\theta$$

The excess of heat input over heat output, which is the accumulation of heat in layer dx, is

$$-kA\,\frac{\partial t}{\partial x}\,d\theta + kA\left(\frac{\partial t}{\partial x} + \frac{\partial^2 t}{\partial x^2}\,dx\right) d\theta = kA\,\frac{\partial^2 t}{\partial x^2}\,dx\,d\theta$$

The accumulation of heat in the layer must increase the temperature of the layer. If c_p and ρ are the specific heat and density, respectively, the accumulation is the product of the mass (volume times density), the specific heat, and the increase in temperature, or $(\rho A\,dx)c_p(\partial t/\partial\theta)d\theta$. Then, by a heat balance,

$$kA\,\frac{\partial^2 t}{\partial x^2}\,dx\,d\theta = \rho c_p A\,dx\,\frac{\partial t}{\partial\theta}\,d\theta$$

or
$$\frac{\partial t}{\partial\theta} = \frac{k}{\rho c_p}\,\frac{\partial^2 t}{\partial x^2} = \alpha\,\frac{\partial^2 t}{\partial x^2} \tag{8-118}$$

The term α in Eq. (8-118) is called the *thermal diffusivity* of the solid and is a property of the material. It has the dimensions of area divided by time (square feet per hour in English units).

The significance of Eq. (8-118) is shown in Fig. 8-33, which is a diagram of temperature conditions in a furnace wall in contact with air at 80°F on one side and gas at 1200°F on the other. The wall is initially all at 80°F; one face of it is suddenly exposed to the hot gas. Line I in Fig. 8-33 shows the temperature pattern in the wall at the instant of exposure to the gas. If there is negligible thermal resistance between the surfaces of the wall and the surroundings, the temperature of the surface contact with the hot gas immediately rises to 1200°F. The rest of the wall, however, is still at 80°F.

Consider now the point A inside the wall at time θ after exposure to the hot gas. The temperature pattern in the wall is shown by line II. The temperature at A is changing with time in accordance with Eq. (8-118).

The rate of temperature change is proportional to the *curvature* of the line, i.e., the degree to which it differs from a straight line. If the curve is strongly flexed, the temperature changes rapidly; if the curvature is slight, the temperature change is slow. Eventually, if the wall is kept in contact with the hot gas for a long enough time, steady-state conditions will be reached, and the temperature at any point in the wall will no longer change. This occurs when the temperature gradient in the wall is uniform and the graph of temperature vs. distance is straight, as shown by line III of Fig. 8-33. Note that the *slope* of the line through any point has no effect on the rate of temperature change with time; it is only the curvature of the line which is significant.

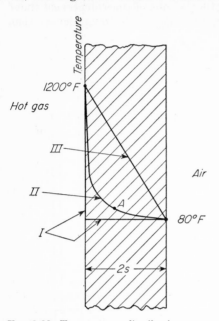

Fig. 8-33. Temperature distributions, unsteady-state heating of furnace wall. I, at instant of exposure of wall to high temperature. II, during heating at time θ. III, at steady state.

General solutions of unsteady-state conduction equations are available only for certain simple shapes such as the infinite slab, the infinitely long cylinder, and the sphere. For example, the integration of Eq. (8-118) for the heating or cooling of an infinite slab of known thickness from both sides by a medium at constant surface temperature gives the following equation.

$$\frac{t_s - \bar{t}_b}{t_s - t_a} = \frac{8}{\pi^2}\left(e^{-a_1 N_{\mathrm{Fo}}} + \frac{1}{9}e^{-9a_1 N_{\mathrm{Fo}}} + \frac{1}{25}e^{-25a_1 N_{\mathrm{Fo}}} + \cdots\right)$$

$$= \Phi(N_{\mathrm{Fo}}) \tag{8-119}$$

where t_s = constant average temperature of surface of slab
t_a = initial temperature of slab
$\bar{t}_b$ = average temperature of slab at time θ_t
N_{Fo} = Fourier number, defined as $\alpha\theta_t/s^2$
α = thermal diffusivity, $\mathrm{ft}^2/\mathrm{hr}$
θ_T = time of heating or cooling, hr
s = one-half slab thickness, ft
a_1 = $(\pi/2)^2$

Equation (8-119) applies only at constant surface temperature, so t_s can be equated to the temperature of the cooling or heating medium only when the temperature difference between the medium and the surface is negligible. This implies that the surface heat-transfer coefficient between surface and medium is large in comparison with k/s. Figure 8-34 is a plot of $\Phi(N_{Fo})$ vs. N_{Fo} and may be used in place of Eq. (8-119).

Equations and graphs are available for the other shapes mentioned above and also are available for the heating and cooling of various shapes when

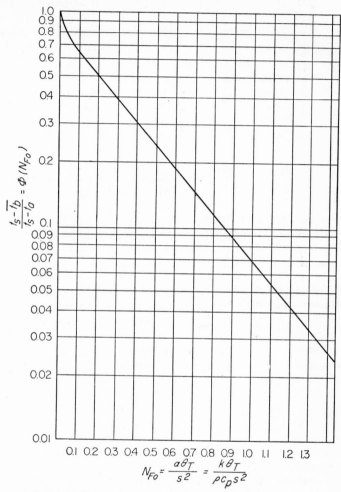

FIG. 8-34. Unsteady-state heating or cooling of large slab.

the heat-transfer coefficient at the surface is sufficiently low to cause variation in the surface temperature.[24b] Temperature distributions during heating and cooling of heterogeneous solids or bodies of complex shape are best found by numerical approximate methods.[13]

Example 8-10. A flat slab of plastic, initially at 70°F, is placed between two platens at 250°F. The slab is 1.0 in. thick. How long will it take to heat the slab to an average temperature of 210°F? The specific gravity of the slab is 0.90, the thermal conductivity is 0.075, and the specific heat is 0.40.

Solution. The quantities for use with Fig. 8-34 are as follows. Standard units are used throughout.

$$k = 0.075 \qquad \rho = 0.90 \times 62.4 = 56.2 \qquad c_p = 0.40$$

$$s = \frac{0.5}{12} = 0.0417 \qquad t_s = 250°F \qquad t_a = 70°F \qquad t_b = 210°F$$

Then $\dfrac{t_s - t_b}{t_s - t_a} = \dfrac{250 - 210}{250 - 70} = 0.222$

$$\alpha = \frac{k}{\rho c_p} = \frac{0.075}{56.2 \times 0.40} = 0.00335$$

From Fig. 8-34, for a temperature difference ratio of 0.222,

$$N_{Fo} = 0.52 = \frac{0.00335\theta_t}{0.0417^2}$$

$$\theta_T = 0.27 \text{ hr, or } 16 \text{ min}$$

Unsteady-state Heat Transfer to Liquids in Agitated Vessels.[4, 10, 18i]

It is often necessary to compute the time required to heat or cool a given amount of liquid in an agitated vessel equipped with a coil or jacket. At any instant of time when the conditions in the vessel and in the coil or jacket are known, the over-all heat-transfer coefficient may be computed by the methods previously discussed. For example, if coooling water is flowing inside a metal coil which is submerged in the agitated liquid, the over-all thermal resistance between the liquid and the water is the sum of: (1) the resistance of the agitated liquid outside the coil, (2) the outside scale resistance, (3) the resistance of the metal wall, (4) the inside scale resistance, and (5) the resistance of the cooling water. Item (1) may be computed from Eq. (8-96) or (8-97), item (5) from Eq. (8-53), and the other quantities as described in previous sections.

In the following derivations it is assumed that the over-all coefficient U is known or can be computed and furthermore that it does not vary with time. In any actual case U is not constant with time, but unless the temperature changes are very great, the assumption of constancy introduces negligible error. Further assumptions are made that the contents of the vessel are so thoroughly agitated that the temperature of the liquid t is the same throughout the vessel at any given time and that the heat capacity of the tank and coil are negligible. Two cases will be discussed: (1) when all of the heating or cooling medium is at a constant temperature, as in the heating of a liquid with steam condensing at a constant pressure; (2) when the heating or cooling medium is a fluid entering at a constant rate and at a constant inlet temperature.

Case 1. Heating Medium at Constant Temperature. A typical agitated vessel with a heat-transfer coil is shown in Fig. 8-35. It contains m lb of liquid having a specific heat of c_p. Steam or other vapor is condensing inside the coil at constant temperature t_h. The surface area of the coil is A ft^2, the over-all coefficient is U Btu/(ft^2)(hr)(°F), and the initial temperature of the liquid is t_a. An equation for calculating the time required to heat the tank from temperature t_a to t_b is desired.

At any instant, when the temperature of the liquid is t, Eq. (8-29) applies in the form

$$q = \frac{dQ}{d\theta} = UA\ \overline{\Delta t_L} \quad (8\text{-}120)$$

Since the temperatures of both fluids are uniform, $\overline{\Delta t_L} = t_h - t$. The rate of heat transfer is also equal to the rate of increase in the enthalpy of the liquid, so

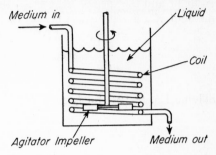

Medium in — Liquid — Coil — Medium out — Agitator Impeller

Fig. 8-35. Vessel for unsteady-state heat transfer.

$$\frac{dQ}{d\theta} = mc_p \frac{dt}{d\theta} \quad (8\text{-}121)$$

Equating the two expressions for $dQ/d\theta$,

$$mc_p \frac{dt}{d\theta} = UA(t_h - t)$$

Separating the variables and integrating between limits $t = t_a$ when $\theta = 0$ and $t = t_b$ when $\theta = \theta_T$,

$$\int_{t_a}^{t_b} \frac{dt}{t_h - t} = \frac{UA}{mc_p} \int_0^{\theta_T} d\theta$$

$$- \ln \frac{t_h - t_b}{t_h - t_a} = \frac{UA\theta_T}{mc_p}$$

and
$$\theta_T = \frac{mc_p}{UA} \ln \frac{t_h - t_a}{t_h - t_b} \quad (8\text{-}122)$$

Case 2. *Liquid Cooling Medium at Constant Inlet Temperature.* This situation is more complicated than that of case 1 because, although the inlet temperature of the cooling medium is constant, its outlet temperature varies with time. Equation (8-120) still applies, but $\overline{\Delta t_L}$ is now, from Eq. (8-30),

$$\overline{\Delta t_L} = \frac{(t_h - t_{ca}) - (t_h - t_c)}{\ln \dfrac{t_h - t_{ca}}{t_h - t_c}} = \frac{t_c - t_{ca}}{\ln \dfrac{t_h - t_{ca}}{t_h - t_c}} \quad (8\text{-}123)$$

where t_{ca} is the constant inlet temperature and t_c is the variable outlet temperature of the cooling liquid. If the flow rate of the cooling liquid is w_c lb/hr and the specific heat is c_{pc}, the rate of heat transfer to the cooling medium at a given instant is $w_c c_{pc}(t_c - t_{ca})$, and

$$\frac{dQ}{d\theta} = w_c c_{pc}(t_c - t_{ca}) = UA \frac{t_c - t_{ca}}{\ln \dfrac{t_h - t_{ca}}{t_h - t_c}} \qquad (8\text{-}124)$$

Solving Eq. (8-124) for t_c gives

$$t_c = t_h - \frac{t_h - t_{ca}}{K} \qquad (8\text{-}125)$$

where

$$K = e^{UA/w_c c_{pc}} \qquad (8\text{-}126)$$

Equating the rate of enthalpy loss of the liquid in the tank to the increase in enthalpy of the cooling medium gives

$$-mc_p \frac{dt_h}{d\theta} = w_c c_{pc}(t_c - t_{ca}) \qquad (8\text{-}127)$$

Eliminating t_c from Eq. (8-127) by Eq. (8-125) gives, after rearranging and integrating with appropriate limits,

$$-\int_{t_{ha}}^{t_{hb}} \frac{dt_h}{t_h - t_{ca}} = \frac{w_c c_{pc}}{mc_p}\left(1 - \frac{1}{K}\right)\int_0^{\theta_T} d\theta$$

and

$$\ln \frac{t_{ha} - t_{ca}}{t_{hb} - t_{ca}} = \frac{w_c c_{pc}}{mc_p}\left(1 - \frac{1}{K}\right)\theta_T \qquad (8\text{-}128)$$

Equation (8-128) may be solved for the time θ_T or the final liquid temperature t_{hb} if the other quantities are known. The area required to transfer a given amount of heat under specified conditions is found from Eq. (8-126) after solving Eq. (8-128) for K.

Example 8-11. An agitated vessel is charged with 5,200 lb of nitrobenzene at 75°F. An internal coil with 40 ft^2 of heat-transfer surface contains steam condensing at 250°F. The average specific heat of the nitrobenzene is 0.61 Btu/(lb)(°F). If $U = 165$ Btu/(ft^2)(hr)(°F), (a) how long will it take to heat the liquid to 200°F, and (b) what will the liquid temperature be 1 hr after the steam is turned on?

Solution. (a) Equation (8-122) is used. The quantities needed are

$$c_p = 0.61 \qquad m = 5,200 \qquad t_h = 250°F$$

$$t_a = 75°F \qquad t_b = 200°F \qquad U = 165 \qquad A = 40$$

$$\theta_T = \frac{0.61 \times 5,200}{165 \times 40} \ln \frac{250 - 75}{250 - 200} = 0.602 \text{ hr}$$

(b) Equation (8-122) can be used to find the liquid temperature at any time. Since $\theta_T = 1$,

$$1 = \frac{0.61 \times 5,200}{165 \times 40} \ln \frac{250 - 75}{250 - t_b}$$

$$\ln \frac{250 - 75}{250 - t_b} = \frac{40 \times 165}{0.61 \times 5,200} = 2.08$$

$$\frac{250 - 75}{250 - t_b} = 8.015$$

From this, $t_b = 229°F$.

Example 8-12. If in Example 8-11 the steam is turned off at the end of 1 hr and 20 gal/min of cooling water admitted at 60°F to the coil, how long will it take to cool the nitrobenzene from 228 to 90°F? Assume $U = 150$.

Solution. Equation (8-128) applies to this case. The new quantities needed are

$$c_{pc} = 1.0 \qquad t_{ca} = 60°F \qquad t_a = 229°F \qquad t_b = 90°F$$

$$w = 20 \times 60 \times 8.33 = 10,000 \text{ lb/hr}$$

$$K = e^{(150 \times 40)/10,000} = e^{0.60} = 1.822$$

Substituting into Eq. (8-128),

$$\ln \frac{229 - 60}{90 - 60} = \frac{10,000}{0.61 \times 5,200} \left(1 - \frac{1}{1.822}\right) \theta_T$$

From this $\theta_T = 1.21$ hr.

HEAT-EXCHANGE EQUIPMENT

The heat-exchange equipment described in this section can be classified into two types. In the first type, latent heat from a condensing vapor is transferred through the tube walls as sensible heat to a second fluid, which is usually a liquid. The simple heater shown in Fig. 8-4 and described on page 425 is an example. In the second type, sensible heat is transferred from one fluid to another. The double-pipe heat exchanger shown in Fig. 8-6 and described on page 426 is a simple example of heat exchanger. An important type of heat-transfer equipment, in which heat is transferred from a condensing vapor to a boiling liquid, is so important that it is given a special name—evaporator—and Chap. 9 is devoted to it.

Heaters

The heater shown in Fig. 8-4 is a single-pass heater, since the entire stream of cooling liquid flows through all the tubes in parallel. In large heaters, this type of flow has a serious limitation. The number of tubes is so large that, in single-pass flow, the velocity through the tubes is too small to yield an adequate heat-transfer coefficient, and the heater is un-economically large. Also, because of the low coefficient, long tubes are needed if the cooling fluid is to be cooled through a reasonably large temperature range, and such long tubes are not practicable.

Multipass Heaters. The limitation of low velocity in a single-pass heater can be removed by using multipass flow. In multipass flow, the tube bundle is divided into sections, and the fluid is forced, by means of baffles placed in the channels, to pass through each section in series. An example of multipass construction is shown in Fig. 8-36. The large plan and section show the baffles in one of the two channels of a large heater. The small diagrams in the figure show how the two channels are divided by baffles to give eight passes. The liquid enters compartment A, flows

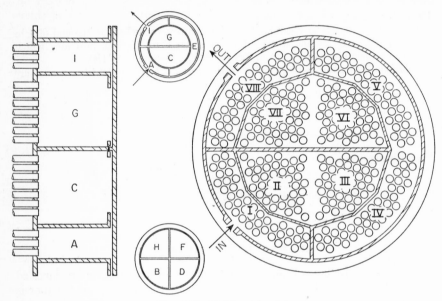

Fig. 8-36. Multipass heater.

to the left into compartment B, back to the right to compartment C, and so on till it finally leaves at I. The tubes thus fall into eight passes, shown by the numerals on the large plan. In the odd-numbered passes the liquid is flowing away from the reader; in the even-numbered ones it is flowing toward him.

Multipass construction decreases the cross section of the fluid path and increases the fluid velocity, with a corresponding increase in the heat-transfer coefficient. The disadvantages are: (1) the heater is slightly more complicated and (2) the friction loss through the equipment is increased because of the larger velocities and the multiplication of exit and entrance losses. For example, the average velocity in the tubes of the heater shown in Fig. 8-36 is eight times that in a single-pass heater having the same number and size of tubes and operated at the same liquid flow rate. The water-side coefficient of the eight-pass heater is approximately $8^{0.8} = 5.2$ times that for the single-pass heater, or even more if the velocity in the single-pass heater is sufficiently low to give laminar flow. The friction loss is approximately $8^{2.8} = 330$ times that in the single heater, not including the additional expansion-and-contraction losses. The most economic design calls for such a velocity in the tubes that the increased cost of power for pumping is offset by the decreased cost of the apparatus. Too low a velocity saves power for pumping but calls for an unduly large (and consequently expensive) heater. Too high a velocity saves on the first cost of the heater but more than makes up for it in the cost of power.

Two-pass Floating-head Heater and Expansion. Because of the differences in temperature existing in heaters, expansion strains may be set up that may be sufficiently severe to buckle the tubes or pull them

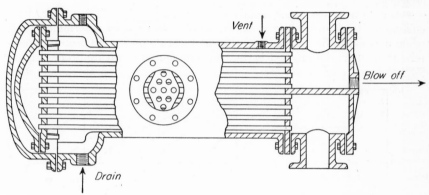

Fig. 8-37. Two-pass floating-head heater.

loose from the tube sheets. Many heaters have a cast-iron shell and relatively thin-walled tubes. When the heater is put into or out of service, the temperature of the shell changes more slowly than that of the tubes, and the resultant strains may result in fluid leakage or mechanical failure. The most common method for avoiding damage from expansion is the use of the floating-head construction, in which one of the tube sheets (and therefore one end of the tubes) is structurally independent of the shell. A two-pass floating-head heater is shown in Fig. 8-37. The construction is obvious, and the figure shows how the tubes may expand or contract, independent of the shell. A perforated plate is placed over the steam inlet to prevent cutting of the tubes by drops of water contained in the steam. A method of allowing for expansion when the heater shell is made of sheet metal is to roll a bulge in the shell, as shown in Fig. 8-4.

Extended-surface Air Heater. In the air heater shown in Fig. 8-18 air is blown or drawn through a rectangular duct past a row of transverse steam-heated tubes. Because the air-film coefficient is much lower than the coefficient of the condensing steam, the outer surface of the tubes is extended by radial fins. This greatly increases the heat-transfer capacity of the unit, as previously discussed. A single row of tubes will typically raise the air temperature by 80°F when the steam is at a gauge pressure of 5 lb force/in.^{2} and the superficial velocity of the air is 8 ft/sec. When a larger temperature rise is needed, the air is passed through several banks of tubes in series.

Correction of Logarithmic Mean Temperature Drop for Crossflow. If a fluid flows perpendicularly to a heated or cooled tube bank, the logarithmic mean temperature drop, as given by Eq. (8-30), applies only if the temperature of one of the fluids is constant. For example, if, in the air heater shown in Fig. 8-18, the heating medium is condensing steam, use of the logarithmic mean temperature drop is correct. If, however, hot water is used as the heating means, the temperature conditions do not correspond to either countercurrent or parallel flow but rather to a type of flow called *crossflow*.

When flow types other than countercurrent or parallel appear, it is customary to define a correction factor F_G, which is so determined that, when it is multiplied by the logarithmic mean temperature drop for countercurrent flow, the product is the true average temperature drop. Figure 8-38 shows a correlation for F_G for crossflow derived on the assumption

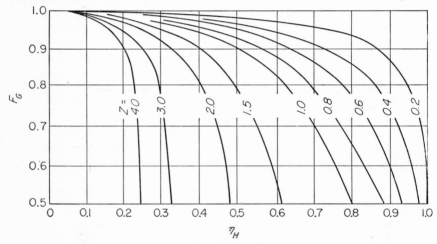

FIG. 8-38. Correction for logarithmic mean temperature difference, cross flow. (*Bowman, Mueller, and Nagle.*[2] *Courtesy of ASME.*)

that neither stream mixes with itself during flow through the exchanger.[2] The temperatures used in Fig. 8-38 are:

Inlet temperature of hot fluid, t_{ha}
Outlet temperature of hot fluid, t_{hb}
Inlet temperature of cold fluid, t_{ca}
Outlet temperature of cold fluid, t_{cb}

Each curved line in the figure corresponds to a constant value of the dimensionless ratio Z, defined as

$$Z = \frac{t_{ha} - t_{hb}}{t_{cb} - t_{ca}} \qquad (8\text{-}129)$$

and the abscissas are values of the dimensionless ratio η_H, defined as

$$\eta_H = \frac{t_{cb} - t_{ca}}{t_{ha} - t_{ca}} \tag{8-130}$$

The factor Z is the ratio of the fall in temperature of the hot fluid to the rise in temperature of the cold fluid. The factor η_H is the "heating effectiveness," or the ratio of the actual temperature rise of the cold fluid to the maximum possible temperature rise obtainable if the warm-end approach were zero (based on countercurrent flow). From the numerical values of η_H and Z the factor F_G is read from Fig. 8-38, interpolating between lines of constant Z where necessary, and multiplied by the logarithmic mean temperature drop for counterflow, to give the true mean temperature drop.

Heat Exchangers

Heat exchangers are so important and so widely used in the process industries that their design has been highly developed. Standards devised and accepted by the Tubular Exchanger Manufacturers Association (TEMA) are available covering in detail materials, methods of construction, technique of design, and dimensions for exchangers.[33] The following sections describe the more important types of exchanger and cover the fundamentals of their engineering, design, and operation. Most exchangers are liquid-to-liquid, but gases and noncondensing vapors can also be treated in them.

Single-pass 1-1 Exchanger. The simple double-pipe exchanger shown in Fig. 8-6 is inadequate for flow rates that cannot readily be handled in a few tubes. If several double pipes are used in parallel, the weight of metal required for the outer tubes becomes so large that the shell-and-tube construction, such as that shown in Fig. 8-39, where one shell serves for many

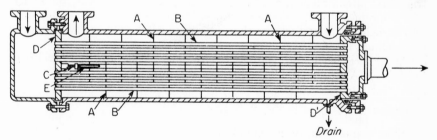

FIG. 8-39. Single-pass 1-1 counterflow heat exchanger: A, baffles. B, tubes. C, guide rods. D, D', tube sheets. E, spacer tubes.

tubes, is more economical. This exchanger, because it has one shell-side pass and one tube-side pass, is a 1-1 exchanger.

In a heater the shell-side coefficient is generally larger than the tube-side coefficient. In an exchanger the shell-side and tube-side coefficients are of comparable importance, and both must be large if a satisfactory over-all coefficient is attained. The velocity and turbulence of the shell-side liquid are as important as those of the tube-side liquid. To prevent weakening of the tube sheets there must be a minimum distance between the tubes, and it is not practicable to space the tubes so closely that the area of the path outside the tubes is as small as that inside the tubes. If the two streams are of comparable magnitude, the velocity on the shell side is low in comparison with that on the tube side. For that reason, baffles are installed in the shell to decrease the cross section of the shell-side liquid and to force the liquid to flow across the tube bank rather than parallel with it. The added turbulence generated in this type of flow further increases the shell-side coefficient.

In the construction shown in Fig. 8-39, the baffles A consist of circular disks of sheet metal with one side cut away. A common practice is to cut away a segment having a height equal to one-fourth the inside diameter of the shell. Such baffles are called *25 per cent baffles*. The baffles are perforated to receive the tubes. To minimize leakage, the clearances between baffles and shell and tubes should be small. The baffles are supported by one or more guide rods C, which are fastened between the tube sheets D and D' by setscrews. To fix the baffles in place short sections of tube E are slipped over the rod C between the baffles. In assembling such a heater it is necessary to assemble the tube sheets, support rods, spacers, and baffles first and then to install the tubes.

The stuffing box shown at the right-hand end of Fig. 8-39 provides for expansion. This construction is practicable only for small shells.

Tubes and Tube Sheets. As described in Chap. 3, tubes are drawn to definite wall thickness in terms of BWG and true outside diameter (OD), and they are available in all common metals. Tables of dimensions of standard tubes are given in Appendix 5. Standard lengths of tubes for heat exchanger-construction are 8, 12, and 16 ft. Tubes are arranged on triangular pitch or square pitch. Unless the shell side tends to foul badly, triangular pitch is used, because more heat-transfer area can be packed into a shell of given diameter than in square pitch. Tubes in triangular pitch cannot be cleaned by running a brush between the rows, as no space exists for cleaning lanes. Square pitch allows cleaning of the outside of the tubes. Also, square pitch gives a lower shell-side pressure drop than triangular pitch.

TEMA standards specify a minimum center-to-center distance 1.25 times the outside diameter of the tubes for triangular pitch and a minimum cleaning lane of $\frac{1}{4}$ in. for square pitch.

Shell and Baffles. Shell diameters are standardized. For shells up to and including 23 in. the diameters are fixed in accordance with ASTM pipe standards. For sizes of 25 in. and above the inside diameter is specified to the nearest inch. These shells are constructed of rolled plate. Minimum shell thicknesses are also specified.

The distance between baffles (center to center) is the *baffle pitch*, or baffle spacing. It should not be less than one-fifth the diameter of the shell or more than the inside diameter of the shell.

Tubes are usually attached to the tube sheets by grooving the holes circumferentially and rolling the tube ends into the holes by means of a rotating tapered mandrel, which stresses the metal of the tube beyond the elastic limit, so the metal flows into the grooves. In high-pressure exchangers, the tubes are welded or brazed to the tube sheet after rolling.

1-2 Parallel-counterflow Heat Exchanger. The 1-1 exchanger has limitations. Higher velocities, shorter tubes, and a more satisfactory

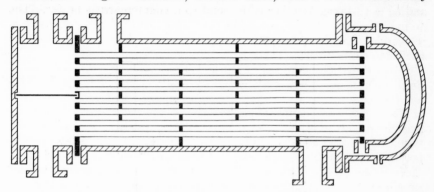

FIG. 8-40. 1-2 parallel-counterflow exchanger.

solution to the expansion problem are realized in multipass construction. An even number of tube-side passes is used in multipass exchangers. The shell side may be either single-pass or multipass. A common construction is the 1-2 parallel-counterflow exchanger, in which the shell-side liquid flows in one pass and the tube-side liquid in two or more passes. Two tube-side passes are common. Such an exchanger is shown in Fig. 8-40. In multipass exchangers, floating heads are frequently used, and the bulge in the shell of the exchanger in Fig. 8-4 and the stuffing box shown in Fig. 8-39

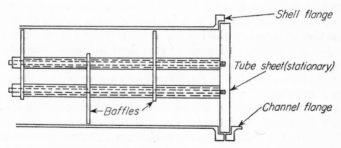

FIG. 8-41. Spacer and baffle details. (*By permission, from "Process Heat Transfer," by D. Q. Kern. Copyright, 1950. McGraw-Hill Book Company, Inc.*)

are unnecessary. The tube-side liquid enters and leaves through the same channel, which is divided by a baffle to separate the entering and leaving tube-side streams. A typical construction of baffles, tie rods, and spacers is shown in Fig. 8-41.

Temperature-drop Correction Factor for 1-2 Exchanger.[18b]   Flow in a 1-2 exchanger is partly parallel and partly countercurrent.   The temperature-length curves for such an exchanger are shown in Fig. 8-42.   Curve $t_{ha} - t_{hb}$ applies to the shell-side fluid, which is assumed to be the hot fluid.   Curve $t_{ca} - t_{ci}$ applies to the first pass of the tube-side liquid, and curve $t_{ci} - t_{cb}$ to the second pass of the tube-side liquid.   Curves $t_{ha} - t_{hb}$ and $t_{ca} - t_{ci}$ taken together are those of a parallel-flow exchanger, and curves $t_{ha} - t_{hb}$ and $t_{ci} - t_{cb}$ taken together correspond to a countercurrent heater.   The

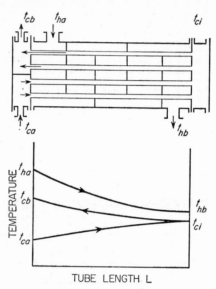

Fig. 8-42. Temperature-length curves, 1-2 exchanger, first nozzle arrangement.

logarithmic mean temperature drop, (LMTD), which applies to either parallel or counterflow but not to a mixture of both, cannot be used without correction to calculate the true mean temperature drop.   The method used in this situation is analogous to that used to correct the LMTD for crossflow: a factor F_G is so defined that, when it is multiplied by the counterflow LMTD, the product is the correct mean temperature drop.

Figure 8-43a shows the factor F_G as a function of the two dimensionless numbers η_H and Z, which have been defined by Eqs. (8-129) and (8-130).

Another arrangement of the inlet and outlet shell nozzles is shown in Fig. 8-44.   The temperature-length curves for this arrangement are also shown.   The factor F_G from Fig. 8-43a can be used for either nozzle arrangement.

Factor F_G is always less than unity.   The mean temperature drop, and, therefore, the capacity, of the exchanger is less than that of a countercurrent exchanger having the same LMTD.   When F_G is less than approximately 0.75, the 1-2 exchanger should not be used, as the actual exchanger no longer follows accurately the assumptions made in the derivation.

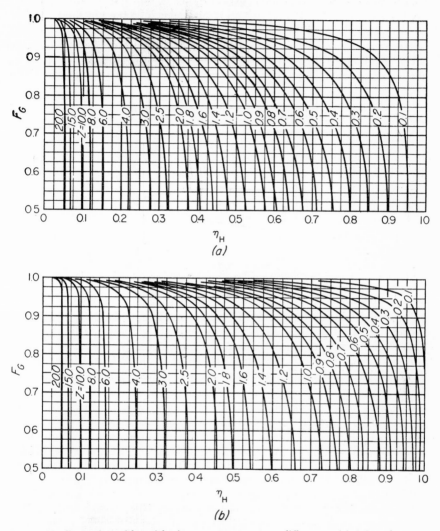

FIG. 8-43. Correction of logarithmic mean temperature difference: (a) 1-2 exchangers. (b) 2-4 exchangers. (*Bowman, Mueller, and Nagle.*[2] *Courtesy of ASME.*)

2-4 Exchanger. The 1-2 exchanger has an important limitation. Because of the parallel-flow pass, the exchanger is unable to bring the exit temperature of one fluid very near to the entrance temperature of the

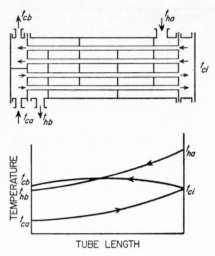

TUBE LENGTH

Fig. 8-44. Temperature-length curves, 1-2 exchangers, second nozzle arrangement.

other. It is clear from Figs. 8-42 and 8-44 that neither can temperature t_{cb} approach closely temperature t_{ha} nor can temperature t_{ca} closely approach temperature t_{hb}. Another way of stating the same limitation is that the heat recovery of a 1-2 exchanger is inherently poor.

A better recovery of heat can be obtained in the 2-4 exchanger, which has two shell-side and four tube-side passes. This type of exchanger also

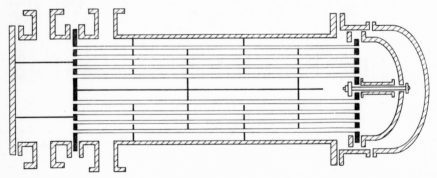

Fig. 8-45. 2-4 exchanger.

gives higher velocities and a larger over-all heat-transfer coefficient than a 1-2 exchanger having two tube-side passes and operating with the same flow rates. An example of a 2-4 exchanger is shown in Fig. 8-45.

The temperature-length relations in a 2-4 exchanger are shown in Fig. 8-46. The dotted lines refer to the shell-side fluid and the full lines to the tube-side fluid. It is assumed that the hotter fluid is in the shell. The hotter pass of the shell-side fluid is in thermal contact with the two hottest tube-side passes and the cooler shell-side pass with the two coolest tube-side passes. The exchanger as a whole approximates a true counter current unit more closely than is possible with a 1-2 exchanger.

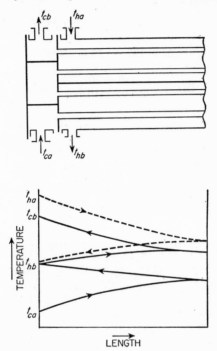

FIG. 8-46. Temperature-length curves, 2-4 exchanger.

The correction factor F_G for 2-4 exchangers is given in Fig. 8-43b. The chart is used in exactly the same manner as Figs. 8-38 and 8-43a. It can be used for two shell-side and four or more tube-side passes.

Other combinations of shell-side passes and tube-side passes are used, but the 1-2 and the 2-4 types are the most common.

Example 8-13. In the 1-2 exchanger sketched in Fig. 8-42, the values of the temperatures are

Tube fluid in: $\qquad\qquad t_{ca} = 70°F$

Tube fluid out: $\qquad\qquad t_{cb} = 130°F$

Shell fluid in: $\qquad\qquad t_{ha} = 240°F$

Shell fluid out: $\qquad\qquad t_{hb} = 120°F$

What is the correct mean temperature drop in this exchanger?

Solution. The correction factor F_G is found from Fig. 8-43a. For this case, from Eqs. (8-129) and (8-130),

$$\eta_H = \frac{130 - 70}{240 - 70} = 0.353 \qquad Z = \frac{240 - 120}{130 - 70} = 2.00$$

From Fig. 8-43a, $F_G = 0.735$.
The temperature drops are

At shell inlet: $\Delta t = 240 - 130 = 110°F$

At shell outlet: $\Delta t = 120 - 70 = 50°F$

$$\overline{\Delta t_L} = \frac{110 - 50}{2.303 \log (110/50)} = 76°F$$

The correct mean is $\overline{\Delta t} = 0.735 \times 76 = 56°F$. These temperatures are marginal for a 1-2 exchanger.

Example 8-14. What is the correct mean temperature difference in a 2-4 exchanger operating with the same inlet and outlet temperatures as in the exchanger in Example 8-13?

Solution. For a 2-4 exchanger, when $\eta_H = 0.353$ and $Z = 2.00$, the correction factor from Fig. 8-43b is $F_G = 0.945$. The $\overline{\Delta t_L}$ is the same as in Example 8-13. The correct mean $\overline{\Delta t} = 0.945 \times 76 = 72°F$.

Calculations for Shell-and-tube Exchanger. The heat-transfer coefficient h_i for the tube-side fluid in a shell-and-tube exchanger can be calculated from Eq. (8-53) or Eq. (8-56). The coefficient for the shell side h_o cannot be so calculated because the direction of flow is not parallel to the tubes but across them and because the cross-sectional area of the stream and the mass velocity of the stream vary as the fluid crosses the tube bundle back and forth across the shell. Also, leakage between baffles and shell and between baffles and tubes short-circuits some of the shell-side liquid and reduces the effectiveness of the heat transfer. No simple general correlation is available for either heat transfer or friction that accounts for all these factors. Data are available for the specialist.[1, 14, 24m, 29c] Typical data for some arrangements are shown in Fig. 8-47. The factors f' and j'_{max} are defined as

$$f' = \frac{2 \, \Delta p g_c}{4 G_{max}^2 N} \tag{8-131}$$

and

$$j'_{max} = \frac{h c_p}{G_{max}} \left(\frac{c_p \mu}{k}\right)^{\frac{2}{3}} \left(\frac{\mu_w}{\mu}\right)^{0.14} \tag{8-132}$$

These factors are plotted against a Reynolds number $D_o G_{max}/\mu$. In these equations G_{max} has the meaning given for Eq. (8-83), N is the number of rows of tubes in the direction of flow, D_o is the tube diameter in feet, and all other symbols have their usual significance.

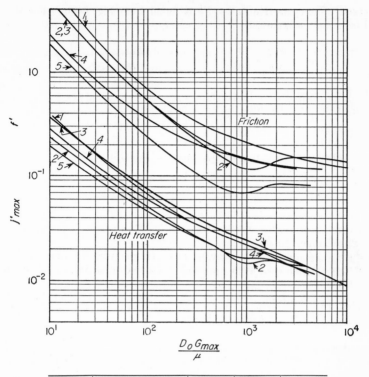

FIG. 8-47. Heat transfer and friction across tube banks. (*Bergelin et al.*[1] *Courtesy of ASME.*)

Curve	Arrangement	Rows	D_o, in.	Pitch D'_o, in.
1	Triangular	10	$\frac{3}{8}$	1.25
2	Square, in line	10	$\frac{3}{8}$	1.25
3	Square, staggered	14	$\frac{3}{8}$	1.25
4	Triangular	10	$\frac{3}{8}$	1.50
5	Square, in line	10	$\frac{3}{8}$	1.50

RADIATION

Radiation, which may be considered to be energy streaming through empty space at the speed of light, may originate in various ways. Some types of material will emit radiation when they are treated by external agencies, such as electron bombardment, electric discharge, or radiation of definite wavelengths. Radiation due to these effects is not of chemical engineering significance, and will not be discussed here. All substances at temperatures above the absolute zero emit radiation that is independent of external agencies. Radiation that is the result of temperature only is called thermal radiation, and the present discussion is restricted to radiation of this type.

Fundamental Facts Concerning Radiation. Radiation moves through space in straight lines, or beams, and only substances in sight of a radiating body can intercept radiation from that body. The fraction of the radiation falling on a body that is reflected is called the *reflectivity*, and is denoted by ρ. The fraction that is absorbed is called the *absorptivity*, and is denoted by α. The fraction that is transmitted is called the *transmissivity*, and is denoted by τ. Since radiation is a form of energy, it cannot be created or destroyed, and the sum of these fractions must be unity, or

$$\alpha + \rho + \tau = 1 \qquad (8\text{-}133)$$

Radiation as such is not heat, and when transformed into heat on absorption, it is no longer radiation. In practice, however, reflected or transmitted radiation usually falls on other absorptive bodies, and is eventually converted into heat, perhaps after many successive reflections.

The radiation emitted by any given mass of substance is independent of that being emitted by other material in sight of, or in contact with, the mass. The *net* energy gained or lost by a body is the difference between the energy emitted by the body and that absorbed by it from the radiation reaching it from the other bodies. Heat flow by conduction and convection may also be taking place independently of the radiation.

When bodies at different temperatures are exposed in sight of one another inside an enclosure, the hotter bodies lose energy by emission of radiation faster than they receive energy by absorption of radiation from the cooler bodies, and the temperatures of the hotter bodies decrease. Simultaneously the cooler bodies absorb energy from the hotter ones faster than they emit energy, and the temperatures of the cooler bodies increase. The process reaches equilibrium when all of the bodies reach the same temperature, just as in heat flow by conduction and convection. The conversion of radiation into heat on absorption and the attainment of temperature equilibrium through the net transfer of radiation justify the usual practice of calling radiation "heat."

Wavelength of Radiation. Known electromagnetic radiations cover an enormous range of wavelengths, from the short cosmic rays having wavelengths of about 10^{-11} cm to long-wave broadcasting waves having lengths of 1,000 m or more.

Radiation of a single wavelength is called *monochromatic*. An actual beam of radiation consists of many monochromatic beams. Although radiation of any wavelength from zero to infinity is, in principle, convertible into heat on absorption by matter, the portion of the electromagnetic spectrum that is of importance in heat flow lies in the wavelength range between 0.5 and 50 microns. Visible light covers a wavelength range of about 0.38 to 0.78 micron, and thermal radiation at ordinary industrial temperatures has wavelengths in the infrared spectrum, which includes waves just longer than the longest visible waves. At temperatures above about 500°C heat radiation in the visible spectrum becomes significant, and the phrases "red heat" and "white heat" refer to this fact. The higher the temperature of the radiating body, the shorter the predominant wavelength of the thermal radiation emitted by it.

For a given temperature, the rate of thermal radiation varies with the state of aggregation of the substance. Elementary gases such as oxygen and nitrogen, which have symmetrical molecules, radiate weakly, even at high temperatures. Under industrial conditions, these gases neither emit nor absorb appreciable amounts of radiation. Compound gases, including water vapor, carbon dioxide, ammonia, sulfur dioxide, and hydrocarbons, have nonsymmetrical molecules and emit and absorb radiation appreciably at furnace temperatures but only in certain bands of wavelength. Solids and liquids emit radiation over the entire spectrum.

Radiation from gases will not be discussed in this text. The following discussion is limited to radiation from solids.

Absorption of Radiation by Opaque Solids. When radiation falls on a solid body, a definite fraction ρ may be reflected, and the remaining fraction $1 - \rho$ enters the solid to be either transmitted or absorbed. Most solids, other than glasses, certain plastics, quartz, and some minerals, absorb radiation of all wavelengths so readily that, except in thin sheets, the transmissivity τ is zero, and all nonreflected radiation is completely absorbed in a thin surface layer of the solid. The absorption of radiation by an opaque solid is, therefore, a surface phenomenon, not a volume phenomenon, and the interior of the solid is not of interest in the absorption of radiation. The heat generated by the absorption can flow into or through the mass of an opaque solid only by conduction.

Reflectivity and Absorptivity of Opaque Solids. Since the transmissivity of an opaque solid is zero, the sum of the reflectivity and the absorptivity is unity, and the factors that influence reflectivity affect absorptivity in the opposite sense. In general, the reflectivity of an opaque solid depends on the temperature and character of the surface, the material of which the surface is made, the wavelength of the incident radiation, and the angle of incidence. Two main types of reflection are encountered, regular and diffuse. The first is characteristic of smooth surfaces such as polished metals; the second is found in reflection from rough surfaces or from dull, or matte, surfaces. In regular radiation, the reflected beam makes a definite angle with the surface, and the angle of incidence equals the angle of reflection. The reflectivity from these surfaces approaches unity, and the absorptivity approaches zero. Matte, or dull, surfaces reflect diffusely in all directions, there is no definite angle of reflection, and the absorptivity can approach unity. Rough surfaces, in which the scale of roughness is large in comparison with the wavelength of the incident radiation, will reflect diffusely even if the radiation from the individual units of roughness is regular. Reflectivities may be either large or small from rough surfaces, depending upon the reflective characteristic of the material itself. Most industrial surfaces of interest to the chemical engineer give diffuse reflection, and in treating practical cases, the important simplifying assumption can be made that reflectivity and absorptivity are independent of angle of incidence. This assumption is equivalent to the *cosine law*, which states that, for a perfectly diffusing surface, the intensity (or brightness, in the case of visible light) of the radiation leaving the surface is independent of the angle from which the surface is viewed.

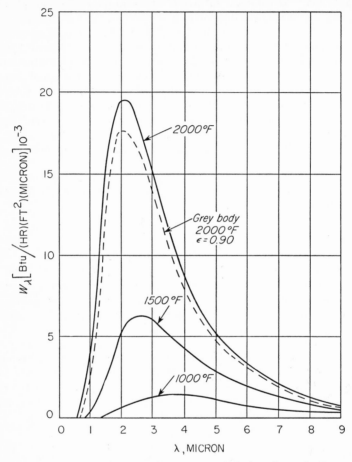

FIG. 8-48. Energy distribution in spectra of black and gray bodies.

The reflectivity may vary with the wavelength of the incident radiation, and the absorptivity of the entire beam is then a weighted average of the monochromatic absorptivities and depends upon the entire spectrum of the incident radiation.

For many industrial surfaces a second simplifying assumption can be made, namely, that the absorptivity of the surface is the same for all wavelengths. A body for which this is true is called a *gray body*. In diffuse radiation from a gray body, the monochromatic absorptivity is independent of both angle of incidence and of wavelength, so the total absorptivity equals the monochromatic absorptivity and is also independent of angle of incidence.

Emissive Power. The monochromatic energy emitted by a radiating surface depends on the temperature of the surface and on the wavelength of the radiation. At constant surface temperature, a curve can be plotted showing the rate of energy emission as a function of the wavelength. Typical curves of this type are shown in Fig. 8-48. Each curve starts at the origin, rises steeply to a maximum, and decreases asymptotically to zero emission at very large wavelengths. The unit chosen for measuring the monochromatic radiation is based on the fact that, from a small area of a radiating surface, the energy emitted is "broadcast" in all directions through any hemisphere centered on the radiation area. The monochromatic radiation emitted in this manner from unit area in unit time is called the *monochromatic radiating power*, and is denoted by W_λ. The units of this quantity are Btu per square foot per hour per micron. The ordinates in Fig. 8-48 are values of W_λ.

For the entire spectrum of the radiation from a surface, the total radiating power, denoted by W, is the sum of all the monochromatic radiations from the surface, or, mathematically,

$$W = \int_0^\infty W_\lambda \, d\lambda \qquad (8\text{-}134)$$

Graphically, W is the entire area under the curve of Fig. 8-48 from zero to infinity. Physically, the total radiating power is the total radiation of all wavelengths emitted by unit area in unit time in all directions through a hemisphere centered in the area. The units of W are Btu per square foot per hour.

Kirchhoff's Law. An important generalization concerning the radiating power of a substance is Kirchhoff's law, which states that, at temperature equilibrium, the ratio of the total radiating power of any body to the absorptivity of that body depends only upon the temperature of the body. Thus, consider any two bodies in temperature equilibrium with common surroundings. Kirchhoff's law states that

$$\frac{W_1}{\alpha_1} = \frac{W_2}{\alpha_2} \qquad (8\text{-}135)$$

where W_1 and W_2 are the total radiating powers and α_1 and α_2 are the absorptivities of the two bodies, respectively. This law applies to both monochromatic and total radiation.

Kirchhoff's law applies to volumes as well as to surfaces. Since absorption by an opaque solid is effectively confined to a thin layer at the surface, the radiation emitted from the surface of the body originates in this same surface layer. Radiating substances absorb their own radiation, and radiation emitted by the material in the interior of the solid is also absorbed in the interior and does not reach the surface.

Because the energy distribution in the incident radiation depends upon the temperature and character of the originating surface, the absorptivity of the receiving surface may also depend upon these properties of the originating surface. Kirchhoff's law does not, therefore, always apply to nonequilibrium radiation. If, however, the receiving surface is gray, a constant fraction, independent of wavelength, of the incident radiation is absorbed by the receiving surface, and Kirchhoff's law applies whether or not the two surfaces are at the same temperature.

Black Bodies. The maximum possible absorptivity is unity, attained only if the body absorbs all radiation incident upon it and reflects or transmits none. A body meeting this requirement is called a black body. By Kirchhoff's law, a black body possesses the maximum attainable radiating power at any given temperature. The black body is used as the standard to which all other radiators are referred. The *emissivity* of a body is then defined as the ratio of the total radiating power of the body to that of a black body at the same temperature. The emissivity is denoted by ϵ.

If the first body referred to in Eq. (8-135) is a black body, $\alpha_1 = 1$, and

$$W_1 = W_b = \frac{W_2}{\alpha_2} \tag{8-136}$$

where W_b denotes the total radiating power of a black body.

By Eq. (8-136), the emissivity of the second body ϵ_2 is

$$\epsilon_2 = \frac{W_2}{W_b} = \alpha_2 \tag{8-137}$$

Thus, when any body is at temperature equilibrium with its surroundings, its emissivity and absorptivity are equal. This relationship may be taken as another statement of Kirchhoff's law. In general, except for black or gray bodies, absorptivity and emissivity are not equal if the body is not in thermal equilibrium with its surroundings.

The absorptivity and emissivity, monochromatic or total, of a black body are, of course, both unity. The cosine law also applies exactly to a black body, as the reflectivity is zero for all wavelengths and all angles of incidence.

Emissivities of Solids. Emissivities of solids are tabulated in standard references.[24s, 29f] Emissivity usually increases with temperature. Emissivities of polished metals are low, in the range 0.03 to 0.08. Those of most oxidized metals range from 0.6 to 0.85, those of nonmetals such as refractories, paper, boards, and building materials, from 0.65 to 0.95, and those of paints, other than aluminum paint, from 0.80 to 0.96.

Practical Source of Black-body Radiation. No actual substance is a black body, although some materials, such as certain grades of carbon black, do approach blackness. An experimental equivalent of a black body is an isothermal enclosure containing a small peephole. If a sight is taken through the peephole on the interior wall of the enclosure, the effect is the same as viewing a black body. The radiation emitted by the interior of the walls or admitted from outside through the peephole is completely absorbed after successive reflections, and the over-all absorptivity of the interior surface is unity.

Laws of Black-body Radiation. A basic relationship for black-body radiation is the Stefan-Boltzmann law, which states that the total emissive power of a black body is proportional to the fourth power of the absolute temperature, or

$$W_b = \sigma T^4 \tag{8-138}$$

where σ is a universal constant depending only upon the units used to measure T and W_b. It has a numerical value [32] 0.1714×10^{-8} Btu/$(ft^2)(hr)(°R^4)$.

The Stefan-Boltzmann law is an exact consequence of the laws of thermodynamics and electromagnetism.[20]

For reasonable numbers in calculations, Eq. (8-138) is often written

$$W_b = 0.1714 \left(\frac{T}{100}\right)^4 \tag{8-139}$$

The distribution of energy in the spectrum of a black body is known accurately. It is given by Planck's law.

$$W_{b,\lambda} = \frac{C_1 \lambda^{-5}}{e^{C_2/\lambda T} - 1} \tag{8-140}$$

where $W_{b,\lambda}$ = monochromatic radiating power of a black body, Btu/$(ft^2)(hr)(micron)$

T = absolute temperature, °R

$C_1 = 1.187 \times 10^8$ Btu/$(ft^2)(hr)(micron^4)$

$C_2 = 2.590 \times 10^4$ micron-°R

λ = wavelength of radiation, microns

The solid lines of Fig. 8-48 are plots of $W_{b,\lambda}$ vs. λ for black-body radiation at temperatures of 1000, 1500, and 2000°F, respectively. The dotted line shows the monochromatic radiating power of a gray body of emissivity 0.9 at 2000°F.

Planck's law can be shown to be consistent with the Stefan-Boltzmann law by substituting $W_{b,\lambda}$ from Eq. (8-140) into Eq. (8-134) and integrating.

At any given temperature, the maximum monochromatic radiating power is attained at a definite wavelength, denoted by λ_{max}. Wien's displacement law states that λ_{max} is inversely proportional to the absolute temperature, or

$$T\lambda_{max} = C \tag{8-141}$$

The constant C is 5,200 when λ_{max} is in microns and T is in degrees Rankine.

Wien's law also can be derived from Planck's law [Eq. (8-140)] by differentiating with respect to λ, equating the derivative to zero, and solving for λ_{max}.

Radiation between Surfaces

The total radiation from a unit area of an opaque body of area A_1 emissivity ϵ_1, and absolute temperature T_1 is

$$\frac{q}{A_1} = \sigma \epsilon_1 T_1^4 \tag{8-142}$$

Practically, however, substances do not emit radiation into empty space that is at the absolute zero of temperature. Even in radiation to a clear night sky, the radiated energy is partially absorbed by the water and carbon dioxide in the atmosphere, and part of the absorbed energy is radiated back to the surface. In the usual situation, the energy emitted by any body is intercepted by other substances in sight of the body, which also are radiators, and their radiation falls on the body, to be absorbed or reflected. For example, a steam line in a room is surrounded by the walls, floor, and ceiling of the room, all of which are radiating to the pipe, and although the pipe loses more energy than it absorbs from its surroundings, the net loss by radiation is less than that calculated from Eq. (8-142).

In furnaces and other high-temperature equipment, where radiation is particularly important, the usual objective is to obtain a controlled rate of net heat exchange between one or more hot surfaces, called "sources" and one or more cold surfaces, called "sinks." In many cases the hot surface is a flame, but exchange of energy between surfaces is common, and a flame can be considered to be a special form of translucent surface. The following treatment is limited to the radiant-energy transfer between opaque surfaces in the absence of any absorbing medium between them.

The simplest type of radiation between two surfaces is where each surface can see only the other, e.g., where the surfaces are very large parallel planes as shown in Fig. 8-49a, and where both surfaces are black. The energy emitted by the first plane is σT_1^4 Btu/(ft^2)(hr). All the radiation from each of the surfaces falls on the other surface and is completely absorbed, so the net loss of energy by the first plane and the net gain of energy by the second (assuming that $T_1 > T_2$) is $\sigma T_1^4 - \sigma T_2^4$, or $\sigma(T_1^4 - T_2^4)$ Btu/(ft^2)(hr).

Actual engineering problems differ from this simple situation in the following ways: (1) One or both of the surfaces of interest see other surfaces. In fact, an element of surface in a concave area sees a portion of its own surface. (2) No actual surface is exactly black, and the emissivities of the surfaces must often be considered.

Angle of Vision. Qualitatively, the interception of radiation from an area element of a surface by another surface of finite size can be visualized in terms of the angle of vision, which is the solid angle subtended by the finite surface at the radiating element. The solid angle subtended by a hemisphere is 2π steradians. This is the maximum angle of vision that can be subtended at any area element by a surface in sight of the element. It will be remembered that the total radiating power of an area element is defined to take this fact into account. If the angle of vision is less than 2π steradians, only a fraction of the radiation from the area element will be intercepted by the receiving area, and the remainder will pass on to be absorbed by other surfaces in sight of the remaining solid angle. Some of the hemispherical angle of vision of an element of a concave surface is subtended by the originating surface itself.

Figure 8-49 shows several typical radiating surfaces. Figure 8-49a shows how, in two large parallel planes, an area element on either plane is sub-

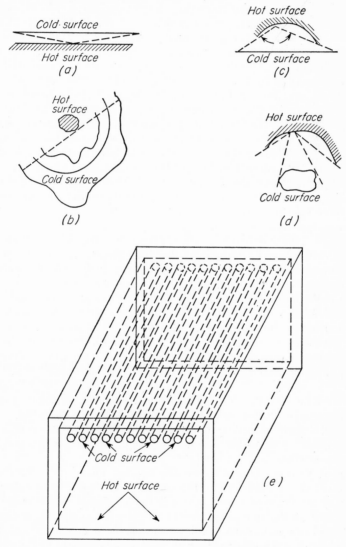

FIG. 8-49. Angle of vision in radiant heat flow.

tended by a solid angle of 2π steradians by the other. The radiation from either plane cannot escape being intercepted by the other. A point on the hot body of Fig. 8-49b sees only the cold surface, and the angle of vision is again 2π steradians. Elements of the cold surface, however, see, for the most part, other portions of the cold surface, and the angle of vision for the hot body is small. This effect of self-absorption is also shown in Fig. 8-49c, where the angle of vision of an element of the hot surface subtended by the cold surface is small. In Fig. 8-49d, the cold surface subtends a small angle at the hot surface, and the bulk of the radiation from the hot surface passes on to some undetermined background. Figure 8-49e shows a simple muffle furnace, in which the radiation from the hot floor, or source, is intercepted partly by the row of tubes across the top of the furnace, which form the sink, and partly by the refractory walls and the refractory ceiling behind the tubes. The refractory in such assemblies is assumed to absorb and emit energy at the same rate, so the net energy effect at the refractory is zero. The refractory ceiling absorbs the energy that passes between the tubes and reradiates it to the backs of the tubes.

If attention is to be focused on the net energy received by the cold surface, the words "hot" and "cold" in Fig. 8-49 may be interchanged, and the same qualitative conclusions hold.

Square-of-the-distance Effect. The energy from a small surface that is intercepted by a large one depends only upon the angle of vision. It is

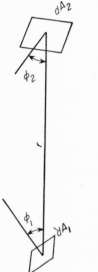

independent of the distance between the surfaces. The energy received per *unit area* of the receiving surface, however, is inversely proportional to the square of the distance between the surfaces. This is the familiar square-of-the-distance law. Also, if the receiving surface is tilted so that it presents a foreshortened appearance at the radiating element, the energy received per unit area is proportional to the cosine of the angle between a normal to the receiving surface and the line of sight to the radiating element. This is the cosine principle, previously mentioned.

Quantitative Calculation of Radiation between Black Surfaces. The above considerations can be treated quantitatively by setting up a differential equation for the net radiation between two elementary areas and integrating the equation for definite types of arrangement of the surfaces. The differential equation for two plane areas dA_1 and dA_2, shown in Fig. 8-50, is

FIG. 8-50. Differential areas for radiation.

$$dq_{12} = \sigma \frac{\cos \phi_1 \cos \phi_2 \, dA_1 \, dA_2}{\pi r^2} (T_1^4 - T_2^4) \quad (8\text{-}143)$$

where dq_{12} = net exchange of radiation between two elementary surfaces of areas A_1 and A_2, respectively

T_1, T_2 = absolute temperatures of surfaces dA_1 and dA_2, respectively

ϕ_1, ϕ_2 = angles between the normals to each of the two surfaces and the straight line connecting the areas

r = length of straight line connecting the areas

σ = Stefan-Boltzmann constant

Equation (8-143) includes application of the cosine law, the square-of-the-distance law, and the Stefan-Boltzmann law. It applies rigorously to black surfaces.

The integration of Eq. (8-143) for a given combination of finite surfaces is usually a laborious multiple integration based on the geometry of the two planes and their relation to each other. The resulting equation for any of these situations can be written in the form

$$q_{12} = \sigma A F (T_1^4 - T_2^4) \qquad (8\text{-}144)$$

where q_{12} = net radiation between two surfaces, Btu/hr

A = area of either of two surfaces, chosen arbitrarily, ft^2

F = dimensionless geometric factor

The factor F is called the view factor; it depends upon the geometry of the two surfaces, their spatial relationship with each other, and the surface chosen for A.

If surface A_1 is chosen for A, Eq. (8-144) can be written

$$q_{12} = \sigma A_1 F_{12} (T_1^4 - T_2^4) \qquad (8\text{-}145)$$

If surface A_2 is chosen,

$$q_{12} = \sigma A_2 F_{21} (T_1^4 - T_2^4) \qquad (8\text{-}146)$$

Comparing Eqs. (8-145) and (8-146),

$$A_1 F_{12} = A_2 F_{21} \qquad (8\text{-}147)$$

If surface A_1 can see only surface A_2, the view factor F_{12} is unity. If surface A_1 sees a number of other surfaces, and if its entire hemispherical angle of vision is filled by these surfaces,

$$F_{11} + F_{12} + F_{13} = \cdots = 1.0 \qquad (8\text{-}148)$$

The factor F_{11} covers the portion of the angle of vision subtended by other portions of body A_1. If the surface of A_1 cannot see any portion of itself, F_{11} is zero. The net radiation associated with an F_{11} factor is, of course, zero.

As a simple example of the use of Eqs. (8-147) and (8-148) consider a small black body of area A_2 ft^2 having no concavities and surrounded by a large black surface of area A_1 ft^2. The factor F_{21} is unity, as area A_2 can see nothing but area A_1. The factor F_{12} is, by Eq. (8-147),

$$F_{12} = \frac{F_{21} A_2}{A_1} = \frac{A_2}{A_1}$$

By Eq. (8-148),

$$F_{11} = 1 - F_{12} = 1 - \frac{A_2}{A_1}$$

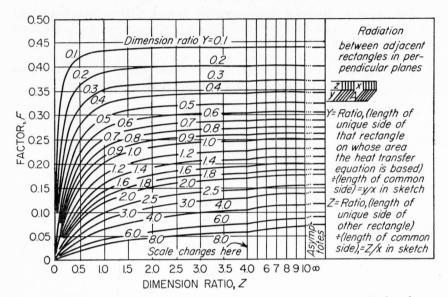

FIG. 8-51. View factor, radiation between adjacent rectangles in perpendicular planes. (*H. C. Hottel, by permission.*)

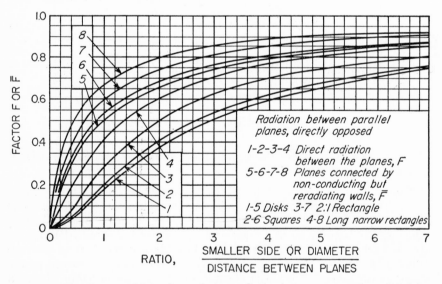

FIG. 8-52. View factor and interchange factor, radiation between opposed parallel disks, rectangles, and squares. (*H. C. Hottel, by permission.*)

The factor F has been determined by Hottel[15] for several important special cases. Figure 8-51 shows the F factor for two mutually perpendicular rectangular planes in contact along one edge. In this figure, Y is the ratio of the unique side of one rectangle to the common side, and Z is the ratio of the unique side of the other to the common side. Figure 8-52 shows the F factor for equal parallel planes directly opposed. Line 1 is

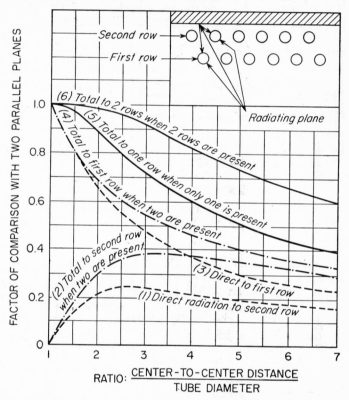

FIG. 8-53. View factor and interchange factor, radiation between parallel plane and rows of tubes. (*H. C. Hottel, by permission.*)

for disks, line 2 for squares, line 3 for rectangles having a ratio of length to width of 2:1, and line 4 is for long, narrow rectangles. In all cases, the factor F is a function of the ratio of the side or diameter of the plane to the distance between them. Figure 8-53 gives factors for radiation to tube banks backed by a layer of refractory which absorbs energy from the rays that pass between the tubes and reradiates the absorbed energy to the backs of the tubes. The factor given in Fig. 8-53 is the radiation absorbed by the tubes calculated as a fraction of that absorbed by a parallel plane of area equal to that of the refractory backing.

Allowance for Refractory Surfaces. When the source and sink are connected by refractory walls in the manner shown in Fig. 8-49e, the factor

F can be replaced by an analogous factor, called the *interchange factor*, denoted by $\bar{F}$, and Eqs. (8-145) and (8-146) written as

$$q_{12} = \sigma A_1 \bar{F}_{12}(T_1^4 - T_2^4) = \sigma A_2 \bar{F}_{21}(T_1^4 - T_2^4) \qquad (8\text{-}149)$$

The interchange factor $\bar{F}$ has been determined accurately for some simple situations.[16] Lines 5 to 8 of Fig. 8-52 give values of $\bar{F}$ for directly opposed parallel planes connected by refractory walls. Line 5 applies to disks, line 6 to squares, line 7 to 2:1 rectangles, and line 8 to long, narrow rectangles.

An approximate equation for $\bar{F}$ in terms of F is

$$\bar{F}_{12} = \frac{A_2 - A_1 F_{12}^2}{A_1 + A_2 - 2A_1 F_{12}} \qquad (8\text{-}150)$$

Equation (8-150) applies where there is but one source and one sink, where neither area A_1 nor A_2 can see itself. It is based on the assumption that the temperature of the refractory surface is constant. This last is a simplifying assumption, as the temperature of the refractory usually varies between those of the source and of the sink.

Nonblack Surfaces. The treatment of radiation between nonblack surfaces becomes, in the general case where absorptivity and emissivity are unequal and both depend upon wavelength and angle of incidence, too complicated for practical calculations. Several important special cases can, however, be treated practically.

A simple example is that where a small nonblack body is surrounded by a black surface. Let the areas of the enclosed and surrounding surfaces be A_1 and A_2 ft^2, respectively, and let their temperatures be T_1 and T_2 °R, respectively. The radiation from surface A_2 falling on surface A_1 is $\sigma A_2 F_{21} T_2^4$ Btu/hr. Of this, the fraction α_1, the absorptivity of area A_1 for radiation from surface A_2, is absorbed by surface A_1. The remainder is reflected back to the black surroundings and completely reabsorbed by the area A_2. Surface A_1 emits radiation in amount $\sigma A_1 \epsilon_1 T_1^4$, where ϵ_1 is the emissivity of surface A_1. All of this radiation is absorbed by the surface A_2, and none is returned by another reflection. The emissivity ϵ_1 and the absorptivity α_1 are not in general equal, because the two surfaces are not at the same temperature. The net energy loss by surface A_1 is

$$q_{12} = \sigma \epsilon_1 A_1 T_1^4 - \sigma A_2 F_{21} \alpha_1 T_2^4 \qquad (8\text{-}151)$$

But, by Eq. (8-147), $A_2 F_{21} = A_1$, and Eq. (8-151), after elimination of $A_2 F_{21}$, becomes

$$q_{12} = \sigma A_1 (\epsilon_1 T_1^4 - \alpha_1 T_2^4) \qquad (8\text{-}152)$$

If surface A_1 is gray, $\epsilon_1 = \alpha_1$, and

$$q_{12} = \sigma A_1 \epsilon_1 (T_1^4 - T_2^4) \qquad (8\text{-}153)$$

If all surfaces are gray, so the absorptivity and emissivity of any one surface are equal, and if the cosine law applies to all surfaces, the following approximate equation can be used to calculate an over-all interchange factor $\mathfrak{F}$, which includes the effect of refractory walls, if present, and accounts for the emissivities of the various surfaces.

$$\mathfrak{F}_{12} = \cfrac{1}{\cfrac{1}{\overline{F}_{12}} + \left(\cfrac{1}{\epsilon_1} - 1\right) + \cfrac{A_1}{A_2}\left(\cfrac{1}{\epsilon_2} - 1\right)} \tag{8-154}$$

where ϵ_1 and ϵ_2 are the emissivities of source and sink, respectively. If no refractory is present, F is used in place of $\overline{F}$. The factor $\mathfrak{F}_{12}$ is used in place of F or $\overline{F}$ in Eqs. (8-145), (8-146), and (8-149).

The following special cases of Eq. (8-154) are useful.

Two Large Parallel Planes. The area ratio A_1/A_2 is unity, there are no refractory surfaces, $\overline{F}_{12} = 1.0$, and Eq. (8-154) becomes

$$\mathfrak{F}_{12} = \cfrac{1}{\cfrac{1}{\epsilon_1} + \cfrac{1}{\epsilon_2} - 1} \tag{8-155}$$

One Surface Completely Surrounded by Another. Let the area of the enclosed body be A_1 and that of the enclosure be A_2. Then $\overline{F}_{12} = 1.0$, and Eq. (8-154) becomes

$$\mathfrak{F}_{12} = \cfrac{1}{\cfrac{1}{\epsilon_1} + \cfrac{A_1}{A_2}\left(\cfrac{1}{\epsilon_2} - 1\right)} \tag{8-156}$$

Equation (8-156) applies strictly to concentric spheres or concentric cylinders, but it can be used without serious error for other shapes.

The case of a gray body surrounded by a black one can be treated as a special case of Eq. (8-156) by setting $\epsilon_2 = 1.0$.

Combined Loss of Heat by Conduction-Convection and Radiation. The total heat loss from a hot body to its surroundings often includes appreciable losses by conduction-convection and radiation. For example, a steam radiator or a hot pipeline in a room loses heat nearly equally by each of the two mechanisms. Since the two types of heat transfer occur in parallel, the total loss is, assuming black surroundings,

$$\begin{aligned}
\frac{q_t}{A} &= \frac{q_c}{A} + \frac{q_r}{A} \\
&= h_c(t_w - t) + 0.1714\epsilon_w \left[\left(\frac{t_w + 460}{100}\right)^4 - \left(\frac{t + 460}{100}\right)^4\right]
\end{aligned} \tag{8-157}$$

where q_t/A = total heat loss, Btu/(ft^2)(hr)

q_c/A = heat loss by conduction-convection, Btu/(ft^2)(hr)

q_r/A = heat loss by radiation, Btu/(ft^2)(hr)

ϵ_w = emissivity of surface

t_w = temperature of surface, °F

t = temperature of surroundings, °F

SYMBOLS

A Area, ft^2; A_b, of bare tube; A_F, of fin; A_i, of inside of tube; $\bar{A}_L$, logarithmic mean; A_o, of outside of tube; A_T, total area of heating surface

a Empirical constant

a_F Fin factor [Eq. (8-88)]

a_1 $(\pi/2)^2$

b Empirical constant

C Constant in Eq. (8-141); equals 5,200 when λ_{max} is in microns and T in degrees Rankine

C_1 Constant in Planck's law, 1.187×10^8 Btu/(ft^2)(hr) (micron^4)

C_2 Constant in Planck's law, 2.590×10^4 micron-°R^4

c_p Specific heat at constant pressure, Btu/(lb)(°F); c_{pc}, for cold fluid; c_{ph}, for warm fluid

D Diameter, ft; D_a, of agitator; D_e, equivalent diameter $4r_H$; D_i, inside diameter; D_o, outside diameter; D_t, diameter of tank; $\bar{D}_L$, logarithmic mean

F View factor for radiation between surfaces; $\bar{F}$, interchange factor; $\mathfrak{F}$, over-all interchange factor

F_G Correction factor for average temperature drop in exchanger, flow different from parallel or countercurrent flow

G Mass velocity, lb/ft^2-sec or lb/ft^2-hr; G_{max}, through minimum cross section in tube bundle

g Acceleration of gravity, ft/sec^2 or ft/hr^2

g_c Newton's-law conversion factor, 32.174 ft-lb/lb force-sec^2, or 4.17×10^8 ft-lb/lb force-hr^2

h Individual or surface heat-transfer coefficient, Btu/(ft^2)(hr)(°F); h_c, coefficient for conduction-convection when radiation is also active; h_i, coefficient for inside of tube; h_{ia}, coefficient for inside of tube, based on arithmetic mean average temperature drop; h_o, coefficient for outside of tube; $\bar{h}$, average coefficient, when local coefficient varies with location on heating surface

h_d Fouling factor, Btu/(ft^2)(hr)(°F); h_{di}, for inside of tube; h_{do}, for outside of tube

i Specific enthalpy, Btu/lb; i_c, for cold stream; i_{ca}, for entering cold stream; i_{cb}, for leaving cold stream; i_{ha}, for entering warm stream; i_{hb}, for leaving warm stream

j'_H Colburn factor, defined by Eq. (8-57); j'_{max}, for crossflow in shell side of tube-and-shell exchanger, defined by Eq. (8-132)

K Constant, defined by Eq. (8-126)

k Thermal conductivity, Btu-ft/(ft^2)(hr)(°F); k_A, k_B, k_C, of layers A, B, C, respectively; k_f, at mean "film" temperature, t_f; k_m, of wall or fin; $\bar{k}_a$, arithmetic average

L Length of tube or cylinder, ft

L_p Perimeter, ft

m Exponent; also, mass of liquid in tank, lb

N Number of horizontal tubes in vertical stack, or number of rows of tubes in tube bank, in direction of flow

n Exponent, also, agitator speed, rph

p Pressure, lb force/ft^2

Q Heat flow, Btu

q Heat flow per unit time, or flux, Btu/hr; equals Q/θ or $dQ/d\theta$; q_A, q_B, q_C, through layers A, B, and C, respectively; q_c, to cold fluid, or from conduction-convection only when radiation is also active; q_h, to warm fluid; q_r, by radiation when radiation and conduction-convection are both active; q_{12}, net heat flow rate by radiation from surface 1 to surface 2

R Resistance, $x_w/(kA)$; R_A, R_B, R_C, of layers A, B, and C, respectively

R_o Gas-law constant

r Radius, ft; r_H, hydraulic radius; r_i, inside radius; r_o, outside radius; $\bar{r}_a$, arithmetic mean; $\bar{r}_L$, logarithmic mean; also, length of straight line connecting area elements of radiating surfaces

S Cross section, ft²; S_c, of annular ring of condensate on vertical tube

s One-half slab thickness, ft

T Absolute temperature, °R; $\overline{T}$, average absolute temperature

t Temperature, °F; t_a, initial temperature; t_b, final temperature; t_c, variable temperature of cooling medium; t_{ca}, of entering cold fluid; t_{cb}, of leaving cold fluid; t_F, temperature at distance y from base of fin; t_f, mean "film" temperature, $(t + t_w)/2$; t_h, boiling or condensing temperature; t_{ha}, temperature of entering warm fluid; t_{hb}, of leaving warm fluid; t_i, of inside of cylinder; t_o, of outside of cylinder; t_s, surface temperature of slab or of boiling tube; t_{wc}, of cold side of tube wall; t_{wh}, of warm side of wall; $\bar{t}_b$, average temperature of slab at time θ_t

U Over-all heat-transfer coefficient, Btu/(ft²)(hr)(°F); U_i, based on inside area; U_o, based on outside area; U_1, U_2, at ends of exchanger

$\overline{V}$ Average linear velocity, ft/sec

v Specific volume, ft³/lb for liquids, ft³/lb mole for gases; $\bar{v}$, average value

W Total radiating power, Btu/(ft²)(hr); W_b, total for black body; $W_{b,\lambda}$, monochromatic radiating power for black body at wavelength λ; W_λ, for monochromatic radiation at wavelength λ

w Mass flow rate of fluid, lb/hr; w_c, for cold fluid; w_h, for warm fluid; also, rate of condensation, lb/hr

x Distance in direction of heat flow, ft; x_A, x_B, x_C, thickness of layers A, B, C, respectively; x_F, length of fin; x_w, tube-wall thickness

y_F Thickness of fin, ft

Greek Letters

α Absorptivity for radiation; also, thermal diffusivity, $k/\rho c_p$, ft²/hr

β Coefficient of volume expansion

Γ Condensate loading, vertical tube, lb/ft of perimeter; Γ' loading for horizontal tube, lb/ft length

Δt Over-all temperature drop $t_h - t_c$, °F; Δt_A, Δt_B, Δt_C, temperature differences over layers A, B, C, respectively; $\overline{\Delta t_F}$, average temperature drop between fluid and fins; Δt_i, temperature drop between wall and fluid in tube; Δt_o, temperature drop between wall and fluid outside tube; $\overline{\Delta t_a}$, arithmetic mean temperature drop; $\overline{\Delta t_L}$, logarithmic mean temperature drop; Δt_1, Δt_2, temperature approaches at ends of exchanger; Δt_w, temperature drop through metal wall

∂ Partial-differential operator

ϵ Emissivity for radiation

η_F Fin efficiency [Eq. (8-86)]

θ Time, hr; θ_T, time required to heat or cool

λ Latent heat of condensation or vaporization, Btu/lb. Also wavelength of radiation, microns

μ Viscosity, lb/ft-sec or lb/ft-hr; μ_f, at mean "film" temperature; μ_w, at wall temperature

ρ Density, lb/ft³; ρ_f, at mean "film" temperature; $\bar{\rho}_a$, arithmetic mean of ρ_1 and ρ_2 at temperatures t_1 and t_2, respectively; also, reflectivity for radiation

σ Stefan-Boltzmann constant, 0.1714×10^{-8} Btu/(ft^2)(hr)($^\circ$R^4)

τ Transmissivity for radiation

Φ Function in Eq. (8-119)

ϕ Angle

ϕ_v Viscosity-correction factor [Eq. (8-52)]

ϕ_n Natural-convection factor [Eq. (8-79)]

χ Function in Eq. (8-69); χ', function in Eq. (8-61)

Ψ_i Function in Eq. (8-48); Ψ_i', function in Eq. (8-47)

Ψ_o Function in Eq. (8-80)

ψ_f Property group, condensation, $(k^3 \rho^2 g / \mu^2)^{\frac{1}{3}}$

Dimensionless Groups

f Fanning friction factor, $2 g_c D \, \Delta p / 4 \rho \overline{V}^2 L$; f', for crossflow in shell side of tube-and-shell exchanger, $2 \, \Delta p g_c / 4 G^2_{\text{ax}} N$

N_{Fo} Fourier number, $\alpha\theta/x_s^2 = k\theta/\rho c_p x_s^2$

N_{Gr} Grashof number, $D^3 \rho^2 \beta g \, \Delta t / \mu^2$

N_{Gz} Graetz number, $w c_p / k L$

N_{Nu} Nusselt number, hD/k

N_{Pe} Peclet number, $D \overline{V} \rho c_p / k$

N_{Pr} Prandtl number, $c\mu/k$

N_{Re} Reynolds number, $D \overline{V} \rho / \mu = DG/\mu$

Z Dimensionless ratio $(t_{ha} - t_{hb})/(t_{cb} - t_{ca})$

η_H Heating effectiveness of exchanger, $(t_{ca} - t_{cb})/(t_{ha} - t_{ca})$

PROBLEMS

8-1. A furnace wall consists of 9 in. of refractory fire-clay brick, $4\frac{1}{2}$ in. of Sil-o-cel brick, and $\frac{1}{4}$ in. of steel plate. The fire side of the refractory is at 2100°F, and the outside surface of the steel is at 90°F. The heat loss from the wall is found, by an accurate heat balance over the furnace, to be 100 Btu/(ft^2)(hr).

It is known that there may be thin layers of air between the layers of brick and steel. To how many inches of Sil-o-cel brick are these air layers equivalent?

8-2. A standard 1-in. Schedule 40 iron pipe carries saturated steam at 250°F. The pipe is lagged (insulated) with a 2-in. layer of 85 per cent magnesia pipe covering, and outside this magnesia there is a 3-in. layer of cork. The inside temperature of the pipe wall is 249°F, and the outside temperature of the cork is 90°F.

Calculate: (*a*) the heat loss from 100 ft of pipe, in Btu per hour, (*b*) the temperatures at the boundaries between metal and magnesia and between magnesia and cork.

8-3. A copper tube, $\frac{1}{2}$ in. OD by 22 BWG wall, is carrying hot water. The tube is to be lagged with an inexpensive insulation material having a k value of 0.18 Btu/(ft)(hr)($^\circ$F). The coefficient of heat transfer between the surface of the lagging and the room is 2.0 Btu/(ft^2)(hr)($^\circ$F).

Calculate and plot, against the thickness of the lagging in inches, the percentage of the heat loss from the bare pipe that is saved by use of the lagging.

8-4. A large sheet of glass 2 in. thick is initially at 300°F throughout. It is plunged into a stream of running water having a temperature of 60°F. How long will it take to cool the glass to an average temperature of 100°F? For glass, $k = 0.40$ Btu-ft/(ft^2)(hr)($^\circ$F); $\rho = 155$ lb/ft^3; and $c_p = 0.20$ Btu/(lb)($^\circ$F).

8-5. Kerosene is heated by hot water in a tube-and-shell heater. The kerosene is inside the tubes, and the water is outside. The flow is countercurrent. The average temperature of the kerosene is 110°F, and the average linear velocity is 5 ft/sec. The properties of the kerosene at 110°F are specific gravity = 0.805, viscosity = 1.5 centipoise, specific heat = 0.583 Btu/(lb)(°F), and thermal conductivity = 0.0875 Btu/(ft)(hr)(°F). The tubes are low-carbon steel $\frac{3}{4}$ in. OD by 16 BWG. The heat-transfer coefficient on the shell side is 300 Btu/(ft²)(hr)(°F). Calculate the over-all coefficient, based on the outside area of the tube.

8-6. Assume that the kerosene of Prob. 8-5 is replaced with water at 110°F and flowing at a velocity of 5 ft/sec. What percentage increase in over-all coefficient may be expected if the tube surfaces remain clean?

8-7. Both surfaces of the tube of Prob. 8-6 become fouled with deposits from the water. The fouling factors are 330 on the inside and 200 on the outside surfaces, both in Btu/(ft²)(hr)(°F). What percentage decrease in over-all coefficient is caused by the fouling of the tube?

8-8. Aniline is to be cooled from 200 to 150°F in a double-pipe heat exchanger having a total outside area of 70 ft². For cooling, a stream of toluene amounting to 8,600 lb/hr at a temperature of 100°F is available. The exchanger consists of Schedule 40 $1\frac{1}{4}$-in. pipe in Schedule 40 2-in. pipe. How much aniline can be cooled?

8-9. Calculate the over-all heat-transfer coefficients based on both inside and outside areas for the following cases. All individual coefficients are given in usual units.

Case 1. Water at 50°F flowing in a $\frac{3}{4}$-in. by 16 BWG condenser tube at a velocity of 15 ft/sec and saturated steam at 220°F condensing on the outside. $h_i = 2150$. $h_o = 2500$. $k_m = 69$.

Case 2. Benzene at atmospheric pressure condensing on the outside of a Schedule 40 1-in. steel pipe and air at 60°F flowing within at 20 ft/sec. $h_i = 5$. $h_o = 225$. $k_m = 26$.

Case 3. Dropwise condensation from steam at a pressure of 50 lb force/in.² gauge on a Schedule 40 1-in. steel pipe carrying oil at a velocity of 3 ft/sec. $h_o = 14,000$. $h_i = 130$. $k_m = 26$.

8-10. It is desired to build a furnace wall, not over 24 in. thick, of one or more of the refractories and insulators listed in Table 8-4. The bricks are to be laid flat, either as

TABLE 8-4. DATA ON REFRACTORIES

Material	Maximum service temp., °F	Thermal conductivity,† Btu-in./(ft²)(hr)(°F)		Unit size, in.
		200°F	1900°F	
Fire clay.................	3000	5.30	10.55	9 × 4.5 × 2.5
Insulating refractory.........	2600	1.85	3.45	9 × 4.5 × 2.5
Insulating brick.............	2000	0.90	1.70	9 × 4.5 × 2.5
Superex insulation.	1900	0.60	1.00	Slab thickness integral inches

† Thermal conductivities are linear with temperature.

headers or spreaders, and to save expense are not to be cut. The Superex comes as slabs in thickness of integral inches. The inside of the furnace wall is to be at 2800°F, and the room temperature may be taken as 100°F. The coefficient of heat transfer from the wall to the room is 2.0 Btu/(ft²)(hr)(°F).

(a) Specify the material and thickness of each layer in the wall in order that the heat loss through the wall is a minimum consistent with the maximum service temperature of the material used. (b) What is the minimum loss in Btu per 100 ft² of wall area? (c) Estimate and plot the temperature gradients through the wall.

8-11. A vertical-tube two-pass heater is used for heating gas oil. Saturated steam at 50 lb force/in.² gauge is used as a heating medium. The tubes are 1 in. OD by 16 BWG and are made of mild steel. The oil enters at 60°F and leaves at 150°F. The viscosity-temperature relation is exponential. The viscosity at 60°F is 5.0 centipoises and at 150°F is 1.8 centipoises. The oil is 37°API (specific gravity = 0.840) at 60°F. The flow of oil is 120 bbl/hr (1 bbl = 42 gal). Assume the steam condenses in film condensation. The thermal conductivity of the oil is 0.078 Btu/(ft)(hr)(°F), and the specific heat is 0.480. The velocity of the oil in the tubes should be considered to be approximately 3 ft/sec.

Calculate the length of tubes needed for this heater.

8-12. By chromium plating the outside of the tubes of Prob. 8-11 and adding mercaptan to the steam, the condensation becomes dropwise, and the condensing steam coefficient becomes 14,000 Btu/(ft²)(hr)(°F). How much oil will the heater heat from 60 to 150°F under these new steam-side conditions?

8-13. Kerosene is heated in a single-pass tube-and-shell heater by hot water. The kerosene is inside the tubes and the hot water outside. The heater is to heat 1,000 gal/min of kerosene (measured at 60°F) from 100 to 180°F. The tubes are mild steel, and are $\frac{5}{8}$ in. OD by 18 BWG wall. The velocity in the tubes is to be about 7 ft/sec. The tubes are arranged on hexagonal centers, and the outside hexagonal periphery of the tubes should be complete. The tubes can be placed on $1\frac{3}{8}$-in. centers. No tube should be within $2\frac{1}{2}$ in. of the shell wall. The water-side coefficient is 350 Btu/(ft²)(hr)(°F). The water enters at 210°F and leaves at 150°F.

The thermal conductivity and viscosity of the kerosene at 140°F are 0.0875 Btu/(ft)(hr)(°F) and 1.0 centipoise, respectively, and the specific gravity is 0.805 at 60°F and 0.772 at 140°F.

Determine the inside diameter of the shell and the length of the tubes.

8-14. Air is flowing through a steam-heated tubular heater under such conditions that the steam and wall resistances are negligible in comparison with the air-side resistance. Assuming that each of the following factors is changed in turn but that all other original factors remain constant, calculate the percentage variation in $q/\overline{\Delta t_L}$ that accompanies each change.

(a) Double the pressure on the gas but keep fixed the mass flow rate of the air. (b) Double the mass flow rate of the air. (c) Double the number of tubes in the heater. (d) Halve the diameter of the tubes.

8-15. Air is blown at a rate of 5,000 ft³/min (measured at 70°F and 760 mm) at right angles to a tube bank 10 pipes and 10 spaces wide and 10 rows deep. The length of each pipe is 8 ft. The tubes are on equilateral centers, and the center-to-center distance is 3 in. It is desired to heat the air from 70 to 100°F at atmospheric pressure. What steam pressure must be used? The pipes which are used are Schedule 40 standard 1-in. steel pipe.

8-16. Oil at 50°F is heated in a horizontal tube 50 ft long having a surface temperature of 100°F. The pipe is Schedule 40 2-in. iron pipe. The oil flow rate is 100 gal/hr at inlet temperature. What will be the oil temperature as it leaves the tubes and after mixing? What is the average heat-transfer coefficient? Properties of the oil are as follows:

	50°F	100°F
Specific gravity............................	0.80	0.75
Thermal conductivity, Btu/(ft)(hr)(°F).....	0.072	0.074
Viscosity, centipoises.....	20	10
Specific heat, Btu/(lb)(°F)...............	0.75	0.75

8-17. A 3-in. Schedule 40 iron pipeline carries steam at 90 lb force/in.^{2} gauge. The line is unlagged and is 200 ft long. The surrounding air is at 80°F. How many pounds of steam will condense per hour? What percentage of the heat loss is from conduction-convection?

8-18. A petroleum oil having the properties given in Table 8-5 is to be heated in a

TABLE 8-5. PROPERTIES OF PETROLEUM OIL

Temp., °F	Thermal conductivity, Btu/(ft)(hr)(°F)	Kinematic viscosity, 10^5 ft^2/sec	Density, lb/ft^3	Specific heat, Btu/(lb)(°F)
100	0.0739	36.6	55.25	0.455
120	0.0737	21.8	54.81	0.466
140	0.0733	14.4	54.37	0.477
160	0.0728	10.2	53.92	0.487
180	0.0724	7.52	53.48	0.498
200	0.0719	5.70	53.03	0.508
220	0.0711	4.52	52.58	0.519
240	0.0706	3.67	52.13	0.530
260	0.0702	3.07	.....	0.540

horizontal multipass heater with steam at 50 lb force/in.^{2} gauge. The tubes are to be steel $\frac{3}{4}$ in. OD by 16 BWG, and their maximum length is 15 ft. The oil enters at 100°F, leaves at 180°F, and enters the tubes at about 3 ft/sec. The total flow rate is 150 gal/min. Assuming complete mixing of the oil after each pass, how many passes are required?

8-19. A large tank of water is heated by natural convection from horizontal steam pipes. The pipes are Schedule 40 3-in. steel. When the steam pressure is atmospheric and the water temperature 80°F, what is the heat transfer to the water in Btu per running foot of pipe?

8-20. Determine the net heat transfer by radiation between two surfaces A and B, expressed as Btu per hour for each square foot of area B, if the temperatures of A and

B are 900 and 400°F, respectively, and the emissivities of A and B are 0.90 and 0.25, respectively. Both surfaces are gray.

(a) Surfaces A and B are infinite parallel planes 10 ft apart.

(b) Surface A is a spherical shell 10 ft in diameter, and surface B is a similar shell concentric with A and 1 ft in diameter.

(c) Surfaces A and B are flat parallel squares 5 by 5 ft, one exactly above the other, 5 ft apart.

(d) Surfaces A and B are concentric cylindrical tubes with diameters of 10 and 9 in. respectively.

(e) Surface A is an infinite plane and surface B is an infinite row of 4-in.-OD tubes set on 8-in. centers.

(f) Same as (e) except 8 in. above the center lines of the tubes is another infinite plane having an emissivity of 0.90, which does not transmit any of the energy incident upon it.

(g) Same as (f) except surface B is a double row of 4-in.-OD tubes set on equilateral 8-in. centers.

8-21. The black flat roof of a building has an emissivity of 0.9 and an absorptivity of 0.8 for solar radiation. The sun beats down at midday with an intensity of 300 Btu/(ft²)(hr). If the temperature of the air and of the surroundings is 80°F, if the wind velocity is negligible, and if no heat penetrates the roof, what is the equilibrium temperature of the roof?

For the rate of heat transfer by conduction-convection use the equation

$$\frac{q}{A} = 0.38 \, (\Delta t)^{1.25}$$

where Δt is the temperature drop between roof and air in degrees Fahrenheit.

8-22. The roof of Prob. 8-21 is painted with an aluminum paint, which has an emissivity of 0.9 and an absorptivity for solar radiation of 0.5. What is the equilibrium temperature of the painted roof?

8-23. Show that for infinite parallel planes of emissivity ϵ_1 and ϵ_2, the factor $\mathfrak{F}$ is

$$\mathfrak{F} = \frac{1}{1/\epsilon_1 + 1/\epsilon_2 - 1}$$

REFERENCES

1. Bergelin, O. P., et al.: *Trans. ASME,* **71:** 27, 369 (1949); **72:** 881 (1950); **74:** 953 (1952); **76:** 841 (1954).
2. Bowman, R. A., A. C. Mueller, and W. M. Nagle: *Trans. ASME,* **62:** 283 (1940).
3. Carpenter, F. G., and A. P. Colburn: International Heat Transfer Conference, sec. I, London, September, 1951.
4. Chaddock, R. E., and M. T. Sanders: *Trans. AIChE,* **40:** 203 (1944).
5. Chilton, T. H., T. B. Drew, and R. H. Jebens: *Ind. Eng. Chem.,* **36:** 510 (1944).
6. Colburn, A. P.: *Trans. AIChE,* **29:** 174 (1933).
7. Colburn, A. P.: *Ind. Eng. Chem.,* **25:** 873 (1933).
8. Colburn, A. P., and O. A. Hougen: *Ind. Eng. Chem.,* **26:** 1178 (1934).
9. Cryder, D. S., and A. C. Finalborgo: *Trans. AIChE,* **33:** 346 (1937).
10. Cummings, G. H., and A. S. West: *Ind. Eng. Chem.,* **42:** 2303 (1950).
11. Drew, T. B., W. M. Nagle, and W. Q. Smith: *Trans. AIChE,* **31:** 605 (1935).
12. Drew, T. B., and W. P. Ryan: *Trans. AIChE,* **26:** 118 (1931).
13. Dusinberre, G. M.: "Numerical Analysis of Heat Flow," McGraw-Hill Book Company, Inc., New York, 1949.

14. Grimison, E. D.: *Trans. ASME,* **59:** 583 (1937); **60:** 381 (1938).
15. Hottel, H. C.: *Mech. Eng.,* **52:** 699 (1930).
16. Hottel, H. C.: "Notes on Radiant Heat Transmission," Cambridge, Mass., 1938.
17. Jakob, M., and W. Fritz: *Forsch. Gebiete Ingenieurw.,* **2:** 434 (1931).
18. Kern, D. Q.: "Process Heat Transfer," (a) p. 130, (b) pp. 139ff., (c) pp. 256ff., p. 261, (d) p. 266, (e) pp. 313ff., (f) pp. 515ff., (g) pp. 538ff., (h) pp. 542ff., (i) pp. 625ff., McGraw-Hill Book Company, Inc., New York, 1950.
19. Kern, D. Q., and D. F. Othmer: *Trans. AIChE,* **39:** 517 (1943).
20. Lewis, G. N., and M. Randall: "Thermodynamics," p. 140, McGraw-Hill Book Company, Inc., New York, 1923.
21. Lyon, R. N.: *Chem. Eng. Progr.,* **47:** 75 (1951).
22. Martinelli, R. C., et al.: *Trans. AIChE,* **38:** 493 (1942).
23. McAdams, W. H.: "Heat Transmission," 2d ed., p. 239, McGraw-Hill Book Company, Inc., New York, 1942.
24. McAdams, W. H.: "Heat Transmission," 3d ed., (a) p. 5, (b) pp. 35ff., (c) p. 167, (d) pp. 172, 180–181, (e) p. 177, (f) pp. 208ff., (g) p. 226, (h) p. 237, (i) p. 241, (j) p. 258, (k) p. 259, (l) p. 268, (m) pp. 271ff., (n) pp. 273–274, (o) pp. 309ff., (p) pp. 330ff., (q) pp. 351ff., (r) p. 393, (s) pp. 472ff., McGraw-Hill Book Company, Inc., New York, 1954.
25. McAdams, W. H., J. N. Addoms, P. M. Rinaldo, and R. S. Day: *Chem. Eng. Progr.,* **44:** 639 (1948).
26. McAdams, W. H., et al.: *Ind. Eng. Chem.,* **41:** 1945 (1949).
27. Nusselt, W.: *Z. Ver. deut. Ing.,* **60:** 541, 569 (1916).
28. Othmer, D. F.: *Ind. Eng. Chem.,* **21:** 576 (1929).
29. Perry, J. H.: "Chemical Engineers' Handbook," 3d ed., (a) p. 131, (b) p. 216, (c) p. 390, (d) p. 464, (e) p. 481, (f) p. 485, McGraw-Hill Book Company, Inc., New York, 1950.
30. Sieder, E. N., and G. E. Tate: *Ind. Eng. Chem.,* **28:** 1429 (1936).
31. Smith, J. C.: *Ind. Eng. Chem.,* **34:** 1248 (1942).
32. Snyder, N. W.: *Trans. ASME,* **76:** 537 (1954).
33. Tubular Exchangers Manufacturers Association: "Standards of the TEMA," 3d ed., p. 87, New York, 1952.

Chapter 9

EVAPORATION

Heat transfer to a boiling liquid has been discussed generally in Chap. 8. A special case occurs so often that it is considered an individual operation. It is called evaporation, and it is the subject of this chapter.

The objective of evaporation is to concentrate a solution consisting of a nonvolatile solute and a volatile solvent. In the overwhelming majority of evaporations the solvent is water. Evaporation is conducted by vaporizing a portion of the solvent to produce a concentrated solution or thick liquor. Evaporation differs from drying in that the residue is a liquid— sometimes a highly viscous one—rather than a solid; it differs from distillation in that the vapor usually is a single component, and even when the vapor is a mixture, no attempt is made in the evaporation step to separate the vapor into fractions; it differs from crystallization in that emphasis is placed on concentrating a solution rather than forming and building crystals. In certain situations, e.g., in the evaporation of brine to produce common salt, the line between evaporation and crystallization is far from sharp. Evaporation sometimes produces a slurry of crystals in a saturated mother liquor. In this book such processes are considered in Chap. 14, which is devoted to crystallization.

Normally, in evaporation the thick liquor is the valuable product and the vapor is condensed and discarded. In one specific situation, however, the reverse is true. Mineral-bearing water often is evaporated to give a solid-free product for boiler feed, for special process requirements, or for human consumption. This technique is often called water distillation, but technically it is evaporation, and it is so considered here.

Equipment for evaporation has changed a great deal during its long history, both as to design and as to methods of supplying the necessary heat. Open, direct-fired pans were used for making salt in the Middle Ages, and this method may still be used in primitive societies. Solar energy has long been used for concentrating brine in open ponds, a technique that is still economical in some parts of the world. Other sources of heat for vaporization are hot gases bubbled through the solution—a method to be discussed later in this chapter—and electrical energy supplied through resistance heaters submerged in the liquid. The energy cost of the latter makes it prohibitive for usual applications.

Most present-day evaporators are heated by steam condensing on metal tubes. Nearly always the material to be evaporated flows inside the tubes. Usually the steam is at a low pressure, below 40 lb force/in.^{2} abs; often the boiling liquid is under a moderate vacuum, up to about 28 in. Hg. Reducing the boiling temperature of the liquid increases the temperature difference between the steam and the boiling liquid and thereby increases the heat-transfer rate in the evaporator.

Liquid Characteristics. The practical solution of an evaporation problem is profoundly affected by the character of the liquor to be concentrated. It is the wide variation in liquor characteristics (which demands judgment and experience in engineering and operating evaporators) that broadens this operation from simple heat transfer to a separate art. Some of the most important properties of evaporating liquids are as follows.

Concentration. Although the thin liquor fed to an evaporator may be sufficiently dilute to have many of the physical properties of water, as the concentration increases, the solution becomes more and more individualistic. The density and viscosity increase with solid content until either the solution becomes saturated or the liquor becomes too sluggish for adequate heat transfer. Continued boiling of a saturated solution causes crystals to form; these must be removed or the tubes clog. The boiling point of the solution may also rise considerably as the solid content increases, so that the boiling temperature of a concentrated solution may be much higher than that of water at the same pressure.

Foaming. Some materials, especially organic substances, foam during vaporization. A stable foam accompanies the vapor out of the evaporator, causing heavy entrainment. In extreme cases the entire mass of liquid may boil over into the vapor outlet and be lost.

Temperature Sensitivity. Many fine chemicals, pharmaceutical products, and foods are damaged when heated to moderate temperatures for relatively short times. In concentrating such materials special techniques are needed to reduce both the temperature of the liquid and the time of heating.

Scale. Some solutions deposit scale on the heating surfaces. The overall coefficient then steadily diminishes, until the evaporator must be shut down and the tubes cleaned. When the scale is hard and insoluble, the cleaning is difficult and expensive.

Materials of Construction. Whenever possible, evaporators are made of cast iron and steel. Many solutions, however, attack ferrous metals, or are contaminated by them. Special materials such as copper, nickel, stainless steel, aluminum, impervious graphite, and lead are then used. Since these materials are expensive, high heat-transfer rates become especially desirable to minimize the first cost of the equipment.

Many other liquid characteristics must be considered by the designer of an evaporator. Some of these are specific heat, heat of concentration, freezing point, gas liberation on boiling, toxicity, explosion hazards, radioactivity, and necessity for sterile operation. Because of the variation in liquor properties, many different evaporator designs have been developed.

The choice for any specific problem depends primarily on the characteristics of the liquid.

Single- and Multiple-effect Operation. When a single evaporator is used, the vapor from the boiling liquid is condensed and discarded. This method is called single-effect evaporation, and although it is simple, it utilizes steam ineffectively. To evaporate 1 lb of water from a solution, from 1000 to 1200 Btu is needed, and this calls for from 1 to 1.3 lb of steam. If the vapor from one evaporator is fed into the steam chest of a second evaporator and the vapor from the second is then sent to a condenser, the operation becomes double-effect. The heat in the original steam is reused in the second effect, and the evaporation achieved by 1 lb of steam fed to the first effect is approximately doubled. Additional effects can be added in the same manner. The general method of increasing the evaporation per pound of steam by using a series of evaporators between the steam supply and the condenser is called multiple-effect evaporation.

TYPES OF EVAPORATORS

The chief types of steam-heated tubular evaporators in use today are:

1. Short-tube evaporators
   (a) Horizontal-tube
   (b) Vertical-tube
2. Long-tube vertical evaporators
   (a) Forced-circulation
   (b) Upward-flow (climbing-film)
   (c) Downward-flow (falling-film)
3. Coil evaporators
4. Agitated-film evaporators

Once-through and Circulation Evaporators. Evaporators may be operated either as once-through or as circulation units. In once-through operation the feed liquor passes through the tubes only once, releases the vapor, and leaves the unit as thick liquor. All the evaporation is accomplished in a single pass. The ratio of evaporation to feed is limited in a single-pass unit; thus these evaporators are well adapted to multiple-effect operation, where the total amount of concentration can be spread over several effects. Agitated-film evaporators are always operated once through; falling-film and climbing-film evaporators can also be operated in this way.

Once-through evaporators are especially useful for heat-sensitive materials. By operating under high vacuum, the temperature of the liquid can be kept low. With a single rapid passage through the tubes the thick liquor is at the evaporation temperature but a short time and can be quickly cooled as soon as it leaves the evaporator.

In circulation evaporators a pool of liquid is held within the equipment. Incoming feed mixes with the liquid from the pool, and the mixture passes through the tubes. Unevaporated liquid discharged from the tubes returns to the pool, so that only part of the total evaporation occurs in one pass. All short-tube and forced-circulation evaporators are operated in this way. Climbing-film evaporators may be either once-through or circulation units.

The thick liquor from a circulation evaporator is withdrawn from the pool. All the liquor in the pool must therefore be at the maximum concentration. Since the liquid entering the tubes may contain several parts of thick liquor for one part of feed, its concentration, density, viscosity, and boiling point are nearly at the maximum. Accordingly, the heat-transfer coefficient tends to be low.

Circulation evaporators are not well suited to concentrating heat-sensitive liquids. With a reasonably good vacuum the temperature of the bulk of the liquid may be nondestructive, but the liquid is repeatedly exposed to contact with hot tubes. Some of the liquid, therefore, may be heated to an excessively high temperature. Although the average residence time of the liquid in the heating zone may be short, part of the liquid is retained in the evaporator for a considerable time. Prolonged heating of even a small part of a heat-sensitive material like a food can ruin the entire product.

Circulation evaporators, however, can operate over a wide range of concentration between feed and thick liquor in a single unit, and are well adapted to single-effect evaporation. They may operate either with natural circulation, with the flow through the tubes induced by density differences, or with forced circulation, with flow provided by a pump.

Short-tube Evaporators. In the older types of evaporators the tubes are "short," 4 to 8 ft long, and fairly large, 2 to 4 in. in diameter. In some units the tubes are horizontal; in some they are vertical. In the horizontal-tube unit shown in Fig. 9-1 the tube bundle is submerged in the liquid; steam condenses inside the tubes, causing the liquid outside to boil. The body of the evaporator contains the pool of liquid. The body is in the form of a vertical cylinder closed top and bottom; usually it has dished heads, although the bottom head may be conical. The lower body ring is provided with steam compartments, which are closed on the outside by cover plates and on the inside by tube sheets. The tubes are held between the tube sheets. Steam is admitted to the steam chest at A, and enters the tubes washing noncondensed gas and condensate ahead of it. Condensate is removed from outlet C through a steam trap; gas is removed at vent B. In normal operation *only* condensate and noncondensed gases are removed from the steam chests. Feed liquid enters through inlet D; thick liquor is withdrawn at E; vapor leaves at G. Sight glasses F are often supplied.

Horizontal-tube evaporators are cheap and are designed for easy replacement of the tubes. Their disadvantages are low heat-transfer coefficients (particularly with viscous materials) because of poor circulation of the liquid and the difficulty of removing scale deposited on the outside of the tubes. These evaporators are seldom built now, but many are still in use.

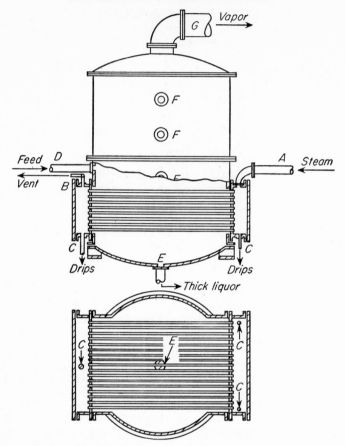

FIG. 9-1. Horizontal-tube evaporator: *A*, steam inlet. *B*, vent for noncondensed gases. *C*, condensate outlet. *D*, feed inlet. *E*, liquor outlet. *F*, sight glasses. *G*, vapor outlet.

In the short-tube vertical evaporator shown in Fig. 9-2 the steam condenses outside the tubes. The tube bundle contains a large central downcomer, the cross-sectional area of which is 25 to 40 per cent of the total cross-sectional area of the tubes. Most of the boiling takes place in the smaller tubes, so that the liquid rises through these tubes and returns through the downcomer. Drops of liquid fall through the vapor in the tall space above the tubes, with their removal from the vapor often assisted by baffle plates over the vapor outlet. Thick liquor is withdrawn from the conical bottom of the shell. In this evaporator the driving force for flow of liquid through the tubes is the difference in density between the liquid in the downcomer and the mixture of liquid and vapor in the tubes.

Short-tube vertical evaporators provide moderately good heat transfer at reasonable cost. They are fairly effective with scaling liquids, for the inside of the tubes can be easily cleaned. Circulation is by natural convec-

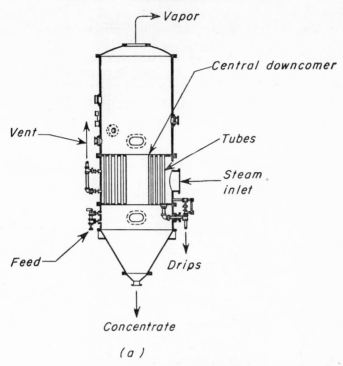

Vapor

Central downcomer

Vent

Tubes

Steam inlet

Feed

Drips

Concentrate

(a)

Fig. 9-2. Short-tube vertical evaporator.

tion but at a much less rapid rate than in long-tube natural-circulation evaporators; the heat-transfer coefficients, therefore, are fairly high with thin liquids but low when the liquid is viscous. Once considered as "standard" evaporators, short-tube vertical units have been largely displaced in new installations by long-tube evaporators and other more specialized designs.

Forced-circulation Evaporators. In a natural-circulation evaporator [4] the liquid enters the tubes at 1 to 3 ft/sec. The linear velocity increases greatly as vapor is formed in the tubes, so that in general the rates of heat transfer are satisfactory. With viscous liquids, however, the over-all coefficient in a natural-circulation unit may be uneconomically low. Higher coefficients are obtained in forced-circulation evaporators, an example of which is shown in Fig. 9-3. Here a centrifugal pump forces liquid through the tubes at an entering velocity of 6 to 18 ft/sec. The tubes are under sufficient static head to ensure that there is no boiling in the tubes; the liquid becomes superheated as the static head is reduced during flow from the heater to the vapor space, and it "flashes" into a mixture of vapor and spray in the outlet line from the exchanger just before entering the body of the evaporator. The mixture of liquid and vapor impinges on a deflector plate in the vapor space. Liquid returns to the pump inlet, where it meets incoming feed; vapor leaves the top of the evaporator body to a condenser or to the next effect. Part of the liquid leaving the separator is continuously withdrawn as concentrate.

In the design shown in Fig. 9-3 the exchanger has horizontal tubes and is two-pass on both tube and shell sides. In others vertical single-pass exchangers are used. In both types the heat-transfer coefficients are high, especially with thin liquids, but the greatest improvement over

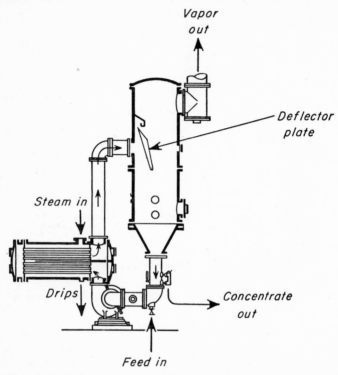

FIG. 9-3. Forced-circulation evaporator with two-pass horizontal separate heating element.

natural-circulation evaporation is with viscous liquids. With thin liquids the improvement with forced circulation does not warrant the added pumping costs over natural circulation, but with viscous material the added costs are justified, especially when expensive metals must be used. An example is caustic soda concentration, which must be done in nickel equipment. In multiple-effect evaporators producing a viscous final concentrate the first effects may be natural-circulation units, and the later ones, handling thick liquid, forced-circulation units. Because of the high velocities in a forced-circulation evaporator, the residence time of the liquid in the tubes is short—about 1 to 3 sec—so that moderately heat-sensitive liquids can be concentrated in them. They are also effective in evaporating salting liquors or those that tend to foam.

Long-tube Evaporators with Upward Flow. Typical long-tube vertical evaporators with upward flow of the liquid are shown in Fig. 9-4. Their essential parts are: (1) a tubular exchanger with steam in the shell and liquid to be concentrated in the tubes, (2) a separator or vapor space for removing entrained liquid from the vapor, and (3) when operated as circulation units, a return leg for the liquid from the separator to the bottom of the exchanger. Inlets are provided for feed liquid and steam,

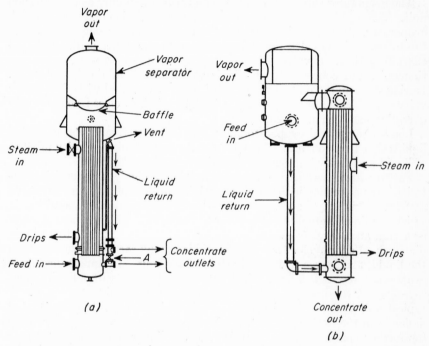

Fig. 9-4. Climbing-film, long-tube vertical evaporators: (a) Integral heating element. (b) Separate heating element.

and outlets are provided for vapor, thick liquor, steam condensate, or "drips," and noncondensable gases from the steam.

The tubes are typically 1 to 2 in. in diameter and 12 to 32 ft long. Liquid and vapor flow upward inside the tubes as a result of the boiling action; separated liquid returns to the bottom of the tubes by gravity. Dilute feed, often at about room temperature, enters the system and mixes with liquid returning from the separator. The mixture enters the bottom of the tubes, on the outside of which steam is condensing. For a short distance the feed to the tubes flows upward as liquid, receiving heat from the steam. Bubbles then form in the liquid as boiling begins, increasing the linear velocity and the rate of heat transfer. Near the top of the tubes the bubbles grow rapidly. In this zone bubbles of vapor alternating with slugs of liquid rise very quickly through the tubes and emerge at high velocity from the top.

From the tubes the mixture of liquid and vapor enters the separator, which may be integral with the exchanger, as in Fig. 9-4a, or away from it, as in Fig. 9-4b. In both designs the diameter of the separator is larger than that of the exchanger, so that the linear velocity of the vapor is greatly reduced. As a further aid in knocking out liquid droplets the vapor impinges on, and then passes around, sets of baffle plates before leaving the separator.

The evaporator shown in Fig. 9-4b can be operated only as a circulation unit. That of Fig. 9-4a can be used either as a circulation or a once-through evaporator. If valve A is closed, the liquid can make but one pass through the tubes, and the thick liquor leaves through the upper of the two concentrate outlets. If valve A is open, the evaporator operates as a circulation unit, and concentrated liquid is withdrawn from the lower outlet. During start-up the once-through unit is usually operated with circulation; valve A, which is open during the initial period, is closed when steady operation has been achieved.

Long-tube vertical evaporators are especially effective in concentrating liquids that tend to foam. Foam is broken when the high-velocity mixture of liquid and vapor impinges against the vapor-head baffle.

Falling-film Evaporators. Concentration of highly heat-sensitive materials such as orange juice requires a minimum time of exposure to heated surface. This may be done in once-through falling-film evaporators, in which the liquid enters at the top, flows downward inside the heated tubes as a film, and leaves from the bottom. The tubes are large, 2 to 10 in. in diameter. Vapor evolved from the liquid is usually carried downward with the liquid, and leaves from the bottom of the unit. In appearance these evaporators resemble long, vertical, tubular exchangers with a liquid-vapor separator at the bottom and a distributor for the liquid at the top. A typical unit is shown in Fig. 9-5. In this design the partly concentrated liquor from the bottom of one tube is pumped to the top of another tube for further concentration. In this way the liquor makes but one pass through each tube, but is given several successive concentrations in the single evaporator.

The chief problem in a falling-film evaporator is that of distributing the liquid uniformly as a film inside the tubes. This is done by a set of perforated metal plates above a carefully leveled tube sheet, by inserts in the tube ends to cause the liquid to flow evenly into each tube, or by "spider" distributors with radial arms from which the feed is sprayed at a steady rate on the inside surface of each tube. Still another way is to use an individual spray nozzle inside each tube.

When recirculation is allowable without damaging the liquid, distribution of liquid to the tubes is facilitated by a moderate recycling of liquid to the tops of the tubes. This provides a larger volume of flow through the tubes than is possible in once-through operation.

For good heat transfer the Reynolds number $4\Gamma/\mu$ of the falling film should be greater than 2,000 at all points in the tube [6] [see Eq. (8-104)]. During evaporation the amount of liquid is continuously reduced as it flows from top to bottom of the tube, so that the amount of concentration that can be done in a single pass is limited. In once-through operation too great a degree of concentration leads to "flooding" of the tube at the inlet or to sticking and build-up of highly concentrated material at the outlet. High concentration ratios must therefore be obtained with multiple stages, as in Fig. 9-5, with the concentrate lifted from the bottom of one tube and evaporated further in another.

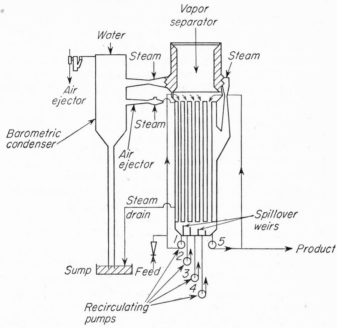

FIG. 9-5. Falling-film evaporator.

Falling-film evaporators, with no recirculation and short residence times, handle sensitive products that can be concentrated in no other way. They are also well adapted to concentrating viscous liquids.

Coil Evaporators. In these evaporators the heating element is a metal coil submerged in the liquid. Steam condenses inside the coil. Large units of this type distill water to make boiler feed; smaller units are employed for batch evaporation of viscous products. The final crystallization in sugar refineries is done in a coil evaporator known as a "strike pan," in which steam is admitted from a manifold to separate heating elements, each consisting of two concentric rings of tubing with an elliptical cross section. The vapor rises past baffles and under a dome into the condensing head, where it is condensed by direct contact with cold water. Noncondensable gas passes out the top to an ejector or vacuum pump.

Agitated-film Evaporator. The principal resistance to over-all heat transfer from the steam to the boiling liquid in an evaporator is on the liquid side. Any method of diminishing this resistance, therefore, gives a considerable improvement in the over-all heat-transfer coefficient. In long-tube evaporators, especially with forced circulation, the velocity of the liquid through the tubes is high. The liquid is highly turbulent, and the rate of heat transfer is large. Another way of increasing turbulence is by mechanical agitation of the liquid film, as in the evaporator shown in Fig. 9-6. This is a modified falling-film evaporator with a single jacketed tube containing an internal agitator. Feed enters at the top of the jacketed section, and is spread out into a thin, highly turbulent, film by the vertical blades of the agitator. Concentrate leaves from the bottom of the jacketed section; vapor rises from the vaporizing zone into an unjacketed separator, which is somewhat larger in diameter than the evaporating tube. In the separator the agitator blades throw entrained liquid outward against stationary vertical plates. The droplets coalesce on these blades and return to the evaporating section. Liquid-free vapor escapes through outlets at the top of the unit.

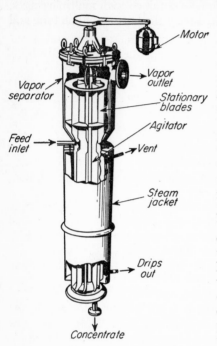

FIG. 9-6. Agitated-film evaporator. (*Rodney-Hunt Machine Co.*)

The chief advantage of an agitated-film evaporator is its ability to give high rates of heat transfer with viscous liquids. The product may have a viscosity as high as 100,000 centipoises at the evaporation temperature. As in other evaporators, the over-all coefficient falls as the viscosity rises, but in this design the decrease is slow. With highly viscous materials the coefficient is appreciably greater than in forced-circulation evaporators and much greater than in natural-circulation units. The agitated-film evaporator is particularly effective with such viscous heat-sensitive products as gelatin, rubber latex, antibiotics, and fruit juices. Its disadvantages are high cost; the internal moving parts, which may need considerable maintenance; and the small capacity of single units, which is far below that of multitubular evaporators.

PERFORMANCE OF TUBULAR EVAPORATORS

The principal measures of the performance of a steam-heated tubular evaporator are the capacity and the economy. *Capacity* is defined as the number of pounds of water vaporized per hour. *Economy* is the number of pounds vaporized per pound of steam fed to the unit. In a single-effect evaporator the economy is nearly always less than 1, but in multiple-effect equipment it may be considerably greater. The *steam consumption*, in pounds of steam per hour, is also important. It equals the capacity divided by the economy.

As shown in Chap. 8, the heat-transfer rate, in Btu per hour, through the heating surface is given by the product of three factors, the area of the heating surface, the over-all temperature drop, and the over-all heat-transfer coefficient.

$$q = UA(t_s - t) = UA \, \Delta t \qquad (9\text{-}1)$$

where q = rate of heat transfer, Btu/hr
A = area of heating surface, ft^2
U = over-all coefficient, Btu/(ft^2)(hr)(°F)
t_s = condensation temperature of steam, °F
t = boiling temperature of solution, °F
$\Delta t = t_s - t$ = over-all temperature drop between steam and solution, °F

If the feed to the evaporator is at the boiling temperature corresponding to the absolute pressure in the vapor space, all of the heat transferred through the heating surface is available for evaporation, and the capacity is proportional to q. Then the capacity depends only on the three factors A, U, and Δt. If the feed is cold, the heat required to heat it to its boiling point may be quite large, and the capacity is reduced accordingly, as heat used in this fashion is not available for evaporation.

The chief factor influencing the economy is the number of effects in the evaporator. By proper design of the equipment the heat from the raw steam can be reused for evaporation one or more times. The economy is also influenced by the heating load, becoming smaller as the heating load increases. Each pound of steam fed to the evaporator yields only a fixed amount of heat. If a large part of this heat is needed to heat the feed to boiling, only a small part is available for evaporation.

TABLE 9-1. FACTORS AFFECTING EVAPORATOR PERFORMANCE

Characteristic	Principally influenced by	Somewhat influenced by
Capacity.......	Area of heating surface Temperature drop Over-all coefficient	Heating load
Economy.......	Number of effects (most important) Heating load	Heat of vaporization of steam Heat of vaporization of liquid Heat of dilution of liquid Amount of superheat or water in steam Heat loss

Minor factors affecting the economy are the heats of vaporization of the steam and the boiling solution, the heat of dilution of the liquid, the amount of superheat or liquid water in the steam, and the amount of heat lost by the evaporator to the surroundings. These are accounted for quantitatively by means of enthalpy balances, which are considered in detail later.

The capacity and economy of a given evaporator thus depend on a number of interrelated factors, some of major and some of minor importance. The relative influence of the various factors is indicated in Table 9-1.

Evaporator Capacity

The temperature drop across the heating surface and the over-all heat-transfer coefficient will be discussed in order. The actual temperature drop depends on the solution being evaporated, the difference in pressure between the steam chest and the vapor space above the boiling liquid, and the depth of liquid over the heating surface. In some evaporators the velocity of the liquid in the tubes also influences the temperature drop, because the frictional loss in the tubes increases the effective pressure of the liquid. When the solution has the characteristics of pure water, its boiling point can be read from steam tables if the pressure is known, as can the temperature of the condensing steam. In actual evaporators, however, the boiling point of a solution is affected by two factors, boiling-point elevation and liquid head.

Boiling-point Elevation and Dühring's Rule. The vapor pressure of most aqueous solutions is less than that of water at the same temperature. Consequently, for a given pressure the boiling point of the solutions is higher than that of pure water. The increase in boiling point over that of water is known as the boiling-point elevation of the solution. It is small for dilute solutions and for solution of organic colloids but may be as large as 150°F for concentrated solutions of inorganic salts. The boiling-point elevation must be subtracted from the temperature drop that is predicted from the steam tables.

For strong solutions the boiling-point elevation is best found from an empirical rule known as Dühring's rule, which states that the boiling point of a given solution is a linear function of the boiling point of pure water at the same pressure. Thus if the boiling point of the solution is plotted against that of water at the same pressure, a straight line results. Different lines are obtained for different concentrations. Over wide ranges of pressure the rule is not exact, but over a moderate range the lines are very nearly straight, though not necessarily parallel. Figure 9-7 is a set of Dühring lines for solutions of sodium hydroxide in water. The use of this figure may be illustrated by an example. If the pressure over a 25 per cent solution of sodium hydroxide is such that water boils at 180°F, by reading up from the x axis at 180°F to the line for 25 per cent solution and then horizontally to the y axis it is found that the boiling point of the solution at this pressure is 200°F. The boiling-point elevation for this solution at this pressure is therefore 20°F.

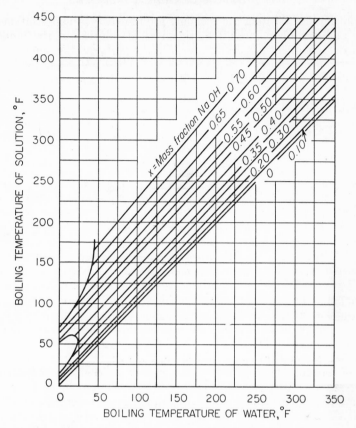

Fig. 9-7. Dühring lines, system sodium hydroxide–water.

Effect of Liquid Head. If the depth of liquid in an evaporator is appreciable, the boiling point corresponding to the pressure in the vapor space is the boiling point of the surface layers of liquid only. A drop of liquid at a distance Z ft below the surface is under a pressure of the vapor space plus a head of Z ft of liquid and therefore has a higher boiling point. In addition, when the velocity of the liquid is large, frictional loss in the tubes further increases the average pressure of the liquid. In any actual evaporator, therefore, the average boiling point of the liquid in the tubes is higher than the boiling point corresponding to the pressure in the vapor space. This increase in boiling point lowers the average temperature drop between the steam and the liquid and reduces the capacity. The amount of reduction cannot be estimated quantitatively with precision, but the qualitative effect of liquid head, especially with high liquor levels and high liquid velocities, should not be ignored.

Apparent and Net Temperature Drops and Apparent Coefficients. It is relatively easy to measure with precision the pressure in the vapor space of an evaporator. It is much less easy to measure the temperature at the same point. Consequently the temperature drop in an evaporator is often found not from measured temperatures but from temperatures calculated from steam tables on the basis of the measured pressures. The temperature drop between the steam and the liquid calculated in this way is called the apparent temperature drop, and over-all coefficients computed from this temperature drop are called apparent coefficients. This method of calculation ignores effects of boiling-point elevation and liquid head.

The boiling-point elevation is often known, or it may be estimated fairly accurately. It is preferable, therefore, to base the calculations on the true

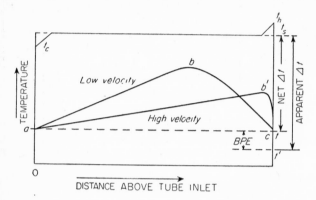

FIG. 9-8. Temperature history of liquor in tubes and temperature drops in long-tube vertical evaporator.

boiling point of the solution under the pressure of the vapor space. The difference between the steam temperature and the true boiling point of the solution at the pressure of the vapor space is known as the net temperature drop. Coefficients based on the net temperature drop are called coefficients corrected for boiling-point elevation. Corrected coefficients, rather than apparent coefficients, are most commonly used.

Figure 9-8 relates the temperatures in an evaporator with the distance along the tube, measured from the bottom. The diagram applies to a long-tube vertical evaporator with upflow of liquid. The steam enters the evaporator at the top of the steam jacket surrounding the tubes and flows downward. The entering steam may be slightly superheated at t_h. The superheat is quickly given up, and the steam drops to saturation temperature t_s. Over the greater part of the heating surface this temperature is unchanged. Before the condensate leaves the steam space, it may be cooled slightly to temperature t_c.

The temperature history of the liquor in the tubes is shown by lines abc and $ab'c$ in Fig. 9-8. The former applies at low velocities, about 3 ft/sec, and the latter at high velocities, above 10 ft/sec, both velocities based on the flow entering the bottom of the tube.[3] It is assumed that the feed enters the evaporator at about the boiling temperature of the liquid at vapor-space pressure, denoted by t. Then the liquid entering the tube is at t, whether the flow is once through or circulatory. At high velocities, the fluid in the tube remains liquid practically to the end of the tube and flashes into a mixture of liquid and vapor in the last few inches of the tube. The maximum liquid temperature occurs at point b', as shown in Fig. 9-8, almost at the exit from the tube.

At lower velocities, the liquid flashes at a point nearer the center of the tube and reaches its maximum temperature, as shown by point b in Fig. 9-8. Point b divides the tube into two sections, a nonboiling section below point b and a boiling section above this point.

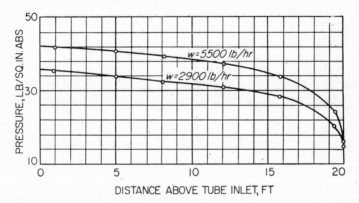

FIG. 9-9. Pressure history of liquor in tubes of long-tube vertical evaporator. (*McAdams.*[7])

At both high and low velocities the vapor and concentrated liquid reach equilibrium at the pressure in the vapor space. If the liquid has an appreciable boiling-point elevation, this temperature t is greater than t', the boiling point of pure water at the vapor-space pressure. The difference between t and t' is the boiling-point elevation (BPE).

The apparent temperature drop is $t_s - t'$, and the net temperature drop, corrected for boiling-point elevation, is $t_s - t$. The true temperature drop, corrected for both boiling-point elevation and static head, is represented by the average distance between t_s and the variable liquid temperature. Although some correlations are available[3] for determining the true temperature drop from the operating conditions, usually this quantity is not available to the designer, and the net temperature drop, corrected for boiling-point elevation only, is used.

The pressure history of the fluid in the tube is shown in Fig. 9-9, in which pressure is plotted against distance from the bottom of the tube.[7] The velocity is such that boiling starts inside the tube. The total pressure drop in the tube, neglecting changes in kinetic energy, is the sum of the static head and the friction loss. The mixture of steam and water in the boiling section has a high velocity, and the friction loss is large in this section. As shown by the curves in Fig. 9-9, the pressure changes slowly in the nonboiling section, where the velocity is low, and rapidly in the boiling section, where the velocity is high.

Heat-transfer Coefficients. As shown by Eq. (9-1), the heat flux and the evaporator capacity are affected by changes both in the temperature drop and in the over-all heat-transfer coefficient. The temperature drop is fixed by the properties of the steam and the boiling liquid, and except for the effect of hydrostatic head is not a function of the evaporator construction. The over-all coefficient, on the other hand, is strongly influenced by the design and method of operation of the evaporator.

As shown in Chap. 8 [Eq. (8-44)], the over-all resistance to heat transfer between the steam and the boiling liquid is the sum of five individual resistances: the steam-film resistance; the two scale resistances, inside and outside the tubes; the tube-wall resistance; and the resistance from the boiling liquid. The over-all coefficient is the reciprocal of the over-all resistance. In most evaporators the fouling factor of the condensing steam and the resistance of the tube wall are very small, and they are usually neglected in evaporator calculations. In an agitated-film evaporator the tube wall is fairly thick, so that its resistance may be a significant part of the total.

Steam-film Coefficients. The steam-film coefficient is characteristically high, even when condensation is filmwise. Promoters are sometimes added to the steam to give dropwise condensation and a still higher coefficient. Since the presence of noncondensable gas seriously reduces the steam-film coefficient, provision must be made to vent noncondensables from the steam chest and, when the steam is at a pressure below atmospheric, to prevent leakage of air inward.

Liquid-side Coefficients. The liquid-side coefficient depends to a large extent on the velocity of the liquid over the heated surface. In most evaporators, and especially those handling viscous materials, the resistance of the liquid side controls the over-all rate of heat transfer to the boiling liquid. In long-tube natural-circulation evaporators the velocity rises to high values near the top of the tubes, much higher than in a short-tube unit. The liquid-side coefficient is correspondingly large. It cannot be estimated with precision in any natural-circulation evaporator. It depends on the characteristics of the liquid, the liquid velocity, the volume of the evolved vapor, and the temperature drop between the tube wall and the boiling liquid.

Forced circulation gives high liquid-side coefficients even though boiling inside the tubes is suppressed by the high static head. The liquid-side coefficient in a forced-circulation evaporator may be estimated by Eq. (8-53) for heat transfer to a nonboiling liquid.

Scale Resistance. The formation of scale on the tubes of an evaporator adds a thermal resistance equivalent to a fouling factor. When true scale is being formed, the over-all coefficient diminishes with time in accordance with the equation [9]

$$\frac{1}{U^2} = \frac{1}{U_0^2} + C\theta \qquad (9\text{-}2)$$

where U = over-all coefficient at any time

U_0 = initial over-all coefficient, with clean tubes

θ = elapsed time

C = constant

A plot of $1/U^2$ vs. θ yields a straight line on arithmetical coordinates. Equation (9-2) is useful in interpreting experimental data so that the optimum time of operation between cleanouts may be calculated.

Over-all Coefficients. Because of the difficulty of measuring the high individual film coefficients in an evaporator, experimental results are usually expressed in terms of over-all coefficients. These may be based on the apparent temperature drop, as discussed earlier, or, preferably, on the net temperature drop corrected for boiling-point elevation. The over-all coefficient, of course, is influenced by the same factors influencing individual coefficients; but if one resistance (say that of the liquid film) is controlling, major changes in the other resistances have almost no effect on the over-all coefficient.

Typical over-all coefficients for various types of evaporators are given in Table 9-2. These coefficients apply to conditions under which the various evaporators are ordinarily used. A horizontal-tube evaporator, for

TABLE 9-2. TYPICAL OVER-ALL COEFFICIENTS IN EVAPORATORS †

Type	Over-all coefficient, Btu/(ft²)(hr)(°F)
Long-tube vertical evaporators:	
Natural circulation..............................	200– 600
Forced circulation...............................	400–2000
Short-tube evaporators:	
Horizontal tube.................................	200– 400
Calandria type..................................	150– 500
Coil evaporators..................................	200– 400
Agitated-film evaporator, Newtonian liquid, viscosity:	
1 centipoise.....................................	400
100 centipoises..................................	300
10,000 centipoises...............................	120

† From G. G. Brown et al., "Unit Operations," p. 484, John Wiley & Sons, Inc., New York, 1950; and E. Lindsey, *Chem. Eng.*, **60**(4): 227 (1953).

example, gives moderately high coefficients with thin liquids but very low ones with viscous materials. A small accumulation of scale reduces the coefficients to a small fraction of the clean-tube values. An agitated-film evaporator gives a seemingly low coefficient with a liquid having a viscosity of 10,000 centipoises, but this coefficient is much larger than would be obtained in any other type of evaporator which could handle such a viscous material at all.

In short-tube natural-circulation evaporators the over-all coefficient is sensitive to the temperature drop and to the boiling temperature of the

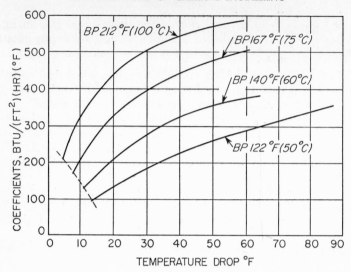

Fig. 9-10. Relation between boiling point, temperature drop, and heat-transfer coefficients in short-tube vertical evaporator. (*Badger and Shepard.*[2])

solution. Typical data for an evaporator with tubes 4 ft long and 2 in. in diameter are shown in Fig. 9-10.

SINGLE-EFFECT CALCULATIONS

There are three relationships available for the quantitative attack on evaporation problems. They are:

1. Material balances
2. Enthalpy balances
3. The capacity equation [Eq. (9-1)]

Material balances, as discussed in Chap. 1, are based on the principle that in equipment operating under steady-state conditions, with no accumulation or depletion of material, whatever enters the system must leave at the same rate. This applies not only to the total amount of material but also to each individual component of the feed. Thus, as shown in Example 1-1, in an evaporator concentrating sodium hydroxide solutions an over-all material balance, a water balance, and a sodium hydroxide balance may be made. Any two of these are independent.

Enthalpy Balances for Single-effect Evaporator. Evaporator economy is entirely a matter of enthalpy balances. In a single-effect evaporator the latent heat of condensation of the steam is transferred through a heating surface to vaporize water from a boiling solution. Two enthalpy balances are needed, one for the steam side and one for the vapor or liquor side.

Figure 9-11 shows diagrammatically a vertical-tube single-effect evaporator. The rate of steam flow and of condensate is w_s lb/hr; that of the thin liquor, or feed, is w_f lb/hr; and that of the thick liquor is w lb/hr. The rate of vapor flow to the condenser, assuming that no solids precipitate from the liquor, is $w_f - w$ lb/hr. Also, let t_s be the condensing temperature of

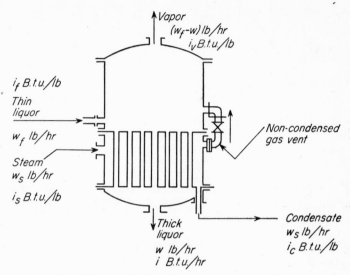

FIG. 9-11. Material and enthalpy balances over single-effect evaporator.

the steam, t the boiling temperature of the liquid in the evaporator, and t_f the temperature of the feed, all in degrees Fahrenheit.

It is assumed that there is no leakage or entrainment, that the flow of noncondensables is negligible, and that heat losses from the evaporator need not be considered. The steam entering the steam chest may be superheated, and the condensate usually leaves the steam chest somewhat subcooled below its boiling point. Both the superheat and the subcooling of the condensate are small, however, and it is acceptable to neglect them in making an enthalpy balance. The small errors made in neglecting them are approximately compensated by neglecting heat losses from the steam chest.

Under these assumptions the difference between the enthalpy of the steam and that of the condensate is simply λ_s, the latent heat of condensation of the steam. The enthalpy balance for the steam side, by Eqs. (1-18) and (1-23), is

$$q_s = w_s(i_c - i_s) = -w_s\lambda_s \qquad (9\text{-}3)$$

where q_s = rate of heat transfer through heating surface to steam, Btu/hr
    i_s = specific enthalpy of steam, Btu/lb
    i_c = specific enthalpy of condensate, Btu/lb
    λ_s = latent heat of condensation of steam, Btu/lb

The quantity q_s is inherently negative since, by convention, heat added to the fluid through the control surface is taken as positive and that abstracted from the fluid as negative.

The enthalpy balance for the liquor side is

$$q = (w_f - w)i_v - w_f i_f + wi \qquad (9\text{-}4)$$

where q = rate of heat transfer from heating surface to liquid, Btu/hr
  i_v = specific enthalpy of vapor, Btu/lb
  i_f = specific enthalpy of thin liquor, Btu/lb
  i = specific enthalpy of thick liquor, Btu/lb

In the absence of heat losses, the heat transferred from the steam to the tubes equals that transferred from the tubes to the liquor, and $-q_s = q$. Thus, by combining Eqs. (9-3) and (9-4),

$$q = w_s \lambda_s = (w_f - w)i_v - w_f i_f + wi \qquad (9\text{-}5)$$

The liquor-side enthalpies i_v, i_f, and i depend upon the characteristics of the solution being concentrated. Most solutions when mixed or diluted at constant temperature do not give much heat effect. This is true of solutions of organic substances and of moderately concentrated solutions of many inorganic substances. Thus sugar, salt, and paper-mill liquors do not possess appreciable heats of dilution or mixing. Sulfuric acid, sodium hydroxide, and calcium chloride, on the other hand, especially in concentrated solutions, evolve considerable heat when diluted and so possess appreciable heats of dilution. An equivalent amount of heat is required, in addition to the latent heat of vaporization, when dilute solutions of these substances are concentrated to high densities. Also, substances having large heats of dilution have large boiling-point elevations, although solutions having little or no heat of dilution also can possess considerable boiling-point elevations.

Enthalpy Balance with Negligible Heat of Dilution. For solutions having negligible heats of dilution, the enthalpies i_f and i can be calculated from the specific heats of the solution. A datum temperature above which the enthalpies can be calculated is chosen. The thick liquor and the vapor leave the vapor head in equilibrium, whether the evaporator is the circulation or the once-through type, and the temperatures of both equal t, the boiling temperature in the evaporator. The most convenient choice of the datum, therefore, is t. If this choice is made, the specific enthalpy i of the thick liquor is zero, and the term wi vanishes. The specific enthalpy i_f of the thin liquor can then be calculated from the specific heat, which is assumed to be constant over the temperature range from t_f to t. Thus

$$i_f = c_{pf}(t_f - t) \qquad (9\text{-}6)$$

where c_{pf} = specific heat of thin liquor, Btu/(lb)(°F)
  t_f = temperature of thin liquor, °F
  t = temperature of thick liquor, °F

The specific heat of a solution having no heat of mixing is a linear function of concentration. Thus if the specific heat c_{p0} is known for some one aqueous solution containing x_0 weight fraction of solute, the specific heat of any other solution of concentration x weight fraction is

$$c_p = 1 - (1 - c_{p0}) \frac{x}{x_0} \tag{9-7}$$

The specific heat of the pure solvent, water, is, of course, 1.

The specific enthalpy i_v to be used in Eq. (9-5) is the enthalpy of the vapor less that of liquid water at the datum temperature t. If the boiling-point elevation of the thick liquor is negligible, this enthalpy difference is simply λ, the latent heat of vaporization of water at the pressure of the vapor space. If the boiling-point elevation of the thick liquor is appreciable, the vapor leaving the solution is superheated by an amount, in degrees, equal to the boiling-point elevation. The specific enthalpy i_v is, accurately, that of the superheated vapor less that of liquid water at the datum temperature t. In practice, however, it is sufficiently accurate, and considerably simpler, to use for i_v the latent heat of vaporization of water λ at the pressure of the vapor space. The error from use of this assumption is such as to underestimate the value of i_v by an amount, in Btu per pound, approximately equal to one-half the boiling-point elevation in degrees Fahrenheit.

Introducing the above assumptions into Eq. (9-5) gives final equations for the enthalpy balances over a single-effect evaporator when the heat of dilution is negligible.

$$q = w_s \lambda_s = (w_f - w)\lambda + w_f c_{pf}(t - t_f) \tag{9-8}$$

If the temperature t_f of the thin liquor is greater than the datum temperature t, the term $c_{pf} w_f(t - t_f)$ is negative. It is the enthalpy brought into the evaporator above the datum temperature by the thin liquor. The enthalpy so introduced into the vapor space is available for evaporation, and it lightens the steam load required for a given evaporation by an equivalent amount. This item is called *flash evaporation* or *self-evaporation*. If the temperature t_f of the thin liquor fed to the evaporator is less than t, the datum temperature, the term $w_f c_{pf}(t_f - t)$ is positive, and, for a given evaporation, additional steam will be required to provide this enthalpy. When the feed must be heated in this manner, the term $w_f c_{pf}(t_f - t)$ is called the *heating load*. In words, Eq. (9-8) states that the heat of vaporization of the steam is utilized: (1) to vaporize water from the solution and (2) to heat the feed to the boiling point; if the feed enters above the boiling point in the evaporator, part of the evaporation is from flash.

Enthalpy Balance with Appreciable Heat of Dilution; Enthalpy-concentration Diagram. If the heat of dilution of the liquor being concentrated is too large to be neglected, the enthalpy is not linear with concentration at constant temperature. The most satisfactory source for the values of i_f and i for use in Eq. (9-5) is then an enthalpy-concentration diagram, in which enthalpy, in Btu per pound of solution, is plotted against concentration, in mass fraction or weight percentage of solute.[8] Isotherms drawn on the diagram show the enthalpy as a function of concentration at constant temperature. In using an enthalpy-concentration diagram it is necessary only to read the ordinate of the intersection of the abscissa for the desired concentration and the isotherm for the given temperature.

Figure 9-12 is an enthalpy-concentration diagram for solutions of sodium hydroxide and water. Concentrations are in mass fraction of sodium hydroxide, temperatures in degrees Fahrenheit, and enthalpies in Btu per pound of solution. The enthalpy of water is referred to the same datum as in the steam tables, namely, liquid water at 32°F, so enthalpies from the figure can be used with those from the steam tables when liquid water or steam is involved in the calculations. In finding data for substitution into Eq. (9-5), values of i_f and i are taken from Fig. 9-12, and the enthalpy i_v of the vapor leaving the evaporator is obtained from the steam tables.

The curved boundary lines on which the isotherms of Fig. 9-12 terminate represent conditions of temperature and concentration under which solid phases form. These are various solid hydrates of sodium hydroxide. The enthalpies of all single-phase solutions lie above this boundary line. The enthalpy-concentration diagram can be extended to include solid phases; a diagram of this type is shown in Fig. 14-2.

The isotherms on an enthalpy-concentration diagram for a system possessing no heat of dilution are straight lines. The curvature of the isotherms in Fig. 9-12 provides a qualitative measure of the effect of the heat of dilution on the enthalpy of solutions of sodium hydroxide and water. Enthalpy-concentration diagrams can be constructed, of course, for solutions having negligible heats of dilution, but they are unnecessary in view of the simplicity of the specific-heat method described in the last section.

There is a close connection between the enthalpy of a solution and its vapor pressure. The vapor pressures of sodium hydroxide solutions, correlated by Dühring lines, are shown in Fig. 9-7, which is consistent with the enthalpy-concentration diagram of Fig. 9-12.

Example 9-1. A solution of organic colloids in water is to be concentrated from 10 to 50 per cent solids in a single-effect evaporator. Steam is available at a gauge pressure of 15 lb force/in.² (249°F). A pressure of 4 in. Hg abs is to be maintained in the vapor space; this corresponds to a boiling point for water of 125°F. The feed rate to the evaporator is 55,000 lb/hr. The over-all heat-transfer coefficient can be taken as 500 Btu/(ft²)(hr)(°F). The solution has a negligible elevation in boiling point and a negligible heat of dilution.

Calculate the steam consumption, the economy, and the heating surface required if the temperature of the feed is: (a) 125°F, (b) 70°F, (c) 200°F.

The specific heat of the feed solution is 0.90, and the latent heat of vaporization of the solution may be taken equal to that of water. Radiation losses may be neglected.

Solution. (a) The flow rate of thin liquor w_f is 55,000 lb/hr. The amount of water evaporated is found by a material balance from the terminal concentrations and the flow rate of the thin liquor. The feed contains $\frac{90}{10} = 9$ lb of water per pound of solid; the thick liquor contains $\frac{50}{50} = 1$ lb of water per pound of solid.

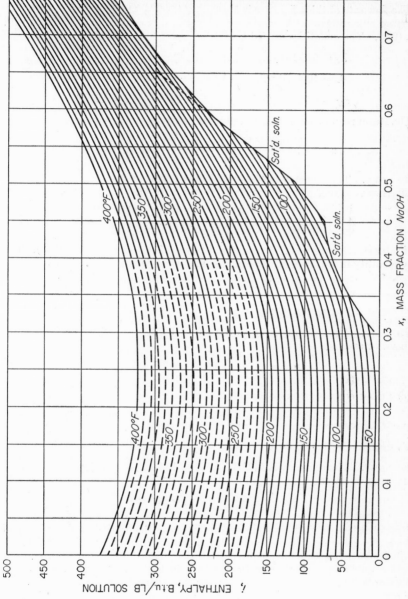

FIG. 9-12. Enthalpy-concentration diagram, system sodium hydroxide–water. (*McCabe.*[8])

The quantity evaporated, therefore, is $9 - 1 = 8$ lb of water per pound of solid, or

$$8 \times 55,000 \times 0.10 = 44,000 \text{ lb/hr}$$

The flow rate of thick liquor w is $55,000 - 44,000 = 11,000$ lb/hr.

STEAM CONSUMPTION. An enthalpy balance is now written in accordance with Eq. (9-8). The quantities needed are

$$\lambda_s = 946 \text{ Btu/lb} \quad \text{(Appendix 6)}$$

$$\lambda = 1023 \text{ Btu/lb} \quad \text{(Appendix 6)}$$

$$c_{pf} = 0.90 \text{ Btu/(lb)(°F)} \quad t_f = 125°F \quad t = 125°F$$

The steam requirement, from Eq. (9-8), is

$$w_s = \frac{(55,000 - 11,000)1023 - 0}{946} = 47,580 \text{ lb/hr}$$

ECONOMY. The economy is $44,000/47,580 = 0.925$.

HEATING SURFACE. This is calculated from Eq. (9-1).

$$A = \frac{q}{U(t_s - t)} = \frac{w_s \lambda_s}{U(t_s - t)} = \frac{47,580 \times 946}{500(249 - 125)} = 725 \text{ ft}^2$$

(b) STEAM CONSUMPTION. The material balance is the same as in part (a). The only quantity in Eq. (9-8) which changes is t_f, which is now 70°F. Thus

$$w_s = \frac{(55,000 - 11,000)1023 - 55,000 \times 0.90(70 - 125)}{946}$$

$$= 50,460 \text{ lb/hr}$$

ECONOMY. The economy is $44,000/50,460 = 0.872$.

HEATING SURFACE

$$A = \frac{50,460 \times 946}{500(249 - 125)} = 770 \text{ ft}^2$$

(c) STEAM CONSUMPTION. t_f is now 200°F.

$$w_s = \frac{(55,000 - 11,000)1023 - 55,000 \times 0.90(200 - 125)}{946}$$

$$= 43,660 \text{ lb/hr}$$

ECONOMY. The economy is $44,000/43,660 = 1.008$.

HEATING SURFACE

$$A = \frac{43,660 \times 946}{500(249 - 125)} = 665 \text{ ft}^2$$

Note that lowering the feed temperature increases the heating area required and reduces the economy. If the feed enters appreciably above its boiling point, the heating area is considerably reduced from that needed with cold feed, and the economy may be greater than unity even in a single-effect evaporator.

Example 9-2. A single-effect evaporator is to concentrate 20,000 lb/hr of a 20 per cent solution of sodium hydroxide to 50 per cent solids. The gauge pressure of the steam is to be 20 lb force/in.²; the absolute pressure in the vapor space is to be 100 mm Hg (1.93 lb force/in.²). The over-all coefficient is estimated to be 250 Btu/(ft²)(hr)(°F). The feed temperature is 100°F.

Calculate the amount of steam consumed, the economy, and the heating surface required.

Solution. The amount of water evaporated is found from a material balance as in Example 9-1. The feed contains $\frac{80}{20} = 4$ lb of water per pound of solid; the thick liquor contains $\frac{50}{50} = 1$ lb of water per pound of solid. The quantity evaporated is $4 - 1 = 3$ lb of water per pound of solid, or

$$3 \times 20{,}000 \times 0.20 = 12{,}000 \text{ lb/hr}$$

The flow rate of thick liquor w is $20{,}000 - 12{,}000 = 8{,}000$ lb/hr.

STEAM CONSUMPTION. Since with strong solutions of sodium hydroxide the heat of dilution is not negligible, the rate of heat transfer is found from Eq. (9-5) and Fig. 9-12. The vaporization temperature of the 50 per cent solution at a pressure of 100 mm Hg is found as follows.

Boiling point of water at 100 mm Hg = 124°F   (Appendix 6)

Boiling point of solution = 197°F   (Fig. 9-7)

Boiling-point elevation = 197 − 124 = 73°F

The enthalpies of the feed and thick liquor are found from Fig. 9-12.

Feed, 20% solids, 100°F: $i_f = $ 55 Btu/lb
Thick liquor, 50% solids, 197°F: $i = $ 221 Btu/lb

The enthalpy of the vapor leaving the evaporator is found from steam tables.[5] The enthalpy of superheated water vapor at 197°F and 1.93 lb force/in.² is 1149 Btu/lb; this is the i_v of Eq. (9-5).

The heat of vaporization of steam λ_s at a gauge pressure of 20 lb force/in.² is, from Appendix 6, 939 Btu/lb.

The rate of heat transfer and the steam consumption may now be found from Eq. (9-5).

$$q = (20{,}000 - 8{,}000)1149 + 8{,}000 \times 221 - 20{,}000 \times 55$$

$$= 14{,}456{,}000 \text{ Btu/hr}$$

$$w_s = \frac{14{,}456{,}000}{939} = 15{,}400 \text{ lb/hr}$$

ECONOMY. The economy is $12{,}000/15{,}400 = 0.78$.

HEATING SURFACE. The condensation temperature of the steam is 259°F. The heating area required is

$$A = \frac{14{,}456{,}000}{250 \, (259 - 197)} = 930 \text{ ft}^2$$

If the enthalpy of the vapor i_v were based on saturated vapor at the pressure in the vapor space, instead of on superheated vapor, the rate of heat transfer would be 14,036,000 Btu/hr, and the heating area would be 906 ft². Thus this approximation would introduce an error of only about 3 per cent.

MULTIPLE-EFFECT EVAPORATION

Multiple-effect evaporation is the most common method of increasing the steam economy of evaporators over that of a single effect. The total pressure difference between the steam and the condenser is spread across two or more effects in the multiple-effect system. The pressure in each effect is lower than that in the effect from which it receives steam and above that of the effect to which it supplies vapor. Each effect, in itself, acts as a single-effect evaporator, and each has a temperature drop across its heating surface corresponding to the pressure drop in that effect. Every statement that has so far been made about a single-effect evaporator applies to each effect of a multiple-effect system. Arranging a series of evaporator bodies into a multiple-effect system is a matter of interconnecting piping, and not of the structure of the individual units.

Principles of Multiple-effect Operation. Figure 9-13 shows three short-tube natural-circulation evaporators arranged to form a multiple-

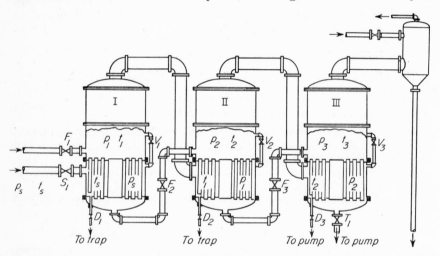

FIG. 9-13. Triple-effect evaporator. I, II, III, first, second, and third effects. D_1, D_2, D_3, condensate valves. F_1, F_2, F_3, feed- or liquor-control valves. S_1, steam valve. T_1, thick-liquor valve. V_1, V_2, V_3, vent valves. p_s, p_1, p_2, p_3, pressures. t_s, t_1, t_2, t_3, temperatures.

effect evaporator. Connections are made so that the vapor from one effect serves as the heating medium for the next. A condenser and air ejector establishes a vacuum in the third effect in the series and withdraws non-condensables from the system. The first effect of a multiple-effect evaporator is the effect to which the raw steam is fed and in which the pressure in the vapor space is the highest. The last effect is that in which the vapor-space pressure is a minimum. The numbering of the effects is independent of the order liquor is fed to them. In Fig. 9-13 dilute feed enters the first effect, where it is partly concentrated; it flows to the second effect for additional concentration, and then to the third effect for final concentration. Thick liquor is pumped out of the third effect.

Imagine that the whole system is cold and at atmospheric pressure and that each effect is filled to an appropriate level with the dilute liquid to be evaporated. All the valves are closed. Imagine that the ejector is turned on and that most of the air and other noncondensable gas is pumped out of the unit. Air flows through the vent lines and vapor lines until the whole apparatus is evacuated.

Now assume that the steam valve S_1 and the condensate valve D_1 are opened until the desired pressure p_s is built up in the steam chest of evaporator I. Let t_s be the temperature of saturated steam at pressure p_s. The steam will first displace any residual air from the steam chest of evaporator I through the vent line. Since the liquid in the tubes is cold, steam will condense on the outside of the tubes. The trap allows condensate to escape as fast as it collects. The liquid in the evaporator body becomes warmer until it reaches the temperature at which it boils. The vapor so generated will gradually displace air from the upper part of evaporator I, from the vapor line, and from the steam chest of evaporator II. The displaced air flows through the vent lines to the ejector.

The steam that is coming off from evaporator I will transmit latent heat to the liquid in evaporator II and be condensed. Condensate valve D_2 will be opened so that this condensate will be removed as fast as it is formed. In condensing, however, it gives up its heat to the liquid in evaporator II, which becomes warmer. As the liquid becomes warmer, the temperature difference between it and the steam becomes smaller, the rate of condensation diminishes, and therefore the pressure in the vapor space of evaporator I gradually builds up, increasing t_1 (the boiling point of the liquid in evaporator I) and cutting down the temperature difference $t_s - t_1$. This continues until the liquid in evaporator II reaches its boiling temperature.

The same process will be repeated in evaporator III. As the liquid in evaporator III becomes warmer and finally begins to boil, the temperature drop between it and the steam from the second effect becomes smaller, pressure begins to build up in the second effect and raises t_2, the boiling point there, so that the temperature difference $t_1 - t_2$ decreases. This decreases the rate of condensation and builds up the pressure in the vapor space of the first effect still more, until finally the evaporator comes to a steady equilibrium with the liquid boiling in all three bodies.

The result of boiling will be to lower the liquid levels. As soon as the level begins to come down in body I, the feed valve F_1 will be opened enough to keep the level constant. As body II begins to boil, feed valve F_2 will be adjusted, and as body III begins to boil, feed valve F_3 will be adjusted. A change in any one valve obviously involves a resetting of the others, but they will be so set that the liquid levels in all three bodies are kept constant. Liquid in body III is becoming more and more concentrated, however, and when it reaches the desired finished concentration the thick-liquor valve T_1 will be opened the proper amount and the thick-liquor pump will be started. Thus when evaporation is proceeding in a steady manner, there will be a continuous feed to the first body, from the first to the second, from the second to the third, and a continuous withdrawal of thick liquor from the third effect. The evaporator is now in continuous operation with a continuous flow of liquid through it, and all the various temperatures and pressures are in equilibrium.

The heating surface in the first effect will transmit per hour an amount of heat given by the equation

$$q_1 = A_1 U_1 \, \Delta t_1 \tag{9-9}$$

If the part of this heat that goes to heat the feed to the boiling point is neglected for the moment, it follows that practically all of this heat must appear as latent heat in the vapor that leaves the first effect. The temperature of the condensate leaving through connection D_2 is very near the temperature t_1 of the vapors from the boiling liquid in the first effect. Therefore, practically all of the heat that was expended in creating vapor in the first effect must be given up when this same vapor condenses in the second effect. The heat transmitted in the second effect, however, is given by the equation

$$q_2 = A_2 U_2 \, \Delta t_2 \tag{9-10}$$

As has just been shown, q_1 and q_2 are nearly equal, and therefore

$$A_1 U_1 \, \Delta t_1 = A_2 U_2 \, \Delta t_2 \tag{9-11}$$

This same reasoning may be extended to show that, roughly,

$$A_1 U_1 \, \Delta t_1 = A_2 U_2 \, \Delta t_2 = A_3 U_3 \, \Delta t_3 \tag{9-12}$$

It should be expressly understood that Eqs. (9-11) and (9-12) are only approximate equations that must be corrected by the addition of terms which are, however, relatively small compared to the quantities involved in the expressions above.

In ordinary practice the heating areas in all the effects of a multiple-effect evaporator are equal. This is to obtain economy of construction. Therefore, from Eq. (9-12) it follows that

$$U_1 \, \Delta t_1 = U_2 \, \Delta t_2 = U_3 \, \Delta t_3 \tag{9-13}$$

From this it follows that the temperature drops in a multiple-effect evaporator are approximately inversely proportional to the heat-transfer coefficients.

A reference to the mechanism by which this imaginary evaporator was started up, as given in the preceding pages, will show that the temperatures t_1 and t_2 were determined as the result of the evaporator's automatically reaching its own equilibrium. Since t_s and t_3 are fixed, this automatically determines Δt_1, Δt_2, and Δt_3. In other words, the equilibrium described by Eqs. (9-12) and (9-13) is automatically and continuously maintained, and cannot be regulated or controlled except by altering U_1, U_2, and U_3.

Example 9-3. A triple-effect evaporator is concentrating a liquid that has no appreciable elevation in boiling point. The temperature of the steam to the first effect is 227°F; the boiling point of the solution in the last effect is 125°F. The overall heat-transfer coefficients, in Btu/(ft²)(hr)(°F), are 500 in the first effect, 400 in the second effect, and 200 in the third effect. At what temperatures will the liquid boil in the first and second effects?

Solution. The total temperature drop is $227 - 125 = 102°F$. As shown by Eq. (9-13), the temperature drops in the several effects will be approximately inversely proportional to the coefficients. Thus

$$\Delta t_1 : \Delta t_2 : \Delta t_3 = \frac{1}{U_1} : \frac{1}{U_2} : \frac{1}{U_3}$$

From this it follows that $\Delta t_1 = 21.5°F$; $\Delta t_2 = 26.8°F$; and $\Delta t_3 = 53.7°F$. Consequently the boiling point in the first effect will be 205.5°F, and that in the second effect, 178.7°F.

Methods of Feeding. The usual method of feeding a multiple-effect evaporator is to pump the thin liquid into the first effect and send it in turn through the other effects, as shown in Fig. 9-14a. This is called *forward feed*. The concentration of the liquid increases from the first effect to the last. This pattern of liquid flow is the simplest. It requires a pump for feeding dilute solution to the first effect, since this effect is often at

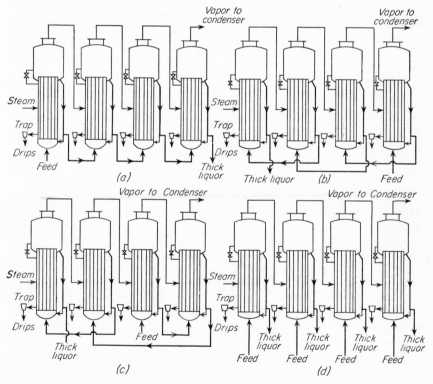

FIG. 9-14. Patterns of liquor flow in multiple-effect evaporators: (*a*) Forward feed. (*b*) Backward feed. (*c*) Mixed feed. (*d*) Parallel feed.

about atmospheric pressure, and a pump to remove thick liquor from the last effect. The transfer from effect to effect, however, can be done without pumps, since the flow is in the direction of decreasing pressure, and control valves in the transfer lines are all that is required.

Another common method is *backward feed*, in which dilute liquid is fed to the last effect and then pumped through the successive effects to the first, as shown in Fig. 9-14b. This method requires a pump between each pair of effects in addition to the thick-liquor pump, since the flow is from low pressure to high pressure. This adds to the investment and operating costs of the evaporator. Care must be taken to avoid leakage of air into the pump casings, which in backward feed are all under partial vacuum.

Backward feed often gives a higher capacity than forward feed when the thick liquor is very viscous. With forward feed the most concentrated liquor is in the last effect, where the temperature is lowest and the viscosity highest. The capacity of the last effect is therefore small because of the low over-all coefficient—so small that it reduces the capacity of the multiple-effect system as a whole. With backward feed the concentrated liquor is in the effect where the temperature is highest. Its viscosity is much lower than at the low temperature of the last effect, and the over-all coefficient can be moderately high.

Backward feed is also advantageous when the feed liquid is cold. If cold feed is sent to the first effect, it must be heated by live steam to the temperature prevailing in the first effect. Since this temperature is the highest in the system, and since the feed is dilute and therefore in large quantity, this means a large consumption of raw steam. The steam so condensed accomplishes no evaporation in either the first or the succeeding effects. A pound of such steam has the potential of evaporating N lb of water, where N is the number of effects, a potential which is now lost. This condensation therefore reduces the economy. With backward feed, on the other hand, the cold feed is heated through a smaller temperature range in the last effect and is heated there by steam each pound of which has already evaporated $N-1$ lb of water. This is partially offset by the fact that there is a heating load on each effect in backward feed, since the feed enters each effect at a temperature lower than the prevailing boiling point. The net result is nevertheless an increase in economy if backward feed is used where the dilute feed liquid is cold. By the same reasoning, forward feed is more economical if the feed liquid is at or above the boiling temperature in the first effect.

Other patterns of feed are sometimes used. In *mixed feed* the dilute liquid enters an intermediate effect, flows in forward feed to the end of the series, and is then pumped back to the first effects for final concentration, as shown in Fig. 9-14c. This eliminates some of the pumps needed in backward feed and yet permits the final evaporation to be done at the highest temperature. In crystallizing evaporators, where a slurry of crystals and mother liquor is withdrawn, feed may be admitted directly to each effect to give what is called *parallel feed*, as shown in Fig. 9-14d. In parallel feed there is no transfer of liquid from one effect to another.

Capacity of Multiple-effect Evaporators. The increase in economy through the use of multiple-effect evaporation is obtained at the cost of reduced capacity. It might be thought that by providing several times as much heating surface the evaporating capacity would be increased, but this is not the case. The total capacity of a multiple-effect evaporator is no greater than that of a single-effect evaporator having a heating surface equal to one of the effects and operating under the same terminal conditions. The amount of water vaporized *per square foot of surface* in an *N*-effect multiple-effect evaporator is approximately $(1/N)$th that in the single effect. This can be shown by the following analysis.

If the heating load and the heat of dilution are neglected, the capacity of an evaporator is directly proportional to the rate of heat transfer. The heat transferred in the three effects in Fig. 9-13 is given by the equations

$$q_1 = U_1 A_1 \,\Delta t_1 \qquad q_2 = U_2 A_2 \,\Delta t_2 \qquad q_3 = U_3 A_3 \,\Delta t_3 \qquad (9\text{-}14)$$

The total capacity is proportional to the total rate of heat transfer q, found by adding these equations.

$$q = q_1 + q_2 + q_3 = U_1 A_1 \,\Delta t_1 + U_2 A_2 \,\Delta t_2 + U_3 A_3 \,\Delta t_3 \qquad (9\text{-}15)$$

Assume that the surface area is A ft^2 in each effect and that the over-all coefficient U is also the same in each effect. Then Eq. (9-15) can be written

$$q = UA(\Delta t_1 + \Delta t_2 + \Delta t_3) = UA \,\Delta t \qquad (9\text{-}16)$$

where Δt is the total temperature drop between the steam in the first effect and the vapor in the last effect.

Suppose now that a single-effect evaporator with a surface area of A ft^2 is operating with the same total temperature drop. If the over-all coefficient is the same as in each effect of the triple-effect evaporator, the rate of heat transfer in the single effect is

$$q = UA \,\Delta t$$

This is exactly the same equation as for the multiple-effect evaporator. No matter how many effects are used, provided the over-all coefficients are the same, the capacity will be no greater than that of a single effect having an area equal to that of each effect in the multiple unit. In actuality, because of boiling-point elevation and other factors, the capacity of a multiple-effect evaporator is almost always less than that of the corresponding single effect.

The cost of an evaporator per square foot of surface is a function of the total area of all effects [1] and decreases with total area, approaching an asymptote for large installations. When the cost of 1 ft^2 of heating surface is constant, regardless of the number of effects, the investment required for an *N*-effect evaporator is *N* times that for a single-effect evaporator of the same capacity. The optimum number of effects must be found from an economic balance between the savings in steam obtained by multiple-effect operation and the added investment required.

Effect of Liquid Head and Boiling-point Elevation. The liquid head and the boiling-point elevation influence the capacity of a multiple-effect evaporator even more than they do that of a single effect. The reduction in capacity caused by the liquid head, as before, cannot be estimated quantitatively. The liquid head, it will be remembered, reduces the temperature drop available in the evaporator. The boiling-point elevation also reduces the available temperature drop in each effect of a multiple-effect evaporator, as follows.

Consider an evaporator that is concentrating a solution with a large boiling-point elevation. The vapor coming from this boiling solution is

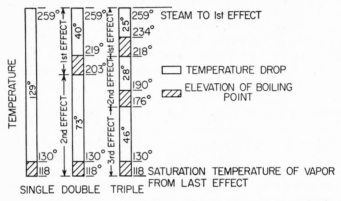

Fig. 9-15. Effect of boiling-point elevation on capacity of evaporators.

at the solution temperature and is therefore superheated by the amount of the boiling-point elevation.

As discussed in Chap. 8, pages 479 to 481, superheated steam is essentially equivalent to saturated steam at the same pressure when used as a heating medium. The temperature drop in any effect, therefore, is calculated from the temperature of saturated steam at the pressure of the steam chest, and not from the temperature of the boiling liquid in the previous effect. This means that the boiling-point elevation in any effect is lost from the total available temperature drop. This loss occurs in every effect of a multiple-effect evaporator, and the resulting loss of capacity is often important.

The influence of these losses in temperature drop on the capacity of a multiple-effect evaporator is shown in Fig. 9-15. The three diagrams in this figure represent the temperature drops in a single-effect, double-effect, and triple-effect evaporator. The terminal conditions are the same in all three; i.e., the steam pressure in the first effect and the saturation temperature of the vapor evolved from the last effect are identical in all three evaporators. Each effect contains a liquid with a boiling-point elevation. The total height of each column represents the total temperature spread from the steam temperature to the saturation temperature of the vapor from the last effect.

Consider the single-effect evaporator. Of the total temperature drop, the shaded part represents the loss in temperature drop due to boiling-point elevation. The remaining temperature drop, the actual driving force for heat transfer, is represented by the unshaded part. The diagram for the double-effect evaporator shows two shaded portions because there is a boiling-point elevation in each of the two effects, and the residual unshaded part is smaller than in the diagram for the single effect. In the triple-effect evaporator there are three shaded portions since there is a loss of temperature drop in each of three effects, and the total net available temperature drop is correspondingly smaller.

Figure 9-15 shows that in extreme cases of a large number of effects or very high boiling-point elevations the sum of the boiling-point elevations in a proposed evaporator could be greater than the total temperature drop available. Operation under such conditions is impossible. The design or the operating conditions of the evaporator would have to be revised to reduce the number of effects or increase the total temperature drop.

The economy of a multiple-effect evaporator is not influenced by boiling-point elevations if minor factors, such as the temperature of the feed and changes in the heats of vaporization, are neglected. A pound of steam condensing in the first effect generates about a pound of vapor, which condenses in the second effect generating another pound there, and so on. The *economy* of a multiple-effect evaporator depends on heat-balance considerations and not on the rate of heat transfer. The *capacity*, on the other hand, is reduced by the boiling-point elevation. The capacity of a double-effect evaporator concentrating a solution with a boiling-point elevation is less than half the capacity of two single effects, each operating with the same over-all temperature drop. The capacity of a triple effect is less than one-third that of three single effects with the same terminal temperatures.

MULTIPLE-EFFECT CALCULATIONS

In designing a multiple-effect evaporator the results usually desired are the amount of steam consumed, the area of the heating surface required, the approximate temperatures in the various effects, and the amount of vapor leaving the last effect. As in a single-effect evaporator, these quantities are found from material balances, enthalpy balances, and the capacity equation [Eq. (9-1)]. In a multiple-effect evaporator, however, a trial-and-error method is used in place of a direct algebraic solution.

Consider, for example, a triple-effect evaporator. There are seven equations which may be written: an enthalpy balance for each effect, a capacity equation for each effect, and the known total evaporation, or the difference between the thin- and thick-liquor rates. If the amount of heating surface in each effect is assumed to be the same, there are seven unknowns in these equations: (1) the rate of steam flow to the first effect, (2) to (4) the rate of flow from each effect, (5) the boiling temperature in the first effect, (6) the boiling temperature in the second effect, and (7) the heating surface per effect. It is possible to solve these equations for the seven unknowns, but the method is tedious and involved. Another method of calculation is as follows:

1. Assume values for the boiling temperatures in the first and second effects.

2. From enthalpy balances find the rates of steam flow and of liquor from effect to effect.

3. Calculate the heating surface needed in each effect from the capacity equations.

4. If the heating areas so found are not nearly equal, estimate new values for the boiling temperatures and repeat items 2 and 3 until the heating surfaces are equal.

Example 9-4. A solution with a negligible boiling-point elevation is to be concentrated from 10 to 50 per cent solids in a triple-effect evaporator. Steam is available at a gauge pressure of 15 lb force/in.^{2} (249°F); the absolute pressure in the third effect is to be 4 in. Hg. This corresponds to a boiling temperature of 125°F. Feed enters at a rate of 55,000 lb/hr and a temperature of 70°F. The specific heat of the solution may be taken as 1.00 Btu/(lb)(°F) at all concentrations. The over-all coefficients, in Btu/(ft^2)(hr)(°F), are estimated to be 550 in the first effect, 350 in the second, and 200 in the third with forward feed; and 450, 350, and 275 with backward feed. Each effect is to have the same amount of heating surface.

Calculate the amount of heating surface needed, the steam consumption, the distribution of temperatures, the economy of each effect, and the over-all economy: (a) with forward feed, (b) with backward feed.

Solution. The total rate of evaporation is the same in both cases, and may be calculated from an over-all material balance, assuming that the solids go through the evaporator without loss.

Material	Flow rate, lb/hr		
	Total	Solid	Water
Feed solution............	55,000	5,500	49,500
Thick liquor.............	11,000	5,500	5,500
Water evaporated......	44,000	.....	44,000

(a) Figure 9-16a shows a flow diagram for this evaporator, with forward feed. The first, second, and third effects are designated as I, II, and III, respectively. For material and heat balances, let

$$w_f = \text{rate of flow of feed}$$
$$w_s = \text{rate of flow of steam}$$
$$w_1, w_2, w_3 = \text{rates of flow of liquor out of I, II, and III}$$
$$\lambda_s = \text{latent heat of steam}$$
$$\lambda_1, \lambda_2, \lambda_3 = \text{latent heats of vaporization in I, II, and III}$$
$$t_f = \text{temperature of feed}$$

t_s = condensing temperature of steam

t_1, t_2, t_3 = boiling temperatures in I, II, and III

c_{pf} = specific heat of feed

c_{p1}, c_{p2}, c_{p3} = specific heats of liquor leaving I, II, and III

q_1, q_2, q_3 = heat-transfer rates in I, II, and III

Since the boiling-point elevations are negligible, the boiling temperature in any one effect equals the condensing temperature in the next succeeding effect; and the

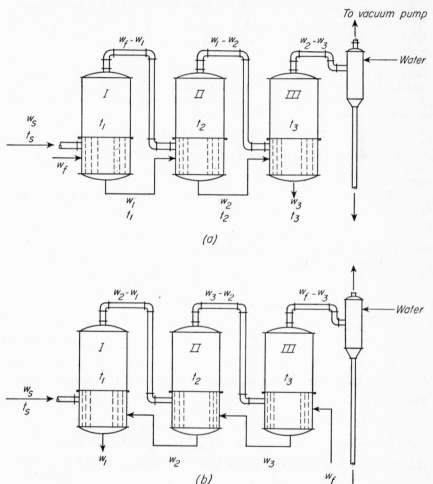

FIG. 9-16. Flow sheets for Example 9-4: (a) Forward feed. (b) Backward feed.

latent heat of condensation in the steam chest of an effect equals the latent heat of vaporization in the next.

The heat-balance equations [Eq. (9-8)] may be written for each of the three effects, aided by reference to Fig. 9-16a.

$$q_1 = w_s\lambda_s = (w_f - w_1)\lambda_1 + w_f c_{pf}(t_1 - t_f)$$

$$q_2 = (w_f - w_1)\lambda_1 = (w_1 - w_2)\lambda_2 + w_1 c_{p1}(t_2 - t_1)$$

$$q_3 = (w_1 - w_2)\lambda_2 = (w_2 - w_3)\lambda_3 + w_2 c_{p2}(t_3 - t_2)$$

Note that each of these equations may be written down by using the verbal rule stated after Eq. (9-8). The "vapor" from any one effect is the "steam" for the next succeeding effect.

As a guide in the first attempt to distribute the total temperature drop among the three effects, the following principles are of value. As discussed on page 559, the temperature drop in any effect is inversely proportional to the over-all heat-transfer coefficient. Also, any effect which has an extra load requires a larger proportion of the total temperature drop than the other effects. In this problem the total temperature drop is $249 - 125$, or $124°F$. The last effect must have the largest drop because of the small coefficient there; because of the large heating load in the first effect, this effect may have a larger temperature drop than the second effect. From these considerations the first assumption is

$$\Delta t_1 = 38°F \qquad \Delta t_2 = 33°F \qquad \Delta t_3 = 53°F$$

The temperatures and latent heats are:

	Temp., °F	Latent heat, Btu/lb
Steam......................	$249 = t_s$	$946 = \lambda_s$
Feed to I....................	$70 = t_f$	
Liquor in I and steam to II:		
$249 - 38$....................	$211 = t_1$	$971 = \lambda_1$
Liquor in II and steam to III:		
$211 - 33$....................	$178 = t_2$	$991 = \lambda_2$
Liquor in III:		
$178 - 53$....................	$125 = t_3$	$1022 = \lambda_3$

In this problem, all specific heats are unity, $w_f = 55,000$ lb/hr, and $w_3 = 11,000$ lb/hr. The three unknowns are w_s, w_1, and w_2. The heat-balance equations are

First effect: $q_1 = 946w_s = 971(55,000 - w_1) + 55,000(211 - 70)$

Second effect: $q_2 = 971(55,000 - w_1) = 991(w_1 - w_2) + (178 - 211)w_1$

Third effect: $q_3 = 991(w_1 - w_2) = 1022(w_2 - 11,000) + (125 - 178)w_2$

Solving the last two of these equations gives $w_3 = $ thick (liquor out)

$$w_1 = 41,380 \text{ lb/hr} \qquad w_2 = 26,660 \text{ lb/hr}$$

From the first equation, w_s is found to be 22,180 lb/hr. The rates of heat transfer are

$$q_1 = 946 \times 22,180 = 20,980,000 \text{ Btu/hr}$$

$$q_2 = 971(55,000 - 41,380) = 13,225,000 \text{ Btu/hr}$$

$$q_3 = 991(41,380 - 26,660) = 14,588,000 \text{ Btu/hr}$$

The areas of the heating surfaces are found from Eq. (9-1).

$$A_1 = \frac{20,980,000}{38 \times 550} = 1,004 \text{ ft}^2$$

$$A_2 = \frac{13,225,000}{33 \times 350} = 1,145$$

$$A_3 = \frac{14,588,000}{53 \times 200} = 1,375$$

$$\text{Average} = \overline{1,175} \text{ ft}^2$$

Since the heating area is to be the same in each effect, the temperature drop must now be adjusted. It may be assumed that the computed areas will change inversely with the temperature drops. The third effect is too large and must have a larger temperature drop; the first two effects are too small and need a smaller temperature drop. Correcting the temperature drops in proportion to the deviations of the computed areas from the average gives

$$\Delta t_1 = 32°\text{F} \qquad \Delta t_2 = 31°\text{F} \qquad \Delta t_3 = 61°\text{F}$$

The revised temperatures and latent heats are:

	Temp., °F	Latent heat, Btu/lb
Steam........................	$249 = t_s$	$946 = \lambda_s$
Feed to I....................	$70 = t_f$	
Liquor in I and steam to II:		
$249 - 32$....................	$217 = t_1$	$967 = \lambda_1$
Liquor in II and steam to III:		
$217 - 31$....................	$186 = t_2$	$987 = \lambda_2$
Liquor in III:		
$186 - 61$....................	$125 = t_3$	$1022 = \lambda_3$

The heat-balance equations are

$$q_1 = 946w_s = 967(55,000 - w_1) + 55,000(217 - 70)$$

$$q_2 = 967(55,000 - w_1) = 987(w_1 - w_2) + (186 - 217)w_1$$

$$q_3 = 987(w_1 - w_2) = 1022(w_2 - 11,000) + (125 - 186)w_2$$

Solving these equations gives

$$w_s = 22,470 \text{ lb/hr} \qquad w_1 = 41,380 \text{ lb/hr} \qquad w_2 = 26,720 \text{ lb/hr}$$

$$q_1 = 21,256,000 \text{ Btu/hr}$$

$$q_2 = 13,171,000 \text{ Btu/hr}$$

$$q_3 = 14,469,000 \text{ Btu/hr}$$

The surface areas required are

$$A_1 = 1,208 \text{ ft}^2$$

$$A_2 = 1,214$$

$$A_3 = 1,185$$

$$\overline{\text{Average} = 1,202 \text{ ft}^2 \text{ per effect}}$$

The economies are

First effect: $\dfrac{55,000 - 41,380}{22,470} = 0.606$ lb evaporated/lb steam

Second effect: $\dfrac{41,380 - 26,720}{55,000 - 41,380} = 1.076$

Third effect: $\dfrac{26,720 - 11,000}{41,380 - 26,720} = 1.072$

Over-all economy: $\dfrac{55,000 - 11,000}{22,470} = 1.96$ lb evaporated/lb steam

(b) The flow sheet for this case is shown in Fig. 9-16b. The heat-balance equations are now

First effect: $q_1 = w_s\lambda_s = (w_2 - w_1)\lambda_1 + w_2 c_{p2}(t_1 - t_2)$

Second effect: $q_2 = (w_2 - w_1)\lambda_1 = (w_3 - w_2)\lambda_2 + w_3 c_{p3}(t_2 - t_3)$

Third effect: $q_3 = (w_3 - w_2)\lambda_2 = (w_f - w_3)\lambda_3 + w_f c_{pf}(t_3 - t_f)$

After a preliminary estimate of temperatures by the method used in the solution of part (a), the revised temperature drops are

$$\Delta t_1 = 35°\text{F} \qquad \Delta t_2 = 42°\text{F} \qquad \Delta t_3 = 47°\text{F}$$

The temperatures and latent heats are:

	Temp., °F	Latent heat, Btu/lb
Steam to I.....................	$249 = t_s$	$946 = \lambda_s$
Liquor in I and steam to II.....	$214 = t_1$	$969 = \lambda_1$
Liquor in II and steam to III....	$172 = t_2$	$995 = \lambda_2$
Liquor in III..................	$125 = t_3$	$1022 = \lambda_3$
Feed to III....................	$70 = t_f$	

The heat-balance equations are, since $w_f = 55,000$ and $w_1 = 11,000$,

$$q_1 = 946 w_s = 969(w_2 - 11,000) + (214 - 172)w_2$$

$$q_2 = 969(w_2 - 11,000) = 995(w_3 - w_2) + (172 - 125)w_3$$

$$q_3 = 995(w_3 - w_2) = 1022(55,000 - w_3) + (125 - 70)55,000$$

Solution of these equations gives

$$w_s = 19{,}150 \text{ lb/hr} \qquad w_2 = 28{,}460 \text{ lb/hr} \qquad w_3 = 43{,}410 \text{ lb/hr}$$

$$q_1 = 18{,}114{,}000 \text{ Btu/hr}$$

$$q_2 = 16{,}919{,}000 \text{ Btu/hr}$$

$$q_3 = 14{,}875{,}000 \text{ Btu/hr}$$

The surface areas required are

$$A_1 = \frac{18{,}114{,}000}{450 \times 35} = 1{,}150 \text{ ft}^2$$

$$A_2 = \frac{16{,}919{,}000}{350 \times 42} = 1{,}151$$

$$A_3 = \frac{14{,}875{,}000}{275 \times 47} = 1{,}151$$

$$\text{Average} = \overline{1{,}151} \text{ ft}^2 \text{ per effect}$$

The economies are

First effect: $\dfrac{28{,}460 - 11{,}000}{19{,}150} = 0.912$ lb evaporated/lb steam

Second effect: $\dfrac{43{,}410 - 28{,}460}{28{,}460 - 11{,}000} = 0.856$

Third effect: $\dfrac{55{,}000 - 43{,}410}{43{,}410 - 28{,}460} = 0.775$

Over-all economy: $\dfrac{55{,}000 - 11{,}000}{19{,}150} = 2.30$ lb evaporated/lb steam

Because of the heating load in each effect, none of the individual economies are greater than unity, as was the case with forward feed. The over-all economy, however, is 2.30 with backward feed compared to 1.96 with forward feed. This is the usual result with backward feed when the feed solution is cold. The heating surface needed for backward feed is slightly smaller than for forward feed, but this is largely due to the particular values assumed for the over-all coefficients and is not a general result for all cases of backward feed.

Example 9-4 is simplified by the absence of boiling-point elevations, heats of dilution, and changes in the specific heat of the solution. In most calculations it is necessary to estimate these quantities, or if adequate data are available, to solve the problem by Eq. (9-5) and an enthalpy-concentration diagram similar to Fig. 9-12. To use the enthalpy-concentration chart it is necessary to estimate the concentrations of the liquors in the various effects. This is easily done by assuming equal evaporation in each effect and correcting, if necessary, the second time through.

VAPOR RECOMPRESSION

The energy in the vapor evolved from a boiling solution can be used to vaporize more water provided there is a temperature drop for heat transfer in the desired direction. In a multiple-effect evaporator this temperature drop is created by progressively lowering the boiling point of the solution in a series of evaporators through the use of lower absolute pressures. The desired driving force can also be obtained by increasing the pressure (and, therefore, the condensing temperature) of the evolved vapor by mechanical or thermal recompression. The compressed vapor is then condensed in the steam chest of the evaporator from which it came.

Mechanical Recompression. The principle of mechanical vapor recompression is illustrated in Fig. 9-17. Cold feed solution enters an ex-

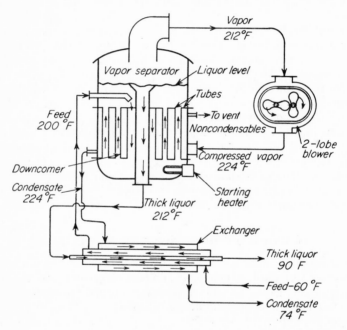

Fig. 9-17. Principle of mechanical vapor-recompression evaporator. (*Arthur D. Little, Inc.*)

changer, where it receives heat from concentrated liquor flowing in a double-pipe heat exchanger. Preheated feed enters the evaporator body nearly at its boiling point at the evaporator pressure, which is 1 atm in the example shown in Fig. 9-17. Since the boiling-point elevation is negligible in this example, the vapor given off on boiling is at 212°F. This vapor is compressed in a two-lobe blower to a pressure of 4 lb force/in.^{2} gauge, corresponding to a saturation temperature of 224°F; it can then give up its heat to the boiling solution, generating more vapor. Condensate flows from the steam chest through the exchanger to preheat the feed. If the solution has an appreciable boiling-point elevation, the vapor must be compressed somewhat more than in this example, though still not very greatly, for the optimum temperature drop for typical operating conditions is only about 10°F.[6] The energy utilization of such a system is very good. The most important application of this equipment is the production of distilled water.

The chief problem in mechanical vapor recompression is the design of a compressor to handle very large volumes of water vapor, especially when the evaporator is to operate under vacuum. Rotary positive-displacement blowers have suitable performance characteristics but are limited to small installations. Centrifugal compressors could handle the vapor from large evaporators but have unsatisfactory performance characteristics, especially if the vapor rate falls much below the design value. Also, they are expensive.

Thermal Recompression. In a thermal recompression system the vapor is compressed by acting on it with high-pressure steam in a jet ejector. This results in more steam than is needed for boiling the solution, so that excess steam must be vented or condensed. The ratio of motive steam to the vapor from the solution depends on the evaporation pressure; for many low-temperature operations the ratio is about 1 lb of steam at 125 lb force/in.^{2} to 2 lb of water evaporated.

Since steam jets can handle large volumes of low-density vapor, thermal recompression is more suited to vacuum evaporation than is mechanical recompression. Jets are cheaper and easier to maintain than blowers and compressors. The chief disadvantages of thermal recompression are the low mechanical efficiency of the jets and lack of flexibility in the system toward changed operating conditions.

EVAPORATOR OPERATION AND AUXILIARIES

In this section, important evaporator auxiliaries and certain operating problems are discussed, namely, the creation of vacuum and the removal of noncondensed gases, the prevention of foam and entrainment, the removal of crystals, and the minimizing of scale.

In a complete installation a number of auxiliaries and controls are needed in addition to the evaporator bodies themselves. Most of these, including pumps, steam traps, steam-jet ejectors, and compressors for vapor recompression, are standard and have been described in Chap. 3. For example, condensate is removed from condensers and steam chests by pumps or traps. Since these must be adapted to function under partial vacuum, self-priming pumps and pressure-return traps are used.

Condensers and Gas Removal. To maintain a low absolute pressure in the vapor space of the last effect, the partial pressures of both water vapor and noncondensable gas must be minimized. The partial pressure of the water equals the vapor pressure of the boiling liquid in the last effect, and should be low to provide a large temperature drop to achieve maximum capacity. Sources of noncondensable gas are the air in the evaporator bodies at start-up, the air leakage from the outside, and the air and gas evolved from the liquid by boiling. As shown in Chap. 8, small concentrations of air in condensing steam seriously reduce the heat-transfer rate, and it is necessary that the partial pressure of noncondensed gases be kept very low by continuously purging gas from all steam chests.

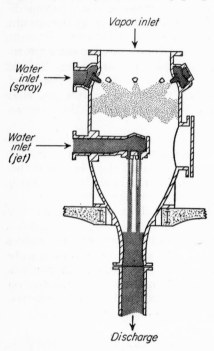

Fig. 9-18. Contact condenser. (*Schutte and Koerting Co.*)

The partial pressure of the water vapor is controlled by a condenser, to which the vapor space of the last effect is directly connected. Water is used as a cooling medium to absorb the latent heat of the vapor. When the condensate from the evaporator cannot be mixed with the cooling water, as in water distillation, a shell-and-tube condenser is used. The vapor space in the shell is carefully designed to minimize friction loss and to facilitate separation and removal of noncondensed gases. The condensate from a surface condenser is usually pumped out, but it may be removed by a barometric leg. This is a vertical tube, about 34 ft long, sealed at the bottom by a condensate-receiving tank. In operation the level of water in the leg automatically adjusts itself so that the difference in head between levels in leg and tank corresponds to the difference in pressure between the atmosphere and the vapor space in the condenser. Then water flows down the leg as fast as it is condensed without breaking the vacuum.

The vapor from most evaporators is condensed by mixing it directly with a stream of cooling water in a contact condenser, an example of which is shown in Fig. 9-18. Contact condensers are much smaller and cheaper than surface condensers. In the design shown in Fig. 9-18, part of the cooling water is sprayed into the vapor stream near the vapor inlet, and the remainder is directed into a discharge throat to complete the condensation and to eject, by the use of a venturi tube, the noncondensable gas. The pressure regain in the downstream cone of the venturi is often sufficient to eliminate the need for a barometric leg.

In either type of condenser the performance of the unit depends on the flow rate and the temperature of the cooling water. In a contact condenser, the partial pressure of the water vapor in the last effect corresponds to the temperature of the mixture of water and condensate leaving the condenser. The cooler the water and the larger the ratio of cooling water to condensate, the lower the boiling point in the evaporator. In a surface condenser the temperature of the condensate, which establishes the boiling point of the liquid, is the temperature of the outlet cooling water plus the warm-end temperature approach in the condenser. The outlet temperature is kept to a minimum if the cooling water enters at a low temperature and at a high rate. These factors are limited, however, by the amount and temperature of the available cooling water.

In a surface condenser, and in some contact condensers, gases are removed by a separate vacuum pump following the condenser. This is necessary if a condensate pump or a barometric leg is used. Steam-jet ejectors are usually used for gas removal. It should be understood that the vacuum pump does *not* establish the boiling point of the liquor in the evaporator. This is done entirely by the condenser.

In multiple-effect evaporators gases from the earlier effects may be vented directly to the condenser, or, more commonly, the steam chest of each effect is vented to the vapor space of the same effect. The cumulative gas from all effects then passes from the last effect through the condenser to the vacuum pump.

Instrumentation. Evaporators were formerly operated manually, by controlling feed pumps, liquor levels, and steam pressure. Modern practice uses instruments and obtains thereby less entrainment, less scaling and salting, and a more uniform product. The minimum instruments are a steam-flow recorder-controller, liquor-level controllers on all effects where liquor levels exist, feed-flow recorder, and absolute-pressure controllers on steam and vacuum. Other instruments, such as recording thermometers, density controllers on the thick liquor, pH recorders, and conductivity recorders on the condensates, are often used.

Foam and Entrainment. These terms, though often used together, are not synonymous. Foaming is the production of stable bubbles in the vapor space of the evaporator. It apparently arises from the formation of a liquid film having a surface tension different from that of the bulk of the liquid, followed by stabilization of this film by finely divided solids or other agents. Various methods of combating foam are used in practice. The mixture of liquid and vapor may be given a high velocity and allowed to impinge against a baffle, which breaks the foam by impact. This is the reason forced-circulation and long-tube evaporators are effective with foamy liquids. Another method is to add a very small quantity of an anti-foaming agent, such as sulfonated castor oil, octyl alcohol, or one of the silicones, which changes the surface tension of the stabilized film.

Entrainment is the actual carry-over of liquid in the vapor. Entrainment may result from foaming or from incomplete separation of the liquid droplets from the vapor. A small amount of foam may have no adverse effect on evaporator performance, but entrainment always means the loss of valuable concentrate and, in multiple-effect units, fouling of the heating surfaces in all effects but the first. Entrainment is avoided by minimizing foaming and by using vapor-liquid separators.

Entrainment Separators. Entrainment is eliminated in two ways: by reducing entrainment in the evaporator body itself and by removing entrained drops from the vapor after leaving the vapor space.

In short-tube evaporators, such as those shown in Figs. 9-1 and 9-2, the vapor space is sufficiently high and the cross section sufficiently large to ensure that most of the drops settle by gravity and return to the liquor. In forced-circulation and long-tube evaporators the velocities are too great to allow gravity separation, and baffles and deflectors are needed. These force the liquid to change its direction and flow as a sheet or curtain,

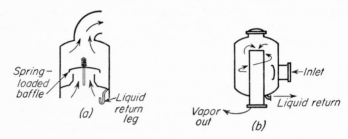

FIG. 9-19. Entrainment separators: (*a*) Deflector type. (*b*) Centrifugal type.

often at a considerable velocity, down toward the wall of the vapor space. These downward-flowing liquids are effective in reducing entrainment and also in breaking foam. A deflector type of internal separator is shown in Fig. 9-3, and a second type in Fig. 9-19*a*.

External separators are often installed in the vapor line from each effect in an evaporator. They may also be built around heating elements and condensers. The liquid recovered from these devices drains back by gravity to the vapor space. By far the most common and effective entrainment separators are the centrifugal cyclones of the type shown in Fig. 9-19*b*. The action in these has been discussed in Chap. 7. It is important in any entrainment separator that the separated liquid not be allowed to dribble back into the vapor stream and thus become reentrained. Centrifugal separators meet this requirement.

Still another way of minimizing entrainment, especially when the pressure drop in the separator must be small, is to insert in the vapor space of the evaporator a "demister" of metal wire mesh.[10]

Crystal Removal. Once a solution being concentrated becomes saturated with dissolved solids at the evaporator temperature, further evaporation results in the precipitation of crystals. The precipitated solids, unless the amount is very small, must be removed if evaporation is to continue. In a small evaporator this may be done by dumping the entire charge. Usually a magma of crystals and mother liquor is pumped continuously from the evaporator to a settling tank, from the bottom of which a thickened mixture of crystals and liquor is taken to a rotary filter or centrifuge and the mother liquor returned to the system. Crystallizing evaporators are described in Chap. 14.

Scale Formation. The solubility of most solutes increases with increasing temperature. Precipitation from solutions of materials having normal solubility curves takes place in the bulk of the fluid rather than at the tube wall. Some inorganic materials, however, including calcium sulfate, anhydrous sodium sulfate, and sodium carbonate monohydrate, have what are known as *inverted solubility curves;* i.e., the solubility of these substances decreases with increasing temperature. When such a solution is concentrated in a tubular evaporator, the solubility of the solute is at a minimum at the tube wall, where the temperature is at a maximum. Precipitation therefore takes place on the tube wall, usually with the formation of a hard, dense, strongly adhering scale. Scale formation is one of the most vexing problems of evaporator operation. Its adverse effect on capacity has already been discussed [see Eq. (9-2)].

When scaling is not too severe, tubes are cleaned by mechanical or chemical means. If the scale is soluble in water, it may be removed by periodic boilouts. Another technique with materials having inverted solubility curves is to preheat the feed by injection of live steam to a temperature well above that at which it boils in the evaporator. This precipitates fine crystals, which may be filtered from the solution before entering the evaporator. On entering the unit, the solution "flashes," cools, and becomes unsaturated. Evaporation thus takes place from an unsaturated solution without deposition of scale on the metal surface.

The forced-circulation evaporator shown in Fig. 9-3 can be effectively operated on some scaling solutions. The suppression of vaporization in the tubes tends to eliminate scaling. Crystals, amounting to 10 to 15 per cent of the circulating solution, provide surfaces for solid to deposit in preference to the surface of the tubes, and if both the temperature drop between tube and solution and the temperature rise of the solution in the tubes are kept low, the supersaturation is sufficiently small to reduce the rate of crystallization to a low point.

Severely scaling liquors may be concentrated in special evaporators. These include evaporators in which the heat-transfer surface can be flexed to break off scale by changing the internal pressure and "switching" evaporators, in which live steam is admitted first to one side of the heat-transfer surface and then to the other side to dissolve scale.

Submerged-combustion Evaporators. The most drastic method of avoiding scale formation is to eliminate all hot metal surfaces from the evaporator and to bring a gaseous heating medium into direct contact with the liquid. Thus there is no surface on which scale can be deposited; as the manufacturers put it, "Scale can't form on a bubble." A submerged-combustion evaporator utilizing this principle consists of a simple tank, often rectangular, to hold the liquid, and one or more burners set to discharge a flame beneath the liquid surface. Air and natural gas are mixed in the burner inlet and ignited by an electrically operated pilot. Elaborate controls are provided to avoid explosions. Heat is transferred by radiation and by direct contact of the bubbles of hot gas with the liquid. Very high rates of heat transfer can thus be obtained in a small volume.

It is difficult to condense the vapors from a submerged-combustion evaporator because they are diluted with a large volume of noncondensable gas. Usually the heat contained in the vapor-gas mixture is wasted. Charring and decomposition of the material being evaporated might be expected to be a problem, because of the high temperatures involved, but they have been shown to be negligible as long as the liquid completely covers the burner. Submerged combustion finds its chief applications in concentrating corrosive inorganic solutions like phosphoric acid, solutions of glauber salt, sulfite waste liquor, and other scale-forming materials.

Submerged-combustion evaporators have several serious disadvantages. The power required to force the combustion gases through the liquid is large; the entrainment losses are severe; the thermal economy is low (equivalent to that of a single effect or less); and the cost of the equipment is high if corrosion-resistant materials are needed.

SYMBOLS

A Area of heating surface, ft^2; A_1, A_2, A_3, in effects I, II, III

C Constant in Eq. (9-2)

c_p Specific heat at constant pressure, Btu/(lb)(°F); c_{pf}, of feed; c_{p0}, of solution of concentration x_0, mass fraction of solid; c_{p1}, c_{p2}, c_{p3}, of liquor from effects I, II, III

i Specific enthalpy, Btu/lb; of liquor leaving single-effect evaporator; i_c, specific enthalpy of condensate; i_f, of feed; i_s, of saturated steam; i_v, of superheated steam

N Number of effects in evaporator

p Pressure; p_s, of steam; p_1, p_2, p_3, in vapor spaces of effects I, II, III

q Rate of heat transfer, Btu/(ft^2)(hr): total heat-transfer rate in all effects; rate of heat transfer from tubes to liquor; q_s, from tubes to steam (inherently negative, by convention); q_1, q_2, q_3, rates of heat transfer in effects I, II, III

t Temperature, °F; boiling temperature in, and of liquor leaving, single-effect evaporator; t_c, temperature of condensate; t_f, of feed; t_s, of steam; t_v, of vapor; t', boiling temperature of liquor at pressure of vapor space; t_1, t_2, t_3, boiling temperature in, and of liquor leaving, effects I, II, III

U Over-all heat-transfer coefficient, Btu/(ft^2)(hr)(°F); U_0, at zero time; U_1, U_2, U_3, in effects I, II, III

w Mass flow rate, lb/hr; mass flow rate of liquor leaving single-effect evaporator; w_f, flow rate of feed to evaporator; w_s, flow rate of steam to steam chest and of condensate out of steam chest; w_1, w_2, w_3, mass flow rate of liquor leaving effects I, II, III

x Mass fraction of solid in liquor

Greek Letters

Γ Loading of condensate on tube, lb/(ft perimeter)(hr)

Δt Net temperature drop, corrected for boiling-point elevation, °F; total over-all corrected temperature drop, all effects; Δt_1, Δt_2, Δt_3, net temperature drop, effects I, II, III

θ Time, hr

λ Latent heat, Btu/lb; λ_s, heat of condensation of steam; λ_1, λ_2, λ_3, heat of vaporization in effects I, II, III

μ Viscosity, lb/ft-hr

PROBLEMS

9-1. A solution of organic colloids is to be concentrated from 20 to 65 per cent solids in a vertical-tube evaporator. The solution has a negligible elevation in boiling point, and the specific heat of the feed is 0.93. Saturated steam is available at 10 lb force/in.^{2} abs, and the pressure in the condenser is 4 in. Hg abs. The feed enters at 60°F. Over-all coefficients may be obtained from Fig. 9-10. The evaporator must evaporate 40,000 lb of water per hour. How many square feet of surface are required, and what is the steam consumption in pounds per hour?

9-2. A forced-circulation evaporator is to be designed to concentrate 50 per cent caustic soda solution to 70 per cent. A throughput of 50 tons of NaOH (100 per cent basis) per 24 hr is needed. Steam at 50 lb force/in.^{2} gauge and 95 per cent quality is available, and the condenser temperature is 100°F. Feed enters at 100°F. Condensate leaves the heating element at a temperature 20°F below the condensing temperature of the steam. Radiation is estimated to be $1\frac{1}{2}$ per cent from the heating element and 2 per cent from the vapor head, both based on the enthalpy difference between steam and condensate. Concentrated liquor leaves at the boiling temperature of the liquid in the vapor head. The over-all coefficient, based on the outside area, is expected to be 350 Btu/(ft^2)(hr)(°F). The boiling-point elevation of the 70 per cent liquor is 150°F.

Calculate: (a) the steam consumption in pounds per hour and (b) the number of tubes required if 8-ft by $\frac{7}{8}$-in. 16 BWG tubes are specified.

9-3. Boarts, Badger, and Meisenburg † show that, for a forced-circulation evaporator operating under conditions where there is negligible boiling in the tubes, the usual equation for heat transfer to nonboiling liquid in turbulent flow is applicable if a constant 0.028 is used in place of 0.023.

A forced-circulation evaporator is constructed with a single-pass vertical heater containing 100 steel tubes 10 ft by 1 in. OD by 16 BWG. The average liquid level is 2 ft above the tops of the tubes. The centrifugal circulating pump maintains a velocity of 8 ft/sec through the tubes. The evaporator is fed with salt brine of 10 per cent concentration, and produces a saturated brine containing 27.8 per cent NaCl by weight. Film condensation may be assumed on the steam side. The feed enters at its own boiling point. Steam is available at 20 lb force/in.^{2} abs and the pressure in the vapor space is 1.5 lb force/in.^{2} abs.

At what rate, in pounds per hour, may this evaporator be fed?

The boiling-point elevation of saturated brine is 12°F. The specific gravity, thermal conductivity, specific heat, and viscosity of saturated brine at the average liquid temperature are 1.20, 0.33 Btu/(ft)(hr)(°F), 0.80 Btu/(lb)(°F), and 1.0 centipoise, respectively.

9-4. An evaporator operating on a scaling liquor gives an over-all coefficient of 500 Btu/(ft^2)(hr)(°F) at the end of 1 hr and 150 at the end of 24 hr. It takes 1 hr to clean the evaporator. How often should the evaporator be cleaned to give maximum daily capacity? Use an integral number of cycles per 24 hr.

9-5. A triple-effect evaporator of the long-tube type is to be used to concentrate 35,000 gal/hr of a 17 per cent solution of dissolved solids to 38 per cent dissolved solids. The feed enters at 60°F and passes through three tube-and-shell heaters, a, b, and c, in series, and then through the three effects in order II, III, I. Heater a is heated by vapor taken from the vapor line between the third effect and the condenser, heater b with vapor from

† *Trans. AIChE*, **33**: 363 (1937).

the vapor line between the second and third effects, and heater c with vapor from the line between the first effect and the second. In each heater the warm-end temperature approach is 10°F. Other data are:

Steam to I, 230°F, dry and saturated.

Vacuum on III, 28 in., referred to a 30-in. barometer.

Condensates leave steam chests at condensing temperatures.

Boiling-point elevations, 1°F in II, 5°F in III, 15°F in I.

Coefficients, corrected for boiling-point elevation, 450 in I, 700 in II, 500 in III.

All effects have equal areas of heating surface.

Concentration solids, %	Sp. gr.	Sp. ht., Btu/(lb)(°F)
10	1.02	0.98
20	1.05	0.94
30	1.10	0.87
35	1.16	0.82
40	1.25	0.75

Calculate: (a) the steam required in pounds per hour, (b) the heating surface per effect, (c) the economy in pounds per pound of steam, (d) the latent heat to be removed in the condenser.

9-6. It is desired to concentrate 1,500,000 lb/day of a liquor containing 3 per cent solids to a concentration of 48 per cent solids. A single evaporator body suitable for this purpose costs $25,000, exclusive of pumps and other accessories common to any number of effects. Fixed charges are 30 per cent per year, and steam costs 40 cents per 1,000 lb. Assume $0.90N$ lb evaporation per pound of steam, where N is the number of effects. The evaporator is to operate 24 hr per day and 300 days per year. Labor costs are independent of the number of effects. How many effects should be used?

9-7. The capacity of a certain quintuple-effect evaporator has fallen considerably below normal. A comparison of the present pressure distribution in the system with that corresponding to normal operation is:

	Normal		Abnormal	
	Pressure, lb force/in.^{2}	Temp., °F	Pressure, lb force/in.^{2}	Temp., °F
Steam.......	45.4	275	45.4	275
Vapor space:				
I........	24.5	239	27.3	245
II........	15.9	216	19.3	226
III........	8.9	188	6.6	174
IV........	4.4	157	3.5	148
V........	1.4	113	1.4	113

The steam space of each effect is vented to the vapor space of that effect. The feed is forward and enters the first effect at 240°F.

(a) What are the possible causes of the low capacity and where are these causes located? (b) What is the present capacity, expressed as per cent of normal?

9-8. A triple-effect standard vertical-tube evaporator each effect of which has 1,500 ft² of heating surface is to be used to concentrate from 5 per cent solids to 40 per cent solids a solution possessing a negligible boiling-point elevation. Forward feed is to be used. Steam is available at 250°F, and the vacuum in the last effect corresponds to a boiling temperature of 105°F. The over-all coefficients can be taken from Fig. 9-10, all specific heats may be taken as unity, and radiation is negligible. Condensates leave at condensing temperature. The feed enters at 200°F.

Calculate: (a) the pounds of 5 per cent liquor that can be concentrated per hour, (b) the steam consumption in pounds per hour.

9-9. A triple-effect forced-circulation evaporator is to be fed with 60,000 lb/hr of 10 per cent caustic soda solution at a temperature of 180°F. The concentrated liquor is to be 50 per cent NaOH.

Saturated steam at 50 lb force/in.² abs is to be used, and the condensing temperature of vapor from the third effect is to be 100°F.

The feed order is II, III, I. Radiation and undercooling of condensate may be neglected. Over-all coefficients, corrected for boiling-point elevation, are 700 in I, 1,000 in II, and 800 in III.

Calculate: (a) the heating surface required in each effect, assuming equal surfaces in each, (b) the steam consumption in pounds per hour, (c) the steam economy in pounds evaporation per pound of steam.

REFERENCES

1. Badger, W. L., in J. H. Perry (ed.): "Chemical Engineers' Handbook," 3d ed., p. 507, McGraw-Hill Book Company, Inc., New York, 1950.
2. Badger, W. L., and P. W. Shepard: *Trans. AIChE*, **13**(I): 101 (1920).
3. Boarts, R. M., W. L. Badger, and S. J. Meisenburg: *Trans. AIChE*, **33**: 363 (1937).
4. Foust, A. S., E. M. Baker, and W. L. Badger: *Trans. AIChE*, **35**: 45 (1939).
5. Keenan, J. H., and F. G. Keyes: "Thermodynamic Properties of Steam," John Wiley & Sons, Inc., New York, 1936.
6. Lindsey, E.: *Chem. Eng.*, **60**(4): 227 (1953).
7. McAdams, W. H.: "Heat Transmission," 3d ed., p. 403, McGraw-Hill Book Company, Inc., New York, 1954.
8. McCabe, W. L.: *Trans. AIChE*, **31**: 129 (1935).
9. McCabe, W. L., and C. S. Robinson: *Ind. Eng. Chem.*, **16**: 478 (1924).
10. York, O. H.: *Chem. Eng. Progr.*, **50**: 421 (1954).

Chapter 10

MASS TRANSFER

A group of operations for separating the components of mixtures is based on the transfer of material from one homogeneous phase to another. These methods, which are covered by the term mass transfer, are differentiated from those of Chap. 7 by the fact that they utilize differences in vapor pressure or solubility rather than variations in density or particle size. The techniques discussed in this and in succeeding chapters are gas absorption, distillation, leaching, liquid extraction, crystallization, air-water contacts, and drying. In this chapter these operations are introduced, some important types of equipment used in conducting them are described, and the principles common to all of them are discussed. Chapters 11 to 16 are devoted to the separate operations and describe the individual characteristics of each.

TYPICAL EXAMPLES OF MASS TRANSFER

Gas Absorption. In gas absorption, a soluble vapor is absorbed, by means of a liquid in which the solute gas is more or less soluble, from its mixture with an inert gas. The washing of ammonia from a mixture of ammonia and air by means of liquid water is a typical example. The solute is subsequently recovered from the liquid by distillation, and the absorbing liquid can be either discarded or reused.

Packed Towers. A common apparatus used in gas absorption, and also in certain other operations, is the packed tower, an example of which is shown in Fig. 10-1. The device consists of a cylindrical column, or tower, equipped with a gas inlet and distributing space at the bottom; a liquid inlet and distributor at the top; liquid and gas outlets at the bottom and top, respectively; and a supported mass of inert solid shapes. The solid shapes are called tower packing or tower filling. The inlet liquid, which may be pure solvent or a dilute solution of solute in the solvent and which is called the weak liquor, is distributed over the top of the packing by the distributor and, in ideal operation, uniformly wets the surfaces of the packing. The solute-containing gas, or rich gas, enters the distributing space below the packing and flows upward through the interstices in the packing countercurrent to the flow of the liquid. The packing provides a large area of contact between the liquid and gas and encourages intimate contact between the phases. The solute in the rich gas is absorbed by the fresh liquid entering the tower, and dilute, or lean, gas leaves the top. The liquid is enriched in solute as it flows down the tower, and concentrated liquid, called strong liquor, leaves the bottom of the tower through the liquid outlet.

Many kinds of tower packing have been invented, and several types are in common use. Perhaps the most common type is the Raschig ring, which

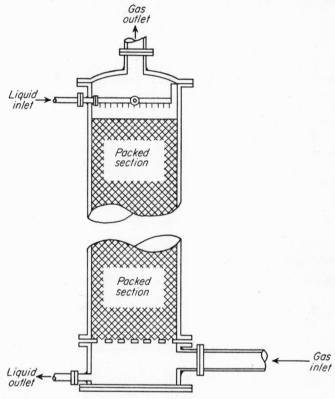

FIG. 10-1. Packed tower.

is a thin-walled cylinder having a diameter equal to the height. A mass of identical rings is dumped at random into the tower, and the rings fill the space above the packing support, which usually is a perforated plate. Details of other tower packings are given in Chap. 11.

Except for the relatively small pressure drop due to fluid friction, the pressure in a packed tower is constant. Absorption of solute by the liquid is accompanied by the evolution of the heats of condensation and of solution of the solute that is absorbed. The absorption rate is usually a maximum at the bottom of the tower, and if the inlet gas is rich, the heat released in that section of the tower causes the temperature to rise appreciably above that in most of the apparatus. Since high temperatures tend to undo the absorption by driving the solute out of solution, external cooling is sometimes provided at the bottom of the tower. More commonly, the entering gas is relatively lean, the heat effect is small, and the temperature in the entire tower is nearly constant.

Distillation. The function of distillation is to separate, by vaporization, a liquid mixture of miscible and volatile substances into individual components or, in some cases, into groups of components. The separation

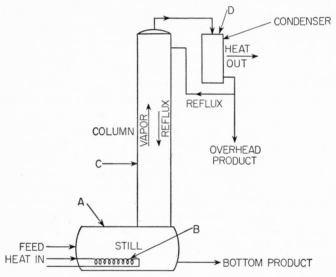

Fig. 10-2. Still with fractionating column: *A*, still. *B*, heating coil. *C*, column. *D*, condenser.

of a mixture of alcohol and water into its components; of liquid air into nitrogen, oxygen, and argon; and of crude petroleum into gasoline, kerosene, fuel oil, and lubricating stock are examples of distillation.

In distillation it is necessary that the vapor formed by boiling the liquid mixture differ in composition from that of the boiling liquid. The component that tends to concentrate in the vapor is called the more volatile component or the low boiler, because that component usually has the lower boiling temperature when pure. The other component is called the less volatile component or the high boiler. In some mixtures, called *azeotropes*, the compositions of the vapor and liquid are the same. Such mixtures act as pure substances in distillation. A mixture of 95 per cent ethanol and 5 per cent water is a typical azeotrope.

A plant for continuous distillation is shown in Fig. 10-2. The still *A* is fed continuously with the liquid mixture to be distilled. The liquid is converted partially into vapor by heat transferred from the heating surface *B*. The vapor formed in the still is richer in low boiler than is the unvaporized liquid, but unless the two components differ greatly in volatility, the vapor contains substantial quantities of both components, and if it were condensed, the condensate would be far from pure. To increase the concentration of the low boiler in the vapor and to remove the high boiler from this stream, the vapor stream from the still is brought into intimate countercurrent contact with a descending stream of boiling liquid in the column or tower C. A packed tower can be used for the purpose. A liquid stream, usually concentrated in low boiler, is introduced at the top of the column. This returned liquid is called *reflux*.

The reflux entering the top of the column is, if cold, immediately heated to its boiling point by the vapor, and throughout the column liquid and vapor are at their boiling and condensing temperatures, respectively. To obtain an increase in the concentration of the low boiler in the vapor, it is necessary that the reflux be richer in low boiler than the equilibrium concentration corresponding to the vapor leaving the still. Then, at all levels in the column, some low boiler spontaneously diffuses from the liquid into the vapor, vaporizing as it passes from one phase to the other. The heat of vaporization of this low boiler is supplied by an equal amount of heat of condensation of high boiler. High boiler diffuses spontaneously from vapor to liquid. The net effect, then, is to transfer high boiler from vapor to liquid and to transfer a thermally equivalent amount of low boiler from liquid to vapor. As the vapor rises in the column, it becomes enriched in low boiler, or more volatile component. As the liquid descends the column, its content of high boiler increases. The flow of the bulk of the low boiler is up the column, and that of the high boiler is down the column. Because of the two-way transfer of material between phases, the change in total quantity of vapor is not great.

The enrichment of the vapor stream as it passes through the column in contact with reflux is called *rectification*. It is immaterial where the reflux originates provided its concentration in low boiler is sufficiently great to give the desired product. The usual source of reflux is the condensate leaving condenser D. Part of the condensate is withdrawn as the product, and the remainder returned to the top of the column. Then the vapor reaching the condenser can be brought as close to complete purity as desired by using a tall tower and a large reflux.

In distillation at atmospheric pressure or above, the pressure losses from friction are relatively small in comparison with the pressure in the unit, so that the pressures in the still, column, and condenser are about the same. In distilling heat-sensitive materials, low pressures are necessary. At absolute pressures of the order of magnitude of several millimeters of mercury, ordinary distillation methods, with some modification in detail, are used, but the pressure drop from friction is an appreciable fraction of the pressure in the still, and the pressure gradient is important. For distillations under high vacuum, at pressures of the order of magnitude of fractions of a millimeter, the usual methods of distillation and rectification become inoperative, because the distances traveled by individual molecules before striking other molecules become so large that the mechanism of vaporization changes radically. Distillations of this type are called molecular distillations. They are not discussed in this book.

From the still, liquid is withdrawn which contains practically all the high boiler, as almost none of this component escapes with the overhead

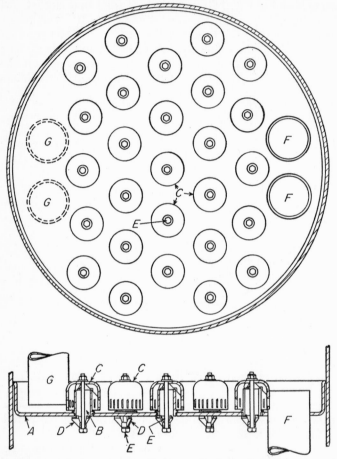

FIG. 10-3. Bubble-cap plate: *A*, tray, or plate. *B*, riser. *C*, bubble caps. *D*, spider. *E*, hold-down bolt. *F*, down pipe to plate below. *G*, down pipe from plate above.

product unless that product is an azeotrope. The liquid from the still, which is called the bottom product or bottoms, is not nearly pure, however, as there is no provision in the equipment of Fig. 10-2 for rectifying this stream. A method for obtaining nearly pure bottom product by rectification is described in Chap. 12.

The Bubble-cap Column. Although a packed tower can be used as a rectifying column, and sometimes is, a more effective and popular device is the bubble-cap column. The column contains a number of bubble-cap plates, or trays, mounted in a stack inside the tower. An example of a single plate is shown in Fig. 10-3. The plate consists of a horizontal tray A carrying a down pipe F, the top of which acts as a weir, and a number of bubble caps C. The down pipe G from the tray above reaches nearly to tray A. Each bubble cap consists of a riser B, a cap C, a spider D, and a hold-down bolt and nut E. The periphery of the cap is slotted. The tops of the slots are slightly below the tops of the risers B. The height of the weir is slightly less than the height of the risers. The construction of the bubble trays leads to the following flow of liquid and vapor. The liquid flows from plate to plate down the column, passing through the down pipes G and F and across the plates. The weir maintains a minimum depth of liquid on the tray, nearly independent of the rate of flow of liquid, and the risers are sufficiently high to prevent liquid flow through them. The vapor flows upward from tray to tray. It enters through the risers, is diverted downward by the caps, and flows out through the slots. Since the slots are completely submerged in the liquid on the plate, the vapor is subdivided by the slots into many small bubbles and passes in intimate contact through the pool of liquid on the tray. Because of the action of the vapor bubbles, the liquid is actually a boiling, frothy mass. Above the froth and below the next tray is fog from collapsing bubbles. This fog for the most part settles back into the liquid, but some is entrained by the vapor and carried to the plate above. Bubble columns are representative of an entire class of equipment called plate columns.

Extraction of Solids, or Leaching. In leaching, soluble material is dissolved from its mixture with an inert solid by means of a liquid solvent.

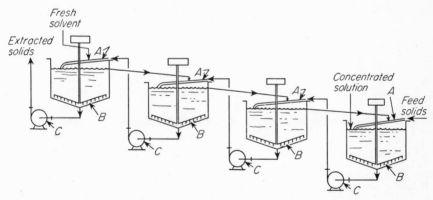

FIG. 10-4. Countercurrent leaching plant: A, launder. B, rake. C, slurry pump.

The dissolved material, or solute, may then be recovered by crystallization or evaporation.

Here, again, the countercurrent method is most effective. A diagrammatic flow sheet of a typical leaching plant is shown in Fig. 10-4. It consists of a series of units, in each of which the solid from the previous unit is mixed with the liquid from the succeeding unit, and the mixture allowed to settle. The solid is then transferred to the next succeeding unit, and the liquid to the previous unit. As the liquid flows from unit to unit, it becomes enriched in solute, and as the solid flows from unit to unit in the reverse direction, it becomes impoverished in solute. The solid discharged from one end of the system is well extracted, and the solution leaving at the other end is strong in solute. The thoroughness of the extraction depends on the amount of solvent and the number of units. In principle, the unextracted solute can be reduced to any desired amount if enough solvent and a sufficient number of units are used.

Any suitable mixer and settler can be chosen for the individual units in a countercurrent leaching system. Those shown in Fig. 10-4 are thickeners of the type shown in Fig. 7-36. Mixing occurs in launders A and in the tops of the tanks, rakes B move solids to the discharge, and slurry pumps C move slurry from tank to tank.

Liquid Extraction. In liquid extraction, a mixture of two components is treated by a solvent that preferentially dissolves one of the components in the mixture. The mixture under treatment is called the raffinate and the solvent-rich phase is called the extract. The component transferred from raffinate to extract is the solute, and the component left in the raffinate is the diluent. The solvent in the extract leaving the extractor is usually recovered and reused.

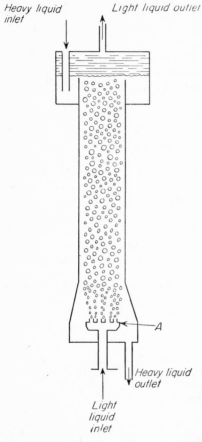

FIG. 10-5. Spray tower. A, nozzle to distribute light liquid. (*Elgin.*[5])

Several kinds of equipment are available for liquid extraction. A series of mixers and settlers arranged to give countercurrent flow can be used. These act in the same manner as the corresponding devices in leaching. Vertical countercurrent towers are also popular, as the lighter of the liquids rises in such a column through the heavier liquid, which is withdrawn at the bottom. Packed towers, with filling especially designed for the purpose, are common. Another device for obtaining countercurrent contact is the spray tower, which is quite suitable for liquid extraction. An example of spray tower is shown in Fig. 10-5. The lighter liquid is introduced at the bottom and distributed as small drops by the nozzles A. The drops of light liquid rise through the mass of heavier liquid, which flows downward as a continuous stream. The drops are collected at the top and form the stream of light liquid leaving the top of the tower. The heavy liquid leaves the bottom of the tower. In Fig. 10-5, light phase is dispersed and heavy phase is continuous. This may be reversed, and the heavy stream sprayed into the light phase at the top of the column, to fall as dispersed phase through a continuous stream of light liquid. Other types of extraction equipment are described in Chap. 13.

Either the raffinate or the extract phase may be the light stream, so the extract may leave at either end of a vertical extractor, and the raffinate then leaves at the other.

The temperature and pressure in a liquid extractor are nearly constant. The only pressure variations are those from friction and gravity. Also, all phases are liquid, and the properties of the liquids are nearly independent of pressure, so pressure effects are important only in controlling the flows of the liquids. Heat effects from transferring solute from phase to phase are small, and the temperature does not vary appreciably in the extractor.

Crystallization. The objective of crystallization is to separate a crystalline solid from its solution. All methods of crystallization depend on creating supersaturation in the solution containing the solute to be crystallized so crystals will form and grow. After growth, the crystals are separated from the residual liquid, which is called the mother liquor, by settling, filtering, or centrifuging.

If the solubility of the solute is considerably higher in the hot solvent than in cold solvent, crystallization can be accomplished by cooling the

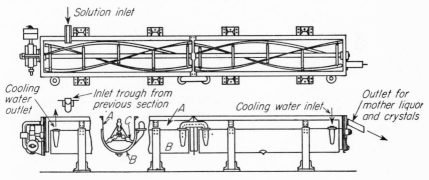

FIG. 10-6. Swenson-Walker crystallizer: A, trough. B, jacket. C, stirrer.

concentrated solution in a special type of heat exchanger. If the decrease in solubility with decrease in temperature is small, or if the solute has an inverted solubility curve, supersaturation is created by evaporation.

A typical example of cooling crystallizer is the Swenson-Walker type, shown in Fig. 10-6. It consists of an open trough A, usually 24 in. wide, with a semicylindrical bottom. A water jacket B is welded to the outside of the trough, and a slow-speed long-pitch helical agitator C is set close to the bottom of the trough. The apparatus is ordinarily built in units 10 ft long; for increased capacity several units are connected in series, 40 ft being the maximum length usually driven from one shaft. If lengths greater than this are needed, several crystallizers are arranged one above the other, with the solution cascading from one crystallizer to the other.

Hot concentrated solution to be crystallized is fed in at one end of the trough, with cooling water flowing in the jacket countercurrent to the solution. When conditions are properly adjusted, crystal nuclei begin to form a short distance from the point where feed is introduced, nuclei which grow steadily as the solution passes down the crystallizer. The primary purpose of the helical stirrer is not to agitate or convey the crystals; it is to prevent an accumulation of crystals on the cooling surface and to lift the crystals that have been formed and shower them down through the solution. In this way the crystals grow while they are freely suspended in the liquid and therefore are usually well-formed individuals, reasonably uniform in size, and free from inclusions or aggregations.

At the end of the trough there may be an overflow weir, where crystals and mother liquor overflow together to a draining table or drain box, from which mother liquor is returned to process and the crystals are raked to a filter or centrifuge. In other designs a short inclined screw conveyor lifts crystals out of the solution and delivers them to the filtering device, while the mother liquor overflows from the trough at a different point.

Other cooling crystallizers and various types of evaporative crystallizers are discussed in Chap. 14.

Air-Water-contact Operations. Several methods of air conditioning and water cooling are carried out by bringing a stream of water into intimate contact with a stream of air. These processes are not used for purposes of separation, but their mechanisms are analogous to those of absorption and distillation.

Air-water-contact operations are conducted for such purposes as cooling air or water and humidifying or dehumidifying air. The air so treated may be used in such processes as drying or for the control of temperature and humidity in air conditioning. Air-water processes and the equipment for conducting them are described in Chap. 15. One example, the cooling of water by direct contact with unsaturated air, is considered here.

A typical natural-draft cooling tower is shown in Fig. 10-7. The purpose of a cooling tower is to conserve cooling water by allowing the cooled water to be reused many times. Warm water, usually from a condenser or other heat-transfer unit, is admitted to the top of the tower and distributed by troughs and overflows to cascade down over slat gratings, which provide large areas of contact between air and water. Flow of air up through the tower is induced by the buoyancy of the warm air in the tower. A cooling tower is, in principle, a special type of packed tower. The usual packing material is cypress wood, which is the most economical tower filling that withstands the combined action of wind and water. In the tower, part of the water evaporates into the air, and sensible heat is transferred from the warm water to the cooler air. Both processes reduce the temperature of the water. Only make-up water, to replace that lost by evaporation and windage loss, is required to maintain the water balance.

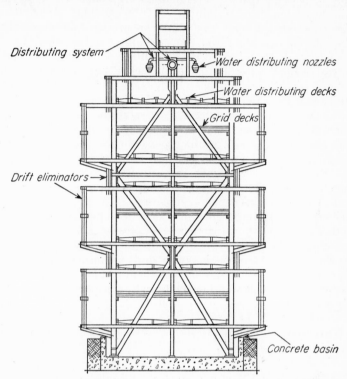

Distributing system

Water distributing nozzles

Water distributing decks

Grid decks

Drift eliminators

Concrete basin

FIG. 10-7. Natural-draft cooling tower. (*Fluor Products Co.*)

Drying. The objective of drying is to remove liquid from a mass of solid. Many methods are available. For example, if the substance is very wet, liquid may be drained, sucked, or squeezed from the solid. The type of drying to be considered in this text is that done by vaporizing water from the substance by applying heat to the material. The solid may be in lumps, cakes, sheets, free-flowing crystals or other particles, or other shapes. Concentrated solutions and slurries may be reduced to dry powders. In several types of dryer, the heat is applied directly to the solid. Common types of dryer are heated by hot air or flue gas, which may either come into direct contact with the solid or which may be separated from the solid by a heat-transfer surface.

A typical example of a countercurrent direct-contact air-heated dryer is shown in Fig. 10-8. A rotating shell *A* made of sheet steel is supported on two sets of rollers *B* and driven by a gear and pinion *C*. At the upper end is a hood *D*, which connects through fan *E* to a stack, and a spout *F*, which brings in wet material from the feed hopper. Flights *G*, which lift the material being dried and shower it down through the current of hot air, are welded inside the shell. At the lower end the dried product discharges into a screw conveyor H. Just beyond the screw conveyor is a set of steam-heated extended-surface pipes which preheat the air. The air is moved through the dryer by a fan, which may, if desired, discharge into the air heater so that the whole system is under a positive pressure. Alternatively, the fan may be placed in the stack as shown, so that it draws the air through the dryer and keeps the system under a slight vacuum. This is desirable when the material tends to dust. Rotary dryers of this kind are widely

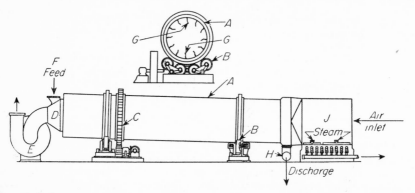

Fig. 10-8. Countercurrent air-heated rotary dryer: A, dryer shell. B, shell-supporting rolls. C, drive gear. D, discharge hood. E, discharge fan. F, feed chute. G, lifting flights. H, product discharge. J, air heater.

used for salt, sugar, and all kinds of granular and crystalline materials which must be kept clean and which may not be directly exposed to very hot flue gas.

Because of the wide variety in the character, moisture content, heat sensitivity, shape, and size of solids that must be dried, the variety of dryers is large. The type shown in Fig. 10-8 is but one of many. Other representative dryers are described in Chap. 16.

In the mass-transfer operations discussed in this text, two streams are in direct contact, one flowing past the other. Usually the flow is counter-current, as shown in Figs. 10-1, 10-2, 10-4, 10-5, 10-7, and 10-8. Parallel flow, as shown in Fig. 10-6, is sometimes used, especially in crystallizers and dryers.

PRINCIPLES OF MASS TRANSFER

The quantitative treatment of mass transfer utilizes material and energy balances, equilibria, and rates of heat and mass transfer. Those portions of these scientific principles that are generally applicable are discussed in this chapter. More specialized topics are treated in the chapters on the individual operations.

Terminology and Symbols. Although the streams may be solid, liquid, or gas, in each process one stream always has a density greater than that of the other. It is convenient to refer generally to the two streams in any one process as the light phase and the heavy phase. Table 10-1 shows how these terms are used in the various operations.

TABLE 10-1. TERMINOLOGY FOR STREAMS IN MASS-TRANSFER OPERATIONS

Operation	Light phase	Heavy phase
Gas absorption.........	Gas	Liquid
Distillation.............	Vapor	Liquid
Leaching...............	Liquid	Solid
Liquid extraction.......	Light liquid	Heavy liquid
Crystallization..........	Mother liquor	Crystals
Air-water contacts......	Air	Water
Drying................	Gas (usually air)	Solid

Note on Concentrations. Strictly, concentration means mass per unit volume, and concentrations are expressed in either pound moles or pounds per cubic foot. It is convenient, however, to extend the use of the word "concentration" to include mole or mass fractions. The relation between concentration and mole or mass fraction is

$$c = \rho x \qquad (10\text{-}1)$$

when x is the mole fraction of a given component, ρ is the molal density of the mixture, in pound moles per cubic foot, and c is then the concentration of that component, also in pound moles per cubic foot. When mass units are chosen, x is the mass fraction, and ρ and c are in pounds per cubic foot. In the following treatment pound molal units are assumed, and the molal density is denoted by ρ_M.

General symbols are needed for flow rates and concentrations. For all operations, use V and L for the flow rates of light and heavy phases, respectively. Use A, B, C, etc., to refer to the individual components. If only one component is transferred between phases, choose component A as that component. Use x for the concentration of a component in the heavy phase, and y for the concentration in the light phase. Thus, y_A is the concentration of component A in a light phase and x_B is that of component B in a heavy phase. When only two components are present in a phase, the concentration of component A is x or y, and that of component B is $1 - y$ or $1 - x$, and the subscripts A and B are unnecessary. Although ordinarily molal units are used, it is sometimes more convenient to use pound units.

Terminal Quantities. Since in mass transfer there are two streams and each must enter and leave, there are four terminal quantities. To identify them, use subscript a to refer to that end of the process where the heavy phase enters and b to refer to that end where the heavy phase leaves. Then, for *countercurrent* flow, the terminal quantities are as shown in Table 10-2.

TABLE 10-2. TERMINAL QUANTITIES FOR COUNTERCURRENT FLOW

Stream	Flow rate	Concentration component A
Heavy, entering plant................	L_a	x_{Aa}
Heavy, leaving plant................	L_b	x_{Ab}
Light, entering plant................	V_b	y_{Ab}
Light, leaving plant................	V_a	y_{Aa}

If there are only two components in a stream, the subscript A can be dropped from the concentration terms.

Equilibria

A limit to mass transfer is reached if the two phases come to equilibrium and the net transfer of material ceases. For a practicable process, which must have a reasonable production rate, equilibrium must be avoided, as the rate of mass transfer at any point is proportional to the driving force, which is the departure from equilibrium at that point. To evaluate driving forces, a knowledge of equilibria between phases is therefore of basic importance. Several kinds of equilibria are important in mass transfer. In all situations, two phases are involved, and all combinations are found except two gas phases or two solid phases. The controlling variables are temperature, pressure, and concentrations. The number of independent concentrations in any one phase is one less than the number of components in the phase. Equilibrium data can be shown in tables, equations, or graphs. For all processes considered in this text, the pertinent equilibrium relationship can be shown graphically.

Classification of Equilibria. To classify equilibria and to establish the number of independent variables or degrees of freedom available in a specific situation, the phase rule is useful. It is

$$\phi = C - P + 2 \tag{10-2}$$

where ϕ = number of degrees of freedom
    C = number of components
    P = number of phases
The equilibria used in mass transfer are analyzed in terms of the phase rule in the following paragraphs. Except in leaching and drying, there are two phases, so $\phi = C$.

Air-Water Equilibrium. The solubility of the air in the water may be neglected. There are two components, air and water, so $\phi = C = 2$. One phase, the water, is pure, and the variables are temperature, pressure, and concentration of water vapor in the light phase. If the pressure is held constant, either the temperature or the vapor-phase concentration may be varied independently, and the value of the other variable follows, so a unique plot of temperature vs. concentration provides an equilibrium relationship. The equilibrium relation between water and air at 1 atm is shown in Fig. 10-9*a*.

Crystallization. There are two components, solvent and solute, and $\phi = C = 2$. The solid phase is pure, and variables are concentration, temperature, and pressure. Fixing one, the pressure, leaves either temperature or concentration as an independent variable. The relation between temperature and concentration is the usual solubility curve.

Distillation; Equilibrium Curve. Assume there are two components, so $\phi = 2$. Both components are found in both phases. There are four variables: pressure, temperature, and the concentrations of component A in the liquid and vapor phases. If the pressure is fixed, only one variable, e.g., liquid-phase concentration, can be changed independently, and both temperature and vapor-phase concentration then follow. The vapor-phase concentration, denoted by y_e, is plotted against the liquid-phase concentration, denoted by x_e. This plot is called the equilibrium curve. A different temperature applies to each point on the curve. An equilibrium curve for the system acetone-ethanol is shown in Fig. 10-9*b*.

If there are more than two components, the equilibrium relationship cannot be represented by a single curve.

Gas Absorption. Assume that only one component is transferred between phases. There are three components and $\phi = 3$. Neglect both the solubility of the inert gas in the liquid and the presence of vapor from the liquid in the gas. Then there are four variables: pressure, temperature, and the concentrations of component A in liquid and gas. The temperature and pressure may be fixed; one concentration remains as an independent variable, which may be varied. The other concentration follows, and an equilibrium curve y_e vs. x_e can be plotted. All points on this curve pertain to the same pressure and temperature. An equilibrium curve for sulfur dioxide–air–water at 1 atm and 30°C is shown in Fig. 10-9*c*.

Liquid Extraction. The number of components is three, so $\phi = 3$. All three components may appear in both phases. The variables are temperature, pressure, and four concentrations. Temperature and pressure may be taken as constant, and one concentration chosen as an independent variable. The relations between that variable and the remaining three are given by special graphical methods described in Chap. 13.

Leaching. Two situations are found in leaching. In the first, the solvent available is more than sufficient to dissolve all the solute, and, at equilibrium, all the solute is in solution. There are, then, two phases, the solid

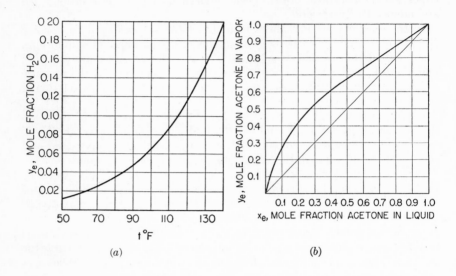

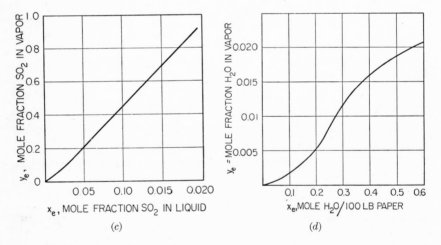

Fig. 10-9. Typical equilibrium relations: (*a*) Air-water at 1 atm. (*b*) Acetone-ethanol at 1 atm. (*c*) Sulfur dioxide–air–water at 1 atm. and 30°C. (*d*) Air–water–paper pulp at 1 atm. and 25°C.

and the solution. The number of components is 3, and $\phi = 3$. The variables are temperature, pressure, and concentration of the solution. All are independently variable.

In the second situation, the solvent available is insufficient to dissolve all the solute, and the excess solute remains as a solid phase at equilibrium. Then the number of phases is 3, and $\phi = 2$. The variables are pressure, temperature, and concentration of the saturated solution. If the pressure is fixed, the concentration depends on the temperature. This relation is the ordinary solubility curve.

Drying. Free liquid water may or may not be present. If it is, there are three phases—vapor, solid, and liquid—and three components, so $\phi = 2$. At constant pressure a unique relation exists between temperature and concentration of water in the vapor, just as in air-water contacts.

The water in hygroscopic solids or in natural materials like wood or leather may be in a loose combination with the solid, with no liquid water present. Then there are two phases and three components, and $\phi = 3$. The variables are temperature, pressure, and the concentrations of water in vapor and solid. If temperature and pressure are fixed, these concentrations can be plotted on an equilibrium curve. The equilibrium curve for air, water, and paper pulp at 1 atm and 25°C is shown in Fig. 10-9d.

Vapor-Liquid Equilibria. In any equilibrium between vapor and liquid, the concept of partial pressure may be used. As shown in Chap. 1, $\bar{p}_A$, the partial pressure of component A in a gas mixture, is related to y, the mole fraction of the component by the equation

$$y_A = \frac{\bar{p}_A}{p} \tag{10-3}$$

where p is the total pressure. In this equation molal units must be used. For a two-component mixture

$$y = \frac{\bar{p}_A}{p} \qquad 1 - y = \frac{\bar{p}_B}{p} \tag{10-4}$$

By definition of partial pressure, the sum of the partial pressures equals the total pressure. Pressures and partial pressures may be expressed in any convenient unit, e.g., in atmospheres.

When a liquid phase having a definite analysis is in equilibrium with a vapor phase at a fixed pressure and temperature, the analysis of the vapor phase is also fixed. From the vapor-phase composition the partial pressures of the vapor-phase components can be calculated by Eq. (10-3). In general, the partial pressure of any one component, such as component A, depends on the temperature, the pressure, and the entire composition of the liquid. For a two-component mixture the composition of the liquid is completely specified by x_e, which is the mole fraction of component A in the liquid at equilibrium. Then $\bar{p}_A = \psi_A(p,t,x_e)$. If the gas phase can be considered an ideal gas, and if the effect of pressure on the properties of the liquid phase is neglected (this assumption is permissible over moderate pressure ranges), the partial pressure is independent of total pressure and depends on temperature and liquid concentration. Then, at constant temperature,

$$\bar{p}_A = \psi_A(x_e) \tag{10-5}$$

The function denoted by $\psi(x_e)$ varies with temperature.

If the liquid phase consists of only one component, e.g., component A, the partial pressure of A equals the vapor pressure of pure A, and

$$\bar{p}_A = p_A$$

where p_A is the vapor pressure of pure A at the temperature of the system.

The usual way of showing the relation of Eq. (10-5) is by a plot of $\bar{p}_A$ vs. x_e or $\bar{p}_B$ vs. either x_e or $1 - x_e$. Such a plot is obtained by analyzing the two phases when in equilibrium and calculating $\bar{p}_A$ or $\bar{p}_B$ by Eq. (10-3) or (10-4). A typical plot of $\bar{p}_A$ and $\bar{p}_B$ vs. x_e is shown in Fig. 10-10, which applies to the system ethyl ether–acetone at 20°C.

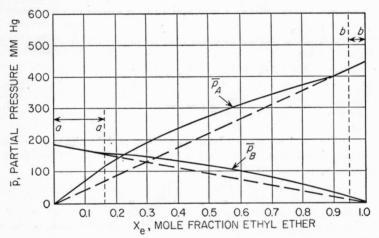

FIG. 10-10. Partial pressure vs. concentration plot, ethyl ether–acetone at 20°C.

Henry's Law. At low concentrations, the graph of $\bar{p}_A$ vs. x_e passes through the origin and is straight for a short distance before it begins to curve. This is shown by the line for ethyl ether from $x_e = 0$ to $x_e \approx 0.15$. Except for substances such as electrolytes, which dissociate in solution, this behavior is general and is a basic characteristic of substances in dilute solutions. The partial pressure is given by the relation

$$\bar{p}_A = H_A x_e \tag{10-6}$$

where H_A is a constant depending on temperature, on the solvent, and, to a minor degree, on pressure. Equation (10-6) is the mathematical expression of Henry's law. Factor H_A is the Henry's law constant. Its value, and the concentration range over which it remains constant, can be found only by experiment.

Raoult's Law. When the solution is nearly pure A, the partial pressure of component A is proportional to its concentration, or

$$\bar{p}_A = K x_e$$

where K is a constant. Since when $x_e = 1.0$, $p_A = \bar{p}_A$, $K = p_A$, and

$$\bar{p}_A = p_A x_e \tag{10-7}$$

The same rule, written for component B, is

$$\bar{p}_B = p_B(1 - x_e) \tag{10-8}$$

Equations (10-7) and (10-8) are mathematical expressions of Raoult's law, which states that the partial pressure of a component over a solution is the product of the vapor pressure of that component and the mole fraction of the component. The dotted lines in Fig. 10-10 represent Raoult's law. It is a corollary of thermodynamics that, if Henry's law applies to one component over a segment of the concentration axis, Raoult's law applies to the other component over the same segment. Thus, in Fig. 10-10 Henry's law applies to the acetone over range bb and to the ether over range aa. Raoult's law applies to acetone over range aa and to ether over range bb.

Ideal Solutions. In some mixtures, Raoult's law applies to each component over the entire concentration range from 0 to 1.0. Such mixtures are called ideal. The partial-pressure–concentration lines are straight, like the dotted lines in Fig. 10-10. Henry's and Raoult's laws are identical for ideal solutions, and the Henry's law constant for a given component is the vapor pressure of that component.

Actually, few solutions are ideal. Mixtures of isotopes are ideal. Mixtures of nonpolar molecules of the same types and of nearly the same size, such as members of some homologous series of hydrocarbons, follow the ideal law closely. Thus, mixtures of benzene, toluene, and xylene and mixtures of the lower paraffins follow Raoult's law at ordinary temperatures. Mixtures of polar substances such as water, alcohol, and electrolytes depart greatly from the ideal-solution law. If one component of a two-component mixture follows the ideal-solution law, so does the other.

The partial-pressure form of the ideal-solution law depends on the assumption that the vapor phase is an ideal gas. Exact thermodynamic relations are available for treating all vapor-liquid equilibria, but they are beyond the scope of this book.

Equilibrium Curves. The use of partial pressures in vapor-liquid equilibria is a means to an end. It is usually more convenient to use equilibrium relationships between y_e, the mole fraction of component A in the vapor, and x_e, the mole fraction of A in the liquid. To convert a partial pressure to a vapor-phase mole fraction it is necessary only to divide by p, the total pressure. Thus Eq. (10-5) becomes

$$y_e = \frac{\bar{p}_A}{p} = \frac{\psi_A(x_e)}{p} \tag{10-9}$$

Henry's law, for components A and B, becomes

$$y_e = \frac{H_A x_e}{p} \qquad 1 - y_e = \frac{H_B(1 - x_e)}{p} \tag{10-10}$$

Raoult's law is

$$y_e = \frac{p_A}{p} x_e \qquad 1 - y_e = \frac{p_B}{p}(1 - x_e) \tag{10-11}$$

For ideal solutions,

$$p_A x_e + p_B(1 - x_e) = p \tag{10-12}$$

Basic Types of Plant: Stage Contact vs. Differential Contact. Inspection of the various countercurrent arrangements shown in Figs. 10-1 to 10-8 shows that two general types of contact devices are used. In the first, represented by the bubble-cap tower (Fig. 10-3) and the countercurrent leaching plant (Fig. 10-4), the contact equipment consists of a series of interconnected individual contact units, or stages, and the study of the assembly as a whole is best made by focusing attention on the streams passing between the individual stages. In the second type, represented by the packed tower (Fig. 10-1), the spray tower (Fig. 10-5), and the countercurrent dryer (Fig. 10-8), the two streams pass each other without the discontinuities characteristic of contact stages. There are no locations at which the equipment is physically separated into units of finite size, as are the contact units, and a differential length or height is the natural basis for analysis.

Stage-contact Plants

An individual unit in a stage-contact plant receives two streams, one light and one heavy, from the two units adjacent to it, brings them into close contact, and delivers heavy and light streams, respectively, to the same adjacent units. The fact that the contact units may be arranged either one above the other, as in the bubble-cap column, or side by side, as in a stage leaching plant, is important mechanically and may affect some of the details of operation of individual stages. The same material-balance equations, however, may be used for either arrangement. A series of contact units of either type is called a cascade.

Terminology for Stage-contact Plants. The individual contact units in a cascade are numbered serially, starting from one end. In this book, the stages are numbered in the direction of flow of the heavy stream, and the last stage is that discharging the heavy stream. A general stage in the system is the nth stage, which is number n counting from the entrance of the heavy phase. The stage immediately ahead of stage n in the sequence is the $(n-1)$st stage, and that immediately following it is the $(n+1)$st stage. Using a plate column as an example, Fig. 10-11 shows how the units in a stage plant are numbered. The total number of stages is N, and the last stage in the plant is therefore the Nth stage.

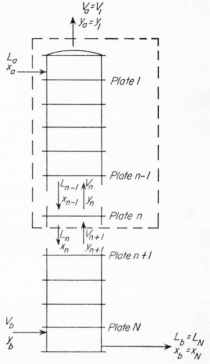

To designate the streams and concentrations pertaining to any one stage, all streams originating in that stage carry the number of the unit as a subscript. Thus, y_{n+1} is the mole fraction of component A in the light stream leaving the $(n+1)$st unit, and L_n is the molal flow rate of the heavy stream leaving the nth unit. The streams entering and leaving the plant and those entering and leaving stage n in a plate tower are shown in Fig. 10-11. Quantities V_a, L_b, y_a, and x_b in Table 10-2 are equal to V_1, L_N, y_1, and x_N, respectively. This may be seen by reference to Fig. 10-11.

Material Balances for Stage-contact Plants. Consider the portion of the cascade that includes stages 1 through n, as shown by the section enclosed by the dotted line

Fig. 10-11. Material-balance diagram for plate column.

in Fig. 10-11. The total input of material to this section is $L_a + V_{n+1}$ moles/hr, and the total output is $L_n + V_a$ moles/hr. Since, under steady flow, there is neither accumulation nor depletion, the input and the output are equal, and

$$L_a + V_{n+1} = L_n + V_a \qquad (10\text{-}13)$$

Equation (10-13) is a total material balance. Another balance can be written by equating input to output for component A. Since the number of moles of this component in a stream is the product of the flow rate and the mole fraction of A in the stream, the input of component A to the section under study is $L_a x_a + V_{n+1} y_{n+1}$ moles/hr, the output is $L_n x_n + V_a y_a$ moles/hr, and

$$L_a x_a + V_{n+1} y_{n+1} = L_n x_n + V_a y_a \qquad (10\text{-}14)$$

A material balance can also be written for component B, but such an equation is not independent of Eqs. (10-13) and (10-14), since if Eq. (10-14) is subtracted from Eq. (10-13), the result is the material-balance equation for component B. Equations (10-13) and (10-14) yield all the information that can be obtained from material balances alone written over the chosen section.

Over-all balances covering the entire plant are found in the same manner. They are

Total material balance: $\quad L_a + V_b = L_b + V_a \qquad (10\text{-}15)$

Component A balance: $L_a x_a + V_b y_b = L_b x_b + V_a y_a \qquad (10\text{-}16)$

Operating Line. Equation (10-14) is a relation between x_n, the concentration of the heavy phase leaving a stage of the column, and y_{n+1}, the concentration of the light phase entering that stage. Equation (10-14) can be written

$$y_{n+1} = \frac{L_n}{V_{n+1}} x_n + \frac{V_a y_a - L_a x_a}{V_{n+1}} \qquad (10\text{-}17)$$

Considering x_n as the abscissa and y_{n+1} as the ordinate, assume that it is possible to plot this equation on rectangular coordinates, as shown in Fig. 10-12. Then all points in the column that have coordinates like

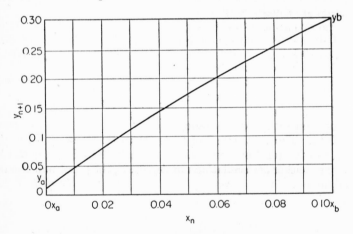

FIG. 10-12. Solution of Example 10-1.

y_{n+1} and x_n must lie on this curve. A line showing such a relation between the concentrations of the light and heavy streams at each point in a countercurrent contact apparatus is called an operating line. Such lines are basic in the graphical treatment of countercurrent contacting devices.

The two ends of the column are represented by two points on the operating line. The coordinates of one point are (x_a,y_a), and those of the other are (x_b,y_b). These points are the terminals of the operating line. This can be shown mathematically as follows. When $x_n = x_a$, $V_{n+1} = V_a$. Substituting into Eq. (10-17) gives

$$y_{n+1} = \frac{L_a x_a}{V_a} + \frac{V_a y_a - L_a x_a}{V_a} = y_a$$

This shows that point (x_a, y_a) lies on the operating line; also, when $x_n = x_b$, $V_{n+1} = V_b$. Substitution into Eq. (10-17) gives

$$y_{n+1} = \frac{L_b x_b}{V_b} + \frac{V_a y_a - L_a x_a}{V_b}$$

or $\qquad V_b y_{n+1} + L_a x_a = V_a y_a + L_b x_b$

Comparing this equation with Eq. (10-16) shows that $y_{n+1} = y_b$, so point (x_b, y_b) also lies on the line.

Construction of Operating Line. To draw an operating line, information must be available on the stream flows V_{n+1} and L_n. Additional data are required for this. Two simple cases cover many of the situations normally encountered in practice. In the first, which is true usually in rectification, neither L nor V changes appreciably from plate to plate. Then L_n and V_{n+1} in Eq. (10-14) are constant, and the operating line is straight. The line is easily drawn for any two points in it. The two terminal points are usually used.

When the operating line is straight, its slope is, from inspection of Eq. (10-14), L/V, or the ratio of the flow of heavy phase to that of the light phase. To emphasize this fact, the over-all balance can be written

$$\frac{y_b - y_a}{x_b - x_a} = \frac{L}{V} \tag{10-18}$$

A second case where the operating line can be drawn is when one component of each stream is not transferred between phases. Then, if L' is the molal flow rate of the inert component in the heavy phase and V' the moles of the other inert component in the light phase, these quantities are constant throughout the cascade. The relationships between V and V' and L and L' are

$$V_{n+1} = \frac{V'}{1 - y_{n+1}} \qquad L_n = \frac{L'}{1 - x_n} \tag{10-19}$$

Substitution of these relationships in Eq. (10-14) gives

$$L'\left(\frac{x_a}{1 - x_a} - \frac{x_n}{1 - x_n}\right) = V'\left(\frac{y_a}{1 - y_a} - \frac{y_{n+1}}{1 - y_{n+1}}\right) \tag{10-20}$$

No subscripts are needed for the quantities V' and L', as these are constant.

When the phases are dilute, x and y are small in comparison with unity, and the operating line becomes practically straight.

Example 10-1. By means of a plate column, acetone is absorbed from its mixture with air in a nonvolatile absorption oil. The entering gas contains 30 mole per cent acetone, and the entering oil is acetone-free. Of the acetone in the air 97 per cent is to be absorbed, and the concentrated liquor at the bottom of the tower is to contain 10 mole per cent acetone. Plot the operating line.

Solution. The concentration of acetone in the leaving air is obtained by an acetone balance. A basis of 100 moles of entering air is chosen. Then

$$V' = 0.70 \times 100 = 70 \text{ moles}$$

The acetone in the entering air is 30 moles, of which 3 per cent, or $0.03 \times 30 = 0.9$ mole, leaves with the gas, and $30 - 0.9 = 29.1$ moles is absorbed to form a 10 per cent solution. Then

$$L' = 29.1 \times \tfrac{90}{10} = 261.9 \text{ moles}$$

The terminal concentrations are

$$y_a = \frac{0.9}{70 + 0.9} = 0.0127$$

$$x_a = 0 \qquad x_b = 0.10 \qquad y_b = 0.30$$

From Eq. (10-20), the equation for the operating line is

$$261.9 \left(0 - \frac{x_n}{1 - x_n} \right) = 70 \left(\frac{0.0127}{1 - 0.0127} - \frac{y_{n+1}}{1 - y_{n+1}} \right)$$

This equation can be solved for either $1 - x$ or $1 - y$. Choosing $1 - y$ gives

$$\frac{y}{1 - y} = 0.0129 + 3.74 \, \frac{x}{1 - x} \qquad \overset{?}{\underset{\circ}{\checkmark}} \, .0127$$

The operating line passes through points ($x = 0$, $y = 0.127$) and ($x = 0.10$, $y = 0.30$). Coordinates for other points are found from the above equation. Thus, when $x = 0.03$,

$$\frac{y}{1 - y} = 0.0129 + \frac{3.74 \times 0.03}{0.97} = 0.1286$$

and

$$y = 0.114$$

Likewise, when $x = 0.05$, $y = 0.173$, and when $x = 0.08$, $y = 0.253$. The operating line is plotted in Fig. 10-12.

Ideal Contact Stages. The ideal stage is a standard to which an actual stage may be compared. In an ideal stage, the light phase leaving the stage is in equilibrium with the heavy phase leaving the same stage. For example, if plate n in Fig. 10-11 is an ideal stage, concentrations x_n and y_n are coordinates of a point on the curve of x_e vs. y_e showing the equilibrium between the phases. In a plate column ideal stages are also called *perfect plates.*

To use ideal stages in design it is necessary to apply a correction factor, called the stage efficiency or plate efficiency, which relates the ideal stage to an actual one. Plate efficiencies are discussed in Chap. 12, and the present discussion is restricted to ideal stages.

Determining the Number of Ideal Stages. A problem of general importance is that of finding the number of ideal stages required in an actual cascade to cover a desired range of concentration x_a to x_b or its equivalent, y_a to y_b. If this number can be determined, and if information on stage efficiencies is available, the number of actual stages can be calculated. This is the usual method of designing cascades.

The most generally satisfactory methods of determining the number of ideal stages in a cascade are graphical ones. The simplest graphical method, and one that is sufficiently precise for most two-component situations, is based on the use of the operating line in conjunction with the equilibrium line, both plotted on the same coordinates.

Figure 10-13 shows the equilibrium curve and the operating line for a typical stage-contact unit, e.g., a gas absorber. The ends of the operating

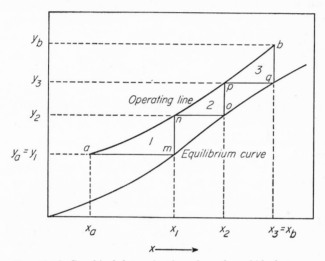

Fig. 10-13. Graphical determination of number of ideal stages.

line are point a, having coordinates (x_a, y_a), and point b, having coordinates (x_b, y_b). It is desired to determine the number of ideal stages needed to accomplish the gas-phase concentration change y_b to y_a and the liquid-phase concentration change x_a to x_b. The problem is solved as follows.

The concentration of the gas leaving the top stage, which is stage 1, is y_a, or y_1. If the stage is ideal, x_1, the concentration of the liquid leaving this stage is, by definition of an ideal stage, such that the point (x_1, y_1) must lie on the equilibrium curve. This fact fixes point m, found by moving horizontally from point a to the equilibrium curve. The abscissa of point m is x_1. The operating line is now used. It passes through all points having coordinates of the type (x_n, y_{n+1}), and since x_1 is known, y_2 is found by moving vertically from point m to the operating line at point n, the coordinates of which are (x_1, y_2). The step, or triangle, defined by points a, m, and n represents one ideal stage, the first one in this column. The second stage is located graphically on the diagram by repeating the same construction, passing horizontally to the equilibrium curve at point o, having coordinates (x_2, y_2), and vertically to the operating line again at point p, having coordinates (x_2, y_3). The third stage is found by again repeating the construction, giving triangle pqb. For the situation shown in Fig. 10-13, the third stage is the last, as the concentration of the gas leaving that stage is y_b, and the liquid leaving it is x_b, which are the desired terminal concentrations. Three ideal stages are required for this separation.

The same construction can be used for determining the number of ideal stages needed in any stage-contact plant, whether it is used for gas absorption, rectification, leaching, or liquid extraction.†

Absorption-factor Method for Calculating the Number of Ideal Stages. When the operating and equilibrium lines are both straight over a given concentration range x_a to x_b, the number of ideal stages can be calculated by formula, and graphical construction is unnecessary. Formulas for this purpose are derived as follows.

Let the equation of the equilibrium line be

$$y_e = mx_e + B \tag{10-21}$$

where, by definition, m and B are constant. If stage n is ideal,

$$y_n = mx_n + B \tag{10-22}$$

Substitution for x_n into Eq. (10-17) gives, for ideal stages and constant L/V,

$$y_{n+1} = \frac{L(y_n - B)}{mV} + y_a - \frac{Lx_a}{V} \tag{10-23}$$

It is convenient to define an absorption factor A by the equation

$$A = \frac{L}{mV} \tag{10-24}$$

The absorption factor is the ratio of the slope of the operating line to that of the equilibrium line. It is a constant when both of these lines are straight. Equation (10-23) can be written

$$y_{n+1} = A(y_n - B) + y_a - Amx_a = Ay_n - A(mx_a + B) + y_a \tag{10-25}$$

The quantity $mx_a + B$ is, by Eq. (10-21), the concentration of the vapor that is in equilibrium with the inlet heavy phase, the concentration of which is x_a. This can be seen from Fig. 10-14. The symbol y^* is used to indicate the concentration of a light phase in equilibrium with a specified

---

† The graphical step-by-step construction utilizing alternately the operating and equilibrium lines to find the number of ideal stages was first applied to the design of rectifying columns, and is known as the McCabe-Thiele method.[15]

heavy phase. Then

$$y_a^* = mx_a + B \tag{10-26}$$

and Eq. (10-25) becomes

$$y_{n+1} = Ay_n - Ay_a^* + y_a \tag{10-27}$$

Equation (10-27) can be used to calculate, step by step, the value of y_{n+1} for each stage starting with stage 1. The method may be followed with the aid of Fig. 10-14.

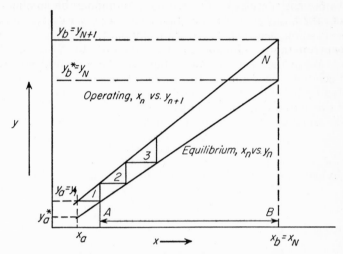

FIG. 10-14. Derivation of absorption-factor equation.

For stage 1, using $n = 1$ in Eq. (10-27) and noting that $y_1 = y_a$,

$$y_2 = Ay_a - Ay_a^* + y_a = y_a(1 + A) - Ay_a^*$$

For stage 2, using $n = 2$ in Eq. (10-27) and eliminating y_2,

$$y_3 = Ay_2 - Ay_a^* + y_a = A[y_a(1 + A) - Ay_a^*] - Ay_a^* + y_a$$
$$= y_a(1 + A + A^2) - y_a^*(A + A^2)$$

These equations may be generalized for the nth stage, giving

$$y_{n+1} = y_a(1 + A + A^2 + \cdots + A^n) - y_a^*(A + A^2 + \cdots + A^n) \tag{10-28}$$

For the entire cascade, $n = N$, the total number of stages, and $y_{n+1} = y_{N+1} = y_b$. Then

$$y_b = y_a(1 + A + A^2 + \cdots + A^N) - y_a^*(A + A^2 + \cdots + A^N) \tag{10-29}$$

The sums in the parentheses of Eq. (10-29) are both sums of geometric series. The sum of such a series is

$$s_n = \frac{a_1(1 - r^n)}{1 - r}$$

where s_n = sum of first n terms of series
   a_1 = first term
   r = constant ratio of each term to preceding term
Equation (10-29) can then be written

$$y_b = y_a \left(\frac{1 - A^{N+1}}{1 - A} \right) - y_a^* A \left(\frac{1 - A^N}{1 - A} \right) \tag{10-30}$$

Equation (10-30) is a form of the Kremser equation.[14] It can be used as such or in the form of a chart relating N, A, and the terminal concentrations.[1,3] It can also be put into a simpler form by the following method.
   Equation (10-27) is, for stage N,

$$y_b = Ay_N - Ay_a^* + y_a \tag{10-31}$$

It can be seen from Fig. 10-14 that $y_N = y_b^*$ and Eq. (10-31) can be written

$$y_a = y_b - A(y_b^* - y_a^*) \tag{10-32}$$

Collecting terms in Eq. (10-30) containing A^{N+1} gives

$$A^{N+1}(y_a - y_a^*) = A(y_b - y_a^*) + (y_a - y_b) \tag{10-33}$$

Substituting $y_a - y_b$ from Eq. (10-32) into Eq. (10-33) gives

$$A^N(y_a - y_a^*) = y_b - y_a^* - y_b^* + y_a^* = y_b - y_b^* \tag{10-34}$$

Taking logarithms of Eq. (10-34) and solving for N gives

$$N = \frac{\log \dfrac{y_b - y_b^*}{y_a - y_a^*}}{\log A} \tag{10-35}$$

and from Eq. (10-32),

$$\frac{y_b - y_a}{y_b^* - y_a^*} = A \tag{10-36}$$

Equation (10-35) can be written

$$N = \frac{\log \dfrac{y_b - y_b^*}{y_a - y_a^*}}{\log \dfrac{y_b - y_a}{y_b^* - y_a^*}} \tag{10-37}$$

The various concentration differences in Eq. (10-37) are shown in Fig. 10-15.

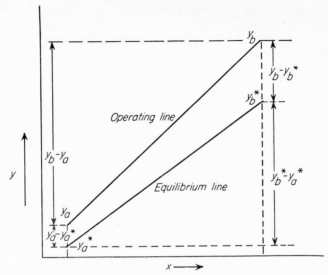

Fig. 10-15. Concentration differences in Eq. (10-37).

Heavy-phase Form of Equation (10-37). The choice of y as the concentration coordinate in the above equations rather than x is arbitrary, and equations analogous to Eqs. (10-35) and (10-37) in x can be derived. They are

$$N = \frac{\log \dfrac{x_b^* - x_b}{x_a^* - x_a}}{\log A} = \frac{\log \dfrac{x_b^* - x_b}{x_a^* - x_a}}{\log \dfrac{x_b^* - x_a^*}{x_b - x_a}} \tag{10-38}$$

where x^* is the equilibrium concentration corresponding to y. The concentration differences in Eq. (10-38) are shown in Fig. 10-16.

When the operating line and the equilibrium line are parallel, A is unity, and Eqs. (10-35), (10-37), and (10-38) are indeterminate. When $A = 1$, Eq. (10-29) becomes

$$y_b = y_a(1 + N) - y_a^* N \tag{10-39}$$

and

$$N = \frac{y_b - y_a}{y_a - y_a^*} = \frac{y_b - y_a}{y_b - y_b^*} \tag{10-40}$$

The corresponding equation in x is

$$N = \frac{x_b - x_a}{x_b^* - x_b} = \frac{x_b - x_a}{x_a^* - x_a} \tag{10-41}$$

Note that the equilibrium line is actually used for steps only over the length shown by line AB in Fig. 10-14. It is necessary only that the actual line be linear over this distance. The quantities x_a^*, x_b^*, y_a^*, and y_b^* do not

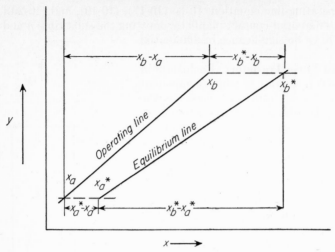

Fig. 10-16. Concentration differences in Eq. (10-38).

need to be actual points on the equilibrium line, and can be used as purely mathematical numbers calculated by the equation of a straight line passing through points A and B.

Material Balances for Differential-contact Plants. In Fig. 10-17 are shown the streams flowing into, through, and out of a differential-contact plant, such as a packed tower. There are no numbered stages, so subscripts pertaining to stage numbers have no meaning and are dropped. Material balances for the portion of the column above an arbitrary section, as shown by the dotted line in Fig. 10-17, are obtained by omitting the subscripts n and $n + 1$ from Eqs. (10-13) and (10-14).

Total material:
$$L_a + V = L + V_a \qquad (10\text{-}42)$$

Component A:
$$L_a x_a + V y = L x + V_a y_a \qquad (10\text{-}43)$$

where V is the molal flow rate of the light phase and L that of the heavy phase at the same point in the tower. The heavy-phase and light-phase concentrations x and y apply to this same location.

The quantities pertaining to the terminal streams carry the usual subscripts, and the over-all material balances given by Eqs. (10-15) and (10-16) apply to a differential-contact column without change.

In the flow of the two phases through a differential length of column, V, L, x, and y change slightly. The differential equation showing this is obtained by differentiating Eq. (10-43).

$$d(Lx) - d(Vy) = 0$$
$$d(Lx) = d(Vy) \qquad (10\text{-}44)$$

This equation is the material balance for component A over a differential section of column.

The operating-line equations [Eqs. (10-17), (10-19), and (10-20)] can be used for differential-contact plants by omitting the subscripts n and $n + 1$, as these have no significance in differential plants.

Rate of Mass Transfer. Each term in Eq. (10-44) is the rate at which component A is transferred from one phase to the other through the area of the interface in the differential section of the column. If N_A is the rate of transfer of component A in moles per hour,

$$d(Lx) = d(Vy) = dN_A \quad (10\text{-}45)$$

The quantity N_A is the flux of component A through the interface between the phases in pound moles per hour. To relate the flux to the properties of the phases and the conditions of operation requires a study of mass transfer from the standpoint of diffusion, which is the subject of the next section.

DIFFUSION

The rate of diffusion of the components in each phase determines the stage efficiency in a stage-contact plant and the flux in a differential-contact plant. Diffusion rates are therefore basic to the design and operation of both types of equipment.

Nature of Diffusion. Diffusion is the movement, under the influence of a physical stimulus, of an individual component

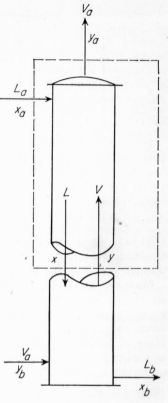

Fig. 10-17. Material-balance diagram for packed column.

through a mixture. The most common cause of diffusion is a concentration gradient of the diffusing component. A concentration gradient tends to move the component in such a direction as to equalize concentrations and destroy the gradient. When the gradient is maintained by constantly supplying the diffusing component to the high-concentration end of the gradient and removing it at the low-concentration end, the flow of the diffusing component is continuous. This movement is exploited in mass-transfer operations. For example, a salt crystal in contact with a stream of water or a dilute solution sets up a concentration gradient in the neighborhood of the interface, and salt diffuses through the liquid layers in a direction perpendicular to the interface. The flow of salt away from the interface continues until the crystal is dissolved. When the salt is intimately mixed with insoluble solid, the process is an example of leaching.

Although the usual cause of diffusion is a concentration gradient, diffusion can also be caused either by a pressure gradient or by a temperature gradient applied over the mixture.   Diffusion induced by total pressure (*not* partial pressure) is called pressure diffusion, and that induced by temperature is thermal diffusion.   Both are rare in chemical engineering, although thermal diffusion is conducted on a large scale in the separation of uranium isotopes.   Only diffusion under a concentration gradient is considered here.

Diffusion is not restricted to molecular transfer through stagnant layers of solid or liquid.   It also takes place in fluid phases by physical mixing and by the eddies of turbulent flow, just as heat flow in a fluid may occur by convection.   Usually the diffusion process is accompanied by bulk flow of the mixture, and it is often associated with heat flow.

Role of Diffusion in Mass Transfer.   In all the mass-transfer operations, diffusion occurs in at least one phase and often in both phases.   In gas absorption, solute diffuses through the gas phase to the interface between the phases and through the liquid phase from the interface.   In rectification, low boiler diffuses through the liquid phase to the interface and away from the interface into the vapor.   The high boiler diffuses in the reverse direction and passes through the vapor into the liquid.   In leaching, diffusion of solute through the solid phase is followed by diffusion into the liquid.   In liquid extraction, the solute diffuses through the raffinate phase to the interface and then into the extract phase.   In crystallization, solute diffuses through the mother liquor to the crystals and deposits on the solid surfaces.   In air-water processes there is no diffusion through the liquid phase, because the liquid phase is pure and no concentration gradient through it can exist; but water vapor diffuses to or from the liquid-gas interface into or out of the gas phase.   In drying, liquid water diffuses through the solid toward the surface of the solid, vaporizes, and diffuses as vapor into the gas.   The zone of vaporization may be either at the surface of the solid or within the solid, depending upon factors that are discussed in Chap. 16.   When the vaporization zone is in the solid, vapor diffusion takes place in the solid between the vaporization zone and the surface, and diffusion of both vapor and liquid occurs within the solid.

Conditions at Interface in Diffusion Processes.   Besides diffusion through the various phases, the action at the actual interface between the two phases (the physical boundary between the two phases) is important. It is difficult to obtain direct evidence of the action at the interface, but it is commonly assumed that, except in crystallization, little or no resistance to mass transfer exists at the interface itself.   If this is true, the two phases are in equilibrium at the boundary surface in spite of the fact that there are concentration gradients in both phases at either side of the interface.[†]

Although equilibrium at the interface has been demonstrated in at least one experimental investigation,[10] other experiments have cast doubt on

[†] It is important to differentiate between two phases in equilibrium only at the interface between them and two phases in equilibrium throughout.   In the latter case, there is no net transfer of material, and the process has stopped.   In the former case, the process is actually in operation.

the accuracy of this assumption.[6] For the usual situation of low and moderate diffusion rates in the individual phases the resistance to mass transfer at the interface is probably small, and will be neglected in all cases except crystal growth.

Theory of Diffusion

In this section quantitative relationships for diffusion are discussed. Attention is focused on diffusion in a direction *perpendicular* to the interface between the phases and at a definite location in the equipment. Assuming a column is being used, the location chosen for analysis is in a definite cross section parallel with the ground. Steady state is assumed, and the concentrations at any point do not change with time.

Comparison of Diffusion and Heat Transfer. An analogy exists between flow of heat and diffusion. In each, a gradient is the cause of the flow. In heat transfer a temperature gradient is the driving force; in diffusion the force is a concentration gradient. In each case the flux per unit area is proportional to the gradient. The analogy cannot be carried further, however, for the reason that heat is not a substance: it is an energy effect. When heat flows from one point to another, it leaves no space behind, nor does it require space in its new location. The velocity of heat flow has no meaning. Diffusion is a physical flow of matter, which occurs at a definite velocity. A diffusing component leaves space behind it, and room must be found for it in its new location.

The fact that diffusion is material flow introduces three complications that are not found in heat transfer: (1) All components in a phase must be considered. In a two-component mixture, to which this discussion is restricted, both components A and B must be considered even if only component A is of interest. There may be a net flow of the entire phase in the direction of diffusion. This net flux is the algebraic sum of the fluxes of the individual components. (2) If the net flux of the entire phase perpendicular to the interface is not zero, the mass of the phase either increases or decreases. The mass of an adjoining phase changes to keep the total mass of both phases constant. Heat, of course, has no mass. (3) Since matter is in motion, the fluxes and velocities are relative to an arbitrary state of rest, which must be chosen before these quantities can be calculated.

Diffusion Velocities. It is desired to derive relationships between flux and velocity on two bases: (1) relative to the interface and (2) relative to the phase as a whole. Consider a section through a fluid phase of two components, as shown in Fig. 10-18a. Let MM represent the interface across which components A and B are diffusing. The area of the interface is A ft^2. Consider, first, velocities relative to the interface. Let the fluxes of components A and B across the interface into the phase be N_A and N_B lb moles/hr, respectively. Since the flow is steady, N_A and N_B are also the fluxes across any other plane parallel with the interface and a fixed distance from it. Consider a thin slice of thickness db ft located b ft from MM. Let the concentrations of components A and B in this slice be c_A

and c_B lb moles/ft³, respectively. Since diffusion is in progress, c_A and c_B vary with b. The fluxes N_A and N_B are constant.

The volume of the elementary slice shown in Fig. 10-18a is $A\ db$ ft³. The amount of component A in the slice is $Ac_A\ db$ lb moles. Since the flow rate of this component is N_A moles/hr, the time of passage through the slice is

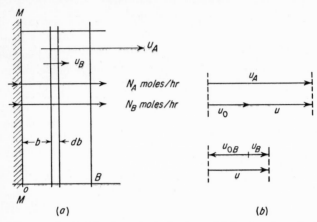

FIG. 10-18. Fluxes and velocities through single phase: (*a*) Section through phase parallel with directions of diffusion. (*b*) Vector diagrams, showing relations among velocities. $u_A = u_{0A} + u$, $u_B = u_{0B} + u$.

$Ac_A\ db/N_A$ hr. The distance traveled by the stream of A during this time is db ft, so the velocity of component A at distance b from the interface is

$$u_A = \frac{db}{c_A A\ db/N_A} = \frac{N_A/A}{c_A} \qquad (10\text{-}46)$$

Likewise, u_B, the velocity of component B at this same location, is

$$u_B = \frac{N_B/A}{c_B} \qquad (10\text{-}47)$$

The total flux of the entire phase is $N_A + N_B$ moles/hr. If the density of the phase is ρ_M moles/ft³, then u, the velocity of the entire phase relative to the interface, is

$$u = \frac{(N_A + N_B)/A}{\rho_M} \qquad (10\text{-}48)$$

If, now, the entire phase is assumed to be at rest, $u = 0$.† The velocity of components A and B relative to the phase as a whole may be denoted by

† This term means that to an observer watching the solution there is no net molal flow in either direction. Suppose that it were possible to color all A molecules red and all B molecules blue and to count them as they pass by. Assume that the red molecules

u_{0A} and u_{0B}, respectively, and the corresponding fluxes of these components by N_{0A} and N_{0B}, also relative to the entire phase. Then,

$$u_{0A} = \frac{N_{0A}/A}{c_A} \tag{10-49}$$

and

$$u_{0B} = \frac{N_{0B}/A}{c_B} \tag{10-50}$$

The velocity of the entire phase relative to itself is, of course, zero. The velocity of the interface relative to that of the entire phase is now $-u$.

The velocities u_A, u_{0A}, and u are related by the equation

$$u_{0A} = u_A - u \tag{10-51}$$

and velocities u_B, u_{0B}, and u by

$$u_{0B} = u_B - u \tag{10-52}$$

These relations are shown in the vector diagrams of Fig. 10-18b. Note that vectors u_{0A} and u_{0B} point in opposite directions.

Elimination of u_{0A}, u_A, and u from Eqs. (10-46), (10-48), and (10-49) gives

$$\frac{N_{0A}}{c_A} = \frac{N_A}{c_A} - \frac{N_A + N_B}{\rho_M} \tag{10-53}$$

Likewise, for component B, from Eqs. (10-47), (10-48), and (10-50),

$$\frac{N_{0B}}{c_B} = \frac{N_B}{c_B} - \frac{N_A + N_B}{\rho_M} \tag{10-54}$$

The concentrations c_A and c_B may be converted to mole fractions by noting that, from Eq. (10-1),

$$c_A = \rho_M y_A \qquad c_B = \rho_M y_B$$

where ρ_M is the molal density of the mixture in pound moles per cubic foot. Then Eqs. (10-53) and (10-54) become, after substituting for c_A and c_B,

$$N_{0A} = N_A - y_A(N_A + N_B) \tag{10-55}$$

$$N_{0B} = N_B - y_B(N_A + N_B) \tag{10-56}$$

Equations (10-55) and (10-56) are independent of the mechanism of transfer and apply in both laminar and turbulent flow.

---

are moving toward the right and the blue molecules toward the left. There would be no net flow of the molecules if the number of red molecules passing the observer during a given time equaled the number of blue ones passing in the other direction during the same time. Note that this refers to *moles*, not *masses*. If the red molecules are, for example, heavier than the blue (higher molecular weight), there is a net *mass* flow toward the right when the net *molal* flow is zero.

Laminar Flow and Molecular Diffusivity. The stream of fluid through which diffusion is taking place is in flow in a direction perpendicular to the direction of diffusion, and laminar flow, turbulent flow, or the usual combination of both may be encountered in the stream. Separation and wake formation also may occur. Consider, first, laminar flow. In the absence of turbulence, diffusion takes place by molecular action only, as there is no physical mixing. Molecular diffusion is, accordingly, analogous to the transfer of heat by thermal conduction.

By analogy with thermal conductivity the diffusivity D_m is defined by the equations

$$\frac{N_{0A}}{A} = -D_{mA}\frac{dy_A}{db} \tag{10-57}$$

and

$$\frac{N_{0B}}{A} = -D_{mB}\frac{dy_B}{db} \tag{10-58}$$

where D_{mA} and D_{mB} are the molal diffusivities of components A and B, respectively. Note that diffusivities are defined in terms of molecular flux relative to the phase as a whole. This specification is necessary for an unambiguous definition.

Comparing molecular diffusion with heat conduction, the molal flux densities N_{0A}/A and N_{0B}/A are analogous to the heat-transfer flux density q/A, the diffusivity D_m to the thermal conductivity k, and the concentration gradients dy_A/db and dy_B/db to the temperature gradient dt/db.

Eliminating N_{0A} from Eqs. (10-55) and (10-57) gives

$$-D_{mA}\frac{dy_A}{db} = \frac{N_A - y_A(N_A + N_B)}{A} \tag{10-59}$$

Likewise, from Eqs. (10-56) and (10-58),

$$-D_{mB}\frac{dy_B}{db} = \frac{N_B - y_B(N_A + N_B)}{A} \tag{10-60}$$

Adding Eqs. (10-57) and (10-58) and noting that $N_{0A} + N_{0B} = 0$, because N_{0A} and N_{0B} refer to no net molecular flow,

$$\frac{N_{0A} + N_{0B}}{A} = -D_{mA}\frac{dy_A}{db} - D_{mB}\frac{dy_B}{db} = 0 \tag{10-61}$$

Since $y_A + y_B = 1$, $dy_A/db = -dy_B/db$, and

$$D_{mA} = D_{mB} = D_m$$

The same molal diffusivity applies to both components of a two-component mixture, and D_m can be used for each of them.

From this point, only the component A is to be followed, and the subscript is dropped from y_A. When reference is made to the concentration of component B, $1 - y$ may be used. Equation (10-59) can be written

$$\frac{dy}{N_A - y(N_A + N_B)} = -\frac{db}{D_m A} \tag{10-62}$$

It is convenient to introduce a quantity z, which is the ratio of the flux of component A to that of the total phase [7] $N_A + N_B$, or

$$z = \frac{N_A}{N_A + N_B} \tag{10-63}$$

Then, from Eq. (10-62),

$$\frac{N_A \, dy}{N_A - y(N_A + N_B)} = -\frac{N_A \, db}{D_m A}$$

$$\frac{[N_A/(N_A + N_B)] \, dy}{N_A/(N_A + N_B) - y} = \frac{z \, dy}{z - y} = -\frac{N_A \, db}{D_m A} \tag{10-64}$$

For use, Eq. (10-64) must be integrated over a finite thickness of the phase, and the diffusivity D_m must be known.

Molal Diffusivity D_m. The dimensions of D_m are, from Eq. (10-57),

$$[D_m] = \left[\frac{Nb}{Ay}\right] = \overline{M}\bar{\theta}^{-1}\overline{L}\overline{L}^{-2} = \overline{M}\bar{\theta}^{-1}\overline{L}^{-1}$$

In engineering units, D_m is conveniently expressed in pound moles per foot per hour.

Diffusivities have been given much attention in the study of the kinetic theories of gases and liquids. Those of gases have been given the more thorough study, and the kinetic theory of diffusion in gases is quite well understood.

Diffusion in Gases. Molecular gas diffusion results from the linear motions of the molecules. At any instant the individual molecules in a gas are moving in random directions at speeds that vary from zero to a very large value. At a given temperature and pressure, the average velocity is constant and of order of magnitude of 1,000 ft/sec. This high speed does not mean, however, that the concentrations equalize rapidly by molecular diffusion. Because of the large molecular population density (approximately 17×10^{21} molecules per cubic foot at 32°F and 1 atm) the frequency of collision is so great that the path of each molecule is often interrupted and the velocity changed in magnitude and direction many times per second. Thus, although the average velocity of the molecules in nitrogen gas at 32°F and 1 atm is 1,490 ft/sec, each molecule suffers on the average 5×10^{11} collisions each second, and the resulting effective speed of bulk movement under a concentration gradient is small.

Small molecules have larger average velocities and lower collision rates than do large molecules. The higher the temperature, the greater the average velocity. The diffusivity increases with increase in temperature and decreases with increase in the molecular weight and the size of the individual molecule. Diffusivities of common gases in air are given in Appendix 17.

Correlations have been found that can be used to estimate gas diffusivities from the characteristics of the individual components. An example is that of Gilliland.[8]

$$D_m = 0.00945 \frac{\sqrt{T}}{(V_A^{\frac{1}{3}} + V_B^{\frac{1}{3}})^2} \sqrt{\frac{1}{M_A} + \frac{1}{M_B}} \qquad (10\text{-}65)$$

where D_m = molal diffusivity, lb moles/ft-hr

T = absolute temperature, °R

M_A, M_B = molecular weights of components A and B, respectively

V_A, V_B = molecular volumes of components A and B, respectively, at their normal boiling points

Rules for calculating the molecular volumes V_A and V_B are given in Table 10-3.

TABLE 10-3. ATOMIC VOLUMES TO BE USED FOR CALCULATING MOLECULAR VOLUMES AT NORMAL BOILING POINT †

Air	29.9	Oxygen, doubly bound	7.4
Carbon	14.8	Coupled to two other elements:	
Chlorine, terminal (as in R—Cl)	21.6	In aldehydes and ketones	7.4
Medial (as in R—CHCl—R')	24.6	In methyl esters	9.1
Fluorine	8.7	In ethyl esters	9.9
Hydrogen, in compounds	3.7	In higher esters and ethers	11.0
In hydrogen molecule	7.15	In acids	12.0
Nitrogen, in primary amines	10.5	In union with sulfur,	
In secondary amines	12.0	phosphorus, nitrogen	8.3
		Phosphorus	27.0
		Silicon	32.0
		Sulfur	25.6
		Water	18.8

† For three-membered ring, as in ethylene oxide, deduct 6.0. For four-membered ring, as in cyclobutane, deduct 8.5. For five-membered ring, as in furan, deduct 11.5. For six-membered ring, as in benzene, deduct 15.0. For naphthalene-ring formation, deduct 30.0. For anthracene-ring formation, deduct 47.5.

Example 10-2. Calculate the diffusivity of gaseous benzoyl chloride (C_6H_5COCl) in air at 120°F.

Solution. The molecular volumes, from Table 10-3, are as follows. For air, $V_A = 29.9$. For benzoyl chloride

$$
\begin{aligned}
C_7 &= 7 \times 14.8 = 103.6 \\
H_5 &= 5 \times 3.7 = 18.5 \\
O &= 1 \times 7.4 = 7.4 \\
Cl &= 1 \times 21.6 = 21.6 \\
\hline
\text{Total} &= 151.1
\end{aligned}
$$

$$
\begin{aligned}
\text{Six-membered ring deduction} &= 15.0 \\
\hline
V_B &= 136.1
\end{aligned}
$$

The molecular weight for air is $M_A = 29.0$, and for benzoyl chloride it is $M_B = 140.6$.

From Eq. (10-65) the diffusivity is

$$D_m = \frac{0.00945\sqrt{460 + 120}}{(29.9^{\frac{1}{3}} + 136.1^{\frac{1}{3}})^2}\sqrt{\frac{1}{29.0} + \frac{1}{140.6}}$$

$$= 6.8 \times 10^{-4} \text{ lb mole/ft-hr}$$

Note that diffusivities refer to a system of two or more components, not to a single component.

Liquid Diffusivities. The theory of diffusion in liquids is not so far advanced nor the experimental data so adequate as in gas diffusion. Diffusivities in liquids are smaller than in gases. Liquid diffusion depends on the energy needed to move the molecules through the liquid, which is a function of the attraction of molecules for each other rather than of molecular velocities. Diffusivities in liquids vary with concentration, approximately linearly, but accurate data showing the extent of the variation are few.

Diffusivities for liquids can be calculated approximately from the equation.[17a, 18]

$$D_m = \frac{4.0 \times 10^{-7} T \rho_M}{\mu(V_B^{\frac{1}{3}} - k_1)} \tag{10-66}$$

where D_m = diffusivity, lb moles/ft-hr

T = absolute temperature, °R

μ = viscosity of mixture, lb/ft-hr

V_B = molecular volume of solute, at normal boiling point, calculated from Table 10-3

ρ_M = molal density, lb moles/ft^3

k_1 = constant

The constant k_1 is 2.0, 2.46, and 2.84, when the solvent is water, methyl alcohol, and benzene, respectively. Equation (10-66) does not apply to electrolytes, and it is valid only at low concentrations.

Volumetric Diffusivity D_v. For both gas and liquid diffusivities, experimental data should be used when available rather than Eqs. (10-65) and (10-67). Such data are found in standard tables of physical and chemical data. Usually data are given not as molecular diffusivities D_m but as volumetric diffusivities D_v. These are related to D_m by the equation

$$D_v = \frac{D_m}{\rho_M} \tag{10-67}$$

where ρ_M is the molal density of the mixture.† The dimensions of D_v are, from Eq. (10-67), $\bar{L}^2\bar{\theta}^{-1}$. Typical values are given in Appendix 17. The

† For ideal gases and liquids, where ρ_M is independent of concentration, $D_{vA} = D_{vB}$. If ρ_M varies with concentration, e.g., as in nonideal gases and liquids, D_{mA} and D_{mB} are still equal, but D_{vA} and D_{vB} are not. Also, D_v varies inversely with pressure, but D_m is independent of pressure.[4]

thermal diffusivity α, used in Chap. 8, has these same dimensions. The usual literature units of D_v are square centimeters per second. To convert from square centimeters per second to square feet per hour multiply by $3,600/30.48^2 = 3.875$. In liquids, values of D_v usually apply to dilute solutions.

Integrated Equation for Molecular Diffusion and Laminar Flow. Returning to Eq. (10-64), this can be integrated if D_m can be considered constant. This assumption is justified for gas diffusion in binary mixtures at all concentrations and for liquid diffusion when the concentration range over the distance of integration B is small. For small changes in D_m, an arithmetic average of the terminal values of D_m can be used as a constant. The temperature should also be constant or nearly so, as diffusivities vary with temperature.

Integration of Eq. (10-64) over the total thickness of the phase gives

$$z \int_{y_i}^{y} \frac{dy}{z-y} = -\frac{N_A}{D_m A} \int_0^B db$$

or

$$z \ln \frac{z-y}{z-y_i} = \frac{N_A B}{D_m A} \tag{10-68}$$

where y = mole fraction of component A where $b = B$
y_i = mole fraction of component A where $b = 0$

Individual Mass-transfer Coefficient. An individual mass-transfer coefficient k, analogous to the individual heat-transfer coefficient h, can be developed from Eq. (10-68) in the following manner.

Define the coefficient k by the equation

$$N_A \phi = kA(y_i - y) \tag{10-69}$$

where ϕ is a factor that takes into account the effect of the quantity z. Eliminating N_A/A from Eqs. (10-68) and (10-69) gives

$$\frac{D_m}{B} z \ln \frac{z-y}{z-y_i} = \frac{k}{\phi}(y_i - y)$$

or

$$\frac{k}{\phi} = \frac{D_m}{B}\left(\frac{z}{y_i - y} \ln \frac{z-y}{z-y_i}\right) = \frac{D_m}{B}\left[\frac{z \ln \dfrac{z-y}{z-y_i}}{(z-y)-(z-y_i)}\right] \tag{10-70}$$

Let

$$k = \frac{D_m}{B} \tag{10-71}$$

and

$$\phi = \frac{(z-y)-(z-y_i)}{z \ln \dfrac{z-y}{z-y_i}} = \frac{\overline{(z-y)_L}}{z} \tag{10-72}$$

where $\overline{(z-y)_L}$ is the logarithmic mean of $z - y_i$ and $z - y$.

Relative-velocity Factor. The factor ϕ, defined by Eq. (10-72), may be called the relative-velocity factor, as it accounts for a possible drift of the entire phase toward or away from the interface. It applies equally to turbulent or laminar flow. It accounts for any combination of the fluxes of the individual components. Two cases are especially important.

Equimolal Diffusion. In rectification, the total diffusional flow $N_A + N_B$ is zero, and $N_A = -N_B$. This type of diffusion is called equimolal.

Since z, by Eq. (10-63), is infinite when $N_A + N_B = 0$, Eq. (10-72) becomes indeterminate. Equation (10-64) can be used, however, by noting that

$$\frac{z}{z - y} = \frac{1}{1 - y/z}$$

and that, when z is infinite, this fraction is unity. Equation (10-64) becomes, after integrating,

$$N_A = \frac{D_m}{B} A(y_i - y) \tag{10-73}$$

When Eqs. (10-69) and (10-73) are compared, it is clear that for equimolal diffusion $\phi = 1$.

Unicomponent Diffusion. In gas absorption, leaching, and extraction, one component, chosen arbitrarily as component B, is stationary with respect to the interface, and only component A diffuses to or from the interface. Then $N_B = 0$, and, by Eq. (10-63), $z = 1$. Equation (10-72) becomes

$$\phi = \frac{(1 - y) - (1 - y_i)}{\ln\left[(1 - y)/(1 - y_i)\right]} = \overline{(1 - y)_L} = \overline{(y_B)_L} \tag{10-74}$$

where $\overline{(y_B)_L}$ is the logarithmic mean of $1 - y$ and $1 - y_i$. Here the relative velocity factor is the logarithmic mean of the boundary concentrations of the inert or nonmoving component.

The concentration gradients for components A and B for a typical case of equimolal diffusion are shown in Fig. 10-19a, and those for unimolal diffusion are shown in Fig. 10-19b. In both cases, concentrations of component A at the interface and at the outer boundary are 0.9 and 0.1, respectively. It is of interest that, although in unimolal diffusion N_B is zero, the concentration gradient of component B is not zero. The gradients are what an observer at rest with respect to the interface would see if the entire phase were moving from left to right just fast enough to prevent component B from moving through the fluid layer to or from the interface. The situation is analogous to a person walking forward on a treadmill the backward velocity of which is just equal to his speed of walking. Component B is diffusing through the stream as a whole, but the stream itself is, in effect, also moving at the same speed in the opposite direction, and component B is at rest with respect to the interface. The entire difference between the situations of Fig. 10-19 is a matter of relative velocities.

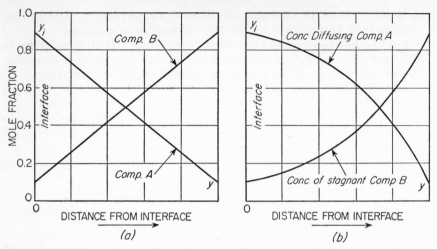

FIG. 10-19. Concentration gradients for equimolal and unimolal diffusion: (a) Components A and B diffusing at same molal rates in different directions. (b) Component A diffusing, component B stationary with respect to interface.

Turbulent Flow. Equation (10-68) is based on steady-state molecular diffusion, and applies only when the flow parallel to the interface is laminar. At Reynolds numbers above the critical, the usual buffer zone and turbulent core appear if the stream is passing through well-defined channels, or if the fluid is passing in contact with solid shapes, as in packed towers, separation and wake formation occur. Transfer to and from bubbles or sprays, as in bubble-cap and spray columns, differs even more greatly from the mechanism of simple molecular diffusion. It is doubtful that steady state is ever reached in such apparatus. An additional complication is that the actual interface area A is not often known. In packed towers, the actual interfacial area is considerably less than the geometric area of the packing, and in sprays and bubbles, the area is unknown.

In the face of these complications, mass transfer is treated by retaining Eq. (10-72) for the definition of ϕ and using Eq. (10-69) to define the coefficient k. This procedure is analogous to the definition of the individual heat-transfer coefficient h by Eq. (8-32). The concentration y is the average concentration of the entire phase and is that reached if the stream is thoroughly mixed. Again, this is analogous to the usage in heat transfer, where the average temperature of a stream is used in defining h. When concentration is defined in this manner, it is identical with that used in the material balances.

The coefficient k is the Drew-Colburn coefficient. Other mass-transfer coefficients are found in the chemical engineering literature, but, for simplicity, k, as defined by Eq. (10-69), is generally used throughout this book. It is used with a subscript y for a light phase and x for a heavy phase. The units of k are lb moles/(ft^2)(hr)(unit mole fraction). Numerically, it is the flux of component A, in moles per hour, through 1 ft^2 of area, under a driving force of unit mole fraction when ϕ is unity.

Application to Liquids. Equations (10-69) and (10-72) apply in principle to liquids as well as to gases.[16] It has been mentioned that, unlike diffusivities in gases, D_m in liquids varies considerably with concentration. Also, experimental data for both D_m and k in liquids are usually determined in such a way that the relative velocity factor is included in them. Furthermore, data for D_m usually refer to dilute solutions, where ϕ is nearly unity. For these reasons, Eq. (10-69) is usually simplified when used for liquids by assuming $\phi = 1$. Thus, if a liquid phase is the heavy phase,

$$N_A = k_x A (x_i - x) \quad (10\text{-}75)$$

Direction of Diffusion. In the above discussion, it is assumed that component A is diffusing from the interface into the phase. It is equally likely that this component diffuses toward the interface, rather than away from it. The same equations can be used by writing $y - y_i$ or $x - x_i$ for $y_i - y$ or $x_i - x$.

Experimental Measurement of Mass-transfer Coefficients. In view of the complexity of mass transfer in actual equipment, fundamental equations for mass transfer in actual equipment are not available, and empirical methods, guided by dimensional analysis and by semitheoretical analogies, are relied upon to give workable equations. The approach to the problem has been made in several steps in the following manner.

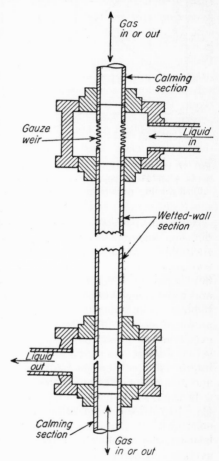

FIG. 10-20. Wetted-wall tower. (*Gilliland and Sherwood.*[9] *Courtesy of Industrial and Engineering Chemistry.*)

1. The coefficient k has been studied in experimental devices in which the area of contact between phases is known and where boundary-layer separation does not take place. The wetted-wall tower shown in Fig. 10-20, which is sometimes used practically, is the usual device of this type. It has given valuable information on mass transfer to and from fluids in turbulent flow. A wetted-wall tower is essentially a vertical tube with means

for admitting liquid at the top and causing it to flow downward along the inside wall of the tube, under the influence of gravity, and means for admitting gas to the inside of the tube, where it flows through the tower in contact with the liquid. Generally the gas enters the bottom of the tower and flows countercurrent to the liquid, but parallel flow can be used. In the wetted-wall tower, the interfacial area, except for some complications from ripple formation, is known, and form drag is absent.

2. Other experiments have been made where the area of contact is known. The principle used in the work is to measure the rate of evaporation or solution of material from solid shapes in a stream of flowing liquid or gas. Evaporation of liquid from wet solids or evaporation or solution of substances from solid masses of the material itself is used. In this technique, the area is known and is simply the area of the evaporating or dissolving solid. Complications from diffusion in the solid itself are eliminated, as the material being transferred originates at the interface between the phases. Data obtained in this manner can be used to extend the correlations obtained in wetted-wall experiments to situations where boundary-layer separation is active.

3. Finally, experiments on actual packed towers, bubble plates, and spray devices are made and the data so obtained correlated along lines suggested by the results from the experiments described in paragraphs 1 and 2. The area effect is taken into account by using not the coefficient k alone but rather by a combined coefficient ka, where a is the (unknown) transfer area per unit volume of equipment. Correlations are then found not for k itself but for the product ka. Some progress has been made in evaluating the factor a independently, and when this is possible, the coefficient k can be isolated and used in the ordinary manner.

The same basic principle is used in all experiments conducted for obtaining numerical values for k or ka. It consists in measuring experimentally the quantities N_A, A, y_i, and y, calculating ϕ from the conditions of the transfer, and calculating k by Eq. (10-69) or its equivalent after averaging over the entire length of the apparatus. If A is not known, the total volume of the equipment is used and ka calculated. Dimensional analysis is used to plan the experiments and to interpret the results in the form of dimensionless groups and equations. Analogies among friction, heat transfer, and mass transfer are useful guides.

Coefficients for Mass Transfer through Known Areas

In this section correlations are given for mass transfer between fluids, or between fluids and solids, where the area A is known. Coefficients for equipment in which the area between the phases is not known are discussed in subsequent chapters.

Dimensional Analysis. From the mechanism of mass transfer, it can be expected that the coefficient k would depend on the diffusivity D_m and on the variables that control the character of the fluid flow, namely, the mass velocity G, the viscosity μ, and some linear dimension D. The shape of the interface can be expected to influence the process, so a different relation should appear from each shape.

For any given shape of transfer surface

$$k = \Psi(D_m, D, G, \mu)$$

Dimensional analysis gives

$$\frac{k\overline{M}}{G} = \Psi_1\left(\frac{DG}{\mu}, \frac{\mu}{D_m\overline{M}}\right) \tag{10-76}$$

Here $\overline{M}$ is the average molecular weight of the entire phase. It is used with k and D_m for consistency, since each is a molal quantity. Equation (10-76) is analogous to the Colburn form of the heat-transfer equation [Eq. (8-50)]. A second dimensionless equation, analogous to the Nusselt form of the heat-transfer equation, is obtained by multiplying Eq. (10-76) by $(DG/\mu)(\mu/D_m\overline{M})$. This gives

$$\frac{kD}{D_m} = \Psi_2\left(\frac{DG}{\mu}, \frac{\mu}{D_m\overline{M}}\right) \tag{10-77}$$

The dimensionless groups in Eqs. (10-76) and (10-77) have been given names and symbols. The group kD/D_m is called the Sherwood number and is denoted by N_{Sh}. This number corresponds to the Nusselt number in heat transfer. The group $\mu/D_m\overline{M}$ is the Schmidt number, denoted by N_{Sc}. It corresponds to the Prandtl number. The Schmidt number is often written as $\mu/\rho D_v$, where ρ is the density of the phase in pounds per cubic foot. Typical values are given in Appendix 17.

Equation (10-76) is also frequently used in the form of a j factor, analogous to the j'_H factor of Eq. (8-57). It is defined by the equation [2]

$$j_M = \frac{k\overline{M}}{G}\left(\frac{\mu}{D_m\overline{M}}\right)^{\frac{2}{3}} \tag{10-78}$$

Wetted-wall Towers. Several correlations are available for wetted-wall towers (Fig. 10-20). The Gilliland-Sherwood equation is [9]

$$N_{\mathrm{Sh}} = 0.023 N_{\mathrm{Re}}^{0.81} N_{\mathrm{Sc}}^{0.44} \tag{10-79}$$

where D in N_{Re} and N_{Sh} is the diameter of the tube. This equation applies over a range of Reynolds numbers from 2,000 to 35,000; of Schmidt numbers from 0.60 to 2.5; and over a pressure range of 0.1 to 3 atm.

A second correlation for wetted-wall towers, somewhat less precise than Eq. (10-79), can be written [17d, 11]

$$j_M = j'_H = \frac{f}{2} = 0.027 N_{\mathrm{Re}}^{-0.2} \tag{10-80}$$

where f is the Fanning friction factor for flow in straight pipes. This correlation is satisfactory for both absorption and rectification in wetted-

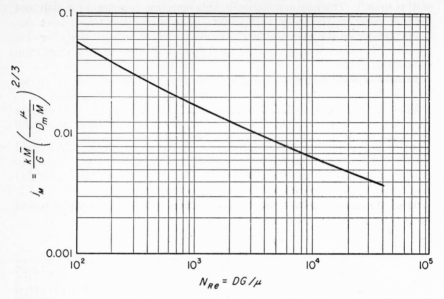

FIG. 10-21. Mass transfer, flow past single cylinders. (*After Sherwood and Pigford.*[17b])

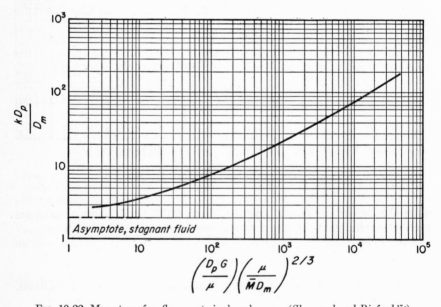

FIG. 10-22. Mass transfer, flow past single spheres. (*Sherwood and Pigford.*[17c])

wall towers.[12] The analogy shown in this equation is general for heat and mass transfer in the same equipment. The analogy between heat and mass transfer on the one hand and friction on the other holds only for skin friction. It does not apply to the total friction if there is form drag from separation of flow.

Flow Perpendicular to Single Cylinders. A correlation of j_M vs. N_{Re} for flow perpendicular to single cylinders is shown in Fig. 10-21.

Flow past Single Spheres. A plot of N_{Sh} vs. $N_{Re}N_{Sc}^{\frac{1}{3}}$ is shown in Fig. 10-22. The line in this figure approaches an asymptote of $N_{Sh} = 2.0$ at low values of $N_{Re}N_{Sc}^{\frac{1}{3}}$. The line is represented by the equation

$$\frac{kD_p}{D_m} = 2 + 0.6 \left(\frac{D_p G}{\mu}\right)^{\frac{1}{2}} \left(\frac{\overline{M} D_m}{\mu}\right)^{\frac{1}{3}} \tag{10-81}$$

Mass Transfer in Beds of Spheres. Many data have been obtained for mass transfer between fluids and beds of spherical particles. The results given by a number of investigators are shown in the form of a plot of

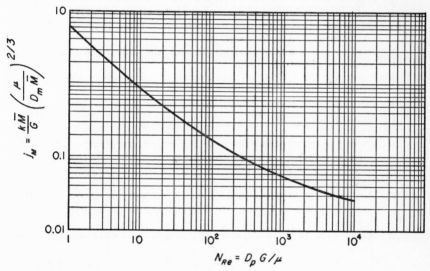

FIG. 10-23. Mass transfer, flow through beds of spheres. (*After Kaufman and Thodos.*[13])

j_M vs. N_{Re} in Fig. 10-23. In Figs. 10-22 and 10-23 D_p is the diameter of the spheres in feet, and G is the superficial mass velocity in pounds per square foot per hour.

SYMBOLS

A Area perpendicular to direction of mass transfer, ft^2, also, absorption factor, L/mV

a Area of interface between phases per unit volume of equipment, ft^{-1}

B Length of path of molecular diffusion through single phase, ft; also, constant in Eq. (10-21)

b Distance from phase boundary in direction of diffusion, ft

C Number of components, phase rule

c Concentration, lb moles/ft^3; c_A, of component A; c_B, of component B

c_p Specific heat at constant pressure, Btu/(lb)(°F)

D Linear dimension or diameter, ft; D_p, diameter of sphere

D_m Molal diffusivity, lb moles/ft-hr; D_{mA}, of component A; D_{mB}, of component B

D_v Volumetric diffusivity, ft^2/hr; D_{vA}, of component A; D_{vB}, of component B

G Mass velocity, lb/ft^2-hr

H Henry's law constant, lb force/(ft^2)(mole fraction); H_A for component A; H_B for component B

K Constant

k Mass-transfer coefficient, lb moles/(ft^2)(hr)(unit mole fraction)

k_1 Constant in Eq. (10-66)

L Flow rate of heavy phase, lb moles/hr; L_a, at entrance; L_b, at exit; L_n, from stage n; L', flow rate of inert component in heavy phase

M Molecular weight; M_A, M_B, of components A and B, respectively; $\overline{M}$ average molecular weight of phase

m Slope of equilibrium curve, dy_e/dx_e

N Total number of ideal stages, also, flux or rate of transfer across phase boundary, lb moles/hr; N_A, of component A; N_B, of component B

N_0 Flux across plane at rest with respect to entire phase, lb moles/hr; N_{0A}, for component A; N_{0B}, for component B

n Serial number of ideal stage, counting from inlet of heavy phase

P Number of phases, phase rule

p Pressure, lb force/ft^2; p_A, p_B, vapor pressures of components A and B, respectively

$\bar{p}$ Partial pressure, lb force/ft^2; $\bar{p}_A$, for component A; $\bar{p}_B$, for component B

T Absolute temperature, °R

t Temperature, °F

u Velocity of entire phase relative to phase boundary, ft/hr; u_A, velocity of component A; u_B, of component B

u_0 Velocity relative to the entire phase, ft/hr; u_{0A}, of component A; u_{0B}, of component B

V Flow rate of light phase, lb moles/hr; V_a, at exit; V_b, at entrance; V_{n+1}, from stage $n+1$; V', flow rate of inert component in light phase

V_A, V_B Molecular volumes of components A and B, respectively

x Mole fraction in heavy phase; used for component A when only two components are present; x_A, mole fraction of component A; x_B, mole fraction of component B; x_{Aa}, of component A in heavy phase at entrance; x_{Ab}, at exit; x_e, mole fraction of component A in heavy phase at equilibrium; x_i, mole fraction of component A in heavy phase at interface between phases; x_n, mole fraction in heavy phase from stage n; x^*, mole fraction in heavy phase in equilibrium with specified stream of light phase; x_a^*, in equilibrium with y_a; x_b^*, in equilibrium with y_b

y Mole fraction in light phase; used for component A when only two components are present; y_A, mole fraction of component A; y_B, mole fraction of component B; y_{Aa}, of component A in light phase at exit; y_{Ab}, at entrance; y_e, mole fraction of component A in light phase at equilibrium; y_i, mole fraction of component A in light phase at interface between phases; y_{n+1}, mole fraction in light phase from stage $n+1$; y^*, mole fraction in light phase in equilibrium with specified stream of heavy phase; y_a^*, in equilibrium with x_a; y_b^*, in equilibrium with x_b

$(\bar{y}_B)_L$ Logarithmic mean mole fraction of component B across path of mass transfer; equals $\overline{(1 - y)}_L$

z Flux of component A relative to phase; equals $N_A/(N_A + N_B)$

Greek Letters

μ Viscosity, lb/ft-hr

ρ Density, lb/ft³; ρ_M, molal density, lb moles/ft³

ϕ Relative-velocity factor, defined by Eq. (10-72); also, number of degrees of freedom, phase rule

Ψ Function; Ψ_1, in Eq. (10-76); Ψ_2, in Eq. (10-77)

ψ Function in Eq. (10-4); ψ_A, for component A; ψ_B, for component B

Dimensionless Groups

j_H' Heat-transfer factor, $(h/c_p G)(c_p \mu/k)^{\frac{2}{3}}$

j_M Mass-transfer factor, $(k\overline{M}/G)(\mu/D_m\overline{M})^{\frac{2}{3}}$

N_{Re} Reynolds number, DG/μ

N_{Sc} Schmidt number, $\mu/D_m\overline{M} = \mu/D_v\rho$

N_{Sh} Sherwood number, kD/D_m

PROBLEMS

10-1. An open circular tank 20 ft in diameter contains benzene at 22°C which is exposed to the atmosphere in such a manner that the liquid is covered with a stagnant air film estimated to be 5 mm thick. The concentration of benzene beyond the stagnant film is negligible. The vapor pressure of benzene at 22°C is 100 mm Hg. If benzene is worth 36 cents per gallon, what is the value of the loss of benzene from this tank, in dollars per day? The specific gravity of benzene is 0.88.

10-2. Alcohol vapor is being absorbed from a mixture of alcohol vapor and water vapor by means of a nonvolatile solvent in which alcohol is soluble but water is not. The temperature is 97°C, and the total pressure is 760 mm Hg. The alcohol vapor can be considered to be diffusing through a film of alcohol-water-vapor mixture 0.1 mm thick. The molal concentration of the alcohol vapor at the outside of the film is 80 per cent, and that on the inside, next to the solvent, is 10 per cent. The volumetric diffusivity of alcohol-water-vapor mixtures at 25°C and 1 atm is 0.15 cm²/sec.

Calculate the rate of diffusion of alcohol vapor in pounds per hour if the area of the film is 100 ft².

10-3. An alcohol-water-vapor mixture is being rectified by contact with an alcohol-water liquid solution. Alcohol is being transferred from gas to liquid and water from liquid to gas. The molal flow rates of alcohol and water are equal but in opposite directions. The temperature is 95°C and the pressure 1 atm. Both components are diffusing through a gas film 0.1 mm thick. The molal concentration of alcohol at the outside of the film is 80 per cent, and that on the inside is 10 per cent.

Calculate the rate of diffusion of alcohol and of water in pounds per hour through a film area of 100 ft².

10-4. A wetted-wall column operating at a total pressure of 518 mm Hg is supplied with water and air, the latter at a rate of 120 g/min. The partial pressure of the water vapor in the air stream is 76 mm, and the vapor pressure of the liquid-water film on the wall of the tower is 138 mm. The observed rate of vaporization of water into the air is 13.1 g/min.

The same equipment, now at a total pressure of 820 mm, is supplied with air at the same temperature as before and at a rate of 100 g/min. The liquid vaporized is *n*-butyl alcohol. The partial pressure of the alcohol is 30.5 mm, and the vapor pressure of the liquid alcohol is 54.4 mm.

What rate of vaporization, in grams per minute, may be expected in the experiment with n-butyl alcohol?

10-5. Air at 100°F and 2.0 atm is passed through a shallow bed of naphthalene spheres $\frac{1}{2}$ in. in diameter at a rate of 5 ft/sec, based on the empty cross section of the bed. The vapor pressure of naphthalene is 117 mm. What is the initial rate of evaporation of naphthalene from 1 ft³ of bed, in pounds per hour, assuming a bed porosity of 40 per cent?

10-6. A solution of SO_2 in H_2O containing 2.5 g of SO_2 per 100 g of H_2O is stripped of SO_2 by countercurrent contact with air in a plate column. The molal ratio of air to water, on a sulfur-dioxide–free basis, is 0.50. The temperature is 20°C, and the equilibrium relationship is

$$Y_e = 2.3X_e$$

where Y_e and X_e are both in moles of SO_2 per mole of SO_2-free stream. The column contains five ideal plates. What fraction of the SO_2 entering with the liquid is stripped from the liquid? What is the concentration of SO_2 in the effluent gas?

REFERENCES

1. Brown, G. G., M. Souders, Jr., and H. V. Nyland: *Ind. Eng. Chem.*, **24**: 522 (1932).
2. Chilton, T. H., and A. P. Colburn: *Ind. Eng. Chem.*, **26**: 1183 (1934).
3. Colburn, A. P., and R. L. Pigford, in J. H. Perry (ed.): "Chemical Engineers' Handbook," 3d ed., p. 554, McGraw-Hill Book Company, Inc., New York, 1950.
4. Drew, T. B.: private communication.
5. Elgin, J. C.: U.S. Pat. 2,364,892 (Dec. 12, 1944).
6. Emmert, R. E., and R. L. Pigford: *Ind. Eng. Chem.*, **50**: 87 (1954).
7. Gaffney, B. J., and T. B. Drew: *Ind. Eng. Chem.*, **42**: 1120 (1950).
8. Gilliland, E. R.: *Ind. Eng. Chem.*, **26**: 681 (1934).
9. Gilliland, E. R., and T. K. Sherwood: *Ind. Eng. Chem.*, **26**: 516 (1934).
10. Goodgame, T. H., and T. K. Sherwood: *Chem. Eng. Sci.*, **3**: 37 (1954).
11. Jackson, M. L., and N. H. Ceaglske: *Ind. Eng. Chem.*, **42**: 1188 (1950).
12. Johnstone, H. F., and R. L. Pigford: *Trans. AIChE*, **38**: 25 (1941).
13. Kaufman, D. J., and G. Thodos: *Ind. Eng. Chem.*, .**43**: 2582 (1951).
14. Kremser, A.: *Natl. Petroleum News*, **22**(21): 42 (May 21, 1930).
15. McCabe, W. L., and E. W. Thiele: *Ind. Eng. Chem.*, **17**: 605 (1925).
16. Tierney, J. W., L. F. Stutzman, and R. L. Daileader: *Ind. Eng. Chem.*, **46**: 1595 (1954).
17. Sherwood, T. K., and R. L. Pigford: "Absorption and Extraction," 2d ed., (a) p. 21, (b) p. 70, (c) p. 74, (d) p. 78, McGraw-Hill Book Company, Inc., New York, 1952.
18. Wilke, C. R.: *Chem. Eng. Progr.*, **45**: 218 (1949).

through the packed bed, give lower pressure drops than dumped packings, in which the gas must frequently change in velocity and direction. This advantage is offset, however, by the poorer contact between liquid and gas in stacked packings.

TABLE 11-1. PHYSICAL CHARACTERISTICS OF TOWER PACKINGS †

Type	Material	Dimensions, in.	Average bulk density, lb/ft³ of tower volume	Surface area a_v, ft²/ft³ of tower volume	Free volume 100ϵ, %
Dumped packings:					
Broken solids............	Coke	3	24	12	50
		6	10	5.5	57
Raschig rings..	Porcelain	$\frac{1}{2} \times \frac{1}{2}$	50	122	64
		1×1	40	58	73
		$1\frac{1}{2} \times 1\frac{1}{2}$	42	35	68
		2×2	37	28	74
Lessing rings............	Porcelain	1×1	50	69	66
		$1\frac{1}{2} \times 1\frac{1}{2}$	58	40	60
		2×2	49	32	68
Berl saddles.............	Porcelain	$\frac{1}{2} \times \frac{1}{2}$	54	142	63
		1×1	45	76	69
		$1\frac{1}{2} \times 1\frac{1}{2}$	38	44	75
Stacked packings:					
Raschig rings............	Porcelain	2×2	..	32	80
Single spiral rings.........	Porcelain	$3\frac{1}{4} \times 3$	52	34	66
		4×4	55	28	67
		6×6	51	19	70
Drip-point grids (No. 6146, General Ceramic Co.)...	....	....	72	18	48

† From L. Clarke, "Manual for Process Engineering Calculations," p. 409, McGraw-Hill Book Company, Inc., New York, 1947; and M. Leva, "Tower Packings and Packed Tower Design," pp. 19–24, U.S. Stoneware Co., Akron, Ohio, 1951.

The requirement of good contact between liquid and gas is the hardest to meet, especially in large towers. Ideally the liquid, once distributed over the top of the packing, flows in thin films over all the packing surface all the way down the tower. Actually the films tend to grow thicker in some places and thinner in others, so that the liquid collects into small rivulets and flows along localized paths through the packing. Especially at low liquid rates much of the packing surface may be dry or, at best, covered by a stagnant film of liquid. This effect is known as *channeling*; it is the chief reason for the poor performance of large packed towers.

Channeling is most severe in towers packed with stacked packing, less severe in dumped packings of crushed solids, and least severe in dumped packings of regular units such as rings.[6] In towers of moderate size channeling can be minimized by having the diameter of the tower at least eight times the packing diameter. If the ratio of tower diameter to packing diameter is less than 8:1, the liquid tends to flow out of the packing and down the walls of the column. Even in small towers filled with packings that meet this requirement, however, liquid distribution and channeling have a major effect on column performance. Figure 11-2 shows a typical

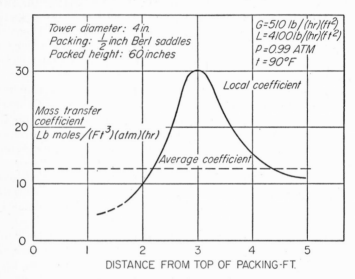

Fig. 11-2. Variation of local mass-transfer coefficients in absorption of carbon tetrachloride from air into dibutyl phthalate.

variation of the local mass-transfer coefficient with distance from the top of the packing in the absorption of carbon tetrachloride from air into dibutyl phthalate in a 4-in. tower packed with $\frac{1}{2}$-in. Berl saddles.[8] The coefficient increases to a maximum 3 ft below the top of the packing as the liquid distribution improves, then falls markedly as channeling begins in the lower part of the tower. The maximum local coefficient is almost $2\frac{1}{2}$ times the average coefficient. Kirschbaum[5] reports a similar variation in the local heat-transfer coefficient in distillations in a 12-in. column packed with 8-mm Raschig rings. In tall towers filled with large packing the effect of channeling is much more pronounced, and the transfer coefficients can be expected to be very low, even when the initial distribution is good and redistributors are included every 10 or 15 ft in the packed section.

Recent studies have shown that regardless of the initial liquid distribution, at low liquid rates much of the packing surface is not wetted by the flowing liquid. As the liquid rate rises, the wetted fraction of the packing surface increases, until at a critical liquid rate, which is usually high, all the packing surface becomes wetted and effective. At liquid rates higher than the critical the effect of channeling is not important. Tentative correlations of the critical liquid rate with the viscosity of the liquid have been published.[18]

DESIGN OF PACKED TOWERS

In the design of a packed tower two major dimensions are first established, the diameter of the tower and the depth of the packed section. The total height of the tower can then be fixed by allowing for the sections carrying the gas outlet, gas inlet, and liquid distributor at the top, and the gas inlet, packing support, and liquid outlet at the bottom. The depth of packing depends on the rate of mass transfer. The diameter, which follows from the cross-sectional area of the tower, depends on the rates of flow of gas and liquid.

Limiting Flow Rates in Packed Towers. In a tower containing a given packing and being irrigated with a definite flow of liquid, there is an upper limit to the rate of gas flow. The gas velocity corresponding to this limit is called the *flooding velocity.* It can be found from an inspection of the relation between the pressure drop through the bed of packing and the gas flow rate, from observation of the holdup of liquid, and by the visual appearance of the packing. The flooding velocity, as identified by these three different effects, varies somewhat with the method of identification and appears more as a range of flow rates than as a sharply defined constant.

Figure 11-3 shows the relation between pressure drop and gas flow rate in a packed tower. The pressure drop per unit packing depth comes from fluid friction; it is plotted on logarithmic coordinates against the gas flow rate. When the packing is dry, the line so obtained is straight and has a slope of about 1.8. The pressure drop therefore increases with the 1.8th power of the velocity, which is consistent with the usual law of friction loss in turbulent flow. If the packing is irrigated with a constant flow of liquid, the relation between pressure drop and gas flow rate follows a line such as *bcdef* in Fig. 11-3. At low and moderate velocities, the pressure drop is proportional to the 1.8th power of the flow rate but is greater than that in dry packing at the same gas velocity. As the gas velocity increases, however, the line curves upward, starting at point *c* in the figure. Then, at a higher velocity, as shown by the line *cde*, the pressure drop increases sharply at nearly constant gas velocity. The rise may follow a smooth curve, as shown by the solid line *cd*, or may show sharp breaks at points *c* and *d*, as shown by the dotted line. While the pressure drop is changing along line *cde*, the holdup, which is constant along line *bc*, increases. Starting at the bottom of the packing, the tower becomes progressively drowned with liquid until, when the point *e* is reached, the entire packed section is flooded. Point *e* represents a flooding point where a layer of liquid is visible on the top of the packing, through which bubbles of gas issue. Then the tower is acting as a gas bubbler rather than as a packed tower. If the gas flow rate is increased beyond the flooding velocity, the pressure drop follows the line *ef*, at a low slope, the increase representing additional friction of the flow of the gas bubbles through the flooded packing.

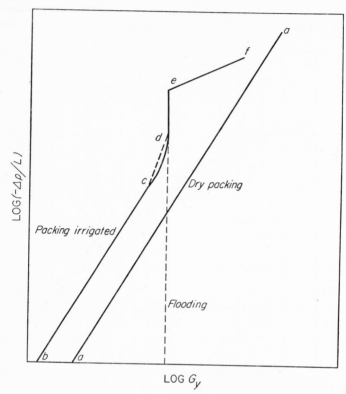

FIG. 11-3. Pressure drop in packed tower: *c*, loading point.   *d*, flooding point.

A packed tower is practical only at velocities below that of point *c*, which is sometimes called the *loading point*. The lower the velocity, the lower the cost of power and the larger the tower. The higher the gas velocity, the larger the power cost and the smaller the tower. Economically, the most favorable gas velocity depends on a balance between the cost of the power and the fixed charges on the equipment. The optimum velocity can be estimated by methods beyond the scope of this book.[13c] It is usually about one-half the flooding velocity.

Figure 11-4 gives a correlation for the estimation of the flooding velocity in packed towers. It consists of a logarithmic plot [14] of the group $G_y^2 a_v (\mu_x')^{0.2}/g_c \epsilon^3 \rho_x \rho_y$ vs. the group $(G_x/G_y)\sqrt{\rho_y/\rho_x}$, where

G_x = mass velocity of liquid, lb/ft^2-hr
G_y = mass velocity of gas, lb/ft^2-hr

ρ_x = density of liquid, lb/ft^3
ρ_y = density of gas, lb/ft^3
a_v = surface area of dry packing per unit packed volume, ft^2/ft^3
μ'_x = viscosity of liquid, centipoises
g_c = Newton's-law conversion factor, 4.17×10^8 ft-lb/lb force-hr^2
ϵ = porosity, or fraction voids, in packed section, dimensionless

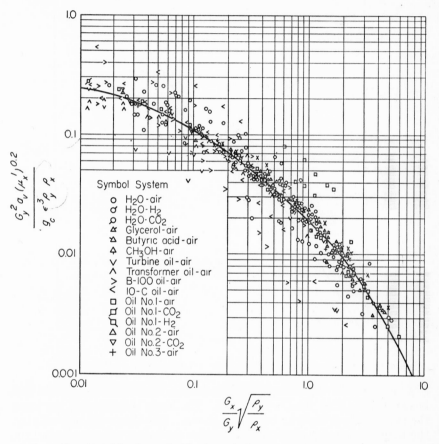

Fig. 11-4. Flooding velocities in packed tower. (*After Lobo et al.*[7])

The ordinates of Fig. 11-4 are not dimensionless, and the stated units must be used. The mass velocities are based on the total tower cross section.

Example 11-1. A tower packed with 1-in. ceramic Raschig rings is to be built to treat 25,000 ft^3 of entering gas per hour. The ammonia content of the entering gas is 2 per cent by volume. Ammonia-free water is used as absorbent. The temperature is 68°F, and the pressure is 1 atm. The ratio of liquid flow to gas flow is 1 lb of gas per pound of liquid. If the gas velocity is to be one-half the flooding velocity, what should be the diameter of the tower?

Solution. The quantities to be used in the groups of Fig. 11-4 are as follows. The average molecular weight of the entering gas is $29 \times 0.98 + 0.02 \times 17 = 28.7$. Then

$$\rho_y = \frac{28.7 \times 492}{359(460 + 68)} = 0.0745 \text{ lb/ft}^3$$

$$\rho_x = 62.3 \text{ lb/ft}^3 \qquad \mu_x' = 1 \text{ centipoise}$$

$$g_c = 4.17 \times 10^8 \qquad \frac{G_x}{G_y} = 1.0$$

For 1-in. ceramic rings, $a_v = 58 \text{ ft}^2/\text{ft}^3$, and $\epsilon = 0.73$ (Table 11-1). Then

$$\frac{G_x}{G_y}\sqrt{\frac{\rho_y}{\rho_x}} = \sqrt{\frac{0.0745}{62.3}} = 0.0346$$

From Fig. 11-4,

$$\frac{G_y^2 a_v (\mu_x')^{0.2}}{g_c \epsilon^3 \rho_x \rho_y} = 0.18$$

The mass velocity at flooding is

$$G_y = \sqrt{\frac{0.18 \times 4.17 \times 10^8 \times 0.73^3 \times 0.0745 \times 62.3}{58 \times 1^{0.2}}} = 1{,}530 \text{ lb/ft}^2\text{-hr}$$

The total gas flow is $25{,}000 \times 0.0745 = 1{,}860$ lb/hr. If the actual velocity is one-half the flooding velocity, the cross-sectional area S of the tower is

$$S = \frac{1{,}860}{1{,}530/2} = 2.43 \text{ ft}^2$$

The diameter of the tower is $\sqrt{2.43/0.7854} = 1\frac{3}{4}$ ft.

Pressure Drop in Wetted Packing. The pressure drop in a bed of solids being irrigated with liquid is considerably greater than that in dry packing at the same gas flow rate. No accurate basic correlation is available for pressure drop through wetted packings, but Fig. 11-5 gives an empirical correction factor for estimating such pressure drops from the corresponding drop through the dry bed. When the water rate G_x, in pounds per square foot per hour, is known, the factor A_L can be read from Fig. 11-5 and used as a multiplier for the pressure drop in dry packing, obtained from Fig. 2-34.[13b] Figure 11-5 applies only for flows less than the loading point and for liquids having approximately the viscosity of water.

Equilibria. The equilibrium relationship of importance in absorption is the plot of x_e, the mole fraction of solute in the liquid, against y_e, the mole fraction in the vapor. The data from which this plot is prepared are ordinarily given in tables of physical constants in the form of partial pressures of solute in millimeters of mercury vs. concentrations in the liquid as grams of solute per 100 g of solvent, equal to pounds of solute per 100 lb of solvent. Thus, for gases that follow Henry's law, such as oxygen, carbon dioxide, and air, a Henry's law constant H_A' is given, as defined by the equation $\bar{p}_A' = H_A' c_A'$, where $\bar{p}_A'$ is the partial pressure of the solute, in millimeters of mercury, and c_A' is the concentration of the solute in the liquid, in grams per 100 g of solvent or pounds per 100 lb of solvent.

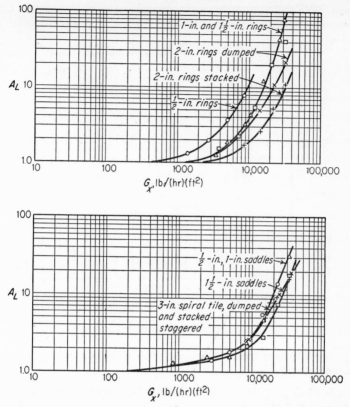

Fig. 11-5. Pressure drop in irrigated packing. (*By permission, from "Absorption and Extraction," 2d ed., by T. K. Sherwood and R. L. Pigford. Copyright, 1952. McGraw-Hill Book Company, Inc.*)

To convert H'_A to m, the slope of the equilibrium line in mole-fraction units, the following equation can be used.

$$m = \frac{100 H'_A M_A}{760 p M_B (1 - x_e)} \tag{11-1}$$

where M_A = molecular weight of solute
M_B = molecular weight of liquid solvent
x_e = mole fraction of solute in liquid at equilibrium
p = total pressure, atm

In Eq. (11-1), the quantity $1 - x_e$ may be taken as unity, since Henry's law applies only to dilute solutions. For more soluble gases, data are required showing partial pressures as functions of concentration and temperature.[10a] Figures 11-6 and 11-7 show such data for ammonia in water and sulfur dioxide in water,[4] respectively. In both figures, partial pressures in millimeters of mercury are plotted against concentrations in pounds

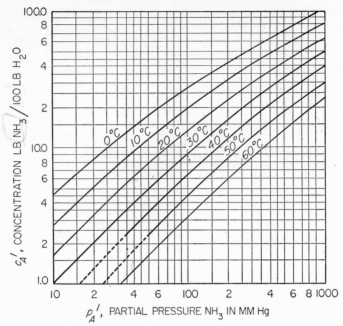

FIG. 11-6. Solubility of ammonia in water.

of solute per 100 lb of water. Points on the curve of x_e vs. y_e can be calculated from these data.

Example 11-2. Plot the equilibrium curve of sulfur dioxide–air–water, in mole-fraction units, at 1 atm and 30°C, over the concentration range covered by Fig. 11-7.

Solution. The y_e coordinates are found by dividing the partial pressures taken from Fig. 11-7 by 760. The x_e coordinates are found by noting that, since the molecular weight of SO_2 is 64, c_A' lb of SO_2 per 100 lb H_2O is $c_A'/64$ moles of SO_2 per 100/18, or 5.55, moles of water. Then, $x_e = (c_A'/64)/(c_A'/64 + 5.55)$. Details of this calculation for several points on the curve of x_e vs. y_e are given in Table 11-2. The equilibrium line is plotted in Fig. 11-8.

TABLE 11-2. DATA FOR EXAMPLE 11-2

c_A', lb SO_2/ 100 lb H_2O	p_A', mm Hg	$y_e = \dfrac{p_A'}{760}$	$x_e = \dfrac{c_A'/64}{c_A'/64 + 5.55}$
1.0	85	0.112	0.0028
2.0	176	0.232	0.0056
3.0	273	0.359	0.0084
4.0	376	0.495	0.0111
5.0	482	0.634	0.0139
6.0	588	0.774	0.0166

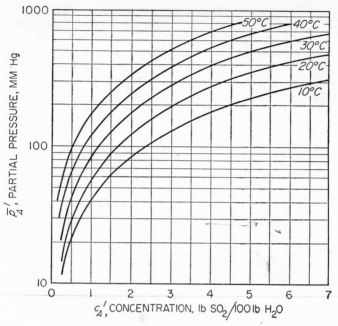

FIG. 11-7. Partial pressure of sulfur dioxide over aqueous solution.

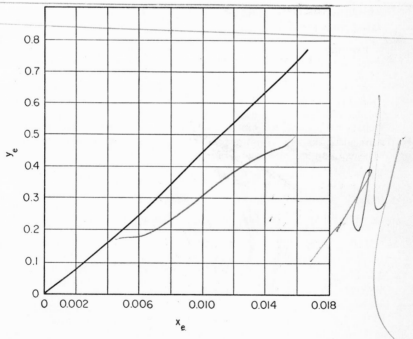

FIG. 11-8. Equilibrium curve, sulfur dioxide in water at 30°C, 1 atm.

Limiting Gas-Liquid Ratio. Equation (10-18) shows that the average slope of the operating line is L/V, the ratio of the molal flows of liquid and gas. Thus, for a given gas flow, a reduction in liquid flow decreases the slope of the operating line. Consider the operating line ab in Fig. 11-9. Assume that the gas rate and the terminal concentrations x_a, y_a, and y_b are held fast and the liquid flow L decreased. The upper end of the operating line then shifts in the direction of the equilibrium line, and x_b, the

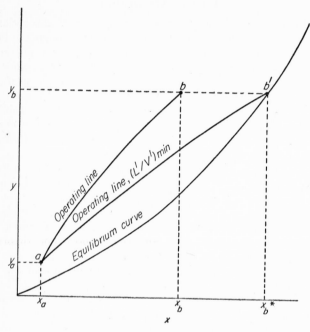

FIG. 11-9. Limiting gas-liquid ratio.

concentration of the strong liquor, increases. The maximum possible liquor concentration and the minimum possible liquid rate are obtained when the operating line just touches the equilibrium line, as shown by line ab' in Fig. 11-9. At this condition, an infinitely deep packed section is necessary, as the concentration difference for mass transfer becomes zero at the bottom of the tower. In any actual tower, the liquid rate must be greater than this minimum if the tower is to operate.

The limiting liquid-gas ratio $(L'/V')_{\min}$ can be computed from Eq. (10-20) by setting $y_{n+1} = y_b$ and $x_n = x_b^*$, where x_b^* is the abscissa of the point on the equilibrium line whose ordinate is y_b.

The L'/V' ratio is important in the economics of absorption in a countercurrent tower. If the liquid-gas ratio is large, the average distance between the operating and equilibrium line is also large, the concentration difference is favorable throughout the tower, and the tower is short. If the solute gas is to be recovered, however, the cost of recovery is high because of the dilution of the strong liquor. If, on the other hand, concentration costs are reduced by using less liquid, the driving forces in the absorber are reduced, and the tower becomes taller and more expensive. The optimum liquid rate is found by balancing recovery costs against fixed costs of the equipment.[2] In general, the operating line should be approximately parallel to the equilibrium line for economical operation.

Rate of Absorption

The height of a packed tower depends on the rate of absorption, which, in turn, is affected by the rate of transfer of mass through the liquid and gas phases. The following treatment has two limitations: no provision is made for possible chemical reactions between the absorbed component and the liquid, and the heat of solution is neglected. Simultaneous mass transfer and heat transfer are considered in Chaps. 15 and 16. Absorption accompanied by chemical reaction in the liquid state is not within the scope of this book.

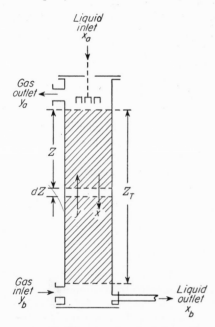

FIG. 11-10. Diagram of packed absorption tower.

Double-resistance (Two-film) Theory. Component A, which is the solute gas transferred from gas to liquid, must pass in series through two diffusional resistances, the gas and the liquid. The theory of this transfer is also called the two-film theory,[16] but in view of the fact that laminar films account for but a part of the resistances, the term double-resistance theory is more descriptive of the actual process.

Consider the packed tower shown in Fig. 11-10. Consider the absorption in the short section of height dZ ft at a level in the tower Z ft from the top of the packed section. Let the rate of absorption be dN_A lb moles/hr. The interfacial area between phases in this section is dA ft^2. The concentrations of the gas and liquid streams, in mole fraction of component A, are y and x, respectively.

The rate of transfer of component A from the bulk of the gas to the interface is, from Eq. (10-69),

$$dN_A = \frac{k_y}{\phi} (y - y_i) \, dA \tag{11-2}$$

where k_y = mass-transfer coefficient, gas phase, lb moles/(ft^2)(hr)(unit mole-fraction difference)

ϕ = relative velocity factor

y_i = mole fraction of A at gas side of interface

The difference $y - y_i$ is the driving force across the gas resistance.

In gas absorption or stripping, component A crosses the interface, while component B is inert and is stationary with respect to the interface. Then $z = 1$, and by Eq. (10-74) ϕ is the logarithmic mean concentration of component B across the gas phase, or

$$\phi = \frac{(1 - y_i) - (1 - y)}{2.303 \log \dfrac{1 - y_i}{1 - y}} = \frac{y - y_i}{2.303 \log \dfrac{1 - y_i}{1 - y}} = \overline{(1 - y)_L} \qquad (11\text{-}3)$$

The rate of transfer of component A in the liquid phase, from the interface to the bulk of the liquid, is, from Eq. (10-75),

$$dN_A = k_x(x_i - x)\, dA \qquad (11\text{-}4)$$

where k_x = mass-transfer coefficient, liquid phase, lb moles/$(\text{ft}^2)(\text{hr})$(unit mole-fraction difference)

x_i = mole fraction of component A at liquid side of interface

x = mole fraction of component A in bulk of liquid

In equations for liquid-phase transfer, ϕ is omitted, for the reasons given in Chap. 10.

If equilibrium at the interface is assumed, x_i and y_i are coordinates of a point on the equilibrium curve, and the curve of x_e vs. y_e is also the relationship between x_i and y_i.

Equations (11-2) and (11-4) are analogous to the heat-transfer equations applicable to two fluid phases in contact. Such equations may be written as

$$q = h_1(t_h - t_i)A \qquad \text{and} \qquad q = h_2(t_i - t_c)A$$

where q = rate of heat transfer

A = area of interface

t_h, t_c = average bulk temperatures of phases 1 and 2

t_i = temperature of interface

h_1, h_2 = individual heat-transfer coefficients for phases 1 and 2

Two complications appear in mass transfer that are not found in heat transfer: (1) The factor ϕ must be taken into account in mass transfer, and no corresponding factor appears in the flow of heat. This is because, in mass transfer, a physical movement of material takes place, while heat transfer is a flow of energy only. (2) The interfacial equilibrium concentrations x_i and y_i are not equal, while in heat transfer the interfacial temperature t_i is the same for both phases.

Because the rate of loss of component A from the gas equals the rate of gain of that component by the liquid, the term dN_A in Eq. (11-2) equals that in Eq. (11-4). Also, by Eq. (10-45), the transfer rate from the gas is $d(Vy)$, and that to the liquid is $d(Lx)$. Then

$$dN_A = \frac{k_y}{\phi}(y - y_i)\, dA = k_x(x_i - x)\, dA = d(Vy) = d(Lx) \qquad (11\text{-}5)$$

In the absence of heat effects and chemical reactions, Eqs. (11-5) provide the basis for the theory of mass transfer in a packed tower.

Equations (11-5) contain terms for both gas and liquid resistances. These can be treated separately. The gas-phase equation is

$$\frac{k_y}{\phi} (y - y_i)\, dA = d(Vy) \tag{11-6}$$

This equation can be transformed for further use. First, since the actual area of transfer in a packed tower cannot readily be measured, the area dA is replaced by the product of the volume of the packed section by the area per unit volume. Also, if S is the cross-sectional area of the tower, the volume of the packing in the section of height dZ ft is $S\, dZ$, and

$$dA = aS\, dZ \tag{11-7}$$

where a is the area of interface per unit volume of packed section, in square feet per cubic foot of packed volume.

The term $d(Vy)$ also can be reduced to a more convenient form. Let V' be the flow rate of component B in moles per hour. Then $V = V'/(1 - y)$. Since component B is not absorbed, V' is constant throughout the tower. Therefore,

$$d(Vy) = V'\, d\left(\frac{y}{1 - y}\right) = V' \frac{dy}{(1 - y)^2} = V \frac{dy}{1 - y} \tag{11-8}$$

Substituting $d(Vy)$ from Eq. (11-8) and dA from Eq. (11-7) and dividing by S gives

$$\frac{k_y a(y - y_i)\, dZ}{\phi} = \frac{V}{S} \frac{dy}{1 - y} = G_{My} \frac{dy}{1 - y} \tag{11-9}$$

where G_{My}, the molal mass velocity of the gas in pound moles per square foot per hour, replaces V/S.

A corresponding treatment of the liquid-phase equation $k_x(x_i - x)\, dA = d(Lx)$ gives

$$k_x a(x_i - x)\, dZ = G_{Mx} \frac{dx}{1 - x} \tag{11-10}$$

where G_{Mx} is the molal liquid mass velocity. Note that G_{Mx} and G_{My} are based on the total cross section of the tower.

Factors k_y and a, and also factors k_x and a, are considered together as single combined quantities. They are evaluated as such by experimental tests on packed towers.

For practical use, Eqs. (11-9) and (11-10) must be integrated over the total depth of packing Z_T. The technique of this integration depends on three considerations: the shape of the equilibrium line, the change in composition of the stream in the tower, and the relative importance of the two resistances. In the most general situation, the equilibrium line is strongly curved, the inlet gas is concentrated and the outlet gas weak, and both resistances are important. In the simplest case the equilibrium line is straight, the changes in concentrations of both liquid and gas are small, and one or the other of the two resistances may be neglected.

General Case: $\Delta x \, \Delta y$ Triangle. In the following construction it is assumed that $k_y a$ and $k_x a$ are known. These coefficients are discussed later. Figure 11-11 shows the operating line CD and the equilibrium curve AB. The operating line is plotted as shown in Example 10-1, from a knowledge of the design conditions. Consider the level in the column where

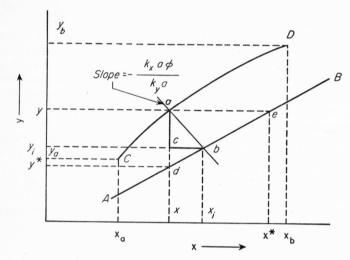

Fig. 11-11. $\Delta x \, \Delta y$ triangle, packed tower.

the gas and liquid concentrations are y and x, respectively. These concentrations are the coordinates of point a lying on the operating line CD.

Since $G_{My} \, dy/(1 - y) = G_{Mx} \, dx/(1 - x)$, it follows from Eqs. (11-9) and (11-10) that $(k_y a/\phi)(y - y_i) \, dZ = k_x a(x_i - x) \, dZ$, or

$$\frac{y - y_i}{x_i - x} = \frac{\phi k_x a}{k_y a} \tag{11-11}$$

Equation (11-11) is that of a straight line, with a slope $-(k_x a\phi/k_y a)$, passing through points (x,y) and (x_i,y_i).[15] Thus, if the factors ϕ, $k_x a$, and $k_y a$ are known, the slope can be calculated and the line ab drawn through point a. The coordinates of point b, which is the intersection of line ab with the equilibrium line, are x_i and y_i. The distance ac is the gas-resistance driving force $y - y_i$, and the distance bc is the liquid-resistance driving force $x_i - x$. The triangle abc may be called the $\Delta x \, \Delta y$ triangle. By constructing several such triangles along the operating line, either Δx or Δy can be determined graphically as a function of y or x.

To use these driving forces, the variables in Eq. (11-9) can be separated and the equation integrated graphically over Z_T, the length of the packed section.

$$\int_{y_a}^{y_b} \frac{\phi \, dy}{(1-y)(y-y_i)} = \overline{\left(\frac{k_y a}{G_{My}}\right)} \int_0^{Z_T} dZ = \overline{\left(\frac{k_y a}{G_{My}}\right)} Z_T \qquad (11\text{-}12)$$

It is assumed that Z is measured from the top down, that y_a is the concentration of the exit gas, and that y_b is the concentration of the inlet gas. The terms ϕ and $1-y$ are retained under the integral sign, as they may vary appreciably with y. The quantity $\overline{k_y a/G_{My}}$ is assumed to be constant. This is an approximation, since G_{My} decreases from bottom to top because of the absorption of component A and $k_y a$ depends on the mass velocity of the gas and also decreases from bottom to top. The effects of changes in these factors tend to compensate, and the ratio $k_y a/G_{My}$ is nearly constant unless the inlet gas is very concentrated. The variation that does occur in $k_y a/G_{My}$ can be taken into account by using an arithmetic mean of the inlet and outlet quantities. This integration also assumes that the effectiveness of the tower is the same at all values of Z. As shown by the data in Fig. 11-2, this assumption is not accurate if channeling is excessive.

If the concentration of the gas phase changes appreciably over the length of the tower, the factors ϕ and $k_y a$ vary also. These factors can be evaluated for the two ends of the tower and their arithmetic average used in calculating the slope of the $\Delta x \, \Delta y$ lines. Experimental data may be given either as $k_y a$ or as $k_y a/\phi$, and must be used accordingly. When ϕ is associated with $k_y a$, it is removed from the integral sign and used on the right-hand side of Eq. (11-12).

The liquid-side equation corresponding to Eq. (11-12) is

$$\int_{x_a}^{x_b} \frac{dx}{(1-x)(x_i-x)} = \overline{\left(\frac{k_x a}{G_{Mx}}\right)} Z_T \qquad (11\text{-}13)$$

In using this equation, $x_i - x$ is read from the $\Delta x \, \Delta y$ triangles and used to carry through a graphical integration of Eq. (11-13).

The choice between Eq. (11-12) and Eq. (11-13) is arbitrary. Both equations give the same result. The precision is better if the larger of the two driving forces is used.

Example 11-3. A tower packed with 1-in. rings is to be designed to absorb sulfur dioxide from air by scrubbing the gas with water. The entering gas is 20 per cent SO_2 by volume, and the leaving gas is to contain not more than 0.5 per cent SO_2 by volume. The entering H_2O is SO_2-free. The temperature is 30°C and the total pressure is 2 atm. The water flow is to be twice the minimum. The air flow rate (SO_2-free basis) is to be 200 lb/ft²-hr.

What depth of packing is required?

The following equations are available for the mass-transfer coefficients for absorption of SO_2 at 30°C in towers packed with 1-in. rings: [17]

$$k_x a = 0.152 G_x^{0.82} \qquad \frac{k_y a}{\varphi} = 0.028 G_y^{0.7} G_x^{0.25}$$

where G_x and G_y are the mass velocities of liquid and vapor, respectively, in pounds per square foot per hour, based on the total tower cross section.

Solution. The first step is to plot the equilibrium curve. Since the pressure is 2 atm, the curve of Example 11-2, which applies at 30°C and 1 atm, can be used if the ordinates are halved. This is because the partial pressure of the SO_2 is unchanged by the increase in pressure, and the mole fraction of SO_2 in the gas phase is now $\bar{p}_A/2$, which is one-half that at 1 atm for the same liquid concentration. The

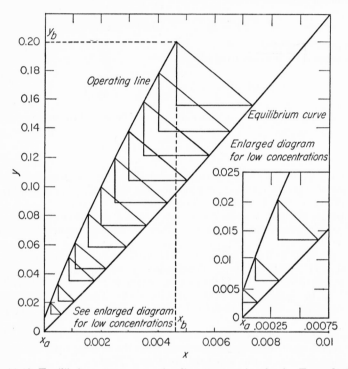

FIG. 11-12. Equilibrium curve, operating line, $\Delta x \, \Delta y$ triangles for Example 11-3.

equilibrium curve is shown in Fig. 11-12. It is slightly curved at the lower end but is straight when $x > 0.001$.

The minimum water rate is next calculated by using the equation of the operating line passing through points (x_b^*, y_b) and (x_a, y_a). Equation (10-20) can be written over the entire column as

$$G'_{Mx}\left(\frac{x_b}{1 - x_b} - \frac{x_a}{1 - x_a}\right) = G'_{My}\left(\frac{y_b}{1 - y_b} - \frac{y_a}{1 - y_a}\right)$$

where G'_{Mx} and G'_{My} are the molal mass velocities of sulfur dioxide–free air and water, respectively. From the conditions of the design:

$$y_b = 0.20 \qquad x_a = 0 \qquad y_a = 0.005 \qquad G'_{My} = \tfrac{200}{29} = 6.90 \text{ moles/ft}^2\text{-hr}$$

From the equilibrium curve, when $y_b = 0.20$, $x_b^* = 0.0092$, and the minimum water rate G'_{Mx} is given by the equation

$$G'_{Mx}\left(\frac{0.0092}{0.9908} - 0\right) = 6.90\left(\frac{0.20}{0.80} - \frac{0.005}{0.995}\right)$$

From this $G'_{Mx} = 182$ moles/ft²-hr.

The actual water rate is twice the minimum, or 364 moles/ft²-hr, and x_b is calculated from the equation

$$364\frac{x_b}{1 - x_b} = 6.90\left(\frac{0.20}{0.80} - \frac{0.005}{0.995}\right)$$

From this, $x_b = 0.00462$.

The equation for the operating line is

$$364\frac{x}{1 - x} = 6.90\left(\frac{y}{1 - y} - \frac{0.005}{0.995}\right)$$

Neglecting x in the term $1 - x$, this equation can be solved for x.

$$x = 0.0189\frac{y}{1 - y} - 0.00010$$

By assigning values to y between 0.005 and 0.200 and calculating the corresponding values of x points on the operating line are obtained. The operating line is plotted in Fig. 11-12. It is slightly concave toward the x axis.

To calculate the mass-transfer coefficients k_x and k_y/ϕ the mass velocities of both gas and liquid streams must be known. They are obtained from the material balances over the tower. The SO₂-free air flow is 6.9 moles/ft²-hr. The SO₂ entering with the gas is $(0.20/0.80)6.9 \times 64.1 = 111$ lb/ft²-hr. The total entering gas is, then, $200 + 111 = 311$ lb/ft²-hr. The SO₂ leaving with the gas is $(0.005/0.995)6.9 \times 64.1 = 2$ lb/ft²-hr, and the total exit gas is $200 + 2 = 202$ lb/ft²-hr. The SO₂ absorbed by the water is $111 - 2 = 109$ lb/ft²-hr. The water fed to the top of the tower is $364 \times 18 = 6,550$ lb/ft²-hr, and the strong liquor is $6,550 + 109 = 6,660$ lb/ft²-hr.

The liquid resistance does not change appreciably from the top to the bottom, and the liquid coefficient can be calculated from the average mass velocity of the liquid, which is $6,550 + 109/2 = 6,610$ lb/ft²-hr. Then

$$k_x a = 0.152 \times 6,610^{0.82} = 206$$

The gas resistance at the bottom of the tower is appreciably greater than that at the top because of the change in G_y. The quantity $k_y a/\phi$ is calculated for both ends of the packed section and the arithmetic average used as a constant. Then

$$\left(\frac{k_y a}{\phi}\right)_b = 0.028 \times 6,660^{0.25} \times 311^{0.7} = 14.1$$

$$\left(\frac{k_y a}{\phi}\right)_a = 0.028 \times 6,550^{0.25} \times 202^{0.7} = 10.4$$

The average of these is 12.3. The slope of the hypotenuse of the $\Delta x\ \Delta y$ triangles is $-206/12.3 = -16.7$.

In Fig. 11-12 are plotted the $\Delta x\ \Delta y$ triangles for a number of positions along the operating line. The intersections of the hypotenuses of the triangles with the equilibrium curve give the values of y_i corresponding to the assumed values of y.

Table 11-3 and Fig. 11-13 show the calculation of the ordinate $1/(1 - y)(y - y_i)$ and the graphical integration of

$$\int_{0.005}^{0.200} \frac{dy}{(1 - y)(y - y_i)}$$

TABLE 11-3. GRAPHICAL INTEGRATION FOR EXAMPLE 11-3

y	$1 - y$	y_i	$y - y_i$	$\dfrac{1}{(1 - y)(y - y_i)}$	Δy	Av. ordinate	Δ area
0.005	0.995	0.0027	0.0023	437			
0.01	0.99	0.0065	0.0035	288	0.005	355	1.77
0.02	0.98	0.0135	0.0065	157	0.01	207	2.07
0.03	0.97	0.0205	0.0095	108	0.01	128	1.28
0.05	0.95	0.0350	0.015	70.2	0.02	84	1.68
0.06	0.94	0.0430	0.017	62.5	0.01	65	0.65
0.08	0.92	0.0580	0.022	49.5	0.02	55.5	1.11
0.10	0.90	0.0730	0.027	41.1	0.02	45.0	0.90
0.12	0.88	0.0885	0.0315	36.0	0.02	38.5	0.77
0.14	0.86	0.105	0.035	33.2	0.02	34.5	0.69
0.16	0.84	0.1215	0.0385	30.9	0.02	32	0.64
0.18	0.82	0.138	0.042	29.1	0.02	29.5	0.59
0.20	0.80	0.1555	0.0445	28.1	0.02	28.5	0.57
Total...	....	....	....	.....	....	.....	12.72

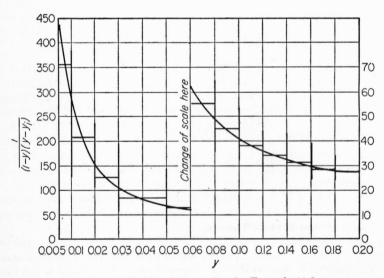

FIG. 11-13. Graphical integration for Example 11-3.

The integral is equal to 12.72. Also,

$$\left(\frac{k_y a}{\phi G_{My}}\right)_a = \frac{10.4}{6.90/0.995} = 1.50$$

$$\left(\frac{k_y a}{\phi G_{My}}\right)_b = \frac{14.1}{6.90/0.80} = 1.63$$

The average of these is 1.56; and, from Eq. (11-12), the height of the packed section is $Z_T = 12.72/1.56 = 8.21$ ft.

Example 11-4. Check the design of the tower of Example 11-3 with respect to flooding.

Solution. For 1-in. rings, $a_v = 58$ ft^2/ft^3, and $\epsilon = 0.73$ (Table 11-1). Also, at the bottom of the tower, $\overline{M}_y = 0.8 \times 29 + 0.2 \times 64 = 36.7$.

$$\rho_y = \frac{36.7 \times 2 \times 273}{359 \times 303} = 0.184 \text{ lb/ft}^3$$

$$\rho_x = 62.3 \text{ lb/ft}^3 \qquad G_x = 6{,}660 \text{ lb/ft}^2\text{-hr}$$

$$G_y = 311 \text{ lb/ft}^2\text{-hr} \qquad \mu_r' = 0.80 \text{ centipoise}$$

$$\frac{G_x}{G_y}\sqrt{\frac{\rho_y}{\rho_x}} = \frac{6{,}660}{311}\sqrt{\frac{0.184}{62.3}} = 1.16$$

From Fig. 11-4,

$$\frac{G_y^2 a_v (\mu_x')^{0.2}}{g_c \epsilon^3 \rho_y \rho_x} = 0.017$$

$$G_y = \sqrt{\frac{0.017 \times 4.17 \times 10^8 \times 0.73^3 \times 62.3 \times 0.184}{58 \times 0.80^{0.2}}} = 755 \text{ lb/ft}^2\text{-hr}$$

Since the maximum mass flow rate in the tower is 311 lb/ft^2-hr, the loading of the tower is $311/755 = 0.41$, or 41 per cent of the flooding velocity. This is safe and is also consistent with sound design.

Curved and Linear Equilibrium Lines. The general method described in the last section is applicable to both curved and straight equilibrium lines. Curved lines are encountered most often when the temperature in the tower varies appreciably from bottom to top. This temperature gradient is found when rich gas is fed to the tower. The rate of absorption is large at the gas inlet, and the heat of condensation and solution of the absorbed constituent may be sufficient to increase considerably the temperature of the liquid. Since the partial pressure of the absorbed component increases with increase in temperature, the concentration of the vapor in equilibrium with a liquid of definite composition also increases with temperature. Even if Henry's law applies, the equilibrium line may be strongly curved in this situation.

The effect on the equilibrium line of a temperature gradient in a tower is shown in Fig. 11-14. Line *OA* is the equilibrium line for isothermal operation, and line *OB* is that where the temperature at the bottom of the tower is larger than that at the top. The top temperature is the same in both situations. If the temperature effect is sufficiently great, the equilibrium line may intersect the operating line, and the process is inoperative near the bottom of the tower. Sometimes cooling coils or other cooling means are installed in the tower to reduce this heat effect.

The calculation of the equilibrium line under nonisothermal conditions can be accomplished by enthalpy balances.[13a]

In most towers fed with dilute or moderately strong gas, the temperature gradient in the column is small and the equilibrium line straight or nearly so. When the curvature of this line can be neglected, the methods described in the following sections can be used.

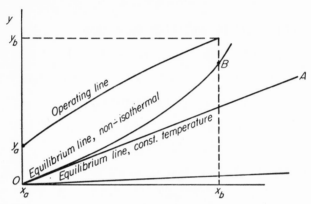

Fig. 11-14. Effect of temperature gradient on equilibrium line. (*Sherwood and Pigford.*[13])

Over-all Coefficients. The experimental determination of the individual coefficients $k_y a$ and $k_x a$ is very difficult, and they are not always known for the system and equipment of specific interest. When the equilibrium line is straight, over-all coefficients, which are more easily determined by experiment, can be used. Over-all coefficients also are simpler to use than individual coefficients because construction of the $\Delta x \, \Delta y$ triangles is unnecessary.

Over-all coefficients are analogous to those used in heat transfer, but because of the change in driving force between phases, they can be defined from the standpoint of either the liquid phase or the gas phase. Each coefficient is based on a calculated over-all driving force. Thus, referring to Fig. 11-11, by continuing the vertical line ac to a point d on the equilibrium curve the quantity y^* is found. This is the equilibrium composition of a gas corresponding to a liquid of concentration x. Since, in an actual tower, equilibrium is not achieved at any location, the quantity y^* has no physical significance in the tower itself but is a mathematical fiction. The over-all driving force is defined as that measured by line segment ad, or $y - y^*$.

Likewise, if a liquid-side equation is to be used, a horizontal line ae, intersecting the equilibrium line at e, defines the liquid composition x^*, which is the composition a liquid would have if it were equilibrated with gas of composition y. The over-all driving force is then represented by line segment ae, or $x^* - x$.

The over-all gas-resistance coefficient K_y is defined by the equation

$$\frac{K_y}{\phi} = \frac{dN_A/dA}{y - y^*} \tag{11-14}$$

and the over-all liquid-resistance coefficient K_x by

$$K_x = \frac{dN_A/dA}{x^* - x} \tag{11-15}$$

Comparison of Eqs. (11-14) and (11-2) shows that

$$\frac{1}{K_y} = \frac{y - y^*}{k_y(y - y_i)} = \frac{(y - y_i) + (y_i - y^*)}{k_y(y - y_i)} = \frac{1}{k_y} + \frac{y_i - y^*}{k_y(y - y_i)}$$

Eliminating $y - y_i$ by use of Eq. (11-11) gives

$$\frac{\phi}{K_y} = \frac{\phi}{k_y} + \frac{y_i - y^*}{k_x(x_i - x)}$$

Reference to Fig. 11-11 shows that $(y_i - y^*)/(x_i - x)$ is the slope of the linear equilibrium line. Call this slope m. Then

$$\frac{\phi}{K_y} = \frac{\phi}{k_y} + \frac{m}{k_x} \tag{11-16}$$

Also, dividing by a gives the over-all coefficient based on unit packed volume $K_y a$.

$$\frac{\phi}{K_y a} = \frac{\phi}{k_y a} + \frac{m}{k_x a} \tag{11-17}$$

A corresponding derivation gives

$$\frac{1}{K_x a} = \frac{1}{k_x a} + \frac{\phi}{m k_y a} = \frac{\phi}{m K_y a} \tag{11-18}$$

where $K_x a$ is the over-all liquid-resistance coefficient, based on unit packed volume. The units of both $K_y a$ and $K_x a$ are the same as those of $k_y a$ and $k_x a$, namely, lb moles/$(\text{ft}^3)(\text{hr})(\text{mole fraction})$.

When changes in ϕ over the length of the tower, which are usually small, are neglected, $K_y a$ and $K_x a$ are constant if $k_y a$ and $k_x a$ are constant. When the equilibrium line is curved, the over-all coefficients are not constant, and they cannot then be used safely in design unless the concentration range covered in the actual tower is nearly the same as that covered by the experiments on which the over-all coefficients are based.

Equations (11-12) and (11-13) become, in terms of over-all coefficients,

$$\int_{y_a}^{y_b} \frac{\phi \, dy}{(1-y)(y-y^*)} = \overline{\left(\frac{K_y a}{G_{My}}\right)} Z_T \tag{11-19}$$

and

$$\int_{x_a}^{x_b} \frac{dx}{(1-x)(x^*-x)} = \overline{\left(\frac{K_x a}{G_{Mx}}\right)} Z_T \tag{11-20}$$

Either Eq. (11-19) or (11-20) may be used, depending on convenience. The integrals can be evaluated graphically, by reading $y - y^*$ or $x^* - x$ from a plot like Fig. 11-12. Alternatively, if the factor ϕ is included in k_y, it is removed from the integral sign and used with $K_y a$ in the right-hand side of Eq. (11-19), as suggested by Eq. (11-17).

Example 11-5. Repeat Example 11-3, assuming a linear equilibrium line and using over-all gas and liquid coefficients.

Solution. The equilibrium line is nearly straight between $y = 0.01$ and $y = 0.200$. The slope of the line between these points is

$$m = \frac{0.200 - 0.010}{0.0092 - 0.0005} = 21.8$$

The over-all gas-side coefficient is, from Eq. (11-17),

$$\frac{\phi}{K_y a} = \frac{\phi}{k_y a} + \frac{m}{k_x a} = \frac{1}{12.3} + \frac{21.8}{206}$$

$$= 0.081 + 0.106 = 0.187$$

$$\frac{K_y a}{\phi} = \frac{1}{0.187} = 5.35 \text{ lb moles}/(\text{ft}^3)(\text{hr})(\text{mole fraction})$$

Here ϕ is lumped with $K_y a$ because it is included in $k_y a$. To use this coefficient the following integral is evaluated.

$$\int_{0.005}^{0.200} \frac{dy}{(1-y)(y-y^*)}$$

The over-all driving force $y - y^*$ is read from the vertical distances between the operating and equilibrium lines of Fig. 11-12 for corresponding values of y and the integration performed graphically. The integral is found to be 5.74. At the top of the tower $(G_{My})_a = 6.90/0.995 = 6.93$ lb moles$/(\text{ft}^2)(\text{hr})$; at the bottom, $(G_{My})_b = 6.90/0.80 = 8.63$ lb moles$/(\text{ft}^2)(\text{hr})$. The average is 7.78. From Eq. (11-19),

$$Z_T = \frac{5.74 \times 7.78}{5.35} = 8.3 \text{ ft}$$

The over-all liquid coefficient is given by Eq. (11-18).

$$\frac{1}{K_x a} = \frac{1}{k_x a} + \frac{\phi}{m k_y a} = \frac{1}{206} + \frac{1}{21.8 \times 12.3} = \frac{1}{0.00858}$$

and

$$K_x a = 117 \text{ lb moles}/(\text{ft}^3)(\text{hr})(\text{mole fraction})$$

To use this coefficient, the following integral is evaluated.

$$\int_0^{0.00462} \frac{dx}{x^* - x}$$

Since x is small, the term $1 - x$ can be considered unity. The over-all driving force $x^* - x$ is read from the horizontal distance between the operating and equilibrium lines and the integral evaluated graphically. The result is an area of 2.35. From Example 11-3, G_{Mx} is 364 lb moles/ft²-hr, and from Eq. (11-20),

$$Z_T = \frac{2.35 \times 364}{117} = 7.3 \text{ ft}$$

HTU Method. A useful method for computing the height of a packed tower is based on the idea of calculating the number of a certain kind of contact units required and multiplying the number of such units by the height of a single unit. This contact unit is called a *transfer unit*,[1,2] and the depth of packing required by a single unit is called the height of one transfer unit, abbreviated to HTU. Then the total height of the packed section is

$$Z_T = N_t H \tag{11-21}$$

where N_t is the number of transfer units and H is the HTU.

The number of transfer units in a column of total height Z_T ft is defined by any one of the following four equations.

$$N_{ty} = \int_{y_a}^{y_b} \frac{\phi \, dy}{(1 - y)(y - y_i)} \tag{11-22a}$$

$$N_{tOy} = \int_{y_a}^{y_b} \frac{\phi \, dy}{(1 - y)(y - y^*)} \tag{11-22b}$$

$$N_{tx} = \int_{x_a}^{x_b} \frac{dx}{(1 - x)(x_i - x)} \tag{11-22c}$$

$$N_{tOx} = \int_{x_a}^{x_b} \frac{dx}{(1 - x)(x^* - x)} \tag{11-22d}$$

Each of these numbers differs from the others in any given situation, and the choice among them is a matter of convenience. The differences among them are compensated for by the magnitudes of the corresponding HTUs. The quantities N_{tOy} and N_{tOx} are based on over-all driving forces and N_{ty} and N_{tx} on individual driving forces.

Substitution of N_{ty} from Eq. (11-22a) in Eq. (11-12) gives

$$N_{ty} = \overline{\left(\frac{k_y a}{G_{My}}\right)} Z_T$$

Comparison of this equation with Eq. (11-21) shows that

$$H_y = \overline{\left(\frac{G_{My}}{k_y a}\right)} \tag{11-23a}$$

Likewise, using Eqs. (11-13), (11-19), and (11-20) with Eqs. (11-22c), (11-22b), and (11-22d), respectively.

$$H_x = \overline{\left(\frac{G_{Mx}}{k_x a}\right)} \qquad (11\text{-}23b)$$

$$H_{Oy} = \overline{\left(\frac{G_{My}}{K_y a}\right)} \qquad (11\text{-}23c)$$

and
$$H_{Ox} = \overline{\left(\frac{G_{Mx}}{K_x a}\right)} \qquad (11\text{-}23d)$$

Quantities H_{Oy} and H_{Ox} are over-all HTUs and H_y and H_x are individual HTUs. Each must be used only with its corresponding N_t.

The over-all HTUs are related to the individual HTUs as follows. Elimination of $K_y a$, $k_y a$, and $k_x a$ from Eq. (11-17) by Eqs. (11-23a), (11-23b), and (11-23c) gives

$$\frac{\phi H_{Oy}}{G_{My}} = \frac{\phi H_y}{G_{My}} + \frac{m H_x}{G_{Mx}}$$

or
$$\phi H_{Oy} = \phi H_y + \frac{m G_{My}}{G_{Mx}} H_x \qquad (11\text{-}24)$$

It is assumed here that $G_{Mx}/k_x a$ and $G_{My}/k_y a$ are constant. Likewise, eliminating $K_x a$, $k_y a$, and $k_x a$ from Eq. (11-18) gives

$$H_{Ox} = H_x + \frac{\phi G_{Mx}}{m G_{My}} H_y = \frac{\phi G_{Mx}}{m G_{My}} H_{Oy} \qquad (11\text{-}25)$$

The over-all HTUs, H_{Oy} and H_{Ox}, are constant when the factors m, G_{Mx}, G_{My}, ϕ, k_y, and k_x are constant throughout the tower. Straight operating and equilibrium lines demand constancy of these same factors, and the straighter the lines, the less the variation in the HTUs. The HTU method is most useful when the equilibrium line is straight and the curvature of the operating line negligible. Equations have been derived for correcting for moderate curvatures in these lines.[3, 9]

Advantages of HTU Method. The HTU is closely related to the mass-transfer coefficient, and the two quantities are essentially equivalent. The HTU is simpler to visualize, as its dimension is simply length and it is measured in feet. The usual order of magnitude of this quantity is 0.5 to 5 ft. The units of the mass-transfer coefficient are more complex, and numerical magnitudes vary over wide limits. Also, since in packed towers $k_y a$ increases with G_{My} and $k_x a$ with G_{Mx}, the ratios $G_{My}/k_y a$ and $G_{Mx}/k_x a$, and therefore H_y and H_x, are nearly independent of the flow rates of liquid and gas. Most modern experimental data on packed towers are given in HTUs rather than in coefficients.

The physical meaning of a transfer unit may be seen by examining the formula for one unit, which, neglecting the effects of ϕ and $1 - y$, is

$$\int_{y}^{y+\Delta y} \frac{dy}{y - y_i} = 1$$

Let the average ordinate under the integral sign be $1/\overline{y - y_i}$, and let Δy be the change in y achieved by one unit. Then $\Delta y/\overline{y - y_i} = 1$. A transfer unit, therefore, is a section of packed tower that achieves a change in composition equal to the average driving force in that section.

Example 11-6. Repeat Example 11-3, using over-all gas and liquid HTUs.

Solution. OVER-ALL GAS-RESISTANCE HTU. Since ϕ is incorporated in $k_y a$, define N_{tOy} by the equation

$$N_{tOy} = \int_{0.005}^{0.200} \frac{dy}{(1 - y)(y - y^*)}$$

From Example 11-5, the above integral, and therefore N_{tOy}, is 5.74 units. Modifying Eq. (11-23c) to include ϕ gives

$$H_{Oy} = \frac{G_{My}\phi}{K_y a}$$

From Example 11-5, $G_{My} = 7.78$; $K_y a/\phi = 5.35$; and

$$H_{Oy} = \frac{7.78}{5.35} = 1.45 \text{ ft}$$

The total packed height is

$$Z_T = 1.45 \times 5.74 = 8.3 \text{ ft}$$

OVER-ALL LIQUID-RESISTANCE HTU. From Example 11-5, $N_{tOx} = 2.35$ units; $G_{Mx} = 364$; $K_x a = 113$; and, from Eq. (11-23d),

$$H_{Ox} = \frac{364}{113} = 3.11 \text{ ft}$$

and

$$Z_T = 3.11 \times 2.35 = 7.3 \text{ ft}$$

Clearly, these results must agree with those of Example 11-5, since the same numbers are used in each example. The HTUs for both gas and liquid are of comparable magnitude, 1.45 and 3.11 ft respectively, while the corresponding coefficients $K_y a/\phi$ and $K_x a$ are of different magnitude, 5.35 and 117, respectively.

Gas-resistance or Liquid-resistance Controlling. With a slightly soluble gas, m, the slope of the equilibrium line, is large, as the concentration in the gas phase at equilibrium is much greater than that in the liquid phase. From Eq. (11-25) it is apparent that unless the ratio G_{Mx}/G_{My} is also large, the term $\phi G_{Mx}/mG_{My}$ is small, and H_{Ox} is nearly equal to H_x. The effect of the gas film can then be ignored, and the liquid resistance is said to "control." The situation is analogous to that in heat transfer, when only one thermal resistance is important in comparison with the others and the over-all coefficient is nearly equal to an individual coefficient.

For soluble gases, the slope m is small, since at equilibrium a low gas concentration corresponds to a large liquid concentration. Then, unless changes in other factors, such as the ratio G_{My}/G_{Mx}, compensate for the small magnitude of m, the term $mG_{My}H_x/G_{Mx}$ in Eq. (11-24) is small, and H_{Oy} is nearly equal to H_y. Then conditions in the liquid are unimportant, and the gas resistance controls.

The conclusion that the gas or liquid resistance controls in a given situation must be drawn cautiously, for the conditions of the operation may be such that a low or high magnitude of m may be compensated by the magnitude of the gas-liquid ratio. For example, when a gas of low solubility is being absorbed, a high liquid rate is used to ensure adequate removal of the solute gas. Then G_{Mx}/G_{My} is large, so the ratio G_{Mx}/mG_{My} may be appreciable in spite of the large value of m, and, as shown by Eq. (11-25), both gas and liquid resistances are important. In most industrial operations both resistances are important, but in experimental work conditions are often deliberately fixed so one resistance is negligible in order that the other resistance can be more easily studied.

Lean Gases. When y is small throughout the tower, the gas is called lean. Several simplifications appear in absorbing such gases. First, the factors $1 - y$ and ϕ are nearly unity, and can be so considered. Equation (11-19), for example, becomes

$$\int_{y_a}^{y_b} \frac{dy}{y - y^*} = \overline{\left(\frac{K_y a}{G_{My}}\right)} Z_T \tag{11-26}$$

Second, the individual quantities $k_y a$, $k_x a$, H_y, and H_x are constant throughout the equipment. Third, the liquid phase is usually also dilute, quantities such as $1 - x$ and $1 - y$ in the equation of the operating line [Eq. (10-20)] become unity, and $L = L'$ and $V = V'$. The equation of the operating line then becomes

$$Vy + x_a L = Vy_a + xL \tag{11-27}$$

Then the operating line is straight, with a slope of L/V. It is readily plotted by drawing a straight line through the two points (x_a, y_a) and (x_b, y_b).

Linear Equilibrium and Logarithmic Mean Driving Force. When both the operating line and the equilibrium line are straight, Eq. (11-26) can be integrated formally. Since y and y^* are both linear with x, their difference is also linear. By using the same method used in Chap. 8 to derive the logarithmic mean temperature difference it is found that

$$G_{My}(y_b - y_a) = K_y a \overline{(y - y^*)}_L Z_T = \frac{N_A}{S} \tag{11-28}$$

where
$$\overline{(y - y^*)}_L = \frac{(y_b - y_b^*) - (y_a - y_a^*)}{2.303 \log \dfrac{y_b - y_b^*}{y_a - y_a^*}} = \overline{\Delta y_L} \tag{11-29}$$

The logarithmic mean over-all concentration difference is used in the same way the logarithmic mean over-all temperature difference is used in heat transfer.

For the liquid phase, the corresponding equation is

$$G_{Mx}(x_b - x_a) = K_x a \overline{(x^* - x)}_L Z_T = \frac{N_A}{S} \tag{11-30}$$

where

$$\overline{(x^* - x)}_L = \frac{(x_b^* - x_b) - (x_a^* - x_a)}{2.303 \log \dfrac{x_b^* - x_b}{x_a^* - x_a}} = \overline{\Delta x_L} \tag{11-31}$$

In terms of over-all HTUs Eqs. (11-28) and (11-30) can be written

$$N_{tOy} = \frac{y_b - y_a}{(y - y^*)_L} = \frac{y_b - y_a}{\overline{\Delta y_L}} \tag{11-32}$$

and

$$N_{tOx} = \frac{x_b - x_a}{\overline{(x^* - x)}_L} = \frac{x_b - x_a}{\overline{\Delta x_L}} \tag{11-33}$$

Example 11-7. Repeat Example 11-3, assuming that the operating and equilibrium lines are straight.

Solution. From Example 11-5, $K_y a = 5.35$ lb moles/(ft^3)(hr)(Δy), and $G_{My} = 7.78$ lb moles/ft^2-hr. Also, from Fig. 11-12, $y_a = 0.005$, $y_b = 0.200$, $y_a^* = 0$, $y_b^* = 0.0955$. From Eq. (11-29),

$$\overline{\Delta y_L} = \frac{(0.200 - 0.0955) - (0.005 - 0)}{2.303 \log \dfrac{0.200 - 0.0955}{0.005 - 0}} = 0.0327$$

$$N_A/S = 7.78(0.200 - 0.005) = 1.52 \text{ lb moles/ft}^2\text{-hr}$$

From Eq. (11-28),

$$Z_T = \frac{N_A/S}{K_{ya}\Delta y_L} = \frac{1.52}{5.35 \times 0.0327} = 8.7 \text{ ft}$$

Coefficients and HTUs in Packed Towers

Although the construction of a packed tower is simple, the action in this equipment is complicated. The process consists of the flow of gas through a bed of solids, the distribution and flow of liquid over the same solids, the interaction of the two fluid streams, and the transfer of the solute component from one stream to the other. Chemical reactions in the liquid also may be important in many situations. Little is known about the details of the process. It is difficult to measure either resistance independently of the other and so to analyze an over-all HTU into the individual ones. The assumption of interfacial equilibrium is questionable, and the two-resistance theory is clearly an oversimplification. Under these circumstances, research on tower coefficients has necessarily been highly empirical, although it has been guided by dimensional analysis.

An individual HTU depends upon both the coefficient k and the area per unit volume a. The variables that control the k of a single phase are, by assumption, the mass velocities of both phases, the viscosity, density, and diffusivity of the phase, and the size and shape of the packing. The same variables, with the exception of the diffusivity, affect the quantity a. It is also known that a is affected by the interfacial tension between the fluid phases, by the presence of surface-active agents in the liquid, and by the wetting characteristics of the packing.[11] The magnitude of a does not bear any simple relationship to the area or shape of the units of packing.

Neglecting the effects of surface factors, the HTU of one phase can be written as

$$H = \psi'(G_y, G_x, \mu, D_m, D_p, b_1, b_2, \ldots, b_n) \tag{11-34}$$

where G_y = mass velocity of gas
$\quad\quad\quad G_x$ = mass velocity of liquid
$\quad\quad\quad \mu$ = viscosity of stream
$\quad\quad\quad D_m$ = molal diffusivity in stream
$\quad\quad\quad D_p$ = size of packing
$b_1, b_2, \ldots, b_n$ = lengths sufficient to define geometry of packing and its arrangement in tower

A dimensional analysis, conducted by the usual method, gives for a gas-phase HTU the result

$$\frac{H_y}{D_p} = \psi_y\left(\frac{D_p G_y}{\mu_y}, \frac{G_x}{G_y}, \frac{\mu_y}{D_{my}\overline{M}_y}, \frac{b_1}{D_p}, \frac{b_2}{D_p}, \ldots, \frac{b_n}{D_p}\right) \tag{11-35}$$

The analogous equation for an HTU based on the liquid is

$$\frac{H_x}{D_p} = \psi_x\left(\frac{D_p G_x}{\mu_x}, \frac{G_y}{G_x}, \frac{\mu_x}{D_{mx}\overline{M}_x}, \frac{b_1}{D_p}, \frac{b_2}{D_p}, \ldots, \frac{b_n}{D_p}\right) \tag{11-36}$$

The Reynolds and Schmidt numbers again appear. The molecular weights of gas and liquid are $\overline{M}_y$ and $\overline{M}_x$, respectively.

In spite of much careful experimental work on coefficients and HTUs, sufficient data are not available to use Eqs. (11-35) and (11-36) completely, and existing correlations account for only some of the variables. For example, the characteristic particle size D_p for packings of different shapes has not been chosen, and the effects of the geometrical shape factors, such as b_1/D_p, are unknown. Some of the more important correlations of HTU data are given in the following paragraphs.

Liquid-resistance HTU. Two kinds of liquid resistances are encountered. For relatively insoluble gases that do not react chemically with the liquid, mass transfer is a physical process. An effective correlation has been found for such gases,[12] which may be written

$$H_x = \frac{1}{\alpha}\left(\frac{G_x}{\mu_x}\right)^n\left(\frac{\mu_x}{D_{mx}\overline{M}_x}\right)^{0.5} \tag{11-37}$$

where α and n are constants, the magnitudes of which are given in Table 11-4.

TABLE 11-4. VALUES OF α AND n OF EQ. (11-37) † FOR VARIOUS PACKING
MATERIALS AT 77°F ‡

Packing type	Packing size, in.	α	n
Rings............	2	80	0.22
	1.5	90	0.22
	1	100	0.22
	0.5	280	0.35
	0.375	550	0.46
Saddles..........	1.5	160	0.28
	1	170	0.28
	0.5	150	0.28
Tile.............	3	110	0.28

† All quantities in Eq. (11-37) must be expressed in fps units if these values of α are used.

‡ From T. K. Sherwood and F. A. L. Holloway, *Trans. AIChE*, 36: 39 (1940).

In the absorption of some common soluble gases in water, ionization and hydrolysis reactions between water and the solute occur which modify the liquid HTU. Chlorine, sulfur dioxide, and probably ammonia show this effect, and Eq. (11-37) does not apply. For example, the following equation for the absorption of sulfur dioxide in water at 77°F is available.[17]

$$k_x a = 0.152 G_x^{0.82} \tag{11-38}$$

where G_x is the liquid mass velocity in pounds per square foot per hour.

The liquid-resistance HTU varies with temperature because of the effect of temperature on the Schmidt number. This temperature effect can be estimated by the equation

$$H_x = H_{x0} e^{-0.013(t - t_0)} \tag{11-39}$$

where H_{x0} = HTU at t_0°F
H_x = HTU at t°F

Gas-resistance HTU. Correlations of data for H_y are not well established. An equation of the following form is usually found in correlating experimental data [10b, 13c]

$$H_y = \beta \frac{G_y^q}{G_x^r} \tag{11-40}$$

where β, q, and r are empirical constants. For different systems treated in the same type of packing, H_y can be assumed to vary with the square root of the Schmidt number. Equation (11-40) is clearly dimensional, and the constants in the equation depend on the choice of units. The gas-phase resistance is not sensitive to temperature, because the Schmidt number does not vary greatly with temperature.

As an example of the application of Eq. (11-40) to a specific situation, the following equation has been found for the absorption of sulfur dioxide in water over a packing of 1-in. rings.[17]

$$H_y = 1.23 \frac{G_y^{0.3}}{G_x^{0.25}} \tag{11-41}$$

where G_y and G_x are the mass velocities of gas and liquid, in pounds per square foot per hour, respectively. This equation, when used with Eq. (11-38), includes the effect of chemical reactions in the liquid phase.

Over-all Coefficients and HTUs. Data on the over-all quantities K_y, K_x, H_{Oy}, and H_{Ox} are given in standard references and handbooks.[10c, 13d] When available, these data are useful, provided they are used under conditions comparable to those under which they were determined. This is especially true when the equilibrium line is strongly curved, because of the importance of m in the relationships of over-all and individual factors.

Absorption in Plate Columns. The calculation of the number of ideal stages in a plate tower is accomplished by the methods of Chap. 10. If either the operating line or the equilibrium line is curved, the McCabe-Thiele method may be used. If both lines are straight, either the McCabe-Thiele or the absorption-factor method [Eq. (10-37)] is applicable. The quotient of the number of ideal stages divided by the plate efficiency gives the number of actual plates. Plate efficiencies are discussed in Chap. 12.

Example 11-8. Assuming a plate efficiency of 20 per cent, how many actual plates are required to accomplish the absorption of sulfur dioxide called for in Example 11-3?

Solution. The McCabe-Thiele construction is shown in Fig. 11-15; 3.8 ideal stages are needed. The actual number of stages is 3.8/0.20 = 19.

The number of ideal stages also can be approximately estimated from Eq. (10-37), although neither the operating nor the equilibrium lines are quite straight. Then

$$y_a = 0.005 \qquad y_b = 0.200$$

$$y_a^* = 0 \qquad y_b^* = 0.0955$$

Substituting these numbers in Eq. (10-37) gives, for the number of ideal stages,

$$N_p = \frac{\log\left[(0.200 - 0.0955)/0.005\right]}{\log\left[(0.200 - 0.005)/0.0955\right]} = 4.26$$

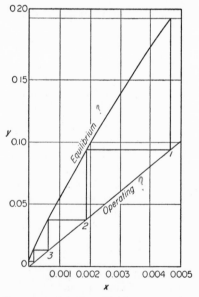

FIG. 11-15. McCabe-Thiele construction for Example 11-8.

The actual number of plates is 4.26/0.20, or 21+. The assumption that the operating and equilibrium lines are straight introduces a positive error of 2+ plates, or 10 per cent.

SYMBOLS

A Area of interface between phases, ft^2

a Area of interface per unit packed volume, ft^2/ft^3; a_v, area of dry packing per unit packed volume

$b_1, b_2, \ldots, b_n$ Defining lengths for packed bed, ft

c'_A Concentration of component A (solute), lb/100 lb of solvent

D_m Molal diffusivity, lb moles/ft-hr; D_{mx}, in liquid; D_{my}, in gas

D_p Linear dimension of unit of packing, ft

G Mass velocity, based on total tower cross section, lb/ft^2-hr; G_x, of liquid stream; G_y, of gas stream

G_M Molal mass velocity based on total tower cross section, lb moles/ft^2-hr; G_{Mx}, of liquid stream; G_{My}, of gas stream

g_c Newton's-law conversion factor, 4.17×10^8 ft-lb/lb force-hr^2

H Height of transfer unit, ft; H_{Ox}, over-all, based on liquid phase; H_{Oy}, over-all, based on gas phase; H_x, individual, based on liquid phase; H_{x0}, at temperature t_0; H_y, individual, based on gas phase

H'_A Henry's law constant, component A, mm Hg/(lb solute/100 lb solvent)

K Over-all mass-transfer coefficient, lb moles/(ft^2)(hr)(unit mole fraction); K_x, based on liquid phase; K_y, based on gas phase

k Individual mass-transfer coefficient, lb moles/(ft^2)(hr)(unit mole fraction); k_x, for liquid phase; k_y, for gas phase

L Molal flow rate of liquid, lb moles/hr; L', molal flow rate of solute-free liquid

$(L'/V')_{min}$ Minimum liquid-gas ratio, based on solute-free streams

M Molecular weight; M_A, of component A; M_B, of component B; $\overline{M}_x$, average molecular weight of liquid; $\overline{M}_y$, average molecular weight of gas

m Slope of equilibrium curve, dy_e/dx_e

N Rate of mass transfer or flux across interface between phases, lb moles/hr; N_A, of component A; N_B, of component B

N_p Number of ideal stages

N_t Number of transfer units; N_{tOx}, over-all, based on liquid phase; N_{tOy}, over-all, based on gas phase; N_{tx}, individual, liquid phase; N_{ty}, individual, gas phase

n Exponent in Eq. (11-37)

p Total pressure, atm

$\bar{p}$ Partial pressure; $\bar{p}_A$, of component A, atm; $\bar{p}'_A$, of component A, mm Hg

q Exponent in Eq. (11-40)

r Exponent in Eq. (11-40)

S Cross-sectional area of tower, ft^2

t Temperature, °F

V Molal flow rate of gas, lb moles/hr; V', molal flow rate of solute-free gas

x Mole fraction component A (solute) in liquid; x_a, at liquid inlet; x_b, at liquid outlet; x_e, at equilibrium; x_i, at gas-liquid interface; x_a^*, equilibrium concentration corresponding to gas-phase concentration y_a; x_b^*, equilibrium concentration corresponding to gas-phase concentration y_b

y Mole fraction component A (solute) in gas; y_a, at gas outlet; y_b, at gas inlet; y_e, at equilibrium; y_i, at gas-liquid interface; y_a^*, equilibrium concentration corresponding to liquid-phase concentration x_a; y_b^*, equilibrium concentration corresponding to liquid-phase concentration x_b

Z Vertical distance below top of packing, ft; Z_T, total depth of packed section

z Ratio, flux of component A to flux of entire phase, in direction toward gas-liquid interface, $N_A/(N_A + N_B)$

Greek Letters

α Constant in Eq. (11-37)

β Constant in Eq. (11-40)

Δx Concentration difference over liquid resistance, $x_i - x$, mole fraction component A; $\overline{\Delta x_L}$, over-all logarithmic mean concentration difference over liquid resistance, defined by Eq. (11-31), mole fraction component A

Δy Concentration difference over gas resistance, $y - y_i$, mole fraction component A; $\overline{\Delta y_L}$, logarithmic mean concentration difference over gas resistance, defined by Eq. (11-29), mole fraction component A

ϵ Porosity, or fraction voids, in packed section

μ Viscosity, lb/ft-hr; μ_x, of liquid; μ_y, of gas; μ_x', of liquid, centipoises

ρ Density, lb/ft^3; ρ_x, of liquid; ρ_y, of gas

ϕ Relative velocity factor, defined by Eq. (10-72)

ψ Function; ψ_x, in Eq. (11-36); ψ_y, in Eq. (11-35); ψ', in Eq. (11-34)

PROBLEMS

11-1. A plant design calls for an absorber which is to recover 95 per cent of the acetone in an air stream, using water as the absorbing liquid. The entering air contains 14 mole per cent acetone. The absorber is to operate at 80°F and 1 atm, and is to produce a product containing 7.0 mole per cent acetone. The water fed to the tower contains 0.02 mole per cent acetone. The tower is to be designed to operate at 50 per cent of the flooding velocity.

(a) How many pounds per hour of water must be fed to the tower if the gas rate is 500 ft^3/min, measured at 1 atm and 32°F? (b) How many transfer units are needed, based on the over-all gas-phase driving force? (c) If the tower is packed with 1-in. Raschig rings, what should be the packed height?

For Equilibrium. Assume that $\bar{p}_A = p_A \gamma_A x$, where $\ln \gamma_A = 1.95(1 - x)^2$.

For Mass Transfer. Use Eq. (11-37) for H_x and calculate H_y from

$$H_y = \frac{1.01 G_y^{0.31}}{G_x^{0.33}}$$

The vapor pressure of acetone at 80°F is 0.33 atm.

11-2. An absorber is to recover 99 per cent of the ammonia in the air-ammonia stream fed to it, using water as the absorbing liquid. The ammonia content of the air is 30 mole per cent. Absorber temperature is to be kept at 30°C by cooling coils; pressure is 1 atm.

(a) What is the minimum water rate? (b) For a water rate 50 per cent greater than the minimum, how many over-all gas-phase transfer units are needed?

11-3. A soluble gas is absorbed in water, using a packed tower. The equilibrium relationship may be taken as

$$Y_e = 0.06 X_e$$

where Y_e and X_e are ratios of moles of solute to moles of inert component. Terminal conditions are:

	Top	Bottom
X	0	0.08
Y	0.001	0.009

If $H_x = 0.8$ ft and $H_y = 1.2$ ft, what is the height of the packed section?

11-4. How many ideal stages are required for the tower of Prob. 11-3?

11-5. A mixture of 5 per cent butane and 95 per cent air is absorbed in a bubble-plate tower containing eight ideal plates. The absorbing liquid is a heavy, nonvolatile oil having a molecular weight of 250 and a specific gravity of 0.90. The absorption takes place at 1 atm and 60°F. The butane is to be recovered to the extent of 95 per cent. The vapor pressure of butane at 60°F is 28 lb force/in.^{2}, and liquid butane has a density of 4.84 lb/gal at 60°F.

(a) Calculate the gallons of fresh absorbing oil per gallon of butane recovered. (b) Repeat, on the assumption that the total pressure is 3 atm and that all other factors remain constant.

Assume that Raoult's and Dalton's laws apply.

11-6. An absorption column is fed at the bottom with a gas containing 5 per cent benzene and 95 per cent air. At the top of the column a nonvolatile absorption oil is introduced, which contains 0.2 per cent benzene by weight. Other data are:

Feed, 4,000 lb of absorption oil per hour
Total pressure, 1 atm
Temperature (constant), 80°F
Molecular weight of absorption oil, 230
Vapor pressure of benzene at 80°F, 106 mm Hg
Volume of entering gas, 40,000 ft^3/hr
Tower packing, Raschig rings, $\frac{1}{2}$ in. nominal size
Fraction of entering benzene absorbed, 0.90
Over-all absorption coefficient based on gas phase $K_y a$, 1.8 lb moles/(hr)(ft^3)(unit mole fraction)
Mass velocity of entering gas, 300 lb/ft^2-hr

Calculate the height and diameter of the packed section of this tower and recommend the total height of the tower.

REFERENCES

1. Chilton, T. H., and A. P. Colburn: *Ind. Eng. Chem.*, **27**: 255 (1935).
2. Colburn, A. P.: *Trans. AIChE*, **35**: 211 (1939).
3. Colburn, A. P.: *Ind. Eng. Chem.*, **33**: 459 (1941).
4. "International Critical Tables," vol. III, p. 302, McGraw-Hill Book Company, Inc., New York, 1929.
5. Kirschbaum, E.: "Distillation and Rectification," Chemical Publishing Company, Inc., New York, 1948.
6. Leva, M.: "Tower Packings and Packed Tower Design," pp. 40–41, U.S. Stoneware Co., Akron, Ohio, 1951.

7. Lobo, W. E., L. Friend, F. Hashmall, and F. A. Zenz: *Trans. AIChE*, **41**: 693 (1945).

8. Millard, W. R.: Ph.D. thesis in chemical engineering, Cornell University, 1950.

9. Othmer, D. F., and E. G. Scheibel: *Trans. AIChE*, **38**: 339 (1942).

10. Perry, J. H.: "Chemical Engineers' Handbook," 3d ed., (*a*) pp. 166–173, 673–680; (*b*) pp. 687–690; (*c*) pp. 693–698, McGraw-Hill Book Company, Inc., New York, 1950.

11. Sherwood, T. K., and F. A. L. Holloway: *Trans. AIChE*, **36**: 21 (1940).

12. Sherwood, T. K., and F. A. L. Holloway: *Trans. AIChE*, **36**: 39 (1940).

13. Sherwood, T. K., and R. L. Pigford: "Absorption and Extraction," 2d ed., (*a*) p. 161; (*b*) p. 244; (*c*) p. 245; (*d*) pp. 278–297, McGraw-Hill Book Company, Inc., New York, 1952.

14. Sherwood, T. K., G. H. Shipley, and F. A. L. Holloway: *Ind. Eng. Chem.*, **30**: 765 (1938).

15. Walker, W. H., W. K. Lewis, W. H. McAdams, and E. R. Gilliland: "Principles of Chemical Engineering," 3d ed., p. 484, McGraw-Hill Book Company, Inc., New York, 1937.

16. Whitman, W. G.: *Chem. & Met. Eng.*, **29**(4): 146 (July 23, 1923).

17. Whitney, R. P., and J. E. Vivian: *Chem. Eng. Progr.*, **45**: 323 (1949).

18. Zenz, F. A.: *Chem. Eng.*, **60**(8): 182 (1953).

Chapter 12

DISTILLATION

The term *distillation* is sometimes employed for those processes where a single constituent is vaporized from a solution, e.g., in "distilling" water. In general, however, this term is properly applied only to those operations where vaporization of a liquid mixture yields a vapor phase containing more than one component and it is desired either to recover one or more of the components in a nearly pure state or to obtain a mixture of controllable composition different from that of the original mixture. Thus, the separation of a mixture of alcohol and water into its constituents is accomplished by distillation, whereas the separation of brine into salt and water is evaporation, even in those cases where the salt is not desired and the condensed water is the valuable product.

EQUILIBRIA

The basic relationship required for the solution of a distillation problem is the constant-pressure equilibrium curve, relating y_e, the concentration of the vapor, with x_e, the concentration of the liquid in equilibrium with it. To follow the temperature change in the process, a second relationship, called the boiling-point curve, is useful.

Boiling-point Diagrams. Figure 12-1 shows the boiling-point diagram at constant pressure for mixtures of component A, boiling at temperature t_A, and component B, boiling at temperature t_B. Component A is the more volatile. Temperatures are plotted as ordinates and concentrations, preferably in mole fractions, as abscissas. The diagram consists of two curves, the ends of which

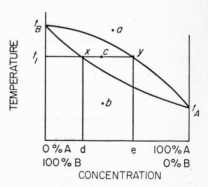

Fig. 12-1. Boiling-point diagram.

coincide. Any point, such as point y, on the upper line represents vapor that will just begin to condense at temperature t_1. The concentration of the first drop of liquid is represented by point d. The upper curve is called

665

the *dew-point curve*. Any point, such as point x, on the lower curve represents liquid that will just begin to boil at temperature t_1. The concentration of the first bubble of vapor is represented by point e. The lower curve is called the *bubble-point curve*. Any two points on the same horizontal line, such as points x and y, represent concentrations of liquid and vapor in equilibrium at the temperature given by their ordinate. The abscissas of such points, therefore, are corresponding values of x_e and y_e. For all points above the top line, such as point a, the mixture is entirely vapor. For all points below the bottom line, such as point b, the mixture is completely liquid. For points between the two curves, such as point c, the system is partly liquid and partly vapor.

Suppose that a liquid mixture of concentration d is heated slowly. It will begin to boil at temperature t_1. Although the first bubble of vapor will have the concentration y, represented by e, as soon as an appreciable amount of vapor has been formed, the concentration of the liquid will no longer correspond to d, since the vapor is richer in the component A than the liquid from which it is evolved and hence the points x and y both tend to move towards t_B.

The boiling-point diagram, like the equilibrium curve, must in general be determined experimentally. The procedure will not be described here.

Determination of Boiling-point Diagram from Raoult's Law. If the mixture of components follows Raoult's law, the boiling-point diagram can be calculated when the vapor-pressure data for the two pure substances are available. It is necessary only to apply Eqs. (10-11) and (10-12) at several temperatures between the boiling points of the pure components and plot the curves of x_e vs. t and y_e vs. t. The equilibrium curve, which is simply a plot of x_e vs. y_e, is obtained from the same data.

Example 12-1. The vapor pressures of benzene and toluene are shown in Table 12-1. Assuming that mixtures of these materials follow Raoult's law, calculate

TABLE 12-1. VAPOR PRESSURES OF BENZENE AND TOLUENE

Temperature, °C	Vapor pressure, mm Hg	
	Benzene	Toluene
80.1	760	
85	877	345
90	1,016	405
95	1,168	475
100	1,344	557
105	1,532	645
110	1,748	743
110.6	1,800	760

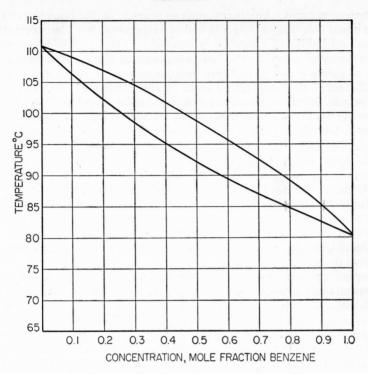

Fig. 12-2. Boiling-point diagram (system benzene-toluene).

and plot the boiling-point and equilibrium curve for the system benzene-toluene at a total pressure of 1 atm.

Solution. Several temperatures between 80.6 and 110.6°C are chosen, and the corresponding data for p_A and p_B are substituted in Eq. (10-12). For example, at 85°, $p_A = 877$ and $p_B = 345$, and

$$877x_e + 345(1 - x_e) = 760$$

From this, $x_e = 0.780$. The vapor concentration is then determined from Eq. (10-11).

$$y_e = \frac{877x_e}{760} = \frac{877 \times 0.780}{760} = 0.900$$

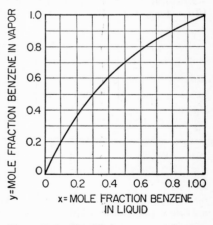

Fig. 12-3. Equilibrium curve (system benzene-toluene).

Other results are found similarly and are given in Table 12-2. The boiling-point diagram is plotted in Fig. 12-2 and the equilibrium curve in Fig. 12-3.

TABLE 12-2. BOILING-POINT CURVES FOR EXAMPLE 12-1

Temperature, °C	Concentration, mole fraction C_6H_6	
	Liquid x_e	Vapor y_e
85	0.780	0.900
90	0.581	0.777
95	0.411	0.632
100	0.258	0.456
105	0.130	0.261
110	0.017	0.039

Azeotropic Mixtures. Many systems do not obey Raoult's law and yet have boiling-point and equilibrium curves qualitatively like those of Figs. 12-2 and 12-3. That is, the boiling-point curves are bow-shaped, the equilibrium curve is concave downward throughout its length, the bubble

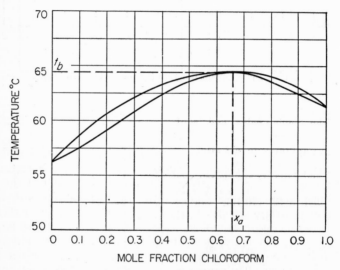

FIG. 12-4. Boiling-point diagram showing maximum-boiling azeotrope (system chloroform-acetone).

and dew points all lie between the boiling temperatures of the pure components, and the vapor in equilibrium with any liquid mixture is richer in component A than the liquid. There are many other important systems that have boiling-point and equilibrium diagrams quite different from those of Figs. 12-2 and 12-3. Figures 12-4 and 12-5 are the boiling-point diagrams for two such mixtures. The first applies to mixtures of chloroform and acetone and the second to mixtures of benzene and ethanol. In Fig. 12-4 the boiling point t_b, corresponding to concentration x_a, is the highest temperature reached by any mixture of these substances and is higher than the boiling temperature of either pure component. In Fig. 12-5 the boiling point t_b, corresponding to concentration x_a, is the lowest boiling temperature reached by any mixture and is lower than that of either pure component. Mixtures of this kind having maximum or minimum boiling points are called azeotropes. Of the two types, minimum-boiling azeotropes are the

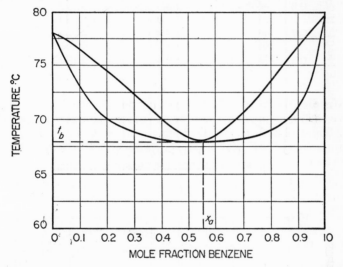

FIG. 12-5. Boiling-point diagram showing minimum-boiling azeotrope (system benzene-ethanol).

more common. At an azeotropic concentration, the liquid and vapor curves touch and have a common horizontal tangent. The composition of the vapor produced from an azeotrope is the same as that of the liquid. An azeotrope, then, may be boiled away at constant pressure without change in concentration in either liquid or vapor. Since, under these conditions, the temperature cannot vary, azeotropes are also called constant-boiling mixtures. An azeotrope cannot be separated by constant-pressure distillation into its components. Furthermore, a mixture on one side of the azeotropic composition cannot be transformed by distillation to a mixture on the other side of the azeotrope. If the total pressure is changed, the azeotropic composition is usually shifted, and this principle can be utilized to obtain separations under pressure or vacuum that cannot be obtained at atmospheric pressure. The same result is more commonly obtained by adding a third component that destroys the azeotrope.

The equilibrium curves for the systems chloroform-acetone and benzene-ethanol are shown in Figs. 12-6 and 12-7, respectively. At an azeotropic concentration, the equilibrium curve crosses the $x = y$ diagonal.

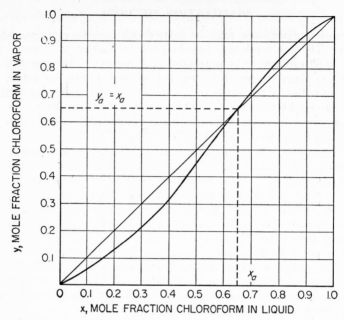

Fig. 12-6. Equilibrium diagram showing maximum-boiling azeotrope (system chloroform-acetone).

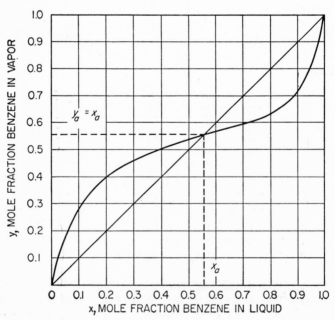

Fig. 12-7. Equilibrium diagram showing minimum-boiling azeotrope (system benzene-ethanol).

DISTILLATION METHODS

In practice, distillation may be carried out by either of two principal methods. The first method is based on the production of a vapor by boiling the liquid mixture to be separated and condensing the vapors without allowing any liquid to return to the still in contact with the vapors. The second method is based on the return of part of the condensate to the still under such conditions that this returning liquid is brought into intimate contact with the vapors on their way to the condenser. This method, called rectification, has been described in Chap. 10.

There are, in turn, two important types of distillation that do not involve rectification. The first of these is flash distillation, and the second is differential distillation. Flash distillation consists of vaporizing a definite fraction of the liquid, keeping all of the residual liquid and all of the vapor in intimate contact so that at the end of the operation the vapor is in equilibrium with the liquid, separating the vapor from the liquid, and condensing the vapor. Flash distillation is usually conducted as a continuous process. In differential distillation, the vapor generated by boiling the liquid is withdrawn from contact with the liquid and condensed as fast as it is formed. Differential distillation is a batch process. A differential distillation may also be carried out in the presence of an inert gas by a method called steam distillation.

Flash Distillation. Figure 12-8 shows the elements of a flash-distillation plant. Feed is pumped by a pump a through heater b, and the pres-

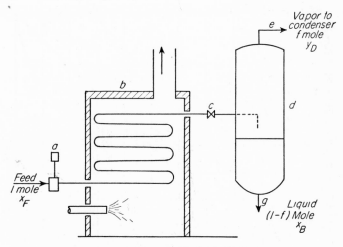

Fig. 12-8. Plant for flash distillation.

sure is reduced through valve c. An intimate mixture of vapor and liquid enters the vapor separator d, in which sufficient time is allowed for the vapor and liquid portions to separate. Because of the intimacy of contact of liquid and vapor before separation, the separated streams are in equilibrium. Vapor leaves through line e and liquid through line g. Consider 1 mole of a two-component mixture † fed to the equipment shown in Fig. 12-8. Let the concentration of the feed be x_F, in mole fraction of the more volatile component. Let f be the molal fraction of the feed that is vaporized and withdrawn continuously as vapor. Then $1 - f$ is the molal fraction of the feed that leaves continuously as liquid. Let y_D and x_B be the concentrations of the vapor and liquid, respectively. By a material balance of the more volatile component, based on 1 mole of feed, all of that component in the feed must leave in the two exit streams, or

$$x_F = fy_D + (1 - f)x_B \qquad (12\text{-}1)$$

There are two unknowns in Eq. (12-1), x_B and y_D. To use the equation a second relationship between the unknowns must be available. Such a relationship is provided by the equilibrium curve, as y_D and x_B are coordinates of a point on this curve. If x_B and y_D are replaced by x and y, respectively, Eq. (12-1) can be written

$$y = -\frac{1 - f}{f}x + \frac{x_F}{f} \qquad (12\text{-}2)$$

Equation (12-2) is the equation of a straight line with a slope of $-(1 - f)/f$ and can be plotted on the equilibrium diagram. The coordinates of the intersection of the line and the equilibrium curve are $x = x_B$ and $y = y_D$. The intersection of this material-balance line and the diagonal $x = y$ can be used conveniently as a point on the line. Thus, substituting x for y in Eq. (12-2) gives

$$x = -\frac{1 - f}{f}x + \frac{x_F}{f}$$

from which $x = y = x_F$. The material-balance line crosses the diagonal at $x = x_F$ for all values of f.

Example 12-2. A mixture of 50 mole per cent benzene and 50 mole per cent toluene is subjected to flash distillation at a separator pressure of 1 atm. Plot the following quantities, all as functions of f, the fractional vaporization: (a) the temperature in the separator, (b) the composition of the liquid leaving the separator, and (c) the composition of the vapor leaving the separator.

Solution. For each of several values of f corresponding quantities $-(1/f - 1)$ are calculated. Using these quantities as slopes, a series of straight lines each passing through point (x_F, x_F) is drawn on the equilibrium curve of Fig. 12-3. These lines, shown on Fig. 12-9, intersect the equilibrium curve at corresponding values

† Flash distillation is used on a large scale in petroleum refining, in which petroleum fractions are heated in pipe stills and the heated fluid flashed into overhead vapor and residual-liquid streams, each containing many components.

of x_B and y_D. The temperature of each vaporization is then found from Fig. 12-2. The results for the lines shown in Fig. 12-9 are shown in Table 12-3 and are plotted

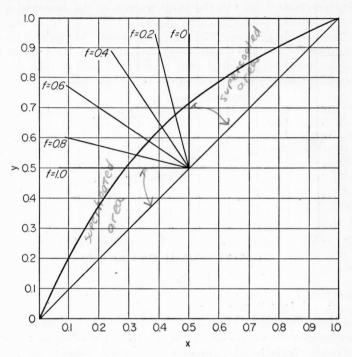

Fig. 12-9. Graphical construction for Example 12-2.

in Fig. 12-10. The lines in Fig. 12-10 are nearly straight. The limits for 0 and 100 per cent vaporization are the bubble and dew points, respectively.

TABLE 12-3. DATA FOR EXAMPLE 12-2

Fraction vaporized f	Slope $-\dfrac{1-f}{f}$	Concentration, mole fraction C_6H_6		Temperature, °C
		Liquid x_B	Vapor y_D	
0	$-\infty$	0.50	0.71	92.2
0.2	-4	0.455	0.67	93.7
0.4	-1.5	0.41	0.63	95.0
0.6	-0.67	0.365	0.585	96.5
0.8	-0.25	0.325	0.54	97.7
1.0	-0	0.29	0.50	99.0

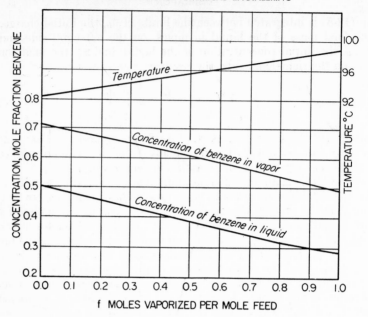

FIG. 12-10. Results for Example 12-2.

Differential Distillation. Consider a batch of n_0 moles † of liquid charged to a differential distillation. Assume that at a given instant during the process there are n moles of liquid left in the still. At this instant let the vapor composition be y and the liquid composition be x. The total amount of component A left in the still will be xn. Assume a small amount of liquid dn moles is vaporized. During the vaporization, the liquid concentration diminishes from x to $x - dx$, and the number of moles of liquid in the still diminishes from n to $n - dn$. There is left in the still $(x - dx)(n - dn)$ moles of A, while the amount $y\,dn$ has been removed as vapor. A material balance on the component A is

$$xn = (x - dx)(n - dn) + y\,dn \qquad (12\text{-}3)$$

Expanding Eq. (12-3) gives

$$xn = xn - x\,dn + dx\,dn - n\,dx + y\,dn \qquad (12\text{-}4)$$

The second-order differential $dx\,dn$ can be neglected, and Eq. (12-4) can be written

$$\frac{dn}{n} = \frac{dx}{y - x} \qquad (12\text{-}5)$$

† Equations (12-3) through (12-6) will hold as well if compositions, including those on the equilibrium curve, are expressed as mass fractions.

If Eq. (12-5) is integrated between the limits of n_0, the initial charge, and n_1, the final mass of the liquid in moles, on the left-hand side and the limits x_0, the initial concentration of the liquid, and x_1, the final concentration, on the right-hand side, the result is

$$\int_{n_1}^{n_0} \frac{dn}{n} = \int_{x_1}^{x_0} \frac{dx}{y - x} = \ln \frac{n_0}{n_1} \qquad (12\text{-}6)$$

Equation (12-6) is known as the *Rayleigh equation*. The function $dx/(y - x)$ can be integrated graphically from the equilibrium curve, since this curve provides a relationship between x and y.

A second relationship for differential distillation is useful when the liquid charge follows Raoult's law and the vapor follows Dalton's law. Thus, assume that the original charge in the still consists of n_{0A} moles of component A and n_{0B} moles of component B. Assume the moles of components A and B remaining in the still at a definite instant during the process are n_A and n_B, respectively. The mole fraction of component A is then $n_A/(n_A + n_B)$, and the partial pressure of this component is, by Raoult's law, $p_A n_A/(n_A + n_B)$, where p_A is the vapor pressure of pure A at the temperature of the liquid. Likewise, the partial pressure of component B is $p_B n_B/(n_A + n_B)$, where p_B is the vapor pressure of pure B. Consider now the vaporization of a small amount of vapor from the liquid. Let the moles of A in this vapor be $-dn_A$ and the moles of B be $-dn_B$. The minus signs denote the fact that n_A and n_B are decreasing during the process. By Dalton's law, the molal ratio of A to B in the vapor is identical with the partial-pressure ratio of the components, or

$$\frac{-dn_A}{-dn_B} = \frac{p_A n_A/(n_A + n_B)}{p_B n_B/(n_A + n_B)} = \frac{p_A n_A}{p_B n_B} = \frac{1}{\alpha_{BA}} \frac{n_A}{n_B} \qquad (12\text{-}7)$$

where $\alpha_{BA} = p_B/p_A$ and is called the *relative volatility*, component B to component A. For a constant-temperature distillation p_A and p_B are constant, and the relative volatility α_{BA} is also constant. Equation (12-7) can be integrated between limits

$$\alpha_{BA} \int_{n_{0A}}^{n_A} \frac{dn_A}{n_A} = \int_{n_{0B}}^{n_B} \frac{dn_B}{n_B}$$

or

$$\log \frac{n_B}{n_{0B}} = \alpha_{BA} \log \frac{n_A}{n_{0A}} \qquad (12\text{-}8)$$

where n_B and n_A are the moles of components B and A, respectively, remaining in the still at any time during the process. Equation (12-8) can also be written

$$\frac{n_B}{n_{0B}} = \left(\frac{n_A}{n_{0A}}\right)^{\alpha_{BA}} \qquad (12\text{-}9)$$

If the pressure is constant, the temperature of the liquid increases during the distillation. Both p_B and p_A also increase, but their ratio α_{BA} remains nearly constant. Equations (12-8) and (12-9) can, therefore, be used for constant-pressure distillations without serious error if an average α_{BA} is used as a constant.

Equation (12-8) shows that, if n_B/n_{OB} is plotted as the ordinate vs. n_A/n_{OA} as the abscissa on logarithmic coordinates, a straight line with a slope of α_{BA} is obtained. Also, since, at the start of the distillation, $n_A = n_{OA}$ and $n_B = n_{OB}$, both n_B/n_{OB} and n_A/n_{OA} initially are unity, and the line passes through the point (1,1). The entire course of any differential distillation at constant α_{BA} is represented, therefore, by a straight line on logarithmic coordinates with a slope of α_{BA} and passing through point (1,1).

Example 12-3. A mixture of 50 mole per cent benzene and 50 mole per cent toluene is subjected to differential distillation at a pressure of 1 atm. Plot the following quantities, each against f, the mole fraction of charge distilled: (a) the temperature in the still, (b) the composition of liquid in the still, (c) the instantaneous composition of vapor leaving the still, and (d) the cumulative average composition of the condensate.

Solution. The initial boiling temperature of the liquid is, from Fig. 12-2, 92.2°C. From the vapor-pressure data in Table 12-1, α_{BA} is 0.41 at 92.2°C. It will be found that the last liquid to be vaporized is pure toluene, and the temperature at the end of the distillation is, therefore, 110.6°C, which is the boiling point of toluene. At

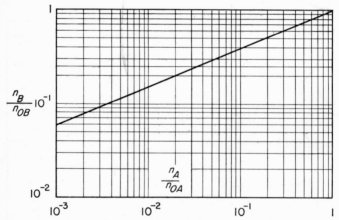

Fig. 12-11. Differential-distillation plot for Example 12-3.

this temperature α_{BA} is 0.42. An average of 0.41 for α_{BA} can be used as a constant. Figure 12-11 is a logarithmic plot of n_A/n_{OA} as the abscissa vs. n_B/n_{OB} as the ordinate. A straight line with a slope of 0.41 and passing through point (1,1) is drawn on Fig. 12-11. Choosing as a basis 1 mole of initial charge and assuming a series of values of n_A, for each choice of n_A the quantity n_A/n_{OA} is calculated, n_B/n_{OB} is read from Fig. 12-11, and n_B is then found. From n_A and n_B the mole fractions of benzene in the liquid are calculated, and the temperatures and instantaneous vapor concentrations are then read from Fig. 12-2. The cumulative average concentrations of the distillate are calculated from the total moles and the moles of benzene vaporized. The results of the calculations are shown in detail in Table 12-4, and the desired curves are plotted in Fig. 12-12.

TABLE 12-4. DATA FOR EXAMPLE 12-3

n_A	$\dfrac{n_A}{0.500}$	$\dfrac{n_B}{0.500}$	n_B	$n_A + n_B$	x	t, °C	Instantaneous y	$0.50 - n_A$	Total condensate	y_{cum}
0.50	1.00	1.00	0.50	1.00	0.50	92.2	0.71	0	0	0.71
0.40	0.80	0.91	0.455	0.855	0.47	93.2	0.68	0.10	0.145	0.69
0.30	0.60	0.80	0.40	0.70	0.43	94.5	0.64	0.20	0.30	0.67
0.20	0.40	0.68	0.34	0.54	0.37	96.2	0.59	0.30	0.46	0.65
0.10	0.20	0.51	0.255	0.355	0.28	99.2	0.49	0.40	0.645	0.62
0.05	0.10	0.38	0.19	0.24	0.21	102.0	0.395	0.45	0.76	0.59
0.03	0.06	0.31	0.155	0.185	0.16	104.7	0.325	0.47	0.815	0.58
0.01	0.02	0.20	0.10	0.11	0.091	106.7	0.205	0.49	0.89	0.55
0.005	0.01	0.15	0.075	0.080	0.062	108.2	0.14	0.495	0.92	0.54
0	0	0	0	0	0	110.6	0	0.50	1.00	0.50

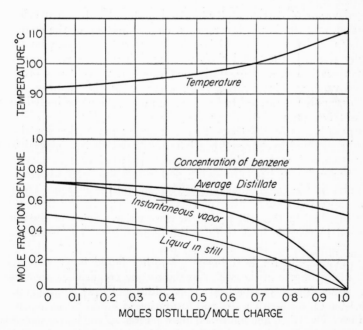

FIG. 12-12. Final curves for Example 12-3.

Differential distillation is approached in commercial batch distillations, where the vapor is removed as fast as it is formed without appreciable condensation. Although as a method of separation this process is not effective, many such stills are used, especially where the components to be separated have widely different boiling points and where sharp separations are not required. Also, most laboratory distillations conducted without reflux are of this type.

Steam Distillation. A special distillation problem is encountered when (1) the boiling point of the material is so high that the substance decomposes at its atmospheric boiling point, (2) the vaporization temperature cannot be reached by steam heat, or (3) the use of direct fire is dangerous or detrimental to the quality of the product. The vaporization of such substances as glycerine, lubricating oil, fatty acids, aniline, and turpentine is impracticable for these reasons. To distill such materials, either to free them from nonvolatile impurities or to separate them into fractions of differing boiling point, it is necessary to lower the distillation temperature to a point where the liquid can be vaporized without reaching excessive temperatures. One method of accomplishing this is to operate under a vacuum. This method, using modern techniques of high-temperature heating and efficient vacuum-producing equipment, is extensively used in both batch and continuous processes. An important older method which utilizes simpler equipment is steam distillation. In this method, the temperature of the liquid is reduced by vaporizing the material into a stream of carrier vapor, the liquid phase of which is immiscible with the material distilled. A mixed vapor of carrier and vaporized material is taken overhead and condensed, and the liquid layers, each containing one of the components, are separated by gravity. Although any inert carrier can be used, for economy steam is the usual choice. Steam distillation can be combined with rectification, usually in continuous fractionating columns, often under a partial vacuum. Lubricating oils are treated in this manner. The following discussion is limited to batch steam distillation conducted to free a material from a nonvolatile impurity.

Theory of Steam Distillation. To conduct a steam distillation the material to be distilled is charged to a batch still, which is equipped with an overhead vapor line, a condenser, a condensate receiver, and gravity separator. Steam is admitted through a perforated pipe in the bottom of the still to give maximum contact between steam and the charge. The process may be operated under either of two conditions. In the first, the only energy supplied to the still is that in the carrier. Some of the carrier is condensed to provide heat to raise the temperature of the still and contents to the operating level, to supply heat of vaporization of the material, and to compensate for heat losses. Another liquid layer, of condensed carrier, then collects in the still. In the second condition, either the inlet carrier is considerably superheated, or additional heat is supplied to the charge by a closed heating coil and formation of liquid thereby prevented. All vapor blown through the liquid then passes out with the product. In either method, when the sum of the partial pressures of carrier and material reaches the total pressure, both substances pass over in the molecular ratio of their partial pressures, and nonvolatile impurities remain behind in the still. The mass ratio of carrier to material is

$$\frac{w_A}{w_B} = \frac{n_A M_A}{n_B M_B} = \frac{\bar{p}_A M_A}{\bar{p}_B M_B} = \frac{(p - \bar{p}_B) M_A}{\bar{p}_B M_B} \tag{12-10}$$

where w_A = mass flow rate of carrier in vapor
w_B = mass flow rate of material in vapor
$\bar{p}_A$ = partial pressure of carrier
$\bar{p}_B$ = partial pressure of material
M_A = molecular weight of carrier
M_B = molecular weight of material
n_A = moles of carrier in vapor
n_B = moles of material in vapor
p = total pressure

Assume that the effect of the nonvolatile matter on the vaporization of the material may be neglected. Then if two liquid layers are allowed to form, there will be three phases (one vapor and two liquids) and two components (material and carrier). By the phase rule [Eq. (10-1)], the system has one degree of freedom. Either the temperature or the pressure —but not both—may be fixed arbitrarily. If the pressure is atmospheric, the temperature so adjusts itself that the sum of the partial pressures of the two components equals the total pressure of 1 atm. This temperature is lower than the boiling point of either pure component. If steam is the inert carrier and the pressure is atmospheric, the temperature is always less than 212°F, and destructive temperatures are not reached.

If the formation of the second liquid layer is prevented, there are two components, two phases, and, therefore, two degrees of freedom. Both pressure and temperature must be fixed independently before the operation is defined. The temperature of the vapor carrier leaving the still is above that corresponding to its condensing point, and this vapor is therefore superheated. In practice, the pressure is either atmospheric or, if below atmospheric, is controlled by the action of the condenser and vacuum pump, just as in an evaporator operated with a surface condenser. The temperature may be fixed either by a temperature controller or by controlling the pressure of the steam in the closed coil used to supply supplementary heat.

Assuming that Dalton's law is applicable, the equilibrium partial pressure of the material is the vapor pressure of the pure material, and if liquid carrier is also present, its partial pressure is the vapor pressure of the carrier. Both vapor pressures are taken at the temperature of the liquid in the still. In practice, contact between carrier and material is not perfect, and the carrier does not reach equilibrium with the liquid, so the actual partial pressure of the material is less than its vapor pressure. A vaporization efficiency η_s defined by Eq. (12-11), accounts for the difference between theory and practice.

$$\eta_s = \frac{\bar{p}_B}{p_B} \quad\leftarrow \text{Partial Pressure} \tag{12-11}$$

where p_B is the vapor pressure of the material. Equation (12-10) may be written

$$\frac{w_A}{w_B} = \frac{(p - \eta_s p_B)M_A}{\eta_s p_B M_B} = \frac{M_A}{\eta_s M_B}\left(\frac{p}{p_B} - \eta_s\right) \tag{12-12}$$

Steam Consumption. The minimum steam consumption (or, more generally, consumption of carrier vapor) is given by Eq. (12-12) when η_s is taken as unity. The actual steam consumption is greater than this because the vaporization efficiency is less than 1, and if supplementary heat is not supplied, steam is also needed to supply the heat requirement of the process. If a second liquid layer forms, the minimum steam consumption per pound of material for the distillation itself is

$$\frac{w_A}{w_B} = \frac{p_A M_A}{p_B M_B} = \alpha_{AB}\frac{M_A}{M_B} \tag{12-13}$$

where p_A = vapor pressure of carrier
      p_B = vapor pressure of material
      $\alpha_{AB} = p_A/p_B$
The total pressure is

$$p = p_A + p_B \tag{12-14}$$

Since both p_A and p_B depend only on temperature, a reduction of temperature reduces the total pressure. The change in steam consumption w_A/w_B with pressure depends on the effect of temperature on the relative-volatility factor α_{AB}, which may either increase, decrease, or remain practically constant as the temperature is changed. For example, the theoretical steam consumption for steam distilling turpentine is shown in Fig. 12-13.

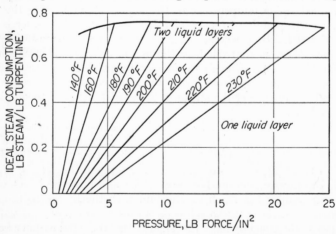

Fig. 12-13. Steam-consumption diagram for steam distillation of turpentine.

The steam consumption for operation with two liquid phases is given by the top line, which shows that in this system α is practically constant and the theoretical steam consumption is nearly independent of pressure. In other systems α may increase or decrease with an increase in pressure, but the change is small, and the theoretical consumption is not sensitive to change in pressure when two liquid layers are present.

If there is no condensation of carrier, Eq. (12-12) shows that, when the total pressure is constant, an increase in temperature always reduces the steam consumption, both theoretical and actual. Likewise, if the temperature is constant and the pressure is lowered, the steam consumption is diminished. For minimum steam consumption, the highest permissible temperature and the lowest practicable pressure should be used. The limit of reduction in pressure is reached when the total pressure is the vapor pressure of the material distilled and the steam consumption vanishes. The process then becomes a vacuum distillation. The effects of changes in temperature and pressure on the steam required to distill 1 lb of turpentine under ideal conditions are shown by the slanting straight lines in Fig. 12-13. In practice, a moderate vacuum readily attainable with the available cooling water often is used combined with the use of steam. In some cases, such as the distillation of fatty acids, high vacuum and steam are both required to obtain vaporization at nondestructive temperatures.

Example 12-4. A batch of 250 gal of turpentine is to be steam-distilled at 1 atm pressure, using saturated steam at a gauge pressure of 5 lb force/in.² The initial temperature of the charge is 70°F. The vaporization efficiency may be taken as 0.85. The following properties of turpentine may be used:

Molecular weight, 140
Latent heat of vaporization, 128 Btu/lb
Specific gravity, 0.87
Specific heat, including heat absorbed by still, 0.50 Btu/(lb)(°F)

TABLE 12-5. VAPOR PRESSURE OF TURPENTINE

Temperature, °F	Vapor pressure, lb force/in.²	Temperature, °F	Vapor pressure, lb force/in.²
140	0.51	190	1.59
150	0.65	200	1.97
160	0.81	210	2.42
170	1.01	220	2.96
180	1.28	230	3.64

Neglecting radiation losses, calculate the pounds of steam required: (a) if no additional heat is supplied, and (b) if the temperature of the batch is brought to, and held at, the temperature of the inlet steam by means of another heat source. Assume that no condensation occurs in the vapor line.

Solution. (a) First, the temperature in the still is calculated. It can be assumed that the partial pressure of the carrier steam leaving the still is p_A, the vapor pressure of water at the temperature of the liquid in the still. The partial pressure of the turpentine is 0.85 p_B, where p_B is the vapor pressure of turpentine at liquid temperature.   Then

$$p = 0.85p_B + p_A = 14.7$$

The vapor pressure of water is given in Appendix 6 and that of turpentine in Table 12-5. Each vapor pressure depends only on temperature. The total pressure p is calculated for several temperatures and plotted against temperature. The temperature at which this line crosses the line $p = 14.7$ is the desired result. A temperature of 205°F is found. The vapor pressure of turpentine at 205°F is 2.18 $lb/in.^2$

The steam consumption is calculated in two increments: that required as a carrier and that required for thermal purposes. The turpentine to be vaporized is $250 \times 8.33 \times 0.87 = 1,810$ lb. The steam required for vaporization is, by Eq. (12-12),

$$\frac{w_A}{w_B} = \frac{18}{0.85 \times 140}\left(\frac{14.7}{2.18} - 0.85\right) = 0.89 \text{ lb/lb charge}$$

The total carrier steam is $0.89 \times 1,810 = 1,610$ lb. The heat required in the process is

Heat charge and still, 60 to 140°F:
$$1,810(205 - 60)0.50 = 131,000 \text{ Btu}$$
Vaporize turpentine:
$$1,810 \times 128 = 232,000$$

$$\text{Total} = 363,000 \text{ Btu}$$

The carrier steam contributes some heat. From Appendix 6, the enthalpy of 1 lb of saturated steam at a gauge pressure of 5 lb force/in.^{2} is 1156.0 Btu, and that of saturated steam at 1 atm is 1147.8 Btu. The heat recovered from the carrier steam is $1,610(1156.0 - 1147.8) = 13,000$ Btu. The heat supplied by heating steam must be $363,000 - 13,000 = 350,000$ Btu.

Each pound of heating steam is condensed to liquid water at 205°F. The enthalpy of water at this temperature is 173.0 Btu/lb. The steam required for heating is

$$\frac{350,000}{1156.0 - 173.0} = 355 \text{ lb}$$

The total steam consumption is $1,610 + 355 = 1,965$ lb. This is $1,965/1,810 = 1.09$ lb per pound of turpentine. The theoretical steam consumption is, from the top line of Fig. 12-13, 0.75 lb per pound of turpentine.

(b) The steam required is only that needed to carry over the vapors, as the thermal requirement is supplied by another heat source. Steam at 5 lb force/in.^{2} gauge condenses at 227°F. At this temperature the vapor pressure of turpentine is 3.5 lb force/in.^{2} The steam requirement, by Eq. (12-12), is

$$\frac{w_A}{w_B} = \frac{18}{0.85 \times 140}\left(\frac{14.7}{3.5} - 0.85\right) = 0.51 \text{ lb/lb turpentine}$$

The theoretical steam consumption, computed by setting $\eta_s = 1$, is

$$\frac{w_A}{w_B} = \frac{(14.7 - 3.5)18}{3.5 \times 140} = 0.41 \text{ lb/lb turpentine}$$

Of course, any fair comparison of operations (a) and (b) should take account of the secondary heat source assumed in part (b).

Rectification. Neither flash nor differential distillation is effective in separating components of comparable volatility. In each process, both the condensate and the residual liquid are far from pure. By many successive redistillations some nearly pure components may finally be obtained. This method, formerly used in the laboratory, is too slow and inefficient for industrial distillations when pure components are wanted. Modern methods used in both laboratory and plant apply the principle of rectification, which has been described on pages 582 to 584.

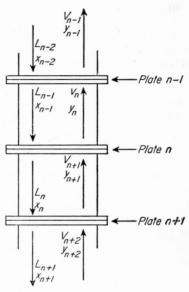

FIG. 12-14. Material-balance diagram of plate n.

Rectification on an Ideal Plate. Consider a single plate in a column or cascade of ideal plates. Assume that the plates are numbered serially from the top down and that the plate under consideration is the nth plate from the top. It is shown diagrammatically in Fig. 12-14. Then the plate immediately above this plate is plate $n - 1$, and that immediately below it is plate $n + 1$. Use subscripts on all quantities showing the point of origin of the quantity.

Two fluid streams enter plate n and two leave it. A stream of liquid, L_{n-1} moles/hr, from plate $n - 1$, and a stream of vapor, V_{n+1} moles/hr, from plate $n + 1$, are brought into intimate contact. A stream of vapor, V_n moles/hr, rises to plate $n - 1$, and a stream of liquid, L_n moles/hr, descends to plate $n + 1$. Since the vapor streams are the light phase, their concentrations are denoted by y, and since the liquid streams are the heavy phase, their concentrations are denoted by x. Then the concentrations of the streams entering and leaving the nth plate are:

Vapor leaving plate, y_n
Liquid leaving plate, x_n
Vapor entering plate, y_{n+1}
Liquid entering plate, x_{n-1}

Figure 12-15 shows the boiling-point diagram for the mixture being treated. The four concentrations given above are shown in this figure. By definition of an ideal plate, the vapor and liquid leaving plate n are in equilibrium, so x_n and y_n represent equilibrium concentrations. This is shown in Fig. 12-15. Since concentrations in both phases increase with the height of the column, x_{n-1} is greater than x_n, and y_n is greater than y_{n+1}. The streams leaving the plate are in equilibrium, and those entering it are not. This can be seen from Fig. 12-15. When the vapor from plate

$n + 1$ and the liquid from plate $n - 1$ are brought into intimate contact, their concentrations tend to move toward an equilibrium state, as shown by the arrows in Fig. 12-15. Some of the more volatile component A is vaporized from the liquid, decreasing the liquid concentration from x_{n+1} to x_n; and some of the less volatile component B is condensed from the vapor, increasing the vapor concentration from y_{n+1} to y_n. Since the liquid streams are at their bubble points and the vapor streams at their dew

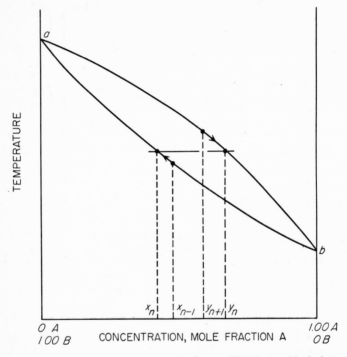

Fig. 12-15. Boiling-point diagram showing rectification on ideal plate.

points, the heat necessary to vaporize component A must be supplied by the heat released in the condensation of component B. Each plate in the cascade acts as an interchange apparatus in which component A is transferred to the vapor stream and component B to the liquid stream. Also, since the concentration in both liquid and vapor increases with column height, the temperature decreases, and the temperature of plate n is greater than that of plate $n - 1$ and less than that of plate $n + 1$.

Combination Rectification and Stripping. The plant described on pages 582 to 584, in which the feed to the unit is supplied to the still, cannot produce a nearly pure bottom product because the liquid in the still is not subjected to rectification. This limitation is removed by admitting the feed to a plate in the central portion of the column. Then the liquid feed flows down the column to the still, which in this type of plant is called the "reboiler," and is subjected to rectification by the vapor rising from the reboiler. Since the liquid reaching the reboiler is stripped of component A, the bottom product can be nearly pure B.

A typical continuous fractionating column equipped with the necessary auxiliaries and containing rectifying and stripping sections is shown in

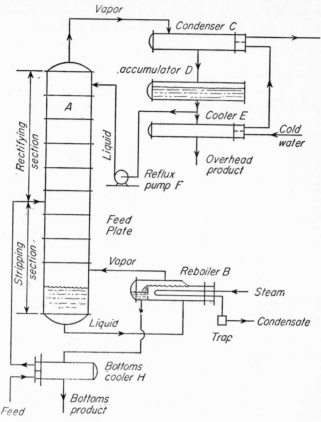

FIG. 12-16. Continuous fractionating column with rectifying and stripping sections.

Fig. 12-16. The column A is fed near its center with a definite flow of feed of definite concentration. Assume that the feed is a liquid at its boiling point. The action in the column is not dependent on this assumption, and other conditions of the feed will be discussed later. The plate on which the feed enters is called the feed plate. All plates above the feed plate constitute the rectifying section, and all plates below the feed, *including the feed plate itself*, constitute the stripping section. The feed flows down the stripping section to the bottom of the column, in which a definite level of liquid is maintained. Liquid flows by gravity to reboiler B. This is a steam-heated vaporizer which generates vapor and returns it to the bottom of the column. The vapor passes up the entire column. At one end of the reboiler is a weir. The bottom product is withdrawn from the pool of liquid on the downstream side of the weir, and flows through the cooler H. This cooler also preheats the feed by heat exchange with the hot bottoms.

The vapors rising through the rectifying section are completely condensed in condenser C, and the condensate is collected in accumulator D, in which a definite liquid level is maintained. Reflux pump F takes liquid from the accumulator and delivers it to the top plate of the tower. This liquid stream is called *reflux*. It provides the down-flowing liquid in the rectifying section that is needed to act on the up-flowing vapor. Without the reflux no rectification would occur in the rectifying section, and the concentration of the overhead product would be no greater than that of the vapor rising from the feed plate. Condensate not picked up by the reflux pump is cooled in heat exchanger E, called the product cooler, and withdrawn as the overhead product. If no azeotropes are encountered, both overhead and bottom products may be obtained in any desired purity if enough plates and adequate reflux are provided.

The plant shown in Fig. 12-16 is often simplified, especially in general chemical operations. In place of the reboiler, a heating coil is placed in the bottom of the column and generates vapor from the pool of liquid there. The condenser is placed above the top of the column and the reflux pump and accumulator omitted. Reflux then returns to the top plate by gravity. A special valve, called a reflux splitter, is used to control the rate of reflux return. The remainder of the condensate forms the overhead product.

DESIGN AND OPERATING CHARACTERISTICS OF PLATE COLUMNS

Important factors in the design and operation of plate columns are the number of plates required to obtain the desired separation, the diameter of the column, the heat input to the reboiler and the heat output from the condenser, the spacing between the plates, the choice of type of plate, and the details of construction of the plates.

In accordance with general principles, the analysis of the performance of plate columns is based on material and energy balances.

Over-all Material Balances. Figure 12-17 is a material-balance diagram for a typical continuous distillation plant. The column is fed with F moles/hr of concentration x_F, and delivers D moles/hr of overhead product of concentration x_D and B moles/hr of bottom product of concentration x_B. Two independent over-all material balances can be written.

Total material balance: $$F = D + B \qquad (12\text{-}15)$$

Component A balance: $$Fx_F = Dx_D + Bx_B \qquad (12\text{-}16)$$

Eliminating B from these equations gives

$$\frac{D}{F} = \frac{x_F - x_B}{x_D - x_B} \qquad (12\text{-}17)$$

Eliminating D gives

$$\frac{B}{F} = \frac{x_D - x_F}{x_D - x_B} \tag{12-18}$$

Equations (12-17) and (12-18) are true for all values of the flows of vapor and liquid within the column.†

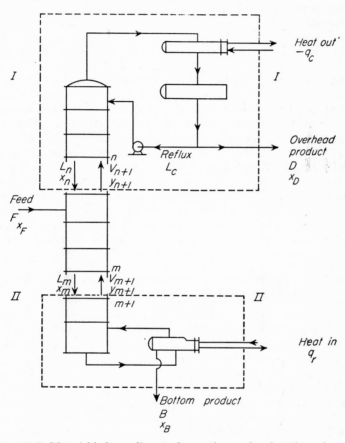

Fig. 12-17. Material-balance diagram for continuous fractionating column.

Operating Lines. Because there are two sections in the column, there are also two operating lines, one for the rectifying section and the other for the stripping section. To derive equations for these lines consider first the part of the plant enclosed by control surface I in Fig. 12-17, which includes the condenser and all plates above $n + 1$. Control surface I re-

† Note the identity of Eqs. (12-17) and (12-18) with Eqs. (7-3) and (7-4) for screening. Unlike as they are physically, both operations are separation operations, and the same over-all material-balance equations apply to them.

ceives V_{n+1} moles/hr of vapor of concentration y_{n+1} from plate $n + 1$ and delivers D moles of product per hour of concentration x_D and L_n moles of liquid per hour of concentration x_n to plate $n + 1$. The following material balances apply.

Total material: $$V_{n+1} = L_n + D \tag{12-19}$$

Component A: $$V_{n+1}y_{n+1} = L_n x_n + Dx_D \tag{12-20}$$

Equation (12-20) can be written

$$y_{n+1} = \frac{L_n}{V_{n+1}} x_n + \frac{Dx_D}{V_{n+1}} \tag{12-21}$$

If y_{n+1} and x_n are considered as mathematical variables, this equation is that of the operating line for the rectifying section. The slope of the line is, as usual, the ratio of the flow of liquid stream to that of the vapor stream. For further analysis it is convenient to eliminate V_{n+1} from Eq. (12-21) by Eq. (12-19), giving

$$y_{n+1} = \frac{L_n}{L_n + D} x_n + \frac{Dx_D}{L_n + D} \tag{12-22}$$

Consider second the part of the plant enclosed by control surface II in Fig. 12-17. To distinguish between quantities in the stripping section and the corresponding quantities in the rectifying section, use subscript m in place of n to denote the plate number, counting from the top down. Control surface II in Fig. 12-17 includes the reboiler and all plates below plate m. Then control surface II receives L_m moles/hr of liquid of concentration x_m from plate m and delivers B moles/hr of bottom product of concentration x_B and V_{m+1} moles/hr of vapor of concentration y_{m+1} to plate m. The material-balance equations are

Total material: $$L_m = V_{m+1} + B \tag{12-23}$$

Component A: $$L_m x_m = V_{m+1}y_{m+1} + Bx_B \tag{12-24}$$

Equation (12-24) can be written

$$y_{m+1} = \frac{L_m}{V_{m+1}} x_m - \frac{Bx_B}{V_{m+1}} \tag{12-25}$$

This is the equation for the operating line in the stripping section. Again the slope is the ratio of the liquid flow to the vapor flow. Eliminating V_{m+1} from Eq. (12-25) by Eq. (12-23) gives

$$y_{m+1} = \frac{L_m}{L_m - B} x_m - \frac{Bx_B}{L_m - B} \tag{12-26}$$

Analysis of Fractionating Columns by McCabe-Thiele Method

When the operating lines represented by Eqs. (12-22) and (12-26) are plotted with the equilibrium curve on the xy diagram, the McCabe-Thiele step-by-step construction can be used to compute the number of *ideal* plates needed to accomplish a definite concentration difference in either the rectifying or the stripping section.[12] It can be seen, however, by inspection of Eqs. (12-22) and (12-26), that unless L_n and L_m are constant, the operating lines are curved and can be plotted only if the change in these reflux streams with concentration is known. That L_n and L_m do not, in fact, vary appreciably in many important situations will be demonstrated in the following paragraphs. Also, because of the unity of the two sections of the column and the means used to obtain both reflux and vapor, a McCabe-Thiele diagram has certain unique features when applied to the entire plant of Fig. 12-17. Special attention is given, therefore, to the graphical construction pertaining to the reboiler, the condenser, and the feed plate.

Constant Molal Overflow. Consider the ideal plate shown in Fig. 12-14. Although the nomenclature in Fig. 12-14 pertains to a plate in the rectifying section, this plate may be any plate, except the feed plate, in either section of the column. A component A balance over plate n is

$$V_{n+1}y_{n+1} + L_{n-1}x_{n-1} = V_ny_n + L_nx_n \qquad (12\text{-}27)$$

A heat balance may be written over the same plate if a datum is chosen. Choose the liquid L_n at the temperature of plate n as this datum. Since an increasing temperature gradient exists in the column from condenser to reboiler, the temperatures on plates $n - 1$, n, and $n + 1$, are in the following order.

$$t_{n-1} < t_n < t_{n+1}$$

The vapor stream V_{n+1} carries both sensible and latent heat; liquid stream L_n carries neither sensible nor latent heat; vapor stream V_n carries latent heat; and liquid stream L_{n-1} carries negative sensible heat. Also, there may be an appreciable heat of mixing and some heat loss by radiation. The various terms in the heat balance will be referred to by letter in accordance with the following scheme:

Heat item	Letter
Latent heat in vapor V_{n+1}	a
Sensible heat in vapor V_{n+1}, above t_n	b
Sensible heat in liquid L_{n-1}, below t_n	c
Latent heat in vapor V_n	d
Heat of mixing	e
Radiation loss	f

Calling all heat-input items positive and all heat-output items negative, the heat balance is

$$a + b - c + d + e - f = 0 \qquad (12\text{-}28)$$

Equation (12-28) may be simplified by taking advantage of an approximation that introduces only a small error in many situations. The two largest items in Eq. (12-28) are, for most systems, the latent heats a and d. The other four items are small: the sensible heats are small because the temperature change from plate to plate is small; the heat of mixing is small or zero unless the two components are considerably different chemically; and the radiation loss is minimized by lagging. Furthermore, items b and e are often balanced against the other two (c and f), so the effect of neglecting all four factors may be less than omitting any one. If the quantity $b + e - c - f$ is neglected in comparison with the greater quantities, a and d, it follows that, approximately,

$$a = d \qquad (12\text{-}29)$$

Then, only latent heats need to be considered. If λ_n and λ_{n+1} are the molal latent heats of streams V_n and V_{n+1}, respectively, Eq. (12-29) can be written

$$V_n\lambda_n = V_{n+1}\lambda_{n+1} \qquad (12\text{-}30)$$

Now a second approximation is in order. Chemically similar materials of nearly the same boiling point have nearly equal latent heats.† If the approximation that the molal latent heats of components A and B are equal is accepted, the heat of vaporization of 1 mole of any mixture of A and B is independent of the concentration of the mixture, and therefore

$$\lambda_n = \lambda_{n+1}$$

Then, from Eq. (12-30),

$$V_n = V_{n+1} \qquad (12\text{-}31)$$

Since plate n is any plate in the column other than the feed plate, this result is general, and the moles of vapor rising to any plate in either the rectifying or stripping section equal the moles of vapor rising from that plate, the feed plate excepted. It is also clear from Eq. (12-27) that

$$L_{n-1} = L_n \qquad (12\text{-}32)$$

An equivalent analysis shows that $L_{m-1} = L_m$.

Equations (12-31) and (12-32) formalize the concepts of *constant molal vaporization* and *constant molal overflow*,[9] respectively. At constant molal overflow, use L and V for the constant liquid and vapor flow rates in the rectifying section and $\overline{L}$ and $\overline{V}$ for the corresponding quantities in the stripping section. When vaporization and overflow are constant in any section of a column, the operating line for that section is straight and can be plotted if two points, or one point and the slope, are known.

A procedure that is free from the assumption of constant molal overflow, and which can be used where such an assumption is unwarranted, is described later in this chapter.

† This generalization is called *Trouton's rule.*

Reflux Ratio. The analysis of fractionating columns is facilitated by the use of a quantity called the reflux ratio. Two such quantities are used. One is the ratio of the reflux to the overhead product, and the other is the ratio of the reflux to the vapor. Both ratios refer to quantities in the rectifying column. The equations for these ratios are

$$R_D = \frac{L}{D} = \frac{V - D}{D} \quad \text{and} \quad R_V = \frac{L}{V} = \frac{L}{L + D} \tag{12-33}$$

In this text only R_D will be used.

If both numerator and denominator of the terms on the right-hand side of Eq. (12-21) are divided by D, the result is, for constant molal overflow,

$$y_{n+1} = \frac{R_D}{R_D + 1} x_n + \frac{x_D}{R_D + 1} \tag{12-34}$$

Equation (12-34) is an equation for the operating line of the rectifying column. The y intercept of this line is $x_D/(R_D + 1)$. The concentration x_D is set by the conditions of the design; and R_D, the reflux ratio, is an operating variable that can be controlled at will by varying the flow rate of the reflux for a given flow rate of the overhead product.

Condenser and Top Plate. Equation (12-21), written for constant molal overflow and dropping the subscripts n and $n + 1$, becomes

$$y = \frac{L}{L + D} x + \frac{D x_D}{L + D} \tag{12-35}$$

The intersection of the operating line represented by Eq. (12-35) and the diagonal represented by the equation $x = y$ is found by simultaneous solution of the two equations. Eliminating y gives

$$x = \frac{L}{L + D} x + \frac{D x_D}{L + D}$$

On simplification this equation gives

$$x = x_D$$

The operating line for the rectifying column, then, intersects the diagonal at point (x_D, x_D). This is true regardless of the performance of the condenser, only provided that but one overhead product is obtained.

The McCabe-Thiele construction for the top plate does depend on the action of the condenser. Figure 12-18 shows material-balance diagrams for the top plate and the condenser. The concentration of the vapor from the top plate is y_1, and that for the reflux to the top plate is x_c. In accordance with the general properties of operating lines, the operating terminus of the line is point a' in Fig. 12-19, having coordinates (x_c, y_1). Triangle $a'b'c'$ in Fig. 12-19 shows the graphical construction for the top plate. The situation represented by this triangle applies, for example, if two con-

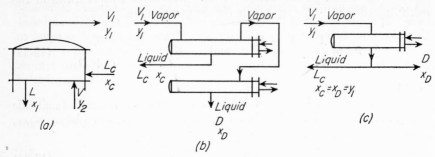

FIG. 12-18. Material-balance diagrams for top plate and condenser: (a) Top plate. (b) Partial and final condensers. (c) Total condenser.

densers operated in series as shown in Fig. 12-18b are used. The first condenser provides reflux and the second, product. In this design y_1, x_c, and x_D are all different.

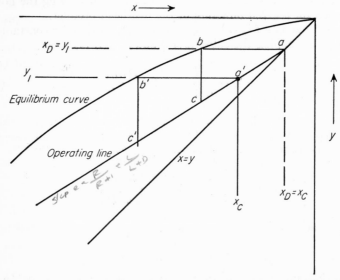

FIG. 12-19. Graphical construction for top plate: Triangle abc, using total condenser. Triangle $a'b'c'$, using partial condenser.

The simplest arrangement for obtaining reflux and liquid product, and one that is frequently used, is the single total condenser shown in Fig. 12-18c, which condenses all the vapor from the column and supplies both reflux and product. When such a single total condenser is used, the concentrations of the vapor from the top plate, of the reflux to the top plate, and of the overhead product are equal, and each can be denoted by x_D. The operating terminus of the operating line becomes point (x_D,x_D), which is the intersection of the operating line with the diagonal. Triangle abc in Fig. 12-19 then represents the top plate.

A second assumption concerning the action of the condenser is usually made. It is assumed that the condenser removes latent heat only and that the condensate is liquid at its boiling point. Then the reflux L is equal to L_c, the reflux from the condenser, and $V = V_1$. If the reflux is cooled below the boiling point, a portion of the vapor coming to plate 1 must condense to heat the reflux; so $V_1 < V$ and $L > L_c$. The assumption of hot reflux is made in the following treatment.

Bottom Plate and Reboiler. The action at the bottom of the column is analogous to that at the top. Thus, Eq. (12-26), written for constant molal overflow, becomes

$$y = \frac{\bar{L}}{\bar{L} - B}\, x - \frac{Bx_B}{\bar{L} - B}$$

Substitution of $y = x$ in this equation and simplifying gives

$$x = x_B$$

The operating line in the stripping column crosses the diagonal at point (x_B,x_B). This is true regardless of the operation of the reboiler, provided only that but one stream leaves the bottom.

The material-balance diagram for the bottom plate and the reboiler is shown in Fig. 12-20. The terminus of the operating line is point (x_b,y_r),

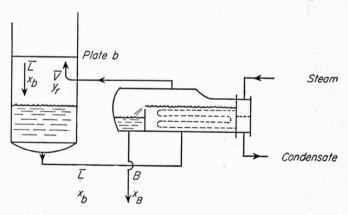

Fig. 12-20. Material-balance diagram for bottom plate and reboiler.

where x_b and y_r are the concentrations of liquid from the bottom plate and of vapor from the reboiler, respectively.

In the most common type of reboiler, shown in Figs. 12-16 and 12-20, the vapor leaving the reboiler is in equilibrium with the liquid leaving as bottom product. Then x_B and y_r are coordinates of a point on the equilibrium curve, and the reboiler acts as an ideal plate. In Fig. 12-21 is shown the graphical construction for the reboiler (triangle *cde*) and the bottom plate (triangle *abc*).

Feed Plate. The addition of the feed increases the reflux in the stripping column, increases the vapor in the rectifying column, or does both. In the rectifying column the vapor flow rate is greater than the liquid flow rate. In the stripping column the liquid flow rate is greater than the vapor flow

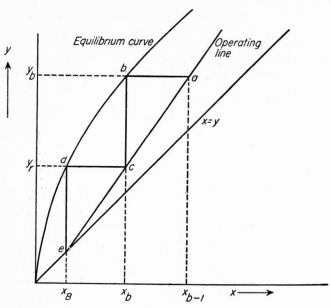

Fig. 12-21. Graphical construction for bottom plate and reboiler: Triangle *cde*, reboiler. Triangle *abc*, bottom plate.

rate. It follows that the slope of the operating line in the stripping column is greater than 1 and that of the operating line in the rectifying column is less than 1.

The quantitative effects of introducing the feed depend on the condition of the feed. Feed may be introduced as a cold liquid, as a saturated liquid at its boiling point, as a mixture of liquid and vapor, as a saturated vapor at its dew point, or as a superheated vapor.

Figure 12-22 shows diagrammatically the liquid and vapor streams into and out of the feed plate for various feed conditions. In Fig. 12-22a cold feed is assumed. The entire feed stream becomes a part of the reflux in the stripping column. Also, to heat the feed to the boiling point some vapor must condense, and this condensate also becomes a part of $\bar{L}$, which consists of: (1) the reflux from the rectifying column, (2) the feed, and (3) the condensate. The vapor to the rectifying column is less, by the amount of condensation, than that in the stripping column.

In Fig. 12-22b the feed is assumed to be at its boiling point. No condensation is required to heat the feed, so $V = \overline{V}$ and $\overline{L} = F + L$. If the feed is partly vapor, as shown in Fig. 12-22c, the liquid portion of the feed becomes part of $\overline{L}$, and the vapor portion becomes part of V. If the feed is saturated vapor, as shown in Fig. 12-22d, the entire feed becomes part of V, so $L = \overline{L}$ and $V = F + \overline{V}$. Finally, if the feed is superheated vapor, as shown in Fig. 12-22e, part of the liquid from the rectifying column

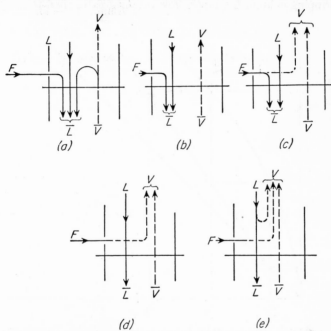

FIG. 12-22. Flow through feed plate for various feed conditions: (a) Feed cold liquid. (b) Feed saturated liquid. (c) Feed partially vaporized. (d) Feed saturated vapor. (e) Feed superheated vapor.

is vaporized to cool the feed to a state of saturated vapor. Then the vapor in the rectifying column consists of: (1) the vapor from the stripping column, (2) the feed, and (3) the vaporization. The reflux to the stripping column is less, by the amount of vaporization, than that in the rectifying column.

All five of the feed types can be correlated by the use of a single factor, denoted by f and defined as the moles of vapor flow in the rectifying section that result from the introduction of each mole of feed. Then f has the following numerical limits for the various feed conditions:

Cold feed, $f < 0$
Feed at bubble point (saturated liquid), $f = 0$
Feed partially vapor, $0 < f < 1$
Feed at dew point (saturated vapor), $f = 1$
Feed superheated vapor, $f > 1$

If the feed is a mixture of liquid and vapor, f is the fraction that is vapor, and $1 - f$ is the fraction that is liquid.

The contribution of the entire feed to V is fF, so

$$V = \overline{V} + fF \quad \text{and} \quad V - \overline{V} = fF \tag{12-36}$$

Likewise, the contribution of the entire feed to $\overline{L}$ is $F(1 - f)$, so the total reflux to the stripping column is the sum of L and $F(1 - f)$, or

$$\overline{L} = L + (1 - f)F \quad \text{and} \quad \overline{L} - L = (1 - f)F \tag{12-37}$$

Equation (12-37) can also be derived from a total material balance around the feed plate, thus

$$F + \overline{V} + L = V + \overline{L}$$

Substituting from Eq. (12-36),

$$F + \overline{V} + L = \overline{V} + fF + \overline{L}$$

$$\overline{L} = L + (1 - f)F$$

Feed Line. Equations (12-36) and (12-37) can be used with material balances to find the locus of all points of intersection of the operating lines. The equation for this line of intersections may be found as follows.

Write Eqs. (12-20) and (12-24) as

$$Vy = Lx + Dx_D \tag{12-38}$$

$$\overline{V}y = \overline{L}x - Bx_B \tag{12-39}$$

Subtract Eq. (12-39) from Eq. (12-38).

$$y(V - \overline{V}) = (L - \overline{L})x + Dx_D + Bx_B \tag{12-40}$$

From Eq. (12-16), the last two terms in Eq. (12-40) can be replaced by Fx_F. Also, substituting for $L - \overline{L}$ from Eq. (12-37) and for $V - \overline{V}$ from Eq. (12-36) and simplifying, the result is

$$Ffy = -(1 - f)Fx + Fx_F$$

or

$$y = -\frac{1 - f}{f} x + \frac{x_F}{f} \tag{12-41}$$

Equation (12-41) represents a straight line, called the feed line, on which all intersections of the operating lines must fall. The position of the line depends only on x_F and f. The slope of the feed line is $-(1 - f)/f$, and, as may be demonstrated by substituting x for y in Eq. (12-41) and simplifying, the line crosses the diagonal at $x = x_F$.

The simplest method of plotting the operating lines is: (1) locate the feed line; (2) calculate the y-axis intercept $x_D/(R_D + 1)$ of the rectifying line, and plot that line through the intercept and the point (x_D, x_D); (3) draw the stripping line through point (x_B, x_B) and the intersection of the rectifying line with the feed line. The operating lines in Fig. 12-23 show this procedure.

In Fig. 12-23 are plotted operating lines for various types of feed, on the assumption that x_F, x_B, x_D, L, and D are all constant. The corresponding feed lines are shown. If the feed is a cold liquid, the feed line slopes upward and to the right; if the feed is a saturated liquid, the line

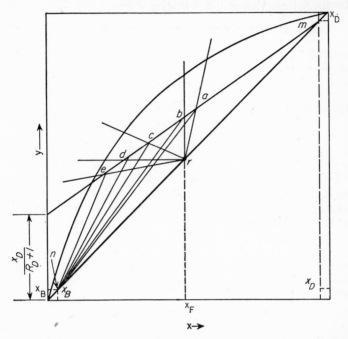

Fig. 12-23. Effect of feed condition on feed line: ra, feed cold liquid. rb, feed saturated liquid. rc, feed partially vaporized. rd, feed saturated vapor. re, feed superheated vapor.

is vertical; if the feed is a mixture of liquid and vapor, the line slopes upward and to the left, and the slope is the negative of the ratio of the liquid to the vapor; if the feed is a saturated vapor, the line is horizontal; and, finally, if the feed is a superheated vapor, the line slopes downward and to the left.

Feed-plate Location. After the operating lines have been plotted, the number of ideal plates is found by the usual step-by-step construction, as shown in Fig. 12-24. The construction can start either at the bottom of the stripping line or at the top of the rectifying line. In the following it is assumed that the construction begins at the top and also that a total condenser is used. As the intersection of the operating lines is approached, a decision must be made as to when the steps should transfer from the rectifying line to the stripping line. The change should be made in such a manner that the maximum enrichment per plate is obtained so that the number of plates is as small as possible. Inspection of Fig. 12-24 shows that this criterion is met if the transfer is made immediately after a value of x is reached that is less than the x coordinate of the intersection of the two operating lines. The feed plate is always represented by the triangle that has one corner on the rectifying line and one on the stripping line. When at the optimum position, the triangle representing the feed plate straddles the intersection of the operating lines.

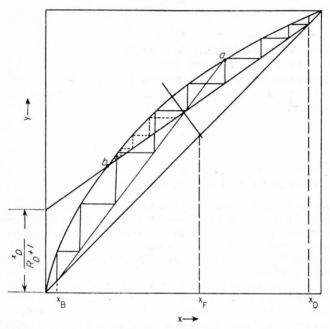

FIG. 12-24. Feed-plate location.

The transfer from one operating line to the other, and hence the feed-plate location, can be made at any location between points a and b in Fig. 12-24, but if the feed plate is placed anywhere but at the optimum point, an unnecessarily large number of plates is called for. It is possible, for example, as shown in Fig. 12-24, to delay transferring to the stripping line until the point b is almost reached. Many small triangles will then be concentrated in the acute angle between the equilibrium and operating lines near point b, and these triangles represent a large number of unnecessary plates in the rectifying section. Likewise, if the transfer from the rectifying line to the stripping line is made nearly at point a, unnecessary plates in the stripping column are called for.

Heating and Cooling Requirements. Radiation from a column is small, and the column itself is essentially adiabatic. The heat effects of the entire plant are confined to the condenser and the reboiler. If the molal latent heat λ is constant (and the assumption of constant molal latent heat requires this assumption), the heat added in the reboiler q_r is $\overline{V}\lambda$, and that removed in the condenser q_c is $-V\lambda$, both in Btu per hour. Heat effects are considered positive when the heat is added to the process and negative when heat is withdrawn; so q_c is negative.

If saturated steam is used as the heating medium, the steam required at reboiler is

$$w_s = \frac{\overline{V}\lambda}{\lambda_s} \qquad (12\text{-}42)$$

where w_s = steam consumption, lb/hr
$\overline{V}$ = vapor from reboiler, lb moles/hr
λ_s = latent heat of steam, Btu/lb
λ = molal latent heat of mixture, Btu/lb mole

If water is used as the cooling medium in the condenser, the cooling-water requirement is

$$w_c = -\frac{-V\lambda}{t_2 - t_1} \qquad (12\text{-}43)$$

where w_c = water consumption, lb/hr
$t_2 - t_1$ = temperature rise of cooling water, °F

Example 12-5. A continuous fractionating column is to be designed to separate 30,000 lb/hr of a mixture of 40 per cent benzene and 60 per cent toluene into an overhead product containing 97 per cent benzene and a bottom product containing 98 per cent toluene. These percentages are by weight. A reflux ratio of 3.5 moles to 1 mole of product is to be used. The molal latent heat of both benzene and toluene is 7240 cal/g mole.

(a) Calculate the moles of overhead product and bottom product per hour.

(b) Determine the number of ideal plates and the positions of the feed plate: (i) if the feed is liquid and at its boiling point; (ii) if the feed is liquid and at 20°C (specific heat = 0.44); (iii) if the feed is a mixture of two-thirds vapor and one-third liquid.

(c) If steam at 20 lb force/in.² gauge is used for heating, how much steam is required per hour for each of the above three cases, neglecting heat losses and assuming the reflux is a saturated liquid?

(d) If cooling water enters the condenser at 80°F and leaves at 150°F, how much cooling water is required, in gallons per minute?

Solution. (a) The molecular weight of benzene is 78 and that of toluene is 92. The concentrations of feed, overhead, and bottoms in mole fraction of benzene are

$$x_F = \frac{\frac{40}{78}}{\frac{40}{78} + \frac{60}{92}} = 0.440$$

$$x_D = \frac{\frac{97}{78}}{\frac{97}{78} + \frac{3}{92}} = 0.974$$

$$x_B = \frac{\frac{2}{78}}{\frac{2}{78} + \frac{98}{92}} = 0.0235$$

The average molecular weight of the feed is

$$\frac{100}{\frac{40}{78} + \frac{60}{92}} = 85.8$$

The feed rate F is $30,000/85.8 = 350$ moles/hr. By an over-all benzene balance, using (Eq. 12-17),

$$D = 350 \frac{0.440 - 0.0235}{0.974 - 0.0235} = 153.4 \text{ moles/hr}$$

$$B = 350 - 153.4 = 196.6 \text{ moles/hr}$$

(i) The first step is to plot the equilibrium diagram and on it erect verticals at x_D, x_F, and x_B. These should be extended to the diagonal of the diagram. Refer to Fig. 12-25.

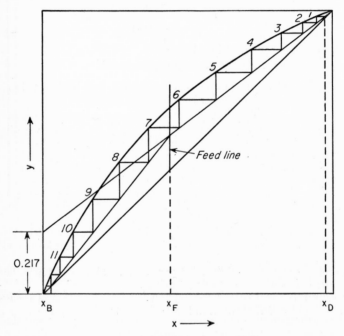

Fig. 12-25. Example 12-5, part (b) (i).

The second step is to draw the feed line. Here, $f = 0$, and the feed line is vertical and is a continuation of line $x = x_F$.

The third step is to plot the operating lines. The intercept of the rectifying line on the y axis is, from Eq. (12-34), $0.974/(3.5 + 1) = 0.216$. From the intersection of this operating line and the feed line the stripping line is drawn.

The fourth step is to draw the rectangular steps between the two operating lines and the equilibrium curve. In drawing the steps, the transfer from the rectifying line to the stripping line is at the seventh step. By counting steps it is found that, besides the reboiler, 11 ideal plates are needed and feed should be introduced on the seventh plate from the top.†

(*ii*) This solution is the same as that of part (*b*)(*i*) except for the feed line. From Fig. 12-2, the boiling point of the feed is 94°C. The specific heat of the feed, in cal/(g mole)(°C) is $0.44 \times 85.8 = 37.7$. To convert 1 g mole of feed to saturated liquid requires $37.7(92 - 20) = 2710$ cal. Since the latent heat is 7240 cal/mole, $2710/7240 = 0.37$ mole of vapor from the stripping section must condense to heat

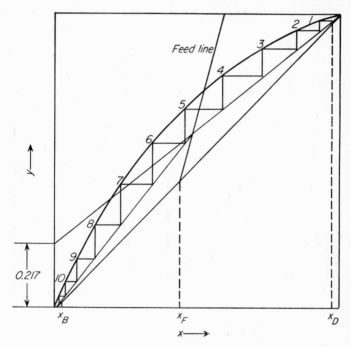

Fig. 12-26. Example 12-5, part (*b*) (*ii*).

1 mole of feed to its boiling point. The value of f is, then, -0.37, and the slope of the feed line is $-(1 + 0.37)/-0.37 = 3.70$. When the steps are drawn for this case, as shown in Fig. 12-26, it is found that a reboiler and 10 ideal plates are needed and that the feed should be introduced on the fifth plate.

(*iii*) From the definition of f it follows that for this case $f = \frac{2}{3}$ and the slope of the feed line is -0.5. The solution is shown in Fig. 12-27. It calls for a reboiler and 12 plates, with the feed entering on the seventh plate.

† To fulfill the conditions of the problem literally, the last step, which represents the reboiler, should reach the concentration x_B exactly. This is nearly true in Fig. 12-25. Usually, x_B does not correspond to an integral number of steps. An arbitrary choice of the four quantities x_D, x_F, x_B, and R_D is not necessarily consistent with an integral number of steps. An integral number can be obtained by a slight adjustment of one of the four quantities, but in view of the fact that a plate efficiency must be applied before the actual number of plates is established, there is little reason for making this adjustment.

(c) The vapor flow in the rectifying column, which must be condensed in the condenser, is 4.5 moles per mole of overhead product, or $4.5 \times 153.4 = 690$ moles/hr. From Eq. (12-36),

$$\overline{V} = 690 - 350f$$

The latent heat of vaporization is 7240 cal/g mole or $7240 \times 1.8 = 13{,}030$ Btu/lb mole. The heat from 1 lb of steam at 20 lb force/in.² gauge is, from Appendix 6,

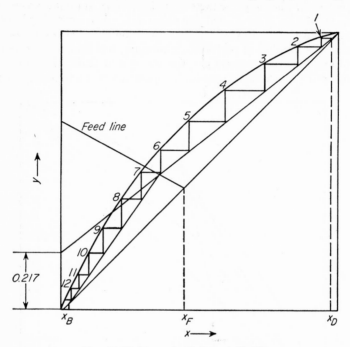

Fig. 12-27. Example 12-5, part (b) (iii).

939 Btu. The steam required is, from Eq. (12-42),

$$w_s = \frac{13{,}030}{939} \overline{V} = \frac{13{,}030}{939} (690 - 350f) \text{ lb/hr}$$

The results are:

Case	f	Steam requirement w_s, lb/hr
(c)(i)	0	9,600
(c)(ii)	-0.36	11,350
(c)(iii)	0.667	6,350

(*d*) The cooling water needed, which is the same in all cases, is, from Eq. (12-43),

$$w_c = -\frac{-13,030 \times 690}{150 - 80} = 128,500 \text{ lb/hr}$$

The water requirement is $128,500/(60 \times 8.33) = 257$ gal/min.

Minimum Number of Plates. Since the slope of the rectifying line is $R_D/(R_D + 1)$, the slope increases as the reflux ratio increases, until, when R_D is infinite, $V = L$, and the slope is 1. The operating lines then both coincide with the diagonal. This condition is called total reflux. At total reflux the number of plates is a minimum, but the feed and both the overhead and bottom products are, for any still of finite size, zero. Total reflux represents one limiting case in the operation of fractionating columns.

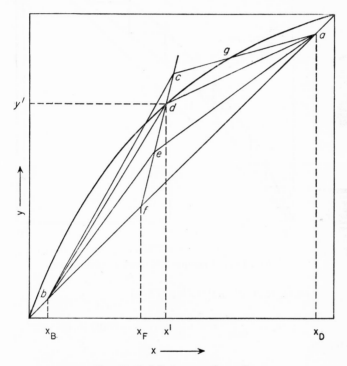

Fig. 12-28. Minimum reflux ratio.

Minimum Reflux Ratio. At any reflux somewhat less than total **a** finite number of plates is required. If the reflux ratio is decreased continuously, however, a second limiting condition is reached at which the reflux ratio becomes a minimum but the number of plates becomes infinite. All actual columns must operate between minimum reflux and total reflux.

The minimum reflux ratio can be found by following the movement of the operating lines as the reflux is reduced. In Fig. 12-28 both operating lines coincide with the diagonal *afb* at total reflux. For an actual operation lines *ae* and *eb* are typical operating lines. As the reflux is further reduced, the intersection of the operating lines moves along the feed line toward the equilibrium curve, the area on the diagram available for steps shrinks, and the number of steps increases. When either one or both of the operating lines touch the equilibrium curve, the number of steps

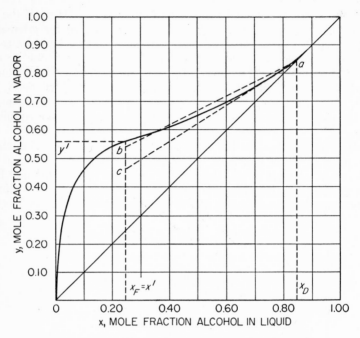

Fig. 12-29. Equilibrium diagram (system ethanol-water).

necessary to cross the point of contact becomes infinite. The reflux ratio corresponding to this situation is, by definition, the minimum reflux ratio.

For the normal type of equilibrium curve, which is concave downward throughout its length, the point of contact, at minimum reflux, of the operating and equilibrium lines is at the intersection of the feed line with the equilibrium curve, as shown by lines *ad* and *db* in Fig. 12-28. A further decrease in reflux brings the intersection of the operating lines outside of the equilibrium curve, as shown by lines *agc* and *cb*. Then even an infinite number of plates cannot pass point *g*, and the reflux ratio for this condition is less than the minimum.

The slope of operating line *ad* in Fig. 12-28 is such that the line passes through the points (x',y') and (x_D,x_D), where x' and y' are the coordinates of the intersection of the feed line and the equilibrium curve. Let the minimum reflux ratio be R_{Dm}. Then

$$\frac{R_{Dm}}{R_{Dm} + 1} = \frac{x_D - y'}{x_D - x'}$$

or
$$R_{Dm} = \frac{x_D - y'}{y' - x'} \qquad (12\text{-}44)$$

Equation (12-44) cannot be applied to all systems. Thus, if the equilibrium curve has a concavity upward, e.g., the curve for ethanol and water shown in Fig. 12-29, it is clear that the rectifying line first touches the equilibrium curve between abscissas x_F and x_D, and line ac corresponds to minimum reflux. Operating line ab is drawn for a reflux less than the minimum, even though it does intersect the feed line below point (x', y'). In such a situation the minimum reflux ratio must be computed from the slope of the operating line ac that is tangent to the equilibrium curve.

Optimum Reflux. As the reflux ratio is increased from the minimum, the number of plates decreases, rapidly at first, and then more and more slowly, until, at total reflux, the number of plates is a minimum. It will be shown later that the cross-sectional area of a column usually is approximately proportional to the flow rate of vapor. As the reflux ratio increases,

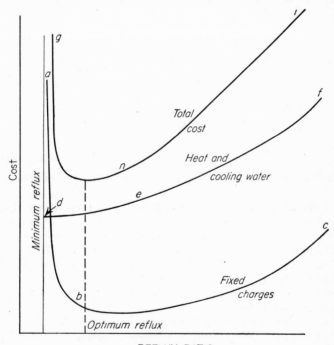

REFLUX RATIO

Fig. 12-30. Optimum reflux ratio.

both V and L increase for a given production, and a point is reached where the increase in column diameter is more rapid than the decrease in number of plates. The cost of the unit is roughly proportional to the total plate area, so the fixed charges first decrease and then increase with reflux. This is shown by line abc in Fig. 12-30.

The reflux is made by supplying heat at the reboiler and withdrawing it at the condenser. The costs of both heat and cooling water increase with reflux, as shown by curve def in Fig. 12-30. The total cost of operation varies with the sum of the fixed charges and the cost of heat and cooling water, as shown by curve ghi. Curve ghi has a minimum, at a definite reflux ratio, which usually is not much greater than the minimum reflux. This is the point of most economical operation, and this reflux ratio is called the optimum reflux ratio. Actually, most plants are operated at refluxes above the optimum, usually in a range of 1.2 to 2.0 times the minimum reflux. The total cost is not very sensitive to reflux ratio in this range, and better operating flexibility is obtained if a reflux greater than the optimum is used.

Example 12-6. What is: (a) the minimum reflux ratio and (b) the minimum number of ideal plates for cases (b)(i), (b)(ii), and (b)(iii) of Example 12-5?

Solution. (a) For minimum reflux ratio use Eq. (12-44). Here $x_D = 0.974$.

Case	x'	y'	R_{Dm}
(b)(i)	0.440	0.658	1.45
(b)(ii)	0.521	0.730	1.17
(b)(iii)	0.300	0.513	2.16

(b) For minimum number of plates, the ratio reflux is infinite, the operating lines coincide with the diagonal, and there are no differences among the three cases. The plot is given in Fig. 12-31. A reboiler and eight ideal plates are needed.

Nearly Pure Products. When either the bottom or overhead product is nearly pure, a single diagram covering the entire range of concentrations is impractical as the steps near $x = 0$ and $x = 1$ become small. This situation may be treated by any one of three methods. In the first, auxiliary diagrams for the ends of the concentration range are prepared, at a large scale, so the individual steps are sufficiently large to be drawn. The operating and equilibrium lines are drawn on the auxiliary plot and the steps transferred to this plot from the main diagram.

The second method for opening up the ends of the diagram is to plot the equilibrium and operating lines on logarithmic coordinates.[18a] The operating line, if straight on rectangular coordinates, is not so on logarithmic coordinates, but calculation of the coordinates of a few points by the material-balance equations allows the curve to be plotted. The steps are distorted on such a diagram, but the number of steps is unaffected by a change from rectangular logarithmic coordinates. This method is especially effective in the portion of the concentration range near an azeotrope.

The third method of treating nearly pure products is based on the principle that Raoult's law applies near $x = 1$ and Henry's law near $x = 0$. At both ends of the curve of x_e vs. y_e, then, both the equilibrium and the oper-

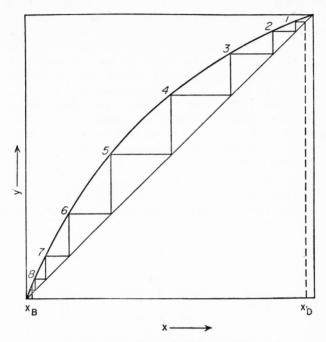

Fig. 12-31. Example 12-6, part (b).

ating lines are straight, so Eq. (10-37) can be used, and no graphical construction is required. The same equation may be used anywhere in the concentration range where both operating line and equilibrium lines are straight or nearly so.

Example 12-7. A mixture of 2 mole per cent ethanol and 98 mole per cent water is to be stripped in a bubble column to a bottom product containing not more than 0.01 mole per cent ethanol. Steam, admitted through an open coil in the liquid on the bottom plate, is to be used as a source of vapor. The feed is at its boiling point. The steam flow is to be 0.2 mole per mole of feed. For dilute ethanol-water solutions, the equilibrium line is straight, and is given by the equation

$$y_e = 9.0x_e$$

How many ideal plates are needed?

Solution. Since both equilibrium and operating lines are straight, Eq. (10-37) rather than a graphical construction may be used. The material-balance diagram is shown in Fig. 12-32. No reboiler is needed, as the steam enters as a vapor. Also, the liquid flow in the tower equals the feed entering the column. By conditions of the problem

$$F = \bar{L} = 1 \qquad \bar{V} = 0.2 \qquad y_b = 0 \qquad x_a = 0.02$$

$$x_b = 0.0001 \qquad m = 9.0 \qquad y_a^* = 9.0 \times 0.02 = 0.18$$

$$y_b^* = 9.0 \times 0.0001 = 0.0009$$

To use Eq. (10-37) y_a, the concentration of the vapor leaving the column, is needed. This is found by an over-all ethanol balance

$$\bar{V}(y_a - y_b) = \bar{L}(x_a - x_b)$$

$$0.2(y_a - 0) = 1(0.02 - 0.0001)$$

from which $y_a = 0.0995$. Substituting into Eq. (10-37),

$$N = \frac{\log \dfrac{0 - 0.0009}{0.0995 - 0.18}}{\log \dfrac{0 - 0.0995}{0.0009 - 0.18}} = \frac{\log \dfrac{0.18 - 0.0995}{0.0009}}{\log \dfrac{0.18 - 0.0009}{0.0995}}$$

$$= \frac{\log 89.4}{\log 1.8} = 7.6 \text{ ideal plates}$$

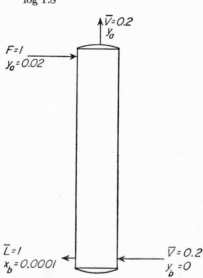

Fig. 12-32. Material-balance diagram for Example 12-7.

Analysis of Fractionating Columns by Ponchon-Savarit Method

The limitation imposed by the assumption of constant molal overflow may be removed if the analysis of fractionating equipment is based on enthalpy balances as well as on material balances. Then the reflux and

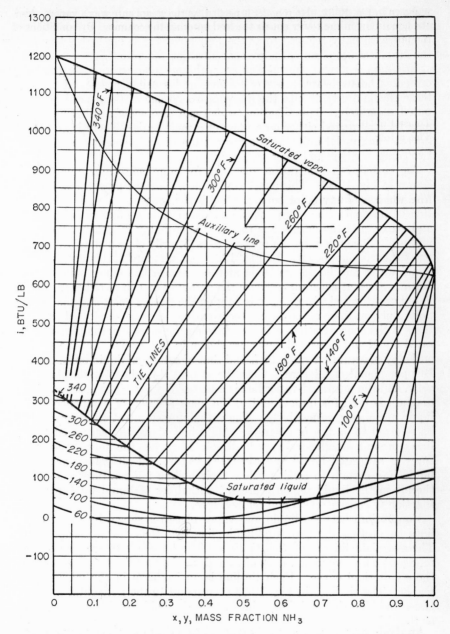

Fig. 12-33. Enthalpy-concentration diagram (system ammonia-water at 10 atm). (*After Bosnjakovic.*)

vapor at any point in the column are found by a rigorous enthalpy balance, which replaces the approximate heat balance shown by Eq. (12-30). A method usually referred to as the Ponchon-Savarit method is based on the graphical use of an enthalpy-concentration diagram.[16, 18b, 22]

In Chap. 9 it was shown that an enthalpy-concentration diagram coordinates latent heats, heats of mixing, and sensible heats and that all these effects are built into the diagram, so no one of them need be computed separately. This diagram, which can also show equilibrium data, may be constructed either on a mass or a mole basis. There is no reason for retaining molal units, as the assumption of equal latent heats for the two components is no longer necessary.

An enthalpy-concentration diagram for ammonia-water mixtures is shown in Fig. 12-33. This chart applies at a constant pressure of 10 atm. The abscissa of any point on Fig. 12-33 is the concentration, in mass fraction of ammonia (the more volatile component) in a mixture of ammonia and water, and the ordinate is the specific enthalpy in Btu per pound of mixture. The enthalpies are referred to arbitrarily chosen standard states, one for each component. In Fig. 12-33 the standard states are the pure liquid water and liquid ammonia at their respective boiling points under 1 atm. Since at constant pressure the enthalpy of 1 lb of either saturated liquid or saturated vapor depends only on the boiling point and the boiling point only on the concentration, these enthalpies may be plotted as functions of the concentration. Curves showing the specific enthalpies of saturated liquid and saturated vapor are shown in Fig. 12-33. Let the specific enthalpy of saturated vapor be i_y and that of saturated liquid be i_x, both in Btu per pound of mixture. Then the saturated-liquid line is a plot of i_x vs. x, and the saturated-vapor line is a plot of i_y vs. y. The abscissa of the diagram is used for both x and y, the mass fraction of component A in liquid and vapor, respectively. The enthalpy-concentration diagram may be referred to as the ix diagram. All points above the saturated vapor line represent superheated vapor; all points between the lines are for mixtures of saturated liquid and saturated vapor; and all points below the liquid line pertain to liquids below their boiling temperatures. The isotherms in the liquid region represent enthalpies of liquid mixtures as a function of concentration and temperature.

Equilibrium between liquid and vapor phases is shown by the straight lines connecting the saturated-liquid line and the saturated-vapor line. These straight lines are called *tie lines*. The terminals of any one tie line represent corresponding values of x_e and y_e, the coordinates of a point on the equilibrium curve. Each tie line is an isotherm in the mixed-phase region. The tie lines in Fig. 12-33 correspond to the horizontal lines between bubble-point and dew-point curves of a boiling-point diagram such as that of Fig. 12-1.

There are an infinite number of tie lines. Several are shown in Fig. 12-33. To obtain other tie lines an interpolation may be made, or, more accurately, the construction shown in Fig. 12-34 can be used. For each tie line a point such as point a is obtained by finding the intersection of the vertical line through x_e and the horizontal line through y_e. The auxiliary line shown in Fig. 12-33 is located in this manner, and the tie line for any desired x_e or y_e is found by reversing this construction. Another good way of obtaining tie lines is to plot an auxiliary x_e vs. y_e diagram and transfer points to the ix diagram.[18c]

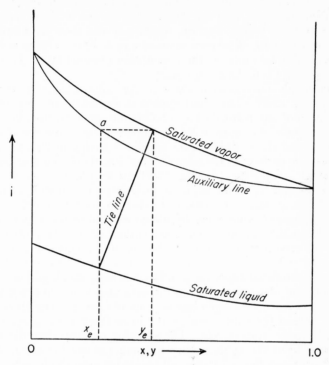

Fig. 12-34. Tie-line construction.

Basic Construction on Enthalpy-concentration Diagram. A number of graphical constructions can be made directly on an ix diagram. These depend on a property that may be demonstrated by considering the simple mixing process shown in Fig. 12-35a. The process consists of mixing w_R lb/hr of a stream R with w_S lb/hr of stream S to form $w_R + w_S$ lb/hr of stream T. The concentrations of streams R, S, and T are x_R, x_S, and x_T, respectively, all in pounds of more volatile component per pound of stream. The specific enthalpies of the same streams are i_R, i_S, and i_T, respectively, all in Btu per pound. Molal units can be used in place of mass units if desired.

Assume, first, that the process is adiabatic. An enthalpy balance [Eq. (1-21)], neglecting mechanical-potential- and kinetic-energy terms, and noting that Q and W_s are zero, is

$$w_R i_R + w_S i_S = (w_R + w_S)i_T \qquad (12\text{-}45)$$

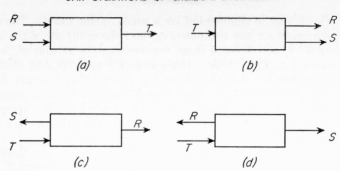

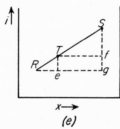

FIG. 12-35. Adiabatic mixing process: (*a*) Mixing R and S to form T. (*b*) Separating T into R and S. (*c*) Removing S from T to give R. (*d*) Removing R from T to give S. (*e*) All processes on *ix* diagram.

A more-volatile-component balance (component A) is

$$w_R x_R + w_S x_S = (w_R + w_S) x_T \tag{12-46}$$

Equations (12-45) and (12-46) can each be solved for the ratio w_R/w_S. The results are

$$\frac{w_R}{w_S} = \frac{i_S - i_T}{i_T - i_R} \tag{12-47}$$

$$\frac{w_R}{w_S} = \frac{x_S - x_T}{x_T - x_R} \tag{12-48}$$

Equating the right-hand sides of Eqs. (12-47) and (12-48) gives

$$\frac{x_S - x_T}{x_T - x_R} = \frac{i_S - i_T}{i_T - i_R} \tag{12-49}$$

Each stream can be represented by a point on the chart. The coordinates of the point are the concentration and the enthalpy of the stream. Plot points R, S, and T, each point representing the stream of the same letter, as shown in Fig. 12-35e. Draw lines RT and TS, and plot points e, f, and g, as shown. It is clear from Eq. (12-49) that triangles TSf and RTe are geometrically similar. Then lines RT and TS have the same slope, and line RTS is a single straight line. In an adiabatic mixing process, therefore, the points on the ix diagram that represent the product stream and the two feed streams lie on the same straight line.

The construction shown in Fig. 12-35e also applies to the three other processes shown in Fig. 12-35b, c, and d. In Fig. 12-35b stream T is separated into streams R and S; in Fig. 12-35c stream S is removed or subtracted from stream T, leaving stream R as a difference; and in Fig. 12-35d stream R is removed from stream T, leaving stream S. Note that for all situations shown in Fig. 12-35 the point representing a single stream entering the process lies between the points representing the two streams leaving the process and that the point for a single stream leaving the process lies between the points for the two streams entering.

Equations (12-47) and (12-48) show that the relative amounts of the three streams are proportional to the lengths of line segments in Fig. 12-35e. Let a vinculum over the letters for the terminals of a line symbolize the length of the line. Then w_R/w_S may be represented by $\overline{Sf}/\overline{Te}$, by $\overline{fT}/\overline{eR}$, or by $\overline{ST}/\overline{TR}$. Also, w_S is proportional to $\overline{Te}$, to $\overline{eR}$, or to $\overline{TR}$, and w_R is proportional to $\overline{Sf}$, to $\overline{fT}$, or to $\overline{ST}$. Again, the ratio of the mass of stream R to that of stream T is given by $\overline{Sf}/\overline{Sg}$, by $\overline{fT}/\overline{gR}$, or by $\overline{ST}/\overline{SR}$.

If Eq. (12-48) is written as $w_R(x_T - x_R) = w_S(x_S - x_T)$, it can be seen that point T represents the center of gravity of points R and S, endowed with masses w_R and w_S, respectively. This will be referred to as the *center-of-gravity principle*.

Nonadiabatic Process. The construction shown in Fig. 12-35e can be extended to nonadiabatic processes by utilizing the following device. Let q be the heat effect accompanying the process, in Btu per hour, as shown in Fig. 12-36a. Assume q is positive for an endothermic process, where the process absorbs heat, and negative for an exothermic process, where the process evolves heat. Imagine the actual process to be divided into two steps, an adiabatic step and a heating step, in which either an incoming stream is preheated by adding the heat q *before* the stream enters the adiabatic step or in which a leaving stream is heated by adding heat q *after* the stream leaves the adiabatic step. Flow charts for these two situations are shown in Fig. 12-36. In Fig. 12-36b the heat effect is absorbed by stream S before that stream enters the adiabatic step, and in Fig. 12-36c the discharge stream T absorbs heat q after it leaves the adiabatic step. The specific enthalpy of stream S, after being heated, is i_S', where

$$i_S' = i_S + \frac{q}{w_S} \tag{12-50}$$

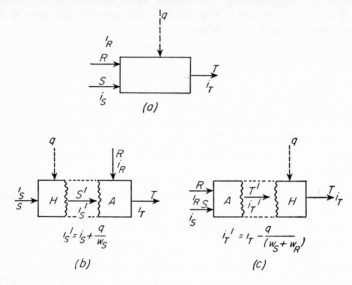

Fig. 12-36. Nonadiabatic mixing processes: (a) Actual process. (b) Equivalent two-step process, heat added to an entering stream. (c) Equivalent two-step process, heat added to leaving stream. A, adiabatic step. H, heating step.

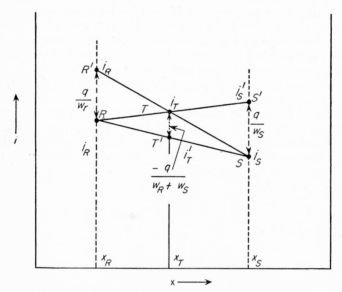

Fig. 12-37. Nonadiabatic process on ix diagram.

Then point S', representing the stream after heating, is collinear with points R and T, as shown in Fig. 12-37. The specific enthalpy of stream T, before being heated, is i'_T, where

$$i'_T = i_T - \frac{q}{w_R + w_S} \tag{12-51}$$

Point T', representing the stream before heating, is collinear with points R and S, representing the feed streams to the process.

If the process is exothermic, q is negative in Eqs. (12-50) and (12-51), and $i'_S < i_S$, $i'_R < i_R$, and $i'_T > i_T$. If the process is endothermic, $i'_S > i_S$; $i'_R > i_R$, and $i'_T < i_T$. The lines in Fig. 12-37 as drawn apply to endothermic processes.

The choice as to which stream is corrected for the heat effect is optional. If two heat effects are involved in a single process, it may be convenient to correct one stream for one heat effect and a second stream for the other.

These principles and constructions can now be applied to the analysis of a fractionating plant with both rectifying and stripping sections on the assumption that an ix diagram is available for the mixture to be processed.

Over-all Enthalpy Balance around Fractionating Column. Consider an over-all enthalpy balance over the plant shown in Fig. 12-17. In addition to the quantities shown in the figure, let i_F, i_D, and i_B represent the specific enthalpies of the feed, overhead product, and bottom product, respectively, all in Btu per pound. Points F, D, and B, representing feed, overhead, and bottoms, are plotted on an enthalpy-concentration diagram in Fig. 12-38. Both products are at their bubble temperatures, so points D and B lie on the saturated-liquid line. The feed may have any thermal condition from cold liquid to superheated vapor. It is shown as a mixture of liquid and vapor in Fig. 12-38. The ratio of vapor to liquid in the feed is, by the center-of-gravity principle, the ratio aF/Fb, where a and b are the intersections of the tie line through point F with the saturated liquid and vapor lines, respectively.

The over-all process is a nonadiabatic splitting of the feed into overhead and bottom products. To use the adiabatic construction on Fig. 12-38, correct stream D for $-q_c$, the heat removed in the condenser, and stream B for q_r, the heat added in the reboiler. Since both streams are product streams, Eq. (12-51) applies, and

$$i'_D = i_D - \frac{-q_c}{D} = i_D + \frac{q_c}{D} \tag{12-52}$$

$$i'_B = i_B - \frac{q_r}{B} \tag{12-53}$$

Points D' and B' represent the corrected product streams. Since the process with corrected streams is adiabatic, points D', F, and B' lie on a single straight line, which may be called the over-all-enthalpy line.

For given feed and product streams, only one of the heat effects, $-q_c$ or q_r, is independent and subject to choice by the designer or operator. The usual choice is $-q_c$, which is fixed when the reflux ratio at the top of the

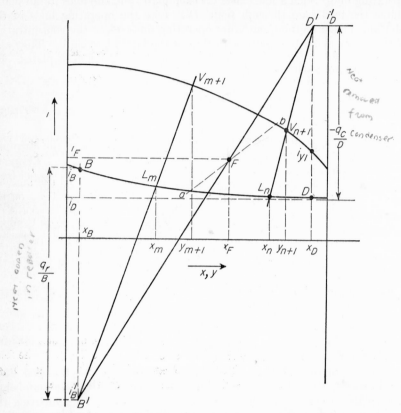

Fig. 12-38. Ponchon-Savarit method. Construction of over-all-enthalpy and enthalpy operating lines.

column is chosen. The reflux ratio is, by the center-of-gravity principle,

$$R_{D1} = \frac{i'_D - i_{y1}}{i_{y1} - i_D} \tag{12-54}$$

where i_{y1} is the specific enthalpy of the vapor leaving plate 1 and entering the condenser.

Enthalpy Balances in Rectifying and Stripping Columns. Let the specific enthalpy of the vapor stream rising from plate $n + 1$ to plate n in the rectifying column shown in Fig. 12-17 be $i_{y(n+1)}$, and let that of the liquid stream leaving plate n be i_{xn}. Let point V_{n+1} on Fig. 12-38 represent the vapor from plate $n + 1$, and let point L_n represent the liquid from plate n. Point V_{n+1} lies on the saturated-vapor line and point L_n on the saturated-liquid line. Since the column is adiabatic, points D', V_{n+1}, and L_n are collinear. The straight line through them is called the enthalpy operating line. Such a line exists for each plate, and the lines for all plates above the feed pass through point D'. Like the operating lines in the McCabe-Thiele method, enthalpy operating lines relate the concentration of the vapor rising to a plate with that of the liquid leaving it. They also show the ratio of the flow rates of vapor and liquid between plates. By the center-of-gravity principle,

$$\frac{V_{n+1}}{L_n} = \frac{\overline{D'L_n}}{\overline{D'V_{n+1}}} = \frac{x_D - x_n}{x_D - y_{n+1}} \qquad (12\text{-}55)$$

Let the specific enthalpy of the vapor rising to plate m in the stripping column be $i_{y(m+1)}$, and let that of the liquid from plate m be i_{xm}. Let point L_m in Fig. 12-38 represent the liquid from plate m and V_{m+1} the vapor from plate $m + 1$. Point L_m is on the saturated-liquid line, and point V_{m+1} is on the saturated-vapor line. Points V_{m+1}, L_n, and B' are the terminal points of an adiabatic mixing process, so line $V_{m+1}L_nB'$ is straight and is an enthalpy operating line. There is an enthalpy operating line for each plate below the feed, and all such lines pass through point B'. The interpretation of these lines is identical with that of the operating lines in the rectifying section.

The ratio of the flow rates of vapor and liquid at any point in the stripping section can be found from the equation

$$\frac{V_{m+1}}{L_m} = \frac{x_m - x_B}{y_{m+1} - x_B} \qquad (12\text{-}56)$$

Number of Ideal Plates. The enthalpy operating lines are used alternately with tie lines to give a step-by-step determination of the number of ideal plates needed to accomplish a given separation. In comparing the Ponchon-Savarit method with the McCabe-Thiele method, the tie lines correspond to the equilibrium curve and the enthalpy operating lines on the ix diagram to the material-balance operating lines on the xy diagram. The details of the Ponchon-Savarit method are shown in Example 12-8.

Example 12-8. A mixture of 35 per cent ammonia and 65 per cent water is to be separated, at a pressure of 10 atm, into overhead and bottom products of 97.5 and 2.5 per cent ammonia, respectively. The feed enters at 60°F. A reflux ratio of 2 is to be used.

(a) Per pound of overhead product, how much heat must be withdrawn at the condenser and how much added at the reboiler? (b) How many ideal plates are needed, and where should the feed be introduced? (c) What are the vapor and liquid flows at each point in the plant?

Solution. The Ponchon-Savarit diagram for this problem is shown in Fig. 12-39. The ix diagram of Fig. 12-33 is used.

(a) Point F is established by the concentration ($x_F = 0.35$) and the temperature (60°F) of the feed. Points D and B are located on the saturated-liquid line at

$x_D = 0.975$ and $x_B = 0.025$, respectively. Point D' is found from the specified reflux ratio by use of Eq. (12-54). From Fig. 12-32, $i_{y1} = 690$, and $i_D = 130$. Then

$$2.0 = \frac{i_D' - 690}{690 - 130}$$

From this, $i_D' = 1810$, which is the abscissa of point D'. The ordinate is $x_D = 0.975$. A straight line through points D' and F is the over-all-enthalpy line. It intersects the line $x_B = 0.025$ at $i_B' = -1000$.

For each pound of product, the feed and bottoms are found by Eqs. (12-17) and (12-15).

$$F = \frac{0.975 - 0.025}{0.35 - 0.025} = 2.92 \text{ lb}$$

$$B = F - D = 2.92 - 1.0 = 1.92 \text{ lb}$$

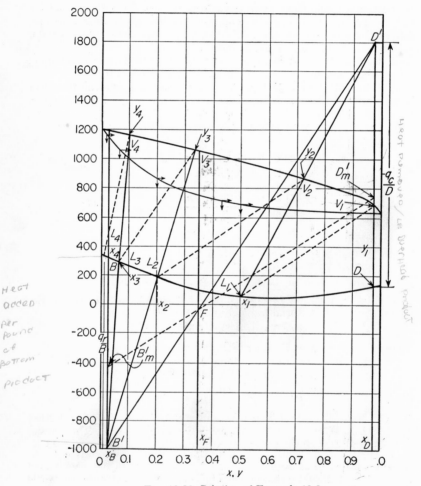

FIG. 12-39. Solution of Example 12-8.

The specific enthalpies of the overhead and bottoms are 130 and 300 Btu/lb, respectively. The heat withdrawn at the condenser is $1810 - 130 = 1680$ Btu, and the heat added at the reboiler is $1.92(300 + 1000) = 2500$ Btu.

(b) Since $x_D = y_1$, the point representing V_1 is on the saturated-vapor line at $x = 0.975$. From point V_1, the appropriate tie line establishes point L_1 on the saturated-liquid line. An operating line through points D' and L_1 intersects the saturated-vapor line at point V_2, and the tie line through this point establishes point L_2.

A decision must be made where the feed plate should be located. The best location for the point of entry of feed is on the first plate where the liquid concentration is less than the abscissa of the intersection of the saturated-liquid line and the over-all-enthalpy line. This means that rectifying operating lines should be used at the right of the over-all-enthalpy line and stripping lines to the left.

Since L_2 is at the left of line $D'FB'$, the next operating line is drawn through point B', thus establishing point V_3 on the saturated-vapor line. From point V_3, point L_3 is located by a tie line; V_4 from an operating line, and L_4 from a tie line. Since x_4 is less than x_B, four steps are sufficient. A reboiler and three ideal plates with feed admitted to the second plate should be specified.

(c) Equations (12-55) and (12-56) can be used to compute the vapor and liquid streams within the column. A basis of 1 lb of overhead product is taken.

By conditions of the design, the liquid to the top plate is 2.0 lb, and the vapor from the top plate is 3.0 lb. From Fig. 12-39, the concentrations within the column are

$$x_1 = 0.50 \qquad x_D = y_1 = 0.975$$

$$x_2 = 0.20 \qquad y_2 = 0.72$$

$$x_3 = 0.06 \qquad y_3 = 0.335$$

$$x_B = x_4 = 0.025 \; \dagger \qquad y_4 = 0.080$$

From Eq. (12-55),

$$\frac{V_2}{L_1} = \frac{x_D - x_1}{x_D - y_2} = \frac{0.975 - 0.50}{0.975 - 0.72} = 1.86$$

Since $V_2 = L_1 + D$, $1.86L_1 = L_1 + 1$, so $L_1 = 1.16$ and $V_2 = 2.16$. From Eq. (12-56),

$$\frac{V_3}{L_2} = \frac{x_2 - x_B}{y_3 - x_B} = \frac{0.20 - 0.025}{0.335 - 0.025} = 0.565$$

$$\frac{V_4}{L_3} = \frac{x_3 - x_B}{y_4 - x_B} = \frac{0.06 - 0.025}{0.080 - 0.025} = 0.636$$

Since $L_2 = V_3 + B$, $L_2 = 0.565L_2 + 1.92$, so $L_2 = 4.42$ and $V_3 = 2.50$. Also, $L_3 = V_4 + B$; $L_3 = 0.636L_3 + 1.92$, so $L_3 = 5.27$ and $V_4 = 3.35$. The concentrations, flows, and heat effects throughout the column are shown in Fig. 12-40.

† Arbitrarily equated to x_B to obtain arithmetic consistency in the material and enthalpy balances. A slight adjustment could be made in the reflux ratio that would give integral steps in Fig. 12-39, but this adjustment is unnecessary.

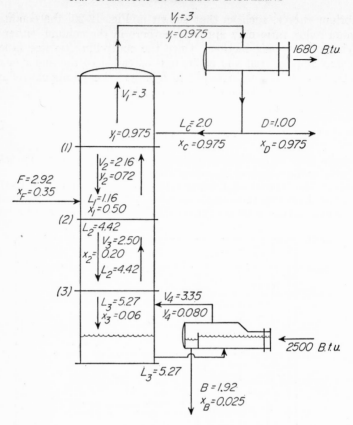

FIG. 12-40. Results for Example 12-8.

Minimum Number of Plates. For total reflux, which corresponds to the minimum number of plates, $-q_c$ and q_r are both very large, and the over-all enthalpy balance and enthalpy operating lines become vertical. Then these vertical lines and the tie lines are used alternately in the usual manner to determine the minimum number of plates.

Minimum Reflux Ratio. When the reflux ratio is reduced from one that calls for a finite number of plates, a situation is reached in which an enthalpy operating line coincides with a tie line. This may occur either above or below the feed. Then the number of steps becomes infinitely large. The reflux ratio corresponding to this condition is the minimum for the given concentrations of feed and products and the given thermal condition of the feed. Usually, the tie line that is encountered first as the reflux is decreased is the one that passes through point F. This is equivalent to the situation on the McCabe-Thiele diagram for normal equilibrium curves, when the operating lines for minimum reflux pass through the intersection of the feed line and the equilibrium curve. For abnormal equilibrium curves, such as that shown in Fig. 12-29, the condition for minimum reflux ratio may appear elsewhere in the column, either in the rectifying or stripping section. Once the controlling tie line is located, which may require trial and error, it is extended to the line $x = x_D$ and point D'_m located, the ordinate of which is i''_{Dm}. The minimum reflux ratio R_{Dm} can then be calculated from the equation

$$R_{Dm} = \frac{i''_{Dm} - i_{y1}}{i_{y1} - i_D} \qquad (12\text{-}57)$$

Example 12-9. What are the minimum number of plates and the minimum reflux ratio for the column of Example 12-8?

Solution. The diagram applying at total reflux is shown in Fig. 12-41. The construction shows that three stages are needed, so a reboiler and two ideal plates are the minimum number that can accomplish the desired separation at total reflux.

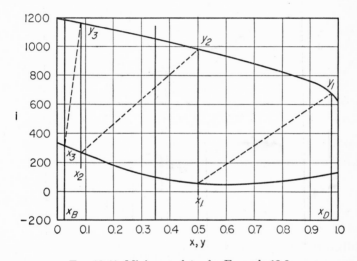

FIG. 12-41. Minimum plates for Example 12-9.

The first tie line that coincides with an enthalpy operating line is found to be that passing through point F in Fig. 12-39. This line intercepts line $x = x_D$ at point D'_m, the ordinate of which is $i''_{Dm} = 740$. The line is plotted as line $D'_m F B'_m$. From Eq. (12-57), the minimum reflux ratio is

$$R_{Dm} = \frac{740 - 690}{690 - 130} = 0.089 \text{ lb/lb overhead}$$

Comparison of Operating Lines on ix and xy Diagrams.[17] If the ix diagram gives straight operating lines on the xy diagram, both methods yield the same results for the same problem. The simpler xy method is usually preferred to the more complicated ix method, provided the error from the assumption of constant overflow is not serious. The question may be asked, "What is the criterion on the ix diagram for straight operating lines on the xy diagram?" The answer is simple. If the saturated-liquid line and saturated-vapor line on the ix diagram are both straight, the xy diagram can be so constructed that operating lines on it are linear.

To demonstrate the fact that straight liquid and vapor lines on the ix diagram can lead to straight operating lines on the xy diagram, consider,

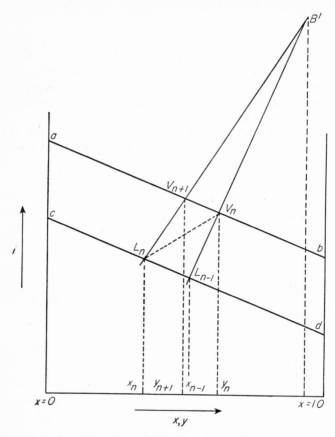

Fig. 12-42. Straight and parallel liquid and vapor lines on ix diagram.

first, the case where the operating and equilibrium lines are not only straight but parallel. Liquid line cd and vapor line ab in Fig. 12-42 are two such lines. Let lines $B'V_nL_{n-1}$ and $B'V_{n+1}L_n$ be any two adjacent enthalpy operating lines in the rectifying section. Since lines ab and cd are parallel, triangles $B'L_{n-1}L_n$ and $B'V_nV_{n+1}$ are geometrically similar. Then

$$\frac{\overline{B'V_n}}{\overline{B'L_{n-1}}} = \frac{\overline{B'V_{n+1}}}{\overline{B'L_n}}$$

By the center-of-gravity principle,

$$\frac{\overline{B'V_n}}{\overline{B'L_{n-1}}} = \frac{L_{n-1}}{V_n} \quad \text{and} \quad \frac{\overline{B'V_{n+1}}}{\overline{B'L_n}} = \frac{L_n}{V_{n+1}}$$

Since $V_n = L_{n-1} + D$ and $V_{n+1} = L_n + D$,

$$\frac{L_{n-1}}{L_{n-1} + D} = \frac{L_n}{L_n + D}$$

and, therefore, $L_n = L_{n-1}$. Since lines $B'V_nL_{n-1}$ and $B'V_{n+1}L_n$ are any two adjacent lines, all enthalpy operating lines in the rectifying section have the same L, and the overflow is constant. An equivalent proof can be given for the stripping section.

If the liquid and vapor lines are straight but not parallel, the above proof fails. These lines can be made parallel, however, by the following

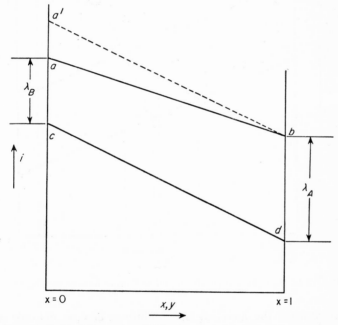

Fig. 12-43. Straight, nonparallel liquid and vapor lines on ix diagram.

device.[14] Consider two straight, nonparallel lines such as lines ab and cd in Fig. 12-43. Assume that the diagram is constructed on a mass basis. As shown in Fig. 12-43, let the enthalpy of vaporization of pure component A be λ_A and that of pure component B, λ_B. Assign a fictitious molecular weight to component B equal to λ_A/λ_B, and imagine that the ix diagram is replotted on a molal basis, using this molecular weight for B

and calling the molecular weight of A unity. Then the "molal" latent heat of pure B, shown by line $a'c$ in Fig. 12-43, equals the "molal" latent heat of A, which remains as λ_A. If a line of i vs. x is straight on a mass basis, it is also straight on a molal basis, so the saturated-liquid and saturated-vapor lines, if plotted on the new basis, would be straight and parallel. If the xy diagram is then plotted on the "molal" basis, the operating lines are straight. Of course, all other quantities and concentrations must also be computed to the same basis.

When Trouton's rule applies to mixtures of A and B, the fictitious molecular weight of B calculated from latent heats is the ratio of the true molecular weight of component B to that of component A. Then a molal xy diagram of the usual type will give straight operating lines, which confirms the approximate analysis of page 689. A glance at the liquid and vapor lines of the ix diagram is sufficient for a decision as to the adequacy of the xy method in any given situation. Unless these lines are strongly curved, the error from assuming constant molal overflow is small, usually less than one ideal plate.

Design of Bubble-cap Columns

To translate ideal plates into actual plates, a correction for the efficiency of the plates must be applied. There are other important decisions, some at least as important as fixing the number of plates, that must be made before a design is complete. A mistake in these decisions results in poor fractionation, lower-than-desired capacity, poor operating flexibility, and, with extreme errors, an inoperative column. Correcting such errors after a plant has been built can be costly. Since many variables that influence plate efficiency depend on the design of the individual plates, the fundamentals of plate design are discussed first.

The extent and variety of rectifying columns and their applications are enormous. The largest units are usually in the petroleum industry, but large and very complicated distillation plants are encountered in fractionating solvents, in treating liquefied air, and in general chemical processing. Tower diameters may range from 1 ft to more than 30 ft and the number of plates from a few plates to scores. Plate spacings may vary from 6 in. or less to several feet. The bubble caps range from less than 1 in. in diameter in liquid-air fractionators to the large single caps used one to a plate in some types of ammonia absorbers and carbonating towers. Many types of liquid distribution and bubble caps are specified. Columns may operate at high pressures or low, from temperatures of liquid gases up to 1600°F reached in the rectification of sodium and potassium vapors. The materials distilled can vary greatly in viscosity, diffusivity, corrosive nature, tendency to foam, and in complexity of composition. Bubble towers are as useful in absorption as in rectification, and the fundamentals of plate design apply to both operations.

Designing fractionating columns, especially large units and those for unusual applications, is best done by experts. Although the number of ideal plates and the heat requirements can be computed quite accurately without much previous experience, other design factors are not precisely calculable, and a number of equally sound designs can be found for the same problem. In common with most engineering activities, sound design of fractionating columns relies on a few principles, on a number of empirical correlations (which are in a constant state of revision), and much experience and judgment.

The following discussion is limited to the usual type of column, equipped with circular bubble caps, operating at pressures not far from atmospheric, and treating mixtures having ordinary properties.

Normal Operation of Bubble Plate. The task of a bubble plate is to bring streams of liquid and vapor into intimate contact. To accomplish

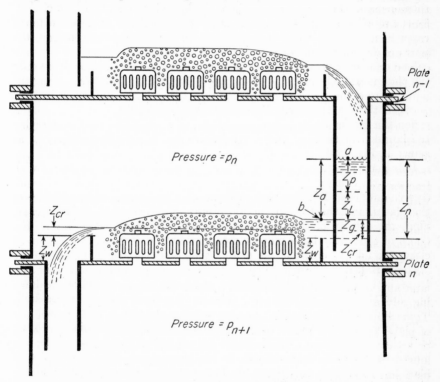

FIG. 12-44. Normal operation of bubble-cap plate: Z_a, height of station a above datum. Z_{cr}, weir crest. Z_g, gradient head. Z_L, liquid-friction head. Z_p, pressure head across plate. Z_n, net head in down pipe. Z_w, weir height.

this definite forces are required to move the fluid streams to, through, and away from the mixing zone. Provision must also be made to obtain a reasonably clean separation between the liquid and vapor leaving the mixing zone.

Figure 12-44 shows a plate in a bubble tower in normal operation. The various pressures and other driving forces are indicated on the figure. The driving force that overcomes the resistance to flow of vapor through the bubble cap and liquid on the plate is pressure. The pressure of the vapor below a plate is greater than that above the plate. This pressure drop across a single plate is usually several inches of water. The pressure drop in the entire column is the product of the number of plates and the pressure drop per plate. The pressure drop over a 30-plate column, then, is of the magnitude of 10 ft of water. Ordinarily this pressure drop is not important, but in vacuum distillation it is the most critical point in the design, because the pressure drop is large with respect to the absolute pressure and the temperature in the base of the column may become too high. A minimum number of plates especially designed for low pressure drop must then be used. The pressure drop in any column is automatically developed by the reboiler, and if the capacity of the reboiler is sufficient to create the required amount of vapor against the back pressure of the column, the pressure drop is not under the control of the operator.

The driving force for the flow of liquid through the down pipe and onto the plate is static head in the down pipe. The driving force for moving the liquid across the plate is gravity, acting through the difference in level between the liquid entering the plate and that leaving it. This head is called the gradient and is denoted by Z_g. Finally, the head for carrying the liquid over the exit weir is the weir-crest head, denoted by Z_{cr}, as shown in Fig. 12-44.

It is good practice to provide a calming space between the last row of caps and the exit weir to allow complete release of vapor, so the down pipes receive only clear liquid. Vapor-free liquid then reaches the inlet sides of the plates, and the frothing action occurs only around the bubble caps. Since the density of the froth is much less than that of clear liquid, the depth of the frothy layer may be greater than the depth of clear liquid at either the inlet or outlet of the plate. This is shown in Fig. 12-44.

Head Balance in Down Pipe. The net static head Z_n, shown in Fig. 12-44, is the difference in level between the surface of the liquid in the down pipe and the top of the exit weir. It is the sum of all driving forces, both liquid and vapor, for the plate. This can be demonstrated as follows.

The pressure drop in the vapor space between plates is negligible, so the entire plate pressure drop can be assumed to occur across the plate, the bubble caps, and the frothy liquid on the plate. The pressure drop across plate n is $p_{n+1} - p_n$, where p_{n+1} is the pressure in the vapor space above plate $n + 1$ and p_n is that above plate n. To relate the pressure drop across the plate to the liquid level in the down pipe, write a Bernoulli equation for the liquid between station a, at the top of the liquid in the down pipe, and station b, at the surface of the clear liquid just entering the bubbling section of plate n. Take the datum of heights through station b and neglect the change in velocity head. Then a Bernoulli relation [Eq. (2-41)] gives

$$\frac{p_a}{\rho_L} + \frac{gZ_a}{g_c} = \frac{p_b}{\rho_L} + H_{fL}$$

or
$$p_b - p_a = \frac{\rho_L Z_a g}{g_c} - H_{fL}\rho_L \qquad (12\text{-}58)$$

where ρ_L = density of liquid

Z_a = height of station a above datum

H_{fL} = friction loss between stations a and b

Pressure drop $p_b - p_a$ is the pressure drop over plate $n - 1$. Since all plates are alike, this is also the pressure drop over plate n, and

$$p_{n+1} - p_n = \frac{\rho_L Z_a g}{g_c} - H_{fL}\rho_L$$

or
$$Z_a = \frac{g_c(p_{n+1} - p_n)}{g\rho_L} + \frac{g_c H_{fL}}{g} \qquad (12\text{-}59)$$

Define Z_p, the pressure head across the plate, and Z_L, the friction head in the liquid, by the equations

$$Z_p = \frac{g_c(p_{n+1} - p_n)}{g\rho_L} \qquad Z_L = \frac{g_c H_{fL}}{g} \qquad (12\text{-}60)$$

Then $Z_a = Z_p + Z_L$, and, from Fig. 12-44 and Eqs. (12-59) and (12-60), the net static head is

$$Z_n = Z_a + Z_g + Z_{cr} = Z_p + Z_L + Z_g + Z_{cr} \qquad (12\text{-}61)$$

These equations are all in consistent units, and the lengths are in feet, but in practice, the various heads in Eq. (12-61) are usually expressed in inches. For simplicity, the conversion from feet to inches is not given here.

Equation (12-61) is a static-head balance in the down pipe. Head Z_p provides the driving force for overcoming the pressure drop through the plate; head Z_L overcomes the friction in the flow of the liquid through and out of the down pipe; head Z_g provides the force for moving liquid across the plate; and head Z_{cr} discharges the liquid over the weir. A major task of the designer of a bubble-plate column is to so proportion his design that a high capacity per square foot of plate area is attained without allowing any of the heads in Eq. (12-61) to become so large that the operation of the column is impaired.

General Approach to Plate Design. The capacity of a column is expressed in terms of the rate of either vapor or liquid flow per square foot of plate area. Once the reflux ratio is chosen, the liquid flow per unit plate area is proportional to that of the vapor, and fixing one establishes the other. The general method of plate design is to choose a plate spacing, estimate the permissible vapor load, and calculate a tentative column cross section. Then a preliminary layout of caps, weirs, and down pipes can be prepared. The designer estimates the various liquid heads and pressure drops, checks the level in the down pipes, and if all is in order, the tentative design can be refined and made final. If abnormal heads are found at any point, it may be necessary to modify the original assumptions of reflux ratio, plate spacing, or vapor velocity and to redesign the plate. In some columns, especially ones of large diameter or ones that handle a large ratio of liquid to vapor, the design is controlled by liquid capacity rather than by vapor rates, and to obtain a safe liquid load the final design may be underloaded with respect to vapor flow.

Construction Details of Plate Columns. To estimate the heads Z_L, Z_p, Z_{cr}, and Z_g several details of construction must be fixed. This discussion is based on the following typical design. Numerous other modifications are used in practice.

The caps are round and have rectangular slots closed at the bottom. The height of the top of the slots above the plate is denoted by Z_{sp}. The caps are arranged on equilateral centers and are staggered with respect to the direction of liquid flow. The caps vary from 3 to 7 in. in diameter. The wall-to-wall distance between caps is 1 to 3 in. Caps are not placed within 1 to 2 in. of the wall of the column or within 2 to 4 in. of down pipes or weirs.

The flow of liquid is across the plate, from the plate above to the outlet weir and down pipe to the plate below. The liquid flows in alternate directions on adjacent plates.

Two common types of weirs are considered. In small columns, the weir is formed by extending the circular down pipes the distance Z_w, the weir height, above the plate. One or more down pipes may be used on each plate. Seal cups are placed at the bottoms of the down pipes to prevent vapor from entering the pipes at the bottom. Flow in a circular down pipe may occur in various manners, depending on the size of the weir and on the influence of streams from the caps and from other nearby weirs. The type of flow assumed here is that obtained when the diameter of the down pipe is more than four times the weir-crest head. Then the liquid flows down the wall of the down pipe as a film until it reaches a definite level of clear liquid in the pipe. For smaller pipes, the pipe may fill with liquid, vapor may be entrained, and a vortex may form. In such situations, the down pipe is reaching a limit of its capacity, and down-pipe area can be the limiting factor in the entire design.

In larger columns, rectangular chord weirs are used. An example is shown in Fig. 12-44. The chord weir is a flat plate that cuts off a sector of the plate. Several circular down pipes, flush with the plate, may be used, or all or nearly all of the sector area may be used for a single large sector-shaped down pipe. An inlet chord weir is often used to provide a liquid seal at the bottom of the down pipes. The area of the plate assigned to the down-pipe area behind the weirs may be 15 to 20 per cent of the entire plate area.

If Z_{sp} denotes the height of the top of the slots above the plate, then $Z_w - Z_{sp}$ is the minimum liquid seal over the slots, corresponding to no liquid flow. This dimension is built into the plate. When the plate is in action, the seal Z_s^o over the outlet caps is $Z_w - Z_{sp} + Z_{cr}$, and that over the inlet caps Z_s^i is $Z_w - Z_{sp} + Z_{cr} + Z_g$. These heads are all based on the clear liquid in the calming sections just before and after the entrance and exit of the bubbling section. The various heads and distances are shown in Fig. 12-45.

Limiting Vapor Velocity. At low velocities the plate efficiency is poor because of poor contact between liquid and vapor. At intermediate velocities in the usual operating range the plate efficiency is at a maximum and is nearly independent of vapor velocity. At high velocities the efficiency drops drastically, and the operating vapor load should be set below this

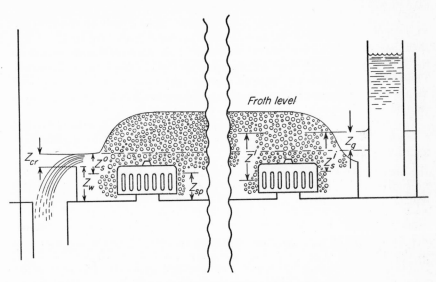

Fig. 12-45. Heads and seals on bubble plate.

point. Poor efficiencies at high velocities are the result of the following phenomena.

Entrainment. At high velocities the drops of liquid in the vapor space above the liquid cannot settle against the upward flow of the vapor, and are swept out and mixed with the liquid on the next plate above. This dilutes the liquid on that plate and undoes some of the rectification performed by the plate. Unless entrainment is unusually severe, its effect on plate efficiency is small, but entrainment adds to the liquid load, and it may be important on the top plates if the overhead product must be free from color or small concentrations of impurity.

Splashing and Foaming. At low spacings and high velocities liquid may geyser or spout clear across the vapor space into bubble caps of the plate above. Or if the liquid tends to form a persistent foam, the foam will also accompany the liquid into the next plate. The effect of either of these happenings is equivalent to heavy entrainment. The foam of this kind is not to be confused with the temporary froth that is a normal and important part of action on the plate. Foam is entirely disadvantageous. In the usual column foaming and splashing are unimportant.

Coning. The most drastic effect of high vapor velocities in reducing plate efficiency is a basic change in the action on the plate. The vapor pushes away the liquid from the vicinity of the bubble caps, passes through the plate in bulk, and fails to contact the liquid adequately. This effect is called coning. Sudden surges of vapor may blow the liquid seals, and vapor then passes up the down pipes.

Calculation of Limiting Vapor Velocity. Formerly it was thought that entrainment was the prime cause of the decrease of plate efficiency at high vapor flow rates. It was assumed that, with droplets of a definite but unspecified size settling under Newton's law, to prevent entrainment it would be necessary to keep the velocity of the vapor below the terminal velocity of the particles. On this assumption, from Eq. (7-60) the maximum permissible vapor velocity would be proportional to $\sqrt{(\rho_L - \rho_v)/\rho_v}$, or

$$u_c = K \sqrt{\frac{\rho_L - \rho_v}{\rho_v}} \qquad (12\text{-}62)$$

where u_c is the maximum permissible velocity, in feet per second, based on the area of the bubbling section of the plate (total area minus the down-pipe area) and K is an empirical constant.[21] In terms of mass velocity, this equation can be written

$$G_c = u_c \rho_v = K \sqrt{\rho_v(\rho_L - \rho_v)} \qquad (12\text{-}63)$$

In spite of the fact that it is now known that entrainment of liquid drops is not the controlling factor in fixing the permissible vapor velocity,

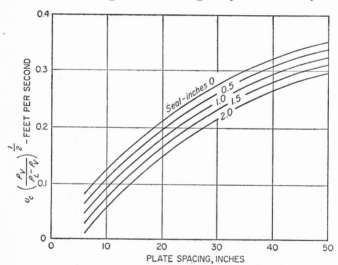

Fig. 12-46. Allowable superficial vapor velocity in bubble-cap columns. (*By permission, from "Elements of Fractional Distillation," 4th ed., by C. S. Robinson and E. W. Gilliland. Copyright, 1950. McGraw-Hill Book Company, Inc.*)

Eqs. (12-62) and (12-63) have survived as empirical relationships for calculating limiting vapor flows. Constant K has been evaluated from plant data, and various correlations of K against operating variables have been made. One such correlation is shown in Fig. 12-46, in which K is shown as a function of plate spacing and liquid seal. The plate spacing is the distance between the top of the clear liquid and the bottom of the next higher plate. The average seal between inlet and outlet is used.

Pressure Head across Plate. The head Z_p across the plate can be divided into two parts. The first, denoted by Z_v, is the friction head of the vapor flowing through the riser, annulus, and slots of the bubble cap. The second, denoted by Z_s, is the static head equal to the seal over the top of the slots. Then

$$Z_p = Z_v + Z_s \qquad (12\text{-}64)$$

Friction Loss in Cap. This loss is from the eddies induced by the changes in direction in flowing through the cap. It is estimated as five to nine velocity heads, depending on the cap design. Seven velocity heads, based on u_s, the maximum velocity through riser, annulus, or slots, is an average factor. Then,

$$Z_v = 7 \frac{u_s^2}{2g_c} \frac{g_c}{g} \frac{\rho_v}{\rho_L} = \frac{3.5 u_s^2 \rho_v}{g \rho_L} \qquad (12\text{-}65)$$

Static Head over Slots. The static head over the slots at the entrance is greater by Z_g than that over the slots at the liquid exit. Since Z_g is small in a normal operation, an average for the inlet and outlet is used. Gradient Z_g is assumed in the first stages of the design and checked later.

Friction Head in Liquid. Friction in the liquid is also composed of two parts, the friction in the straight pipe, and the reversal loss at the bottom of the down pipe. For the free flow assumed for down pipes, the friction in the straight pipe is usually negligible. It can be estimated on the basis that 40 diameters of pipe give a friction loss of 1 velocity head.

The reversal loss at the bottom of the down pipe can be taken as three velocity heads, based on the maximum velocity at the base of the down pipe. This maximum velocity may be either that in the down pipe or that at the end of the pipe in the seal cup, depending on which cross section is the smaller. Assuming that the maximum velocity is through the pipe, the friction loss Z_L is

$$Z_L = \left(3 + \frac{L_d}{40 D_d}\right) \frac{u_d^2}{2 g_c} \frac{g_c}{g} = \left(1.5 + \frac{L_d}{80 D_d}\right) \frac{u_d^2}{g} \qquad (12\text{-}66)$$

where u_d = velocity in down pipe, ft/sec
  L_d = length of down pipe, ft
  D_d = diameter of down pipe, ft
  g = acceleration of gravity, ft/sec²

If the maximum velocity is elsewhere than in the down pipe, use the maximum velocity instead of u_d and neglect $L_d/80 D_d$. The liquid velocity in the down pipe below the liquid level should not be more than about 0.5 ft/sec.

When the net head required in the downpipe exceeds the plate spacing, liquid accumulates on the plates, and the column floods.

Weir Crest. The head over the weir Z_{cr} is calculated from the weir formulas given in Chap. 2. For a circular weir at the top of a down pipe, from Eq. (2-119),

$$Z_{cr} = 0.46 \left(\frac{q_L}{b}\right)^{0.71} = 0.20 \left(\frac{q_L}{D_d}\right)^{0.71} \tag{12-67}$$

where q_L = liquid flow rate, ft³/sec
  b = perimeter of pipe, ft
  D_d = diameter of pipe, ft

For a chord weir, Eq. (2-118) gives

$$Z_{cr} = 0.45 \left(\frac{q_L}{b}\right)^{\frac{2}{3}} \tag{12-68}$$

Equation (12-68) applies where the velocity of approach to the weir is negligible. Because of the influence of the wall of the column, this is not

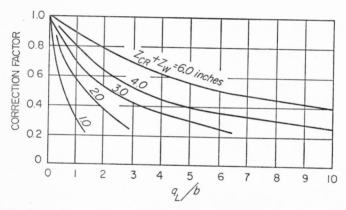

FIG. 12-47. Correction to chord-weir equation. (*By permission, from "Elements of Fractional Distillation," 4th ed., by C. S. Robinson and E. W. Gilliland. Copyright, 1950. McGraw-Hill Book Company, Inc.*)

true in a bubble plate. Figure 12-47 gives a correction, to be multiplied into the right-hand side of Eq. (12-68), that allows for velocity of approach.[18d] The factor depends on q_L/b and $Z_{cr} + Z_w$.

Gradient Head. Accurate calculation of Z_g is difficult. The following dimensionless equation is recommended for estimation purposes.[18e]

$$Z_g = 25{,}000 \left(\frac{Z_o}{Z_{sp}} - 1\right)^{\frac{2}{3}} \frac{(w_L/b_b)^2 L_b}{g r_H \rho_L^2 Z_o^2} \left[\frac{Z_o \mu_L}{r_H(w_L/b_b)}\right]^{1.2} \tag{12-69}$$

where Z_g = gradient head, ft

Z_o = total static head at outlet, $Z_w + Z_{cr}$, ft

Z_{sp} = height of top of slots above plate, ft

w_L = flow rate of liquid, lb/sec

b_b = average width of bubbling section, perpendicular to direction of flow, ft

L_b = length of bubbling section, parallel to direction of flow, ft

g = acceleration of gravity, ft/sec^2

μ_L = viscosity of liquid, lb/ft-sec

ρ_L = density of liquid, lb/ft^3

r_H = hydraulic radius, calculated from Eq. (12-70)

$$r_H = \frac{2b_b Z_o}{b_b + 4Z_o} \tag{12-70}$$

For viscosities of 1 centipoise or less, assume $\mu_L = 0.00067$. For viscosities greater than 1 centipoise, use the actual viscosity of liquid at flowing temperature.

Velocity Distribution across Plate; Stable and Unstable Operation. Ideally, the vapor flows through all caps on a plate should be equal. Since the liquid level on the liquid-entrance side is higher by height Z_g than that on the exit side, vapor flows must be somewhat different at the ends, and do vary across the plate. Let superscript o refer to the plate outlet and superscript i to the inlet. Since the pressure head Z_p is the same for all caps on the same plate, it follows from Eq. (12-64) that

$$Z_v^o + Z_s^o = Z_v^i + Z_s^i = Z_p \tag{12-71}$$

Substituting from Eq. (12-65),

$$\frac{3.5(u_s^o)^2}{g\rho_L} + Z_s^o = \frac{3.5(u_s^i)^2}{g\rho_L} + Z_s^i \tag{12-72}$$

Since $Z_s^i = Z_s^o + Z_g$,

$$Z_g = \frac{3.5[(u_s^o)^2 - (u_s^i)^2]}{g\rho_L} \tag{12-73}$$

In columns of small diameter, Z_g is small, usually of the order of a few tenths of an inch, the difference between u_s^i and u_s^o is unimportant, and each cap receives nearly the same flow of vapor. In larger columns, Z_g increases rapidly with plate diameter. This can be seen from Eq. (12-69). The factors in this equation that vary with diameter are w_L, b_b and L_b. Using an exponent of 1 instead of 1.2, the gradient head is proportional to $w_L L_b/b_b$. When the column diameter is increased, the permissible loads of vapor and liquid increase in proportion to the cross-sectional area of the column; so w_L is proportional to the square of the diameter if the plate is fully loaded. Since L_b/b_b is independent of diameter, $w_L L_b/b_b$ is proportional to the square of the diameter, and the gradient increases with diameter in accordance with the same rule. If the column diameter is doubled, for example, the gradient is approximately quadrupled. The larger the gradient, the poorer the vapor distribution.

If Z_g becomes so large that $Z_s^i = Z_p$, from Eq. (12-71) $Z_v^i = 0$. Then, by Eq. (12-72) it is seen that u_s^i becomes zero, and no vapor flows through the caps near the liquid inlet. A plate on which some of the caps do not bubble is called an unstable plate. If all caps bubble, the plate is called stable, even though the vapor flow through some of the caps is small. The action on an unstable plate is shown in Fig. 12-48. When operation is unstable, the vapor flows mainly through the caps near the liquid outlet. In extreme situations, the liquid may accumulate at the liquid inlet to such

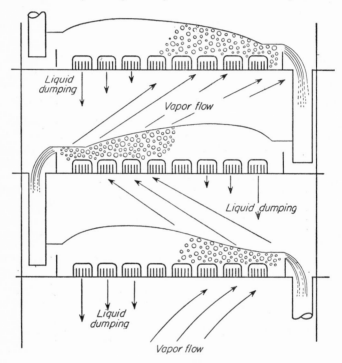

FIG. 12-48. Action on unstable plate. (*Kemp and Pyle.*[8])

a depth that liquid flows back through the risers of the bubble caps near the inlet. This is called *dumping*. The liquid that dumps through bubble caps bypasses two plates. Associated with dumping may be a vapor flow so large at the liquid outlet that liquid is blown away from the caps near the outlet, and vapor instead of liquid flows through the down pipes, so the contact between liquid and vapor over the entire·plate is nearly destroyed. Such a plate may be inoperable.

Plate stability can be obtained in large columns by various methods,[8, 18f] one of which is a complete change in the design of the plates so the liquid flows across only a part of the diameter of the plate. These methods are not discussed here. All plate designs should be checked for stability, and if Z_s^i in a given design approaches Z_p, the design should be modified to reduce the gradient head. If the approximate calculation shows that the gradient head is critical, more elaborate methods available in the literature should be used to check the conclusion.[4, 6, 8]

Example 12-10. The stripping column of Example 12-8 is to treat 400 gal/min of feed. Design a suitable plate for this column using 4-in. caps and 24-in. plate spacing.

Solution. Dimensions of a typical 4-in. bubble cap are:

Cap: ID, $3\frac{3}{4}$ in.
 OD, 4 in.
 Height, $2\frac{1}{4}$ in. from surface of plate
 Slot size, $\frac{3}{4}$ by $\frac{3}{32}$ in.
 Number of slots, 48
 Height of top of slots above plate, 1.00 in.
Riser: ID, $1\frac{7}{8}$ in.
 OD, 2.0 in.
 Height, $1\frac{13}{16}$ in.

PLATE DIAMETER. A tentative decision on weir height and seal is made. Assume a minimum seal at no flow of $\frac{1}{4}$ in. The weir height Z_w is then 1.25 in. Assume a weir crest Z_{cr} of 1.65 in. Assume also that Z_g will be about 0.5 in. The seal at the liquid inlet is then $1.25 + 1.65 + 0.50 - 1.00 = 2.4$ in., and that at the outlet is 1.9 in. For estimating the permissible vapor velocity assume an average seal of 2.15 in. The total distance between plates is 24 in. To allow for liquid depth and plate construction discount this by 4 in., and use a net spacing of 20 in. From Fig. 12-46, a 20-in. spacing and a 2.15-in. seal gives $u_c\sqrt{\rho_v/(\rho_L - \rho_v)} = 0.145$. Since both vapor and liquid at the bottom plate are essentially water, ρ_v and ρ_L are obtained from Appendixes 6 and 14. To allow for pressure drop in the column, assume the pressure in the bottom of the column is 16 lb force/in.^{2} and the temperature is 216°F. The densities of liquid and vapor are

$$\rho_L = 59.7 \text{ lb/ft}^3 \qquad \rho_v = \frac{1}{24.8} = 0.0403 \text{ lb/ft}^2$$

The permissible velocity through the bubbling area of the plate is

$$u_c = 0.145 \sqrt{\frac{59.7 - 0.04}{0.0403}} = 5.56 \text{ ft/sec}$$

The average molecular weight of the feed is

$$\frac{100}{\frac{98}{18} + \frac{2}{46}} = 18.2$$

The specific gravity of the feed can be taken as that of water, and the feed and the liquid flow down the column are

$$F = L = \frac{400 \times 8.33}{18.2} = 183 \text{ moles/min}$$

This is $(183 \times 18)/60 = 55.0$ lb/sec, or $55.0/59.7 = 0.92$ ft³/sec. The vapor flow is $183 \times 0.20 = 36.6$ moles/hr, or

$$\frac{36.6 \times 18}{60 \times 0.0403} = 272 \text{ ft}^3/\text{sec}$$

The cross-sectional area of the bubbling portion of the plate is $272/5.56 = 49$ ft². Allowing 4 ft² for down-pipe area, the total area is 53 ft², and the diameter of the column is $\sqrt{53/0.7854} = 8.2$ ft.

LENGTH OF WEIRS. Assume that identical chord weirs are used at each side of the plate, one for liquid entrance and the other for liquid exit. Neglecting the velocity of approach as a first approximation, Eq. (12-68) gives

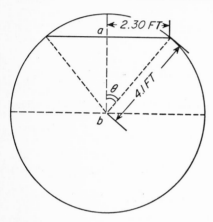

$$\frac{1.65}{12} = 0.45 \left(\frac{0.92}{b}\right)^{\frac{2}{3}}$$

From this $0.92/b = 0.17$. From Fig. 12-47, for $q_L/b = 0.17$ and $Z_{cr} + Z_w = 1.65 + 1.25 = 2.9$ in., the correction factor for velocity of approach is 0.90, and

$$\frac{1.65}{12} = 0.45 \times 0.90 \left(\frac{0.92}{b}\right)^{\frac{2}{3}}$$

From this $b = 4.6$ ft. Each weir is a chord, 4.6 ft long, across an 8.2-ft circle. The distance of each weir from the center of the plate and the area available for down pipes are calculated with the aid of Fig. 12-49. The distance ab is

FIG. 12-49. Distances on plate for Example 12-10.

$$\sqrt{4.1^2 - 2.3^2} = 3.4 \text{ ft}$$

The center of the weir is $4.1 - 3.4 = 0.7$ ft from the wall of the column. The sine of angle θ is $2.3/4.1 = 0.56$, and $\theta = 34.1°$. The area for down pipes is

$$\frac{52.0 \times 2 \times 34.1}{360} - 2.3 \times 3.4 = 2.0 \text{ ft}^2$$

The area behind both weirs is 4.0 ft², which agrees satisfactorily with the assumption.

PLATE LAYOUT. The layout of caps on the plate can now be prepared. Assuming a 2-in. spacing between caps, the center-to-center distance is 6 in. One quadrant of the plate layout is shown in Fig. 12-50. A center row and seven other rows on each side of the center can be accommodated. The total number of caps is 197.

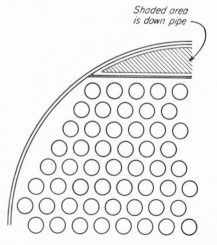

FIG. 12-50. Plate layout for Example 12-10.

FRICTION HEAD THROUGH CAPS Z_v. To calculate the friction loss through the caps the maximum velocity in the bubble cap is needed. The areas of the passages through the cap are

Riser: $1.875^2 \times 0.7854 = 2.76$ in.^{2}

Annulus $0.7854(3.75^2 - 2^2) = 7.9$ in.^{2}

Slots: $48 \times \frac{3}{4} \times \frac{3}{32} = 3.37$ in.^{2}

The velocity through the risers controls, and this velocity is

$$u_s = \frac{272 \times 144}{197 \times 2.76} = 72 \text{ ft/sec}$$

From Eq. (12-65),

$$Z_v = \frac{3.5 \times 72^2 \times 0.0403}{32.2 \times 59.7} = 0.38 \text{ ft, or } 4.6 \text{ in.}$$

GRADIENT HEAD Z_g. The gradient head is estimated from Eq. (12-69). The quantities to be used in this equation are

$$Z_o = Z_w + Z_{cr} = 1.25 + 1.65 = 2.90 \text{ in., or } 0.242 \text{ ft}$$

$$Z_{sp} = 1.0 \text{ in.} \qquad w_L = 55.0 \text{ lb/sec} \qquad b_b = \frac{8.2 + 4.6}{2} = 6.4 \text{ ft}$$

$$L_b = 2 \times 3.4 = 6.8 \text{ ft} \qquad r_H = \frac{2 \times 6.4 \times 0.242}{6.4 + 4 \times 0.242} = 0.42 \text{ ft}$$

$$\rho_L = 59.7 \text{ lb/ft}^3$$

Since the viscosity of water at 216°F is 0.2, well below 1 centipoise, use 0.00067 for μ_L. Substituting into Eq. (12-69),

$$Z_g = \frac{25,000 \left(\dfrac{2.90}{1.0} - 1\right)^{\frac{2}{3}} \left(\dfrac{55.0}{6.4 \times 59.7 \times 0.242}\right)^2 \dfrac{6.8}{32.2 \times 0.42}}{\left(\dfrac{0.42 \times 55.0}{0.00067 \times 0.242 \times 6.4}\right)^{1.2}}$$

$$= 0.041 \text{ ft, or } 0.50 \text{ in.}$$

PRESSURE HEAD THROUGH PLATE Z_p. The average static head, or seal, above the tops of the slots is $1.9 + 0.50/2 = 2.15$ in. The total pressure head over the plate is $Z_p = 4.6 + 2.15 = 6.75$ in. The pressure-head distribution is:

	Inlet, in.	Average, in.	Outlet, in.
Friction through caps Z_b.....	4.35	4.60	4.85
Static head Z_s..............	2.40	2.15	1.90
Total...................	6.75	6.75	6.75

Since Z_p is considerably greater than Z_s^i, the plate should be stable. The ratio of vapor flows at the inlet and the outlet is, approximately,

$$\frac{u_s^o}{u_s^i} = \sqrt{\frac{4.85}{4.35}} = 1.06$$

The vapor distribution is excellent.

DOWN-PIPE AND LIQUID-HEAD LOSS Z_L. Assume that the cross section of the down pipe is 90 per cent of the area of the sector behind the down pipe. The velocity in the down pipe, then, is

$$\frac{0.92}{0.90 \times 2.0} = 0.51 \text{ ft/sec}$$

The friction loss in the down pipes is negligible at this velocity. Assume that the down pipe reaches to 1.0 in. of the plate. Then the area available for flow out of the down pipe is $4.6/12 = 0.383$ ft^2, and the velocity at the bottom of the down pipe is $u_d = 0.92/0.383 = 2.4$ ft/sec. From Eq. (12-66),

$$Z_L = \frac{1.5 \times 2.4^2}{32.2} = 0.268 \text{ ft, or } 3.22 \text{ in.}$$

TOTAL DEPTH OF LIQUID IN DOWN PIPES. The total head, and its distribution in the down pipes, is

	Inches
Weir height Z_w	1.25
Weir crest Z_{cr}	1.65
Gradient head Z_g	0.50
Liquid-friction head Z_L	3.22
Pressure head across plate Z_p	6.75
Total	13.37

A vapor space of $24 - 13 = 11$ in. is left above the level of liquid in the down pipe. This provides a satisfactory margin of safety for surges in flow or for the temporary presence of foam in the down pipe.

Plate Efficiency

To translate ideal plates into actual plates the plate efficiency must be known. The following discussion applies to both absorption and fractionating columns.

Types of Plate Efficiency. Three kinds of plate efficiency are used: (1) over-all efficiency, which concerns the entire column; (2) Murphree efficiency, which has to do with a single plate; and (3) local efficiency, which pertains to a specific location on a single plate.

The *over-all efficiency*, denoted by η_o, is simple to use but is the least fundamental. It is defined as the ratio of the number of ideal plates needed in an entire column to the number of actual plates.[9] For example, if six ideal plates are called for and the plate efficiency is 60 per cent, the number of actual plates is $6/0.60 = 10$.

The *Murphree efficiency*,[13] denoted by η_M, is defined by the equation

$$\eta_M = \frac{y_n - y_{n+1}}{y_n^* - y_{n+1}} \tag{12-74}$$

where y_n = actual concentration of vapor leaving plate n
y_{n+1} = actual concentration of vapor entering plate n
y_n^* = concentration of vapor in equilibrium with liquid leaving down pipe from plate n
The *local efficiency*, denoted by η', is defined by the equation

$$\eta' = \frac{y_n' - y_{n+1}'}{y_{en}' - y_{n+1}'} \tag{12-75}$$

where y_n' = concentration of vapor leaving specific location in plate n
y_{n+1}' = concentration of vapor entering plate n at same location
y_{en}' = concentration of vapor in equilibrium with liquid at same location
Since y_n' cannot be greater than y_{en}', a local efficiency cannot be greater than 1.00, or 100 per cent.

Relation between Murphree and Local Efficiencies. In small columns, the liquid on a plate is sufficiently agitated by vapor flow through the bubble caps, so that there are no measurable concentration gradients in the liquid as it flows across the plate. The concentration of the liquid in the down pipe x_n is that of the liquid on the entire plate. Since the concentration of the liquid on the plate is constant, that of the vapor from the plate is also constant, and no gradients exist in the vapor streams. A comparison of the quantities in Eqs. (12-74) and (12-75) shows that $y_n' = y_n$, $y_{n+1} = y_{n+1}'$, and $y_n^* = y_{en}'$. Then $\eta_M = \eta'$, and the local and Murphree efficiencies are equal.

In larger columns, liquid mixing in the direction of flow is not complete, and a concentration gradient does exist in the liquid on the plate. The maximum possible variation is from a concentration of x_{n-1} at the liquid inlet to a concentration of x_n at the liquid outlet. To trace the effect of such a concentration gradient on the Murphree efficiency, consider two situations: plate a, where mixing is complete, and plate b, where no liquid mixing occurs. Assume that y_{n+1}, x_n, and x_{n-1} are identical in the two cases. The quantity y_n^* is also the same for both plates, because this is the equilibrium concentration corresponding to liquid concentration x_n. In plate b, however, y_n is greater than that in plate a, because some of the vapor in plate b has the benefit of contact with liquid of concentration x_{n-1}, which is greater than the concentration of any of the liquid on plate a. The Murphree efficiency, by Eq. (12-74), is higher for plate b than for plate a. Also, when mixing does not occur, the Murphree efficiency is greater than the local efficiency. It is possible, on an unmixed plate, for

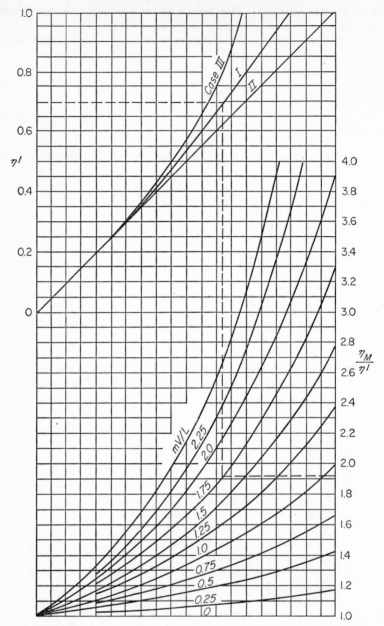

FIG. 12-51. Relation between local and Murphree efficiencies. (*By permission, from "Elements of Fractional Distillation,"* 4th ed., *by C. S. Robinson and E. W. Gilliland. Copyright, 1950. McGraw-Hill Book Company, Inc.*)

y_n to be greater than y_n^*. The Murphree efficiency is then greater than 100 per cent. This has been observed in large columns.[2]

The relation between η_M and η' has been established mathematically for the following three situations: [10]

Case 1. Vapor completely mixed, liquid unmixed.

Case 2. Vapor unmixed, liquid unmixed, with the flow of liquid in the same direction on all plates.

Case 3. Vapor unmixed, liquid unmixed, with flow of liquid in opposite directions on adjacent plates.

The usual bubble-cap column acts in a manner intermediate between cases 1 and 3. Special construction is needed to obtain conditions specified in case 2.

The results of this analysis are shown in Fig. 12-51, in which the ratio η_M/η' is shown as a function of η' and mV/L for each of the three cases. The slope of the equilibrium curve m, the slope of the operating line L/V, and the Murphree efficiency are assumed to be constant.

Relation between Murphree and Over-all Efficiencies. There is a definite relation between the over-all efficiency η_o and the Murphree efficiency η_M

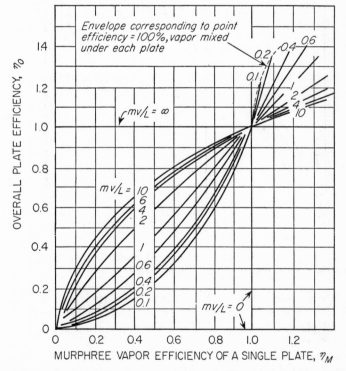

Fig. 12-52. Relation between Murphree and over-all efficiencies. (*By permission, from "Chemical Engineers' Handbook," 3d ed., by J. H. Perry. Copyright, 1950. McGraw-Hill Book Company, Inc.*)

when the equilibrium and the operating line are both straight. This is [10]

$$\eta_o = \frac{\log \left[1 + \eta_M (mV/L - 1)\right]}{\log \left(mV/L\right)} \tag{12-76}$$

Equation (12-76) is plotted in Fig. 12-52. When the operating and equilibrium lines are parallel, or where both efficiencies are nearly 1.0, $\eta_M = \eta_o$. Otherwise there is a considerable difference between η_o and η_M, the magnitude of which depends on the ratio of the slopes of the two lines. There is no simple relationship between these efficiencies when either the equilibrium line or the operating line is strongly curved.

Use of Murphree Efficiency. When the Murphree efficiency is known, it can readily be used in the McCabe-Thiele diagram. The diagram for an

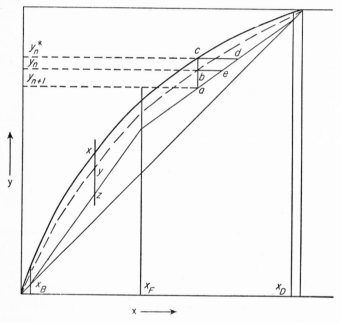

FIG. 12-53. Use of Murphree efficiency on xy diagram. Dotted line is effective-equilibrium curve, y'_e vs. x_e for $\eta_M = 0.60$. $ba/ca = yz/xz = 0.60$.

actual plate as compared with that for an ideal plate is shown in Fig. 12-53. Triangle acd represents the ideal plate and triangle abe the actual plate. The actual plate, instead of enriching the vapor from y_{n+1} to y^*_n, shown by line segment ac, accomplishes a lesser enrichment, $y_n - y_{n+1}$, shown by line segment ab. By definition of η_M, the Murphree efficiency is given by the ratio ab/ac. To apply a known Murphree efficiency to an entire column, it is necessary only to replace the true equilibrium curve y_e vs. x_e by an effective equilibrium curve y'_e vs. x_e, whose ordinates are calculated from the equation

$$y'_e = y + \eta_M (y_e - y) \tag{12-77}$$

In Fig. 12-53 an effective equilibrium curve for $\eta_M = 0.60$ is shown. Note that the position of the y'_e vs. x_e curve depends on both the operating line and the true equilibrium curve. Once the effective equilibrium curve has been plotted, the usual step-by-step construction is made and the number of actual plates determined. The reboiler is not subject to a discount for

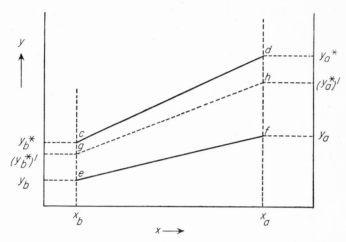

Fig. 12-54. Linear equilibrium and operating lines for use with η_M: cd, true equilibrium line, y_e vs. x_e. ef, operating line, y_{n+1} vs. x_n. gh, effective-equilibrium line, y'_e vs. x_e.

plate efficiency, and the true equilibrium curve is used for the last step in the stripping column.

An analogous construction using η_M can be used in the Ponchon-Savarit method.[3]

When the operating and equilibrium lines are both straight, the number of actual plates can be calculated by either of two equivalent methods. In the first method, the number of ideal plates is calculated by Eq. (10-37) and η_o by Eq. (12-76). Dividing the number of ideal plates by η_o gives the number of actual plates. The second method is to correct Eq. (10-37) for η_M and obtain the number of actual plates directly. To obtain positive factors Eq. (10-37) may be written for the fractionating column diagrammed in Fig. 12-54 as

$$N = \frac{\log \left[(y_a^* - y_a)/(y_b^* - y_b) \right]}{\log \left[(y_a^* - y_b^*)/(y_a - y_b) \right]} \tag{12-78}$$

For actual plates, when η_M is constant, the logarithmic numerator in Eq. (12-78) is unchanged since both $y_a^* - y_a$ and $y_b^* - y_b$ are multiplied by η_M and their ratio is unchanged. Also, $y_a - y_b$, the vapor-concentration range covered by the operating line, is unchanged. The correction of

Eq. (12-78) for plate efficiency affects only the term $y_a^* - y_b^*$. As shown by Fig. 12-54, this difference is corrected by replacing y_a^* and y_b^* in the denominator by

$$(y_a^*)' = y_a + \eta_M(y_a^* - y_a)$$

and

$$(y_b^*)' = y_b + \eta_M(y_b^* - y_b)$$

When Eq. (12-78) is so corrected, it gives the number of actual plates directly.

Factors Influencing Plate Efficiency. Although complete quantitative knowledge of the effect of all design and operating variables on plate efficiency is not available, sufficient data are at hand to show the major factors involved and to provide a basis for estimating the efficiencies for conventional types of columns operating on mixtures of common substances.

The most important requirement for obtaining satisfactory efficiencies is that the plates operate properly. Adequate and intimate contact between vapor and liquid is essential. Any misoperation of the column, such as excessive foaming or entrainment, poor vapor distribution or coning, or short-circuiting or dumping of liquid, lowers the plate efficiency.

Plate efficiency is a function of the mass transfer between liquid and vapor. On a bubble plate mass transfer occurs in three successive zones: at the point of formation of the bubbles, during the contact between the vapor and liquid on the plate, and during contact between fog and vapor in the space above the liquid. Mass transfer is rapid during bubble formation. The second stage—the transfer of mass between vapor and the pool of liquid on the plate—has been given the most theoretical and experimental attention. It has been found that the concepts of the two-resistance theory of mass transfer apply and that each locality on a plate acts much like a miniature absorber. The over-all transfer rate in this zone depends on the relative importance of liquid-phase and vapor-phase resistances, which in turn depends largely on the slope of the equilibrium curve. When the liquid resistance is large, the viscosity of the liquid is an important variable. Efficiencies in absorbers are usually less than those in fractionating columns because of the low solubility (low slope of the equilibrium curve) and high liquid viscosity.

The velocity of the vapor does not affect plate efficiencies greatly provided it is high enough to give adequate contact and low enough to prevent excessive entrainment or coning. Details of plate and bubble-cap construction affect the efficiency, especially slot width and depth of liquid.

The following empirical dimensional equations summarize many of the data on the point efficiency of conventional bubble columns, and have been recommended for estimating purposes.[18i]

$$\eta' = 1 - e^{-n'} \tag{12-79}$$

Exponent n' is given by the equation

$$n' = \frac{(Z')^{0.5}}{(2.5 + 0.37\alpha_{AB}M/\rho_L)\mu_L^{0.68}b_s^{0.33}} \tag{12-80}$$

where Z' = vertical distance between center of slots and top of liquid level over weir (see Fig. 12-45)

α_{AB} = relative volatility, component A to component B †

η' = point plate efficiency

ρ_L = density of liquid, lb/ft³

M = molecular weight of liquid

μ_L = viscosity of liquid, centipoises

b_s = width of slots, in.

Example 12-11. How many actual plates should be specified for the column of Example 12-10?

Solution. The point plate efficiency is estimated from Eqs. (12-79) and (12-80). The quantities for substitution are

$$Z' = 1.90 + \frac{0.75}{2} = 2.275 \text{ in.}$$

$$m = 9 \qquad \rho_L = 59.7 \text{ lb/ft}^3 \qquad M = 18$$

$$\mu = 0.28 \text{ centipoise} \qquad b_s = \tfrac{3}{32} = 0.094 \text{ in.}$$

$$n' = \frac{2.275^{0.5}}{\left[2.5 + \dfrac{0.37 \times 9 \times 18}{59.7}\right] 0.28^{0.68} \times 0.094^{0.33}} = 2.23$$

$$\eta' = 1 - e^{-2.23} = 0.892$$

If the liquid and vapor are unmixed, Fig. 12-52 shows that, for case 3, at $\eta' = 0.892$ and $mL/V = \tfrac{9}{5} = 1.8$, $\eta_M/\eta' = 2.05$. Since considerable mixing is inevitable on an 8-ft plate, assume that one-fourth of the improvement from the ideal situation of no mixing is realized. Then $\eta_M/\eta' = 1 + \tfrac{1}{4}(2.05 - 1) = 1.26$, and

$$\eta_M = 1.26 \times 0.892 = 1.12$$

The actual number of plates may be calculated from the over-all efficiency, which is found from Eq. (12-76).

$$\eta_o = \frac{\log\left[1 + 1.12(\tfrac{9}{5} - 1)\right]}{\log \tfrac{9}{5}} = 1.09$$

The number of theoretical plates was found to be 7.7 in Example 12-8, and the number of actual plates is $7.7/1.09 = 7.0$. Seven plates should be installed.

The actual plates can also be calculated from Eq. (12-78). The concentrations are

$$y_a = 0.0995 \qquad y_b = 0$$

$$y_a^* = 0.18 \qquad y_b^* = 0.0009$$

Then

$$(y_a^*)' = 0.0995 + 1.12(0.18 - 0.0995) = 0.1900$$

$$(y_b^*)' = 0 + 1.12(0.0009 - 0) = 0.00101$$

$$N = \frac{\log \dfrac{0.18 - 0.0995}{0.0009 - 0}}{\log \dfrac{0.1900 - 0.00101}{0.0995 - 0}} = 7.0$$

† For absorbers, $m = dy_e/dx_e$ is recommended in place of α_{AB}.

Sieve-plate and Other Distillation Towers

Within recent years the design of plates for distillation towers has been given much study. Many modified bubble-cap plates have been devised, and sieve-plate columns, formerly restricted almost entirely to plates making alcohol or liquid air, have found applications in many other services.

Modified Bubble-cap Tower. A modified bubble-cap tray is illustrated in Fig. 12-55. This Uniflux tray [1] is made up of interlocking S-shaped

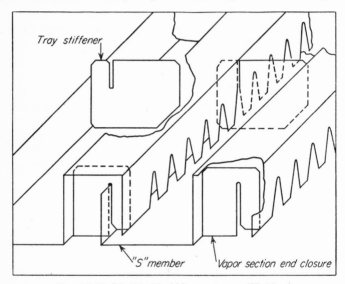

FIG. 12-55. Modified bubble-cap tray. (*Uniflux.*)

sections which form, when assembled, a series of transverse tunnel caps with vapor outlets on one side only. The vapor is directed toward the liquid outlet of each plate, to minimize the gradient head. The construction members are light but strong and are cheaper than bell caps. All that must be kept in stock at the plant for repair purposes are lengths of the slotted S sections, instead of the many component parts of the usual bubble-cap column. This reduces the inventory of replacement parts and simplifies maintenance. Directing the vapor horizontally into the liquid reduces entrainment, resulting in high plate efficiencies and large capacities.

Sieve-plate Towers. In many towers a flat plate perforated with a multitude of small holes is used in place of a bubble-cap tray. The liquid passes from plate to plate over weirs and through downcomers as in a bubble-cap tower; it is kept from flowing through the holes in the plates by the upward flow of vapor. Formerly it was thought that the operating range of a sieve-plate tower was narrow, but this has been shown to be untrue. On the contrary, properly designed sieve-plate towers have a somewhat greater effective operating range than bubble-cap towers; they give higher plate efficiencies and less entrainment; they permit considerably higher flow rates of vapor at the same pressure drop (or lower pressure drops at the same flow rate); and they are cheaper.[7,11] For these reasons they are finding increasing favor over bubble-cap columns, especially in chemical manufacturing.

Typical sieve plates carry $\frac{1}{8}$-in. holes on $\frac{3}{8}$-in. triangular centers, punched or drilled in $\frac{1}{4}$- or $\frac{1}{8}$-in. plate. The combined area of the perforations equals 6 to 15 per cent of the total cross-sectional area of the tower. The height of the liquid-overflow weir on each tray Z_w is usually between $\frac{3}{4}$ and 3 in. The pressure drop per plate may be computed from the static head of liquid above the plate $Z_w + Z_{cr} + Z_g$ and the pressure drop through the perforations, using a discharge coefficient C of 0.85.[11] The gradient head Z_g is usually negligible, and is sometimes eliminated by tilting the plates slightly toward the outlet.

Example 12-12. Design a sieve-plate tower for the same duty as in Example 12-10, using the same tower diameter. The weir height Z_w is to remain at 1.25 in. and the weir crest Z_{cr} at 1.65 in. Estimate the pressure drop per plate in the sieve-plate tower, and compare with that in the bubble-cap tower of Example 12-10.

Solution. The diameter of the perforations will be $\frac{3}{16}$ in. The combined cross-sectional area of the holes will be set at 10 per cent of the column area.

$$\text{Cross-sectional area of column (Example 12-10)} = 49 \text{ ft}^2$$

$$\text{Total area of holes} = 0.10 \times 49 = 4.9 \text{ ft}^2$$

$$\text{Area of one } \tfrac{3}{16}\text{-in. hole} = 1.92 \times 10^{-4} \text{ ft}^2$$

$$\text{Number of holes} = \frac{4.9}{1.92 \times 10^{-4}} = 25{,}520$$

PRESSURE DROP PER PLATE. The gradient head may be neglected. The depth of liquid on each plate, therefore, is $1.25 + 1.65 = 2.90$ in.

The pressure drop across the holes is found from the velocity through the holes, using the simplified orifice equation. The quantities needed are

$$\text{Vapor velocity in column (Example 12-10)} \; u_c = 5.56 \text{ ft/sec}$$

$$\text{Vapor velocity through holes } V_o = \frac{5.56}{0.10} = 55.6 \text{ ft/sec}$$

$$C = 0.85 \qquad \rho_L = 59.7 \text{ lb/ft}^3 \qquad \rho_v = 0.0403 \text{ lb/ft}^3$$

From Eqs. (2-110) and (2-111), assuming $\beta = 0$,

$$p_a - p_b = \frac{\rho_v V_0^2}{2 g_c C^2} = \frac{0.0403 \times 55.6^2}{2 \times 32.17 \times 0.85^2} = 2.68 \text{ lb force/ft}^2$$

$$Z_n = \frac{2.68 \times 12}{59.7} = 0.54 \text{ in.}$$

The total pressure head across the plate is $2.90 + 0.54 = 3.44$ in.† The pressure head across the plate in the bubble-cap tower (Example 12-10) is 6.75 in.

† The actual pressure head across the working sieve plate may differ from this by 10 to 25 per cent. A more accurate and elaborate method of calculation is given in Ref. 7.

The pressure drop in a sieve-plate column is usually much less than that in a bubble-cap column operating under the same conditions. The permissible vapor throughput is also larger in the sieve-plate column, and the plate efficiency is somewhat higher. Sieve plates are less expensive than bubble-cap trays and because of their simplicity are easier to maintain and keep clean. Sieve plates, however, must be leveled more carefully than bubble-cap plates and cannot be used where corrosion by the fluids tends to enlarge the holes. At low vapor velocities there is a tendency for liquid to "weep" through the holes, lowering the plate efficiency.

FIG. 12-56. Turbogrid tray. (*Courtesy of Shell Development Company.*)

Turbogrid Towers. The Turbogrid plates [19] illustrated in Fig. 12-56 are modified perforated plates. They are simple slotted plates set horizontally in the tower. Normally downcomers are not used. The slot width and total slot area are selected to give the proper dynamic balance, so that at any instant the liquid descends through a fraction of the slots while vapor passes into the froth region above the plate through the remaining slots. Fouling is avoided by proper choice of slot width. Because all the tray area is available for contact between liquid and vapor, and because the liquid need not flow through downcomers, a Turbogrid tower has considerably more capacity than a bubble-cap tower of the same diameter. The plates are said to be as cheap as any other type of plate available. The height of a theoretical stage in columns of small or moderate diameter is equivalent to that of bubble-cap trays, but the plate efficiency declines as the column diameter is increased. The operating range, although narrower than that of a bubble-cap or sieve-plate tower, is adequate for many distillation problems.

Rectification in Packed Towers. Packed towers are sometimes used for rectification, usually where the required separation is fairly easy. The diameter of packed distillation columns rarely exceeds 30 in. The principles of their operation are the same as those of absorption columns. Equations (11-9), (11-10), (11-22a), and (11-22c) are as applicable to rectification as to absorption.[5] The relative velocity factor ϕ is unity because of countercurrent diffusion of the components, and $1 - y$ is unity because of constant molal overflow. Over-all coefficients and HTUs are, however, difficult to use because m, the slope of the equilibrium curve, varies greatly over the concentration range $x = 0$ to $x = 1$, and over-all coefficients are reasonably constant only at the ends of the diagram. Adequate data for the individual coefficients k_y and k_x and the HTUs H_y and H_x are too few to be of much use. At present only empirical over-all factors are available [15b, 18j] for estimating packing depths in packed distillation columns.

SYMBOLS

B Bottom product, moles/hr or lb/hr

b Perimeter of down pipe or length of weir, ft; b_b, average width of bubbling section on plate; b_s, width of slot, in.

C Discharge coefficient, flow through perforations of sieve plate

D Overhead product, moles/hr or lb/hr

D_d Diameter of down pipe, ft

F Feed, moles/hr or lb/hr

f Fraction of feed that is vaporized, also, moles of vapor to rectifying column per mole of feed

G_c Maximum permissible mass velocity of vapor in column, based on area of bubbling section, lb/ft²-sec

g Acceleration of gravity, ft/sec²

g_c Newton's-law conversion factor, 32.174 ft-lb/lb force-sec²

H_{JL} Friction loss in liquid between stations a and b, ft-lb force/lb

i Specific enthalpy, Btu/lb; i_B, of bottom product; i_D, of overhead product; i_R, i_S, i_T, of streams R, S, and T, respectively; i_x, of saturated liquids; i_{xm}, of liquid from plate m of stripping column; i_{xn}, of liquid from plate n of rectifying column; i_y, of saturated vapor; $i_{y(m+1)}$, of vapor from plate $m + 1$ in stripping column; $i_{y(n+1)}$, of vapor from plate $n + 1$ in rectifying column; i_{y1}, of vapor from top plate; i'_B, of bottom product, corrected for heat effect q_r; i'_D, of overhead product, corrected for heat effect $-q_c$; i'_{Dm}, of overhead product for minimum reflux, corrected for heat effect $-q_c$

K Constant in Eq. (12-62)

L Molal liquid flow rate in rectifying column, lb moles/hr; L_c, reflux from condenser; L_m, liquid flow rate from plate m in stripping column, lb moles/hr or lb/hr; L_n, liquid flow rate from plate n in rectifying column, lb moles/hr or lb/hr; $\bar{L}$ molal liquid flow rate in stripping column, lb moles/hr; also, length, ft; L_b, of bubbling section; L_d, of down pipe

M Molecular weight; M_A, of component A; M_B, of component B

m Serial number of plate in stripping column, counting from feed plate; also, slope of equilibrium curve, dy_e/dx_e

N Number of ideal plates

n Moles of liquid in batch; n_0, at start of distillation; n_{0A}, of component A at start; n_{0B}, of component B at start; n_1, moles of liquid remaining at end of distillation; also, serial number of plate in rectifying column, counting from top

n' Exponent in Eq. (12-79)

p Pressure, atm; p_A, vapor pressure of component A; p_B, of component B; p_a, pressure at upstream station a; p_b, at downstream station b; p_{n-1}, p_n, p_{n+1}, pressures in vapor space above plates $n - 1$, n, and $n + 1$, respectively

$\bar{p}$ Partial pressure, atm; $\bar{p}_A$, of component A; $\bar{p}_B$, of component B

q Rate of heat flow, Btu/hr (positive for endothermic process and negative for exothermic process); q_r, heat added at reboiler; $-q_c$, heat rejected at condenser

q_L Volumetric flow rate of liquid in down pipe, ft³/sec

R Reflux ratio; $R_D = L/D$; $R_V = L/V$; R_{Dm}, minimum reflux ratio

r_H Hydraulic radius, ft

t Temperature, °F or °C; t_A, boiling temperature of component A; t_B, of component B; t_{n-1}, t_n, t_{n+1}, temperatures on plates $n - 1$, n, and $n + 1$, respectively; t_1, inlet temperature of cooling water to condenser; t_2, outlet temperature of cooling water from condenser

u Linear velocity, ft/sec; u_{cr} maximum permissible vapor velocity, based on area of bubbling section; u_d, liquid velocity in down pipe; u_s, vapor velocity in cap, annulus, or slot of bubble cap, whichever is largest

V Molal vapor flow rate in rectifying column, lb moles/hr; V_{m+1}, vapor flow rate from plate $m + 1$ in stripping column, lb moles/hr or lb/hr; V_{n+1}, vapor flow rate from plate $n + 1$ in rectifying column, lb moles/hr or lb/hr; V_1, vapor flow rate from top plate to condenser; $\overline{V}$, molal vapor flow rate in stripping column, lb moles/hr

V_o Vapor velocity through holes in sieve plate, ft/sec

w Mass flow rate, lb/hr or lb/sec; w_A, w_B, of components A and B, respectively; w_c, of cooling water to condenser; w_L, of liquid on and across bubble plate; w_R, w_S, w_T, of streams R, S, and T, respectively; w_s, of steam to reboiler

x Mole fraction or mass fraction of component A in liquid; x_a, in liquid entering single section column; x_B, in bottom product; x_b, in liquid leaving single section column; x_c, in reflux from condenser; x_D, in overhead product; x_e, in equilibrium with vapor of concentration y_e; x_F, in feed to column; x_m, in liquid from plate m in stripping column; x_{n-1}, x_n, in liquid from plates $n - 1$ and n, respectively, in rectifying column; x_0, initial concentration in batch; x_R, x_S, x_T, in streams R, S, T, respectively

y Mole or mass fraction of component A in vapor; y_a, of vapor leaving single-section column; y_b, of vapor entering single-section column; y_D, in vapor overhead product; y_e, at equilibrium with liquid of concentration x_e; y_{m+1}, of vapor from plate $m + 1$ in stripping column; y_n, y_{n+1}, in vapor from plates n and $n + 1$, respectively, in rectifying column; y^*, in vapor in equilibrium with specific stream of liquid; y_a^*, in equilibrium with x_a; y_b^*, in equilibrium with x_b; y_n^*, in equilibrium with x_n; y_e', effective equilibrium concentration in vapor corresponding to Murphree efficiency [Eq. (12-77)]; y_{en}', in equilibrium with liquid at specific location on plate n; y_{n+1}', in vapor entering plate n from plate $n + 1$, at specific location on plate n

Z Distance above datum, or head, ft; Z_a, height of station a above datum; Z_{cr}, weir-crest head; Z_g, gradient across plate; Z_L, friction head in liquid; Z_n, net static head, from level in down pipe to top of weir; Z_o, total static head at outlet, $Z_{cr} + Z_w$; Z_p, pressure head across plate, in vapor stream; Z_s, liquid seal above top of slot, $Z_w - Z_{sp} + Z_{cr}$; Z_{sp}, height of top of slot above plate; Z_v, friction head, vapor flow through riser, annulus, and slots; Z_w, weir height; Z', distance between center of slots and liquid level over weir

Greek Letters

α Relative volatility; $\alpha_{AB} = p_A/p_B$; $\alpha_{BA} = p_B/p_A$

η Efficiency; η_M, Murphree plate efficiency; η_o, over-all plate efficiency; η_s, efficiency of vaporization, steam distillation; η', local plate efficiency

λ Latent heat of vaporization, Btu/lb mole; λ_n, of vapor from plate n; λ_{n+1}, of vapor from plate $n + 1$; λ_s, latent heat of steam, Btu/lb

μ_L Viscosity of liquid, lb/ft-sec

ρ Density, lb/ft^3; ρ_L, of liquid; ρ_v, of vapor

Superscripts

i Liquid-inlet side of plate

o Liquid-outlet side of plate

PROBLEMS

12-1. The vapor pressures of n-pentane and n-hexane are given in Table 12-6. It may be assumed that Raoult's law applies to mixtures of these two substances.

TABLE 12-6. VAPOR PRESSURES OF n-PENTANE AND n-HEXANE

Temp., °C	Vapor pressure, atm		Temp., °C	Vapor pressure, atm	
	n-Pentane	n-Hexane		n-Pentane	n-Hexane
17.5	0.500	0.138	46	1.355	0.460
20	0.552	0.154	48.5	1.49	0.500
25	0.665	0.195	55	1.79	0.631
30	0.802	0.242	60	2.08	0.750
36	1.00	0.316	65	2.41	0.875
40	1.13	0.362	69	2.70	1.00

(a) Construct a temperature-concentration diagram for a pressure of 1.0 atm. (b) Construct a pressure-concentration diagram for a temperature of 55°C. (c) Construct an xy diagram for a total pressure of 1.0 atm.

12-2. A liquid containing 30 mole per cent toluene, 40 mole per cent ethylbenzene, and 30 mole per cent water is subjected to a continuous flash distillation at a total pressure of 0.5 atm. Vapor-pressure data for these substances are given in Table 12-7.

TABLE 12-7. VAPOR PRESSURES OF ETHYLBENZENE, TOLUENE, AND WATER

Temp., °C	Vapor pressure, mm Hg		
	Ethylbenzene	Toluene	Water
50	53.8	. . . . .	92.5
60	78.6	139.5	149.4
70	113.0	202.4	233.7
80	160.0	289.4	355.1
90	223.1	404.6	525.8
100	307.0	557.2	760.0
110	414.1		
110.6	. . . . .	760	
120	545.9		

Assuming that mixtures of ethylbenzene and toluene obey Raoult's law and that the hydrocarbons are completely immiscible in water, calculate the temperature and compositions of liquid and vapor phases: (a) at the bubble point, (b) at the dew point, (c) at the 50 per cent point (one half of the feed leaves as vapor and the other half as liquid).

12-3. It is desired to produce an overhead product containing 80 mole per cent benzene from a feed mixture of 70 mole per cent benzene and 30 mole per cent toluene. The following methods are considered for this operation. All are to be conducted at atmospheric pressure. For each method calculate the moles of product per 100 moles of feed and the number of moles vaporized per 100 moles feed.

(a) Continuous equilibrium distillation.

(b) Differential distillation.

(c) Continuous distillation in a still fitted with a partial condenser, in which 50 mole per cent of the entering vapors are condensed and returned to the still. The partial condenser is so constructed that vapor and liquid leaving it are in equilibrium, and holdup in it is negligible.

(d) Batch distillation, with 50 mole per cent of the vapors entering the partial condenser returned to the still. The liquid and vapor leaving the condenser are in equilibrium.

12-4. The boiling-point–equilibrium data for the system acetone-methanol at 760 mm Hg are given in Table 12-8. A column is to be designed to separate a feed analyzing 25

TABLE 12-8. SYSTEM ACETONE-METHANOL

Temp., °C	Mole fraction acetone		Temp., °C	Mole fraction acetone	
	Liquid	Vapor		Liquid	Vapor
64.5	0.00	0.000	56.7	0.50	0.586
63.6	0.05	0.102	56.0	0.60	0.656
62.5	0.10	0.186	55.3	0.70	0.725
60.2	0.20	0.322	55.05 †	0.80	0.80
58.65	0.30	0.428	56.1	1.00	1.00
57.55	0.40	0.513			

† Azeotrope.

mole per cent acetone and 75 mole per cent methanol into an overhead product containing 78 mole per cent acetone and a bottom product containing 1.0 mole per cent acetone. The feed enters as an equilibrium mixture of 30 per cent liquid and 70 per cent vapor. A reflux ratio equal to twice the minimum is to be used. An external reboiler is to be used. Bottom product is removed from the reboiler. The condensate (reflux and overhead product) leaves the condenser at 25°C, and the reflux enters the column at this temperature.

The molal latent heats of both components are 7700 g cal/g mole. The Murphree plate efficiency is 70 per cent.

Calculate: (a) the number of plates required above and below the feed; (b) the heat required at the reboiler, in Btu per pound mole of overhead product; (c) the heat removed in the condenser, in Btu per pound mole of overhead product.

12-5. A plant must distill a mixture containing 80 mole per cent methanol and 20 per cent water. The overhead product is to contain 99.99 mole per cent methanol and

the waste 0.005 mole per cent. The feed is cold, and for each mole of feed 0.4 mole of vapor is condensed at the feed plate. The reflux ratio at the top of the column is 1.35, and the reflux is at its boiling point.

Calculate: (a) the minimum number of plates; (b) the minimum reflux ratio; (c) the number of plates using a total condenser and a still, assuming an average Murphree plate efficiency of 70 per cent; (d) the number of plates using a still and a partial condenser operating with the reflux in equilibrium with the vapor going to a final condenser. Equilibrium data are given in Table 12-9.

TABLE 12-9. EQUILIBRIUM DATA FOR METHANOL-WATER

x	0.1	0.2	0.3	0.4	0.5	0.6	0.7	0.8	0.9	1.0
y	0.417	0.579	0.669	0.729	0.780	0.825	0.871	0.915	0.959	1.0

12-6. An equimolal mixture of benzene and toluene is to be separated in a bubble-plate tower at the rate of 100 lb moles/hr. The overhead product must contain at least 98 mole per cent benzene. The feed is saturated liquid.

A tower is available containing 24 plates. Feed may be introduced either on the eleventh or the seventeenth plate from the top. The maximum vaporization capacity of the reboiler is 120 lb moles/hr. The plates are about 50 per cent efficient.

How many moles per hour of overhead product can be obtained from this tower?

12-7. An aqueous solution of a volatile component A containing 7.94 mole per cent A preheated to its boiling point is to be fed to the top of a continuous stripping column operated at atmospheric pressure. Vapor from the top of the column is to contain 11.25 mole per cent A. No reflux is to be returned. Two methods are under consideration, both calling for the same expenditure of heat, namely, a vaporization of 0.562 mole per mole feed in each case. Method 1 is to use a still at the bottom of a plate column, generating vapor by use of steam condensing inside a closed coil in the still. In method 2 the still and heating coil are omitted, and live steam is injected directly below the bottom plate. Equilibrium data are given below. The usual simplifying assumptions may be made.

What are the advantages of each method?

EQUILIBRIUM DATA IN MOLE FRACTION A

x	0.0035	0.0077	0.0125	0.0177	0.0292	0.0429	0.0590	0.0784
y	0.0100	0.0200	0.0300	0.0400	0.0600	0.0800	0.1000	0.1200

12-8. A rectifying column containing the equivalent of three ideal plates is to be supplied continuously with a feed consisting of 0.5 mole per cent ammonia and 99.5 mole per cent water. Before entering the column the feed is converted wholly into saturated vapor, and it enters between the second and third plates from the top of the column. The vapors from the top plate are totally condensed but not cooled. Per mole of feed, 1.3 moles of condensate is returned to the top plate as reflux, and the remainder of the distillate is removed as overhead product. The liquid from the bottom plate overflows to a reboiler, which is heated by closed steam coils. The vapor generated in the reboiler enters the column below the bottom plate, and bottom product is continuously removed from the reboiler. The vaporization in the reboiler is 0.6 mole per mole of feed.

Over the concentration range involved in this problem, the equilibrium relation is given by the equation

$$y = 12.6x$$

Calculate the mole fraction of ammonia in: (a) the bottom product from the reboiler, (b) the overhead product, (c) the liquid reflux leaving the feed plate.

12-9. A plant has available 80 per cent by weight acetic acid which must be separated into two fractions, one containing 99 weight per cent acid and the other 35 weight per cent. This is done in a continuous column operating with a reflux ratio of 4.0. The feed temperature is 68°F. Data on the system acetic acid–water are given in Table 12–10.

TABLE 12-10. DATA ON SYSTEM ACETIC ACID–WATER

Temp., °C	Weight H_2O, %	
	Liquid	Vapor
100.0	100	100
100.2	70.0	77.8
100.5	55.0	66.7
100.8	45.0	58.5
101.2	35.0	48.5
102.2	25.0	36.9
103.4	15.0	23.6
104.8	10.0	16.3
108.1	5.0	8.45
114.7	1.0	2.17
118.1	0.0	0.0

	Water	Acetic acid
Specific heat.............	1.00	0.53
Latent heat, Btu/lb......	965	153

Calculate: (a) the heat input to the still, in Btu per pound of feed; (b) the heat removed in the condenser, in Btu per pound of feed; (c) the total number of ideal plates and the position of the feed plate.

12-10. A tower containing six ideal plates, a reboiler, and a total condenser is used to separate, partially, oxygen from air at 65 lb force/in.² gauge. It is desired to operate at a reflux ratio (reflux to product) of 2.5 and to produce a bottom product containing 45 weight per cent oxygen. The air is fed to the column at 65 lb force/in.² gauge and 25 per cent vapor by mass. The enthalpy of oxygen-nitrogen mixtures at this pressure is given in Table 12-11. Compute the composition of the overhead if the vapors are just condensed but not cooled.

TABLE 12-11. ENTHALPY OF OXYGEN-NITROGEN AT 65 LB FORCE/IN.^{2} GAUGE

Temp., °C	Liquid		Vapor	
	N_2, wt %	i_x, cal/g mole	N_2, wt %	i_y, cal/g mole
−163	0.0	420	0.0	1840
−165	7.5	418	19.3	1755
−167	17.0	415	35.9	1685
−169	27.5	410	50.0	1625
−171	39.0	398	63.0	1570
−173	52.5	378	75.0	1515
−175	68.5	349	86.0	1465
−177	88.0	300	95.5	1425
−178	100.0	263	100.0	1405

REFERENCES

1. Bowles, V. O.: *Chem. Eng.*, **61**(5): 174 (1954).
2. Brown, G. G., M. Souders, Jr., H. V. Nyland, and W. H. Hessler: *Ind. Eng. Chem.*, **27**: 383 (1935).
3. Brown, G. G., and associates: "Unit Operations," p. 343, John Wiley & Sons, Inc., New York, 1950.
4. Davies, J. A.: *Ind. Eng. Chem.*, **39**: 774 (1947).
5. Deed, D. W., P. W. Schutz, and T. B. Drew: *Ind. Eng. Chem.*, **39**: 766 (1947).
6. Good, A. J., M. H. Hutchinson, and W. C. Rousseau: *Ind. Eng. Chem.*, **34**: 1445 (1942).
7. Jones, J. B., and C. Pyle: *Chem. Eng. Progr.*, **51**: 424 (1955).
8. Kemp, H. S., and C. Pyle: *Chem. Eng. Progr.*, **45**: 435 (1949).
9. Lewis, W. K.: *Ind. Eng. Chem.*, **14**: 492 (1922).
10. Lewis, W. K., Jr.: *Ind. Eng. Chem.*, **28**: 399 (1936).
11. Mayfield, F. D., W. L. Church, A. C. Green, D. C. Lee, and R. W. Rasmussen: *Ind. Eng. Chem.*, **44**: 2238 (1952).
12. McCabe, W. L., and E. W. Thiele: *Ind. Eng. Chem.*, **17**: 605 (1925).
13. Murphree, E. V.: *Ind. Eng. Chem.*, **17**: 747 (1925).
14. Peters, W. A., Jr.: *Ind. Eng. Chem.*, **14**: 476 (1922).
15. Pigford, R. L., and A. P. Colburn, in J. H. Perry (ed.): "Chemical Engineers' Handbook," 3d ed., (*a*) p. 551, (*b*) p. 620, McGraw-Hill Book Company, Inc., New York, 1950.
16. Randall, M., and B. Longtin: *Ind. Eng. Chem.*, **30**: 1063, 1188 (1938).
17. Randall, M., and B. Longtin: *Ind. Eng. Chem.*, **30**: 1311 (1938).
18. Robinson, C. S., and E. R. Gilliland: "Elements of Fractional Distillation," 4th ed., (*a*) p. 127, (*b*) p. 146, (*c*) p. 149, (*d*) p. 411, (*e*) pp. 416ff., (*f*) pp. 423ff., (*g*) p. 430, (*h*) p. 449, (*i*) p. 462, (*j*) p. 466, McGraw-Hill Book Company, Inc., New York, 1950.
19. Shell Development Co. Engineering Staff: *Chem. Eng. Progr.*, **50**: 57 (1954).
20. Sherwood, T. K., and F. A. L. Holloway: *Trans. AIChE*, **36**: 39 (1940).
21. Souders, M., and G. G. Brown: *Ind. Eng. Chem.*, **26**: 98 (1934).
22. Thiele, E. W.: *Ind. Eng. Chem.*, **27**: 392 (1932).

Chapter 13

LEACHING AND EXTRACTION

This chapter discusses the methods of removing one constituent from a solid or liquid by means of a liquid solvent. These techniques fall into two categories. The first, called leaching, is used to dissolve soluble matter from its mixture with an insoluble solid. The second, called liquid extraction, is used to separate two miscible liquids by the use of a solvent which preferentially dissolves one of them. Although the two processes have certain common fundamentals, the differences in equipment and, to some extent, in theory are sufficient to justify separate treatment.

LEACHING

Leaching differs very little from the washing of filtered solids, as discussed in Chap. 7, and leaching equipment strongly resembles the washing section of various filters. In leaching, the amount of soluble material removed is often rather greater than in ordinary filtration washing, and the properties of the solids may change considerably during the leaching operation. Coarse, hard, or granular feed solids may disintegrate into pulp or mush when their content of soluble material is removed.

When the solids form an open, permeable mass throughout the leaching operation, solvent may be percolated through an unagitated bed of solids. With impermeable solids, or materials which disintegrate during leaching, the solids are dispersed into the solvent and are later separated from it. Both methods may be either batch or continuous.

Leaching by Percolation through Solid Beds; Stationary Solid Beds. Stationary solid-bed leaching is done in a tank with a perforated false bottom to support the solids and permit drainage of the solvent. Typical tank bottoms are shown in Fig. 13-1. Solids are loaded into the tank, sprayed with solvent until their solute content is reduced to the economical minimum, and excavated. In some cases the rate of solution is so rapid that one passage of solvent through the material is sufficient; but countercurrent flow of solvent through a battery of tanks is more common. In this method, fresh solvent is fed to the tank containing the solid that is most nearly extracted; it flows through the several tanks in series, and is finally withdrawn from the tank that has been freshly charged. Such a series of tanks is called an *extraction battery*. The solid in any one tank is stationary until it is completely extracted. The piping is arranged so that fresh solvent can be introduced to any tank and strong solution withdrawn from any tank, making it possible to charge and discharge one tank at a time. The other tanks in the battery are kept in countercurrent operation by advancing the inlet and drawoff tanks one at a time as the material is charged and removed. Such a process is sometimes called a *Shanks process*.

Each unit of an extraction battery is similar to the gravity nutsch filter described in Chap. 7. This type of equipment can be used only with coarse,

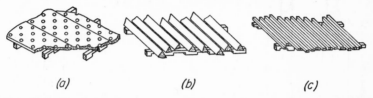

(a)   (b)   (c)

Fig. 13-1. Tank filter-bottom constructions: (*a*) Perforated boards on bearers. (*b*) Support for gravel filter. (*c*) Support for coarse material or filter cloth.

free-draining solids. In ore-dressing, where such tanks are most common, the units are very large and operating cycles are long. In leaching copper ore, for example, the batteries consist of 13 tanks in series, each 175 by $67\frac{1}{2}$ by 18 ft in size and charged with 9,000 tons of ore. The tanks operate on a 13-day cycle, and are automatically loaded and excavated in the proper sequence.[1]

In some solid-bed leaching the solvent is volatile, necessitating the use of closed vessels operated under pressure. Pressure is also needed to force solvent through beds of some less permeable solids. A series of such pressure tanks operated with countercurrent solvent flow is known as a *diffusion battery*.

Moving-bed Leaching.[3] In the machines illustrated in Fig. 13-2 the solids are moved through the solvent with little or no agitation. The Bollman extractor (Fig. 13-2*a*) contains a bucket elevator in a closed casing. There are perforations in the bottom of each bucket. At the top right-hand corner of the machine, as shown in the drawing, the buckets are loaded with flaky solids such as soybeans and are sprayed with appropriate amounts of "half miscella" as they travel downward. Half miscella is the intermediate solvent containing some extracted oil and some small solid particles. As solids and solvent flow concurrently down the right-hand side of the machine, the solvent extracts more oil from the beans. Simultaneously the fine solids are filtered out of the solvent, so that clean "full miscella" may be pumped from the right-hand sump at the bottom of the casing. As the partially extracted beans rise through the left side of the machine, a stream of pure solvent percolates countercurrently through them. It collects in the left-hand sump and is pumped to the half-miscella storage tank. Fully extracted beans are dumped from the buckets at the top of the elevator into a hopper, from which they are removed by paddle conveyors. The capacity of typical units is 50 to 500 tons of beans per 24-hr day.

The Hildebrandt extractor shown in Fig. 13-2b consists of a U-shaped screw conveyor with a separate helix in each section. The helices turn at different speeds to give considerable compaction of the solids in the horizontal section. Solids are fed to one leg of the U and fresh solvent to the other, to give countercurrent flow.

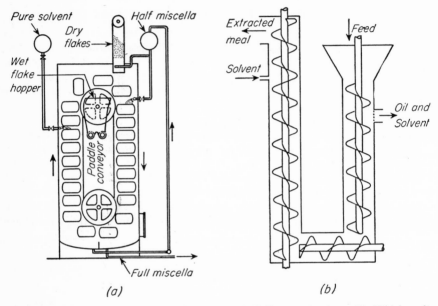

FIG. 13-2. Moving-bed leaching equipment: (a) Bollman extractor. (b) Hildebrandt extractor.

Another moving-bed leaching unit is the Rotocel extractor, which contains several compartments moving over a perforated horizontal disk. The compartments are successively charged with solids, passed under sprays of solvent, and emptied through an opening in the stationary disk. This device bears the same relation to a stationary-bed leaching tank as a continuous horizontal vacuum filter (page 335) bears to a vacuum nutsch. Single-deck and multiple-deck rake classifiers are also used for leaching coarse solids.

Dispersed-solid Leaching. Solids which form impermeable beds, either before or during leaching, are treated by dispersing them in the solvent by mechanical agitation in a tank or flow mixer. The leached residue is then separated from the strong solution by settling or filtration.

Small quantities may be leached in this way in a single agitated vessel with a bottom drawoff for settled residue. Continuous countercurrent leaching is obtained with several gravity thickeners connected in series, as in Fig. 10-4, or when the contact in a thickener is inadequate, by placing an agitator tank in the equipment train between each pair of thickeners. A still further refinement, used when the solids are too fine to settle out by gravity, is to separate the residue from the miscella in continuous solid-bowl helical-conveyor centrifuges. Many other leaching devices have been developed for special purposes, such as the solvent extraction of various oilseeds, with their specific design details governed by the properties of the solvent and of the solid to be leached.[3]

Continuous Countercurrent Leaching

The most important method of leaching is the continuous countercurrent method using stages. Even in a diffusion battery or a Shanks system, where the solid is not moved physically from stage to stage, the charge in any one cell is treated by a succession of liquids of constantly decreasing concentration as if it were being moved from stage to stage in a countercurrent system.

Because of its importance, only the continuous countercurrent method is discussed here. Also, since the stage method is normally used, the differential-contact method is not considered. In common with other stage cascade operations, leaching may be considered, first, from the standpoint of ideal stages and, second, from that of stage efficiencies.

Ideal Stages in Countercurrent Leaching. Figure 13-3 shows a material-balance diagram for a continuous countercurrent plant. The stages are numbered in the direction of flow of the solid. The light phase

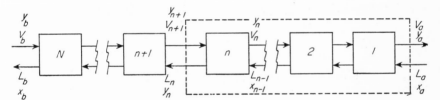

Fig. 13-3. Countercurrent leaching plant.

is the liquid that overflows from stage to stage in a direction counter to that of the flow of the solid, dissolving solute as it moves from stage N to stage 1. The heavy phase is the solid flowing from stage 1 to stage N. Exhausted solids leave stage N, and concentrated solution overflows from stage 1.

It is assumed that the solute-free solid is insoluble in the solvent and that the flow rate of this solid is constant throughout the plant. Let V refer to the overflow solution and L to the liquid retained by the solid, both based on a definite flow of dry, solute-free solid. Also, in accordance with standard nomenclature, the terminal concentrations are: of the solution on entering solid x_a, of solution on leaving solid x_b, of fresh solvent entering the system y_b, and of concentrated solution leaving the system y_a.

As in absorption and distillation, the quantitative performance of a countercurrent system can be analyzed by utilizing an equilibrium line and an operating line, and, as before, the method to be used depends on whether these lines are straight or curved.

Equilibrium. In leaching, provided sufficient solvent is present to dissolve all the solute in the entering solid and there is no adsorption of solute by the solid, equilibrium is attained when the solute is completely dissolved, and the concentration of the solution so formed is uniform. Such a condition may be obtained simply or with difficulty, depending on the structure of the solid. These factors are considered when stage efficiency is discussed. At present, it is assumed that the requirements for equilibrium are met. Then the concentration of the liquid retained by the solid leaving any stage is the same as that of the liquid overflow from the same stage. The equilibrium relationship is, simply, $x_e = y_e$.

Operating Line. The equation for the operating line is obtained by writing material balances for that portion of the plant consisting of the first n units, as shown by the control surface indicated by the dotted lines in Fig. 13-3. These balances are

Total solution:
$$V_{n+1} + L_a = V_a + L_n \tag{13-1}$$

Solute:
$$V_{n+1}y_{n+1} + L_a x_a = L_n x_n + V_a y_a \tag{13-2}$$

Eliminating V_{n+1} and solving for y_{n+1} gives

$$y_{n+1} = \frac{1}{1 + (V_a - L_a)/L_n} x_n + \frac{V_a y_a - L_a x_a}{L_n + V_a - L_a} \tag{13-3}$$

This is the equation of the operating line. As usual, the line passes through points (x_a, y_a) and (x_b, y_b).

Constant and Variable Underflow. Two cases are to be considered. If the density and viscosity of the solution change considerably with solute concentration, the solids from the lower-numbered stages may retain more liquid than those from the higher-numbered stages. Then, as shown by Eq. (13-3), the slope of the operating line varies from unit to unit. If, however, the mass of solution retained by the solid is independent of concentration, L_n is constant, and the operating line is straight. This condition is called constant solution underflow. If the underflow is constant, so is the overflow. Constant and variable overflow are given separate consideration.

Number of Ideal Stages for Constant Underflow. When the operating line is straight, a McCabe-Thiele construction may be used to determine the number of ideal stages, but since in leaching the equilibrium line is always straight, Eq. (10-37) can be used directly for constant underflow. The use of this equation is especially simple here because $y_a^* = x_a$ and $y_b^* = x_b$.

Equation (10-37) cannot be used for the entire cascade if L_a, the solution entering with the unextracted solids, differs from L, the underflows within the system. Equations have been derived for this situation,[2, 9] but it is easy to calculate, by material balances, the performance of the first stage separately and then to apply Eq. (10-37) to the remaining stages.

Example 13-1.[2] By extraction with kerosene, 2 tons of waxed paper per day is to be dewaxed in a continuous countercurrent extraction plant which contains a number of ideal stages. The waxed paper contains, by weight, 25 per cent paraffin wax and 75 per cent paper pulp. The extracted pulp is put through a dryer to evaporate the kerosene. The pulp, which retains the unextracted wax after evaporation, must not contain over 0.2 lb of wax per 100 lb of wax-free pulp. The kerosene used for the extraction contains 0.05 lb of wax per 100 lb of wax-free pulp. Experiments show that the pulp retains 2.0 lb of kerosene per pound of kerosene-and wax-free pulp as it is transferred from cell to cell. The extract from the battery is to contain 5 lb of wax per 100 lb of wax-free kerosene. How many stages are required?

Solution. Any convenient units may be used in Eq. (10-37) as long as the units are consistent and as long as the overflows and underflows are constant. Thus, mole fractions, mass fractions, or mass of solute per mass of solvent are all permissible choices for concentration. The choice should be made that gives constant underflow. In this problem, since it is the ratio of kerosene to pulp that is constant, flow rates should be expressed in pounds of kerosene. Then, all concentrations must be in pounds of wax per pound of wax-free kerosene. The unextracted paper has no kerosene, so the first cell must be treated separately. Equation (10-37) can then be used for calculating the number of remaining units.

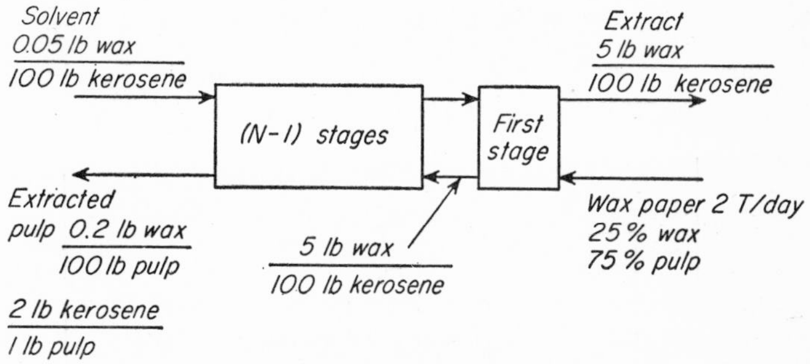

Fig. 13-4. Material-balance diagram for Example 13-1.

The flow quantities and concentrations for this plant are shown in Fig. 13-4. The kerosene in with the fresh solvent is found by an over-all wax balance. Take a basis of 100 lb of wax- and kerosene-free pulp, and let s be the pounds of kerosene fed in with the fresh solvent. The wax balance is, in pounds:

Wax in with pulp, $100 \times \frac{25}{75} = 33.33$
Wax in with solvent, $0.0005s$
  Total wax input, $33.33 + 0.0005s$
Wax out with pulp, $100 \times 0.002 = 0.200$
Wax out with solution, $(s - 200)0.05 = 0.05s - 10$
  Total wax output, $0.05s - 9.80$

Therefore

$$33.33 + 0.0005s = 0.05s - 9.80$$

From this $s = 871$ lb. The kerosene in the exhausted pulp is 200 lb, and that in the strong solution is $871 - 200 = 671$ lb. The wax in this solution is $671 \times 0.05 =$

33.55 lb. The concentration in the underflow to the second unit equals that of the overflow from the first stage, or 0.05 lb of wax per pound of kerosene. The wax in the underflow to unit 2 is $200 \times 0.05 = 10$ lb. The wax in the overflow from the second cell to the first is, by a wax balance over unit 1,

$$10 + 33.55 - 33.33 = 10.22 \text{ lb}$$

The concentration of this stream is, therefore, $10.22/871 = 0.0117$. The quantities for substitution in Eq. (10-37) are

$$x_a = y_a^* = 0.05 \qquad y_a = 0.0117$$

$$x_b = y_b^* = \frac{0.2}{200} = 0.001 \qquad y_b = 0.0005$$

Equation (10-37) gives, since stage 1 has already been taken into account,

$$N - 1 = \frac{\log \dfrac{0.0005 - 0.001}{0.0117 - 0.05}}{\log \dfrac{0.0005 - 0.0117}{0.001 - 0.050}} = \frac{\log \dfrac{0.05 - 0.0117}{0.001 - 0.0005}}{\log \dfrac{0.050 - 0.001}{0.0117 - 0.0005}} = 3$$

The total number of ideal stages is $N = 1 + 3 = 4$.

Number of Ideal Stages for Variable Underflow. When the underflow and overflow vary from stage to stage, a modification of the Ponchon-Savarit graphical method may be used for leaching calculations. The modification

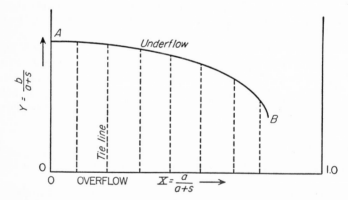

Fig. 13-5. XY diagram for leaching: Line AB, underflow line. Dotted lines, vertical tie lines. a, solute. b, solid. s, solvent.

consists (1) in considering each stream to be a mixture of solid and solution and (2) in using the ratio of solid to solution in place of specific enthalpy. The solution, in turn, is a mixture of solute and solvent. Let a represent the solute, b the solid, and s the solvent. These letters may refer to either concentration or to amounts. Then the abscissa of the Ponchon-Savarit diagram becomes $a/(a + s)$. Denote this by X. The ordinate becomes $b/(a + s)$. Denote this by Y. The XY diagram for a typical system of solid and solution is shown in Fig. 13-5. The curved line corresponds to the saturated-liquid line in the ix chart. It is a plot of the ratio of the solid to the retained solution as a function of the concentration of the solution. It is the curvature of this line that makes Eq. (10-37) inapplicable. Unlike the corresponding curve on the ix diagram, this line is not a property of the solution itself but depends on the conditions of drainage in the leaching plant. Data for plotting this line must be determined experimentally under the conditions of the extraction.

The XY diagram for leaching is simpler than the ix diagram for distillation in two ways: (1) Since the solid is assumed to be insoluble in the solvent, all points representing overflow streams lie on the axis of abscissas, where $Y = 0$, and there is no line corresponding to the enthalpy of saturated vapor. (2) Since at equilibrium the solution concentrations for both underflow and overflow for the same stage are equal, all tie lines are vertical. This is shown in Fig. 13-5.

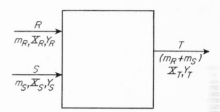

FIG. 13-6. Mixing process, $R + S = T$.

The basic construction for the adiabatic mixing process on the ix diagram carries over without change to the XY diagram: the point for a mixture of two streams lies on the straight line connecting the points for the streams that are mixed, and the center-of-gravity principle applies to the line segments between the points. This may be shown as follows.

Figure 13-6 is a flow diagram showing the mixing of streams R and S to form stream T. Let the masses of the solutions in streams R, S, and T be m_R, m_S, and $m_R + m_S$. Let the corresponding solution concentrations be X_R, X_S, and X_T and the corresponding values of the mass of solid per unit mass of solution be Y_R, Y_S, and Y_T. Two independent material balances may be written.

Solute balance: $$m_R X_R + m_S X_S = (m_R + m_S)X_T \qquad (13\text{-}4)$$

Solid balance: $$m_R Y_R + m_S Y_S = (m_R + m_S)Y_T \qquad (13\text{-}5)$$

Comparison of these equations with Eqs. (12-45) and (12-46) shows that the only difference is the use of X in Eq. (13-4) in place of x in Eq. (12-46) and Y in Eq. (13-5) instead of i in Eq. (12-45). All conclusions and constructions deduced from Eq. (12-49) for the ix diagram are therefore applicable to the XY diagram. It is important to note that the line segments on the XY diagram represent not the masses of the total streams but only the masses of the solution in the streams. With this understanding, the center-of-gravity principle applies as before.

Over-all Balances and Point J. The over-all balance for a leaching plant shows four streams, the inlet solvent V_b, the outlet concentrated liquid V_a, the inlet feed L_a, and the outlet exhausted solid L_b. The simple mixing construction shown in Fig. 13-6, which applies only to three streams, must be modified. This may be done, for calculation purposes, by assuming the over-all process to be replaced by two fictitious processes in series, in the first of which streams V_b and L_a are mixed to form stream J, and in the second of which stream J is divided into streams V_a and L_b. These

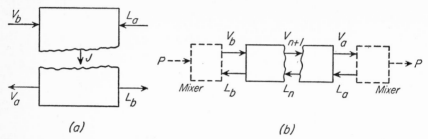

FIG. 13-7. Definition of points J and P: (a) Processes $V_b + L_a = J$ and $J = V_a + L_b$. (b) Processes $P = V_a - L_a$ and $P = V_b - L_b$.

processes are shown in Fig. 13-7a and are represented by the equations

$$V_b + L_a = J = V_a + L_b \tag{13-6}$$

Five points, labeled V_b, L_a, J, V_a, L_b, each representing a corresponding stream, are plotted in Fig. 13-8. A straight line through points V_a and L_b intersects a straight line through points V_b and L_a at point J. By the center-of-gravity principle, the mass of solution in stream L_a is inversely proportional to the line segment $\overline{JL_a}$; the solution in stream V_a is inversely proportional to segment $\overline{JV_a}$; and so on. If, for example, the mass ratio of the solution in the incoming solvent to that in the feed is known, this ratio equals $\overline{JL_a}/\overline{JV_b}$, and point J can be plotted. Then if the analysis of one other stream, e.g., V_a, is known, the line L_bJV_a can be plotted. Since point L_b is on the underflow line, it is also the intersection of the underflow line with line L_bJV_a, and L_b is thereby determined.

The Operating Line and Point P. Over-all balances can also be treated by assuming a fictitious mixer added to either end of the cascade, as shown by the dotted rectangles in Fig. 13-7b. Imaginary stream P is then assumed to be formed by subtracting L_a from V_a or by subtracting L_b from V_b. These processes are represented by the equations

$$V_a = L_a + P \qquad L_b + P = V_b$$

or
$$P = V_a - L_a = V_b - L_b \tag{13-7}$$

These equations are represented on the XY diagram in Fig. 13-8. Since stream P is a mathematical quantity that does not exist physically, either or both of its coordinates may be negative.

A material balance over the first n stages and the fictitious mixer ahead of stage 1 is

$$V_{n+1} = L_n + P \tag{13-8}$$

The operating line represented by this equation is a straight line passing through points P, V_{n+1}, and L_n. All other individual operating lines pass through point P, and these operating lines, used alternately with the verti-

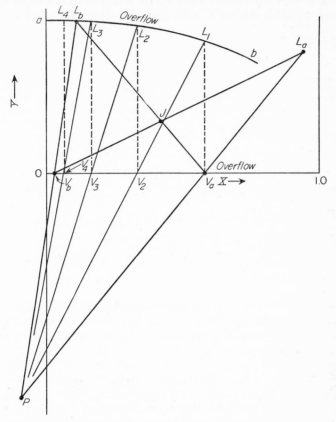

FIG. 13-8. Leaching on Ponchon-Savarit diagram.

cal equilibrium lines, give the Ponchon-Savarit construction shown in Fig. 13-8.

The fictitious quantity P is the net flow of all material through the cascade, considered as a single stream, from stage N to stage 1. From Eqs. (13-7) and (13-8),

$$P = V_{n+1} - L_n = V_a - L_b$$

and P is the same for all stages. It is constant throughout the cascade. If $V_a < L_a$, the net flow is negative, and if $V_a > L_a$, P is positive.

Example 13-2. Oil is to be extracted from meal by means of benzene, using a continuous countercurrent extractor. The unit is to treat 2,000 lb of meal (based on completely exhausted solid) per hour. The untreated meal contains 800 lb of oil and 50 lb of benzene. The fresh solvent mixture contains 20 lb of oil and 1,310 lb of benzene. The exhausted solids are to contain 120 lb of unextracted oil. Experiments carried out under conditions identical with those of the projected battery show that the solution retained depends on the concentration of the solution, as shown in Table 13-1.

TABLE 13-1. DATA FOR EXAMPLE 13-2

Concentration, lb oil/lb solution	Solution retained, lb/lb solid
0.0	0.500
0.1	0.505
0.2	0.515
0.3	0.530
0.4	0.550
0.5	0.571
0.6	0.595
0.7	0.620

Find: (a) the concentration of the strong solution, or extract, (b) the concentration of the solution adhering to the extracted solids, (c) the mass of solution leaving with the extracted meal, (d) the mass of extract, (e) the number of stages required.

Solution. The data in Table 13-1 provide the coordinates of the Y vs. X line for all underflows. Values of X are given in the first column of the table, and corresponding values of Y are the reciprocals of the numbers in the second column. The YX line for the underflow is plotted as curve ab in Fig. 13-9. From the conditions of the problem, the coordinates of points L_a and V_b are

$$\text{Point } L_a: \quad X = \frac{800}{800 + 50} = \frac{800}{850} = 0.941 \qquad Y = \frac{2,000}{850} = 2.35$$

$$\text{Point } V_b: \quad X = \frac{20}{1,310 + 20} = \frac{20}{1,330} = 0.015 \qquad Y = 0$$

Points L_a and V_b are plotted on Fig. 13-9, and line $L_a V_b$ drawn. Point J lies on this line, a distance $850/(1,330 + 850) = 0.39$ times the distance between points V_b and L_a, measured from V_b. Point J is plotted. The ratio of Y to X for point L_b is the ratio of the solid to the solute in this stream, or $2,000/120 = 16.67$. Point L_b lies on line ab. A straight line through the origin with a slope of 16.67 intersects line ab at point L_b, and another straight line through points L_b and J intersects the $X = 0$ line at point V_a. The X coordinate of point L_b is 0.12, and that of point V_a is 0.592.

The total solution input is $1,330 + 850 = 2,180$ lb, and this equals the total solution leaving in streams V_a and L_b. This flow is divided between the two streams in proportion to the line segments on line $V_a J L_b$. The X coordinate of point J is 0.372, and, by the center-of-gravity principle,

$$L_b = \frac{0.592 - 0.372}{0.592 - 0.120} \, 2,180 = 1,020 \text{ lb}$$

$$V_a = 2,180 - 1,020 = 1,160 \text{ lb}$$

The answers to parts (a) to (d) are: part (a), 0.592; part (b), 0.12; part (c) 1,020 lb; part (d) 1,160 lb.

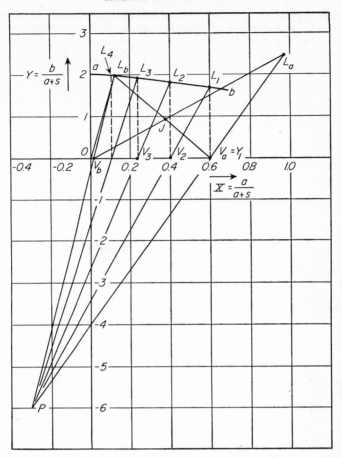

Fig. 13-9. Solution of Example 13-2.

To determine the number of ideal stages, point P is established as the intersection between straight lines through points L_a and V_a and L_b and V_b, as shown in Fig. 13-9. The Ponchon-Savarit construction is shown in the figure, and from this the number of ideal stages is found to be slightly less than four. The answer to part (e), then, is four stages.

Saturated Concentrated Solution. A special case of leaching is encountered when the solute is of limited solubility and the concentrated solution reaches saturation. This situation can be treated by the above methods.[5] The solvent input to stage N should be the maximum that is consistent with a saturated overflow from stage 1, and all liquids except that adhering to the underflow from stage 1 should be unsaturated. If saturation is attained in stages other than the first, all but one of the "saturated" stages are unnecessary, and the solute concentration in the underflow from stage N is higher than it needs to be.

If the carrier solid is partially soluble in the solvent, the Ponchon-Savarit method can be extended to take that fact into account. A second line above the X axis is drawn, which is the Y vs. X relationship for the overflow, including its content of dissolved solid. Otherwise, the construction is unchanged.

Stage Efficiencies. The stage efficiency in a leaching process depends on the time of contact between the solid and the solution and on the rate of diffusion of the solute through the solid and into the liquid. Consider a particle of solid containing soluble extractable material and in contact with an unsaturated solution. At a given instant θ hr from the time of initial contact with the liquid, assume that the concentration of the solute in the bulk of the liquid is X_1 and that the average concentration in the solution retained in the solid is $\overline{X}$. The concentration at the interface may lie between X_1 and $\overline{X}$ if both the resistance to diffusion in the solid and that from the interface into the liquid are of comparable magnitude. A general solution, then, requires that both resistances be calculable. This is not possible with present knowledge, but certain special cases may be considered.

First, consider a solid entirely impervious and inert to the action of the solvent with a film of strong solution on its surface. For such a case the process would involve simply the equilization of concentrations in the bulk of the extract and in the adhering film. Such a process is rapid, and any reasonable time of contact will bring about equilibrium. The counter-current leaching process described on page 585 is of this type, and the stage efficiency is taken as unity in calculations for such processes.

A second case is where the soluble material is uniformly distributed throughout the interior of a permeable solid, e.g., black ash or *caliche*. Then the rate of diffusion of the solute to the interface may be the rate-controlling factor. Agitation is of little help, but the process is hastened if the solid is finely ground.

A third important situation is that where the solid is impermeable and contains the solute distributed through it in fine particles. Then the surface of the particles of solute may be an extremely small proportion of the total surface of the solid, and the rate of extraction is small. Here the best method is to grind the solid to such a size that the individual particles of solute are released.

Still another case is encountered in the leaching of natural materials such as sugar beets, tanbark, or soybeans. The solute is contained in vegetable cells, and the process requires the diffusion of solute by osmosis through the cell walls. Subdividing the material is ineffective because of the small size of the individual cells, and fairly large particles and long contact times are used. The resistance to diffusion from the interface into the bulk of the solution is small in comparison with that of diffusion in the solid itself, and the concentration at the interface can be assumed equal to that of the bulk of the liquid.

Under certain idealized conditions, the stage efficiency in extracting some (but not all) cellular materials can be estimated from experimental diffusion data obtained under the conditions of temperature and agitation to be used in the plant. The assumptions are as follows: [7, 10, 16]

1. The diffusion rate is represented by the equation

$$\frac{\partial X}{\partial \theta} = D'_v \frac{\partial^2 X}{\partial^2 z}$$ (13-9)

where D'_v is the diffusivity,† in length squared per time, and z is distance, measured in the direction of diffusion, through which diffusion occurs.

2. The diffusivity is constant.

3. The solid can be considered to be equivalent to thin slabs of constant density, size, and shape.

4. The concentration X_1 of the liquid in contact with the solid is constant.

5. The initial concentration in the solid is uniform throughout the solid.

Under these assumptions, Eq. (13-9) can be integrated in exactly the same manner as Eq. (8-118), for conduction of heat through a slab. The result may be written

$$\frac{\overline{X} - X_1}{X_0 - X_1} = \frac{8}{\pi^2} \left(e^{-a_1\beta} + \frac{1}{9} e^{-9a_1\beta} + \frac{1}{25} e^{-25a_1\beta} \cdots \right)$$

$$= \phi(\beta)$$ (13-10)

where $\overline{X}$ = average concentration of solute in solid at time θ

X_0 = constant concentration of solute in solid at zero time

X_1 = concentration of solute in solid at equilibrium and in bulk of solution at all times

and

$$\beta = \frac{D'_v \theta}{r_p^2} \qquad a_1 = \left(\frac{\pi}{2}\right)^2$$ (13-11)

where $2r_p$ is the thickness of the particle.

Note that Eq. (13-10) is Eq. (8-119) with t replaced by X, and N_{Fo} by β. Figure 8-34, therefore, also represents Eq. (13-10).

The Murphree stage efficiency for leaching is given by the equations

$$\eta_M = \frac{X_0 - \overline{X}}{X_0 - X_1} = 1 - \frac{\overline{X} - X_1}{X_0 - X_1} = 1 - \phi(\beta) = 1 - \phi\left(\frac{D'_v \theta}{r_p^2}\right)$$ (13-12)

Equation (13-12) follows from the fact that $X_0 - \overline{X}$ is the actual concentration change for the stage and $X_0 - X_1$ is the concentration change that would be obtained if equilibrium were reached.

† The diffusivity D'_v differs from the volumetric diffusivity D_v of Chap. 10: (1) it is based on a gradient expressed in terms of pounds of solute per pound of solute-free solvent rather than mole fraction of solute; (2) the transfer is in pounds rather than pound moles; and (3) the relative-velocity factor ϕ is contained in D'_v. Since the diffusivity D'_v can be found only by experiment on the material to be extracted, it is actually used as an empirical constant, and these differences are unimportant in practice.

Example 13-3. Assuming, for the meal in Example 13-2, $D'_v = 5 \times 10^{-9}$ cm²/ sec, that the thickness of the flakes is 0.04 cm, and that the time of contact in each stage is 3 hr, estimate the actual number of stages required.

Solution. Since $2r_p = 0.04$, $r_p = 0.02$ cm. Also, $\theta = 3 \times 3,600 = 1.08 \times 10^4$ sec. Then

$$\beta = \frac{D'_v \theta}{r_p^2} = \frac{5 \times 10^{-9} \times 1.08 \times 10^4}{0.02^2} = 0.135$$

From Fig. 8-34, $\phi(\beta) = 0.59$, and the Murphree efficiency is

$$\eta_M = 1 - 0.59 = 0.41$$

The average efficiency is nearly equal to the Murphree efficiency, as m is unity and V/L nearly so. The actual number of stages is $4/0.41 = 9+$. Either 9 or 10 stages should be used.

LIQUID EXTRACTION

Liquid extraction is used when distillation and rectification are difficult or ineffective. Close-boiling mixtures or substances that cannot withstand the temperature of vaporization, even under a vacuum, may often be separated by extraction. Extraction utilizes differences in the solubilities of the components rather than differences in their volatilities. Since solubility depends on chemical properties, extraction exploits chemical differences instead of vapor-pressure differences.

When either distillation or extraction may be used, the choice is usually distillation, in spite of the fact that heat and cooling are needed. In extraction the solvent must be recovered for reuse, which also requires a distillation step, and the combined operation is more complicated and often more expensive than a corresponding distillation. Also, in extraction, a third component, the solvent, must be processed and its losses replaced. Extraction is a relatively new operation and as experience with it accumulates, it is being used more and more for separations where distillation was formerly chosen. In many problems, the choice between these methods should be based on a comparative study of both.

Extraction may be used to separate more than two components; and mixtures of solvents, instead of a single solvent, are needed in some applications. These more complicated methods are not treated in this text.

Extraction Equipment

In liquid-liquid extraction, as in gas absorption and distillation, two phases must be brought into good contact to permit transfer of material and then be separated. In absorption and distillation the mixing and separation are easy and rapid. In extraction, however, the two phases have comparable densities, so that the energy available for mixing and separation—if gravity flow is used—is small, much smaller than when one phase is a liquid and the other is a gas. The two phases are often hard to mix and harder to separate. The viscosities of both phases, also, are relatively high, and linear velocities through most extraction equipment are low. In some types of extractors, therefore, energy for mixing and separation is supplied mechanically.

Extraction equipment may be operated batchwise or continuously. A quantity of feed liquid may be mixed with a quantity of solvent in an agitated vessel, after which the layers are settled and separated into extract and raffinate. This gives about one theoretical contact, which is adequate in simple extractions. The operation may of course be repeated if more than one contact is required, but when the quantities involved are large and several contacts are needed, continuous flow becomes economical. Most extraction equipment is continuous, with either successive stage contacts or differential contacts. Representative types are mixer-settlers, vertical towers of various kinds which operate by gravity flow, agitated tower extractors, and centrifugal extractors. The characteristics of various types of extraction equipment are listed in Table 13-2.

TABLE 13-2. PERFORMANCE OF COMMERCIAL EXTRACTION EQUIPMENT †

Type	Liquid capacity of combined streams, ft^3/ft^2-hr	HTU, ft	Plate or stage efficiency, %	Spacing between plates or stages, in.	Typical applications
Mixer-settler....	. . . . . .	. . . . .	75–100	. . . . .	Duo-Sol lube-oil process
Spray column...	50–250	10–20	. . . . . .	. . . . .	Ammonia extraction of salt from caustic soda
Packed column..	20–150	5–20	. . . . . .	. . . . .	Phenol recovery
Perforated-plate column	10–200	1–20	6–24	30–70	Furfural lube-oil process
Baffle column...	60–105	4–6	5–10	4–6	Acetic acid recovery
York-Schiebel column	50–100	1–2	80–100	12–24	Pharmaceuticals and organic chemicals
Luwesta........	3,800 gal/hr	. . . . .	Approx. 3 stages/ unit		Penicillin extraction
Podbelniak......	To 20,000 gal/hr	. . . . .	Low: 3–5 stages/ unit		Antibiotics
	1,000–5,000 gal/hr	. . . . .	High: 8–20 stages/unit		

† From R. L. Von Berg and H. F. Wiegandt, *Chem. Eng.*, **59**(6): 189 (1952).

Mixer-Settlers. For batchwise extraction the mixer and settler may be the same unit. A tank containing a turbine or propeller agitator is most common. At the end of the mixing cycle the agitator is shut off, the layers allowed to separate by gravity, and extract and raffinate drawn off to separate receivers through a bottom drain line carrying a sight glass. The mixing and settling times required for a given extraction can be determined only by experiment; 5 min for mixing and 20 min for settling are typical, but both shorter and much longer times are common.

For continuous flow the mixer and settler must be separate pieces of equipment. The mixer may be a small agitated tank provided with inlets and a drawoff line and baffles to prevent short-circuiting; or it may be a centrifugal pump or other flow mixer. The settler is often a simple continuous decanter of the type described in Chap. 7, page 377. With liquids

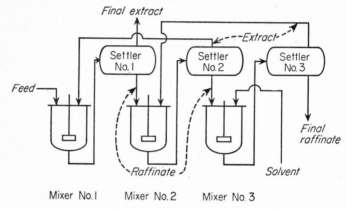

Fig. 13-10. Mixer-settler extraction system.

which emulsify easily and which have nearly the same density it may be necessary to pass the mixer discharge through a screen or pad of glass fiber to coalesce the droplets of the dispersed phase before gravity settling is feasible. For even more difficult separations tubular or disk-type centrifuges are employed.

If, as is usual, several contact stages are required, a train of mixer-settlers is operated with countercurrent flow, as shown in Fig. 13-10. The raffinate from each settler becomes the feed to the next mixer, where it meets intermediate extract or fresh solvent. The principle is identical with that of continuous countercurrent stage leaching system in Fig. 10-4.

Spray and Packed Extraction Towers. These tower extractors give differential contacts, not stage contacts, and mixing and settling proceed simultaneously and continuously. In the spray tower shown in Fig. 10-5, for example, drops of one liquid are dispersed into, and rise through, a slowly falling continuous phase of the other liquid. In some extractions the dispersed phase may be the heavy liquid. In either case each drop is constantly being "mixed," i.e., brought into fresh contact with the other phase, and is constantly being separated from it. There is continuous transfer of material between phases, and the composition of each phase changes as it flows through the tower. At any given level, of course, equilibrium is not reached; indeed, it is the departure from equilibrium that provides the driving force for material transfer. By using a tall tower, however, a number of ideal contacts can theoretically be obtained.

In actual spray towers the contact between the drops and continuous phase is not highly effective except where the drops are initially dispersed. As much as 40 to 45 per cent of the total transfer may occur at the point of dispersion.[20] Thus while spray towers are simple to build and easy to operate, they are not highly effective. Adding more height does not always help much; it is much more effective to redisperse the drops at frequent intervals throughout the tower. This may be done by filling the tower with packing, such as rings, saddles, crushed coke, or wood grids. The packing causes the drops to coalesce and re-form and, as shown in Table 13-2, appreciably increases the effectiveness of the tower. Packed towers approach spray towers in simplicity, and can be made to handle almost any problem of corrosion or pressure at a reasonable cost. Their chief disadvantage is that solids tend to collect in the packing and cause channeling.

Perforated-plate Towers. Redispersion of liquid drops is also done by transverse perforated plates like those in the sieve-plate distillation tower described in Chap. 12. The perforations in an extraction tower are $\frac{1}{16}$ to $\frac{3}{8}$ in. in diameter. Plate spacings are 6 to 24 in. Usually the light liquid is

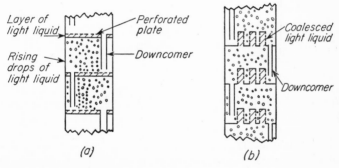

FIG. 13-11. Perforated-plate extraction towers: (a) Perforations in horizontal plates. (b) Perforations in vertical sections.

the dispersed phase, and downcomers carry the heavy continuous phase from one plate to the next. As shown in Fig. 13-11a, light liquid collects in a thin layer beneath each plate and jets into the thick layer of heavy liquid above. In many perforated-plate towers the plates are made as shown in Fig. 13-11b, with the perforations in vertical sections of the tray.[15] In this design the light liquid jets horizontally into the continuous phase.

Bubble-cap towers may be used for extraction, though with low plate efficiencies of 5 to 10 per cent. They offer no advantages over perforated-plate towers, however, and are rarely if ever used industrially.

Baffle Towers. These extraction towers contain sets of horizontal baffle plates, as shown in Fig. 13-12. Heavy liquid flows over the top of each baffle and cascades to the one beneath; light liquid flows under each baffle and sprays upward from the edge through the heavy phase. The most common arrangements of baffles are the "disk-and-doughnut" baffles of Fig. 13-12a and the segmental, or "side-to-side," baffles of Fig. 13-12b. In both types the spacing between baffles is 4 to 6 in.

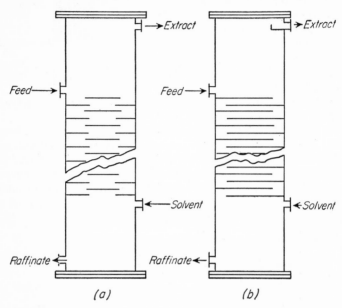

(a)   (b)

FIG. 13-12. Baffle extraction towers: (a) Disk-and-doughnut baffles. (b) Side-to-side baffles.

Baffle towers contain no small holes to clog or be enlarged by corrosion. They can handle dirty solutions containing suspended solids; one modification of the disk-and-doughnut towers even contains scrapers to remove deposited solids from the baffles. Because the flow of liquid is smooth and even, with no sharp changes in velocity or direction, baffle towers are valuable for liquids that emulsify easily. For the same reason, however, they are not effective mixers, and each baffle is equivalent to only 0.05 to 0.1 ideal stage.[24]

Agitated Tower Extractors: York-Scheibel Extractor. Mixer-settlers supply mechanical energy for mixing the two liquid phases, but the tower extractors so far described do not. They depend on gravity flow both for mixing and for separation. In some tower extractors, however, mechanical energy is provided by internal turbines or other agitators, mounted on a central rotating shaft. Sets of agitators are often separated by partitions or calming sections to give, in effect, a stack of mixer-settlers one above the other. The York-Scheibel tower illustrated in Fig. 13-13 is an example. Here the calming sections between the agitators are packed with wire mesh to encourage coalescence and separation of the phases. Most of the extraction takes place in the mixing sections, but some also occurs in the calming sections, so that the efficiency of each mixer-settler unit is sometimes greater than 100 per cent. Typically each mixer-settler is 1 to 2 ft high, which means that several theoretical contacts can be provided in a reasonably short col-

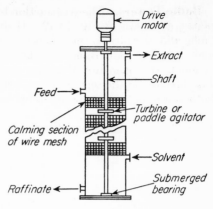

FIG. 13-13. York-Scheibel extraction tower.

umn. The problem of maintaining the internal moving parts, however, particularly where the liquids are corrosive, may be a serious disadvantage.

Pulse Columns. Agitation may also be provided by external means as in the pulse column shown in Fig. 13-14. A bellows or other reciprocating pump "pulses" the entire contents of the column at frequent intervals, so that a reciprocating motion is superimposed on the usual flow of the

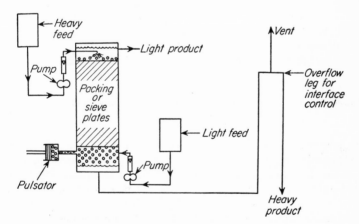

FIG. 13-14. Schematic layout of pulse extraction tower.

liquid phases. The tower may contain ordinary packing or special sieve plates. In a packed tower the pulsation thoroughly disperses the liquids and eliminates channeling, and the contact between the phases is greatly improved. In sieve-plate pulse towers the holes are smaller than in non-pulsing towers; they range from $\frac{1}{16}$ to $\frac{1}{8}$ in. in diameter, with a total open area in each plate 6 to 23 per cent of the cross-sectional area of the tower. No downcomers are used. Ideally the pulsation causes light liquid to be dispersed into the heavy phase on the upward stroke and the heavy phase to jet into the light phase on the downward stroke. Under these conditions the stage efficiency may reach 70 per cent. This is possible, however, only when the volumes of the two phases are nearly the same and where there is almost no volume change during extraction. In the more usual case the successive dispersions are less effective, and there is back mixing of one phase in one direction. The plate efficiency then drops to about 30 per cent. Nevertheless, in both packed and sieve-plate pulse columns the height required for a given number of transfer units or theoretical contacts is often less than one-third that required in an unpulsed column.[19, 26]

Centrifugal Extractors. The dispersion and separation of the phases may be greatly accelerated by centrifugal force, and several commercial extractors make use of this. The Luwesta extractor contains three mixer-settlers, one above the other, in separate compartments in a tall centrifuge bowl. As the bowl turns at high speed, the liquids are sprayed outward into each compartment from a hollow central shaft; there they separate and flow to the next spray distributor or to the outlet. Solids thrown down by the centrifugal force collect inside the bowl, from which they are periodically removed.[24]

In the Podbelniak extractor a perforated spiral ribbon inside a heavy metal casing is wound about a hollow horizontal shaft through which the liquids enter and leave. Light liquid is pumped to the outside of the spiral at a pressure between 50 and 200 lb force/in.^{2} to overcome the centrifugal force; heavy liquid is pumped to the center. The liquids flow countercurrently through the passage formed by the ribbon and the casing walls. Heavy liquid moves outward along the outer face of the spiral; light liquid is forced by displacement to flow inward along the inner face. The high shear at the liquid-liquid interface results in rapid mass transfer. In addition, some liquid sprays through the perforations in the ribbon and increases the turbulence. Up to 20 theoretical contacts may be obtained in a single machine, although 3 to 10 contacts are more common. Operating characteristics of these machines are listed in Table 13-2. Centrifugal extractors are complex, expensive devices and find relatively limited use. They have the advantages of providing many theoretical contacts in a small space and of very short holdup times—about 4 sec. Thus they are valuable in the extraction of sensitive products such as vitamins and antibiotics.

Auxiliary Equipment. The dispersed phase in an extraction tower is allowed to coalesce at some point into a continuous layer from which one product stream is withdrawn. The interface between this layer and the predominant continuous phase is set in an open section at the top or bottom of a packed tower; in a sieve-plate tower it is commonly near the middle. The interface level is most often automatically controlled by a vented overflow leg for the heavy phase, as in the continuous gravity decanter shown in Fig. 7-39. In large columns the interface is sometimes held at the desired point by a level controller actuating a valve in the heavy-liquid discharge line.

In liquid-liquid extraction the solvent which is added to separate one component from another nearly always must be removed from the extract or raffinate or both. Thus auxiliary stills, evaporators, heaters, and condensers form an essential part of most extraction systems and often cost much more than the extraction device itself. As mentioned at the beginning of this section, if a given separation can be done either by extraction or distillation, economic considerations usually favor distillation. Extraction provides a solution to problems that cannot be solved by distillation alone but does not usually eliminate the need for distillation or evaporation in some part of the separation system.

Principles of Extraction

Since most continuous extraction methods use countercurrent contacts between two phases, one a light liquid and the other a heavier one, many of the fundamentals of countercurrent gas absorption and of rectification carry over into the study of liquid extraction. Thus questions about ideal stages, stage efficiency, minimum ratio between the two streams, and size of equipment have the same importance in extraction as in gas absorption or rectification. The equilibrium relationships in liquid extraction are, however, sometimes more complicated than in the other two operations.

Triangular Coordinates. Although either the raffinate or the extract may be the heavy phase and the other the light, for convenience in the following discussion it is assumed that the solvent phase is the light liquid, so its magnitude is denoted by V and its concentration by y. Then the amount of the raffinate phase is denoted by L and its concentration by x. Also, identify the three components as follows: the solute, which is the substance extracted from the raffinate, is component a; the diluent, which is primarily retained in the raffinate, is component b; and the pure solvent is component s.

Equilibrium data show how the component a is distributed between the raffinate and extract phases at equilibrium. Also, since frequently the diluent is somewhat soluble in the solvent and the solvent in the diluent, all three components may be found in both raffinate and extract streams. It is convenient to consider, first, the situation where such mutual solubility between components b and s is negligible and, second, that where this solubility cannot be ignored.

When the extract stream is practically free of diluent and the raffinate of solvent, the equilibrium becomes a simple relation between the concentrations of solute in the two phases. This situation is identical with that of gas absorption, and the equilibrium curve is the usual x_e vs. y_e plot of the kind used in Chap. 11. In dilute solutions, Henry's law is followed. In some systems the equilibrium line is straight over a wide range of concentrations. As in the solubility of gases in liquids, the equilibrium is sensitive to temperature; but, in contrast to gas-liquid equilibria, liquid-liquid equilibria are nearly independent of pressure.

When the mutual solubility of diluent and solvent cannot be neglected, the solubility and equilibrium relationships are often shown on triangular coordinates. In this method of plotting, the composition of any three-component mixture can be shown by a point lying inside an equilateral triangle, as shown in Fig. 13-15. The triangular diagram has certain important characteristics.

Concentrations represented by triangular diagrams are based on the entire mixture, not on one or two components. Then the concentrations of the three components must add to unity, or

$$x_A + x_B + x_S = 1 \quad \text{and} \quad y_A + y_B + y_S = 1 \quad (13\text{-}13)$$

Either mole fractions or mass fractions may be used. Mass fractions are the more common in these diagrams.

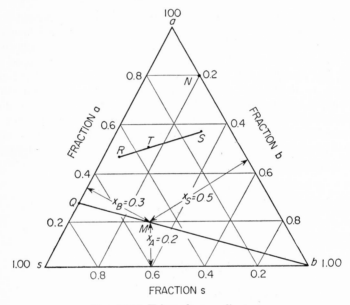

Fig. 13-15. Triangular coordinates.

The fact that the sum of the concentrations is unity is correlated with the diagram by the geometric principle that the sum of the perpendicular distances from any point to the three sides of an equilateral triangle equals the altitude of the triangle. Thus, if the altitude is taken as unity, the perpendicular distances from any point to the sides automatically add to unity, and these distances can be used to represent the individual concentrations of the three components. Any apex represents a pure component. Let the apices of the triangle of Fig. 13-15 represent the components a, b, and s, as shown. Then point M, for example, is the point for a mixture consisting of 20 per cent a, 30 per cent b, and 50 per cent s.

One side of the triangle represents mixtures of the two components represented by the ends of that side; e.g., point N corresponds to a mixture of 80 per cent a and 20 per cent b.

As in the ix and the XY diagrams already described, a point for a mixture of two streams lies on the straight line connecting the points for the two streams, and the center-of-gravity principle applies. This can be demonstrated by the usual method of correlating material-balance equations with similar triangles on the diagram. Point T in Fig. 13-15 corresponds to a mixture of the two liquids represented by points R and S, combined in the ratio of m_R lb of R to m_S lb of S, where

$$\frac{m_S}{m_R} = \frac{\overline{RT}}{\overline{TS}}$$

As a special case of the mixing rule, when one pure component is added to a given mixture, the locus of all points for the resulting solution is the straight line connecting the point for the original mixture with the apex for the added component. Likewise, if one component is stripped from a mixture, the point for the two-component mixture so formed is found by projecting, to the side of the triangle, the straight line connecting the original point and the apex for the component removed. For example, if the mixture represented by point M is diluted with component b, the points for all mixtures so formed lie on line Mb. If component b is stripped from the original mixture, the b-free solution obtained is represented by point Q.

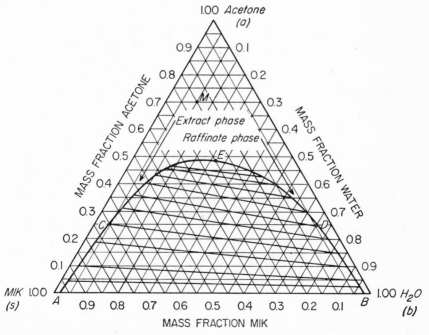

Fig. 13-16. System acetone-MIK-water at 25°C. (*Othmer, White, and Trueger.*[17])

Liquid-Liquid Equilibria with Partial Miscibility of Solvent and Diluent. Figure 13-16 is a triangular diagram showing equilibria in the system acetone, methyl isobutyl ketone (MIK), and water at 25°C. The MIK may be considered the solvent for extracting the solute, acetone, from the diluent, water. This figure is an example of the first of two kinds of system important in extraction.

MIK is soluble in water to a concentration of approximately 2 per cent MIK, and water is soluble in MIK to about 2 per cent water. This is shown by points B and A in Fig. 13-16. Any mixture of MIK and water of composition lying between points A and B forms two liquid layers, represented by these points. The mass ratio of the two layers is found by the center-of-gravity principle.

If acetone is added to a two-layer mixture of MIK and water, the acetone distributes between the layers, and the compositions of the layers follow two solubility curves, one for the water layer, or raffinate phase, and the other for the MIK layer, or extract phase. Line BDE shows compositions of the saturated water layer, and line ACE compositions of the saturated MIK layer. As the over-all acetone content of the mixture increases, the

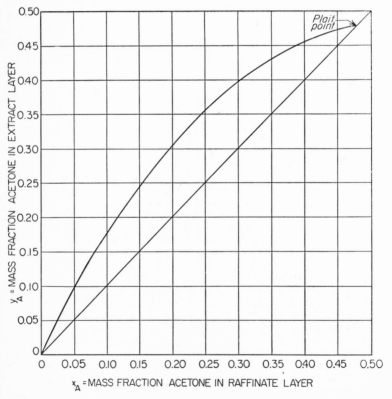

FIG. 13-17. $x_A y_A$ plot, system acetone-MIK-water at 25°C.

concentrations of acetone in both layers increase, and the solubility curves approach each other. The two phases become identical at point E in Fig. 13-16. Such a common point on both curves is called a *plait point*.

The area below the dome-shaped curve $ABEDC$ is a composition field within which are located the points for all mixtures that form two liquid layers at equilibrium. All points outside the dome are for single-phase mixtures. Point M, for example, is that for a homogeneous mixture of 70 per cent acetone, 20 per cent MIK, and 10 per cent water.

Tie Lines and Distribution Curves. Just as in enthalpy-concentration diagrams, equilibria in triangular diagrams are shown by tie lines. These are straight lines connecting points on the two solubility curves. The ends of a tie line are the points for two phases in equilibrium. Several tie lines are shown in Fig. 13-16. Tie lines decrease in length as the plait point is approached and vanish at the plait point.

Each point within the dome lies on a tie line, and an infinite number of such lines exists. Graphical methods for interpolating between known tie lines are available.[21a] To avoid cluttering the diagram with construction lines, a distribution curve showing x_A, the acetone concentration in one phase, plotted against y_A, the acetone concentration in the other, can be used as source of tie-line data. The distribution curve terminates on the $x = y$ diagonal at the concentration of the plait point. This curve is analogous to the xy curve used in distillation and absorption. Such a curve for the system acetone-MIK-water is shown in Fig. 13-17. To

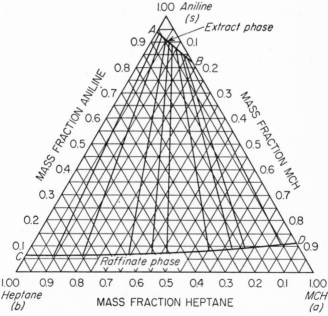

FIG. 13-18. System aniline–n-heptane–MCH at 25°C: a, solute, MCH. b, diluent, n-heptane. s, solvent, aniline. (*Varteressian and Fenske.*[23])

locate a tie line when the acetone concentration of one phase is known, that for the other is found from the distribution curve and used to locate the point on the other solubility curve. To locate the tie line for a given point within the dome trial and error is required.

A second type of equilibrium diagram is shown in Fig. 13-18, which applies to the system aniline, heptane, and methyl cyclohexane (MCH) at 25°C. Aniline is the solvent, MCH the solute, and heptane the diluent.

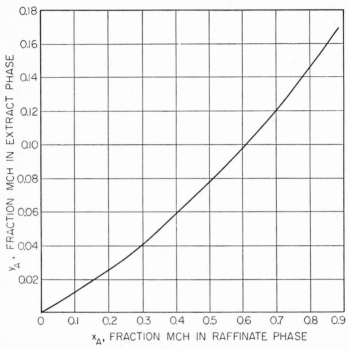

Fig. 13-19. $x_A y_A$ plot, system aniline–n-heptane–MCH at 25°C.

Line AB gives the solubility relationships for the extract phase and line CD that for the raffinate phase. Tie lines, as usual, provide phase-equilibrium data. To interpolate between tie lines it is convenient to plot, as a distribution curve, the concentration of MCH in the extract phase against that in the raffinate, as shown in Fig. 13-19.

Systems like that for acetone-MIK-water are called type I, and those like aniline-heptane-MCH are called type II.

Ponchon-Savarit Method on Triangular Coordinates. Since the graphical mixing construction and the center-of-gravity principle both apply to triangular diagrams, the complete Ponchon-Savarit method carries over unchanged to these coordinates. Typical constructions for both type I and type II systems are shown in Fig. 13-20. Points P and J have the usual significance. In using triangular coordinates it must be remembered that stream magnitudes and concentrations are based on total streams. The constructions shown in Fig. 13-20 apply to the flow diagrams of Figs. 13-3 and 13-7. In the situation shown in Fig. 13-20a, three ideal stages are required, and in that of Fig. 13-20b, five stages are called for.

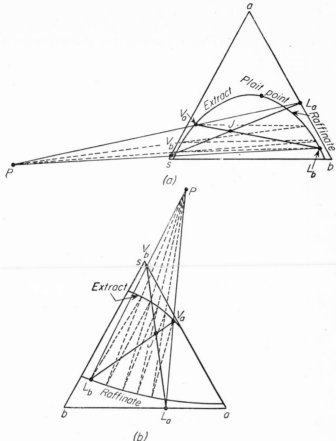

Fig. 13-20. Ponchon-Savarit construction on triangular diagram: (a) type I system; (b) type II system.

Rectangular Coordinates. Equilateral triangular diagrams have some practical disadvantages. They require special plotting paper, their scales cannot be changed, and they cannot be enlarged when it is necessary to follow changes in a narrow concentration range. For these reasons, various other kinds of diagrams, which can be plotted on ordinary plotting paper, and the scales of which can be set individually at will, are used. One excellent method is to use a right-angle triangle in place of the equilateral triangle, by making one of the apices of the latter into a right angle.[11] Concentrations of two of the components are then plotted along the rectangular axes and the third along the hypotenuse. Another method is to use rectangular coordinates and to choose the sum of two of the components as a base mixture for calculations and graphical constructions. This is the method used above in treating leaching. It is also consistent with the ix diagram for rectification, where enthalpy is used as a "component" and the diagram based on the sum of the two real components.

In principle, any pair of components may be chosen as the base mixture. In type I systems, a useful choice is the combination of diluent and solvent, or $s + b$. Then the abscissa, denoted by X, is defined as the ratio of the diluent to the base mixture, or

$$X = \frac{b}{b + s}$$

The ordinate, denoted by Y, is defined as the ratio of the solute to the base mixture, or

$$Y = \frac{a}{b + s}$$

Here a, b, and s refer to either masses or concentrations.

The distribution diagram for finding tie lines is a plot of Y_r for the raffinate vs. Y_e for the extract. The rectangular diagram and distribution curve for the system acetone-water-MIK are given in Fig. 13-21.

In type II systems, the usual choice of base mixture is the combination of solute and diluent, or $a + b$. The abscissa X is the ratio of solute to base mixture, or $X = a/(a + b)$. The ordinate Y is the ratio of the solvent

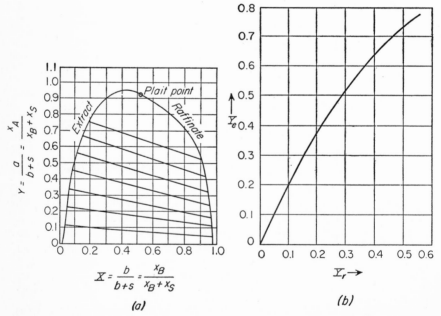

Fig. 13-21. System acetone-MIK-water at 25°C: (a) XY diagram. a, solute, acetone. b, diluent, water. s, solvent, MIK. (b) Y_e vs. Y_r.

to the base mixture, or $Y = s/(a + b)$. The distribution curve is a plot of X_r for the raffinate vs. X_e for the extract. Plots for the system aniline-heptane-MCH are given in Fig. 13-22. An auxiliary line, shown in Fig. 13-22a, may also be used for locating tie lines.

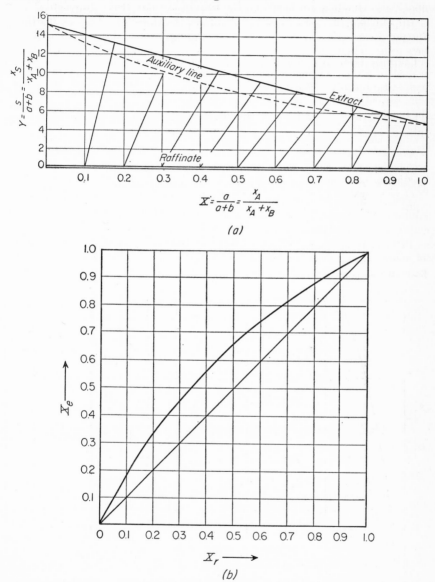

(a)

(b)

FIG. 13-22. System aniline–n-heptane–MCH at 25 °C: (a) XY diagram. a, solute, MCH. b, diluent, n-heptane. s, solvent, aniline. (b) Equilibrium, raffinate and extract, on solvent-free basis, X_e vs. X_r.

Note that in both type I and type II diagrams the two components chosen for the base mixtures in the rectangular plots are the same as those represented by the base in the triangular diagrams.

The usual Ponchon-Savarit constructions can be used on all rectangular diagrams. It must be remembered, however, that all concentrations and stream magnitudes refer to the base mixture in the stream, not to the total stream.

Example 13-4. A countercurrent extraction plant containing three ideal stages is used to extract acetone from its mixture with water by means of MIK at a temperature of 30°C. The feed consists of 40 per cent acetone and 60 per cent water. Pure solvent equal in mass to the feed is used as the extracting liquid. What is the fraction of the inlet acetone remaining unextracted in the final raffinate? What is the analysis of the final extract? What are the analyses of extract and raffinate if all the solvent is removed from them?

Solution. Choose the rectangular coordinates as shown in Fig. 13-21a. Let acetone be component a, water component b, and MIK component s. The combination of water and MIK is the base mixture. Per 100 lb of feed, there are 60 lb of base mixture $(s + b)$ and 40 lb of solute a. The coordinates of point L_a, representing the feed, are

$$X = \tfrac{60}{60} = 1.00 \qquad Y = \tfrac{40}{60} = 0.667$$

The fresh solvent consists entirely of MIK, and the coordinates of point V_b, representing this stream, are $X = 0$ and $Y = 0$. Points L_a and V_b are plotted on Fig. 13-23, which is an enlargement of the lower part of Fig. 13-21a. Since the ratio

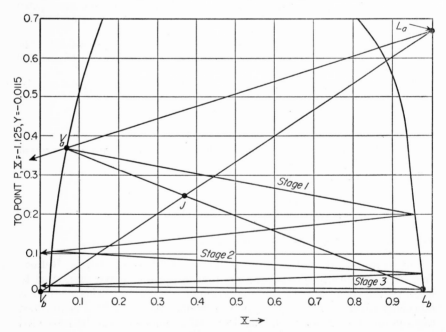

FIG. 13-23. Solution of Example 13-4.

of stream L_a to stream V_b is known from the conditions of the problem, point J can be plotted on the straight line connecting these points. The base mixture in the feed is 60 lb, and that in the entering solvent equals the entire feed, or 100 lb. By the center-of-gravity principle, point J divides the line V_bL_{a}, into segments $\overline{V_bJ}$ and $\overline{JL_a}$ in the ratio

$$\frac{\overline{V_bJ}}{\overline{JL_a}} = \frac{60}{100} = 0.60$$

The analyses of the final extract and raffinate are unknown. A straight line through points L_b and V_a, which represent these streams, must pass through point J. For any such line there is a point P, which is the intersection of lines L_aV_a and L_bV_b. For each P point there is a definite number of ideal stages, found by the Ponchon-Savarit construction. Tie lines are found by the use of Fig. 13-21b. The solution to this problem requires a trial-and-error determination of the line V_aJL_b, pivoting around the point J that leads to three stages. The final result is shown in Fig. 13-23.

The coordinates of point V_a are $X = 0.070$ and $Y = 0.37$. Those of point L_b are $X = 0.975$ and $Y = 0.010$.

The base mixture in each of the exit streams is found by splitting the entire input of base mixture in proportion to the line segments $\overline{V_aJ}$ and $\overline{JL_b}$. The total input of $s + b$ is $100 + 60 = 160$ lb, and the base mixture in the extract is

$$\frac{0.975 - 0.375}{0.975 - 0.070} \, 160 = 106 \text{ lb}$$

The base mixture in the raffinate is $160 - 106 = 54$ lb, and the unextracted acetone is $54 \times 0.010 = 0.54$ lb, so the fraction of unextracted acetone is $0.54/40 = 0.0135$.

In the final extract, $y_A = 0.37/1.37 = 0.270$, and $y_B = 0.070/1.37 = 0.051$. The analysis of this stream is, then, 27.0 per cent acetone, 5.1 per cent water, and 67.9 per cent MIK.

For the raffinate L_b, $X = 0.975$, and $Y = 0.010$. Then, on a solvent-free basis,

$$\frac{Y}{X} = \frac{x_B}{x_A} = \frac{0.010}{0.975} = \frac{x_B}{1 - x_B}$$

From this, $x_B = 0.010$, and $x_A = 0.990$. The solvent-free raffinate is 1.0 per cent acetone and 99.0 per cent water.

For the solvent-free extract

$$x_A = \frac{0.270}{0.270 + 0.051} = 0.84 \quad \text{and} \quad x_B = 0.16$$

The solvent-free extract is 84 per cent acetone and 16 per cent water.

Countercurrent Extraction of Type II Systems Using Reflux. Just as in rectification, reflux can be used in countercurrent extraction to improve the separation of the components in the feed. This method is especially effective in treating type II systems, because with a center-feed cascade and the use of reflux the two feed components can be separated into nearly pure products.

A flow diagram for countercurrent extraction with reflux is shown in Fig. 13-24. To emphasize the analogy between this method and fractionation it is assumed that the cascade is a plate column. Any other kind of

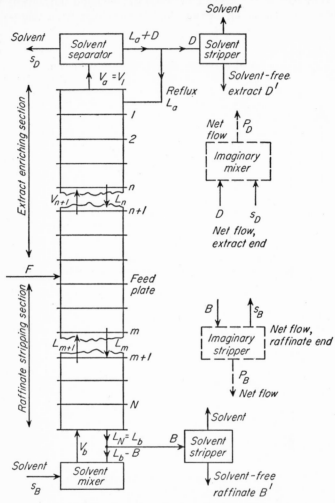

FIG. 13-24. Countercurrent extraction with reflux.

cascade, however, may be used. The method requires that sufficient solvent be removed from the extract leaving the cascade to form a raffinate, part of which is returned to the cascade as reflux, and the remainder is withdrawn from the plant as a product. Also, a portion of the raffinate leaving the cascade is withdrawn as a product, and the remainder is diluted with solvent to form an extract which is returned to the cascade. The solvent separator, which is ordinarily a still, and the solvent mixer are shown in Fig. 13-24. As also shown in Fig. 13-24, both products may be stripped of solvent in solvent strippers to give solvent-free products.

In Fig. 13-24, the feed is denoted by F. It may or may not contain solvent. The part of the cascade between the feed and the solvent separator is the extract-enriching section, and the product from this section is

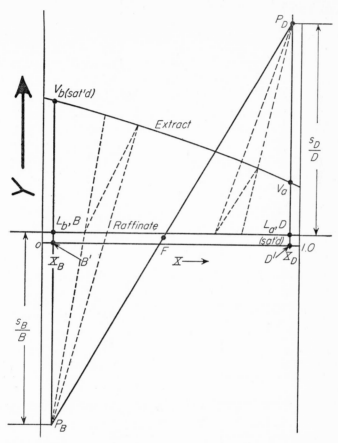

Fig. 13-25. Extraction with reflux. (Ponchon-Savarit construction.)

denoted by D. The part of the cascade between the feed and the solvent mixer is the raffinate-stripping section, and the product from this section is denoted by B. Ideal stages are numbered serially from the extract end. The total number of stages is N. The rectangular diagram of Fig. 13-25 shows the graphical analysis of the process of Fig. 13-24. A type II system is postulated, and the coordinates of Fig. 13-22a are chosen. As before, in considering ix, xy, and XY diagrams, the same letter is used to designate both the flow rate of a stream and the point on the diagram representing that stream. Also, since the diagram is on a solvent-free basis, all flow rates are in pounds per hour of $a + b$. Any solvent that accompanies a stream must be computed separately. For example, assume that an extract stream V is flowing at a rate of 150 lb/hr, total stream, and that the stream is 20 per cent solute, 30 per cent diluent, and 50 per cent solvent. Then, for this stream, $V = 150(0.20 + 0.30) = 75$ lb/hr. The solvent accompanying this is also 75 lb/hr. The coordinates of point V are

$$X = \frac{0.20}{0.20 + 0.30} = 0.40 \quad \text{and} \quad Y = \frac{0.50}{0.20 + 0.30} = 1.00$$

On the diagram of Fig. 13-25, the coordinates for pure solvent are $X = 0$, $Y = \infty$. Any mixing or stripping process involving pure solvent is represented by a vertical straight line connecting the points for the initial and final streams.

The graphical method for countercurrent extraction follows closely that for a center-feed fractionating column.[18, 23] Let V_a denote the extract leaving stage 1, and let s_D be the solvent removed from this stream in the solvent separator. Also, assume that the solvent removed in the separator is just sufficient to convert stream V_a into a saturated raffinate. Then product D and reflux L_a have the same composition, and the point representing these streams lies on the saturated-raffinate line. Let L_b denote the raffinate leaving stage N of the cascade. Assume that, after product B has been removed from this stream, the remainder is diluted, in the solvent mixer, with just sufficient solvent, denoted by s_B, to convert it to a saturated extract, denoted by V_b. Then point V_b lies on the saturated-extract line.

The net flow, denoted by P_D, leaving the extract end of the plant is an imaginary mixture of product D and solvent s_D. If the coordinates of point D are X_D and Y_D, those of point P_D are X_D and $Y_D + s_D/D$. Likewise, the net flow out of the plant at the raffinate end is an imaginary flow of product B less solvent s_B. If the coordinates of point B are X_B and Y_B, those of point P_B are X_B and $Y_B - s_B/B$. Points P_D and P_B are plotted in Fig. 13-25.

Since the feed F is split into two net flows P_D and P_B, points P_D, F, and P_B lie on a common straight line, the over-all-balance line, as shown in Fig. 13-25. The number of ideal stages is found by the Ponchon-Savarit method, exactly as in the design of a fractionating column, as shown in Fig. 12-38. Points D' and B', representing solvent-free products leaving the solvent strippers, are also shown in Fig. 13-25.

The following balances, on a solvent-free basis, can be written.

$$D = P_D \qquad B = P_B$$

$$V_a = L_a + D \qquad L_b = V_b + B \qquad (13\text{-}14)$$

$$F = P_D + P_B = D + B$$

By the center-of-gravity principle, the following equalities between ratios of streams and of line segments can be written.

$$\frac{L_a}{P_D} = \frac{L_a}{D} = \frac{\overline{P_D V_a}}{\overline{V_a L_a}} \qquad \frac{V_b}{P_B} = \frac{V_b}{B} = \frac{\overline{L_b P_B}}{\overline{V_b L_b}}$$

(13-15)

$$\frac{D}{F} = \frac{\overline{F P_B}}{\overline{P_D P_B}} \qquad \frac{B}{F} = \frac{\overline{P_D F}}{\overline{P_B P_D}}$$

The quantity L_a/D is the ratio of the reflux to the extract product. It may be denoted by R_D, and it corresponds to the reflux ratio R_D in a fractionating column.

The close analogy between rectification and extraction, both using reflux, is shown in Table 13-3. Note that the solvent plays the same part in extraction that heat does in rectification.

TABLE 13-3. COMPARISON OF EXTRACTION WITH RECTIFICATION, BOTH USING REFLUX

Rectification	Extraction
Vapor flow in cascade V	Extract flow in cascade V
Liquid flow in cascade L	Raffinate flow in cascade L
Overhead product D	Extract product D
Bottom product B	Raffinate product B
Condenser	Solvent separator
Reboiler	Mixer
Bottom-product cooler	Raffinate solvent stripper
Overhead-product cooler	Extract solvent stripper
Heat to reboiler q_r	Solvent to mixer s_B
Heat removal in condenser $-q_c$	Solvent removal in separator s_D
Reflux ratio $R_D = L_a/D$	Reflux ratio $R_D = L_a/D$
Rectifying section	Extract-enriching section
Stripping section	Raffinate-stripping section

Limiting Reflux Ratios. Just as in rectification, two limiting cases exist in operating a countercurrent extractor with reflux. As the reflux ratio R_D becomes very great, the number of stages approaches a minimum, and as R_D is reduced, a minimum value of the reflux ratio is reached where the number of stages becomes infinite. The minimum number of stages and the minimum reflux ratio are found by exactly the same methods used to determine the same quantities in rectification.

Application of the McCabe-Thiele Method to Extraction. The distribution relation shown in Fig. 13-22b is analogous to the xy diagram used in distillation. By considering this line as an equilibrium curve, the McCabe-Thiele method can be applied to countercurrent extraction with reflux, provided the operating lines can be located.[14] In extraction, the extract and raffinate streams vary from stage to stage, and the McCabe-Thiele operating lines are generally far from straight. The Ponchon-

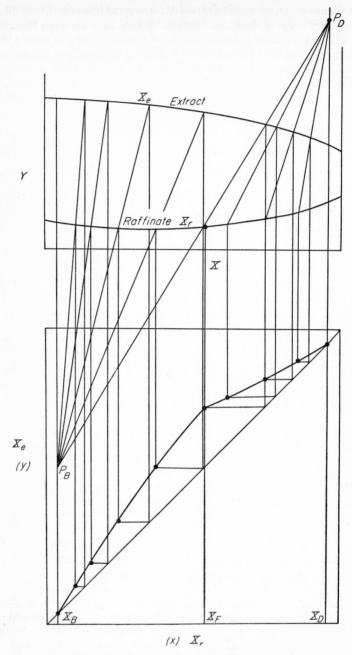

F$_{\text{IG}}$. 13-26. Construction of operating lines on xy plane.

Savarit diagram can be used to establish operating lines on the xy diagram. It is necessary only to draw, at random, several lines through the net flow points P_D and P_B and transfer the intersections of these lines with the saturated-raffinate and extract lines to the xy diagram, as shown in Fig. 13-26. The operating lines are then drawn through these points. The usual McCabe-Thiele construction for ideal stages can then be applied.†

Whenever both the equilibrium and operating lines on the xy plane are straight, Eq. (10-37) or (10-38) can, of course, be used.

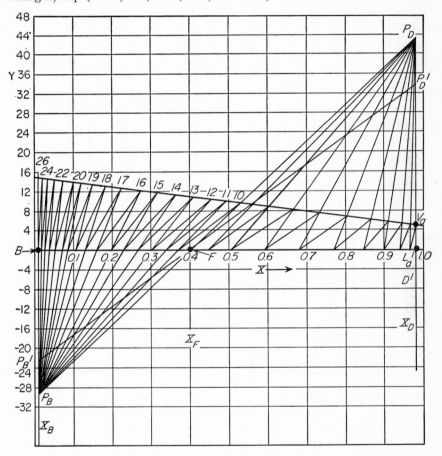

FIG. 13-27. Example 13-5. Ponchon-Savarit diagram.

Example 13-5. A mixture of 40 per cent MCH and 60 per cent heptane is extracted with pure aniline at 25°C in a countercurrent extractor with reflux. The raffinate product is to contain 1 per cent MCH and the extract product 98 per cent MCH, both on a solvent-free basis. A reflux ratio R_D of 7.7 is to be used. What

† Many other graphical constructions have been developed for extraction calculations. For an exhaustive and comprehensive treatment of these see Ref. 22.

is the minimum reflux ratio, and how many ideal stages are needed? Per 100 lb of feed, how much solvent is added in the mixer, and how much is removed in the solvent separator?

Solution. The rectangular diagram of Fig. 13-22a is used. Refer to Fig. 13-27. The solvent content of the raffinates is low, and Y_r is considered zero. From the conditions of the problem, $X_F = 0.40$, $X_D = 0.98$, and $X_B = 0.01$. The minimum reflux ratio R'_D is found by locating the over-all balance line, collinear with a tie line, intersecting line $X = X_D$ at the point farthest from the X axis. By trial this is found to be the line collinear with the tie line through the point F. Any other line through a tie line to be used in the Ponchon-Savarit construction intersects the vertical line through X_D at a point nearer the X axis than does line $P'_D F P'_B$, and this line establishes the condition for minimum reflux ratio. The ordinates of points P'_D, V_a, and L_a are, respectively, 34.0, 5.0, and 0. Then the minimum reflux ratio is

$$R'_D = \frac{L'_a}{D} = \frac{34.0 - 5.0}{5.0} = 5.8$$

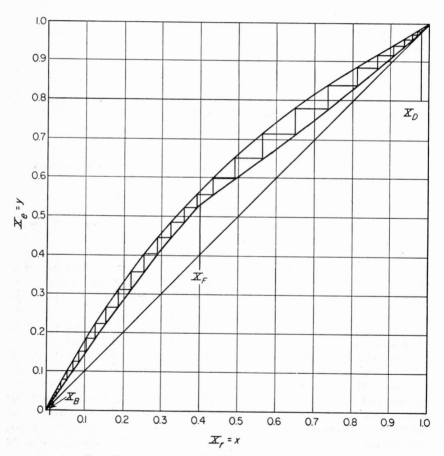

Fig. 13-28. Example 13-5. McCabe-Thiele construction

Since the operating reflux ratio is 7.7, Y, the ordinate of point P_D, is given by the center-of-gravity equation

$$R_D = \frac{Y - 5.0}{5.0} = 7.7 \quad \text{or} \quad Y = 43.5$$

Line $P_D F P_B$, drawn through points P_D and F, establishes the over-all-balance line. The Ponchon-Savarit construction is then drawn, as in Fig. 13-27. Tie-line terminals are found from Fig. 13-22b. The total number of ideal stages is 26. The feed should be placed on the eleventh stage.

The solution of this problem by the McCabe-Thiele method is shown in Fig. 13-28. The operating lines in this figure are located by the method of Fig. 13-26.

Per 100 lb of feed, the solvent-free raffinate product is

$$B' = \frac{0.98 - 0.40}{0.98 - 0.01} 100 = 59.8 \text{ lb}$$

The solvent-free extract product is $D' = 100 - 59.8 = 40.2$ lb.

The solvent added in the mixer, per pound of product D, is the Y coordinate of point P_B, which is -29.2. Then

$$s_D = 29.2 \times 59.8 = 1,750 \text{ lb}$$

The solvent removed in the solvent separator, per pound of extract product D, is the ordinate of point P_D, which is 43.5. Then

$$s_B = 43.5 \times 40.2 = 1,750 \text{ lb}$$

Since the raffinates are nearly solvent-free, it is assumed that solvent strippers are not necessary.

Extraction in Packed and Spray Columns

The principles of liquid extraction in countercurrent packed columns are identical with those for absorption and rectification in the same kind of equipment. Extraction in spray columns is also treated in the same way. The mass-transfer coefficients $K_x a$, $K_y a$, $k_x a$, and $k_y a$, the HTUs H_{Ox}, H_{Oy}, H_x and H_y, and the techniques for integrating over the length of the packed section are all the same in extraction as in the other operations. If the operating line is curved, it can be located by the method of Fig. 13-26. When the equilibrium and operating lines are straight, logarithmic mean driving forces, as used in Eqs. (11-28) to (11-33), are applicable.

Flooding Velocities. If the flow rate of either the dispersed phase or the continuous phase is held constant and that of the other phase gradually increased, a point is reached where the dispersed phase coalesces, the holdup of that phase increases, and finally both phases leave together through the continuous phase outlet. The effect, like the corresponding action in an absorption column, is called flooding. The larger the flow rate of one phase at flooding, the smaller is that of the other. A column obviously should be operated at flow rates below the flooding point.

Flooding velocities in packed columns may be estimated from Fig. 13-29. In this figure the abscissa is the group

$$\frac{(\sqrt{u_c} + \sqrt{u_d})^2 \rho_c}{a_p \mu_c}$$

The ordinate is the group

$$\frac{\mu_c}{\Delta \rho}\left(\frac{\sigma}{\rho_c}\right)^{0.2}\left(\frac{a_p}{\epsilon}\right)^{1.5}$$

where u_c, u_d = average linear velocities of continuous and dispersed phases, respectively, based on cross section of empty column, ft/hr

μ_c = viscosity of the continuous phase, lb/ft-hr

σ = interfacial tension between phases, dynes/cm

ρ_c = density of the continuous phase, lb/ft^3

$\Delta \rho$ = density difference between phases, lb/ft^3

a_p = specific surface area of packing, ft^2/ft^3

ϵ = fraction voids or porosity of packed section

The groups in Fig. 13-29 are not dimensionless, and the proper units must be used.

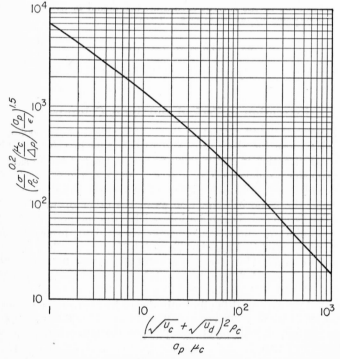

Fig. 13-29. Flooding velocities in packed extraction towers. (*After Crawford and Wilke.*[4])

Coefficients in Extraction Columns. Mass-transfer coefficients and HTUs in extraction equipment depend on the same variables as in liquid-gas contact equipment, and, in addition, they are also sensitive to the details of distribution of the dispersed phase throughout the mass of the continuous one. It is even more difficult to separate the over-all coefficients or HTUs into the individual factors than in gas-liquid contacts, and the over-all coefficients $K_x a$ and $K_y a$ and the over-all HTUs H_{Ox} and H_{Oy}

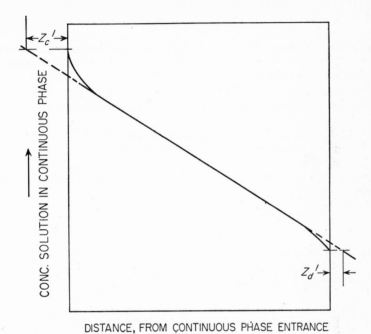

Fig. 13-30. End effects in extraction column. (*After Kreager and Geankopolis.*[12])

are usually reported. Customarily, these over-all quantities are based on the aqueous phase if either the solvent or the diluent is water.

Because of the complexity of the action in extraction equipment and the many factors involved, no basic correlations of extraction coefficients are available. Many data have been reported. They are given in standard references,[6, 21b, 22a] which should be consulted for coefficients for specific systems and designs. The following general statements may be made.

1. Coefficients in spray columns are more sensitive to variations in a than in K or k, and this parameter depends on many factors. The size of the drops, the number of drops per unit volume of continuous phase, and the time of residence of the drops in the extractor are especially important. Drops rise (or fall) in general agreement with the laws of terminal velocities discussed in Chap. 7.

2. The relative importance of the two individual resistances depends,

as in gas-liquid contacts, on the slope of the equilibrium curve, in accordance with Eqs. (11-17) and (11-18), but this influence of solubility is often obscured by changes in factor a. HTUs vary linearly with the factor G_{Mx}/mG_{My} as shown by Eqs. (11-24) and (11-25).[13]

3. End effects are important, especially in short columns. Extraction in long columns is less effective per unit of total height than in short ones. Data on short columns should not, therefore, be extrapolated to tall columns. Figure 13-30 shows how end effects in towers are found and evaluated.[12] In this figure the concentration of solute in the continuous phase is plotted against the distance from the continuous-phase entrance. Near this entrance the rate of extraction is abnormally large, and this is also true, to a lesser extent, near the dispersed-phase inlet. In the middle of the column, the rate, as measured by the slope of the concentration-length line, is constant. If the straight line applying at the center of the column is extrapolated at both ends until it intersects horizontal lines through the terminal concentrations, two lengths Z'_c and Z'_d are determined, as shown in Fig. 13-30. These lengths are fictitious column heights equivalent to the inlet end effects of the continuous and dispersed phases.

4. In packed columns the coefficients are larger for small packings than for large ones, and packed columns give much larger coefficients than do unpacked spray columns operated under the same conditions.[13]

5. In packed columns it is important which phase is chosen as the dispersed phase. Preferably, this phase should not wet the packing. If it does so, it tends to flow along the surface of the packing as rivulets rather than as small, discrete droplets, and the area factor a is small.

6. Except for the condition of paragraph 5, a general rule is to choose the phase flowing in larger volume as the dispersed phase.

7. Especially in spray columns there may be considerable recirculation in the continuous phase between top and bottom.[8] This flattens the operating line and distorts the material balances. If recirculation is neglected in evaluating coefficients, apparent, rather than true, coefficients are found, and these should not be used for columns of sizes or shapes much different from those used in measuring the coefficients.

8. Small quantities of impurities may concentrate at the interface between the two phases and form an additional resistance to mass transfer.[25] There is no way of predicting such effects.

RECAPITULATION OF MASS-TRANSFER-CALCULATION METHODS

In this and previous chapters countercurrent contacting devices have been discussed from the standpoint of the individual operations of absorption, desorption, rectification, leaching, and liquid extraction. General methods, based on a common approach, have been applied in turn to each of these operations. The methods for calculating the number of stages or the height of the unit may now be summarized as follows.

Two types of contacting plant are important, stage contacts and differential contacts. Each is treated by its own methods. The choice of method within each class of unit depends on the curvature, or lack of curvature, of two lines, the equilibrium line and the operating line.

In stage contacts, the absorption-factor method, as given by Eqs. (10-37) and (10-38), is recommended when both lines are straight; the McCabe-Thiele construction is suggested when the equilibrium line is curved and the operating line is straight or when it may be easily plotted by use of material balances; and the Ponchon-Savarit method is available when both lines are curved. In the last situation, an alternative method may be used, namely, the McCabe-Thiele method, utilizing operating lines located by the Ponchon-Savarit construction.

In differential-contact plants, when both lines are straight, over-all coefficients or over-all HTUs, in combination with logarithmic mean driving forces, as given by Eqs. (11-28), (11-30), (11-32), and (11-33), are used. When the equilibrium line is straight and the operating line curved, graphical integration based on over-all mass-transfer coefficients K_x and K_y or the over-all HTUs H_{Ox} and H_{Oy} are used. If the operating line is strongly curved, the over-all mass-transfer coefficient is, in principle, the more accurate. When the equilibrium line is strongly curved, graphical integration based on individual coefficients k_x and k_y or individual HTUs H_x and H_y should be used whenever data on these individual factors are available or can be estimated. Otherwise K or H_O factors, determined under conditions approaching those for the design, must be used.

If in one part of the diagram the lines are straight and in another part of the same diagram they are curved, the calculation may be made in two parts, using the simpler methods where the lines are straight and more complicated methods where the lines are curved. The results from the two calculations are then added.

SYMBOLS

a Solute in stream, concentration or mass

a_p Specific surface of packing, ft^2/ft^3

a_1 $(\pi/2)^2$ [Eq. (13-11)]

B Bottom product, lb base mixture/hr; B', bottom product leaving stripper

b Solid in stream, concentration or mass, leaching; diluent in stream, concentration or mass, extraction

D Overhead product, lb base mixture/hr; D', overhead product leaving solvent stripper

D'_v Volumetric diffusivity of solute through stationary solution contained in solid, ft^2/hr

F Feed to extraction cascade, lb base mixture/hr

G_M Molal mass velocity, based on tower cross section, lb moles/ft^2-hr; G_{Mx}, of heavy phase; G_{My}, of light phase

H_O Height of transfer unit, over-all, ft; H_{Ox}, based on heavy phase; H_{Oy}, based on light phase

J Total input to, and output from, cascade, lb base mixture/hr

Ka Over-all volumetric mass-transfer coefficient, lb moles/(ft^3)(hr)(unit mole fraction); K_xa, based on heavy phase; K_ya, based on light phase

ka Volumetric mass-transfer coefficient, lb moles/(ft^3)(hr)(unit mole fraction); k_xa, based on heavy phase; k_ya, based on light phase

L Underflow, or heavy phase, lb base mixture/hr; L_a, entering cascade; L_b, leaving cascade; L_n, leaving stage n

m Mass flow, lb; m_R, m_S, m_T, for streams R, S, T, respectively; also, slope of equilibrium curve

N Number of ideal stages

P Net flow through cascade, in at stage N and out at stage 1, lb base mixture/hr; P_B, net flow leaving extract-enriching section; P_D, net flow leaving raffinate-stripping section

q Heat added, Btu/lb; $-q_c$, to condenser; q_r, to reboiler

R_D Reflux ratio; R_D', minimum

r_p One-half particle thickness, ft

s Solvent in stream, concentration or mass; s_B, solvent added to bottom product, lb/hr; s_D, solvent removed from overhead product, lb/hr

u Linear velocity, based on cross section of column, ft/hr; u_c, of continuous phase; u_d, of dispersed phase

V Overflow, or light phase, lb base mixture/hr; V_a, leaving cascade; V_b, entering cascade; V_{n+1}, leaving stage $n + 1$

X Abscissa on XY diagram; for leaching, $X = a/(a + s)$; for type I extraction, $X = b/(b + s)$; for type II extraction, $X = a/(a + b)$; X_e for saturated extract in equilibrium with X_r; X_r, for saturated raffinate in equilibrium with X_e; X_R, X_S, X_T, for underflow streams R, S, T, respectively; X_0, initial concentration of solute in solution within solid; X_1, concentration of solute in solution in contact with solid; $\overline{X}$, average concentration of solute in solution within solid at time θ

x Mass fraction of solute in underflow or heavy liquid; x_a, at entrance; x_b, at exit; x_A, x_B, x_S, mass fraction of solute, diluent, solvent, respectively, based on entire heavy phase

Y Ordinate on XY diagram; for leaching, $Y = b/(a + s)$; for type I extraction, $Y = a/(b + s)$; for type II extraction, $Y = s/(a + b)$; Y_e, for saturated extract in equilibrium with Y_r; Y_r, for saturated raffinate in equilibrium with Y_e; Y_R, Y_S, Y_T, for overflow streams R, S, T, respectively

y Mass fraction of solute in overflow or heavy liquid; y_a, at exit; y_b, at entrance; y_A, y_B, y_S, mass fraction of solute, diluent, solvent, respectively, based on entire light phase

y^* Concentration of overflow solution in equilibrium with specific underflow solution; y_a^*, in equilibrium with x_a; y_b^*, in equilibrium with x_b

Z' Equivalent column height for end effects, ft; Z_c', at continuous-phase entrance; Z_d', at dispersed-phase entrance

z Distance measured in direction of diffusion, ft

Greek Letters

$\Delta\rho$ Density difference between phases, lb/ft^3

∂ Partial-differential operator

ϵ Fraction voids in packed section

η_M Murphree stage efficiency

θ Time, hr

μ_c Viscosity of continuous phase, lb/ft-hr

ρ_c Density of continuous phase, lb/ft^3

σ Interfacial tension, dynes/cm

ϕ Function in Eq. (13-10)

Dimensionless Group

β Diffusion group, $D_v'\theta/r_p^2$

PROBLEMS

13-1. Roasted copper ore containing the copper as $CuSO_4$ is to be extracted in a countercurrent stage extractor. Each hour a charge consisting of 10 tons of gangue, 1.2 tons of copper sulfate, and 0.5 ton of water is to be treated. The strong solution produced is to consist of 90 per cent H_2O and 10 per cent $CuSO_4$ by weight. The recovery of $CuSO_4$ is to be 98 per cent of that in the ore. Pure water is to be used as the fresh solvent. After each stage, 1 ton of inert gangue retains 2 tons of water plus the copper sulfate dissolved in that water. Equilibrium is attained in each stage.

How many stages are required?

13-2. A five-stage countercurrent extraction battery is used to extract the sludge from the reaction

$$Na_2CO_3 + CaO + H_2O \rightarrow CaCO_3 + 2NaOH$$

The $CaCO_3$ carries with it 1.5 times its weight of solution in flowing from one unit to another. It is desired to recover 98 per cent of the NaOH. The products from the reaction enter the first unit with no excess reactants, but with 6.5 lb of H_2O per pound of $CaCO_3$.

(a) How much wash water must be used per pound of calcium carbonate? (b) What is the concentration of the solution leaving each unit, assuming that $CaCO_3$ is completely insoluble? (c) Using the same quantity of wash water, how many units must be added to recover 99.5 per cent of the sodium hydroxide?

13-3. In Prob. 13-2 it is found that the sludge retains solution varying with the concentration as follows:

NaOH, weight %	Lb solution/lb $CaCO_3$
0	1.50
5	1.75
10	2.05
15	2.70
20	3.60

It is desired to produce a 10 per cent solution of NaOH. How many stages must be used to recover 95 per cent of the NaOH?

13-4. A solution of oxalic acid containing 0.50 lb of acid per gallon of solution is to be treated countercurrently with ethyl ether in a packed tower. It is necessary to reduce the concentration of the acid to 0.05 lb/gal. The ratio of aqueous solution to ether is to be 0.0735 gal of solution per gallon of ether. The equilibrium data for this system are:

$C_e \times 10^5$	1.92	8.33	16.7	25.0	33.4	50.0
K	19	17	14.9	13.8	13.1	12.1

where $K = C_w/C_e$ and C_w and C_e are the concentrations of oxalic acid in the water and ether phases, respectively, in pound moles per gallon. The water and ether are mutually insoluble.

How many transfer units (over-all, based on the water phase) are needed? What will be the concentration of acid in the extract?

13-5. Oil is to be extracted from halibut livers by means of ether in a countercurrent extraction battery. The entrainment of solution by the granulated liver mass was found by experiment to be:

Solution retained by 1 lb exhausted livers, gal	Solution concentration, gal oil/gal solution
0.035	0
0.042	0.1
0.050	0.2
0.058	0.3
0.068	0.4
0.081	0.5
0.099	0.6
0.120	0.68

In the extraction battery, the charge per cell is to be 100 lb, based on completely exhausted livers. The unextracted livers contain 0.043 gal of oil per pound of exhausted material. A 95 per cent recovery of oil is desired. The final extract is to contain 0.65 gal of oil per gallon of extract. The ether fed to the system is oil-free.

(a) How many gallons of ether are needed per charge of livers? (b) How many extractors are needed?

13-6. Two liquids A and B which have nearly identical boiling points are to be separated by extraction with a solvent C. The following data represent the equilibrium between the two liquid phases at 95°C.

Determine the minimum amount of reflux that must be returned from the extract product and from the raffinate product to produce an extract containing 83 per cent A and 17 per cent B (on a solvent-free basis) and a raffinate product containing 10 per cent A and 90 per cent B (solvent-free). The feed contains 35 per cent A and 65 per cent B on a solvent-free basis and is a saturated raffinate. Determine the number of ideal stages on both sides of the feed required to produce the same end products from the same feed when the reflux ratio of the extract, expressed as pounds of extract reflux per pound of extract product (including solvent), is twice the minimum. Calculate the masses of the various streams per 1,000 lb of feed, all on a solvent-free basis.

EQUILIBRIUM DATA

Extract layer			Raffinate layer		
A, %	B, %	C, %	A, %	B, %	C, %
0	7	93.0	0	92.0	8.0
1.0	6.1	92.9	9.0	81.7	9.3
1.8	5.5	92.7	14.9	75.0	10.1
3.7	4.4	91.9	25.3	63.0	11.7
6.2	3.3	90.5	35.0	51.5	13.5
9.2	2.4	88.4	42.0	41.0	17.0
13.0	1.8	85.2	48.1	29.3	22.6
18.3	1.8	79.9	52.0	20.0	28.0
24.5	3.0	72.5	47.1	12.9	40.0
31.2	5.6	63.2	Plait point		

REFERENCES

1. Aldrich, H. W., and W. G. Scott: *Trans. AIMME*, **106:** 650 (1933).
2. Baker, E. M.: *Trans. AIChE*, **32:** 62 (1936).
3. Cofield, E. P., Jr.: *Chem. Eng.*, **58**(1): 127 (1951).
4. Crawford, J. W., and C. R. Wilke: *Chem. Eng. Progr.*, **47:** 423 (1951).
5. Elgin, J. C.: *Trans. AIChE*, **32:** 451 (1936).
6. Elgin, J. C., in J. H. Perry (ed.): "Chemical Engineers' Handbook," 3d ed., pp. 748–753, McGraw-Hill Book Company, Inc., New York, 1950.
7. Fan, H. P., J. C. Morris, and H. Wakeman: *Ind. Eng. Chem.*, **40:** 195 (1948).
8. Gier, T. E., and J. E. Hougen: *Ind. Eng. Chem.*, **45:** 1362 (1953).
9. Grosberg, J. A.: *Ind. Eng. Chem.*, **42:** 154 (1950).
10. King, C. O., D. L. Katz, and J. C. Brier: *Trans. AIChE*, **40:** 533 (1944).
11. Kinney, G. F.: *Ind. Eng. Chem.*, **34:** 1102 (1942).
12. Kreager, R. M., and C. J. Geankopolis: *Ind. Eng. Chem.*, **45:** 2156 (1953).
13. Leibson, I., and R. B. Beckmann: *Chem. Eng. Progr.*, **49:** 405 (1953).
14. Maloney, J. O., and E. A. Schubert: *Trans. AIChE*, **36:** 741 (1940).
15. Morello, V. S., and N. Poffenberger: *Ind. Eng. Chem.*, **42:** 1021 (1950).
16. Osburn, J. O., and D. L. Katz: *Trans. AIChE*, **40:** 511 (1944).
17. Othmer, D. F., R. E. White, and E. Trueger: *Ind. Eng. Chem.*, **33:** 1240 (1941).
18. Randall, M., and B. Longtin: *Ind. Eng. Chem.*, **30:** 1188 (1938).
19. Sege, G., and F. W. Woodfield: *Chem. Eng. Progr.*, **50:** 396 (1954).
20. Sherwood, T. K., J. E. Evans, and J. V. A Longcor: *Ind. Eng. Chem.*, **31:** 1144 (1939).
21. Sherwood, T. K., and R. L. Pigford: "Absorption and Extraction," 2d ed., (*a*) p. 402, (*b*) pp. 427–753, McGraw-Hill Book Company, Inc., New York, 1952.
22. Treybal, R. E.: "Liquid Extraction," (*a*) pp. 312–342, McGraw-Hill Book Company, Inc., New York, 1951.
23. Varteressian, K. A., and M. R. Fenske: *Ind. Eng. Chem.*, **29:** 270 (1937).
24. Von Berg, R. L., and H. F. Wiegandt: *Chem. Eng.*, **59**(6): 189 (1952).
25. West, F. B., et al.: *Ind. Eng. Chem.*, **44:** 625 (1952).
26. Wiegandt, H. F., and R. L. Von Berg: *Chem. Eng.*, **61**(7): 183 (1954).

Chapter 14

CRYSTALLIZATION

Crystallization is important industrially because of the variety of materials that are marketed in the crystalline form. Its wide use is based on the fact that a crystal formed from even an impure solution is itself pure (unless mixed crystals occur) and that it affords a practical method of obtaining pure chemical substances in a satisfactory condition for packaging and storing.

Commercial Importance of Crystal Size and Shape. It is obvious that yield and purity are important in operating a crystallization process, but these two factors are not the only ones to be considered. The size, and often the shape, of the crystals is important; it is especially necessary that the crystals be of uniform size. Uniform size is desirable for appearance, for ease in filtering and washing, and for consistent behavior in use. Uniformity of size also minimizes caking in the package. Large crystals are often demanded, although such demands are not always justified by any real advantage. Sometimes a definite shape is also required, e.g., needles rather than plates or cubes. Strong, single (nonaggregated) crystals not easily broken are needed for some purposes.

CRYSTAL GEOMETRY

A crystal is the most highly organized type of nonliving matter. It is characterized by the fact that its constituent particles, which may be atoms or ions, are arranged in orderly three-dimensional arrays called *space lattices*. The distances between the particles in a space lattice can be measured by X-ray methods, and these lattice constants have been determined for most common crystalline materials.

Crystal Forms: Law of Haüy. As a result of the space-lattice arrangement of the particles, when crystals are allowed to form without hindrance from other crystals or outside bodies, they appear as polyhedrons having sharp corners and flat sides, or faces. Although the relative sizes of the faces and edges of various crystals of the same material may be widely different, the angles made by corresponding faces of all crystals of the same material are equal and are characteristic of that material.

Crystallographic Systems. Since all crystals of a definite substance have the same interfacial angles in spite of wide differences in the extent of development of individual faces, crystal forms are classified on the basis of these angles. For example, consider a definite crystal. Choose three important faces and call them axial planes. When the crystal is a symmetrical solid, such as a cube, the axial faces are taken parallel with planes of symmetry. The three intersections of the axial planes determine three nonparallel lines, and the three lines parallel with such lines drawn through a single arbitrary point are called the axes of the crystal. The axes may all be mutually perpendicular; two of them may be perpendicular to the third but not to each other; they may be equally inclined to each other, with a common angle different from either 60 or 90°; or they may be mutually inclined with three different angles, all differing from 60 or 90°.

In addition to the three axial faces, a fourth fundamental face, intersecting all three axes, is chosen. The lengths of the segments so cut off from the three axes are expressed as ratios, the length of one of the segments being taken as unity. The lengths of the axes so determined may be equal for all axes, equal for two axes but unequal for the third, or unequal for all three.

One class of crystals, showing a hexagonal cross section with 60° angles between the hexagonal sides, is most conveniently referred to four axes instead of the usual three. Three of the axes are at 60° to each other and in the same plane, and the fourth is perpendicular to the plane of the other three.

The combinations of angles and lengths of the axes give rise to seven classes of crystals. These are:

1. *Triclinic System.* Three mutually inclined and unequal axes, all angles unequal and other than 30, 60, or 90°.

2. *Monoclinic System.* Three unequal axes, two of which are inclined but perpendicular to the third.

3. *Orthorhombic System.* Three unequal rectangular axes.

4. *Tetragonal System.* Three rectangular axes, two of which are equal and different in length from the third.

5. *Trigonal System.* Three equal and equally inclined axes.

6. *Hexagonal System.* Three equal coplanar axes inclined 60° to each other, and a fourth axis different in length from the other three and perpendicular to them.

7. *Cubic System.* Three equal rectangular axes.

PRINCIPLES OF CRYSTALLIZATION

Crystallization may be analyzed from the standpoints of purity, yield, energy requirements, and rates of formation and growth.

Purity of Product. A crystal itself is pure. It retains, however, mother liquor when removed from the final magma, and if the crop contains crystalline aggregates, considerable amounts of mother liquor may be occluded within the solid mass. When retained mother liquor of low purity is dried on the product, contamination results, the extent of which depends on the amount and degree of impurity of the mother liquor retained by the crystals.

In practice, much of the retained mother liquor is separated from the crystals by filtration or centrifuging, and the balance is removed by washing with fresh solvent. The effectiveness of these purification steps depends on the size and uniformity of the crystals. Small mushlike crystals or loose crystal agglomerates cannot be filtered and washed as effectively as a crop consisting of single crystals of uniform size.

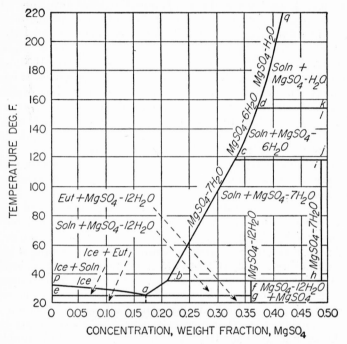

FIG. 14-1. Phase diagram, system MgSO₄-H₂O. (*By permission, from "Chemical Engineers' Handbook," 3d ed., by J. H. Perry. Copyright, 1950. McGraw-Hill Book Company, Inc.*)

Purity may also be improved by recrystallization. This is common in laboratory practice but is usually more expensive and less satisfactory in plant operation than filtering, centrifuging, and washing.

Equilibria and Yields. Equilibrium is attained in crystallization when the mother liquor is saturated. Solubility data are given in standard tables.[7,14] Many important inorganic substances crystallize with water of crystallization. In some systems, several different hydrates are formed, depending on concentration and temperature, and phase equilibria in such systems can be quite complicated.

The phase diagram for the system magnesium sulfate–water is shown in Fig. 14-1. Concentration in mass fraction of anhydrous magnesium sulfate is plotted against temperature in degrees Fahrenheit. The entire area above and to the left of the broken solid line represents undersaturated solutions of magnesium sulfate in water. The broken line *eagfhij* represents complete solidification of the liquid solution to form various solid phases. The area *pae* represents mixtures of ice and saturated solution. Any solution containing less than 16.5 per cent $MgSO_4$ precipitates ice when the temperature reaches line *pa*. Broken line *abcdq* is the solubility curve. Any solution more concentrated than 16.5 per cent precipitates, on cooling, a solid when the temperature reaches this line. The solid formed at point *a* is called eutectic. It consists of an intimate mechanical mixture of ice and $MgSO_4 \cdot 12H_2O$. Between points *a* and *b* the crystals are $MgSO_4 \cdot 12H_2O$; between *b* and *c* the solid phase is $MgSO_4 \cdot 7H_2O$ (epsom salt); between *c* and *d* the crystals are $MgSO_4 \cdot 6H_2O$; and above point *d* they are $MgSO_4 \cdot H_2O$. In the area *cihb*, the system at equilibrium consists of mixtures of saturated solution and crystalline $MgSO_4 \cdot 7H_2O$. In area *dljc*, the mixture consists of saturated solution and crystals of $MgSO_4 \cdot 6H_2O$. In area *qdk*, the mixture is saturated solution and $MgSO_4 \cdot H_2O$. The graphical mixing rule and the center-of-gravity principle apply in these phase diagrams for *isothermal* mixing processes.

In many industrial crystallization processes, the crystals and mother liquor are in contact long enough to reach equilibrium, and the mother liquor is saturated at the final temperature of the process. The yield of the process can then be calculated from the concentration of the original solution and the solubility at the final temperature. If appreciable evaporation occurs during the process, this must be known or estimated.

When the rate of crystal growth is slow, considerable time is required to reach equilibrium. This is especially true when the solution is viscous or where the crystals collect in the bottom of the crystallizer so there is little crystal surface exposed to the supersaturated solution. In such situations, the final mother liquor may retain appreciable supersaturation, and the actual yield will be less than that calculated from the solubility curve.

If the crystals are anhydrous, calculation of the yield is simple, as the solid phase contains no solvent. When the crop contains water of crystallization, account must be taken of the water accompanying the crystals, since this water is not available for retaining solute in solution. Solubility data are usually given either in parts by mass of anhydrous material per hundred parts by mass of total solvent or in mass per cent anhydrous solute. These data ignore water of crystallization. The key to calculations of yields of hydrated solutes is to express all masses and concentrations in terms of hydrated salt and free water. Since it is this latter quantity that remains in the liquid phase during crystallization, concentrations or amounts based on free water may be subtracted to give a correct result.

Example 14-1. A solution consisting of 30 per cent $MgSO_4$ and 70 per cent H_2O is cooled to 60°F. During cooling, 5 per cent of the total water in the system evaporates. How many pounds of crystals are obtained per ton of original mixture?

Solution. From Fig. 14-1 it is noted that the crystals are $MgSO_4 \cdot 7H_2O$ and that the concentration of the mother liquor is 24.5 per cent anhydrous $MgSO_4$ and 75.5 per cent H_2O. Per 2,000 lb of original solution, the total water is $0.70 \times 2,000 = 1,400$ lb. The evaporation is $0.05 \times 1,400 = 70$ lb. The molecular weights of $MgSO_4$ and $MgSO_4 \cdot 7H_2O$ are 120.4 and 246.5, respectively, so the total $MgSO_4 \cdot 7H_2O$ in the batch is $2,000 \times 0.30(246.5/120.4) = 1,228$ lb, and the free water is $2,000 - 70 - 1,228 = 702$ lb. In 100 lb of mother liquor, there is $24.5(246.5/120.4) = 50.16$ lb of $MgSO_4 \cdot 7H_2O$ and $100 - 50.16 = 49.84$ lb of free water. The $MgSO_4 \cdot 7H_2O$ in the mother liquor, then, is $(50.16/49.84)702 = 706$ lb. The final crop is $1,228 - 706 = 522$ lb.

The problem can also be solved by use of the center-of-gravity principle. The over-all concentration of the batch after evaporation, is, in mass fraction of $MgSO_4$,

$$\frac{0.30 \times 2,000}{2,000 - 70} = \frac{600}{1,930} = 0.311$$

Since the concentration of the mother liquor is 0.245 and that of $MgSO_4 \cdot 7H_2O$ is 0.488 (the abscissas of the ends of line hi in Fig. 14-1), the yield is, by the center-of-gravity principle,

$$1,930 \frac{0.311 - 0.245}{0.488 - 0.245} = 524 \text{ lb}$$

Enthalpy Balances. In heat-balance calculations for crystallizers, the heat of crystallization is important. This is the latent heat evolved when solid forms from a solution. Ordinarily, crystallization is exothermic, and the heat of crystallization varies with both temperature and concentration. Rigorously, the heat of crystallization is related to the heat of solution of the crystals in a large amount of solvent and to the heat of dilution of the solution from saturation to high dilution. Data on heats of solution are available,[1,12] but heats of dilution are few and, when available, are inconvenient to use directly. A common approximation is to use the negative of the heat of solution as the heat of crystallization and ignore the heat of dilution.

Application of the Enthalpy-concentration Diagram to Solid Phases. The enthalpy-concentration diagram, discussed in Chap. 9, may be extended to account for the enthalpies of solid phases. This diagram is useful in following crystallization processes. An ix diagram, showing enthalpies of solid phases, for the system $MgSO_4$ and H_2O is given in Fig. 14-2. This diagram is consistent with the phase diagram of Fig. 14-1. As before, enthalpies are given in Btu per pound. They refer to 1 lb of total mixture regardless of the number of phases in the mixture. The area above line $pabcdq$ represents enthalpies of undersaturated solutions of $MgSO_4$ in H_2O, and the isotherms in this area have the same significance as those in Fig. 9-12. The area eap in Fig. 14-2 represents all equilibrium mixtures of ice and freezing $MgSO_4$ solutions. Point n represents pure ice. The isothermal (25°F) triangle age gives the enthalpies of all combinations of ice with partly solidified eutectic or of partly solidified eutectic with $MgSO_4 \cdot 12H_2O$. Area $abfg$ gives the enthalpy-concentration points for all magmas consisting of $MgSO_4 \cdot 12H_2O$ crystals and mother liquor. The isothermal (35.7°F) triangle bhf shows the transformation of $MgSO_4 \cdot 7H_2O$ to $MgSO_4 \cdot$

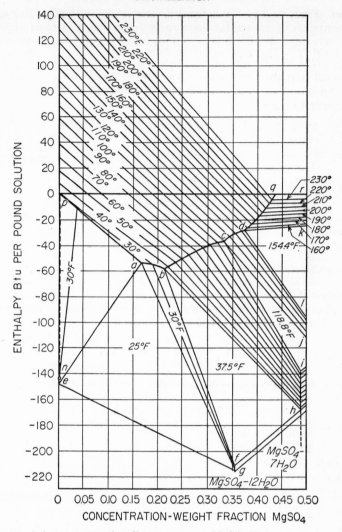

FIG. 14-2. Enthalpy-concentration diagram, system MgSO₄-H₂O. (*By permission, from "Chemical Engineers' Handbook," 3d ed., by J. H. Perry. Copyright, 1950. McGraw-Hill Book Company, Inc.*)

12H₂O, and this area represents mixtures consisting of a saturated solution of concentration 21 per cent, solid MgSO₄·7H₂O, and solid MgSO₄·12H₂O. The area *cihb* represents all magmas of MgSO₄·7H₂O and its mother liquor. Isothermal (118.8°F) triangle *cji* represents mixtures consisting of a saturated solution containing 33 per cent MgSO₄, solid MgSO₄·6H₂O, and solid MgSO₄·7H₂O. Area *dljc* gives enthalpies of MgSO₄·6H₂O and its mother liquor. The isothermal (154.4°F) triangle *dkl* represents mixtures of a saturated solution containing 37 per cent MgSO₄, solid MgSO₄·H₂O, and solid $MgSO_4 \cdot 6H_2O$. Area $qrkd$ is part of the field representing saturated solutions in equilibrium with $MgSO_4 \cdot H_2O$. Except for the isotherms in the liquid-phase field and the solubility and freezing curves, all lines on the ix diagram are straight.

The usual constructions for adiabatic and nonadiabatic mixing processes may be made on Fig. 14-2, including use of points in the areas representing more than one phase.

Example 14-2. A 32.5 per cent solution of $MgSO_4$ at 120°F is cooled, without appreciable evaporation, to 70°F in a Swenson-Walker crystallizer. How much heat must be removed from the solution per ton of crystals?

Solution. The initial solution is represented by the point on Fig. 14-2 at a concentration of 0.325 in the undersaturated-solution field on the 120°F isotherm. The enthalpy coordinate of this point is -33.0 Btu/lb. The point for the final magma lies on the 70°F isotherm in area *cihb* at concentration 0.325. The enthalpy coordinate of this point is -78.4. Per 100 lb of original solution the heat *absorbed* by the solution is

$$100(-78.4 + 33.0) = -4540 \text{ Btu}$$

This is a heat *evolution* of 4540 Btu.

The split of the final magma between crystals and mother liquor may be found by the center-of-gravity principle applied to the 70°F isotherm in either Fig. 14-1 or 14-2. The concentration of the mother liquor is 0.259, and that of the crystals is 0.488. Then, per 100 lb of magma, the crystals are

$$100 \frac{0.325 - 0.259}{0.488 - 0.259} = 28.8 \text{ lb}$$

The heat evolved per ton of crystals is $(4540/28.8)2000 = 315,000$ Btu.

Crystal Formation

Crystallization from a solution takes place in two steps, formation and growth. Crystal formation is also called nucleation. Although in modern theories nucleation and growth are considered to be fundamentally much alike, it is convenient to consider them, first, separately and, second, in combination.

Both nucleation and growth require that the solution be supersaturated, but the effect of supersaturation is different in the two processes.

Nucleation. The usual solubility relation is an equilibrium between saturated solution and crystals of at least moderate size (larger than, say, 150-mesh). As the particle size is reduced, the solubility increases. This increase in solubility becomes appreciable at about 200-mesh and below. It is a result of the large increase in surface achieved by subdividing a solid mass into fine particles. The surface energy so created changes the equilibrium in the direction of greater solubility. In a given system, the effect of particle size on solubility is given approximately by the equation

$$\log \frac{y_r}{y_e} = \frac{C_1}{D_p} \tag{14-1}$$

where y_r = solubility of crystal of size D_p

$\quad\ y_e$ = normal solubility of large crystals

$\quad\ C_1$ = constant

When the D_p of Eq. (14-1) is extrapolated to zero, the equation predicts that no crystal could ever form because it would be infinitely soluble at birth. Actually, however, chance encounters between high-energy solute molecules lead to the formation of nuclei large enough to survive.[8] The formation of fog or rain in supercooled vapor is another example of the same kind of phenomenon. The magnitude of the smallest nucleus that is just large enough to exist in equilibrium with the supersaturated solution in which it is born is called the critical size. Nuclei smaller than this critical size immediately dissolve, and those larger, grow. A critical nucleus contains in the order of 10 to several hundred molecules. One for liquid water, for example, contains about 80 molecules. The larger the supersaturation the smaller the critical size, and the more rapid is the nucleation, measured in the number of nuclei per unit time per unit volume.

FIG. 14-3. Effect of supersaturation on rate of nucleation.

Miers Supersolubility Curve. The general shape of the curve showing the rate of nucleation as a function of the supersaturation is shown in Fig. 14-3. The rate is practically zero until a fairly definite supersaturation, shown by point a, is reached. A small increase in supersaturation beyond this point brings about nucleation, the rate of which increases rapidly with supersaturation. This is shown by curve ab.

When curves such as that of Fig. 14-3 are determined for several saturation temperatures, the locus of points showing the start of nucleation may be plotted along with the usual solubility curve as shown in Fig. 14-4. Line AB is the solubility curve, and line CD, known as the Miers supersolubility curve, shows the temperatures and concentrations where nucleation begins. Lines AB and CD divide the entire concentration-temperature field into three zones: the undersaturated region to the left of the solubility curve, the metastable region between the two curves, and the labile region at the right of the supersolubility curve. A solution represented by a point, such as a, in the undersaturated zone does not crystallize. A supersaturated solution represented by a point, such as c, in the metastable zone does not spontaneously crystallize, but one represented by a point, such as d, in the labile region crystallizes spontaneously until the solution reaches saturation at point b.

The supersolubility curve differs from the solubility curve in that its position is not a property of the system only but also depends on other factors. There is some evidence that in very carefully purified solutions completely free from all solid particles and at rest the position of the supersolubility curve is fixed and reproducible. It is difficult, however, to locate the curve accurately, as nucleation often occurs before the curve is reached, and the so-called supersaturation curve is probably a range rather than a line.

The width of the metastable zone is reduced by strong agitation, by mechanical shock,[19] by sonic or ultrasonic irradiation,[18] or by the presence of solid particles. Variation in cooling rate may also change its location.

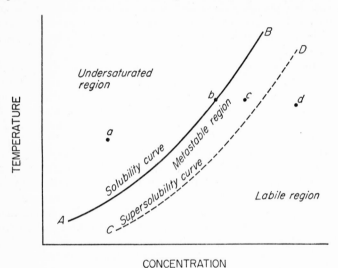

FIG. 14-4. Miers solubility and supersolubility curves.

In commercial practice, the maximum supersaturation represented by the Miers curve is seldom, if ever, reached. Solid particles, including dust, considerably reduce the width of the metastable zone. Some act as catalysts and are especially effective in starting nucleation. The most common material of this kind is the crystalling substance itself. Crystals of other materials, especially those having crystal lattices like that of the crystallizing substance, are also effective. Solutions containing small crystals of the crystallizing solute are called *seeded solutions*. Once nucleation starts, the solution is, of course, seeded, and crystal growth necessarily occurs only in a seeded solution.

The position of the supersaturation curve of a seeded solution depends on the size and concentration of the seeds, the cooling rate, and the degree of agitation.[17] Given sufficient time, any seeded solution will form new nuclei even at low supersaturations. Under specific conditions, however, the supersaturation curves for a seeded solution are definite and reproducible. Within this operating metastable zone, crystals may grow for considerable time without forming new nuclei. In the labile zone nucleation and growth occur together. In an undersaturated solution crystals dissolve, and in a saturated solution neither nucleation nor growth can take place, nor does the crystal dissolve.

The width of the zone for either seeded or unseeded solutions varies greatly from system to system. In very general terms, soluble substances and substances that crystallize with water of crystallization have wide metastable regions where growth without substantial nucleation is possible, and large crystals may be made from aqueous solutions. Unseeded potassium chloride solutions, for example, may be supercooled by about 10°C without entering the labile zone, and seeded solutions of this substance can support 2 to 3°C of supersaturation. Unseeded solutions of hydrated salts such as sodium thiosulfate or magnesium sulfate may be supersaturated by 20°C or more.

Less soluble substances or those crystallizing in the anhydrous form do not usually support appreciable supersaturations. Sodium chloride, for example, has a narrow metastable zone. It is difficult to grow, from aqueous solution, really large crystals of such substances. Large, perfect crystals of these substances are best produced by very slow cooling of the molten, solvent-free material.

Methods of Forming Nuclei in Solution. In any crystallization process, both nucleation and growth should be under control. In a batch process, if uniform crystals are desired, as large a proportion as possible of the nuclei should be formed at the same time. Otherwise, a nonuniform crop is obtained. In a continuous process, the average rate of nucleation equals the rate at which crystals are removed from the unit. Nuclei may be obtained either by adding the desired number of seeds, usually in the form of crushed crystals, to the crystallizer or, more commonly, by forming nuclei *in situ*. Nuclei are created *in situ* by one or more of the following methods:

1. *Spontaneous Nucleation from Unseeded Solutions.* This occurs only when the solution becomes sufficiently supersaturated to enter the labile zone.

2. *Mechanical Impact.* Vigorous stirring or collisions of crystals with each other or with the wall of the crystallizer lead to nucleation. Small crystalline fragments may also be formed by attrition. When the magma is circulated through a pump, the corners and edges of the crystals may be knocked off. Such fragments act as nuclei and grow into normal crystals.

3. *Catalytic Effect of Existing Crystals.* As concentration increases, the rate of nucleus formation induced by the presence of existing crystals sharply increases.

4. *Local Variations in Concentration.* It is unnecessary that nucleation occur throughout the liquid, as nuclei formed locally are distributed by agitation throughout the entire crystallizing solution. In most crystallizers, nucleation is localized. Thus, the transfer of heat through a crystallizer wall requires temperature gradients which lead to local increases in supersaturation, and the nucleation rate near the wall is increased. Evaporation

from the surface of the solution causes concentration gradients near the surface of the solution which also result in nucleation. Nuclei also tend to form in the small region where the solid wall, the vapor, and the liquid meet. Liquid splashed against the wall dries and forms nuclei. If the liquid is boiling, droplets of solution ejected into the vapor space form nuclei which drop back into the solution.

Crystal Growth

Crystal growth is a diffusional process, modified by the effect of the solid surfaces on which the growth occurs. Solute molecules or ions reach the growing faces of a crystal by diffusion through the liquid phase. The usual mass-transfer coefficient k_y applies to this step. On reaching the surface, the molecules or ions must be accepted by the crystal and organized into the space lattice. The reaction occurs at the surface at a finite rate, and the over-all process consists of two steps in series. Neither the diffusional nor the interfacial step will proceed unless the solution is supersaturated.

Geometry of Crystal Growth: Invariant Crystals. Under ideal conditions, a growing crystal maintains geometric similarity during growth. Such a crystal is called invariant. Figure 14-5 shows cross sections of an

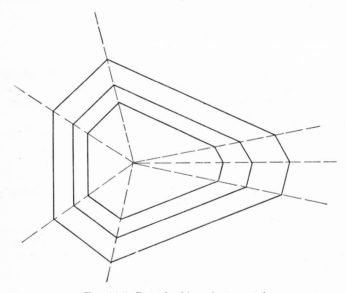

Fig. 14-5. Growth of invariant crystal.

invariant crystal during growth. Each of the polygons in the figure represents the outline of the crystal at a different time. Since the crystal is invariant, these polygons are geometrically similar, and the dotted lines connecting the corners of the polygons with the center of the crystal are straight. The rate of growth of any face is measured by the velocity of translation of the face away from the center of the crystal, in a direction perpendicular to the face. It may be seen that, unless the crystal is a regular polyhedron, the rates of growth of the various faces of an invariant crystal are not equal. The smaller faces grow faster than the larger ones.

Over-all and Individual Growth Coefficients. Consider a single crystal face, e.g., face i of a growing crystal having a total of n faces. Let the area of the face be A_i. Let y be the concentration of the solution at a distance from the face, y_e the saturation concentration, and y' the concentration of the solution at the interface between crystal and liquid. If there were no interfacial reaction, y' would be equal to y_e, but since such a reaction does exist, $y' > y_e$. The rate of mass transfer from the bulk of the liquid to the interface is given by the equation

$$\frac{dm_i}{A_i\,d\theta} = k_y(y - y') \qquad (14\text{-}2)$$

where dm_i is the moles of solute transferred to face i in time $d\theta$ hr and k_y is the usual mass-transfer coefficient as defined by Eq. (10-69).†

Mass-transfer coefficients for crystal growth have been correlated by the typical equation

$$\frac{k_y D_e}{D_m} = C\left(\frac{D_e G}{\mu}\right)^{0.6}\left(\frac{\mu}{\overline{M}D_m}\right)^{0.3} \qquad (14\text{-}4)$$

where k_y = mass-transfer coefficient, lb moles/(ft^2)(hr)(unit mole fraction)
D_m = molal diffusivity, lb moles/ft-hr
D_e = diameter of sphere having same surface area as crystal (equivalent diameter), ft
G = mass velocity of flow of solution past crystal, lb/ft^2-hr
μ = viscosity of solution, lb/ft-hr
$\overline{M}$ = average molecular weight of solution
C = constant

For copper sulfate pentahydrate and magnesium sulfate heptahydrate the constant C is 0.29 and 0.48, respectively.

† In crystallization, the relative-velocity factor ϕ is neither the logarithmic mean solvent concentration nor unity. This is because of the movement of the face of the growing crystal into the solution, which introduces another velocity component for the interface. The correct equation is [6]

$$\phi = \frac{(R - y')\ln\dfrac{R - y}{R - y'}}{(1 - y')\ln\dfrac{1 - y}{1 - y'}} \qquad (14\text{-}3)$$

where $R = M_B/(M_B - M_A)$ and M_A and M_B are the molecular weights of the solute and solvent, respectively. Fortunately, ϕ calculated from this equation is so nearly unity that it can be ignored. It is omitted, therefore, from Eq. (14-2).

Assume that the rate of transfer at the interface is proportional to the supersaturation at the interface, or

$$\frac{dm_i}{A_i \, d\theta} = k_{si}(y' - y_e) \tag{14-5}$$

where k_{si} is the coefficient of the interfacial reaction at face i. Elimination of y' from Eqs. (14-2) and (14-5) gives

$$\frac{dm_i}{A_i \, d\theta} = \frac{y - y_e}{1/k_y + 1/k_{si}} = K_i(y - y_e) \tag{14-6}$$

The surface coefficient k_{si} varies from face to face in the same crystal. Modern theories of crystal growth hold that growth occurs layer by layer. Each new layer of particles on a given face begins as a two-dimensional nucleus attached to that face. Once formed, the nucleus grows rapidly into a completed layer, which covers the entire face. Then another two-dimensional nucleus forms, grows to another layer, and so on.[2a] The energy required to form a stable two-dimensional nucleus on an existing face is less than that needed for forming the stable three-dimensional nucleus of a new crystal, so growth occurs at supersaturations too low to provide driving force for three-dimensional nucleation. Crystals can grow, therefore, within the metastable region where three-dimensional nucleation is rare. The coefficient k_{si} is sensitive to temperature but is independent of velocity.[6]

Growth of Entire Crystal. Let the ratio of the area of face i to A, the total area of the crystal, be α_i. Then

$$A_i = \alpha_i A \tag{14-7}$$

Equation (14-6) may be written for each of the n faces

Face 1: $\qquad\qquad dm_1 = A\alpha_1 K_1(y - y_e) \, d\theta$

Face 2: $\qquad\qquad dm_2 = A\alpha_2 K_2(y - y_e) \, d\theta$

$$\cdots\cdots\cdots\cdots\cdots \tag{14-8}$$

Face n: $\qquad\qquad dm_n = A\alpha_n K_n(y - y_e) \, d\theta$

It is assumed that the over-all supersaturation $y - y_e$ is the same for all faces and for all crystals in the batch.

Adding Eqs. (14-8) gives for dm, the total molal growth of the crystal during time $d\theta$,

$$dm = dm_1 + dm_2 + \cdots + dm_n$$
$$= A(y - y_e)(\alpha_1 K_1 + \alpha_2 K_2 + \cdots + \alpha_n K_n) \, d\theta \tag{14-9}$$

Since the crystal is a polygon, its mass and area may be expressed in terms of shape factors and the arbitrarily chosen linear dimension used to measure size. This is the method used on page 203 for defining the geometry and size of comminuted particles. The shape factors a and b are given by the equations

$$A = 6bD_p^2 \tag{14-10}$$

$$m = \frac{a\rho_s D_p^3}{M_A} \tag{14-11}$$

where ρ_s is the density of the crystal and M_A is the molecular weight of the crystal, including that of the water of hydration if the solute is hydrated. Differentiating Eq. (14-11) on the assumption that a is constant gives

$$dm = \frac{3a\rho_s D_p^2 \, dD_p}{M_A} \tag{14-12}$$

Substituting A from Eq. (14-10) and dm from Eq. (14-12) into Eq. (14-9) gives

$$\frac{3a\rho_s D_p^2 \, dD_p}{M_A} = 6bD_p^2(y - y_e)(\alpha_1 K_1 + \alpha_2 K_2 + \cdots + \alpha_n K_n) \, d\theta$$

On simplification this becomes

$$dD_p = \frac{2(y - y_e)\overline{K} M_A}{\lambda \rho_s} \, d\theta \tag{14-13}$$

where $\lambda = a/b$ and

$$\overline{K} = \alpha_1 K_1 + \alpha_2 K_2 + \cdots + \alpha_n K_n \tag{14-14}$$

An average interfacial coefficient for the entire crystal, denoted by $\bar{k}_s$, may be defined by the equation

$$\overline{K} = \frac{1}{1/k_y + 1/\bar{k}_s} \tag{14-15}$$

Coefficient $\bar{k}_s$ can be determined experimentally.[6]

Example 14-3. Crystals of epsom salt having an equivalent diameter of 0.10 in. are growing in a solution at a supersaturation of 0.002 mole fraction of $MgSO_4 \cdot 7H_2O$ and a temperature of 86°F. The velocity of the solution relative to the crystal is 1.0 ft/sec. The following data are available for saturated solutions of $MgSO_4$ at 86°F.[6]

$\bar{k}_s = 0.33$ lb mole $MgSO_4 \cdot 7H_2O/(ft^2)(hr)$(unit mole fraction)

$\mu = 7.66$ centipoises $\rho_L = 82.5$ lb/ft³   (density of solution)

$D_m = 3.56 \times 10^{-4}$ lb mole/ft-hr

$\rho_s = 105$ lb/ft³   (density of crystals)

The crystals, although rhombic, may be assumed to be essentially cubical.
What growth rate may be expected, in inches per hour? What growth rate should be expected when the crystals have reached an equivalent diameter of 0.20 in.?

Solution. The mass-transfer coefficient k_y is calculated by Eq. (14-4). From Fig. 14-1, the solubility of $MgSO_4$ at 86°F is 28.5 per cent, based on anhydrous $MgSO_4$. The molecular weights of $MgSO_4 \cdot 7H_2O$ and $MgSO_4$ are 246.5 and 120.4, respectively. The moles of $MgSO_4 \cdot 7H_2O$ in 100 lb of solution are $28.5/120.4 = 0.237$. The mass of $MgSO_4 \cdot 7H_2O$ is $0.237 \times 246.5 = 58.5$ lb, and the free water is $100 - 58.5 = 41.5$ lb, or $41.5/18 = 2.31$ moles. The average molecular weight of the solution is

$$\overline{M} = \frac{100}{0.237 + 2.31} = 39.2$$

Substituting in Eq. (14-4),

$$k_y = \frac{0.48 \times 3.56 \times 10^{-4}}{0.10/12} \left[\frac{(0.10/12)1.0 \times 82.5}{7.66 \times 6.72 \times 10^{-4}} \right]^{0.6} \left[\frac{2.42 \times 7.66}{39.2 \times 3.56 \times 10^{-4}} \right]^{0.3}$$

$$= 3.34 \text{ lb moles/(ft}^2)(\text{hr})(\text{unit mole fraction})$$

From Eq. (14-15),

$$\overline{K} = \frac{1}{1/3.34 + 1/0.33} = \frac{1}{0.30 + 3.0} = 0.30 \text{ lb mole/(ft}^2)(\text{hr})(\text{unit mole fraction})$$

Since the crystals are considered cubes, $\lambda = 1$. From Eq. (14-13),

$$\frac{dD_p}{d\theta} = \frac{2 \times 0.002 \times 246.5 \times 0.30}{105} = 2.8 \times 10^{-3} \text{ ft/hr}$$

$$= 12 \times 2.8 \times 10^{-3} = 0.034 \text{ in./hr}$$

For $D_e = 0.20$ in.,

$$k_y = 3.34 \left(\frac{0.20}{0.10} \right)^{-0.4} = 2.53$$

$$\overline{K} = \frac{1}{1/2.53 + 1/0.33} = 0.29$$

$$\frac{dD_p}{d\theta} = \frac{0.29}{0.30} 0.034 = 0.033 \text{ in./hr}$$

Ideal Crystal Growth. An ideal batch process of crystal growth, to which an actual growth process may be compared, conforms to the following requirements: (1) all crystals in the batch have the same shape, and remain invariant during growth, and (2) no new nuclei form.

A crystal remains invariant if the over-all coefficients K_1, K_2, ..., K_n are constant during the entire growth of the crystal from its formation as a nucleus. These coefficients are constant if the individual coefficients k_y, k_{s1}, k_{s2}, ..., k_{sn} are constant.

When the crystal is invariant, the geometrical factors a, b, α_1, α_2, ..., α_n are constant, and, therefore, the average over-all coefficient $\overline{K}$ and the average interfacial coefficient $\overline{k}_s$ are also constant.

For an ideal process, the total growth of a crystal during time θ_T is found by integrating Eq. (14-13), giving

$$D_{p2} - D_{p1} = \Delta D_p = \frac{2\overline{K}M_A}{\lambda \rho_s} \int_0^{\theta_T} (y - y_e)\, d\theta \qquad (14\text{-}16)$$

where D_{p1} is the crystal size when $\theta = 0$, and D_{p2} is the size when $\theta = \theta_T$. The total growth is ΔD_p. Equation (14-16) shows that, in an ideal growth, the total growth of a crystal is independent of the initial size of the crystal and is the same for all crystals in the batch.[9]

If the over-all supersaturation $y - y_e$ is also constant, Eq. (14-16) becomes

$$\Delta D_p = \frac{2(y - y_e)\overline{K}M_A\theta_T}{\lambda \rho_s} \qquad (14\text{-}17)$$

N vs. D_p Curves for Ideal Growth. From the screen analysis of a mixture of crystals a curve showing the cumulative population N as a function of the particle size D_p can be determined by using Eq. (4-12) and plotting N against D_p. Plots of this type for the seeds and product of an ideal process are shown in Fig. 14-6. The ΔD_p of all crystals is, by Eq. (14-16), independent of size. This is shown by the horizontal ΔD_p vs. N line in Fig.

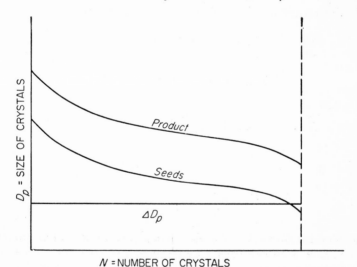

FIG. 14-6. N vs. D_p curves, ideal process.

14-6. The curve for the product is found, then, by vertically displacing the curve of the seeds a distance equal to the growth ΔD_p of the process.

When Eq. (14-16) applies, the screen analysis of the product may be estimated from that of the seeds if the mass ratio of product to seeds is also known.[9]

Example 14-4. Copper sulfate pentahydrate crystals having the screen analysis shown in Table 14-1 are used as seeds in an ideal process. The mass of the product is 10 times that of the seeds. Estimate the screen analysis of the product.

TABLE 14-1. SCREEN ANALYSIS OF SEEDS FOR EXAMPLE 14-4

| Meshes/in. | Opening, in. | Analysis, % | | Meshes/in. | Opening, in. | Analysis, % | |
		Diff.	Cum.			Diff.	Cum.
On 14	0.046	0	0	On 28	0.0232	27	60
16	0.039	3	3	32	0.0195	20	80
20	0.0328	10	13	35	0.0164	15	95
24	0.0276	20	33	42	0.0138	5	100

Solution. Consider the unit mass of the seeds. The cumulative screen analysis of the seeds ϕ_s vs. D_{ps} is plotted from Table 14-1 as curve A in Fig. 14-7. The number of crystals in a small mass of seeds $d\phi_s$ having a size D_{ps} is

$$dN = \frac{d\phi_s}{a\rho_s D_{ps}^3}$$

where ϕ_s is the cumulative mass fraction of seeds at least as large as D_{ps}. Under ideal conditions these crystals grow to size $D_{ps} + \Delta D_p$, where ΔD_p is the linear growth of all crystals in the batch. If dm is the mass of these same crystals after growth, it follows that, since the number of crystals has not changed,

$$dN = \frac{dm}{a\rho_s(D_{ps} + \Delta D_p)^3} = \frac{d\phi_s}{a\rho_s D_{ps}^3}$$

From this, by solving for dm and integrating,

$$m = \int_0^{\phi_s} (1 + \Delta D_p/D_{ps})^3 d\phi_s \quad \text{and} \quad m_p = \int_0^{1.0} (1 + \Delta D_p/D_{ps})^3 d\phi_s \quad (14\text{-}18)$$

where m_p is the total mass of the product from unit mass of seeds. By the conditions of the problem, when $\phi_s = 1.0$, $m_p = 10$. By trial and error, a value of ΔD_p may be found consistent with $m_p = 10$ and giving this result when used in a graphical integration of Eq. (14-18).

The result of such a trial and error shows that, when ΔD_p is 0.027 in., the integral of Eq. (14-18) is 10. The details of the calculation are given in Table 14-2, and

TABLE 14-2. COORDINATES OF CURVE OF ϕ_s VS. $(1 + 0.027/D_{ps})^3$ FOR EXAMPLE 14-4

ϕ_s	D_{ps}	$\left(1 + \dfrac{0.027}{D_{ps}}\right)^3$	ϕ_s	D_{ps}	$\left(1 + \dfrac{0.027}{D_{ps}}\right)^3$
0	0.046	4.00	0.60	0.0232	10.2
0.03	0.039	4.85	0.80	0.0195	13.5
0.13	0.0328	6.05	0.95	0.0164	18.5
0.33	0.0276	7.74	1.00	0.0138	25.8

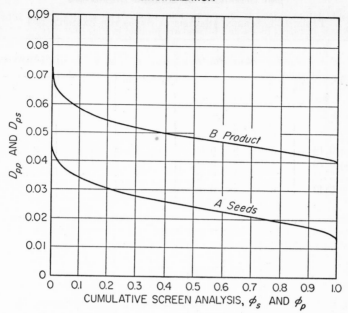

FIG. 14-7. Screen analysis of seeds and product for Example 14-4.

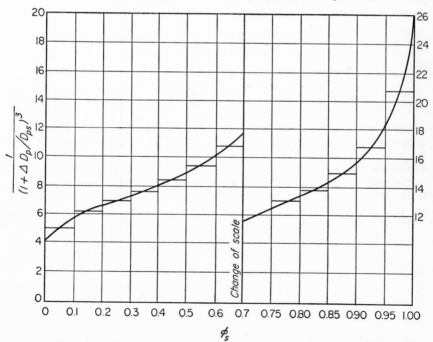

FIG. 14-8. Graphical integration for Example 14-4.

the graphical integration is shown in Fig. 14-8. The results of the integration are shown, as m vs. ϕ_s, in Table 14-3 and are plotted in Fig. 14-9.

TABLE 14-3. COORDINATES OF POINTS ON CURVE OF ϕ_s VS. m FOR EXAMPLE 14-4

ϕ_s	m	ϕ_s	m	ϕ_s	m
0	0	0.50	3.41	0.85	7.37
0.10	0.50	0.60	4.35	0.90	8.12
0.20	1.12	0.70	5.43	0.95	8.96
0.30	1.81	0.75	6.03	1.00	10.00
0.40	2.57	0.80	6.68		

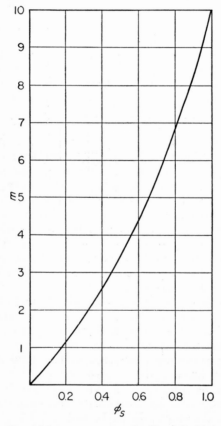

FIG. 14-9. m vs. ϕ_s curve for Example 14-4.

The results shown in Table 14-3 and Fig. 14-9 are translated into the screen analysis of the product by the following method. The largest crystal in the product has a size of $0.046 + 0.027 = 0.073$ in., and the smallest crystal has a size of $0.0138 + 0.027 = 0.0408$ in. Reference to Appendix 18 shows that all the product passes a 9-mesh screen, and is retained on a 16-mesh screen. To obtain the size distribution of the product, assign to D_{pp}, the size of the product, values equal to the actual mesh sizes of the screens; read the corresponding values of ϕ_s from curve A of Fig. 14-7; and then read the values of m from Fig. 14-9. The cumulative and differential analyses are then calculated by expressing the values of m on a percentage basis. This calculation is shown in Table 14-4, and the cumulative analysis of the product ϕ_p vs. D_{pp} is shown as curve B in Fig. 14-7.

TABLE 14-4. CALCULATION OF SCREEN ANALYSIS OF PRODUCT FOR EXAMPLE 14-4

Meshes/in.	D_{pp}	$D_{ps} = D_{pp} - 0.027$	ϕ_s	m	Screen analysis	
					Cum.	Diff.
On 9	0.078	0.051	0	0	0	0
10	0.065	0.038	0.035	0.15	1.5	1.5
12	0.055	0.028	0.290	1.75	17.5	16.0
14	0.046	0.019	0.81	6.80	68.0	50.5
16	0.039	0.012	1.00	10.00	100.0	32.0

Actual Crystal Growth. In actual growth in a batch of seeded solution the process may differ considerably from the ideal. The first difference is the appearance of new nuclei. Unwanted nuclei are called *false grain*. False grain may be minimized by careful control of supersaturation, but it cannot always be eliminated. When growth is small, e.g., if ΔD_p is not greater than the average size of the seeds, the crystals remain essentially invariant, and ΔD_p is independent of crystal size.[9] In large growths, however, a second difference between actual and ideal growth is that the crystals do not remain invariant [10] and ΔD_p is not, therefore, independent of size. Experiment shows that the larger crystals in a batch grow more than do the smaller ones.

Typical curves of N vs. D_p for the seeds and product of an actual batch process with large growth are shown on Fig. 14-10. The new nuclei, or false grain, are shown as the extension of the curve for the product beyond the terminus of that for the seeds. The plot of ΔD_p vs. N shows the preferential growth of the large crystals over that of the small.

The failure of the actual growth process to follow the first requirement of an ideal process, namely, that the crystals remain invariant, is probably caused by variations in the individual coefficients; e.g., k_y depends on both velocity and particle size, as shown in Eq. (14-4). Under the usual agitation conditions in a crystallizer, it is probable that the relative velocities of the larger crystals are greater than those of the smaller ones. Although the effect of particle size is counter to that of velocity, the velocity effect predominates, and the larger the crystal, the faster it grows.   The individual interfacial constants, such as k_{si}, may also vary haphazardly because of fortuitous changes in the structure of the surface.[3]

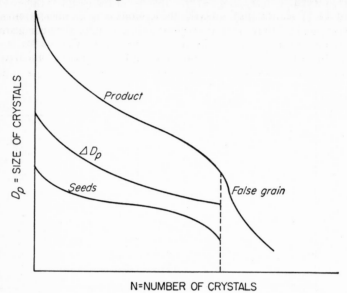

Fig. 14-10. N vs. D_p, actual process.

Combined Nucleation and Growth.   Except when, in batch operation, all the nuclei are either added as seeds or are formed at once *in situ*, growth and nucleation occur simultaneously.   The result of this combined process on the size distribution of the product depends on the relative rates of nucleation and growth.   As the solute crystallizes, the precipitating solid must either appear as new nuclei or must attach itself to existing crystals, and a competition exists between the two processes.   The mass represented by new nuclei is insignificant, but if large numbers of nuclei do form, the solid available for each nucleus is small, and, for a given yield, the final crop consists of a few large and many small crystals.   This is why suppression of false grain is so important when uniform crystals of reasonable size are wanted.

Actual data for rates of nucleation and growth are almost nonexistent, and only qualitative reasoning can be used.   Such reasoning, however, is useful in analyzing the effects of operating variables on an industrial crystallization process.   To aid in this analysis reference is made to the hypothetical curves shown in Fig. 14-11.   This figure shows two rate curves, each plotted against supersaturation.   Curve A shows the growth rate, measured as increase in linear dimension per unit time, and curve B shows the nucleation rate, measured in number of new nuclei formed per unit time in unit volume of solution.   On the assumption that the growth rate is approximately proportional to the supersaturation, curve A is straight;

on the other hand, the nucleation rate, as shown by curve B, is low at the start, is small in the metastable region, and rises rapidly as the labile region is entered.

Figure 14-11 shows that growth predominates over nucleation at low concentrations but that nucleation predominates strongly over growth at higher supersaturations, especially beyond the break in curve B.

The major difficulty in controlling crystallization lies in the control of nucleation rather than growth. The rate of growth increases uniformly

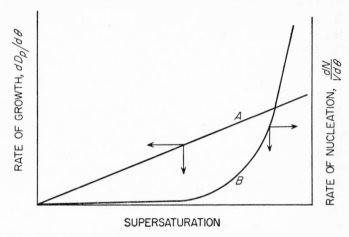

FIG. 14-11. Rate-of-growth and rate-of-nucleation curves.

with increase in supersaturation, but that of nucleation changes so rapidly when the labile zone is entered that it is too low to be significant at a supersaturation just short of the labile zone and too great at a slightly larger supersaturation just inside the labile zone.

The positions of the curves of Fig. 14-11 vary with operating variables and with the characteristics of the substance. The position of curve A depends on the conditions of agitation and the viscosity, density, and diffusivity of the solution. It is also a function of temperature. The position of curve B depends on the concentration of nuclei, the agitation, and the size and area of the crystals present in the solution.[17] When nucleation occurs from attrition or from a vapor-liquid interface or from local concentration and temperature gradients, curve B is displaced to the left. An important variable in controlling nucleation is the ratio of crystals to magma. This ratio, which may be expressed either as mass ratio or volume ratio, is called the magma density. High magma densities tend to suppress nucleation and encourage growth. Most modern crystallizers use high magma densities for this reason. A range of from 20 to 35 per cent suspended solids is usual.

In the operation of a continuous unit, the rate of nucleation may cycle from a minimum to a maximum and back again. When the rate of nucleation is low, the crystal population decreases, the total rate of precipitation in pounds per hour drops, because of the scarcity of surface for growth, and the supersaturation builds up. When the labile zone is entered, copious nucleation occurs for a short time, new area for growth is quickly created, the rate of precipitation increases, and the supersaturation decreases. This completes the cycle.

Usually, the objective of crystallization is to obtain a maximum growth rate consistent with a low nucleation rate. Obviously, this is attained by operating at an average supersaturation just short of that corresponding to rapid nucleation. It is not always satisfactory, however, to operate under such conditions, as control at this point is obviously difficult. Also, the quality of the crystal product sometimes suffers.[11] Capacity must then be sacrificed by reducing the supersaturation. Also, at low supersaturations, the required nuclei may have to be added as seeds or created by means other than entering the labile zone. Balancing these conflicting factors is an art rather than a science.

Effects of Impurities. Small concentrations of impurities or of components other than the solvent and solute may strongly influence crystallization. The solubility may be modified and the yield changed. If impure solutions are used, it is advisable to determine the solubility curve on the actual solution rather than to accept data from the literature based on pure solutions. Impurities also affect both nucleation and growth. Small concentrations of high-molecular-weight substances may retard or even stop nucleation, while other substances act catalytically to reduce the width of the metastable zone and to accelerate nucleation.[5]

The relative growth of the various faces of the same crystal may be greatly modified by small amounts of soluble substances, especially dyes. Certain faces may be completely suppressed in favor of the growth of other faces not found at all in crystals grown in pure solution. Except for angles, the entire geometry of the crystal may be changed thereby. These variations are called changes in crystal habit. Many examples are known.[2b]

CRYSTALLIZATION EQUIPMENT

Commercial crystallizers may operate either continuously or batchwise. Except for special applications, continuous operation is preferred. The first requirement of any crystallizer is to create a supersaturated solution, as crystallization cannot occur without supersaturation. The means used to produce supersaturation depends primarily on the solubility curve of the solute. Solutes like common salt have solubilities nearly independent of temperature, and others, such as anhydrous sodium sulfate and sodium carbonate monohydrate, have inverted solubility curves and become more soluble as the temperature is lowered. To crystallize these materials, supersaturation must be developed by evaporation. Other materials, such as glauber salt, epsom salt, copperas, and hypo, are much less soluble at low temperatures than at high, and cooling without evaporation produces supersaturation. In intermediate cases, a combination of evaporation and cooling is effective. Sodium nitrate, for example, may be satisfactorily crystallized by cooling without evaporation, evaporation without cooling, or a combination of both.[16]

A classification of crystallization equipment based on the method used to create supersaturation is the following:

1. Supersaturation produced by cooling without appreciable evaporation, e.g., tank crystallizers, Swenson-Walker crystallizer, Krystal cooling crystallizer.

2. Supersaturation produced by evaporation without appreciable cooling, e.g., crystallizing evaporators, Krystal evaporator-crystallizer, strike pans.

3. Evaporation combined with adiabatic cooling, e.g., Swenson vacuum crystallizer, Krystal vacuum crystallizer.

Tank Crystallizers. The oldest method of producing crystals in quantity is to prepare hot, nearly saturated solutions and to cool them, by natural convection, in open rectangular tanks. Little or no attempt is made to seed these tanks, to provide agitation, or to control supersaturation. When the tanks have cooled sufficiently, which usually is a matter of several days, remaining mother liquor is drained, and the crystals removed by hand. This requires much labor. The crop is usually aggregated and occludes impurities. Floor space and inventory in process are large. The method is now used only in crude operations and has been largely superseded by more modern and efficient methods.

Crystallizers Cooled by Liquid Cooling Medium. To speed crystallization and to obtain control of the process, heat may be removed from the wall of the crystallizer. The Swenson-Walker unit described in Chap. 10 is typical of this type. Another crystallizer of this kind is the Krystal cooling crystallizer shown in Fig. 14-12. A basic principle of this unit is the creation of supersaturation without nucleation in one step and the relief of supersaturation by both nucleation and growth in a second step. The supersaturation developed in the first step is sufficiently small to ensure that the labile region is not entered. In the unit shown in Fig. 14-12 a circulating stream of solution passes up through the tubes of cooler

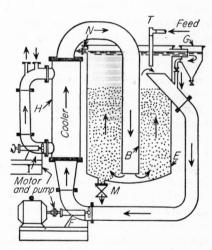

FIG. 14-12. Krystal cooling crystallizer. (*Svanoe.*[15] *Courtesy of Industrial and Engineering Chemistry.*)

H, where supersaturation is produced by cooling provided by water or brine circulating through the shell side of the exchanger. The supersaturated solution flows through pipe N to the bottom of pipe B and upward through the mass of growing crystals in tank E. The flow rate is such that the bed of crystals is fluidized and so kept in agitation at high magma density. Solution, partially relieved of its supersaturation by contact with the crystals in tank E, is mixed with incoming feed entering through pipe T, and the mixture of feed and recycle returns to the cooler to be supersaturated. Nucleation occurs by attrition, by action of the existing crystals, or by areas of local concentration.

The crystals are retained by hindered-settling classification in the crystal bed until they are sufficiently large to be removed as product through discharge M. The small separator G eliminates excess nuclei by removing part of the liquor from the top of the tank, settling out the nuclei, which represent a very small mass, and returning the settled liquor to the system. Sufficient mother liquor is removed with the product to maintain the overall material balance. The magma density is high.

Crystallizing Evaporators. The vertical-tube evaporators described in Chap. 9 may be used to concentrate salting liquors, and so function as crystallizing evaporators. The crystals are removed continuously or periodically as described in Chap. 9. Although these units are designed largely from the standpoint of heat transfer and evaporation and control of crystallization is rather incidental, the classification action of the currents in them is such that the crystal size of product is small but quite uniform. Circulating pumps or internal agitators are often used to maintain adequate velocity to keep the crystals in suspension. Also, the use of basic crystallization principles to prevent salting on the tubes and bodies or the evaporator is important. These principles are to use a high magma density, to maintain adequate flow, and to keep the supersaturation low.

Strike Pan. An example of crystallization in a batch evaporator utilizing sound crystallization principles is the sugar strike pan. This is the classic method of crystallizing both cane and beet sugar. A strike pan is a special coil-tube single-effect evaporator-crystallizer, in which the steam pressure and vacuum are under the accurate control of the operator. Concentrated solution is admitted to the pan and further concentrated, with the addition of more feed to maintain level. When the supersaturation approaches the labile region, the operator either adds seeds or, by careful control of steam pressure and vacuum, "shocks" sufficient nuclei to seed the batch. Then, by remaining within the metastable zone and avoiding formation of false grain the operator builds the seeds to the required size. This may be done by automatic control of steam pressure, using the boiling-point elevation of the solution as the control measurement. When the desired size of grain has been achieved, the entire charge is quickly dumped, and the mother liquor, or molasses, is removed by centrifuging. Since all nuclei were formed at the same time and were grown under identical conditions for the same length of time, the product is uniform in size.

Krystal Evaporator-Crystallizer. Figure 14-13 shows another form of the Krystal equipment. In this unit supersaturation is created by evaporation rather than by cooling. Circulating liquid flows downward through the tubes of heater H, where it is heated by steam condensing in the shell side of the heater. The static head on the solution during heating is sufficient to prevent evaporation. The heated solution then flows up to the vapor head A, where flash evaporation occurs and supersaturation is created. Vapor leaves through connection U to a condenser. Circulation is maintained by pump F. Heater H, vapor head A, and pump F consti-

tute a forced-circulation evapora- tor, the function of which is to in- crease the supersaturation of the liquid. Supersaturated solution flows down through the leg B and then passes upward through the bed of forming and growing crys- tals in tank E. The action here is identical with that in tank E of the unit shown in Fig. 14-12. Classified product leaves through discharge M, and feed is mixed at connection T with the circulating solution just before it enters heater H.

Vacuum Crystallizers. The most important type of crystallizer in use today is the vacuum crys- tallizer, which utilizes adiabatic evaporation and cooling to create supersaturation. Assume that a warm saturated solution is admitted to a closed lagged vessel in which a vacuum is maintained that corre- sponds to a boiling point of the solu- tion lower than the feed tempera- ture. The solution must then flash and spontaneously cool by adiabatic evaporation to its equilibrium tem- perature at the absolute pressure in the vessel. The sensible heat from the cooling is automatically con-

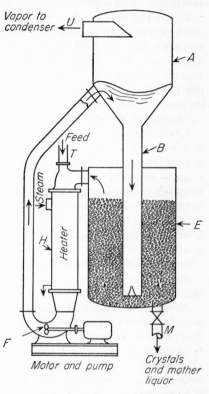

Fig. 14-13. Krystal evaporator-crystal- lizer. (Svanoe.[15] Courtesy of Industrial and Engineering Chemistry.)

verted into latent heat of evaporation, and the loss of both temperature and solvent creates supersaturation in the liquid, provided the solubility relation is normal.

A vacuum crystallizer is simple, contains a minimum of moving parts, and requires no heating or cooling surface. It can, therefore, be built of acid-resisting metals or be rubber- or lead-lined. Its capacity can be large. By the use of steam-jet ejectors of the kind described in Chap. 3 it is pos- sible to produce low temperatures and therefore obtain large yields, thus returning a minimum amount of material to the process as mother liquor.

Vacuum crystallizers cannot be used when refrigeration temperatures are required or where the solution has a large boiling-point elevation. For such applications, cooling crystallizers are required.

Two major types of vacuum crystallizer are available, the Swenson type, in which supersaturation and crystallization occur in the same locations, and the Krystal type, in which supersaturation and crystallization are separated. The Swenson unit may be operated either batchwise or continuously. The Krystal unit is a continuous crystallizer.

Under the low absolute pressure existing in a vacuum crystallizer, the effect of static head on boiling point is important; e.g., water at 45° has a vapor pressure of 0.30 in. Hg. This is a pressure readily obtainable by steam jets. A static head of 1 ft increases the absolute pressure to 1.18 in. Hg, where the boiling point of water is 84°F. Thus, a feed at 84°F entering 1 ft below the surface of the liquid would not flash at the point of entrance. This effect of static head is even greater with heavy magmas.

Because of the effect of static head, evaporation and cooling occur only in liquid layers near the surface, and concentration and temperature gradients near the surface are formed. Also, the crystals tend to settle to the bottom of the mass, where there is naturally no supersaturation. A vacuum crystallizer, therefore, is not operable unless the charge is well agitated, to equalize concentration and temperature gradients and to suspend the crystals uniformly throughout the magma. In batch operation agitation is provided by internal stirrers; in continuous units, forced flow of either magma or of liquid, by outside pumps, is used to maintain circulation.

Swenson Vacuum Crystallizer. Figure 14-14 shows a batch vacuum crystallizer of the Swenson type. Vessel A is charged with warm, nearly saturated solution through feed inlet F. Propellers B generate a centrifugal swirling in the liquid to equalize gradients and suspend the crystals. Vacuum is established by the steam-jet booster C and the condenser D. The booster is needed whenever the final temperature of the magma is below that reachable by the available cooling water. An appropriate vacuum pump, which is usually a two-stage steam jet, follows the condenser to remove air and noncondensed gases.

As the pressure in the unit decreases, the temperature drops, and crystals form and grow. When the final temperature is reached, the batch is dumped through valve E, and the crystals and mother liquor are separated by a filter or centrifuge. Typical cooling curves of a batch vacuum crystallizer are shown in Fig. 14-15. Two curves are shown, one where the batch is seeded at saturation and the other where nuclei were formed spontaneously *in situ*. Better control was obtained when the batch was seeded.

The crystal size obtainable in a batch vacuum crystallizer is small, not larger than about 60-mesh. Batch units have two advantages, however, over continuous units. It is easy to build them of corrosion-resistant metals or to line them with rubber or lead. Also, their steam consumption is lower than that of continuous units operating over the same temperature range. The steam required by a booster increases with the ratio of the pressure of the discharge to that of the feed vapor. The rise is especially large at high compression ratios. In batch operation, much of the vapor is removed at a relatively high pressure, and only at the end is the full

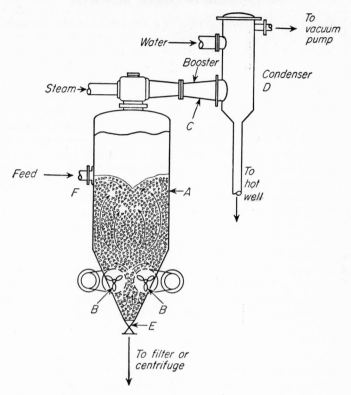

FIG. 14-14. Batch vacuum crystallizer. (*After Seavoy and Caldwell.*)

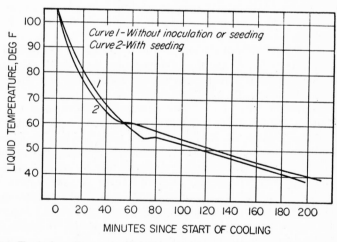

FIG. 14-15. Typical cooling curves, batch vacuum crystallizer. (*Seavoy and Caldwell.*[13] *Courtesy of Industrial and Engineering Chemistry.*)

pressure ratio needed. The average steam consumption, therefore, is considerably less than for continuous operation at the same terminal temperature, where all the vapor must be pumped out at the maximum compression ratio.

Most vacuum crystallizers are operated continuously. Two examples are shown in Fig. 14-16. The essential operating feature of these units is

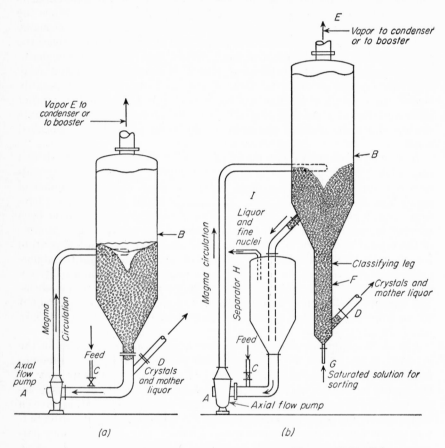

Fig. 14-16. Continuous Swenson vacuum crystallizers. (*Caldwell.*[4])

that the entire mass of magma in the unit is circulated by the outside pump *A*, which takes suction from the cone bottom of the crystallizer body *B* and discharges tangentially at a level just below the liquor level in the body. The tangential flow of the large stream discharged from the pump generates the swirling action and agitation necessary for temperature and concentration uniformity and for crystal suspension. Feed enters continuously through connection *C*, immediately before the pump suction, and product crystals and mother liquor leave continuously through product outlet D. Vapor leaves the body through vapor line E and flows directly to a booster or a condenser.

The temperature of the combined feed and recycle stream returning to the body from the pump should be only about 2°F above the magma temperature in the body.[4] Exceeding this differential leads to excessive nucleation near the point of entrance. This temperature differential is regulated by controlling the ratio of recycled magma to feed. Since the feed is necessarily considerably warmer than the magma if a reasonable yield is to be obtained, a small amount of feed is sufficient to increase the temperature of a large amount of recirculation, and the ratio of magma to feed is large. The crystal discharge pipe D is slanted upward and away from the crystallizer in order that, if the discharge line is temporarily closed, the crystals in the line will settle back into the crystallizer rather than continue to grow and plug the line. For good growth, the magma density should be high, of the order of 20 to 30 per cent of suspended solids.

In a continuous crystallizer nuclei are continuously formed as seeds and continuously removed as product. For continuity, the number of crystals removed during a given time must, on the average, equal the number of nuclei born during the same time. The magma in the unit contains, therefore, crystals of all ages, from those just born to some that may have been recirculated for considerable time. The product is a representative sample of the crystal inventory in the unit and contains crystals of all sizes, from zero to quite large individuals. The mass of fines is small, and the abnormally large crystals are few, so most of the product lies within a relatively narrow size range, as much as 80 per cent on one screen in a screen analysis. The usual product from a vacuum unit of the type of Fig. 14-16a is about 20- to 30-mesh. It is not practicable to grow crystals larger than this in this equipment.

Use of Classification and Nuclei Removal. The smaller crystals in the product, although low in mass, are often undesirable, as their presence tends to favor caking in the package. These smaller crystals may be eliminated by using a classifying leg, as shown in Fig. 14-16b. A supplementary flow of sorting liquid, in the form of a saturated solution, gives a hindered-settling classification in the leg F. The solution enters the bottom of the leg through connection G. Product, free of the smaller particles, which are washed back into crystallizer body B by the sorting liquid, is withdrawn through connection D near the bottom of the leg F. The use of classification does not affect the size distribution of particles larger than the average.

Classification is effective only when the number of crystals being grown can be regulated. If excessive nucleation occurs, the unit is overloaded with small crystals that cannot be grown to the desired minimum size, and continuity of operation becomes impossible. To remove unwanted nuclei, the circulation magma is passed through separator H, where the velocity is reduced to allow the larger crystals to settle, so the smaller nuclei can be removed in a stream of liquid withdrawn through connection I from the top of the separator. By controlling the flow through this connection, the number of nuclei remaining in the unit may be fixed at the proper level to give the product desired. The number of nuclei removed from separator H may be many times the number of crystals removed as product through discharge D.

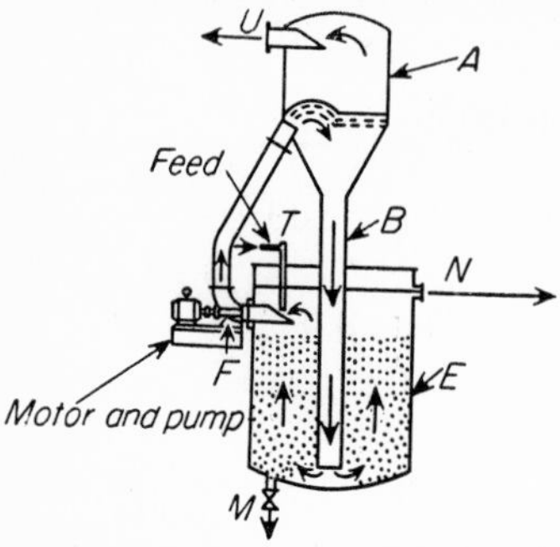

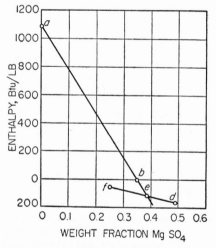

FIG. 14-17. Krystal vacuum crystallizer. (*Svanoe*.[15] *Courtesy of Industrial and Engineering Chemistry*.)

The *Krystal vacuum crystallizer* is shown in Fig. 14-17. Hot feed enters through connection T and is mixed with recycle liquid; the combined stream is sent by pump F to the vapor head A, where adiabatic flash evaporation occurs. Supersaturated solution flows downward through pipe B to the crystallizing tank E in the usual manner. Excess nuclei may be removed through pipe N and product leaves through M.

Yield of Vacuum Crystallizers. The yield from a vacuum crystallizer may be calculated by enthalpy and material balances. A graphical calculation based on the ix chart is shown in Fig. 14-18. Since the process is an adiabatic split of the feed into magma and vapor, points b, for the feed, a for the vapor, and e for the magma lie on a straight line. The isotherm connecting point d for the crystals with point f for the mother liquor also passes through point e, and this point is located by the intersection of lines ab and df. From line segments and the center-of-gravity principle, the ratios of the various streams are calculated.

Example 14-5. A continuous vacuum crystallizer is fed with 100,000 lb/hr of a 35 per cent $MgSO_4$ solution at 183°F. An absolute pressure of 0.2 lb force/in.² is maintained by a jet booster. The solution has a boiling-point elevation of 10°F. Calculate the yield and the evaporation.

Solution. Figure 14-18 is the graphical solution of this problem. The temperature of boiling water at 0.2 lb force/in.² is 53°F, and the equilibrium temperature of the magma in the crystallizer is $53 + 10 = 63$°F. From steam tables, the enthalpy of 1 lb of steam at 0.2 lb force/in.² and 63°F is 1089.5 Btu. The coordinates of point a are $i = 1089.5$ Btu/lb, $x = 0$. The enthalpy of the

FIG. 14-18. Solution of Example 14-5. (*By permission, from "Chemical Engineers' Handbook," 3d ed., by J. H. Perry. Copyright, 1950. McGraw-Hill Book Company, Inc.*)

feed solution, at a concentration of 0.35 and a temperature of 183°F, is zero. The coordinates of point b are $i = 0$ Btu/lb, $x = 0.35$. The straight line df is the 63°F isotherm in the area $bcih$ of Fig. 14-2. The coordinates of its terminals are, for point f, $i = -49.5$ Btu/lb, $x = 0.25$, and for point d, $i = -157.5$ Btu/lb, $x = 0.488$. Lines fd and ab intersect at point e, which has coordinates $i = -111.2$ Btu/lb, $x = 0.386$. By the center-of-gravity principle, the evaporation is

$$100,000 \, \frac{0.386 - 0.35}{0.386 - 0} = 9,320 \text{ lb}$$

and the final magma is $100,000 - 9,320 = 90,680$ lb. The yield is

$$90,680 \, \frac{0.386 - 0.25}{0.488 - 0.25} = 51,800 \text{ lb}$$

SYMBOLS

A Area of crystal, ft^2; A_i, of face i; A_1, A_2, ..., A_n, of face 1, 2, ..., n, respectively

a Shape factor for mass of crystal, defined by Eq. (14-11)

b Shape factor for area of crystal, defined by Eq. (14-10)

C Constant in Eq. (14-4); C_1, constant in Eq. (14-1)

D_e Equivalent diameter of crystal, defined as diameter of sphere having volume of crystal, ft

D_m Molal diffusivity through solution, lb moles/ft-hr

D_p Linear size of crystal, ft; D_{pp}, of product; D_{ps}, of seed; D_{p1}, initial size of crystal at zero time; D_{p2}, final size of crystal at time θ_T

G Mass velocity of solution past crystal, lb/ft^2-hr

i Specific enthalpy, Btu/lb

K Over-all mass-transfer coefficient, lb moles/(ft^2)(hr)(unit mole fraction); K_i, for face i; K_1, K_2, ..., K_n, for faces 1, 2, ..., n, respectively; $\bar{K}$, average value for entire crystal

k_s Coefficient of interfacial reaction, lb moles/(ft^2)(hr)(unit mole fraction); k_{si}, at face i; k_{s1}, k_{s2}, ..., k_{sn}, at faces 1, 2, ..., n, respectively; $\bar{k}_s$, average value for entire crystal

k_y Mass-transfer coefficient from solution to face of crystal, lb moles/(ft^2)(hr)(unit mole fraction)

M Molecular weight; M_A, M_B, of solute and solvent, respectively; $\bar{M}$, average molecular weight of solution

m Mass, lb; m_i, mass deposited on face i; m_1, m_2, ..., m_n, mass deposited on faces 1, 2, ..., n, respectively

N Number of crystals; N_s, number of seeds

n Number of faces on single crystal

R $M_B/(M_B - M_A)$

x Mole or mass fraction of anhydrous solute, in general or in heavy phase (crystals)

y Mole or mass fraction of solute (hydrated basis) in light phase (liquid); y_e, normal solubility of large crystals; y_r, solubility of crystals of size D_p; y', concentration of solution in contact with crystal face

Greek Letters

α Ratio, area of face to total area of crystal; α_i, for face i; α_1, α_2, ..., α_n, for faces 1, 2, ..., n, respectively

ΔD_p Increase in size of crystal during growth, ft, $D_{p2} - D_{p1}$
θ Time, hr; θ_T, time of growth
λ Shape factor, a/b
μ Viscosity, lb/ft-hr
ρ Density, lb/ft^3; ρ_L, of liquid; ρ_s, of crystal
ϕ Relative velocity factor [Eq. (14-3)]; also, cumulative mass fraction of crystals larger than size D_p; ϕ_p, of product crystals; ϕ_s, of seed crystals

PROBLEMS

14-1. Sal soda ($NaCO_3 \cdot 10H_2O$) is to be made by dissolving soda ash in a mixture of mother liquor and water to form a 30 per cent solution by weight at 45°C and then cooling to 15°C. The wet crystals removed from the mother liquor consist of 90 per cent sal soda and 10 per cent mother liquor by weight. This mother liquor is to be dried on the crystals as additional sal soda. The remainder of the mother liquor is to be returned to the dissolving tanks. At 15°C the solubility of Na_2CO_3 is 14.2 parts per 100 H_2O.

Crystallization is to be done in a Swenson-Walker crystallizer. This is to be supplied with water at 10°C, and sufficient cooling water is to be used to ensure that the exit water will not be over 20°C. The Swenson-Walker crystallizer is built in units 10 ft long, containing 3 ft^2 of heating surface per foot of length. An over-all heat-transfer coefficient of 35 Btu/(ft^2)(hr)(°F) is expected.

The latent heat of crystallization of sal soda at 15°C is approximately 25,000 g cal/g mole. The specific heat of the solution is 0.85. A production of 1.0 ton/hr of dried crystals is desired. Radiation losses and evaporation from the crystallizer are negligible.

(a) What amounts of water and of soda ash are to be added to the dissolver per hour? (b) How many units of crystallizer are needed? (c) What is to be the capacity of the refrigeration plant, in tons of refrigeration, if the cooling water is to be cooled and recycled? One ton of refrigeration is equivalent to 12,000 Btu/hr.

14-2. $CuSO_4 \cdot H_2O$ containing 3.5 per cent of a soluble impurity is dissolved continuously in sufficient water and recycled mother liquor to make a saturated solution at 80°C. The solution is then cooled to 25°C and crystals of $CuSO_4 \cdot 5H_2O$ thereby obtained. These crystals carry 10 per cent of their dry weight as adhering mother liquor. The crystals are then dried to zero free water ($CuSO_4 \cdot 5H_2O$). The allowable impurity in the product is 0.6 per cent.

Calculate: (a) the weight of water and of recycled mother liquor required per 100 lb of impure copper sulfate; (b) the percentage recovery of copper sulfate, assuming that the mother liquor not recycled is discarded.

The solubility of $CuSO_4 \cdot 5H_2O$ at 80°C is 120 g per 100 g of free H_2O and at 25°C is 40 g per 100 g of free H_2O.

14-3. A solution of $MgSO_4$ containing 43 g of solid per 100 g of water is fed to a Swenson vacuum crystallizer at 220°F. The vacuum in the crystallizer corresponds to a H_2O boiling temperature of 43°F, and a saturated solution of $MgSO_4$ has a boiling-point elevation of 7°F. How much solution must be fed to the crystallizer to produce 1 ton of epsom salt ($MgSO_4 \cdot 7H_2O$) per hour?

14-4. A Swenson-Walker crystallizer is fed with a saturated solution of $MgSO_4$ at 110°F. The solution and its crystalline crop are cooled to 40°F. The inlet solution contains 1 g of seed crystals per 100 g of solution. The seeds are 80-mesh. Assuming ideal growth, what is the mesh size of the crystals leaving with the cooled product? Evaporation may be neglected.

14-5. The process of Prob. 14-3 is to be conducted in a Krystal vacuum crystallizer of the type shown in Fig. 14-17. The feed temperature and concentration and pressure in the vapor head are unchanged. For each pound of feed, 5 lb of liquid is pumped to the vapor head for flashing. It may be assumed that the mother liquor leaving the crystallizer is saturated at 50°F.

(a) What is the temperature and flow per 100 lb of feed of the supersaturated solution returning from the vapor head to the crystallizing vessel? (b) What is the yield of crystals, in pounds of $MgSO_4 \cdot 7H_2O$ per 100 lb of feed?

REFERENCES

1. Bichowsky, F. R., and F. D. Rossini: "Thermochemistry of Chemical Substances," Reinhold Publishing Corporation, New York, 1936.
2. Buckley, H. E.: "Crystal Growth," (a) pp. 169–222; (b) pp. 529–559, John Wiley & Sons, Inc., New York, 1951.
3. Bunn, C. W.: *Discussions Faraday Soc.*, **5**: 132 (1949).
4. Caldwell, H. B.: private communication.
5. Egli, P. H., and S. Zerfoss: *Discussions Faraday Soc.*, **5**: 61 (1949).
6. Hixson, A. W., and K. L. Knox: *Ind. Eng. Chem.*, **43**: 2144 (1951).
7. "International Critical Tables," vol. IV, McGraw-Hill Book Company, Inc., New York, 1928.
8. La Mer, V. K.: *Ind. Eng. Chem.*, **44**: 1270 (1952).
9. McCabe, W. L.: *Ind. Eng. Chem.*, **21**: 30, 112 (1929).
10. McCabe, W. L., and R. P. Stevens: *Chem. Eng. Progr.*, **47**: 168 (1951).
11. Miller, P., and W. C. Saeman: *Chem. Eng. Progr.*, **43**: 667 (1947).
12. Perry, J. H.: "Chemical Engineers' Handbook," 3d ed., p. 246, McGraw-Hill Book Company, Inc., New York, 1950.
13. Seavoy, G. E., and H. B. Caldwell: *Ind. Eng. Chem.*, **32**: 627 (1940).
14. Seidell, A.: "Solubilities," 3d ed., D. Van Nostrand Company, Inc., New York, 1940.
15. Svanoe, H.: *Ind. Eng. Chem.*, **32**: 636 (1940).
16. Svanoe, H.: private communication.
17. Ting, H. H., and W. L. McCabe: *Ind. Eng. Chem.*, **26**: 1201 (1934).
18. Van Hook, A., and F. Frulla: *Ind. Eng. Chem.*, **44**: 1305 (1952).
19. Young, S. W.: *J. Am. Chem. Soc.*, **33**: 148 (1911).

Chapter 15

AIR-WATER-CONTACT OPERATIONS

Processes depending on contact between air and water have been introduced in Chap. 10. Calculations on such operations require a knowledge of the amount of water vapor carried by air under various conditions, of the enthalpy of mixtures of air and water vapor, of the volume of air-water mixtures, and of the change in enthalpy, water content, and temperature when water is brought into contact with air. The special situation of contact of air with wet solid is discussed in Chap. 16, on drying. The present chapter deals with the properties of mixtures of air and water, the interactions of air with water, and the equipment in which these operations are conducted. Although the discussion is concerned primarily with water and air, the fundamentals apply to any mixture of a fixed gas and a condensable vapor, and some attention is given to such other systems.

Heat or mass transfer have been basic to many operations considered in previous chapters, but temperature and concentration changes have been treated separately. In air-water contacts and in drying, however, heat transfer and mass transfer occur together, and concentration and temperature change simultaneously. The principles of these combined processes are discussed in this chapter.

Definitions. For engineering use the simplest basis for computing the various quantities is unit mass of dry air.[2] Another choice is 1 mole of dry air, but a basis of 1 lb of dry air is used throughout this chapter. Because the properties of a gas-vapor mixture vary with total pressure, the pressure must be fixed. Unless otherwise specified, a total pressure of 1 atm is assumed. Also, it is assumed that mixtures of air and water vapor follow the ideal-gas laws.

Humidity H is the pounds of water vapor carried by 1 lb of dry air. So defined, humidity depends only on the partial pressure of the water vapor in the mixture when the total pressure is fixed. If the partial pressure of the water vapor is $\bar{p}_A$ atm, the molal ratio of water to dry air at 1 atm is $\bar{p}_A/(1 - \bar{p}_A)$. Since the molecular weights of air and water are 29 and 18, respectively, the humidity is

$$H = \frac{18\bar{p}_A}{29(1 - \bar{p}_A)} \tag{15-1}$$

Saturated air is air in which the water vapor is in equilibrium with liquid water at air temperature. If Dalton's law is accepted, the partial pressure of water in saturated air equals the vapor pressure of water at air temperature. If H_s is the saturation humidity, and if p_A is the vapor pressure of water,

$$H_s = \frac{18p_A}{29(1 - p_A)} \tag{15-2}$$

Relative humidity H_R is defined as the ratio of the partial pressure of the water vapor to the vapor pressure of water at air temperature. It is usually expressed on a percentage basis, so 100 per cent humidity means saturated air, and 0 per cent humidity means dry air. By definition

$$H_R = 100 \frac{\bar{p}_A}{p_A} \tag{15-3}$$

Percentage humidity H_A is the ratio of the actual humidity H to the saturation humidity H_s at air temperature, or

$$H_A = 100 \frac{H}{H_s} = 100 \frac{\bar{p}_A/(1 - \bar{p}_A)}{p_A/(1 - p_A)} = H_R \frac{1 - p_A}{1 - \bar{p}_A} \tag{15-4}$$

At all humidities other than 0 or 100 per cent, the percentage humidity is less than the relative humidity.

Humid heat c_s is the Btu necessary to increase the temperature of 1 lb of dry air, plus whatever water it may contain, by 1°F. For the usual temperature range, the specific heats of dry air and water vapor may be taken as constants, equal to 0.24 and 0.46 Btu/(lb)(°F), respectively. Then, by definition,

$$c_s = 0.24 + 0.46H \tag{15-5}$$

Humid volume v_H is the total volume, in cubic feet, of 1 lb of dry air, plus whatever water vapor it may contain, at 1 atm and air temperature. By using the gas laws, v_H is calculated from humidity and temperature by the equation

$$v_H = \frac{359(t + 460)}{29 \times 492} + \frac{359(t + 460)H}{18 \times 492}$$

$$= (0.730t + 335.7)\left(\frac{1}{29} + \frac{H}{18}\right)$$

$$= (0.730t + 335.7)\left(\frac{1}{29} + \frac{H_A H_s}{1,800}\right) \tag{15-6}$$

where t is the temperature in degrees Fahrenheit. For dry air $H = 0$, and v_H is the specific volume of dry air. For saturated air, $H_A = 100$, $H = H_s$, and v_H becomes the *saturated volume*.

Dew point is the temperature to which a mixture of air and water vapor must be cooled (at constant humidity) to become saturated. The dew point of saturated air equals the air temperature.

Total enthalpy of air i_y is the enthalpy of 1 lb of air, plus whatever water it may contain. To calculate i_y, two reference states must be chosen, one for air and one for water. Let t_0 be the datum temperature chosen for both components, and base the enthalpy of the water on liquid water at t_0. Let the temperature of the air be t °F and the humidity be H. The total enthalpy is the sum of three items: the sensible heat of the water vapor, $0.46(t - t_0)H$; the latent heat of the water at t_0, $H\lambda_0$, where λ_0 is the latent heat of water at t_0; and the sensible heat of the dry air, $0.24(t - t_0)$. Then

$$i_y = 0.24(t - t_0) + H[\lambda_0 + 0.46(t - t_0)]$$

$$= c_s(t - t_0) + H\lambda_0 \tag{15-7}$$

Adiabatic-saturation Temperature. Consider the process in Fig. 15-1. Air, with an initial humidity H and temperature t, flows continuously through the spray chamber A. The chamber is lagged, so the process

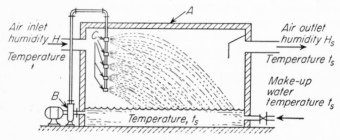

FIG. 15-1. Adiabatic saturator: A, spray chamber. B, circulating pump. C, sprays.

is adiabatic. Water is circulated by pump B from the reservoir in the bottom of the spray chamber through sprays C and back to the reservoir. The air passing through the chamber is cooled and humidified. The temperature of the water reaches a definite steady-state temperature, denoted by t_s and called the adiabatic-saturation temperature. Unless the entering air is saturated, the adiabatic-saturation temperature is lower than the temperature of the entering air. If contact between water and air is sufficient to bring the water and the exit air into equilibrium, the air leaving the chamber is saturated at temperature t_s. Since the water evaporated into the air is lost from the chamber, make-up water is needed. This is supplied to the reservoir as liquid water at temperature t_s.

An enthalpy balance may be written over this process. Pump work is neglected, and the enthalpy balance is based on temperature t_s as a datum. Then the enthalpy of the make-up water is zero, and the total enthalpy of the entering air equals that of the leaving air. Since the latter is at datum temperature, its enthalpy is simply $H_s\lambda_s$, where H_s is the saturation humidity and λ_s is the latent heat, both at t_s. From Eq. (15-7) the total enthalpy of the entering air is $c_s(t - t_s) + H\lambda_s$, and the enthalpy balance is

$$c_s(t - t_s) + H\lambda_s = H_s\lambda_s$$

or $$\frac{H - H_s}{t - t_s} = -\frac{c_s}{\lambda_s} = -\frac{0.24 + 0.46H}{\lambda_s} \qquad (15\text{-}8)$$

Humidity Chart. A convenient diagram showing the properties of mixtures of a permanent gas and a condensable vapor is the humidity chart. A chart for mixtures of air and water at 1 atm is shown in Fig. 15-2, to be found at the end of the book. Many forms of such charts have been proposed. Figure 15-2 is based on the Grosvenor [2] chart.

On Fig. 15-2 temperatures are plotted as abscissas and humidities as ordinates. Any point on the chart represents a definite mixture of air and water. The curved line marked "100%" gives the humidity of saturated air as a function of air temperature. By using the vapor pressure of water the coordinates of points on this line are found by use of Eq. (15-2). Any point above and to the left of the saturation line represents a mixture of saturated air and liquid water. This region is important only in checking fog formation. Any point below the saturation line represents undersaturated air, and a point on the temperature axis represents dry air. The curved lines between the saturation line and the temperature axis marked in even per cents represent mixtures of air and water of definite percentage humidities. As shown by Eq. (15-4), linear interpolation between the saturation line and the temperature axis may be used to locate the lines of constant percentage humidity.

The slanting lines running upward and to the left to the saturation line are called *adiabatic-cooling lines*. They are plots of Eq. (15-8), each drawn for a given constant value of the adiabatic-saturation temperature. For a given value of t_s, both H_s and λ_s are fixed, and the line of H vs. t can be plotted by assigning values to H and calculating corresponding values of t. Inspection of Eq. (15-8) shows that the slope of an adiabatic-cooling line, if drawn on truly rectangular coordinates, is $-c_s/\lambda_s$, and, by Eq. (15-5), this slope depends on the humidity. On rectangular coordinates, then, the adiabatic-cooling lines are neither straight nor parallel. In Fig. 15-2 the ordinates are sufficiently distorted to straighten the adiabatics and render them parallel, so interpolation between them is easy. The ends of the adiabatics are identified with the corresponding adiabatic-saturation temperatures.

Lines are shown on Fig. 15-2 for the specific volume of dry air and the saturated volume. Both lines are plots of volume against temperature. Volumes are read on the scale at the right. Coordinates of points on these lines are calculated by use of Eq. (15-6), calling $H_A = 0$ for dry air and $H_A = 100$ for saturated air. Linear interpolation between the two lines, based on percentage humidity, gives the humid volume of unsaturated air. Also, the relation between the humid heat c_s and humidity is shown as a line on Fig. 15-2. This line is a plot of Eq. (15-5) The scale for c_s is at the top of the chart.

Use of Humidity Chart. The utility of the humidity chart as a source of data on a definite air-water mixture may be shown by reference to Fig. 15-3, which is a portion of the chart of Fig. 15-2. Assume, for example, that a given stream of undersaturated air is known to have a temperature t_1 and a percentage humidity H_{A1}. Point a represents this air on the chart. This point is the intersection of the constant-temperature line for

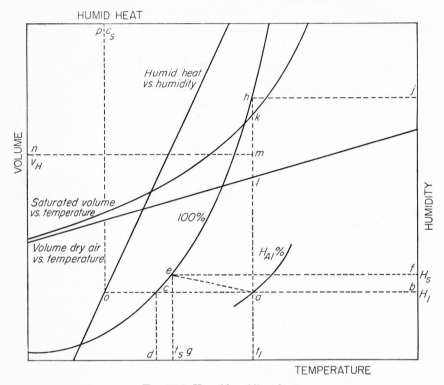

FIG. 15-3. Use of humidity chart.

t_1 and the constant-percentage-humidity line for H_{A1}. The humidity H_1 of the air is given by point b, the humidity coordinate of point a. The dew point is found by following the constant-humidity line through point a to the left to point c on the 100 per cent line. The dew point is then read at point d on the temperature axis. The adiabatic-saturation temperature is the temperature applying to the adiabatic-cooling line through point a. The humidity at adiabatic saturation is found by following the adiabatic line through point a to its intersection e on the 100 per cent line, and reading humidity H_s at point f on the humidity scale. Interpolation between the adiabatic lines may be necessary. The adiabatic-saturation temperature t_s is given by point g. If the original air is subsequently saturated at constant temperature, the humidity after saturation is found by following the constant-temperature line through point a to point h on the 100 per cent line, and reading the humidity at point j.

The humid volume of the original air is found by locating points k and l on the curves for saturated and dry volumes, respectively, corresponding to temperature t_1. Point m is then found by moving along line lk a distance $(H_A/100)\overline{kl}$ from point l, where $\overline{kl}$ is the line segment between points l and k. The humid volume v_H is given by point n on the volume scale. The humid heat of the air is found by locating point o, the intersection of the constant humidity line through point a and the humid-heat line, and reading the humid heat c_s at point p on the scale at the top.

Example 15-1. The temperature and dew point of the air entering a certain dryer are 150 and 60°F, respectively. What additional data for this air can be read from the humidity chart?

Solution. The dew point is the temperature coordinate on the saturation line corresponding to the humidity of the air. The saturation humidity for a temperature of 60°F is 0.011 lb of water per pound of dry air, and this is the humidity of the air. From the temperature and humidity of the air, the point on the chart for this air is located. At $H = 0.011$ and $t = 150$°F, the percentage humidity H_A is found by interpolation to be 5.9 per cent. The adiabatic-cooling line through this point intersects the 100 per cent line at 85°F, and this is the adiabatic-saturation temperature. The humidity of saturated air at this temperature is 0.026 lb of water per pound of dry air. The humid heat of the air is 0.245 Btu/(lb dry air) (°F). The saturated volume at 150°F is 20.7 ft³ per pound of dry air, and the specific volume of dry air at 150°F is 15.35 ft³/lb. The humid volume is, then,

$$v_H = 15.35 + 0.059(20.7 - 15.35) = 15.67 \text{ ft}^3/\text{lb dry air}$$

Humidity Charts for Systems Other than Air–Water. A humidity chart may be constructed for any system at any desired total pressure. The data required are the vapor pressure and latent heat of vaporization of the condensable component as a function of temperature, the specific heats of dry gas and vapor, and the molecular weights of both components. If a chart on a molal basis is desired, all equations may easily be modified to the use of molal units. If a chart at a pressure other than 1 atm is wanted, obvious modifications in the above equations may be made. Charts for several common systems besides air-water have been published.[1,8]

Wet-bulb Temperature and Measurement of Humidity

The air-water properties discussed above and those shown on the humidity chart are static or equilibrium quantities. Equally important are the rates at which mass and heat are transferred between water and air not in equilibrium. A useful quantity depending on both of these rates is the wet-bulb temperature.

Wet-bulb Temperature. The wet-bulb temperature is the steady-state, nonequilibrium temperature reached by a small mass of water immersed under adiabatic conditions in a continuous stream of air. The mass of the water is so small in comparison with the air stream that there is only a negligible change in the properties of the air, and the effect of the process is confined to the water. The method of measuring the wet-bulb temperature is shown in Fig. 15-4. A thermometer, or an equivalent temperature-measuring device such as a thermocouple, is covered by a wick, which is saturated with liquid water and immersed in a stream of air having a definite temperature t and humidity H. Assume that initially the temperature of the water is about that of the air. Since the air is not saturated, water evaporates, and because the process is adiabatic, the

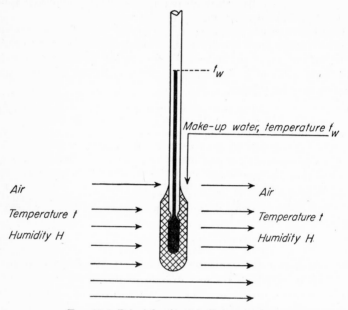

FIG. 15-4. Principle of wet-bulb temperature.

latent heat is supplied at first by cooling the water. As the temperature of the water decreases below that of the air, sensible heat is transferred to the water. Ultimately a steady state is reached at such a water temperature that the heat needed to evaporate the water and heat the vapor to air temperature is exactly balanced by the sensible heat flowing from the air to the water. It is this steady-state temperature, denoted by t_w, that is called the wet-bulb temperature. It is a function of both t and H.

To measure the wet-bulb temperature with precision, three precautions are necessary:[13] (1) the wick must be completely wet so no dry areas of the wick are in contact with the air; (2) the velocity of the air should be large enough to ensure that the rate of heat flow by radiation from warmer surroundings to the bulb is negligible in comparison with the rate of sensible heat flow by conduction and convection from the air to the bulb; (3) if make-up water is supplied to the bulb, it should be at the wet-bulb temperature. When these precautions are taken, the wet-bulb temperature is independent of air velocity over a wide range of flow rates.

The wet-bulb temperature superficially resembles the adiabatic-saturation temperature t_s. Indeed, for air-water mixtures the two temperatures are nearly equal. This is fortuitous, however, and is not true of mixtures other than air and water. The wet-bulb temperature differs fundamentally from the adiabatic-saturation temperature. In the latter the temperature and humidity of the air are changed, and the end point of the process is a true equilibrium, rather than a dynamic steady state.

Commonly, an uncovered thermometer is used along with the wet-bulb to measure t, the actual air temperature, and the air temperature is usually called the dry-bulb temperature.

Theory of Wet-bulb Temperature. At the wet-bulb temperature the rate of heat transfer from the air to the water may be equated to the product of the rate of vaporization and the sum of the latent heat of evaporation and the sensible heat of the vapor. Since radiation may be neglected, this balance may be written

$$q = 18N_A[\lambda_w + 0.46(t - t_w)] \tag{15-9}$$

where N_A = molal rate of vaporization of water
 q = rate of sensible heat transfer to water
 t_w = temperature of water, °F
 t = temperature of air, °F
 λ_w = latent heat at wet-bulb temperature

The rate of heat transfer may be expressed in terms of the area, the temperature drop, and the heat-transfer coefficient in the usual way, or

$$q = h_y(t - t_i)A \tag{15-10}$$

where h_y = heat-transfer coefficient between air and surface of drop, Btu/(ft^2)(hr)(°F)
 t_i = temperature at interface, °F
 A = surface area of water, ft^2

The rate of mass transfer may be expressed in terms of the mass transfer coefficient, the area, and the driving force in mole fraction of water vapor, or

$$N_A = \frac{k_y}{\phi}(y_i - y)A \tag{15-11}$$

where N_A = molal rate of transfer of water vapor
 y_i = mole fraction of water vapor at interface
 y = mole fraction of water in air stream
 k_y = mass-transfer coefficient, lb moles water/(ft^2)(hr)(unit mole fraction)
 ϕ = relative-velocity factor

If the wick is completely wet and no dry spots show, the entire area of the wick is available for both heat and mass transfer, and the areas in Eqs. (15-10) and (15-11) are equal. Since the temperature of the water is constant, no temperature gradients are necessary in the liquid to act as driving forces for heat transfer within the water, the surface of the liquid is at the same temperature as the interior, and the surface temperature of the water is t_w. Since the water is pure, no concentration gradients exist, and, granting interfacial equilibrium, y_i is the mole fraction of water vapor in saturated air at temperature t_w. It is convenient to replace the mole fraction terms in Eq. (15-11) by humidities. By the definition of humidity

$$y = \frac{H/18}{1/29 + H/18} \qquad y_i = \frac{H_w/18}{1/29 + H_w/18} \tag{15-12}$$

where H_w is the saturation humidity at the wet-bulb temperature. Substitution of q from Eq. (15-10), of N_A from Eq. (15-11), and of y and y_i from Eqs. (15-12) into Eq. (15-9) gives

$$h_y(t - t_w) = \frac{k_y}{\phi}\left(\frac{H_w}{1/29 + H_w/18} - \frac{H}{1/29 + H/18}\right)[\lambda_w + 0.46(t - t_w)] \tag{15-13}$$

Equation (15-13) may be simplified without introducing serious error in the usual range of temperatures and humidities as follows: (1) the relative-velocity factor ϕ is nearly unity and can be deleted; (2) the sensible-heat item $0.46(t - t_w)$ is small in comparison with λ_w and can be neglected; (3) the terms $H_w/18$ and $H/18$ are small in comparison with $1/29$ and may be dropped from the denominators of the humidity terms. With these simplifications, Eq. (15-13) becomes

$$h_y(t - t_w) = 29k_y\lambda_w(H_w - H)$$

or
$$\frac{H - H_w}{t - t_w} = -\frac{h_y}{29k_y\lambda_w} = -\frac{h_y}{k_y'\lambda_w} \tag{15-14}$$

where $29k_y = k_y'$.

For a given wet-bulb temperature, both λ_w and H_w are fixed. The relation between H and t then depends on the ratio h_y/k_y. The close analogy between mass transfer and heat transfer provides considerable information on the magnitude of this ratio and the factors that affect it.

It has been shown in Chap. 8 that heat transfer by conduction and convection between a stream of fluid and a solid or liquid boundary depends on the Reynolds number DG/μ and the Prandtl number $c_p\mu/k$, or

$$\frac{h_y}{c_pG} = \psi\left(\frac{DG}{\mu}, \frac{c_p\mu}{k}\right) \tag{15-15}$$

Also, as shown in Chap. 10, the mass-transfer coefficient depends on the Reynolds number and the Schmidt number $\mu/D_m\overline{M}$.

$$\frac{k_y\overline{M}}{G} = \psi\left(\frac{DG}{\mu}, \frac{\mu}{D_m\overline{M}}\right) \tag{15-16}$$

where $\bar{M}$ is the average molecular weight of the air stream and D_m is the molal diffusivity. Other symbols have their usual meanings. Since both processes in a given situation are under the control of the same boundary layer, whether or not separation occurs, the functions ψ of Eqs. (15-15) and (15-16) may be taken as identical. The identity of the equations for the Chilton-Colburn j factors for heat and mass transfer, discussed on page 623, is a special case of this principle. If, now, it is assumed that this function can be written as a product of the powers of the dimensionless groups,

$$\frac{h_y}{c_p G} = b \left(\frac{DG}{\mu}\right)^n \left(\frac{c_p \mu}{k}\right)^{-m} \tag{15-17}$$

and

$$\frac{\bar{M} k_y}{G} = b \left(\frac{DG}{\mu}\right)^n \left(\frac{\mu}{D_m \bar{M}}\right)^{-m} \tag{15-18}$$

where b, n, and m are constants. Substitution of h_y from Eq. (15-17) and of k_y from Eq. (15-18) in Eq. (15-14), assuming that $\bar{M}/29$ is unity, gives

$$\frac{H - H_w}{t - t_w} = -\frac{h_y}{k_y' \lambda_w} = -\frac{c_p}{\lambda_w}\left(\frac{\mu/D_m \bar{M}}{c_p \mu/k}\right)^m \tag{15-19}$$

and

$$\frac{h_y}{k_y'} = c_p \left(\frac{\mu/D_m \bar{M}}{c_p \mu/k}\right)^m \tag{15-20}$$

Experiment shows that Eq. (15-20) is reasonably accurate if m is taken as $\frac{2}{3}$. This is consistent with the Chilton-Colburn analogy. The experimental value of h_y/k_y' for water is 0.26.[14] That for organic liquids in air is larger, usually in the range 0.4 to 0.5. The difference is, as shown by Eq. (15-20), the result of the differing ratios of Prandtl and Schmidt numbers for water and for organic vapors. Note that Eq. (15-19) predicts that velocity should not influence the wet-bulb temperature.

Psychrometric Line and Lewis Relation. For a given wet-bulb temperature, Eq. (15-19) may be plotted on the humidity chart as a straight line having a slope of $-h_y/k_y' \lambda_w$ and intersecting the 100 per cent line at t_w. This line is called the psychrometric line. When both a psychrometric line, from Eq. (15-19), and an adiabatic-cooling line, from Eq. (15-8), are plotted for the same point on the 100 per cent curve, the relation between the lines depends on the relative magnitudes of c_s and h_y/k_y'. For air-water these quantities are nearly equal, and the two lines become essentially the same. The assumption that $h_y/k_y' = c_s$ is called the Lewis relation.[6] When this relation is true, the adiabatic-saturation lines can be used as the psychrometric lines with reasonable precision. The Lewis relation provides an important simplification in studying rate processes in the air-water system. It cannot, of course, be used in other systems, where separate lines must be used for psychrometric lines.

Measurement of Humidity. The humidity of a stream or mass of air may be found by measuring either the dew point or the wet-bulb temperature, or by direct absorption methods.

Dew-point Methods. If a water-cooled, polished disk is inserted in the air of unknown humidity and the temperature of the water gradually lowered, the disk reaches a temperature at which mist condenses on the polished surface. The temperature at which this mist just forms is the temperature of equilibrium between the water in the air and liquid water. It is therefore the dew point. A check on the reading is obtained by slowly increasing the water temperature and noting the temperature at which the mist just disappears. From the observed temperatures of mist formation and disappearance, the humidity may be read from Fig. 15-2.

Psychrometric Methods. A very common method of measuring the humidity of air is to determine simultaneously the wet-bulb and dry-bulb temperatures. From these readings the humidity is found by locating the psychrometric line intersecting the saturation line at the observed wet-bulb temperature and following the psychrometric line to its intersection with the ordinate of the observed dry-bulb temperature.

The *sling psychrometer* is a useful instrument for measuring the wet- and dry-bulb temperatures of atmospheric air. It consists of the two thermometers fastened in a metal frame that may be whirled about a handle. The psychrometer is whirled for several seconds and the reading of the wet bulb quickly taken. The operation is repeated until successive readings of the wet-bulb temperature show that it has reached its steady state. The dry-bulb temperature is also read. The humidity is found from these readings in the usual manner.

In process equipment, the wet-bulb and dry-bulb thermometers are permanently installed in the equipment. A current of air at sufficient velocity (usually above 15 ft/sec) to minimize the radiation error is forced over the bulbs. The two readings may be taken visually or recorded on charts.

Direct Methods. The water content in air may be determined by direct absorption, in which a known volume of air is drawn through a drying agent such as sulfuric acid, phosphorus pentoxide, or other moisture-absorbing substance. The water absorbed from the measured volume of air is weighed and the moisture content of the air calculated.

EQUIPMENT FOR AIR-WATER PROCESSES

The use of unsaturated air for cooling water in a cooling tower has been described in Chap. 10. Air-water-contact processes are also used to humidify or to dehumidify air. In a humidifier water is sprayed into a stream of warm, dry air, and sensible-heat and mass transfer take place in the manner described in the discussion of the adiabatic-saturation temperature. The air is humidified and cooled adiabatically. It is not necessary that final equilibrium be reached, and the air may leave the spray chamber at less than full saturation. In dehumidifying air, warm humid air is brought into contact with cold water. The temperature of the air is reduced to a level such that the air cannot retain all its water vapor, and water condenses, thus reducing the moisture content of the air. After dehumidification, the air may be reheated to its original dry-bulb temperature.

Humidifiers. A typical air humidifier consists of a rectangular chamber 6 to 8 ft long through which air is drawn at 8 to 12 ft/sec. Within the chamber centrifugal spray nozzles create a finely divided spray of warm water, nearly all of which evaporates into the air. Baffles at the intake of the chamber distribute the air; other baffles, or "eliminators," at the outlet remove any unevaporated moisture. Humidifiers are nearly always placed on the inlet side of a ventilating fan and are used in connection with ventilating systems. As mentioned above, the water is neither heated nor cooled during recirculation.

Dehumidifiers. In processes demanding an atmosphere of constant humidity, it may be necessary to humidify the air in winter and to dehumidify it in summer. Equipment for dehumidification may utilize a direct spray of cold water into the air, a spray of water on refrigerated coils, or refrigerated coils alone.

Spray units are similar to spray humidifiers, except that since the function of the water is to condense and remove moisture from the air, a coarse spray is used, and the air velocity is only 7 to 9 ft/sec. Much more liquid is sprayed into the air than in a humidifier, and more care is taken to ensure complete removal of liquid water from the discharged air.

In a typical coil-type dehumidifier the air is first passed through a dry filter, then over cold metal coils, and finally over heating coils when warm dry air is required. A refrigerant is expanded in the cooling coils, or cold water or brine is pumped through them. Drip pans under the coils collect the condensate.

Cooling Towers and Spray Ponds

The chief industrial use of air-water contacts is to cool large quantities of water. Many processes require a coolant at a temperature below the prevailing summer temperatures of the surface water. In a few localities cold well water is available in unlimited amounts, but it is becoming increasingly scarce, and recourse must be had to cooling the surface water by exposing it to undersaturated air.

The driving force for evaporation is, very nearly, the difference between the vapor pressure of the water and the vapor pressure it would have at the wet-bulb temperature of the air. Obviously the water cannot be cooled to below the wet-bulb temperature. In practice the discharge temperature of the water must differ from the wet-bulb temperature by at least 4 or 5°F. This difference in temperature is known as the *approach*. The change in temperature in the water from inlet to outlet is known as the *range*. Thus if water were cooled from 95 to 80°F by exposure to air with a wet-bulb temperature of 70°F, the range would be 15°F and the approach 10°F.

The lowest temperature to which water may be cooled throughout the year depends not on the summer dry-bulb temperature but on the maximum wet-bulb temperature. Tables of maximum wet-bulb temperatures have been published for various points in the United States and many other parts of the world.[7, 12d]

The loss of water by evaporation during cooling is small. Since roughly 1000 Btu is required to vaporize 1 lb of water, 100 lb must be cooled 10°F to give up enough heat to evaporate 1 lb. Thus for a change of 10°F in the water temperature there is an evaporation loss of 1 per cent. In addition there are mechanical spray losses, but in a well-designed tower these amount to only about 0.2 per cent. Under the conditions given above, then, the total loss of water during passage through the cooler would be approximately $\frac{15}{10} \times 1 + 0.2 = 1.7$ per cent.

Relatively small amounts of water are cooled by spray ponds; larger amounts, up to 100,000 gal/min, are cooled in cooling towers.

Spray Ponds. In these devices water is cooled by spraying it upward into the air and allowing it to fall back into a shallow pond. Spray nozzles set about 2 ft above the pond are arranged so that the spray pattern from one nozzle does not overlap that from any other. Otherwise the humid air would not readily escape. The pond is about 3 ft deep and often very large in area. Since there is only a single dispersion of water and no positive circulation of the air, the cooling capacity of even a sizable pond is limited. Charts are available for estimating the capacity of spray ponds under various conditions.[12c]

Cooling Towers. In a cooling tower water cascades downward over a filling of wood slats set to impede the direct fall of the water streams and to break them up repeatedly into drops. Open passages are provided in the filling for the flow of air. The air flow may be crosswise, upward and countercurrent to the water flow, or a combination of both. It may be by natural circulation of the prevailing wind, in what is called an "atmospheric" tower, or by mechanical draft.

Atmospheric Towers. A typical atmospheric tower with crosswise flow of the air is described in Chap. 10 and illustrated in Fig. 10-7. Such towers must be located in an exposed area, broadside to the prevailing wind. A minimum wind velocity of $2\frac{1}{2}$ to 3 mph is necessary for good performance. Where these conditions can be met, an atmospheric tower is usually the proper choice; but in congested localities or where winds are light and variable, a mechanical-draft tower is more economical.

Mechanical-draft Towers. In most mechanical-draft cooling towers air is drawn through the filling by large fans set at the top of the unit, as illustrated in Fig. 15-5. Such towers are known as "induced-draft towers." The older "forced-draft towers," in which air was blown through the filling by a fan at the bottom of the tower, have lost favor, since they encouraged recirculation of the moist air from the top of the unit. In induced-draft towers of medium size, cooling from 1,000 to 5,000 gal/min of water, the air enters through the side of the tower near the bottom, flows upward through the filling, and passes between zigzag baffles, or "drift eliminators," to the fan intake. A stack surrounding the fan discharge carries the moist air upward, preferably to a height of 50 ft above ground, to discourage recirculation to the air inlet. Water is sprayed upward into the air stream near the top of the tower. The towers are often multiple, with two or more fans.

In large towers, cooling from 5,000 to 100,000 gal/min of water, crossflow of the air is used. A central fan draws air inward and upward through

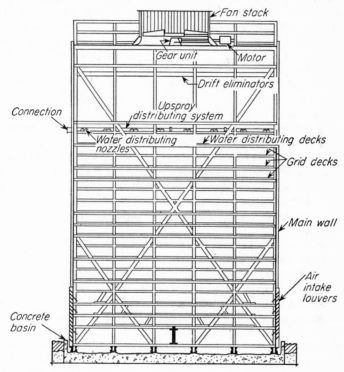

FIG. 15-5. Induced-draft cooling tower. (*Fluor Products Co.*)

two sets of filling, one on each side, and past inclined drift eliminators. The entire side of the tower is left open for intake of air. As in smaller towers, the ends of the tower are covered with a casing of wood slats. The tower filling is assembled without nails or other metal fittings.

The height of a tower required depends on the range and the approach. Sufficient time must be available for contact of air and water. If it is not, no increase in the ratio of air to water will result in the desired cooling. Empirically it has been found that a certain minimum height of filling is necessary, depending on conditions. For example, with a cooling range of 25 to 35°F and an approach of 15 to 20°F, a tower 15 to 20 ft high is adequate. With the same range but an approach of 4 to 8°F, the water must fall a minimum of 35 to 40 ft.[12a]

The cross-sectional area of the tower is determined by the amount of air and the superficial air velocity, which is usually set at about 6 ft/sec, or by the quantity of water to be cooled and by the "water concentration," i.e., the volume of water flowing per square foot of cross section. Too little water results in poor distribution over the filling; too much leads to interference with the flow of air. The range of allowable water concentrations extends only from 1 to 3 gal/ft²-min. The variation of the water

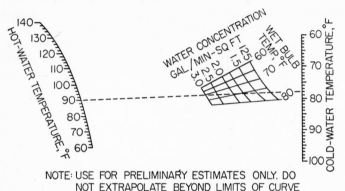

NOTE: USE FOR PRELIMINARY ESTIMATES ONLY. DO
NOT EXTRAPOLATE BEYOND LIMITS OF CURVE

To find required size of cooling tower, place straight edge on points
representing (1) HOT WATER, (2) COLD WATER, and Wet Bulb
Temperatures. Then read the water concentration. The quantity of
water to be cooled divided by water concentration, gives effective
ground area of the cooling tower required.

Fig. 15-6. Capacity of induced-draft cooling tower.[12b]   (*Fluor Products Co.*)

concentration with the cooling range, for different values of approach in a typical induced-draft tower, is shown in Fig. 15-6.

Example 15-2. What cross-sectional area would be required to cool 1,000 gal/ min of water from 90 to 78°F in an induced-draft cooling tower if the wet-bulb temperature is 75°F?

Solution. The water concentration (from Fig. 15-6) is 1.8 gal/ft²-min, and the area required is $1,000/1.8 = 556$ ft².

THEORY AND CALCULATION OF AIR-WATER PROCESSES

The mechanism of the reaction of unsaturated air and water at the wet-bulb temperature of the air has been discussed under the description of wet- and dry-bulb thermometry. The process has been shown to be controlled by the flow of heat and the diffusion of water vapor through the air at the interface between the air and the water. Although these factors are sufficient for the discussion of the adiabatic humidifier, where the water is at constant temperature, it is necessary, in the case of dehumidifiers and water coolers, where the water is changing temperature, to consider heat flow in the water phase also.

In an adiabatic humidifier, where the water remains at constant adiabatic-saturation temperature, there is no temperature gradient through the water. In dehumidification and in water cooling, however, where the temperature of the water is changing, sensible heat flows into or from the water, and a temperature gradient is thereby set up. This introduces a liquid-phase resistance to the flow of heat. On the other hand, it is apparent that there can be no liquid-phase diffusion resistance in any case, since there can be no concentration gradient in pure water.

Mechanism of Interaction of Air and Water. It is important to obtain a correct picture of the relationships of the transfer of heat and of water vapor in all situations of air-water contact. In Figs. 15-7, 15-8, 15-10, and 15-11, distances measured perpendicular to the interface are plotted as abscissas, and temperatures and humidities as ordinates. In all figures let

t_x = temperature of bulk of water
t_i = temperature at interface
t_y = temperature of bulk of air
H_i = humidity at interface
H = humidity of bulk of air

Broken arrows represent the diffusion of water vapor through the gas phase, and full arrows represent the flow of heat (both latent and sensible) through air and water phases. In all processes, t_i and H_i represent equilibrium conditions and are therefore coordinates of points lying on the saturation line on the humidity chart.

The simplest case, that of adiabatic humidification with the water at constant temperature, is shown diagrammatically in Fig. 15-7. The latent-heat flow from water to air just balances the sensible-heat flow from air to water, and there is no temperature gradient in the water. The air temperature t_y must be higher than the interface temperature t_i in order that sensible heat may flow to the interface; and H_i must be greater than H in order that the air be humidified.

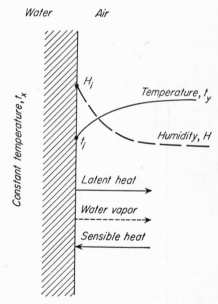

FIG. 15-7. Conditions in adiabatic humidifier.

Conditions at one point in a dehumidifier are shown in Fig. 15-8. Here H is greater than H_i, and therefore water vapor must diffuse to the interface. Since t_i and H_i represent saturated air, t_y must be greater

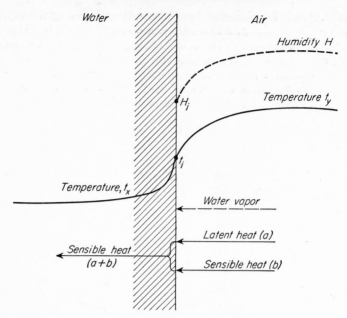

Fig. 15-8. Conditions in dehumidifier.

than t_i; otherwise the bulk of the air would be supersaturated with water vapor.

This reasoning leads to the conclusion that moisture can be removed from unsaturated air by direct contact with sufficiently cold water without first bringing the bulk of the air to saturation. This operation is shown in

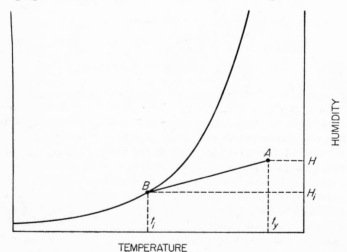

Fig. 15-9. Dehumidification by cold water.

Fig. 15-9. Point A represents the air that is to be dehumidified. Suppose that water is available at such a temperature that saturation conditions at the interface are represented by point B. It has been shown experimentally[4] that in such a process the path of the air on the humidity chart is nearly a straight line between points A and B. As a result of the humidity and temperature gradients, the interface receives both sensible heat and water vapor from the air. The condensation of the water liberates

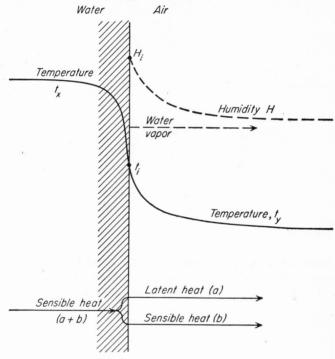

FIG. 15-10. Conditions in top of cooling tower.

latent heat, and both sensible and latent heat are transferred to the water phase. This requires a temperature difference $t_i - t_x$ through the water.

The conditions in a countercurrent cooling tower depend on whether the air temperature is below or above the temperature at the interface. In the first case, e.g., in the upper part of the cooling tower, the conditions are shown diagrammatically in Fig. 15-10. Here the flow of heat and of water vapor (and hence the direction of temperature and humidity gradients) are exactly the reverse of those shown in Fig. 15-8. The water is being cooled both by evaporation and by transfer of sensible heat, the humidity and temperature of the air decrease in the direction of interface to air, and the temperature drop $t_x - t_i$ through the water must be sufficient to give a heat-transfer rate high enough to account for both heat items.

In the lower part of the cooling tower, where the temperature of the air is above that of the interface temperature, the conditions shown in Fig. 15-11 prevail. Here the water is being cooled, hence the interface must be cooler than the bulk of the water, and the temperature gradient through the water is toward the interface (t_i is less than t_x). On the other hand, there must be a flow of sensible heat from the bulk of the air to the interface (t_y is greater than t_i). The flow of water vapor away from the interface

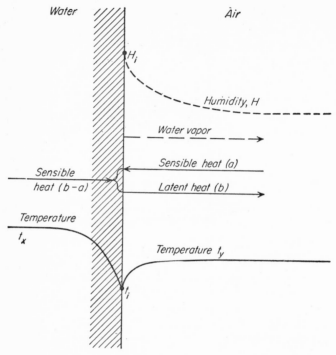

Fig. 15-11. Conditions in bottom of cooling tower.

carries, as latent heat, all the sensible heat supplied to the interface from both sides. The resulting temperature profile $t_x t_i t_y$ has a striking V shape, as shown in Fig. 15-11.

Equations for Air-Water Contacts. Consider the countercurrent air-water contactor shown diagrammatically in Fig. 15-12. Air at humidity H_b and temperature t_{yb} enters the bottom of the contactor and leaves at the top with a humidity H_a and temperature t_{ya}. Water enters the top at temperature t_{xa} and leaves at the bottom at temperature t_{xb}. The mass velocity of the air is G'_y lb of dry air per square foot of tower cross section per hour. The mass velocities of the water at inlet and outlet are, respectively, G_{xa} and G_{xb} lb per square foot of tower cross section per hour. Let dZ be the height of a small section of the tower at distance Z ft from the bottom of the contact zone. Let the mass velocity of the water at height

Z be G_x, the temperatures of air and water be t_y and t_x, respectively, and the humidity be H. At the interface between the air and the water phases, let the temperature be t_i and the humidity be H_i. The cross section of the tower is S ft^2, and the height of the contact section is Z_T ft. Assume that the water is warmer than the air, so the conditions at height Z are those

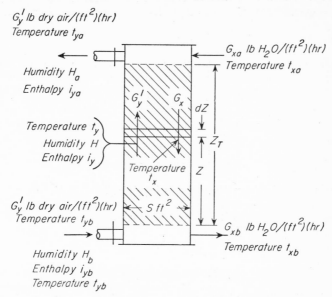

FIG. 15-12. Countercurrent water cooler, flow diagram.

shown in Fig. 15-10. The following equations may be written over the small volume $S\ dZ$.

The enthalpy balance is

$$G_y'\ di_y = d(G_x i_x) \tag{15-21}$$

where i_y and i_x are the total enthalpies of air and water, respectively.

The rate of heat transfer from water to interface is

$$d(G_x i_x) = h_x(t_x - t_i)a_H\ dZ \tag{15-22}$$

where h_x is the heat-transfer coefficient from water to interface and a_H is the heat-transfer area, in square feet per cubic foot of contact volume.

The rate of heat transfer from interface to air is

$$G_y'c_s\ dt_y = h_y(t_i - t_y)a_H\ dZ \tag{15-23}$$

The rate of mass transfer of water vapor from interface to air is

$$G_y'\ dH = k_y'(H_i - H)a_M\ dZ \tag{15-24}$$

where a_M is the mass-transfer area, in square feet per cubic foot of contact volume. The factors a_M and a_H are not necessarily equal. If the contactor is packed with solid packing, the water may not completely wet the packing, and the area available for heat transfer, which is the entire area of the packing, is larger than that for mass transfer, which is limited to the surface that is actually wet.

The Lewis relation (for air-water) is

$$\frac{h_y}{k_y'} = c_s \tag{15-25}$$

These equations may be simplified and rearranged. First, neglect the change of G_x with height and write for the enthalpy of the water

$$i_x = c_L(t_x - t_0) \tag{15-26}$$

where c_L is the specific heat of the liquid (unity for water) and t_0 is the base temperature for computing enthalpy. Then

$$d(G_x i_x) = G_x \, di_x = G_x c_L \, dt_x \tag{15-27}$$

Substituting $d(G_x i_x)$ from Eq. (15-27) into Eq. (15-22) gives

$$G_x c_L \, dt_x = h_x(t_x - t_i)a_H \, dZ$$

This may be written

$$\frac{dt_x}{t_x - t_i} = \frac{h_x a_H}{G_x c_L} dZ \tag{15-28}$$

Second, Eq. (15-23) may be rearranged to read

$$\frac{dt_y}{t_i - t_y} = \frac{h_y a_H}{c_s G_y'} dZ \tag{15-29}$$

Third, Eq. (15-24) may be written

$$\frac{dH}{H_i - H} = \frac{k_y' a_M}{G_y'} dZ \tag{15-30}$$

Finally, using Eq. (15-21), Eq. (15-27) may be written

$$\frac{di_y}{dt_x} = \frac{G_x c_L}{G_y'} \tag{15-31}$$

Three working equations, which are applied later, may now be derived. First, multiply Eq. (15-24) by λ_0 and add Eq. (15-23) to the product, giving

$$G_y'(c_s \, dt_y + \lambda_0 \, dH) = [\lambda_0 k_y'(H_i - H)a_M + h_y(t_i - t_y)a_H] \, dZ \tag{15-32}$$

If the packing is completely wet with liquid, so that the area for mass transfer equals that for heat transfer, $a_M = a_H = a$. Neglecting the change of c_s with H, Eq. (15-7) gives, on differentiation,

$$di_y = c_s\, dt_y + \lambda_0\, dH \tag{15-33}$$

From Eq. (15-33), di_y may be substituted in the left-hand side of Eq. (15-32). This gives

$$G'_y\, di_y = [\lambda_0 k'_y(H_i - H)a + h_y(t_i - t_y)a]\, dZ$$

Also, using Eq. (15-25), h_y can be eliminated. These simplifications give

$$G'_y\, di_y = k'_y a[(\lambda_0 H_i + c_s t_i) - (\lambda_0 H + c_s t_y)]\, dZ \tag{15-34}$$

If i_i is the enthalpy of the air at the interface,

$$i_i = \lambda_0 H_i + c_s(t_i - t_0)$$

From this definition of i_i and the expression for i_y given by Eq. (15-7), the bracketed term in Eq. (15-34) is, simply, $i_i - i_y$. Then Eq. (15-34) becomes

$$G'_y\, di_y = k'_y a(i_i - i_y)\, dZ$$

or

$$\frac{di_y}{i_i - i_y} = \frac{k'_y a}{G'_y}\, dZ \tag{15-35}$$

Second, from Eqs. (15-21) and (15-22),

$$G'_y\, di_y = h_x(t_x - t_i)a\, dZ \tag{15-36}$$

Eliminating dZ from Eqs. (15-35) and (15-36) and rearranging gives, with the aid of Eq. (15-25),

$$\frac{i_i - i_y}{t_i - t_x} = -\frac{h_x}{k'_y} = -\frac{h_x c_s}{h_y} \tag{15-37}$$

Third, Eq. (15-29) is divided by Eq. (15-35), giving

$$\frac{dt_y/(t_i - t_y)}{di_{y'}/(i_i - i_y)} = \frac{h_y}{k'_y c_s}$$

From Eq. (15-25), the right-hand side of this equation is unity, and

$$\frac{dt_y}{di_y} = \frac{i_i - i_y}{t_i - t_y} \tag{15-38}$$

Note that, since the Lewis relation is used in their derivation, Eqs. (15-35), (15-37), and (15-38) apply only to the air-water system and also for situations where $a_M = a_H$. In the following sections the discussion is limited to air-water contacts.

Adiabatic Humidification. Adiabatic humidification corresponds to the process shown in Fig. 15-1 except that the air leaving the humidifier is not necessarily saturated, and, for design, rate equations must be used to calculate the size of the contact zone. The inlet and outlet water temperatures are equal. It is assumed in the following that the make-up water enters at adiabatic-saturation temperature and that the volumetric-area factors a_M and a_H are identical. The wet-bulb and adiabatic-saturation temperatures are equal and constant. Then,

$$t_{xa} = t_{xb} = t_i = t_x = t_s = \text{const}$$

where t_s is the adiabatic-saturation temperature of the inlet air. Also, for water, $c_L = 1.0$. Equation (15-29) then becomes

$$\frac{dt_y}{t_s - t_y} = \frac{h_y a}{c_s G_y'} dZ \qquad (15\text{-}39)$$

Using $\bar{c}_s$ as the average humid heat over the humidifier, Eq. (15-39) may be integrated

$$\int_{t_{yb}}^{t_{ya}} \frac{dt_y}{t_s - t_y} = \frac{h_y a}{\bar{c}_s G_y'} \int_0^{Z_T} dZ$$

$$\ln \frac{t_{yb} - t_s}{t_{ya} - t_s} = \frac{h_y a Z_T}{\bar{c}_s G_y'} = \frac{h_y a S Z_T}{\bar{c}_s G_y' S} = \frac{h_y a V_T}{\bar{c}_s V'} \qquad (15\text{-}40)$$

where $V_T = S Z_T = $ total contact volume, ft^3
$V' = G_y' S = $ total flow of dry air, lb/hr

An equivalent equation, based on mass transfer, may be derived from Eq. (15-30), which is written for adiabatic humidification as

$$\frac{dH}{H_s - H} = \frac{k_y' a}{G_y'} dZ$$

Since H_s, the saturation humidity at t_s, is constant, this equation may be integrated in the same manner as Eq. (15-39) to give

$$\ln \frac{H_s - H_b}{H_s - H_a} = \frac{k_y' a}{G_y'} Z_T = \frac{k_y' a V_T}{V'} \qquad (15\text{-}41)$$

Application of HTU Method. The HTU method is applicable to adiabatic humidification. Thus, by definition,

$$N_t = \ln \frac{H_s - H_b}{H_s - H_a} \qquad (15\text{-}42)$$

where N_t is the number of humidity transfer units. From the definition of H_t and Eqs. (15-41) and (15-42),

$$H_t = \frac{Z_T}{N_t} = \frac{G_y'}{k_y' a} \qquad (15\text{-}43)$$

where H_t is the height of one humidity transfer unit.

The number of transfer units may also be defined on the basis of heat transfer, as follows.

$$N_t = \ln \frac{t_{yb} - t_s}{t_{ya} - t_s} \qquad H_t = \frac{G'_y \bar{c}_s}{h_y a} \qquad (15\text{-}44)$$

That H_t given by Eq. (15-43) is the same as that given by Eq. (15-44) can be shown by dividing Eq. (15-43) by the value of H_t from Eq. (15-44). The result is Eq. (15-25). Also, the values of N_t calculated from Eqs. (15-42) and (15-44) are the same, because, in both instances, $N_t = Z_T/H_t$.

The heat-transfer method, using Eqs. (15-40) and (15-44), is equivalent to the mass-transfer method using Eqs. (15-41) to (15-43). The two methods give the same result.

Example 15-3. For a certain process requiring air at controlled temperature and humidity there is needed 15,000 lb of dry air per hour at 20 per cent humidity and 130°F. This air is to be obtained by conditioning air at 20 per cent humidity and 70°F by first heating, then humidifying adiabatically to the desired humidity, and finally reheating the humidified air to 130°F. The humidifying step is to be conducted in a spray chamber. Assuming the air leaving the spray chamber is to be 4°F warmer than the adiabatic-saturation temperature, to what temperature should the air be preheated, at what temperature should it leave the spray chamber, how much heat will be required for pre- and reheating, and what should be the volume of the spray chamber? Take $h_y a$ as 85 Btu/(ft³)(hr)(°F).

Solution. The temperature-humidity path of the air through the heaters and spray chamber is plotted on the section of the humidity chart shown in Fig. 15-13.

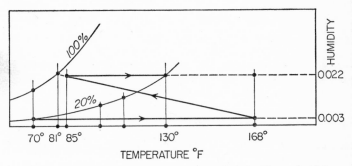

Fig. 15-13. Temperature-humidity path of air for Example 15-3.

Air at 20 per cent humidity and 130°F has a humidity of 0.022. The air leaving the spray chamber is at this same humidity, and the point representing it on the humidity chart is located by finding the point where the coordinate for $H = 0.022$ is 4°F from the end of an adiabatic-cooling line. By inspection, it is found that the adiabatic line for the process in the spray chamber corresponds to an adiabatic-saturation temperature of 81°F and that the point on this line at $H = 0.022$ and $t = 85$°F represents the air leaving the chamber. The original air has a humidity of 0.0030. To reach the adiabatic cooling line for $t_s = 81$°F, the temperature of the air leaving the preheater must be 168°F.

The humid heat of the original air is, from Fig. 15-2, 0.241. The heat required to preheat the air is, then,

$$0.241 \times 15,000(168 - 70) = 354,000 \text{ Btu/hr}$$

The humid heat of the air leaving the spray chamber is 0.250, and the heat required in the reheater is

$$0.250 \times 15,000(130 - 85) = 169,000 \text{ Btu/hr}$$

The total heat required is $354,000 + 169,000 = 523,000$ Btu/hr.

To calculate the volume of the spray chamber, Eq. (15-40) may be used. The average humid heat is

$$\bar{c}_s = \frac{0.241 + 0.250}{2} = 0.2455 \text{ Btu/(lb dry air)(°F)}$$

Substituting in Eq. (15-40),

$$2.303 \log \frac{168 - 81}{85 - 81} = \frac{85 V_T}{15,000 \times 0.2455}$$

From this, the volume of the spray chamber is $V_T = 134$ ft³.

Countercurrent Adiabatic Water Cooling. In a water cooler such as a countercurrent forced-draft cooling tower, heat transfer takes place through the water to the interface, and $t_x - t_i$ cannot be neglected. By

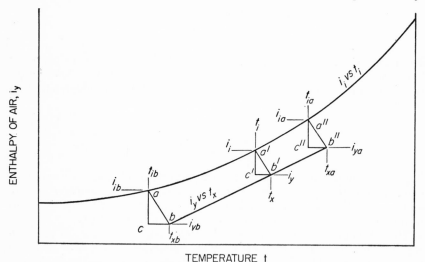

FIG. 15-14. Graphical design of cooling tower. (*After Merkel.*[10])

using a graphical method due to Merkel,[10] the size of the tower may be calculated if the individual coefficients $h_x a$ and $k_y' a$ are known and if the simplifying assumptions made in the analysis of the adiabatic humidifier are retained. The method is analogous to the general method of treating countercurrent differential-contact equipment in that it utilizes an operating line, an equilibrium curve, and sloping lines to isolate the individual driving forces over the water and air resistances. The method is unique in that the coordinates are not phase concentrations for the two streams but are the enthalpy and temperature of the air.

In Fig. 15-14 is plotted a curve of i_i vs. t_i. Points on this curve are calculated from Eq. (15-7), using $t_0 = 32°F$. With this choice of base temperature, $\lambda_0 = 1076$ Btu/lb, and

$$i_{yi} = c_s(t_i - 32) + 1076H_i$$

where t_i and H_i are coordinates of points on the saturation curve of the humidity chart. It is this curve that is used as an equilibrium curve. Interfacial equilibrium between water and air is assumed. From Eq. (15-31),

$$di_y = \frac{G_x c_L}{G_y'} dt_x$$

This equation may be integrated over the bottom of the tower.

$$\int_{i_{yb}}^{i_y} di_y = \frac{G_x c_L}{G_y'} \int_{t_{xb}}^{t_x} dt_x$$

$$i_y = \frac{G_x c_L}{G_y'} (t_x - t_{xb}) + i_{yb} \qquad (15\text{-}45)$$

Equation (15-45), plotted on the diagram of i_y vs. t, gives an operating line. Since, under the assumption that G_x is constant the slope of the operating line is also constant, this line is straight. By integrating over the entire tower, the over-all enthalpy balance is

$$i_{ya} - i_{yb} = \frac{G_x c_L}{G_y'} (t_{xa} - t_{xb})$$

The operating line has the usual properties of such relations: it has a slope equal to the ratio of the liquid flow to the vapor flow (since $c_L = 1.0$), and it passes through points (t_{xa}, i_{ya}) and (t_{xb}, i_{yb}), representing the terminals of the process. An operating line $bb'b''$ is shown in Fig. 15-14.

The temperature difference over the water-phase resistance $t_x - t_i$ and the enthalpy difference over the air-phase resistance $i_i - i_y$ may be found by a construction like that used to establish the $\Delta x \, \Delta y$ triangles in absorption. Equation (15-37) shows that if a line is drawn with a slope of $-h_x/k_y'$ through point (t_i, i_{yi}) on the equilibrium curve, it intersects the operating line at point (t_x, i_y) corresponding to (t_i, i_i). Such sloping lines appear in Fig. 15-14 for both ends of the column and for one point within the tower, as lines ab, $a'b'$, and $a''b''$. Line segments ac, $a'c'$, and $a''c''$, each equal to a value of $i_i - i_y$, are read from these lines just as values of Δy are read from the $\Delta x \, \Delta y$ triangles in the diagram for an absorption column. From the values of $i_i - i_y$ the height of the packed section can be calculated by integrating Eq. (15-35).

$$\int_{i_{yb}}^{i_{ya}} \frac{di_y}{i_i - i_y} = \frac{k_y'a}{G_y'} Z_T = \frac{h_y a}{G_y' \bar{c}_s} Z_T \qquad (15\text{-}46)$$

The integral in this equation may be evaluated graphically.

Alternately, values of $t_i - t_x$ may be read from horizontal segments, such as cb, $c'b'$, and $c''b''$, and used in a graphical integration of Eq. (15-28). The results from the two methods should agree.

Example 15-4. A countercurrent packed tower is used to cool water by direct contact with air. The water enters at 105°F and leaves at 75°F. The air enters at a wet-bulb temperature of 55°F and a dry-bulb of 70°F. A liquid rate of 800 lb/ft²-hr is used. The coefficients may be calculated from the equations [9]

$$h_x a = 0.8(G_y')^{0.7} G_x^{0.5} \qquad h_y a = 2.2(G_y)^{0.7} G_x^{0.07}$$

where G_y' and G_x are the air and water mass velocities, respectively, in pounds per square foot of tower cross section per hour.

(a) What is the minimum air flow that can accomplish this cooling? (b) If the air flow is 1,000 lb/ft²-hr, dry basis, how deep should the packed section be?

Solution. (a) The equilibrium and operating lines for this problem are plotted in Fig. 15-15. The equilibrium line is the plot of i_i vs. t_i. From the humidity

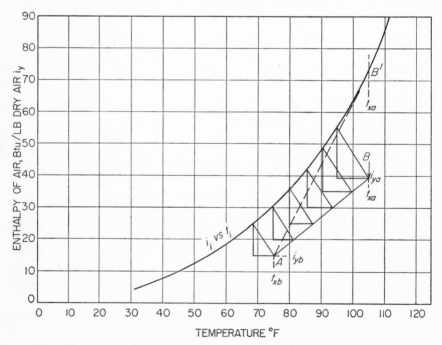

FIG. 15-15. Solution of Example 15-4.

chart (Fig. 15-2), the humidity of the entering air is 0.0057, and the humid heat is 0.243. Its enthalpy i_{yb} is, from Eq. (15-7),

$$0.243(70 - 32) + 1076 \times 0.0057 = 15.4 \text{ Btu/lb dry air}$$

The minimum air flow is found by determining the slope of a line through point A, for the inlet air and the outlet water, and the point on the equilibrium line at 105°F. This line is the dotted line AB' of Fig. 15-15. The slope of this line is 1.9, and the minimum air flow G'_{min} is given by $800/G'_{min} = 1.9$, from which $G'_{min} = 420$ lb/hr.

(b) For $G' = 1,000$, the slope of the operating line is $800/1,000 = 0.80$. The operating line is shown as line AB in Fig. 15-15. The enthalpy of the leaving air is $i_{ya} = 39.2$ Btu/lb.

From the above equations,

$$h_x a = 0.8 \times 1,000^{0.7} \times 800^{0.5} = 2840 \text{ Btu/(ft}^2)(\text{hr})(°\text{F})$$

$$h_y a = 2.2 \times 1,000^{0.7} \times 800^{0.07} = 440 \text{ Btu/(ft}^2)(\text{hr})(°\text{F})$$

The air leaving the tower will be nearly saturated, and its humid heat may be taken as 0.250. The average humid heat $\bar{c}_s$ is

$$\bar{c}_s = \frac{0.250 + 0.243}{2} = 0.246$$

The slope of the $\Delta t \, \Delta i$ lines is, by Eq. (15-37),

$$-\frac{2840 \times 0.246}{440} = -1.6$$

The $\Delta t \, \Delta i$ triangles are plotted in Fig. 15-15, and a graphical integration of

$$\int_{15.4}^{39.2} \frac{di_y}{i_i - i_y}$$

as called for by Eq. (15-46), using the vertical sides of the $\Delta t \, \Delta i$ triangles for values of $i_i - i_y$, gives a result of 1.98.†

Then, by Eq. (15-46),

$$1.98 = \frac{440}{1,000 \times 0.246} Z_T$$

and $Z_T = 1.1$ ft.‡

Finding the Air Temperature. The construction shown in Fig. 15-14 does not show the bulk temperature of the air stream in the process. If the inlet air temperature t_{yb} is known, the air temperature throughout the tower may be found by a graphical method proposed by Mickley.[11] This method is shown for the tower of Example 15-4 in Fig. 15-16. The enthalpy of the air at any level in the column is given by the ordinate of the point on the operating line for that level. The entering air is represented

† This is also the number of gas-phase transfer units needed.

‡ This height is small in comparison with that of an actual tower operating with the same range and approach (see page 851). The contact area in a cubic foot of packed tower is much greater than that in an equal volume of a conventional cooling tower.

on Fig. 15-16 by point A, which is located from the known enthalpy i_{yb} and the known temperature t_{yb}. Line BC is the $\Delta t\,\Delta i$ line for the bottom of the tower, and the coordinates of point C are (t_{ib}, i_{ib}). Draw straight line AC. Then, by Eq. (15-38), the slope of this line is dt_y/di_y at point A. Line AC establishes the trend of the t_y vs. i_y line at the air entrance. Although t_y is not linear with i_y, it may be assumed to be so over a short

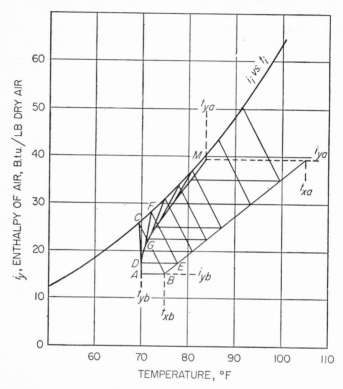

Fig. 15-16. Construction for determining air temperature in cooling tower. (*After Mickley.*[11])

distance such as the line segment AD. Point D is chosen arbitrarily, and a horizontal line intersecting the operating line at point E is drawn. A second $\Delta t\,\Delta i$ line is drawn, intersecting the equilibrium curve at point F. A straight line FD provides a trend of the t_y vs. i_y line at point D. Another arbitrary segment DG is chosen and the construction repeated until the horizontal line $i = i_{ya}$ is reached at point M. The locus of all points such as A, D, G, ..., M is the desired relation of t_y vs. i_y. The shorter the assumed segments, such as AD and DG, the more precise the result.

Fog Formation. The Mickley construction shows whether or not fog formation is likely in the tower. If the t_y vs. i_y line crosses the equilibrium line, the air must either become supersaturated or form fog. Fog formation is incompatible with the rate equations [Eqs. (15-22) to (15-24)], and calculations based on them are invalid. It is advisable to change the design conditions to eliminate the possibility of fog formation.

The principles used above may be adapted to other air-water processes, such as dehumidification by direct contact with cold water, or by indirect contact with refrigerant through tubular or extended-surface heat exchangers.[11]

Systems Other than Air-Water. Calculations on systems other than air-water are complicated by the fact that the Lewis relation does not hold. Methods are available for such situations,[5, 15] but they are not considered in this book.

Coefficients

Satisfactory experimental data on the coefficients $h_x a$, $h_y a$, and $k'_y a$ are few. These data are difficult to obtain. Another complication comes from the fact that, except at large liquid and gas flow rates, the area factors a_H and a_M are not equal. The simplest correlations for individual coefficients are in the form [9, 16]

$$ha = b(G'_y)^n G_x^m \tag{15-47}$$

where b, n, and m are constant. Some investigations have given data that do not follow a simple equation, and which are best read from plots.[3]

SYMBOLS

A Surface area of water, ft^2

a Transfer area, ft^2/ft^3; a_H, for heat transfer; a_M, for mass transfer

b Constant

c_p Specific heat, Btu/(lb)(°F); c_L, specific heat of liquid

c_s Humid heat, Btu/(lb)(°F); $\bar{c}_s$, average humid heat

D Diameter, ft

D_m Diffusivity, lb moles/ft-hr

G Mass velocity, lb/ft^2-hr; G_x, of water at any point; G_{xa}, of water at entrance; G_{xb}, of water at exit; G'_y, of air, lb dry air/(ft^2 tower cross section) (hr)

H Humidity, lb water vapor/lb dry air; H_a, at top of contactor; H_b, at bottom of contactor; H_i, at air-water interface; H_s, saturation humidity; H_w, saturation humidity at wet-bulb temperature

H_A Percentage humidity, $100H/H_s$

H_R Relative humidity, $100\bar{p}_A/p_A$

H_t Height of humidity transfer unit, ft

h Heat-transfer coefficient, Btu/(ft^2)(hr)(°F); h_x, water to interface; h_y, air to interface

i Enthalpy, Btu/lb; i_i, of air at interface; i_{ib}, at interface in Fig. 15-16; i_x, of water; i_y, of air; i_{ya}, of air at top of contactor; i_{yb}, of air at bottom of contactor

k Thermal conductivity, Btu-ft/(ft^2)(hr)(°F)

k_y Mass-transfer coefficient, lb moles/(ft^2)(hr)(unit mole fraction); $k'_y = 29k_y$

$\bar{M}$ Average molecular weight of air stream

m Exponent

N_A Rate of transfer or vaporization of water, moles/hr

N_t Number of humidity transfer units

n Exponent

p Pressure, atm; p_A, vapor pressure of water; $\bar{p}_A$, partial pressure of water vapor

q Rate of sensible heat transfer to water, Btu/hr

S Cross-sectional area of tower, ft^2

t Temperature, °F; t_i, at air-water interface; t_{ib}, at interface in Fig. 15-16; t_s, adiabatic saturation temperature; t_w, wet-bulb temperature; t_x, of bulk of water; t_{xa}, of water at top of contactor; t_{xb}, of water at bottom of contactor; t_y, of bulk of air; t_{ya}, of air at top of contactor; t_{yb}, of air at bottom of contactor; t_0, datum for computing enthalpy; t_1, temperature in Fig. 15-3

V_T Total contact volume, ft^3

V' Total flow rate of dry air, lb/hr

v_H Humid volume, ft^3/lb

y Mole fraction of water in air stream; y_i, at air-water interface

Z Distance from bottom of contact zone, ft; Z_T, total height of contact section

Greek Letters

λ Latent heat of vaporization, Btu/lb; λ_s, at t_s; λ_w, at t_w; λ_0, at t_0

μ Viscosity, lb/ft-hr

ρ Density, lb/ft^3

ϕ Relative-velocity factor

ψ Operator, meaning "function of"

PROBLEMS

15-1. One method of removing acetone from cellulose acetate is to blow an air stream over the cellulose acetate fibers. To know the properties of the air-acetone mixtures, the process control department requires a humidity chart for air-acetone. After investigation, it was found that an absolute humidity range of 0 to 6.0 and a temperature range of 40 to 130°F would be satisfactory.

Construct the following portions of such a humidity chart for air-acetone at a total pressure of 760 mm Hg: (*a*) percentage humidity lines for 50 and 100 per cent, (*b*) saturated volume vs. temperature, (*c*) latent heat of acetone vs. temperature, (*d*) humid heat vs. humidity, (*e*) adiabatic-cooling lines for adiabatic-saturation temperatures of

TABLE 15-1. PROPERTIES OF ACETONE

Temp., °C	Vapor pressure, mm Hg	Latent heat, joules/g
0	. . . .	564
10	115.6	
20	179.6	552
30	281.0	
40	420.1	536
50	620.9	
56.1	760.0	521
60	860.5	517
70	1,189.4	
80	1,611.0	495

70 and 100°F, (f) wet-bulb temperature (psychrometric) lines for wet-bulb temperatures of 70 and 100°F.

The necessary data are given in Table 15-1, above. For acetone vapor, $c_p = 0.35$ and $h/k_y' = 0.41$.

15-2. It is desired to install a humidifier with a capacity of 10,000 lb of dry air per hour from 15 per cent humidity and 60°F to a wet-bulb temperature of 80°F and a dry-bulb temperature of 100°F. The process is to consist of preheating in tempering coils which cost $6 per square foot, including all accessories, then adiabatically partially saturating with water, and then reheating to the desired final temperature in tempering coils of the same cost per square foot as the preheater. The over-all coefficient of heat transfer from steam at 5 lb force/in.² gauge to the air in the coils is 10 Btu/(ft²)(hr)(°F). The heat-transfer coefficient in the spray chamber is 100 Btu of sensible heat per cubic foot of spray-chamber volume per hour per degree Fahrenheit. The spray chamber, including all accessories, costs $2 per cubic foot. If the steam cost is 50 cents per 1,000 pounds and the investment costs, including taxes, etc., are 25 per cent of the first cost per year, what size spray chamber should be used? Neglect pumping cost.

15-3. The following data were obtained during a test run of a packed cooling tower operating at atmospheric pressure:

> Height of packed section, 6 ft
> Inside diameter, 12 in.
> Average temperature of entering air, 100°F
> Average temperature of leaving air, 103°F
> Average wet-bulb temperature of entering air, 80°F
> Average wet-bulb temperature of leaving air, 96°F
> Average temperature of entering water, 115°F
> Average temperature of leaving water, 95°F
> Rate of entering water, 2,000 lb/hr
> Rate of entering air, 480 ft³/min

(a) Using the entering-air conditions, calculate the humidity of the exit air by means of an enthalpy balance. Compare the result with the humidity calculated from the wet- and dry-bulb readings.

(b) Assuming that the water-air interface is at the same temperature as the bulk of the water (water-phase resistance to heat transfer negligible), calculate the driving force $H_i - H$ at the top and at the bottom of the tower. Using an average $H_i - H$, estimate the average value of $k_y'a$.

15-4. Air is to be cooled and dehumidified by countercurrent contact with water in a packed tower. The tower is to be designed for the following conditions:

> Dry-bulb temperature of inlet air, 82°F
> Wet-bulb temperature of inlet air, 76.5°F
> Flow rate of inlet air, 1,500 lb of dry air per hour
> Inlet water temperature, 50°F
> Outlet water temperature, 65°F

(a) For the entering air, find: (i) the humidity, (ii) the per cent relative humidity, (iii) the dew point, (iv) the enthalpy, based on air and liquid water at 32°F.

(b) What is the maximum water rate which may be used to meet design requirements, assuming a very tall tower?

(c) Calculate the number of transfer units required for a tower that meets design specifications when 1,000 lb/hr of water is used and if the liquid-phase resistance to heat transfer is negligible.

15-5. It is desired to cool 10,000 gal/min of water from 110 to 85°F in an induced-draft cooling tower. The maximum summer wet-bulb temperature is 77°F.

(a) What should be the cross-sectional area of the tower?

(b) Approximately how high should the tower be?

(c) If mechanical losses are 0.2 per cent of the water fed to the tower, how much make-up water will be required?

REFERENCES

1. Brown, G. G., and associates: "Unit Operations," p. 545, John Wiley & Sons, Inc., New York, 1950.
2. Grosvenor, W. M.: *Trans. AIChE*, **1**: 184 (1908).
3. Hensel, S. L., Jr., and R. E. Treybal: *Chem. Eng. Progr.*, **48**: 362 (1952).
4. Keevil, C. S., and W. K. Lewis: *Ind. Eng. Chem.*, **20**: 1058 (1928)
5. Lewis, J. G., and R. R. White: *Ind. Eng. Chem.*, **45**: 486 (1953).
6. Lewis, W. K.: *Trans. AIME*, **44**: 325 (1922).
7. Marley Company: "Summer Weather Data," Kansas City, Mo., 1944.
8. Marshall, W. R., and S. J. Friedman, in J. H. Perry (ed.): "Chemical Engineers' Handbook," 3d ed., pp. 812–816, McGraw-Hill Book Company, Inc., New York, 1950.
9. McAdams, W. H., J. B. Pohlenz, and R. C. St. John: *Chem. Eng. Progr.*, **45**: 241 (1949).
10. Merkel, F.: *Z. ges. Kälte-Ind.*, **34**: 117 (1927).
11. Mickley, H. S.: *Chem. Eng. Progr.*, **45**: 739 (1949).
12. Perry, J. H.: "Chemical Engineers' Handbook," 3d ed., (a) p. 790, (b) p. 791, (c) pp. 795, 797, (d) p. 796, McGraw-Hill Book Company, Inc., New York, 1950.
13. Sherwood, T. K., and E. W. Comings: *Trans. AIChE*, **28**: 88 (1932).
14. Sherwood, T. K., and R. L. Pigford: "Absorption and Extraction," 2d ed., pp. 97–101, McGraw-Hill Book Company, Inc., New York, 1952.
15. Sherwood, T. K., and C. E. Reed: "Applied Mathematics in Chemical Engineering," pp. 134–137, McGraw-Hill Book Company, Inc., New York, 1939.
16. Yoshida, F., and T. Tanaka: *Ind. Eng. Chem.*, **43**: 1469 (1951).

Chapter 16

DRYING

A precise definition of drying that differentiates it from other operations such as evaporation is difficult to formulate. In general, drying means the removal of relatively small amounts of water from solids, liquids, or gases. Drying a solid is usually the final step in a series of operations, and the product from a dryer is often ready for final packaging. Drying is differentiated from evaporation in that the latter is used to remove relatively large quantities of water from solutions. Although a solid product, in the form of a magma, may be formed in an evaporator, such a product is too wet to be packaged and must be further dehydrated in a dryer before it is a finished product.

Moisture may be removed from liquids or gases by adsorption on a solid such as silica gel or alumina, by absorption in liquid such as sulfuric acid, by refrigeration, or by distillation. Water may be removed from solids mechanically by presses or centrifuges or thermally by vaporization. This chapter is restricted to drying by thermal vaporization. It is generally cheaper to remove water mechanically than thermally, and it is advisable to so reduce the moisture content as much as practicable before feeding the material to a heated dryer.

The moisture content of a dried substance varies from product to product. Occasionally the product contains no water, and is called "bone-dry." More commonly, the product does contain some water. Dried table salt, for example, contains about 0.5 per cent water, dried coal about 4 per cent, and dried casein about 8 per cent. Drying is a relative term and means merely that there is a reduction in moisture content from an initial value to a final one.

DRYING EQUIPMENT

Many types of commercial dryers are described in standard references.[11c, 12] Here only a small number of important types are considered. They are listed in Table 16-1 and are classified as to whether the material is a rigid solid, a granular solid or semisolid paste, a continuous sheet, or a liquid solution or slurry; as to whether or not the material is agitated during drying; and as to whether the operation is batch or continuous. Still another division is between direct-contact dryers, in which the solids

TABLE 16-1. TYPICAL DRYERS

Material dried	Batch or cyclic operation	Continuous operation
Rigid or preformed shapes or cakes..	Tray dryer	Tunnel dryer
Granular solids or pastes:		
Material not agitated............	Tray dryer	Screen-conveyor dryer
Material agitated...............	Pan dryer	Rotary dryer
		Screw-conveyor dryer, flash dryer, tower dryer
Continuous flexible sheets.........	....	Festoon dryer, cylinder dryer
Solutions or slurries..............	....	Drum dryer, spray dryer

are dried by exposure to hot air or flue gas, and indirect dryers, in which heat is transferred to the material from a heating medium through a metal wall.

Dryers for Solids and Pastes

The nature of the product usually dictates the type of dryer to be used. Coarse, granular inorganic solids may be dried in an agitated dryer by direct contact with hot flue gases; fragile crystals of heat-sensitive organic materials must be dried in unagitated equipment by indirect heating or by contact with warm air at carefully controlled temperature and humidity. Operation under vacuum is often used to reduce drying temperatures. As in other unit operations, batch equipment is favored when production rates are low, holdup times in the unit are long, or when many different products are to be dried in the same unit. Continuous dryers are employed when tonnages are large and the drying rate is so rapid that the time of drying is not excessive.

Typical dryers for rigid or granular solids, as listed in Table 16-1, include tray, tunnel, and screen-conveyor dryers for materials that cannot be agitated and pan, tower, rotary, screw-conveyor, and flash dryers for solids that can be agitated. Some of these dryers are batch, some continuous; some are direct-contact, others indirect. Continuous flexible sheets are dried in special continuous units, such as festoon and cylinder dryers.

Tray Dryers. A typical batch tray dryer is illustrated in Fig. 16-1. It consists of a rectangular chamber of sheet metal containing two trucks which support racks H. Each rack carries a number of shallow trays, perhaps 30 in. square and 2 to 6 in. deep, which are loaded with the material to be dried. Heated air is circulated at 7 to 15 ft/sec between the trays by fan C and motor D and passes over heaters E. Baffles G distribute the air uniformly over the stack of trays. Some moist air is continuously vented through exhaust duct B; make-up fresh air enters through inlet A. The racks are mounted on truck wheels I, so that at the end of the drying cycle the trucks may be pulled out of the chamber and taken to a tray-dumping station.

Tray dryers are useful when the production rate is less than 50 to 100 lb/hr of dry product. They can dry almost anything, but because of the labor required for loading and unloading are expensive to operate. They find most frequent application on valuable products like dyes and pharmaceuticals. Drying by circulation of air across stationary layers of solid is slow, and drying cycles are long: 4 to 48 hr per batch. In some tray dryers

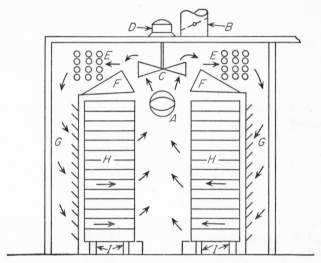

Fig. 16-1. Tray dryer. (*Proctor and Schwarz Co.*)

coarse-grained solids are supported on screen-bottom trays, and the air is passed vertically through them, giving what is known as *through-circulation* drying. Drying in a through-circulation unit is much more rapid than in a cross-circulation unit, but through circulation is usually neither economical nor necessary in batch dryers, because shortening the drying cycle does not reduce the labor required for each batch. As mentioned above, when production rates are high and short cycles can be used, a continuous dryer should be selected.

Tray dryers may be operated under vacuum, often with indirect heating. The trays may rest on hollow metal plates supplied with steam or hot water or may themselves have spaces for a heating medium. Water vapor from the solid is removed by an injector or vacuum pump. *Freeze-drying* is the sublimation of water vapor from ice under high vacuum at temperatures below 30°F. This is done in special vacuum tray dryers for drying vitamins and other extremely heat-sensitive products.

Tunnel Dryers. A tray dryer may be made continuous or semicontinuous by elongating the drying chamber into a tunnel, through which the racks or trays, mounted on trucks, are pushed or drawn. Such a dryer is shown in Fig. 16-2. A truck of wet material is admitted to one end of the tunnel; simultaneously a truck of dry product emerges from the other end. Drying conditions at any point in the tunnel do not vary with time, as they do in a tray dryer, but differ from point to point in the tunnel. Tunnel dryers are effective with solid units of complex shape, as in pottery

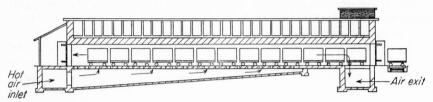

Hot air inlet — Air exit

FIG. 16-2. Tunnel dryer.

making, and with solids which must remain in the drying chamber for prolonged periods. When drying cycles are short (say 2 hr or less), the continuous screen-conveyor dryer is favored.

Screen-conveyor Dryers. A typical through-circulation screen-conveyor dryer is shown in Fig. 16-3. A layer 1 to 6 in. thick of material to be dried is slowly carried on a traveling metal screen through a long drying

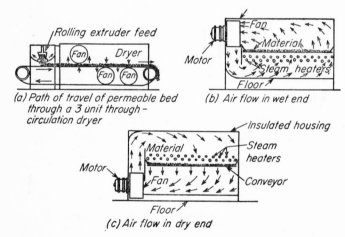

(a) Path of travel of permeable bed through a 3 unit through - circulation dryer

(b) Air flow in wet end

(c) Air flow in dry end

FIG. 16-3. Through-circulation screen-conveyor dryer. (*Perry*.[11])

chamber or tunnel. The chamber consists of a series of separate sections, each with its own fan and air heater. At the inlet end of the dryer the air usually passes upward through the screen and the solids; near the discharge end, where the material is dry and may be dusty, air is passed downward through the screen. The air temperature and humidity may differ in the various sections, to give optimum conditions for drying at each point.

Screen-conveyor dryers are typically 6 ft wide and 12 to 150 ft long, giving drying times of 5 to 120 min. The minimum screen size is about

30-mesh. Coarse granular, flaky, or fibrous materials can be dried by through circulation without any pretreatment and without loss of material through the screen. Pastes and filter cakes of fine particles, however, must be preformed before they can be handled on a screen-conveyor dryer. Preforming is done by finned drums, extruders, granulators, and other devices which aggregate the solid.[11d] The aggregates usually retain their shape while being dried and do not dust through the screen except in small amounts. Provision is sometimes made for recovering any fines which do sift through the screen.

Screen-conveyor dryers handle a variety of solids continuously and with a very gentle action; their cost is reasonable; and their steam economy is high. A typical value is 2 lb of steam per pound of water evaporated. Air may be recirculated through, and vented from, each section separately or passed from one section to another countercurrently to the solid. These dryers are particularly applicable when the drying conditions must be appreciably changed as the moisture content of the solid is reduced.

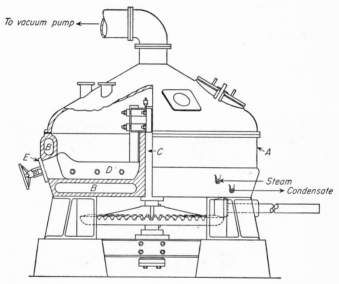

FIG. 16-4. Pan dryer for batch operation under vacuum: A, dryer pan. B, steam jackets. C, shaft. D, stirrer arms. E, discharge door. (*Buflovac Equipment Division, Blaw-Knox Co.*)

Pan Dryers. A typical batch dryer in which the solid is agitated during drying is the pan dryer shown in Fig. 16-4. It consists of a shallow pan A with a steam jacket B and a central shaft C carrying scraper blades D. Wet material is dumped into a manhole in the top of the dryer. The agitator is started, the steam turned on, and vacuum applied through the duct in the top of the unit. At the end of the drying cycle the vacuum is released and dry solid discharged through door E by the action of the scraper blades.

Pan dryers are sometimes open to the air and operate at atmospheric pressure. Closed pans of the type shown in Fig. 16-4 are usually operated under vacuum and are especially valuable for removing nonaqueous solvents from solids. The solvents can be recovered by condensing the vapors in a surface condenser.

Rotary Dryers. A rotary dryer consists of a revolving cylindrical shell, horizontal or slightly inclined toward the outlet. Wet feed enters one end of the cylinder; dry material discharges from the other. As the shell ro-

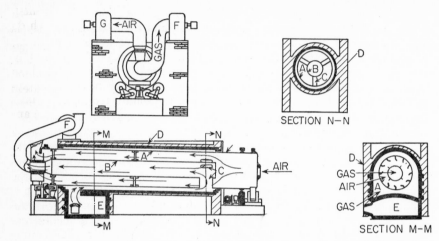

SECTION N–N

SECTION M–M

FIG. 16-5. Indirect fire-heated reversed-current rotary dryer: *A*, dryer shell. *B*, flue-gas duct. *C*, connecting nozzles. *D*, brick setting. *E*, firebox. *F*, flue-gas-discharge fan. *G*, air-discharge fan.

tates, internal flights lift the solids and shower them down through the interior of the shell. Rotary dryers are heated by direct contact of air or gas with the solids, by hot gases passing through an external jacket on the shell, or by steam condensing in a set of longitudinal tubes mounted on the inner surface of the shell. The last of these types is called a steam-tube rotary dryer. A direct-contact rotary dryer using heated air is shown in Fig. 10-8.

Figure 16-5 shows an *indirect fire-heated reversed-current rotary dryer*. It consists of an outer shell *A* and an inner tube *B* concentric with the shell. At the right end of the shell are nozzles *C* connecting the interior of the inner tube with the outside of the shell. The entire shell is mounted in a brick chamber *D*, which contains a firebox *E* under the left end. Flue gases from the firebox pass around the outside shell, traveling from left to right, through nozzles *C* into the inner tube *B*, and then through the inner tube from right to left. The cooled gases are exhausted by fan *F* or by a stack. The gases do not come into contact with the material at any point. Material is fed at the left end, passes through the annular space between the shell and the central tube, and is discharged at the right end. The feed and outlet connections are not shown in Fig. 16-5. Air enters the right-hand end, passes through the annular space of the dryer in countercurrent contact with the material, and is removed by the separate fan G. A dryer of this type would be used for materials, such as high-grade clays and similar inert materials, which can withstand the high temperature of the flue gas but which must be protected from dirt.

The allowable mass velocity of the gas in a direct-contact rotary dryer depends on the dusting characteristics of the solid being dried and ranges from 200 lb/ft^2-hr for fine particles to 10,000 lb/ft^2-hr for coarse, heavy particles. Dryer diameters vary from 1 to 10 ft.

Rotary dryers are effective with free-flowing granular materials. When the properties of the material permit, countercurrent flow of solid and gas is used for economy. The dry solid then comes into contact with the hottest gases. With heat-sensitive materials this can damage the product. With such materials, therefore, indirect heat or parallel flow of solid and gas is used. The temperature of the material in a direct-heated continuous dryer approximates the wet-bulb temperature of the air so long as the solid retains appreciable amounts of water. At low moisture contents the stock temperature approximates the dry-bulb temperature of the air. In countercurrent operation, therefore, the maximum stock temperature approaches the temperature of the hot inlet air. In parallel-current operation, the hottest gases are in contact with the wet solid, the temperature of which is limited by the wet-bulb effect, and the coolest gases are in contact with the dry solid, thus limiting the maximum stock temperature to a nondestructive level. The reversed-current flow used in the dryer of Fig. 16-5 is intermediate between parallel and counterflow.

Screw-conveyor Dryers. A screw-conveyor dryer is a continuous indirect-heat dryer, consisting essentially of a horizontal screw conveyor (or paddle conveyor) enclosed in a cylindrical jacketed shell. Solid fed in one end is conveyed slowly through the heated zone, and discharges from the other end. The vapor evolved is withdrawn through pipes set in the roof of the shell. The shell is 3 to 24 in. in diameter and up to 20 ft long; when more length is required, several conveyors are set one above another in a bank. Often the bottom unit in such a bank is a cooler, not a dryer; water or another coolant in the jacket lowers the temperature of the dried solids before they are discharged.

The rate of rotation of the conveyor is slow, from 2 to 30 rpm. Heat-transfer coefficients are based on the entire inner surface of the shell, even though the shell runs only 10 to 60 per cent full. The coefficient depends on the loading in the shell and on the conveyor speed. It ranges, for many solids, between 3 and 10 $Btu/(ft^2)(hr)(°F)$.

Screw-conveyor dryers handle solids that are too fine and too sticky for rotary dryers. They are completely enclosed and permit recovery of solvent vapors with little or no dilution by air. When they are provided with star feeders, as described in Chap. 5, they may be operated under moderate vacuum. Thus they are adaptable to the continuous removal and recovery of volatile solvents from solvent-wet solids, such as spent meal, from leaching operations. For this reason they are sometimes known as *desolventizers*.

Tower Dryers. A tower dryer contains a series of circular trays mounted one above the other on a central rotating shaft. Solid feed dropped on the topmost tray is exposed for a short time to a stream of hot air or gas which passes across the tray. The solid is then scraped off and dropped to the tray below. It travels in this way through the dryer, discharging as dry

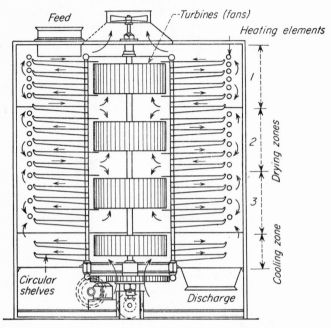

Fig. 16-6. Turbodryer. (*Wyssmont Co.*)

product from the bottom of the tower. The flow of solids and gas may be either parallel or countercurrent.

The *turbodryer* illustrated in Fig. 16-6 is a tower dryer with internal recirculation of the heating gas. Turbine fans circulate the air or gas outward between some of the trays, over heating elements, and inward between other trays. Gas velocities are commonly 2 to 8 ft/sec. The bottom two trays of the dryer shown in Fig. 16-6 constitute a cooling section for dry solids. Preheated air is usually drawn in the bottom of the tower and discharged from the top, giving countercurrent flow. A turbodryer may be considered as a continuous tray dryer, with, however, some agitation of the solids as they tumble from one tray to another.

Flash Dryers. In a flash dryer a wet pulverized solid is transported for a few seconds in a hot gas stream. A dryer of this type is shown in Fig. 16-7. Drying takes place during transportation. The rate of heat transfer from the gas to the suspended solid particles is high, and drying is rapid, so that no more than 3 or 4 sec is required to evaporate substantially all the

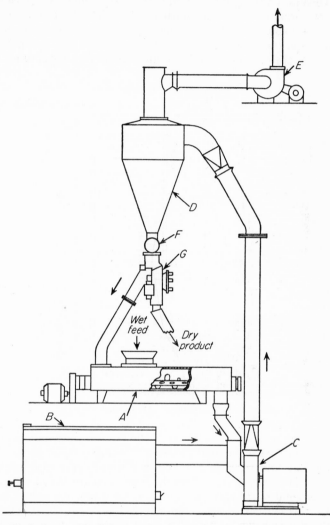

FIG. 16-7. Flash dryer with disintegrator: *A*, paddle-conveyor mixer. *B*, oil-fired furnace. *C*, hammer mill. *D*, cyclone separator. *E*, vent fan. *F*, star feeder. *G*, solids flow divider and timer.

moisture from the solid. The temperature of the gas is high—often about 1200°F at the inlet—but the time of contact is so short that the temperature of the solid rarely rises more than 100°F during drying. Flash drying may therefore be applied to sensitive materials that in other dryers would have to be dried indirectly by a much cooler heating medium.

Sometimes a pulverizer is incorporated in the flash-drying system to give simultaneous drying and size reduction. Such a system is shown in Fig. 16-7. Wet feed enters mixer A, where it is blended with enough dry material to make it free-flowing. Mixed material discharges into hammer mill C, which is swept with hot flue gas from furnace B. Pulverized solid is carried out of the mill to the gas stream through a fairly long duct, in which drying takes place. Gas and dry solid are separated in cyclone D, with the cleaned gas passing to vent fan E. Solids are removed from the cyclone through star feeder F, which drops them into solids divider G. This divider is nearly always needed to recycle some dry solid for mixing with the wet feed. It operates by a timer which moves a flapper valve so that for a fixed period of time dry solid returns to the mixer and for another fixed period it is removed as product. Usually more solid is returned to the mixer than is withdrawn. Typical recirculation ratios are 3 to 4 lb of solid returned per pound of product removed from the system.

Festoon Dryers. A festoon dryer resembles a screen-conveyor dryer except that the screen is replaced by the material that is being dried. The path of travel is much longer than in a screen-conveyor dryer, and the rate of travel is considerably greater. Festoon dryers are used for continuous wide sheets of material such as films, plastics, and textiles. Usually the material passes alternately upward and downward over a number of metal rolls set across the drying chamber. Drying is done directly by heating the rolls with steam or hot water, by direct contact with hot air or gas, or by both methods simultaneously. As in screen-conveyor dryers, the drying conditions may differ from point to point in the dryer. Positive control of humidity is especially important in most festoon dryers.

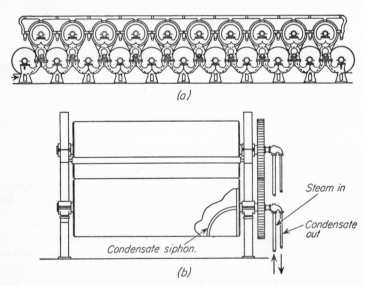

(a)

(b)

FIG. 16-8. Cylinder dryer: (*a*) General view. (*b*) Detail of roll construction.

Cylinder Dryers. In the manufacture of paper, roofing felt, and other materials that are produced in continuous sheets, it is desirable to dry them at the same rate at which they are made. The dryer on a paper machine is an example of this method, and is illustrated in Fig. 16-8. After the sheet has been formed, it is carried in a zigzag manner over the surfaces of a series of hollow steam-heated rolls. These rolls are driven mechanically at the same speed at which the sheet of paper is produced. The number of rolls, their temperature, and the speed of the sheet are so adjusted that the sheet is dry when it leaves the last roll.

Dryers for Solutions and Slurries

Tray dryers and pan dryers can handle liquid materials, though usually with very long drying cycles. Some special pan dryers have a false bottom covered with filter medium through which some of the liquid in a slurry is drawn by vacuum before thermal drying is begun. Some top-feed filters, as described in Chap. 7, also function as dryers by drawing hot air through the filter cake before it is discharged. A few dryers, which are discussed in this section, evaporate solutions and slurries to dryness entirely by thermal means. These are drum dryers and spray dryers.

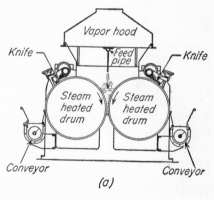

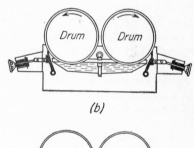

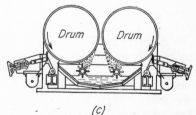

Fig. 16-9. Drum dryers: (*a*) Double-drum with center feed. (*b*) Twin-drum with dip feed. (*c*) Twin-drum with splash feed. (*Buflovac   Equipment   Division,   Blaw-Knox Co.*)

Drum Dryers. A drum dryer consists of one or more heated metal rolls on the outside of which a thin layer of liquid is evaporated to dryness. The dried solid is scraped off the rolls as they slowly revolve.

Various types of atmospheric drum dryers are illustrated in Fig. 16-9. In the *double-drum dryer with center feed*, shown in Fig. 16-9*a*, liquid is fed from a trough or perforated pipe into a pond in the narrow space above and between two rolls. The pond is confined there by stationary end plates. Heat is transferred by conduction to the liquid, which is partly concentrated in the space between the rolls.

Concentrated liquid issues from the bottom of the pond as a viscous layer covering the remainder of the drum surfaces. Substantially all the liquid is vaporized from the solid as the drums turn, leaving a thin layer of dry material which is scraped off by doctor blades into conveyors below. Vaporized moisture is collected and removed through a vapor head above the drums.

Double-drum dryers are effective with dilute solutions, concentrated solutions of highly soluble materials, and moderately heavy slurries. They are not suitable for solutions of salts with limited solubility or for slurries of abrasive solids that settle out and create excessive pressure between the drums.

Materials not suitable for a double-drum dryer may be handled by a *twin-drum dryer with dip feed* (Fig. 16-9b) or with *splash feed* (Fig. 16-9c). In these dryers the two drums operate independently and revolve in directions opposite to those in a double-drum dryer. The film of liquid adhering to the drums is, in general, thinner and more uniform than in a double-drum dryer, and in drying slurries the particle size of the solids is better preserved. Dip feed is used when the liquid spontaneously forms a sufficiently thick layer on the submerged drum; splash feed is necessary with liquids which do not otherwise adhere well enough to the drum surface. Single-drum dryers are also available, utilizing either dip feed or splash feed. Fully enclosed drum dryers operating under vacuum find occasional applications.

The rolls of a drum dryer are 2 to 5 ft in diameter and 2 to 12 ft long, revolving at 4 to 10 rpm. The time that the solid is in contact with hot metal is 6 to 15 sec, which is short enough to result in little decomposition even of heat-sensitive products. The heat-transfer coefficient is high, from 220 to 360 Btu/(ft^2)(hr)(°F) under optimum conditions, although it may be as low as 30 when conditions are adverse.[11h] The drying capacity is proportional to the active drum area; it is usually between 1 and 10 lb of dry product per square foot of drying surface per hour.

Spray Dryers. In a spray dryer a slurry or liquid solution is dispersed into a stream of hot gas in the form of a mist of fine droplets. Moisture is rapidly vaporized from the droplets, leaving residual particles of dry solid, which are then separated from the gas stream. The flow of liquid and gas may be cocurrent, countercurrent, or a combination of both in the same unit.

The droplets are formed inside a cylindrical drying chamber, either by spray nozzles or by high-speed spray disks. In either case it is essential to prevent the droplets or wet particles of solid from striking solid surfaces before drying has taken place, so that the drying chambers are necessarily large. Diameters of 8 to 30 ft are common.

Typical spray dryers are shown in Figs. 16-10 and 16-11. In the design shown in Fig. 16-10 the chamber is a cylinder with a short conical bottom. Liquid feed is pumped into a spray-disk atomizer set in the roof of the chamber. In this dryer the spray disk is about 12 in. in diameter and rotates at 5,000 to 10,000 rpm. It atomizes the liquid into tiny drops, which

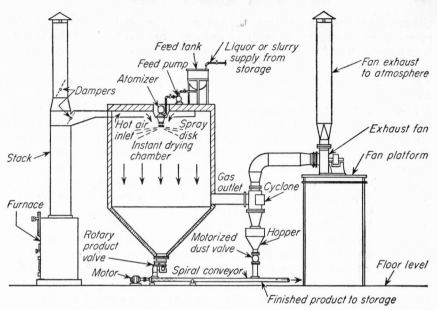

Fig. 16-10. Spray dryer with parallel flow. (*Instant Drying Company.*)

are thrown radially into a stream of hot gas, which enters near the top of the chamber. Cooled gas is drawn by an exhaust fan through a horizontal discharge line set in the side of the chamber at the bottom of the cylindrical section. The gas passes through a cyclone separator, where any entrained particles of solid are removed. Much of the dry solid settles out of the gas into the bottom of the drying chamber, from which it is removed by a rotary valve and screw conveyor and combined with any solid collected in the cyclone.

Figure 16-11 shows a "mixed-flow" spray dryer utilizing both parallel and countercurrent flow of solids and gas. The drying chamber has a short cylindrical upper section and a long bottom cone. The chamber is typically 22 ft in diameter and 35 ft high. Hot gas, which is a mixture of flue gas from an air heater and air from a primary air fan, is admitted to the drying chamber tangentially near the top. The gas spirals gently downward near the walls, then reverses direction and spirals upward more rapidly near the axis of the chamber, leaving through the outlet duct at the top. Feed liquid enters through a spray pipe held by the spray-pipe guide in the roof of the drying chamber. The feed pipe ends in a spray nozzle about halfway down the chamber. Liquid is pumped to the nozzle at a pressure of 1,200 to 3,000 lb force/in.^{2} and is atomized into the central rising spiral gas stream. The liquid evaporates, leaving particles of dry solid, which pass outward into the outer spiral gas stream. This sweeps them gently toward the walls and downward to the bottom of the cone. From here they are removed, with some gas, to a dry-solids collector.

Most of the gas leaves from the top of the drying chamber and passes to a wet collector. This is a simple wet-gas scrubber in which the gas, which is still fairly hot, is brought into contact with a thin liquid feed. Thin feed is thus preconcentrated and is withdrawn from the bottom of the collector at a controlled rate for further concentration in a separate evapora-

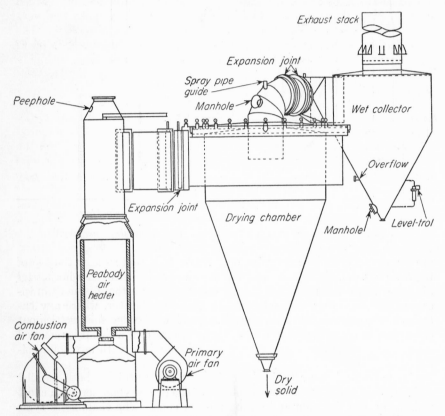

FIG. 16-11. Spray dryer with combined parallel and countercurrent flow. (*Swenson Evaporator Company.*)

tor or for use directly as feed to the spray dryer. In the wet collector, any entrained solids are also scrubbed out of the discharged gas, which then passes to the stack.

Many other designs of spray dryers are also in use.

The chief advantages of spray dryers are the very short drying time, of the order of 2 to 20 sec, which permits drying of highly heat-sensitive materials, and the production of solid or hollow spherical particles. The desired consistency, bulk density, appearance, and flow properties of some products, such as foods or synthetic detergents, may be difficult or impossible to obtain in any other type of dryer. Spray dryers also have the advantage of yielding from a solution, slurry, or thin paste, in a single step, a dry product that is ready for the package. A spray dryer may combine the functions of an evaporator, a crystallizer, a dryer, a size-reduction unit, and a classifier. Where one can be used, the resulting simplification of the over-all manufacturing process may be considerable.

Considered as dryers alone, spray dryers are not highly efficient. Much heat is ordinarily lost in the discharged gases. They are bulky and very large, often 80 ft or more high, and are not always easy to operate. The bulk density of the dry solid—a property of especial importance in packaged products—is often difficult to keep constant, for it may be highly sensitive to changes in the solids content of the feed, to the inlet gas temperature, and to other variables.[11f]

In spray drying solutions the evaporation from the surface of the drops leads to initial deposition of solute at the surface before the interior of the drop reaches saturation. The rate of diffusion of solute back into the drop is slower than the flow of water from the interior to the surface, and the entire solute content accumulates at the surface. The final dry particles are, then, usually hollow, and the product from a spray dryer is very porous.

The performance of a spray dryer depends on the time the drops spend in the drying chamber. This in turn depends on many factors, including the size and shape of the chamber, the size and terminal velocities of the drops, and the flow pattern and velocity of the air. Since the drops formed in commercial atomization equipment cover a wide range of sizes, the time required to reach dryness varies from drop to drop and is, of course, longer for large drops than for small. It is possible to overdry the small drops and at the same time to underdry the large ones, which then leave the dryer in a sticky condition. Although a great deal has been learned about the action in spray dryers,[8, 11g] their design should be left to specialists.

PRINCIPLES OF DRYING

Because of the wide variety of materials that are dried in commercial equipment and the many types of equipment that are used, there is no single theory of drying that covers all materials and dryer types. Variations in phase, in shape and size of stock, in moisture equilibria, in the mechanism of flow of moisture through the solid, and the method of providing the heat required for the vaporization—all prevent a unified treatment. General principles, used in a semiquantitative way, are relied upon. Dryers are seldom designed by the user but are bought from companies that specialize in the engineering and fabrication of drying equipment.

Equilibria. A typical equilibrium relationship between solid and moist air is discussed in Chap. 10 and illustrated in Fig. 10-9d. In this figure moisture contents are expressed in molal units. More commonly, the moisture equilibria for solids are given as relationships between the relative humidity of the air or gas and the moisture content of the solid in pounds of water per pound (or 100 lb) of bone-dry solid. Examples of such relations are given in Fig. 16-12. Curves of this type are nearly independent of dry-bulb temperature. The ordinates of such curves are readily converted to humidities in pounds of water per pound of dry air.

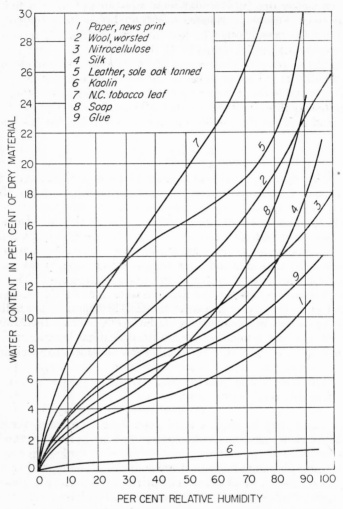

FIG. 16-12. Equilibrium-moisture curves (temperature = 25°C).[7]

When a wet solid is brought into contact with air of lower humidity than that corresponding to the moisture content of the solid, as shown by the humidity-equilibrium curve, the solid tends to lose moisture and dry to equilibrium with the air. When the air is more humid than the solid in equilibrium with it, the solid absorbs moisture from the air until equilibrium is attained.

For ease in calculations all moisture contents are expressed in pounds of water per pound (or 100 lb) of bone-dry solid. This practice is followed throughout the present chapter.

Equilibrium Moisture and Free Moisture. The air entering a dryer is seldom completely dry but contains some moisture and has a definite relative humidity. For air of definite humidity, the moisture content of the solid leaving the dryer cannot be less than the equilibrium-moisture content corresponding to the humidity of the entering air. That portion of the water in the wet solid that cannot be removed by the inlet air, because of the humidity of the latter, is called the equilibrium moisture.

The free water is the difference between the total-water content of the solid and the equilibrium-water content. Thus, if X_T is the total moisture content, and if X^* is the equilibrium-moisture content, the free moisture X is

$$X = X_T - X^* \tag{16-1}$$

It is X, rather than X_T, that is of interest in drying calculations.

Bound and Unbound Water. If an equilibrium curve like those in Fig. 16-12 is continued to its intersection with the axis for 100 per cent humidity, the moisture content so defined is the minimum moisture this material can carry and still exert a vapor pressure at least as great as that exerted by liquid water at the same temperature. If such a material contains more water than that indicated by this intersection, it can still exert only the vapor pressure of water at the stock temperature. This makes possible a distinction between two types of water held by a given material. The water up to the lowest concentration that is in equilibrium with saturated air (given by the intersection of the curves of Fig. 16-12 with the line for 100 per cent humidity) is called bound water because it exerts a vapor pressure less than that of liquid water at the same temperature. Substances containing bound water are often called hygroscopic substances. The condition at which wood, textiles, and other cellular materials is in equilibrium with saturated air is called the *fiber-saturation* point.

Bound water may exist in several conditions. Liquid water in fine capillaries exerts an abnormally low vapor pressure because of the highly concave curvature of the surface; moisture in cell or fiber walls may suffer a vapor-pressure lowering because of solids dissolved in it; water in natural organic substances is in physical and chemical combination, the nature and strength of which vary with the nature and moisture content of the solid. Unbound water, on the other hand, exerts its full vapor pressure and is largely held in the voids of the solid.

The terms employed in this discussion may be clarified by reference to Fig. 16-12. Consider, for instance, curve 2 for worsted yarns. This intersects the curve for 100 per cent humidity at 26 per cent moisture; consequently, any sample of wool that contains less than 26 per cent water contains only bound water. Any moisture that a sample may contain above 26 per cent is unbound water. If the sample contains 30 per cent water, for example, 4 per cent of this water is unbound and 26 per cent bound. Assume, now, that this sample is to be dried with air of 30 per cent relative humidity. Curve 2 shows that the lowest moisture content that can be reached under these conditions is 9 per cent. This, then, is the equilibrium-moisture content for this particular set of conditions. If a sample containing 30 per cent total moisture is to be dried with air at 30 per cent relative humidity, it contains 21 per cent free water and 9 per cent equilibrium moisture. The distinction between bound and unbound water depends on the material itself, while the distinction between free and equilibrium moisture depends on the drying conditions.

Rate of Drying

The capacity of a thermal dryer depends on both the rate of heat transfer and the rate of mass transfer. Since water vapor must be vaporized, heat of vaporization must be supplied to the vaporization zone, which, depending on the type of material and the conditions of the process, may be either at or near the surface of the solid or within the solid. The moisture must flow either as a liquid or a vapor through the solid and as a vapor from the surface of the solid into the gas or air stream.

Heat Transfer in Dryers. Several methods of heat transfer are used in the dryers described in the last section. When the stock is in the form of sheets, lumps, or cakes and is not agitated, heat may be supplied either directly by contact with hot gas or indirectly by contact with hot surfaces. In drying sheet materials in tunnel and festoon dryers the first method is used, and in cylinder dryers, the second. In tray dryers, where the cakes of solid are held in trays, a combination of both methods is used.

When all the heat for vaporizing the water is supplied by direct contact with hot gases and heat transfer by conduction from contact with hot boundaries or from radiation from solid walls is negligible, the process is called *adiabatic drying*. Then the latent heat is obtained entirely from the sensible heat of the gas. From the standpoint of the gas, the process is much like adiabatic humidification. In actual dryers, radiation is active to some extent, and in high-temperature equipment, heated by flue gas or flames, radiation may be of controlling importance.

So long as unbound water is present, the temperature at the zone of vaporization is that corresponding to the temperature of equilibrium between air and liquid water. If only bound water is present, as in the drying of hygroscopic materials, the partial pressure of the water vapor at the zone of vaporization is less than the vapor pressure of water at the temperature of the vaporization zone, so the water vapor is superheated. When the last traces of water are being removed from a hygroscopic solid, the temperature of the solid may be close to that of the air.

In through-circulation drying, as in screen-conveyor dryers, where hot gas is blown through a bed of solid particles or a porous sheet, another kind of heat transfer—that between hot gas and a bed of solids—occurs. Simultaneously, mass transfer between the same phases takes place. The basic mechanism is like that of humidification in a packed tower except that the water diffuses through the solid particles instead of flowing over them.

In spray dryers and rotary dryers the material is in the form of particles or drops. Heat is supplied to them by allowing them to fall through, or be carried by, a stream of air or gas. In spray dryers and in direct rotary dryers heat transfer is primarily by conduction and convection, and adiabatic conditions prevail. At high temperatures, especially in indirect rotaries, radiation is important and may be controlling.

Rotary dryers are designed on the basis of heat transfer. Since the particles are falling through the hot gas, the heat-transfer rate in direct rotaries is a function of the surface area of the particles rather than the surface of the shell. This is also true of spray dryers. In indirect rotaries, heat transfer both from the walls and to the surface of the particles must be considered. Empirical equations are available for direct rotary dryers.[11e]

The laws and procedures for heat transfer given in Chap. 8 may be used in dryer calculations. Experimental data on the heat-transfer coefficients are sometimes needed for accurate design.

Mass transfer from the solid to the gas or air is covered by the relations discussed in Chap. 15. The flow of moisture in the solid itself, however, depends on the structure and character of the solid. To understand this process, knowledge of the internal mechanism of drying is required. This mechanism is the subject of the next section.

Mechanism of Drying of Solids. Two quite different kinds of solids are found in drying practice. Although many solids are intermediate between the two extremes, it is convenient to assume that the solid is either porous or nonporous. Also, either type may be either hygroscopic or nonhygroscopic.

The details of drying a given solid may be examined from the standpoints of how the drying rate varies with air conditions and what happens in the interior of the solid during drying. For this purpose, consider a slab of stock which is being uniformly dried from both sides but which is protected against moisture loss from the edges. Assume that the temperature, humidity, and velocity and direction of flow of the air across the drying surface are constant. This type of drying is called drying under *constant drying conditions*. Note that only the conditions in the air stream are constant, as the moisture content and other factors in the solid are changing.

Rate-of-drying Curve. To study the mechanism of drying under constant conditions, a plot of the instantaneous rate of drying, in pounds of water removed per square foot of drying area per hour, as a function of the instantaneous free-moisture content, in pounds of water per pound of bone-dry solid, is useful. Rate-of-drying curves for four different materials are given in Figs. 16-13, 16-15, 16-17, and 16-18. Figure 16-13 is the curve for a nonporous clay slab; Fig. 16-15 is that for a porous, nonhygroscopic ceramic plate having small pores; Fig. 16-17 shows the rate of drying of a bed of sand particles; and Fig. 16-18 one for sheets of paper pulp, which is both hygroscopic and porous. The pronounced differences in the shapes of these curves are the result of differences in the mechanism of moisture

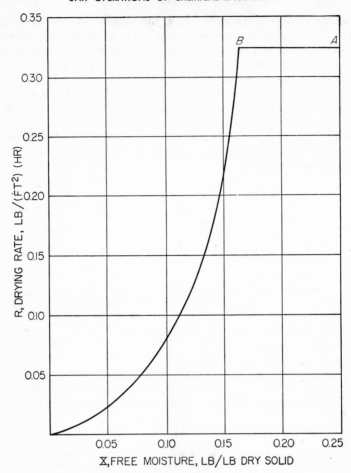

F$_{IG}$. 16-13. Drying-rate curves for nonporous clay slab. (*After Sherwood.*[15])

flow in the various materials. Experimental study of the distribution of moisture during drying is also illuminating.

Constant-rate Period. It is apparent that each drying-rate curve has at least two distinct segments. After a preliminary period (not shown in the various rate-of-drying figures), during which the temperature of the material adjusts itself to the drying conditions, each curve has a horizontal segment *AB*, which pertains to the first major drying period. This period, which may be absent if the initial moisture content of the solid is less than a certain minimum, is called the constant-rate period. It is characterized by a rate of drying independent of moisture content. During this period, the solid is so wet that a continuous film of water exists over the entire drying surface, and this water acts as if the solid were not there. If the solid is nonporous, the water removed in this period is mainly superficial water on the surface of the solid. In a porous solid, most of the water removed in the constant-rate period is supplied from the interior of the solid. The evaporation from a porous material is by the same mechanism as that from a wet-bulb thermometer, and the process occurring at a wet-bulb thermometer is essentially one of constant-rate drying. In the absence of radiation or of heat transfer by conduction through direct contact of the solid with hot surfaces, the temperature of the solid during the constant-rate period is the wet-bulb temperature of the air.

To calculate the rate of drying during the constant-rate period either the mass-transfer equation [Eq. (16-2)] or the heat-transfer equation [Eq. (16-3)] may be used.

$$w = k_y'(H_i - H)A \qquad (16\text{-}2)$$

$$w = \frac{h_y(t - t_i)A}{\lambda_i} \qquad (16\text{-}3)$$

where w = rate of evaporation, lb/hr
A = drying area, ft^2
h_y = heat-transfer coefficient, Btu/(ft^2)(hr)(°F)
k_y' = mass-transfer coefficient, lb/(ft^2)(hr)(unit humidity difference)
H_i = humidity of air at interface, lb water/lb dry air
H = humidity of air, lb water/lb dry air
t = temperature of air, °F
t_i = temperature at interface, °F
λ_i = latent heat at temperature t_i, Btu/lb

When the air is flowing parallel with the surface of the solid, the heat-transfer coefficient may be estimated by the dimensional equation

$$h_y = 0.128G^{0.8} \qquad (16\text{-}4)$$

where G is the mass velocity in pounds per square foot per hour.

When the flow is perpendicular to the surface, the equation is

$$h_y = 0.37G^{0.37} \qquad (16\text{-}5)$$

When radiation from hot surroundings and conduction from solid surfaces in contact with the stock are not negligible, the temperature of the interface is greater than the wet-bulb temperature of the air, and the rate of drying is, by Eq. (16-3), increased accordingly. Methods for estimating these effects are available.[11a, 14]

Critical Moisture Content and Falling-rate Period. As the moisture content decreases, the constant-rate period ends at a definite moisture content, and during further drying the rate decreases. The point terminating the constant-rate period, shown by point B in Figs. 16-13, 16-15, 16-17, and 16-18, is called the *critical point*.[14] This point marks the instant when the liquid water on the surface is insufficient to maintain a continuous film covering the entire drying area. In nonporous solids, the critical point occurs at about the time when the superficial moisture is evaporated. In porous solids, the critical point is reached when the rate of moisture flow to the surface no longer equals the rate of evaporation called for by the wet-bulb evaporative process.

If the initial moisture content of the solid is below the critical, the constant-rate period is not found.

The critical moisture content varies with the thickness of the material and with the rate of drying. It is therefore not a property of the material itself. The critical moisture content is best determined experimentally, although some approximate data are available.[11b]

The period subsequent to the critical point is called the falling-rate period. It is clear from Figs. 16-13, 16-15, 16-17, and 16-18 that the drying-rate curve in the falling-rate period varies from one type of material to another. The shape of the curve also depends on the thickness of the material and on the external variables.

Drying of Nonporous Solids and Diffusion Theory. The moisture distribution in a solid giving a falling-rate curve like that in Fig. 16-13 is shown as the dotted line in Fig. 16-14, where the local-moisture content is plotted against distance from the surface. The curve is concave downward throughout its length. Rate curves like Fig. 16-13 and moisture-distribution curves like Fig. 16-14 are usually found in the drying of nonporous materials like soap, glue, and plastic clay. These substances are

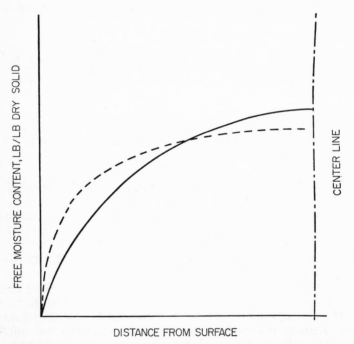

Fig. 16-14. Moisture distribution in slab dried from both faces. Moisture flow by diffusion. Full curve, theoretical distribution. Dotted curve, actual distribution.

essentially colloidal gels of solids and water. They retain considerable amounts of bound water. Dense cellular solids, such as wood and leather, give curves of the same types when they are being dried at moisture contents below the fiber-saturation point.

The shape of the moisture-distribution curves of Fig. 16-14 is qualitatively consistent with that called for by assuming that the moisture is flowing by diffusion through the solid, in accordance with Eq. (13-9). This equation has long been used as a basis for quantitative calculations of the rate of drying of nonporous solids.[10, 13] Materials drying in this way are said to be drying by diffusion, although the actual mechanism is probably considerably more complicated than that of simple diffusion.

Diffusion is characteristic of slow-drying materials. The resistance to the mass transfer of water vapor to the air is usually negligible, and the diffusion in the solid controls the over-all drying rate. The velocity of the air has little or no effect, and the humidity of the air influences the process primarily through its effect on the equilibrium-moisture content. Since diffusivity increases with temperature, the rate of drying increases with the temperature of the solid.

Equations for Diffusion. Assuming that the diffusion law given by Eq. (13-9) does apply, the integrated forms of this equation may be used to relate the time of drying with the initial and final moisture contents. Thus, if the assumptions made in integrating Eq. (13-9) are all accepted, Eq. (13-10) is obtained on integration. For drying, the result may be written

$$\frac{X_T - X^*}{X_{T1} - X^*} = \frac{X}{X_1} = \frac{8}{\pi^2}\left(e^{-a_1\beta} + \frac{1}{9}e^{-9a_1\beta} + \frac{1}{25}e^{-25a_1\beta} + \cdots\right) \quad (16\text{-}6)$$

where $\beta = D'_v\theta/s^2$
  $a_1 = (\pi/2)^2$
  X_T = average total moisture content at time θ hr
  X = average free-moisture content at time θ hr
  X^* = equilibrium-moisture content
  X_{T1} = initial moisture content at start of drying when $\theta = 0$
  X_1 = initial free-moisture content
  D'_v = diffusivity of moisture through solid, ft^2/hr
  s = one-half slab thickness, ft

All moisture contents are in pounds of water per pound of bone-dry solid.

Figure 8-34, with β in place of N_{Fo} and X/X_1 in place of $(t_s - \bar{t}_b)/(t_s - t_a)$, is a plot of Eq. (16-6).

The accuracy of the diffusion theory for drying suffers from the fact that the diffusivity usually is not constant but varies with moisture content. It is especially sensitive to shrinkage. The value of D'_v is less at small moisture contents than at large and may be very small near the drying surface. Thus, the moisture distribution called for by the diffusion theory with constant diffusivity is like that shown by the full line in Fig. 16-14.

In practice, an average value of D'_v established experimentally on the material to be dried is used.

When β is greater than about 0.1, only the first term on the right-hand side of Eq. (16-6) is significant, and the remaining terms of the infinite series may be dropped. By differentiating the resulting equation with respect to time it is found that

$$-\frac{dX}{d\theta} = \left(\frac{\pi}{2}\right)^2 \frac{D'_v}{s^2} X \tag{16-7}$$

Assuming constant diffusivity, an approximate equation may be derived for θ_f, the time necessary to dry, by diffusion only, the solid from an initial free moisture X_1 to a final moisture content of X_2. Thus, by integrating Eq. (16-7),

$$-\int_{X_1}^{X_2} \frac{dX}{X} = \left(\frac{\pi}{2}\right)^2 \frac{D'_v}{s^2} \int_0^{\theta_f} d\theta$$

or
$$\theta_f = \frac{4s^2}{\pi^2 D'_v \ln(X_1/X_2)} \tag{16-8}$$

Equation (16-8) may be used in place of the more accurate relation of Fig. 8-34 in most practical situations.

Equation (16-7) shows that, when diffusion controls, the rate of drying is inversely proportional to the square of the thickness. Equation (16-8) shows that, if time is plotted against the logarithm of the free-moisture content, a straight line should be obtained.

Example 16-1. Planks of wood 1 in. thick are dried from an initial moisture content of 25 per cent to a final moisture of 5 per cent, using air of negligible humidity. If D'_v for the wood is 3.2×10^{-5} ft²/hr, how long should it take to dry the wood?

Solution. Since the air is dry, the equilibrium-moisture content is zero, and $X_1 = 0.25$ and $X_2 = 0.05$. Then $X_2/X_1 = 0.05/0.25 = 0.20$. From Fig. 8-34, β is read on the abscissa corresponding to the known ordinate of 0.20 and found to be 0.57. Since $s = 0.5/12$ ft,

$$\frac{3.2 \times 10^{-5}\theta_f}{(0.5/12)^2} = 0.57$$

Solving for θ_f gives a drying time of 31.0 hr.

Another, more approximate solution can be found by Eq. (16-8). Substituting the appropriate values in this equation gives

$$\theta_f = \frac{4(0.5/12)^2 2.303 \log(25/5)}{\pi^2 3.2 \times 10^{-5}} = 35.5 \text{ hr}$$

Shrinkage and Casehardening. When bound moisture is removed from a colloidal nonporous solid, the material shrinks. In small pieces this effect may not be important, but in large units improper drying may lead to serious product difficulties which are basically a result of shrinkage.

Since the outer layers necessarily lose moisture before the interior portions, the concentration of moisture in these layers is less than that in the interior, and the surface layers shrink against an unyielding, constant-volume core. This surface shrinkage causes checking, cracking, and warping. Also, since the diffusivity is sensitive to moisture concentration and decreases with concentration, the resistance to diffusion in the outer layers is increased by surface dehydration. This accentuates the effect of shrinkage by impeding the flow of moisture to the surface and so increasing the moisture gradient near the surface.

In extreme cases, the shrinkage and drop in diffusivity may combine to give a skin, practically impervious to moisture, which encloses the bulk of the material so the interior moisture cannot be removed. This is called casehardening.

Warping, checking, cracking, and casehardening may be minimized by reducing the rate of drying, thereby flattening the concentration gradients in the solid. Then the shrinkage of the surface is reduced, and the diffusivity throughout the solid is more nearly constant. The moisture gradient at the surface is flattened, and the entire piece is protected against shrinkage.

The rate of drying is controlled most readily by controlling the humidity of the drying air. Since the equilibrium-moisture concentration at the surface is fixed by the humidity of the air, increasing the latter also increases the former. For a given total moisture concentration X_T, the free-moisture concentration X is reduced by increasing the equilibrium moisture X^*, as shown by Eq. (16-1). The over-all concentration gradient of free water is then reduced, the drying slowed, and the skin effect minimized.

Drying of Porous Solids and Flow of Water by Capillarity. The flow of water through porous solids does not conform to the laws of diffusion given by Eq. (13-9). This may be seen by comparing the moisture distribution in a solid of this type during drying with that for diffusion. A typical moisture-distribution curve for a porous solid is shown in Fig. 16-16. A point of inflection divides the curve into two parts, one concave upward and one concave downward. This is completely contrary to the distribution called for by the diffusion law and shown in Fig. 16-14.

Moisture flows through porous solids by capillarity.[3-5] A porous material contains a complicated network of interconnecting pores and channels, the cross sections of which vary greatly. At the surface are the mouths of pores of various sizes. As water is removed by vaporization, a meniscus across each pore is formed, which sets up capillary forces by the interfacial tension between the water and the solid. The capillary forces possess components in the direction perpendicular to the surface of the solid. It is these forces that provide the driving force for the movement of water through the pores toward the surface.

The strength of capillary forces at a given point in a pore depends on the curvature of the meniscus, which is a function of the pore cross section. Small pores develop greater capillary forces than large ones, and small pores, therefore, can pull water out of the large pores. As the water at the surface is depleted, the large pores tend to empty first. Air must displace the water so removed. This air enters either through the mouths of the larger pores at the drying surface or from the sides and back of the material if drying is from one side only.

The forces developed in fine pores by capillarity are surprisingly large. The flow of sap to the tops of tall trees is accomplished by capillarity. In drying unbound water from large pieces of wood, internal forces are developed inside the piece in locations not accessible to air, and the structure of

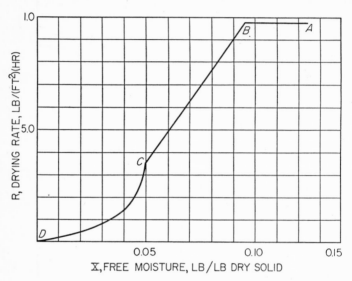

Fig. 16-15. Drying-rate curve for porous ceramic plate. (*After Sherwood and Comings.*[16])

the wood collapses. If the solid is plastic, considerable shrinkage occurs when water is removed by capillarity.

The rate-of-drying curve for a typical porous solid having small pores is shown in Fig. 16-15. As long as the delivery of water from the interior to the surface is sufficient to keep the surface completely wet, the drying rate is constant. The pores are progressively depleted of water, and at the critical point the surface layer of water begins to recede into the solid. This starts with the larger pores. The high points on the surface of the solid emerge like Mt. Ararat after the flood, and the area available for mass transfer from the solid into the air decreases. Then, although the rate of evaporation per unit *wetted* area remains unchanged, the rate based on the *total* surface, including both wet and dry areas, is less than that in the constant-rate period. The rate continues to decrease as the fraction of dry surface increases.

The first section of the falling-rate period is shown by line *BC* in Fig. 16-15. The rate of drying during this period depends on the same factors that are active during the constant-rate period, since the mechanism of evaporation is unchanged and the vaporization zone is at or near the surface. The state of the water during this period is called the *funicular state*. The water in the pores is the continuous phase and the air the dispersed phase. In the first falling-rate period, the rate-of-drying curve is usually linear.

As the water is progressively removed from the solid, the fraction of the pore volume that is occupied by air increases. When the fraction reaches

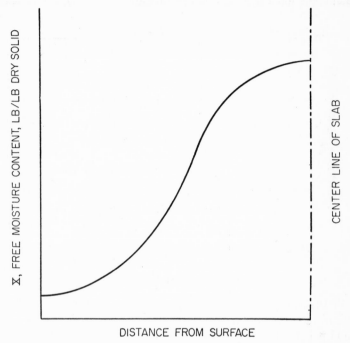

FIG. 16-16. Moisture distribution in slab dried from both faces. Moisture flow by capillarity.

a certain limit, there is insufficient water left to maintain continuous films across the pores, the interfacial tension in the capillaries breaks, and the pores fill with air, which now becomes the continuous phase. The remaining water is relegated to small isolated pools in the corners and interstices of the pores. This state is called the *pendular state*. When this state appears, the rate of drying again suddenly decreases as shown by line *CD* in Fig. 16-15. The moisture content at which this break appears, shown by point *C* in Fig. 16-15, is called the second critical point, and the period that it initiates is called the second falling-rate period.

In the pendular state, the vaporization zone is in the form of small isolated areas dispersed throughout the solid, and the vaporization rate is practically independent of the velocity of the air. The water vapor must diffuse through the solid, and the heat of vaporization must be transmitted to the vaporization zones by conduction through the solid. Temperature gradients are set up in the solid, and the temperature of the solid surface approaches the dry-bulb temperature of the air. For fine pores, the rate-of-drying curve during the second falling-rate period conforms to the diffusion law, and the drying-rate curve is concave upward.

Effect of Gravity. In quite porous solids, such as beds of sand, the pores are large, the resistance to moisture flow is low, and the capillary forces are small. The force of gravity is then large in comparison with the capillary forces, and there is a directional effect due to gravity.[6]

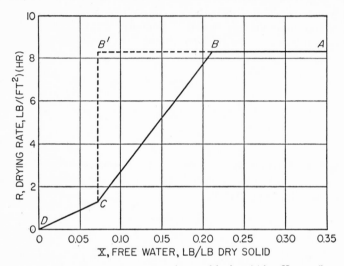

Fig. 16-17. Drying-rate curves for sand beds. (*After Hougen.*[6])

Drying-rate curves for a horizontal bed of sand particles are shown in Fig. 16-17. Two broken lines are shown. The full line *ABCD* is obtained for a sample dried from the top and the dotted line *AB'CD* for a sample dried from the bottom. In top drying gravity opposes capillarity, and the first critical point is reached early, as shown by point *B*. Then the usual two falling-rate periods, represented by segments *BC* and *CD*, are found. In bottom drying capillary and gravity forces act in the same direction to move water to the drying surface, and the constant-rate period continues until the pendular state appears. The constant-rate period is represented by segment *AB'*. At the critical point *B'* the rate drops suddenly to that corresponding to the second critical point found in top drying. Only one critical point and one falling-rate period are obtained.

In very porous solids, the rate-of-drying curve in the second falling-rate period is often straight, and the diffusion equations do not apply.

Calculation of Drying Time under Constant Drying Conditions. In the design of dryers, an important quantity is the time required for drying the material under the conditions existing in the dryer, as this fixes the size of the equipment needed for a given capacity. For drying under constant drying conditions, the time of drying may be determined from the rate-of-drying curve if this can be constructed. Often the only source of this curve is an experiment on the material to be dried, and this gives the drying time directly. Drying-rate curves for one set of conditions often may be modified to other conditions, and then working back from the drying-rate curve to drying time is useful. For example, consider the drying of a slab of solid of half-thickness s ft and face area S ft^2, dried from both sides. The ordinate of the rate-of-drying curve is $-dm/2S \, d\theta$, where m is the total moisture content in the solid, in pounds and θ is the drying time in hours. Call this ordinate R. The abscissa of the curve is X, the pounds of free moisture per pound of bone-dry solid. If ρ_s is the density of the solid, in pounds per cubic foot of bone-dry material, and if shrinkage is neglected so that ρ_s is constant,

$$dm = 2sS\rho_s \, dX \qquad (16\text{-}9)$$

Then
$$R = -\frac{dm}{2S \, d\theta} = \frac{2sS\rho_s \, dX}{2S \, d\theta} = \frac{s\rho_s \, dX}{d\theta} \qquad (16\text{-}10)$$

Since the rate-of-drying curve provides a relation between R and X, Eq. (16-10) may be integrated between X_1 and X_2, the initial and final free-moisture contents, respectively.

$$\theta_T = s\rho_s \int_{X_2}^{X_1} \frac{dX}{R} \qquad (16\text{-}11)$$

where θ_T is the time of drying. This equation may be integrated graphically from the rate-of-drying curve or analytically if equations are available giving R as a function of X.

Equations for Constant Drying Conditions. In two situations equations are available for formally integrating Eq. (16-11). These are during the constant-rate period and during a falling-rate period in which the rate-of-drying curve is linear.

During the constant-rate period, R is constant, and it may be denoted by R_c. Equation (16-11) gives, for θ_c, the time of drying in a constant-rate period,

$$\theta_c = s\rho_s \int_{X_2}^{X_1} \frac{dX}{R_c} = \frac{s\rho_s(X_1 - X_2)}{R_c} \qquad (16\text{-}12)$$

When the rate of drying is linear in X during a falling-rate period,

$$R = aX + b \qquad (16\text{-}13)$$

where a and b are constants. Differentiating Eq. (16-13) gives

$$dR = a \, dX$$

Substituting for dX in Eq. (16-11) gives, for the time required in the falling-rate period,

$$\theta_f = \frac{s\rho_s}{a}\int_{R_2}^{R_1}\frac{dR}{R} = \frac{s\rho_s}{a}\ln\frac{R_1}{R_2} \tag{16-14}$$

where R_1 and R_2 are the ordinates for the initial and final moisture contents, respectively. The constant a is the slope of the rate-of-drying curve and may be written as

$$a = \frac{R_c - R'}{X_c - X'} \tag{16-15}$$

where R_c = rate at first critical point
R' = rate at second critical point
X_c = free-moisture content at first critical point
X' = free-moisture content at second critical point
Substitution of a from Eq. (16-15) into Eq. (16-14) gives

$$\theta_f = \frac{s\rho_s(X_c - X')}{R_c - R'}\ln\frac{R_1}{R_2} \tag{16-16}$$

When the drying process covers both a constant-rate period and a falling-rate period, the X_2 of Eq. (16-12) equals X_c, and the R_1 of Eq. (16-16) equals R_c. The total time of drying θ_T is then

$$\theta_T = \theta_c + \theta_f = s\rho_s\left(\frac{X_1 - X_c}{R_c} + \frac{X_c - X'}{R_c - R'}\ln\frac{R_c}{R_2}\right) \tag{16-17}$$

Here X_1 is the moisture content at the start of the entire process, and R_2 is the drying rate at the end of the process.

In some situations, a single straight line passing through the origin adequately represents the entire falling-rate period. The point (X_c, R_c) lies on this line. When this approximation may be made, Eq. (16-17) may be simplified by noting that b, R', X' drop, out, that $a = R_c/X_c$, and that $R_c/R_2 = X_c/X_2$. Equation (16-17) then becomes

$$\theta_T = \frac{s\rho_s}{R_c}\left[(X_1 - X_c) + X_c\ln\frac{X_c}{X_2}\right] \tag{16-18}$$

Here X_2 is the moisture content at the end of the entire process.

Also, R_c, the rate of drying at the critical point, may be estimated from either Eq. (16-2) or (16-3). Choosing the heat-transfer form and assuming drying from both sides of the sheet,

$$R_c = -\frac{dm}{2S\,d\theta} = \frac{w}{A} = \frac{h_y(t - t_i)}{\lambda_i} \tag{16-19}$$

Equation (16-18) may then be written

$$\theta_T = \frac{s\rho_s\lambda_i}{h_y(t - t_i)}\left[(X_1 - X_c) + X_c \ln\frac{X_c}{X_2}\right] \tag{16-20}$$

Note that $t - t_i$ must be obtained during the constant-rate period; this temperature difference may decrease greatly during the falling-rate period, especially toward the end of the process.

Equations (16-12), (16-17), and (16-18) show that the rate of drying of materials following these equations is inversely proportional to the thickness of the solid, except as the thickness affects the critical moisture contents. This is in contrast to the rate of drying when diffusion controls, where the rate is inversely proportional to the square of the thickness.

Example 16-2. A filter cake 24 in. square and 2 in. thick is dried from both sides with air at a wet-bulb temperature of 80°F and a dry-bulb temperature of 120°F. The air flows parallel with the faces of the cake at a velocity of 2.5 ft/sec. The dry density of the cake is 120 lb/ft³. The equilibrium-moisture content is negligible. Under the conditions of drying the critical moisture is 9 per cent, dry basis. Assuming the rate-of-drying in the falling-rate period is proportional to the moisture content, how long will it take to dry this material from an initial moisture content of 20 per cent to a final moisture of 2 per cent?

Solution. Since the initial moisture content is greater than the critical, and the final moisture less, both constant-rate and falling-rate periods are involved. Also, since the rate-of-drying curve passes through the origin, Eq. (16-20) may be used. The interface temperature t_i is the wet-bulb temperature of the air, 80°F. Then λ_i is, by Appendix 6, 1049 Btu/lb. The mass velocity of the air is

$$\frac{2.5 \times 29 \times 492 \times 3,600}{359(460 + 120)} = 618 \text{ lb/ft}^2\text{-hr}$$

The half-thickness is $s = \frac{1}{12}$ ft, and the coefficient h_y is, by Eq. (16-4),

$$h_y = 0.12 \times 618^{0.8} = 20.5 \text{ Btu/(ft}^2)(\text{hr})(°\text{F})$$

Substituting in Eq. (16-20),

$$\theta_T = \frac{\frac{1}{12} \times 120 \times 1049}{20.5(120 - 80)}\left[(0.20 - 0.09) + 0.09 \times 2.303 \log\frac{0.09}{0.02}\right]$$

$$= 3.1 \text{ hr}$$

Drying of Hygroscopic Porous Solids. When the solid is both porous and hygroscopic, as in a slab of paper pulp, only the unbound water can form the funicular and pendular states. When the unbound water has been removed, considerable bound water is left. This is then removed by progressive vaporization below the surface of the solid, which is accompanied by diffusion of water vapor through the solid.[9] The second critical point is suppressed, and only one critical point, that at the end of the constant-rate period, is found. The drying-rate curves for several thicknesses of porous paper pulp are shown in Fig. 16-18. It is clear that there is no simple relation between drying rate and slab thickness in drying materials of this character.

Drying Time under Variable Drying Conditions. Under certain restricted conditions, equations may be derived for calculating the drying time in a continuous or batch dryer when the humidity and temperature of the air are varying.[1,17] This involves an integration over the entire process that takes account of the variations in these variables. It is gen-

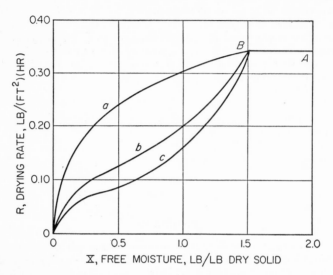

FIG. 16-18. Drying-rate curves for paper-pulp slabs: (a) Thickness 0.108 cm. (b) Thickness 0.648 cm. (c) Thickness 2.37 cm. (*After McCready and McCabe.*[9])

erally necessary that the solid be porous and that the drying fall either in a constant-rate period or in the first falling-rate period.

When diffusion controls, no satisfactory method of calculation is known for variable-condition drying. The basic difficulty is that the instantaneous rate is not a function of the average moisture content of the solid but depends also on the previous history of the material during drying. Also, the change in solid temperature during drying cannot be predicted. The most satisfactory method for obtaining drying times experimentally in such situations is to conduct a small-scale drying experiment on an actual sample of the material under the same variations in temperature, humidity, and air velocity that are to be used in the commercial dryer.[2] This gives the drying time directly.

SYMBOLS

A Area for drying, ft^2

a Slope of drying-rate curve [Eq. (16-15)]

a_1 $(\pi/2)^2$

D_v' Volumetric diffusivity of moisture through solid, ft^2/hr

G Mass velocity of air, lb/ft^2-hr

H Humidity, lb water vapor/lb dry air; H_i, saturation value at t_i

h_y Heat-transfer coefficient, from air to surface of solid, $Btu/(ft^2)(hr)(°F)$

k_y' Mass-transfer coefficient, from surface of solid to air, $lb/(ft^2)(hr)(unit\ humidity$ difference)

R Rate of drying, lb/ft^2-hr; R_c, for first critical point; R_1, for start of drying; R_2, for end of drying; R', for second critical point

S Area of face of slab, ft^2

s One-half slab thickness, ft

t Temperature, °F; t_i, for air-solid interface

w Rate of evaporation, lb/hr

X Average moisture content, lb water/lb bone-dry solid; X, free moisture; X_c, free moisture at first critical point; X_T, total moisture; X_1, free moisture at start of drying; X_2, free moisture at end of drying; X^*, equilibrium moisture; X', free moisture at second critical point

Greek Letters

θ Time of drying, hr; θ_c, time in constant-rate period; θ_f, time in falling-rate period; θ_T, total time in both periods

λ_i Latent heat at t_i, Btu/lb

ρ_s Dry density of solid, lb bone-dry solid/ft^3

Dimensionless Group

β $D_v'\theta/s^2$

PROBLEMS

16-1. Wood chips are to be dried in a continuous rotary countercurrent dryer. The chips entering the dryer contain 40 per cent moisture (wet basis), and the chips discharged from the dryer contain 15 per cent moisture (wet basis). The dryer is to deliver 2,000 lb of dry (15 per cent moisture) product per hour. Air enters at 230°F and 0.010 humidity and leaves at 120°F. The chips must remain in the dryer 2 hr. Small-scale experiments show that air velocities must not exceed 8 ft/sec or chips will be blown from the dryer. The average chip size is 1 by 2 by $\frac{1}{4}$ in. The ratio of length to diameter of the dryer is to be 5:1. The chips occupy 0.17 ft^3 per pound of moisture-free chips. The average pressure in the dryer is 1 atm.

What should be the length and diameter of the dryer?

16-2. A porous solid is dried in a batch dryer under constant drying conditions. Six hours is required to reduce the moisture content from 30 to 10 per cent. The critical moisture content was found to be 16 per cent and the equilibrium moisture 2 per cent. All moisture contents are on the dry basis.

Assuming that the rate of drying during the falling-rate period is proportional to the free-moisture content, how long should it take to dry a sample of the same solid from 35 to 6 per cent under the same drying conditions?

16-3. A countercurrent dryer is operating on a crystalline material having water of crystallization. In order that the crystals may be dried without losing any of their water of crystallization, the air entering the dryer must have a dry-bulb temperature of 90°F and a dew point of 85°F.

The air leaving the dryer is dehumidified by countercurrent contact with a solution of lithium salts, which enters the dehumidifier under thermostatic control at 110°F. The air leaving the dehumidifier and entering the dryer is in substantial equilibrium with the solution entering the dehumidifier.

The boiling-point elevations of solutions of the salts used in the dehumidifier at a constant solution temperature of 110°F are given in the following table. The crystals leave the dryer with no excess water over that of crystallization. They enter the dryer containing 50 per cent of their total weight as excess water. The feed to the dryer is 2 tons/hr. The weight of solution fed to the dehumidifier is 4,400 lb/hr.

Calculate the concentration of the solution leaving the dehumidifier.

Total solids, weight %	Boiling-point elevation, °F
45	65
43	55
40	44
38	38
35	30
30	19
25	10
20	6

16-4. A slab with a wet weight of 10 lb originally contains 50 per cent moisture (wet basis). The slab is 2 by 3 ft by 3 in. thick. The equilibrium-moisture content is 5 per cent of the total weight when in contact with air of 70°F and 20 per cent humidity. The drying rate is given in the table below for contact with air of the above quality at a definite velocity. Drying is from one face.

Wet-slab weight, lb	Drying rate, lb/ft²-hr
20.0	1.0
15.8	1.0
11.6	0.9
9.3	0.8
7.2	0.7
6.1	0.4
5.5	0.2

How long will it take to dry the slab to 15 per cent moisture content (wet basis)?

16-5. A continuous countercurrent dryer is to be designed to dry 500 lb of wet porous solid per hour from 60 per cent moisture to 10 per cent, both on the wet basis. Air at 120°F dry bulb and 70°F wet bulb is to be used. The exit humidity is to be 0.012. The average equilibrium-moisture content is 5 per cent of the dry weight. The total moisture content (wet basis) at the critical point is 30 per cent. The stock may be assumed to remain at a temperature 3°F above that of the wet-bulb temperature of the air throughout the dryer. The mass-transfer coefficient k_y' is 50 lb/(ft²)(hr)(unit humidity). The stock has 1.0 ft² surface exposed to the air per pound of dry stock.

How long must the stock remain in the dryer?

REFERENCES

1. Badger, W. L., and W. L. McCabe: "Elements of Chemical Engineering," 2d ed., pp. 308–316, McGraw-Hill Book Company, Inc., New York, 1936.
2. Broughton, D. B., and H. S. Mickley: *Chem. Eng. Progr.*, **49**: 319 (1953).

3. Ceaglske, N. H., and O. A. Hougen: *Trans. AIChE,* **33:** 283 (1937).
4. Ceaglske, N. H., and F. C. Kiesling: *Trans. AIChE,* **36:** 211 (1940).
5. Comings, E. W., and T. K. Sherwood: *Ind. Eng. Chem.,* **26:** 1096 (1934).
6. Hougen, O. A.: paper on Dryer Calculations and Design, in "Teaching of Chemical Engineering," *Proc. Chem. Eng. Div. Soc. Prom. Eng. Educ.,* Second Summer School in Chemical Engineering, June, 1939.
7. "International Critical Tables," vol. II, pp. 322–325, McGraw-Hill Book Company, Inc., New York, 1928.
8. Marshall, W. R., Jr.: Atomization and Spray Drying, *Chem. Eng. Progr. Monograph Ser.,* **50**(2), 1954.
9. McCready, D. W., and W. L. McCabe: *Trans. AIChE,* **29:** 131 (1933).
10. Newman, A. B.: *Trans. AIChE,* **27:** 203, 310 (1931).
11. Perry, J. H.: "Chemical Engineers' Handbook," 3d ed., (*a*) pp. 803–804, (*b*) p. 808, (*c*) pp. 813–875, (*d*) p. 823, (*e*) p. 831, (*f*) p. 841, (*g*) pp. 845–848, (*h*) p. 865, McGraw-Hill Book Company, Inc., New York, 1950.
12. Riegel, E. R.: "Chemical Process Machinery," 2d ed., chap. 17, Reinhold Publishing Corporation, New York, 1953.
13. Sherwood, T. K.: *Ind. Eng. Chem.,* **21:** 12 (1929).
14. Sherwood, T. K.: *Ind. Eng. Chem.,* **21:** 976 (1929).
15. Sherwood, T. K.: *Ind. Eng. Chem.,* **24:** 307 (1932).
16. Sherwood, T. K., and E. W. Comings: *Trans. AIChE,* **28:** 118 (1932).
17. Walker, W. H., W. K. Lewis, W. H. McAdams, and E. R. Gilliland: "Principles of Chemical Engineering," 3d ed., pp. 666–680, McGraw-Hill Book Company, Inc., New York, 1937.

Appendix 1

SYMBOLS, DIMENSIONS, AND UNITS OF IMPORTANT QUANTITIES

Symbol	Name	Typical unit	Dimension
L	Length	ft	$\bar{L}$
m	Mass	lb or lb mole	$\bar{M}$
θ	Time	sec or hr	$\bar{\theta}$
F	Force	lb force	$\bar{F}$
t, T †	Temperature	°F or °R †	$\bar{T}$
Q	Heat	Btu	$\bar{H}$

† Absolute temperature.

Sym-bol	Name	Typical unit	Dimensions					
			$\bar{L}$	$\bar{M}$	$\bar{\theta}$	$\bar{F}$	$\bar{T}$	$\bar{H}$
A	Area	ft^2	2					
a	Acceleration	ft/sec^2	1		-2			
	Surface per volume	1/ft	-1					
B	Heavy-product flow rate	lb/hr or lb moles/hr		1	-1			
C	Coefficient	Dimensionless						
c	Concentration	lb/ft^3 or lb moles/ft^3	-3	1				
c_p	Specific heat, constant pressure	Btu/(lb)(°F)		-1			-1	1
D	Diameter	ft	1					
	Light-product flow rate	lb/hr or lb moles/hr		1	-1			

SECONDARY QUANTITIES (*Continued*)

Symbol	Name	Typical unit	Dimensions					
			L	$\bar{M}$	$\bar{\theta}$	$\bar{F}$	$\bar{T}$	$\bar{H}$
D_m	Diffusivity, molal	lb moles/ft-hr	-1	1	-1			
D_v	Diffusivity, volumetric	ft²/hr	2		-1			
E	Energy, mechanical	ft-lb force	1			1		
G	Mass velocity	lb/ft²-hr	-2	1	-1			
g	Acceleration of gravity	ft/sec² or ft/hr²	1		-2			
g_c	Newton's-law conversion factor	ft-lb/lb force-sec²	1	1	-2	-1		
H	Height of transfer unit	ft	1					
	Humidity	Dimensionless						
H_f	Friction loss	ft-lb force/lb	1	-1		1		
h	Heat-transfer coefficient, individual	Btu/(ft²)(hr)(°F)	-2		-1		-1	1
i	Specific enthalpy	Btu/lb		-1				1
J	Mechanical equivalent of heat	ft-lb force/Btu	1			1		-1
K	Mass-transfer coefficient, over-all	$\dfrac{\text{lb moles/ft}^2\text{-hr}}{\text{mole fraction}}$	-2	1	-1			
k	Mass-transfer coefficient, individual	$\dfrac{\text{lb moles/ft}^2\text{-hr}}{\text{mole fraction}}$	-2	1	-1			
	Thermal conductivity	Btu-ft/(ft²)(hr)(°F)	-1		-1		-1	1
L	Heavy-phase flow rate	lb/hr or lb moles/hr		1	-1			
M	Molecular weight	Dimensionless						
n	Number of moles	lb moles		1				
	Speed of rotation	rpm			-1			
P	Power	hp	1		-1	1		
p	Pressure	lb force/ft²	-2			1		
q	Rate of heat flow	Btu/hr			-1			1
	Volumetric flow rate	ft³/hr	3		-1			
R_o	Gas constant	ft-lb force/ (lb mole)(°R)	1	-1		1	-1	
S	Cross-sectional area	ft²	2					
U	Heat-transfer coefficient, over-all	Btu/(ft²)(hr)(°F)	-2		-1		-1	1
u	Linear velocity	ft/sec	1		-1			
V	Light-phase flow rate	lb/hr or lb moles/hr		1	-1			
	Volume	ft³	3					
$\bar{V}$	Volumetric average velocity	ft/sec	1		-1			
v	Specific volume	ft³/lb	3	-1				
	Linear velocity	ft/sec	1		-1			

SECONDARY QUANTITIES (*Continued*)

Sym-bol	Name	Typical unit	Dimensions					
			L	$\bar{M}$	$\bar{\theta}$	$\bar{F}$	T	$\bar{H}$
W	Work	ft-lb force	1			1		
	Total emissive power	Btu/ft²-hr	-2		-1			1
w	Mass flow rate	lb/hr or lb/sec		1	-1			
X	Mole or mass ratio, heavy phase	Dimensionless						
x	Mole or mass fraction, heavy phase	Dimensionless						
Y	Mole or mass ratio, light phase	Dimensionless						
y	Mole or mass fraction, light phase	Dimensionless						
Z	Head, or height above datum plane	ft	1					
Z_t	Height of transfer unit	ft	1					
α	Absorptivity, radiation	Dimensionless						
	Kinetic-energy factor	Dimensionless						
	Specific cake resistance	lb/ft	-1	1				
	Thermal diffusivity	ft²/sec	2		-1			
β	Coefficient of volumetric expansion	1/°F or 1/°C					-1	
Γ	Condensate loading	lb/ft-hr	-1	1	-1			
γ	Specific weight	lb force/ft³	-3			1		
ϵ	Emissivity, radiation	Dimensionless						
	Porosity	Dimensionless						
η	Efficiency	Dimensionless						
κ	Specific-heat ratio	Dimensionless						
λ	Latent heat	Btu/lb or Btu/lb mole		-1				1
	Wavelength	Microns	1					
	Particle shape factor	Dimensionless						
μ	Viscosity, absolute	lb/ft-sec or lb/ft-hr	-1	1	-1			
μ_F	Viscosity, gravitational	lb force-sec/ft²	-2		1	1		
ν	Kinematic viscosity	ft²/sec	2		-1			
ρ	Density	lb/ft³	-3	1				
	Reflectivity, radiation	Dimensionless						
σ	Stefan-Boltzmann constant	Btu/(ft²)(hr)(°R⁴)	-2		-1		-4	1
	Interfacial tension	lb force/ft	-1			1		
τ	Shear stress	lb force/ft²	-2			1		
ϕ	Fraction of particles above cut size	Dimensionless						
	Relative-velocity factor	Dimensionless						
ω	Angular velocity	radians/sec			-1			

Appendix 2

CONVERSION FACTORS

Quantity	Symbol	Factor
Density...........................	ρ	$\dfrac{1 \text{ lb/ft}^3}{1 \text{ g/cm}^3} = 62.438$
Heat..............................	Q	1 Btu/1 cal † = 251.996
Length...........................	L	1 yd/1 cm = 3,600/3,937 ‡
		1 in./1 cm = 2.54
		1 ft/1 cm = 30.48
Mass.............................	m	1 lb/1 g = 453.5924277 ‡
Mechanical energy..................	E_m	1 joule/1 erg = 10^7 ‡
		1 joule/1 wattsec = 1 ‡
Mechanical equivalent of heat..........	J	1 cal †/1 joule = 4.1873
		1 Btu/1 ft-lb force = 778.26
		1 kwhr/1 Btu = 3,412.75
Newton's-law conversion factor.........	g_c	1 g force-sec²/1 g-cm = 980.665 ‡
		1 lb force-sec²/1 ft-lb = 32.174
Pressure...........................	p	$\dfrac{1 \text{ atm}}{1 \text{ lb force/in.}^2} = 14.696$
		1 atm/1 mm Hg ¶ = 760 ‡
		1 atm/1 in. Hg ¶ = 29.92
Power.............................	P	$\dfrac{1 \text{ hp}}{1 \text{ ft-lb force/sec}} = 550$ ‡
		1 hp/1 kw = 0.74548
Specific heat......................	c	$\dfrac{1 \text{ cal/(g)(°C)}}{1 \text{ Btu/(lb)(°F)}} = 1$ ‡
Temperature difference................	Δt	1°C/1°F = 1.8 ‡
Viscosity..........................	μ	$\dfrac{1 \text{ centipoise}}{1 \text{ lb/ft-sec}} = 6.72 \times 10^{-4}$
		$\dfrac{1 \text{ centipoise}}{1 \text{ lb/ft-hr}} = 2.42$
		$\dfrac{1 \text{ centipoise}}{1 \text{ lb force-sec/ft}^2} = 2.089 \times 10^{-5}$
Volume............................	V	1 ft³/1 liter = 28.316
		1 U.S. gal/1 in.³ = 231 ‡
		1 ft³/1 gal = 7.48

† International steam-table (IT) calorie.
‡ Exact value, by definition.
¶ At density of 13.5951 g/cm³.

Appendix 3

DIMENSIONLESS GROUPS

Symbol	Name	Definition
C_D	Drag coefficient	$2g_cF_D/u^2\rho A_p$
f	Fanning friction factor	$-\Delta p\, g_cD/2L\rho\overline{V}^2$
j_H'	Heat-transfer factor	$(h/c_pG)(c_p\mu/k)^{\frac{2}{3}}(\mu_w/\mu)^{0.14}$
j_M	Mass-transfer factor	$(k/G_m)(\mu/D_m\overline{M})^{\frac{2}{3}}$
N_{Fo}	Fourier number	$k\theta/c_p\rho s^2$
N_{Fr}	Froude number	u^2/gL
N_{Gr}	Grashof number	$L^3\rho^2\beta g\,\Delta t/\mu^2$
N_{Gz}	Graetz number	wc_p/kL
N_{Nu}	Nusselt number	hD/k
N_{Pe}	Peclet number	LGc_p/k
N_{po}	Power number	$Pg_c/\rho n^3D^5$
N_{Pr}	Prandtl number	$c_p\mu/k$
N_{Re}	Reynolds number	DG/μ
N_{Sc}	Schmidt number	$\mu/D_m\overline{M}$
N_{Sh}	Sherwood number	kD/D_m

Appendix 4

DIMENSIONS, CAPACITIES, AND
WEIGHTS OF STANDARD STEEL PIPE †

Nominal pipe size, in.	Outside diameter, in.	Schedule No.	Wall thickness, in.	Inside diameter, in.	Cross-sectional area of metal, in.^{2}	Inside sectional area, ft^2	Circumference, ft, or surface, ft^2/ft of length — Outside	Circumference, ft, or surface, ft^2/ft of length — Inside	Capacity at 1 ft/sec velocity — U.S. gal/min	Capacity at 1 ft/sec velocity — Water, lb/hr	Pipe weight/ ft, lb
$\frac{1}{8}$	0.405	40	0.068	0.269	0.072	0.00040	0.106	0.0705	0.179	89.5	0.25
		80	0.095	0.215	0.093	0.00025	0.106	0.0563	0.112	56.0	0.32
$\frac{1}{4}$	0.540	40	0.088	0.364	0.125	0.00072	0.141	0.0954	0.323	161.5	0.43
		80	0.119	0.302	0.157	0.00050	0.141	0.0792	0.224	112.0	0.54
$\frac{3}{8}$	0.675	40	0.091	0.493	0.167	0.00133	0.177	0.1293	0.596	298.0	0.57
		80	0.126	0.423	0.217	0.00098	0.177	0.1110	0.440	220.0	0.74
$\frac{1}{2}$	0.840	40	0.109	0.622	0.250	0.00211	0.220	0.1630	0.945	472.5	0.85
		80	0.147	0.546	0.320	0.00163	0.220	0.1430	0.730	365.0	1.09
$\frac{3}{4}$	1.050	40	0.113	0.824	0.333	0.00371	0.275	0.2158	1.665	832.5	1.13
		80	0.154	0.742	0.433	0.00300	0.275	0.1942	1.345	672.5	1.48
1	1.315	40	0.133	1.049	0.494	0.00600	0.344	0.2745	2.690	1,345	1.68
		80	0.179	0.957	0.639	0.00499	0.344	0.2505	2.240	1,120	2.17
$1\frac{1}{4}$	1.660	40	0.140	1.380	0.669	0.01040	0.435	0.362	4.57	2,285	2.28
		80	0.191	1.278	0.881	0.00891	0.435	0.335	3.99	1,995	3.00
$1\frac{1}{2}$	1.900	40	0.145	1.610	0.799	0.01414	0.498	0.422	6.34	3,170	2.72
		80	0.200	1.500	1.068	0.01225	0.498	0.393	5.49	2,745	3.64
2	2.375	40	0.154	2.067	1.075	0.02330	0.622	0.542	10.45	5,225	3.66
		80	0.218	1.939	1.477	0.02050	0.622	0.508	9.20	4,600	5.03
$2\frac{1}{2}$	2.875	40	0.203	2.469	1.704	0.03322	0.753	0.647	14.92	7,460	5.80
		80	0.276	2.323	2.254	0.02942	0.753	0.609	13.20	6,600	7.67
3	3.500	40	0.216	3.068	2.228	0.05130	0.917	0.804	23.00	11,500	7.58
		80	0.300	2.900	3.016	0.04587	0.917	0.760	20.55	10,275	10.3
$3\frac{1}{2}$	4.000	40	0.226	3.548	2.680	0.06870	1.047	0.930	30.80	15,400	9.11
		80	0.318	3.364	3.678	0.06170	1.047	0.882	27.70	13,850	12.5
4	4.500	40	0.237	4.026	3.173	0.08840	1.178	1.055	39.6	19,800	10.8
		80	0.337	3.826	4.407	0.07986	1.178	1.002	35.8	17,900	15.0
5	5.563	40	0.258	5.047	4.304	0.1390	1.456	1.322	62.3	31,150	14.7
		80	0.375	4.813	6.112	0.1263	1.456	1.263	57.7	28,850	20.8
6	6.625	40	0.280	6.065	5.584	0.2006	1.734	1.590	90.0	45,000	19.0
		80	0.432	5.761	8.405	0.1810	1.734	1.510	81.1	40,550	28.6
8	8.625	40	0.322	7.981	8.396	0.3474	2.258	2.090	155.7	77,850	28.6
		80	0.500	7.625	12.76	0.3171	2.258	2.000	142.3	71,150	43.4
10	10.75	40	0.365	10.020	11.90	0.5475	2.814	2.620	246.0	123,000	40.5
		80	0.593	9.564	18.92	0.4989	2.814	2.503	224.0	112,000	64.4
12	12.75	40	0.406	11.938	15.77	0.7773	3.338	3.13	349.0	174,500	53.6
		80	0.687	11.376	26.03	0.7058	3.338	2.98	317.0	158,500	88.6

† Based on ASA Standards B36.10—1939.

Appendix 5

CONDENSER AND HEAT-EXCHANGER TUBE DATA †

Outside diameter, in.	Wall thickness BWG No.	Wall thickness In.	Inside diameter, in.	Cross-sectional area metal, in.²	Inside sectional area, ft²	Circumference, ft, or surface, ft²/ft of length — Outside	Circumference — Inside	Velocity, ft/sec for 1 U.S. gal/min	Capacity at 1 ft/sec velocity U.S. gal/min	Capacity — Water, lb/hr	Weight/ft, lb ‡
$\frac{5}{8}$	12	0.109	0.407	0.1767	0.000903	0.1636	0.1066	2.468	0.4053	202.7	0.657
	14	0.083	0.459	0.1460	0.00115	0.1636	0.1202	1.938	0.5161	258.1	0.526
	16	0.065	0.495	0.1143	0.00134	0.1636	0.1296	1.663	0.6014	300.7	0.425
	18	0.049	0.527	0.0887	0.00151	0.1636	0.1380	1.476	0.6777	338.9	0.330
$\frac{3}{4}$	12	0.109	0.532	0.2195	0.00154	0.1963	0.1393	1.447	0.6912	345.6	0.817
	14	0.083	0.584	0.1739	0.00186	0.1963	0.1529	1.198	0.8348	417.4	0.647
	16	0.065	0.620	0.1398	0.00210	0.1963	0.1623	1.061	0.9425	471.3	0.520
	18	0.049	0.652	0.1079	0.00232	0.1963	0.1707	0.962	1.041	520.5	0.401
$\frac{7}{8}$	12	0.109	0.657	0.2623	0.00235	0.2291	0.1720	0.948	1.055	527.5	0.976
	14	0.083	0.709	0.2065	0.00274	0.2291	0.1856	0.813	1.230	615.0	0.768
	16	0.065	0.745	0.1654	0.00303	0.2291	0.1950	0.735	1.360	680.0	0.615
	18	0.049	0.777	0.1271	0.00329	0.2291	0.2034	0.678	1.477	738.5	0.473
1	10	0.134	0.732	0.3654	0.00292	0.2618	0.1916	0.763	1.310	655.0	1.36
	12	0.109	0.782	0.3051	0.00334	0.2618	0.2048	0.667	1.499	750.0	1.14
	14	0.083	0.834	0.2391	0.00379	0.2618	0.2183	0.588	1.701	850.5	0.890
	16	0.065	0.870	0.1909	0.00413	0.2618	0.2277	0.538	1.854	927.0	0.710
$1\frac{1}{4}$	10	0.134	0.982	0.4698	0.00526	0.3271	0.2572	0.424	2.361	1,181	1.75
	12	0.109	1.032	0.3907	0.00581	0.3271	0.2701	0.384	2.608	1,304	1.45
	14	0.083	1.084	0.3042	0.00641	0.3271	0.2839	0.348	2.877	1,439	1.13
	16	0.065	1.120	0.2419	0.00684	0.3271	0.2932	0.326	3.070	1,535	0.900
$1\frac{1}{2}$	10	0.134	1.232	0.5750	0.00828	0.3925	0.3225	0.269	3.716	1,858	2.14
	12	0.109	1.282	0.4763	0.00896	0.3925	0.3356	0.249	4.021	2,011	1.77
	14	0.083	1.334	0.3694	0.00971	0.3925	0.3492	0.229	4.358	2,176	1.37
2	10	0.134	1.732	0.7855	0.0164	0.5233	0.4534	0.136	7.360	3,680	2.92
	12	0.109	1.782	0.6475	0.0173	0.5233	0.4665	0.129	7.764	3,882	2.41

† Condensed from J. H. Perry (ed.), "Chemical Engineers' Handbook," 3d ed., pp. 425–426, McGraw-Hill Book Company, Inc., 1950.

‡ For brass; for copper, multiply by 1.05; for steel, multiply by 0.91.

Appendix 6

PROPERTIES OF SATURATED STEAM AND WATER †

Temp. t, °F	Vapor press. p_A, lb force/in.^{2}	Specific vol., ft^3/lb		Enthalpy, Btu/lb		
		Liquid v_x	Sat. vapor v_y	Liquid i_x	Vaporization λ	Sat. vapor i_y
32	0.08854	0.01602	3,306	0.00	1075.8	1075.8
35	0.09995	0.01602	2,947	3.02	1074.1	1077.1
40	0.12170	0.01602	2,444	8.05	1071.3	1079.3
45	0.14752	0.01602	2,036.4	13.06	1068.4	1081.5
50	0.17811	0.01603	1,703.2	18.07	1065.6	1083.7
55	0.2141	0.01603	1,430.7	23.07	1062.7	1085.8
60	0.2563	0.01604	1,206.7	28.06	1059.9	1088.0
65	0.3056	0.01605	1,021.4	33.05	1057.1	1090.2
70	0.3631	0.01606	867.9	38.04	1054.3	1092.3
75	0.4298	0.01607	740.0	43.03	1051.5	1094.5
80	0.5069	0.01608	633.1	48.02	1048.6	1096.6
85	0.5959	0.01609	543.5	53.00	1045.8	1098.8
90	0.6982	0.01610	468.0	57.99	1042.9	1100.9
95	0.8153	0.01612	404.3	62.98	1040.1	1103.1
100	0.9492	0.01613	350.4	67.97	1037.2	1105.2
110	1.2748	0.01617	265.4	77.94	1031.6	1109.5
120	1.6924	0.01620	203.27	87.92	1025.8	1113.7
130	2.2225	0.01625	157.34	97.90	1020.0	1117.9
140	2.8886	0.01629	123.01	107.89	1014.1	1122.0
150	3.718	0.01634	97.07	117.89	1008.2	1126.1
160	4.741	0.01639	77.29	127.89	1002.3	1130.2
170	5.992	0.01645	62.06	137.90	996.3	1134.2
180	7.510	0.01651	50.23	147.92	990.2	1138.1
190	9.339	0.01657	40.96	157.95	984.1	1142.0
200	11.526	0.01663	33.64	167.99	977.9	1145.9

PROPERTIES OF SATURATED STEAM AND WATER (*Continued*)

Temp. t, °F	Vapor press. p_A, lb force/in.^{2}	Specific vol., ft^3/lb		Enthalpy, Btu/lb		
		Liquid v_x	Sat. vapor v_y	Liquid i_x	Vaporization λ	Sat. vapor i_y
210	14.123	0.01670	27.82	178.05	971.6	1149.7
212	14.696	0.01672	26.80	180.07	970.3	1150.4
220	17.186	0.01677	23.15	188.13	965.2	1153.4
230	20.780	0.01684	19.382	198.23	958.8	1157.0
240	24.969	0.01692	16.323	208.34	952.2	1160.5
250	29.825	0.01700	13.821	218.48	945.5	1164.0
260	35.429	0.01709	11.763	228.64	938.7	1167.3
270	41.858	0.01717	10.061	238.84	931.8	1170.6
280	49.203	0.01726	8.645	249.06	924.7	1173.8
290	57.556	0.01735	7.461	259.31	917.5	1176.8
300	67.013	0.01745	6.466	269.59	910.1	1179.7
310	77.68	0.01755	5.626	279.92	902.6	1182.5
320	89.66	0.01765	4.914	290.28	894.9	1185.2
330	103.06	0.01776	4.307	300.68	887.0	1187.7
340	118.01	0.01787	3.788	311.13	879.0	1190.1
350	134.63	0.01799	3.342	321.63	870.7	1192.3
360	153.04	0.01811	2.957	332.18	862.2	1194.4
370	173.37	0.01823	2.625	342.79	853.5	1196.3
380	195.77	0.01836	2.335	353.45	844.6	1198.1
390	220.37	0.01850	2.0836	364.17	835.4	1199.6
400	247.31	0.01864	1.8633	374.97	826.0	1201.0
410	276.75	0.01878	1.6700	385.83	816.3	1202.1
420	308.83	0.01894	1.5000	396.77	806.3	1203.1
430	343.72	0.01910	1.3499	407.79	796.0	1203.8
440	381.59	0.01926	1.2171	418.90	785.4	1204.3
450	422.6	0.0194	1.0993	430.1	774.5	1204.6

† Abstracted from abridged edition of "Thermodynamic Properties of Steam," by Joseph H. Keenan and Fredrick G. Keyes, John Wiley & Sons, Inc., New York, 1937, with the permission of the authors and publisher.

Appendix 7

VISCOSITIES OF GASES †

No.	Gas	X	Y	No.	Gas	X	Y
1	Acetic acid	7.7	14.3	29	Freon-113	11.3	14.0
2	Acetone	8.9	13.0	30	Helium	10.9	20.5
3	Acetylene	9.8	14.9	31	Hexane	8.6	11.8
4	Air	11.0	20.0	32	Hydrogen	11.2	12.4
5	Ammonia	8.4	16.0	33	$3H_2 + 1N_2$	11.2	17.2
6	Argon	10.5	22.4	34	Hydrogen bromide	8.8	20.9
7	Benzene	8.5	13.2	35	Hydrogen chloride	8.8	18.7
8	Bromine	8.9	19.2	36	Hydrogen cyanide	9.8	14.9
9	Butene	9.2	13.7	37	Hydrogen iodide	9.0	21.3
10	Butylene	8.9	13.0	38	Hydrogen sulfide	8.6	18.0
11	Carbon dioxide	9.5	18.7	39	Iodine	9.0	18.4
12	Carbon disulfide	8.0	16.0	40	Mercury	5.3	22.9
13	Carbon monoxide	11.0	20.0	41	Methane	9.9	15.5
14	Chlorine	9.0	18.4	42	Methyl alcohol	8.5	15.6
15	Chloroform	8.9	15.7	43	Nitric oxide	10.9	20.5
16	Cyanogen	9.2	15.2	44	Nitrogen	10.6	20.0
17	Cyclohexane	9.2	12.0	45	Nitrosyl chloride	8.0	17.6
18	Ethane	9.1	14.5	46	Nitrous oxide	8.8	19.0
19	Ethyl acetate	8.5	13.2	47	Oxygen	11.0	21.3
20	Ethyl alcohol	9.2	14.2	48	Pentane	7.0	12.8
21	Ethyl chloride	8.5	15.6	49	Propane	9.7	12.9
22	Ethyl ether	8.9	13.0	50	Propyl alcohol	8.4	13.4
23	Ethylene	9.5	15.1	51	Propylene	9.0	13.8
24	Fluorine	7.3	23.8	52	Sulfur dioxide	9.6	17.0
25	Freon-11	10.6	15.1	53	Toluene	8.6	12.4
26	Freon-12	11.1	16.0	54	2,3,3-trimethylbutane	9.5	10.5
27	Freon-21	10.8	15.3	55	Water	8.0	16.0
28	Freon-22	10.1	17.0	56	Xenon	9.3	23.0

Coordinates for use with figure opposite.

† By permission, from "Chemical Engineers' Handbook," 3d ed., by J. H. Perry Copyright, 1950. McGraw-Hill Book Company, Inc.

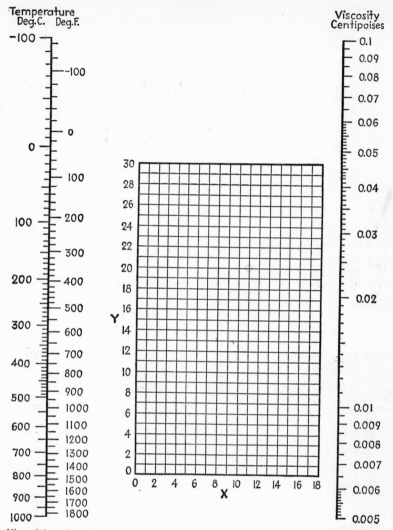

Viscosities of gases and vapors at 1 atm; for coordinates, see table opposite.

Appendix 8

VISCOSITIES OF LIQUIDS †

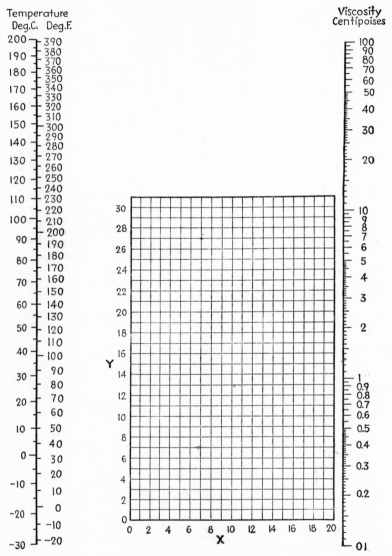

Viscosities of liquids at 1 atm. For coordinates, see table opposite.

† By permission, from "Chemical Engineers' Handbook," 3d ed., by J. H. Perry. Copyright, 1950. McGraw-Hill Book Company, Inc.

No.	Liquid	X	Y	No.	Liquid	X	Y
1	Acetaldehyde	15.2	4.8	56	Freon-22	17.2	4.7
2	Acetic acid, 100%	12.1	14.2	57	Freon-113	12.5	11.4
3	Acetic acid, 70%	9.5	17.0	58	Glycerol, 100%	2.0	30.0
4	Acetic anhydride	12.7	12.8	59	Glycerol, 50%	6.9	19.6
5	Acetone, 100%	14.5	7.2	60	Heptane	14.1	8.4
6	Acetone, 35%	7.9	15.0	61	Hexane	14.7	7.0
7	Allyl alcohol	10.2	14.3	62	Hydrochloric acid, 31.5%	13.0	16.6
8	Ammonia, 100%	12.6	2.0	63	Isobutyl alcohol	7.1	18.0
9	Ammonia, 26%	10.1	13.9	64	Isobutyric acid	12.2	14.4
10	Amyl acetate	11.8	12.5	65	Isopropyl alcohol	8.2	16.0
11	Amyl alcohol	7.5	18.4	66	Kerosene	10.2	16.9
12	Aniline	8.1	18.7	67	Linseed oil, raw	7.5	27.2
13	Anisole	12.3	13.5	68	Mercury	18.4	16.4
14	Arsenic trichloride	13.9	14.5	69	Methanol, 100%	12.4	10.5
15	Benzene	12.5	10.9	70	Methanol, 90%	12.3	11.8
16	Brine, CaCl₂, 25%	6.6	15.9	71	Methanol, 40%	7.8	15.5
17	Brine, NaCl, 25%	10.2	16.6	72	Methyl acetate	14.2	8.2
18	Bromine	14.2	13.2	73	Methyl chloride	15.0	3.8
19	Bromotoluene	20.0	15.9	74	Methyl ethyl ketone	13.9	8.6
20	Butyl acetate	12.3	11.0	75	Naphthalene	7.9	18.1
21	Butyl alcohol	8.6	17.2	76	Nitric acid, 95%	12.8	13.8
22	Butyric acid	12.1	15.3	77	Nitric acid, 60%	10.8	17.0
23	Carbon dioxide	11.6	0.3	78	Nitrobenzene	10.6	16.2
24	Carbon disulphide	16.1	7.5	79	Nitrotoluene	11.0	17.0
25	Carbon tetrachloride	12.7	13.1	80	Octane	13.7	10.0
26	Chlorobenzene	12.3	12.4	81	Octyl alcohol	6.6	21.1
27	Chloroform	14.4	10.2	82	Pentachloroethane	10.9	17.3
28	Chlorosulfonic acid	11.2	18.1	83	Pentane	14.9	5.2
29	Chlorotoluene, ortho	13.0	13.3	84	Phenol	6.9	20.8
30	Chlorotoluene, meta	13.3	12.5	85	Phosphorus tribromide	13.8	16.7
31	Chlorotoluene, para	13.3	12.5	86	Phosphorus trichloride	16.2	10.9
32	Cresol, meta	2.5	20.8	87	Propionic acid	12.8	13.8
33	Cyclohexanol	2.9	24.3	88	Propyl alcohol	9.1	16.5
34	Dibromoethane	12.7	15.8	89	Propyl bromide	14.5	9.6
35	Dichloroethane	13.2	12.2	90	Propyl chloride	14.4	7.5
36	Dichloromethane	14.6	8.9	91	Propyl iodide	14.1	11.6
37	Diethyl oxalate	11.0	16.4	92	Sodium	16.4	13.9
38	Dimethyl oxalate	12.3	15.8	93	Sodium hydroxide, 50%	3.2	25.8
39	Diphenyl	12.0	18.3	94	Stannic chloride	13.5	12.8
40	Dipropyl oxalate	10.3	17.7	95	Sulphur dioxide	15.2	7.1
41	Ethyl acetate	13.7	9.1	96	Sulphuric acid, 110%	7.2	27.4
42	Ethyl alcohol, 100%	10.5	13.8	97	Sulphuric acid, 98%	7.0	24.8
43	Ethyl alcohol, 95%	9.8	14.3	98	Sulphuric acid, 60%	10.2	21.3
44	Ethyl alcohol, 40%	6.5	16.6	99	Sulphuryl chloride	15.2	12.4
45	Ethyl benzene	13.2	11.5	100	Tetrachloroethane	11.9	15.7
46	Ethyl bromide	14.5	8.1	101	Tetrachloroethylene	14.2	12.7
47	Ethyl chloride	14.8	6.0	102	Titanium tetrachloride	14.4	12.3
48	Ethyl ether	14.5	5.3	103	Toluene	13.7	10.4
49	Ethyl formate	14.2	8.4	104	Trichloroethylene	14.8	10.5
50	Ethyl iodide	14.7	10.3	105	Turpentine	11.5	14.9
51	Ethylene glycol	6.0	23.6	106	Vinyl acetate	14.0	8.8
52	Formic acid	10.7	15.8	107	Water	10.2	13.0
53	Freon-11	14.4	9.0	108	Xylene, ortho	13.5	12.1
54	Freon-12	16.8	5.6	109	Xylene, meta	13.9	10.6
55	Freon-21	15.7	7.5	110	Xylene, para	13.9	10.9

Coordinates for use with figure opposite.

Appendix 9

THERMAL CONDUCTIVITIES OF METALS [†]

Metal	Thermal conductivity k [‡]		
	32°F	64°F	212°F
Aluminum...................	117	.....	119
Antimony....................	10.6	.....	9.7
Brass (70 copper, 30 zinc).......	56	.....	60
Cadmium.....................	.....	53.7	52.2
Copper (pure).................	224	.....	218
Gold.........................	.....	169.0	170.0
Iron (cast)....................	32	.....	30
Iron (wrought)................	.....	34.9	34.6
Lead..........................	20	.....	19
Magnesium...................	92	92	92
Mercury (liquid)..............	4.8		
Nickel........................	36	.....	34
Platinum.....................	.....	40.2	41.9
Silver........................	242	.....	238
Sodium (liquid)...............	.....	.....	49
Steel (mild)..................	.....	.....	26
Steel (1% carbon).............	.....	26.2	25.9
Steel (stainless, type 304)........	.....	.....	9.4
Steel (stainless, type 316)........	.....	.....	9.4
Steel (stainless, type 347)........	.....	.....	9.3
Tantalum.....................	.....	32	
Tin..........................	36	.....	34
Zinc.........................	65	.....	64

[†] Based on W. H. McAdams, "Heat Transmission," 3d ed., pp. 445–447, McGraw-Hill Book Company, Inc., New York, 1954.

[‡] k = Btu-ft/(ft^2)(hr)(°F).

Appendix 10

THERMAL CONDUCTIVITIES OF GASES AND VAPORS †

Substance	Thermal conductivity k ‡	
	32°F	212°F
Acetone.................................	0.0057	0.0099
Acetylene..............................	0.0108	0.0172
Air....................................	0.0140	0.0184
Ammonia..............................	0.0126	0.0192
Benzene...............................	0.0052	0.0103
Carbon dioxide........................	0.0084	0.0128
Carbon monoxide......................	0.0134	0.0176
Carbon tetrachloride..................	......	0.0052
Chlorine..............................	0.0043	
Ethane................................	0.0106	0.0175
Ethyl alcohol.........................	......	0.0124
Ethyl ether...........................	0.0077	0.0131
Ethylene..............................	0.0101	0.0161
Helium................................	0.0818	0.0988
Hydrogen..............................	0.0966	0.1240
Methane...............................	0.0176	0.0255
Methyl alcohol........................	0.0083	0.0128
Nitrogen..............................	0.0139	0.0181
Nitrous oxide.........................	0.0088	0.0138
Oxygen................................	0.0142	0.0188
Propane...............................	0.0087	0.0151
Sulfur dioxide........................	0.0050	0.0069
Water vapor (at 1 atm abs pressure)....	......	0.0136

† Based on W. H. McAdams, "Heat Transmission," 3d ed., pp. 457–458, McGraw-Hill Book Company, Inc., New York, 1954.

‡ k = Btu-ft/(ft^2)(hr)(°F).

Appendix 11

THERMAL CONDUCTIVITIES OF LIQUIDS OTHER THAN WATER [†]

Liquid	Temp., °F	k [‡]
Acetic acid...................	68	0.099
Acetone......................	86	0.102
Ammonia (anhydrous)..........	5–86	0.29
Aniline......................	32–68	0.100
Benzene.....................	86	0.092
n-Butyl alcohol...............	86	0.097
Carbon bisulfide..............	86	0.093
Carbon tetrachloride..........	32	0.107
Chlorobenzene................	50	0.083
Ethyl acetate.................	68	0.101
Ethyl alcohol (absolute)........	68	0.105
Ethyl ether..................	86	0.080
Ethylene glycol...............	32	0.153
Gasoline.....................	86	0.078
Glycerine....................	68	0.164
n-Heptane...................	86	0.081
Kerosene....................	68	0.086
Methyl alcohol...............	68	0.124
Nitrobenzene.................	86	0.095
n-Octane....................	86	0.083
Sulfur dioxide................	5	0.128
Sulfuric acid (90%)...........	86	0.21
Toluene.....................	86	0.086
Trichloroethylene.............	122	0.080
o-Xylene....................	68	0.090

[†] Based on W. H. McAdams, "Heat Transmission," 3d ed., pp. 455–456, McGraw-Hill Book Company, Inc., New York, 1954.

[‡] $k = \text{Btu-ft}/(\text{ft}^2)(\text{hr})(°\text{F})$.

Appendix 12

SPECIFIC HEATS OF GASES †

C = Specific heat = Btu/(Lb)(Deg F) = Pcu/(Lb)(Deg C)

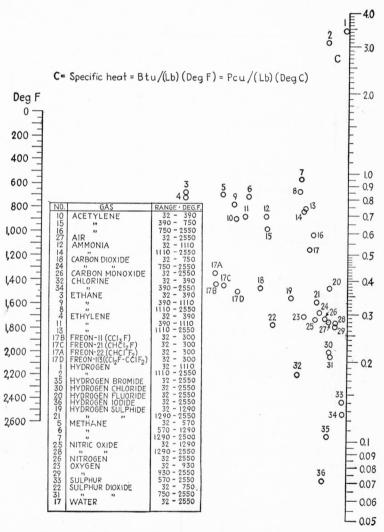

NO.	GAS	RANGE · DEG.F.
10	ACETYLENE	32 – 390
15	"	390 – 750
16	"	750 – 2550
27	AIR	32 – 2550
12	AMMONIA	32 – 1110
14	"	1110 – 2550
18	CARBON DIOXIDE	32 – 750
24	" "	750 – 2550
26	CARBON MONOXIDE	32 – 2550
32	CHLORINE	32 – 390
34	"	390 – 2550
3	ETHANE	32 – 390
9	"	390 – 1110
8	"	1110 – 2550
4	ETHYLENE	32 – 390
11	"	390 – 1110
13	"	1110 – 2550
17B	FREON-11 (CCl$_3$F)	32 – 300
17C	FREON-21 (CHCl$_2$F)	32 – 300
17A	FREON-22 (CHCl F$_2$)	32 – 300
17D	FREON-113(CCl$_2$F–CClF$_2$)	32 – 300
1	HYDROGEN	32 – 1110
2	"	1110 – 2550
35	HYDROGEN BROMIDE	32 – 2550
30	HYDROGEN CHLORIDE	32 – 2550
20	HYDROGEN FLUORIDE	32 – 2550
36	HYDROGEN IODIDE	32 – 2550
19	HYDROGEN SULPHIDE	32 – 1290
21	" "	1290 – 2550
5	METHANE	32 – 570
6	"	570 – 1290
7	"	1290 – 2500
25	NITRIC OXIDE	32 – 1290
28	" "	1290 – 2550
26	NITROGEN	32 – 2550
23	OXYGEN	32 – 930
29	"	930 – 2550
33	SULPHUR	570 – 2550
22	SULPHUR DIOXIDE	32 – 750
31	" "	750 – 2550
17	WATER	32 – 2550

True specific heats c_p of gases and vapors at 1 atm pressure.

† Courtesy of T. H. Chilton.

Appendix 13

SPECIFIC HEATS OF LIQUIDS †

Specific heat = Btu/(Lb)(Deg F) = Pcu/(Lb)(Deg C)

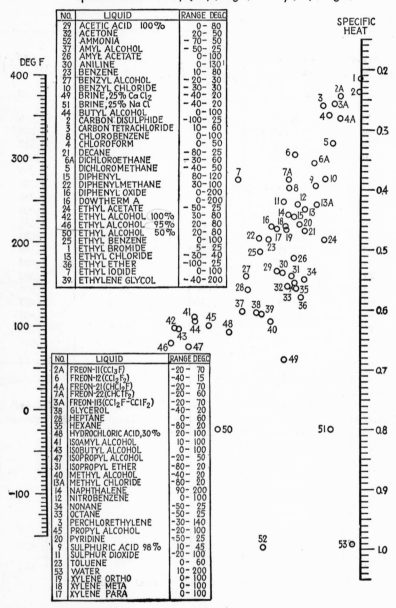

NO.	LIQUID	RANGE DEG.C
29	ACETIC ACID 100%	0 – 80
32	ACETONE	0 – 50
52	AMMONIA	– 70 – 50
37	AMYL ALCOHOL	– 50 – 25
26	AMYL ACETATE	0 – 100
30	ANILINE	0 – 130
23	BENZENE	10 – 80
27	BENZYL ALCOHOL	– 20 – 30
10	BENZYL CHLORIDE	– 30 – 30
49	BRINE, 25% CaCl₂	– 40 – 20
51	BRINE, 25% Na Cl	– 40 – 20
44	BUTYL ALCOHOL	0 – 100
2	CARBON DISULPHIDE	– 100 – 25
3	CARBON TETRACHLORIDE	10 – 60
8	CHLOROBENZENE	0 – 100
4	CHLOROFORM	0 – 50
21	DECANE	– 80 – 25
6A	DICHLOROETHANE	– 30 – 60
5	DICHLOROMETHANE	– 40 – 50
15	DIPHENYL	80 – 120
22	DIPHENYLMETHANE	30 – 100
16	DIPHENYL OXIDE	0 – 200
16	DOWTHERM A	0 – 200
24	ETHYL ACETATE	– 50 – 25
42	ETHYL ALCOHOL 100%	30 – 80
46	ETHYL ALCOHOL 95%	20 – 80
50	ETHYL ALCOHOL 50%	20 – 80
25	ETHYL BENZENE	0 – 100
1	ETHYL BROMIDE	5 – 25
13	ETHYL CHLORIDE	– 30 – 40
36	ETHYL ETHER	– 100 – 25
7	ETHYL IODIDE	0 – 100
39	ETHYLENE GLYCOL	– 40 – 200

NO.	LIQUID	RANGE DEG.C
2A	FREON-11 (CCl₃F)	– 20 – 70
6	FREON-12 (CCl₂F₂)	– 40 – 15
4A	FREON-21 (CHCl₂F)	– 20 – 70
7A	FREON-22 (CHClF₂)	– 20 – 60
3A	FREON-113 (CCl₂F–CClF₂)	– 20 – 70
38	GLYCEROL	– 40 – 20
28	HEPTANE	0 – 60
35	HEXANE	– 80 – 20
48	HYDROCHLORIC ACID, 30%	20 – 100
41	ISOAMYL ALCOHOL	10 – 100
43	ISOBUTYL ALCOHOL	0 – 100
47	ISOPROPYL ALCOHOL	– 20 – 50
31	ISOPROPYL ETHER	– 80 – 20
40	METHYL ALCOHOL	– 40 – 20
13A	METHYL CHLORIDE	– 80 – 20
14	NAPHTHALENE	90 – 200
12	NITROBENZENE	0 – 100
34	NONANE	– 50 – 25
33	OCTANE	– 50 – 25
3	PERCHLORETHYLENE	– 30 – 140
45	PROPYL ALCOHOL	– 20 – 100
20	PYRIDINE	– 50 – 25
9	SULPHURIC ACID 98%	10 – 45
11	SULPHUR DIOXIDE	– 20 – 100
23	TOLUENE	0 – 60
53	WATER	10 – 200
19	XYLENE ORTHO	0 – 100
18	XYLENE META	0 – 100
17	XYLENE PARA	0 – 100

† Courtesy of T. H. Chilton.

Appendix 14

PROPERTIES OF LIQUID WATER

Temperature t, °F	Viscosity † μ', centipoises	Thermal conductivity ‡ k, Btu/(ft)(hr)(°F)	Density ¶ ρ, lb/ft^3	$\psi_f = \left(\dfrac{k^3 \rho^2 g}{\mu^2}\right)^{\frac{1}{3}}$
32	1.794	0.320	62.42	1,410
40	1.546	0.326	62.43	1,590
50	1.310	0.333	62.42	1,810
60	1.129	0.340	62.37	2,050
70	0.982	0.346	62.30	2,290
80	0.862	0.352	62.22	2,530
90	0.764	0.358	62.11	2,780
100	0.682	0.362	62.00	3,020
120	0.559	0.371	61.71	3,530
140	0.470	0.378	61.38	4,030
160	0.401	0.384	61.00	4,530
180	0.347	0.388	60.58	5,020
200	0.305	0.392	60.13	5,500
220	0.270	0.394	59.63	5,960
240	0.242	0.396	59.10	6,420
260	0.218	0.396	58.51	6,830
280	0.199	0.396	57.94	7,210
300	0.185	0.396	57.31	7,510

† From "International Critical Tables," vol. 5, p. 10, McGraw-Hill Book Company, Inc., New York, 1929.

‡ From E. Schmidt and W. Sellschopp, *Forsch. Gebiete Ingenieurw.*, **3:** 277 (1932).

¶ Calculated from J. H. Keenan and F. G. Keyes, "Thermodynamic Properties of Steam," John Wiley & Sons, Inc., New York, 1937.

Appendix 15

PRANDTL NUMBERS FOR GASES AT 1 ATM AND 212°F [†]

Gas	$N_{\mathrm{Pr}} = \dfrac{c_p \mu}{k}$
Air	0.69
Ammonia	0.86
Argon	0.66
Carbon dioxide	0.75
Carbon monoxide	0.72
Helium	0.71
Hydrogen	0.69
Methane	0.75
Nitric oxide, nitrous oxide	0.72
Nitrogen	0.70
Oxygen	0.70
Water vapor	1.06

[†] Based on W. H. McAdams, "Heat Transmission," 3d ed., p. 471, McGraw-Hill Book Company, Inc., New York, 1954.

Appendix 16

PRANDTL NUMBERS FOR LIQUIDS †

Liquid	$N_{Pr} = \dfrac{c_p \mu}{k}$	
	60°F	212°F
Acetic acid..............	14.5	10.5
Acetone.................	4.5	2.4
Aniline.................	69	9.3
Benzene................	7.3	3.8
n-Butyl alcohol..........	43	11.5
Carbon tetrachloride......	7.5	4.2
Chlorobenzene...........	9.3	7.0
Ethyl acetate............	6.8	5.6
Ethyl alcohol............	15.5	10.1
Ethyl ether..............	4.0	2.3
Ethylene glycol...........	350	125
n-Heptane...............	6.0	4.2
Methyl alcohol...........	7.2	3.4
Nitrobenzene............	19.5	6.5
n-Octane................	5.0	3.6
Sulfuric acid (98%).......	149	15.0
Toluene.................	6.5	3.8
Water..................	7.7	1.5

† Based on W. H. McAdams, "Heat Transmission," 3d ed., p. 470, McGraw-Hill Book Company, Inc., New York, 1954.

Appendix 17

DIFFUSIVITIES AND SCHMIDT NUMBERS FOR GASES IN AIR AT 32°F AND 1 ATM [†]

	Diffusivity		$N_{Sc} = \dfrac{\mu}{\rho D_v}$ [‡]
Gas	Volumetric D_v, ft²/hr	Molal D_m, lb moles/ft-hr $\times 10^3$	$= \dfrac{\mu}{D_m \overline{M}}$
Acetic acid	0.413	1.15	1.24
Acetone	0.32 [¶]	0.89	1.60
Ammonia	0.836	2.33	0.61
Benzene	0.299	0.83	1.71
n-Butyl alcohol	0.273	0.76	1.88
Carbon dioxide	0.535	1.49	0.96
Carbon tetrachloride	0.24 [¶]	0.67	2.13
Chlorine	0.36 [¶]	1.00	1.42
Chlorobenzene	0.24 [¶]	0.67	2.13
Ethane	0.42 [¶]	1.17	1.22
Ethyl acetate	0.278	0.77	1.84
Ethyl alcohol	0.396	1.10	1.30
Ethyl ether	0.302	0.84	1.70
Hydrogen	2.37	6.60	0.22
Methane	0.61 [¶]	1.70	0.84
Methyl alcohol	0.515	1.43	1.00
Naphthalene	0.199	0.55	2.57
Nitrogen	0.52 [¶]	1.45	0.98
n-Octane	0.196	0.546	2.62
Oxygen	0.690	1.92	0.74
Phosgene	0.31 [¶]	0.86	1.65
Propane	0.34 [¶]	0.95	1.51
Sulfur dioxide	0.40 [¶]	1.11	1.28
Toluene	0.275	0.77	1.86
Water vapor	0.853	2.38	0.60

[†] By permission, from "Absorption and Extraction," 2d ed., p. 20, by T. K. Sherwood and R. L. Pigford. Copyright, 1952. McGraw-Hill Book Company, Inc.

[‡] The value of μ/ρ is that for pure air, 0.512 ft²/hr.

[¶] Calculated by Eq. (10-65).

Appendix 18

TYLER STANDARD SCREEN SCALE

This screen scale has as its base an opening of 0.0029 in., which is the opening in 200-mesh 0.0021-in. wire, the standard sieve, as adopted by the National Bureau of Standards.

Mesh	Clear opening, in.	Clear opening, mm	Approx. opening, in.	Wire diameter, in.
	1.050	26.67	1	0.148
†	0.883	22.43	$\frac{7}{8}$	0.135
	0.742	18.85	$\frac{3}{4}$	0.135
†	0.624	15.85	$\frac{5}{8}$	0.120
	0.525	13.33	$\frac{1}{2}$	0.105
†	0.441	11.20	$\frac{7}{16}$	0.105
	0.371	9.423	$\frac{3}{8}$	0.092
$2\frac{1}{2}$ †	0.312	7.925	$\frac{5}{16}$	0.088
3	0.263	6.680	$\frac{1}{4}$	0.070
$3\frac{1}{2}$ †	0.221	5.613	$\frac{7}{32}$	0.065
4	0.185	4.699	$\frac{3}{16}$	0.065
5 †	0.156	3.962	$\frac{5}{32}$	0.044
6	0.131	3.327	$\frac{1}{8}$	0.036
7 †	0.110	2.794	$\frac{7}{64}$	0.0328
8	0.093	2.362	$\frac{3}{32}$	0.032
9 †	0.078	1.981	$\frac{5}{64}$	0.033
10	0.065	1.651	$\frac{1}{16}$	0.035
12 †	0.055	1.397	. .	0.028
14	0.046	1.168	$\frac{3}{64}$	0.025
16 †	0.0390	0.991	. .	0.0235
20	0.0328	0.833	$\frac{1}{32}$	0.0172
24 †	0.0276	0.701	. .	0.0141
28	0.0232	0.589	. .	0.0125
32 †	0.0195	0.495	. .	0.0118
35	0.0164	0.417	$\frac{1}{64}$	0.0122
42 †	0.0138	0.351	. .	0.0100

[Continued on next page

TYLER STANDARD SCREEN SCALE (*Continued*)

Mesh	Clear opening, in.	Clear opening, mm	Approx. opening, in.	Wire diameter, in.
48	0.0116	0.295	. .	0.0092
60 †	0.0097	0.246	. .	0.0070
65	0.0082	0.208	. .	0.0072
80 †	0.0069	0.175	. .	0.0056
100	0.0058	0.147	. .	0.0042
115 †	0.0049	0.124	. .	0.0038
150	0.0041	0.104	. .	0.0026
170 †	0.0035	0.088	. .	0.0024
200	0.0029	0.074	. .	0.0021

For coarser sizing—3- to $1\frac{1}{2}$-in. opening

			Approx. opening, in.	Wire diameter, in.
			3	0.207
			2	0.192
			$1\frac{1}{2}$	0.148

† These screens, for closer sizing, are inserted between the sizes usually considered as the standard series. With the inclusion of these screens the ratio of diameters of openings in two successive screens is as $1:\sqrt[4]{2}$ instead of $1:\sqrt{2}$.

Appendix 19

CAPACITIES AND SPEEDS OF
BELT CONVEYORS †

Belt width, in.	Normal operating speed, ft/min	Capacity for various bulk densities at 100 ft/min, tons/hr					
		30 lb/ft³	50 lb/ft³	75 lb/ft³	100 lb/ft³	125 lb/ft³	150 lb/ft³
14	200	9	16	24	32	40	47
16	200	13	21	31	42	52	63
18	250	16	27	40	53	67	80
20	250	20	33	50	67	83	100
24	300	29	49	73	98	120	145
30	350	47	79	120	160	195	235
36	400	70	115	175	235	290	350
42	400	98	165	245	325	410	490
48	400	130	220	330	440	550	655
54	450	170	285	425	570	710	855
60	450	215	360	540	720	900	1,080

† By permission, from H. L. Strube, *Chem. Eng.*, **61**(4):195 (1954).

Appendix 20

APPROXIMATE CAPACITIES OF HORIZONTAL SCREW CONVEYORS †

Screw diameter, in.	Standard length, ft	Light nonabrasive material, e.g., grain		Heavy nonabrasive material, e.g., coal		Heavy abrasive material, e.g., ashes	
		Capacity, ft^3/hr	Max. rpm	Capacity, ft^3/hr	Max. rpm	Capacity, ft^3/hr	Max. rpm
3	8	74	250	37	125		
4	8	171	220	86	110	46	90
5	8	304	210	150	105	85	85
6	10	500	200	255	100	135	80
7	10	820	190	410	95	200	75
8	10	1,180	180	590	90	300	75
9	10	1,600	175	780	85	400	70
10	10	2,050	160	1,030	80	516	65
12	12	3,300	150	1,660	75	820	60
14	12	4,000	140	2,000	70	1,200	55
16	12	7,000	130	3,400	65	1,630	50
18	12	9,000	120	4,500	60	2,100	45
20	12	12,000	115	5,800	55	2,860	40

† By permission of authors and publisher, from G. G. Brown and associates, "Unit Operations," p. 53, John Wiley & Sons, Inc., New York, 1950; and E. R. Riegel, "Chemical Process Machinery," 2d ed., p. 97, Reinhold Publishing Corporation, New York, 1953.

INDEX

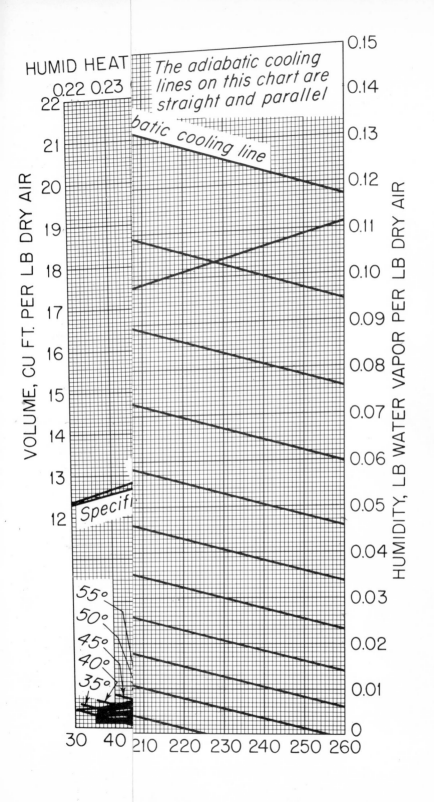